WILEY SERIES ON PHARMACEUTICAL SCIENCE AND BIOTECHNOLOGY: PRACTICES, APPLICATIONS AND METHODS

Series Editor:

Mike S. Lee
Milestone Development Services

Mike S. Lee • *Integrated Strategies for Drug Discovery Using Mass Spectrometry*

Birendra Pramanik, Mike S. Lee, and Guodong Chen • *Characterization of Impurities and Degradants Using Mass Spectrometry*

Mike S. Lee and Mingshe Zhu • *Mass Spectrometry in Drug Metabolism and Disposition: Basic Principles and Applications*

MASS SPECTROMETRY IN DRUG METABOLISM AND DISPOSITION

MASS SPECTROMETRY IN DRUG METABOLISM AND DISPOSITION

Basic Principles and Applications

Edited by

Mike S. Lee
Milestone Development Services

Mingshe Zhu
Bristol-Myers Squibb

A JOHN WILEY & SONS, INC., PUBLICATION

Published by John Wiley & Sons, Inc., Hoboken, New Jersey
Published simultaneously in Canada

For general information on our other products and services or for technical support, please contact our Customer Care Department within the United States at (800) 762-2974, outside the United States at (317) 572-3993 or fax (317) 572-4002.

Wiley also publishes its books in a variety of electronic formats. Some content that appears in print may not be available in electronic formats. For more information about Wiley products, visit our web site at www.wiley.com.

Library of Congress Cataloging-in-Publication Data:

Mass spectrometry in drug metabolism and disposition: basic principles and applications / edited by Mike S. Lee, Mingshe Zhu.

p. ; cm.

Includes bibliographical references and index.

ISBN 978-0-470-40196-5 (cloth) 1. Drugs—Metabolism—Analysis. 2. Metabolites—Spectra. 3. Mass spectrometry. 4. Mass spectrometry. I. Lee, Mike S., 1960- II. Zhu, Mingshe. [DNLM: 1. Pharmaceutical Preparations—metabolism. 2. Biopharmaceutics—methods. 3. Drug Design. 4. Mass Spectrometry—methods. 5. Pharmacokinetics. QV 38]

RM301.55.M367 2011

615′.7—dc22

2010028341

CONTENTS

FOREWORD

Studies in the areas of drug metabolism and pharmacokinetics have assumed progressively greater importance in pharmaceutical research over the past two decades, reflecting an increased awareness of the critical impact on successful drug development of the absorption, metabolism, distribution, elimination, and toxicity (ADMET) properties of candidate therapeutic agents. Indeed, the role of drug metabolism studies in the pharmaceutical industry, formerly limited to later phases of the development process, now spans the continuum from early discovery efforts through lead optimization, preclinical development, clinical trials, and postmarketing surveillance. Information on the identities and exposure levels of drug metabolites, first in animals and subsequently in human subjects, represents an essential component of preclinical and clinical safety assessment programs and, in those cases where circulating metabolites are pharmacologically active, provides the basis for assessing their pharmacokinetic/pharmacodynamic (PK/PD) relationships and contribution to the effects of the parent drug. Chemically reactive drug metabolites, which can be detected and characterized through specialized in vitro "trapping" techniques, generally are viewed as risk factors in drug development in light of their association with several forms of drug-induced toxicities, and early information on their identities is key to the design of optimized new chemical entities that lack this potential liability.

The detection, structural characterization, and quantitative analysis of drug metabolites in complex biological matrices often is a challenging endeavor, given the low levels that derive from highly potent parent compounds that are administered at doses of a few milligrams per day or less. As a result, stringent demands are placed on the analytical methodology employed for drug metabolism studies conducted either in vitro or in vivo, in terms of sensitivity and specificity of detection, and of wide dynamic range. In this regard, mass spectrometry, which always has been an important technique in drug metabolism studies, rapidly became the dominant technology in the field following the introduction, in the early 1990s, of the first commercial LC–MS/MS systems. Over the past decade, remarkable technical advances have been made in ion source design and ionization methods, rapid scanning and highly sensitive mass analyzers, efficient methods for inducing fragmentation of parent ions, rapid-response detectors with wide dynamic range, and powerful data acquisition and processing systems with sophisticated software packages and expert systems designed specifically for investigations in drug metabolism. The evolution of hybrid mass analyzers for MS/MS studies, and ancillary techniques such as ion mobility spectrometry, have added new dimensions to the mass spectrometry experiment, while the advent of high mass resolution capabilities on an LC time scale (even with "fast" chromatography) is having a truly revolutionary impact on the utility of LC–MS/MS in this field.

Mass Spectrometry in Drug Metabolism and Disposition: Basic Principles and Applications addresses each of these areas through a series of chapters authored by eminent scientists well versed in the application of contemporary mass spectrometry techniques to problems in drug metabolism and pharmacokinetics, with an emphasis on issues in drug discovery and development. The reader cannot help but be impressed by the capabilities of the current generation of LC–MS/MS instruments, which provide a combination of sensitivity, specificity, versatility, and speed of analysis that was difficult to envisage only a few years ago, and which have transformed the way drug metabolism studies are conducted. One can only wonder what lies in the years ahead!

Thomas A. Baillie
Seattle, WA
August, 2010

PREFACE

Two decades ago, drug metabolism research in the pharmaceutical industry was limited to radiolabeled in vivo drug disposition studies conducted in late stages of development. Drug metabolite identification was accomplished via a long and tedious process: metabolite separation and isolation followed by mass spectrometric and nuclear magnetic resonance analysis. Now, drug metabolism plays a critical role in the drug discovery and development process from lead optimization to clinical drug–drug interaction studies. Commercialized liquid chromatography/mass spectrometry (LC/MS) platforms have become the dominant analytical instrument employed in drug metabolism and pharmacokinetics (DMPK) studies and revolutionized the productivity of drug metabolism research. Certainly, the need for fast, sensitive and accurate measurements of drugs and metabolites in complex biological matrices has driven the continued development of novel LC/MS technology. Drug metabolism science and mass spectrometry technology have been integrated into an inseparable arena in drug discovery and development as well in related academic research activities. This book provides a unique thesis on mass spectrometry in drug metabolism with specific emphasis on both principles and applications in drug design and development. We believe that this integration will provide a unique contribution to the field that details both drug metabolism and analytical perspectives. Therefore, this book can be used as a teaching and/or reference tool to delineate and understand the "why" and "how" with regard to the many creative uses of mass spectrometry in drug metabolism and disposition studies. This work, authored by internationally renowned researchers, represents a combination of complementary backgrounds to bring technical and cultural awareness to this very important endeavor while serving to address needs within academia and industry.

The book is organized into three parts. Part I provides the reader with the basic concepts of drug metabolism and disposition. These concepts are intended to build a unique foundation of knowledge for drug metabolism and the subsequent studies performed during drug discovery and drug development endeavors. The book begins with an elegant perspective on drug metabolism. This review or perhaps "state of the union" provides unique insight into where we are, how we got there, and where we appear to be headed. Next we delve into the details of drug metabolism with a chapter on common biotransformation reactions. Further detail is provided in Chapter 3 from a medicinal chemistry perspective as the metabolic activation of organic functional groups is described along with considerations on how to address the reactive metabolite issues in drug design. Chapter 4 provides an overview on metabolizing enzymes, transporters, and their involvement with drug–drug interactions. In vitro experimental approaches to assess and predict drug–drug interactions in humans are elaborated. Chapter 5 starts with DMPK strategies in a drug discovery setting followed by a comprehensive overview of various experimental models applied in drug metabolism and disposition studies. Selection and data interpretation of the appropriate model are also discussed. Prediction of human pharmacokinetics is the focus of Chapter 6. Basic concepts and principles are discussed along with the use of mathematical models to predict pharmacokinetics. The actual use of drug metabolism information within the pharmaceutical industry is described in Chapter 7. In this chapter, the reader will obtain insight into the strategies used to design experiments for characterizing drug metabolism properties and addressing drug metabolism related issues from drug discovery to regulatory registrations.

Part II of the book highlights the principles and common practices of mass spectrometry in drug metabolism. The basic concepts and theory of mass spectrometry are presented in Chapter 8. In this chapter the reader will be able to obtain an updated thesis on the major components of this enabling instrumentation as well as the various mass analyzer platforms in use today. Some of the most common LC/MS-based methods used for metabolite identification is described in Chapter 9. Strategies for identification are reviewed and include a variety of mass spectrometry formats. Chapter 10 provides a review of common fragmentation reactions in the gas phase that are the foundation for mass spectral interpretation. Detailed examples provide the reader with the necessary tools for metabolite structure eludication. We dedicate an entire chapter to techniques that facilitate the performance of mass spectrometry during metabolite studies. And so, Chapter 11 provides concise background and industrial use of liquid chromatographic techniques as well as other detection techniques that are used to enhance the analytical performance of the mass spectrometer.

Part III of the book focuses on LC/MS techniques in drug metabolism and disposition. The application of quantitative LC/MS in drug metabolism and disposition is highlighted in Chapter 12. Critical studies that are routinely performed in drug discovery that involve metabolic stability, enzyme kinetics, metabolizing enzyme inhibition and induction, permeability and absorption, and in vitro transporter experiments are described. Chapter 13 provides both background and advantages of modern high-resolution mass spectrometry along with the use of newly developed data-mining tools for in vitro and in vivo drug metabolite identification. The understanding of the tissue distribution of a drug and the corresponding metabolites is illustrated in Chapter 14. The recent applications of imaging mass spectrometry for these studies are described. Novel instrumentation and mass spectrometry scan functions of hybrid triple quadrupole–linear ion trap mass spectrometry are discussed in Chapter 15. The applications for both bioanalysis and metabolite identification are highlighted. Quantitative drug metabolism studies using accelerator mass spectrometry are introduced and described in Chapter 16. The utility of this powerful technique for microdosing and DMPK studies in early clinical studies is described. Chapter 17 provides a provocative description of novel approaches, typically used in the field of protein identification, to obtain standard-free estimation of metabolite levels using nanospray mass spectrometry. A unique perspective on the profiling and characterization of Chinese herbal medicine and their metabolites using LC/MS is provided in Chapter 18. The approaches to determine the chemical composition of these medicines and their subsequent metabolites are discussed. The emerging area of bioanalysis of protein therapeutics in biological matrices is discussed in Chapter 19. Unique perspectives on digestion efficiency, internal standards, and biomarker validation are discussed in detail. Biomarker analysis is an exciting and emerging area of interest. Chapter 20 provides the reader with real-world application of mass spectrometry for the analysis of DNA, protein, peptide, and lipid biomarkers of oxidative stress. Part III of the book concludes with an updated perspective on endogenous metabolite profiling and small-molecule biomarker discovery (Chapter 21). In this chapter, the relationship between a perturbation and effected biochemical pathways is described within the context of biomarker discovery.

So, what criteria will emerge as the most desirable analytical figure of merit for high-performance LC/MS analysis in drug metabolism? It is our sincere hope that this book will provide an updated perspective on mass spectrometry in drug metabolism and disposition with recent applications, novel technologies, and innovative workflows.

Finally, the authors wish to acknowledge the contributions of many who transformed this book from idea to passion to reality. Specifically, the contributions from the authors and their respective collaborators, the editorial staff at John Wiley & Sons, and, of course, the families of the editors.

Mike S. Lee

Mingshe Zhu

CONTRIBUTORS

ZEPER ABLIZ
Key Laboratory of Bioactive Substances and Resource Utilization of Chinese Herbal Medicine, Ministry of Education
Institute of Materia Medica
Chinese Academy of Medical Sciences and Peking Union Medical College
Beijing, 100050 China

MARIA D. BACOLOD
Department of Chemistry
Lundbeck Research USA,
Paramus, New Jersey

TOM BAILLIE
School of Pharmacy
University of Washington
Seattle, Washington

KEVIN P. BATEMAN
Drug Metabolism and Pharmacokinetics
Merck Frosst Canada Ltd.
Quebec, H9H 3L1 Canada

IAN A. BLAIR
Centers for Cancer Pharmacology and Excellence in Environmental Toxicology
University of Pennsylvania
Philadelphia, Pennsylvania

MANUEL CAJINA
Department of Chemistry
Lundbeck Research USA,
Paramus, New Jersey

DONGMEI DAI,
Key Laboratory of Bioactive Substances and Resource Utilization of Chinese Herbal Medicine, Ministry of Education
Institute of Materia Medica
Chinese Academy of Medical Sciences and Peking Union Medical College
Beijing, 100050 China

Xinxin Ding
Wadsworth Center,
New York State Department of Health
Albany, New York

Stephen R. Dueker
Vitalea Science
Davis, California

Timothy J. Garrett
Department of Medicine
University of Florida
Gainesville, Florida

Stacy L. Gelhaus
Centers for Cancer Pharmacology and Excellence in Environmental Toxicology
University of Pennsylvania
Philadelphia, Pennsylvania

Ping Geng
Key Laboratory of Bioactive Substances and Resource Utilization of Chinese Herbal Medicine, Ministry of Education
Institute of Materia Medica
Chinese Academy of Medical Sciences and Peking Union Medical College
Beijing, 100050 China

Qing Ping Han
Department of Chemistry
Lundbeck Research USA,
Paramus, New Jersey

Mark Hayward
Analytical, Automation, and Formulation Laboratories
Department of Chemistry
Lundbeck Research USA
Paramus, New Jersey

Jiuming He
Key Laboratory of Bioactive Substances and Resource Utilization of Chinese Herbal Medicine, Ministry of Education
Institute of Materia Medica
Chinese Academy of Medical Sciences and Peking Union Medical College
Beijing, 100050 China

Akihiro Hisaka
Department of Pharmacy
The University of Tokyo Hospital
Faculty of Medicine
The University of Tokyo
Tokyo, 113-8655 Japan

Serhiy Hnatyshyn
Bioanalytical and Discovery Analytical Sciences
Applied and Investigational Metabonomics
Bristol-Myers Squibb
Princeton, New Jersey

Gérard Hopfgartner
Life Sciences Mass Spectrometry
School of Pharmaceutical Sciences
University of Geneva
University of Lausanne
Geneva, Switzerland

W. Griffith Humphreys
Department of Biotransformation
Bristol-Myers Squibb Research and Development
Princeton, New Jersey

Qin C. Ji
Bioanalytical Sciences, Analytical Research & Development
Bristol-Myers Squibb
Princeton, New Jersey

Wenying Jian
BA/DMPK
Pharmaceutical Research and Development
Johnson & Johnson
Raritan, New Jersey

Elliott B. Jones
Applied Biosystems Inc.
Foster City, California

Amit S. Kalgutkar
Pharmacokinetics, Dynamics, and Metabolism Department,
Pfizer Global Research and Development
Eastern Point Road, Groton, Conneticut

Brad Keck
Vitalea Science
Davis, California

Lewis J. Klunk
Department of Drug Metabolism and Pharmacokinetics
Biogen Idec
Cambridge, Massachusetts

Walter Korfmacher
Exploratory Drug Metabolism
Merck Research Laboratories
Kenilworth, New Jersey

Mike Lee
Milestone Development Services
Newtown, PA

Fumin Li
Bioanalytical Department
Covance Laboratories
Madison, Wisconsin

Jian Liu
Key Laboratory of Bioactive Substances and Resource Utilization of Chinese Herbal Medicine, Ministry of Education
Institute of Materia Medica
Chinese Academy of Medical Sciences and Peking Union Medical College
Beijing, 100050 China

Peter Lohstroh
Vitalea Science
Davis, California

Steven W. Louie
Department of Pharmacokinetics and Drug Metabolism
Amgen, Inc.
Thousand Oaks, California

Chuang Lu
Drug Metabolism and Pharmacokinetics
Millennium Pharmaceuticals, Inc
Cambridge, Massachusetts

Gang Luo
Covance Laboratories
Madison, Wisconsin

Daniel P. Magparangalan
Department of Chemistry
University of Florida
Gainesville, Florida

Russell J. Mortishire-Smith
Janssen Pharmaceutical Companies of Johnson & Johnson
B-2340 Beerse, Belgium

Sidney D. Nelson
Department of Medicinal Chemistry, School of Pharmacy
University of Washington
Seattle, Washington

Chandra Prakash
Department of Drug Metabolism and Pharmacokinetics
Biogen Idec
Cambridge, Massachusetts

Birendra Pramanik
Chemical Research
Merck Research Laboratories
Kenilworth, New Jersey

Richard F. Reich
Department of Chemistry
University of Florida
Gainesville, Florida

Michael D. Reily
Bioanalytical and Discovery Analytical Sciences
Applied and Investigational Metabonomics
Bristol-Myers Squibb
Princeton, New Jersey

Petia Shipkova
Bioanalytical and Discovery Analytical Sciences
Applied and Investigational Metabonomics
Bristol-Myers Squibb
Pennington, New Jersey

MAGANG SHOU
Department of Pharmacokinetics and Drug Metabolism
Amgen, Inc.
Thousand Oaks, California

HIROSHI SUZUKI
Department of Pharmacy
The University of Tokyo Hospital
Faculty of Medicine
The University of Tokyo
Tokyo, 113-8655 Japan

JOHN S. VOGEL
Vitalea Science
Davis, California

BO WEN
Department of Drug Metabolism and Pharmacokinetics
Hoffmann-La Roche
Nutley, New Jersey

RONALD E. WHITE
White Global Pharma Consultants
Cranbury, New Jersey

JING-TAO WU
Drug Metabolism and Pharmacokinetics
Millennium Pharmaceuticals, Inc.
Cambridge, MA

LIN XU
Department of Drug Metabolism and Pharmacokinetics
Biogen Idec
Cambridge, Massachusetts

MING YAO
Department of Biotransformation
Bristol-Myers Squibb Research and Development
Princeton, New Jersey

TAKEHITO YAMAMOTO
Department of Pharmacy,
The University of Tokyo Hospital
Faculty of Medicine
The University of Tokyo
Tokyo, 113-8655 Japan

Richard A. Yost
Department of Chemistry
University of Florida
Gainesville, Florida

Donglu Zhang
Department of Biotransformation
Bristol-Myers Squibb
Princeton, New Jersey

Haiying Zhang
Department of Biotransformation
Bristol-Myers Squibb Research & Development
Pennington, New Jersey

Li-Kang Zhang
Global Analytical Chemistry
Merck Research Laboratoriess
Kenilworth, New Jersey

Ruiping Zhang
Key Laboratory of Bioactive Substances and Resource Utilization of Chinese Herbal Medicine, Ministry of Education
Institute of Materia Medica
Chinese Academy of Medical Sciences and Peking Union Medical College
Beijing, 100050 China

Mingshe Zhu
Department of Biotransformation
Bristol-Myers Squibb Research and Development
Princeton, New Jersey

Zack Zou
Department of Chemistry
Lundbeck Research USA,
Paramus, New Jersey

PART I
Basic Concepts of Drug Metabolism and Disposition

1 Progression of Drug Metabolism

RONALD E. WHITE
White Global Pharma Consultants, Cranbury, New Jersey

1.1 INTRODUCTION

In a certain sense, the field of drug metabolism (DM) is standing still. More specifically, the basic experiment of drug metabolism (i.e., administering a new drug to an animal or human and determining the structures, amounts, and disposition of the metabolites) has changed very little over a period of decades. Remarkably, the experimental design and resulting data set from a typical absorption, distribution, metabolism, and excretion (ADME) study conducted today would be instantly recognized and understood by DM scientists from 50 years ago. This is not the case with most other disciplines in the life sciences.

Mass Spectrometry in Drug Metabolism and Disposition: Basic Principles and Applications, First Edition. Edited by Mike S. Lee and Mingshe Zhu.

ɹstance, 20 years ago protein sequencing was based on peptide chemistry, ɹ the basic experiment was the assay of amino acids released from tryptic peptides, a process that required months or years to complete. Now, protein sequencing is based on nucleotide chemistry, and the basic experiment is the automated assay of oligonucleotides from partial hydrolysis of complementary deoxyribonucleic acid (cDNA) a process that takes about a day. The reason that ADME experiments have not evolved much is that we have never devised a surrogate for the whole human body for metabolism studies. All systems tried up to this point (e.g., animals, transgenic animals, perfused organs, in vitro incubations, three-dimensional (3D) microfluidic cell culture devices, in silico calculations) fail to reliably predict the actual metabolic fate of NCEs. To be sure, the technology we use for the ADME experiment has advanced greatly, but even with the newest methods, DM is still essentially a chemical exercise at its core, and the backbone technology remains mass spectrometry (MS).

However, if we consider the more complete picture, DM has expanded enormously in scope and level of understanding in those 50 years. While the chemistry-based core remains intact and actively growing, several other kinds of DM studies have layered over the core, all coexisting and relevant in contemporary DM science. Thus, in addition to the purely chemical description of the structures of metabolites and probable chemical mechanisms of their formation, we now have a very good biochemical understanding of the various enzymes that catalyze these biotransformation processes, as well as a cellular and genetic understanding of the expression and regulation of those enzymes. We are even making progress toward reliable prediction of the fates of xenobiotic substances in human beings (Anderson et al., 2009), although this goal remains out of reach for the present. The current level of understanding of DM is presented by several experts in Part I of this book.

1.2 HISTORICAL PHASES OF DRUG METABOLISM

Near the end of the twentieth century, I suggested that there had been four overlapping phases of DM in industrial drug discovery and development (White, 1998). These can be summarized as follows.

1.2.1 The "Chemistry" Phase (1950–1980)

During this period, only a descriptive account of the disposition of a new chemical entity (NCE) was provided, largely consisting of chemical information. Major urinary and fecal metabolites were isolated and identified by classic chemical techniques including column and thin-layer chromatography, crystallization, and derivatization. Eventually, spectroscopy was used for the structural elucidation, including mass spectroscopy, infrared, and nuclear magnetic resonance (NMR). These techniques required much smaller quantities to be isolated and allowed high-performance liquid chromatography (HPLC) to replace column chromatography. Interestingly, early in this period R.T. Williams published his monograph *Detoxication Mechanisms*, which can be considered the first identification of DM as a discrete field of study (Murphy, 2008). The publication of that book also showed that academic researchers were beginning to think about the biological basis and implications of DM, although this way of thinking took some time to make an impact in the industrial world.

1.2.2 The "Biochemistry" Phase (1975–Present)

Starting in the mid-1970s, we began to determine the underlying biochemical processes responsible for the disposition of xenobiotics (e.g., which enzymes were involved). Illustrating the indistinct separation of these phases, the pioneers in this phase often came from a chemical background, and they sought to describe the enzymes in chemical terms. The proteins were isolated so that they could be treated as discreet chemical reagents, describable in classical chemical terms of composition, reaction stoichiometry, thermodynamics, and reaction mechanism. However, after about a decade or so, the chemical approach to studying the enzymes transitioned to a biochemical and cell biology approach in which enzyme kinetics and protein–protein and membrane–protein interactions became the "hottest" topics. All of the important DM enzymes were characterized, named, and even made commercially available to industrial researchers. The latest advance in the biochemistry phase was the realization that even the *exposure* of drugs to the drug-metabolizing enzymes was a biochemical event, mediated by physical enzymes called *transporters* (Wu and Benet, 2005). Advances in our understanding of the biochemistry of DM in academic laboratories were reflected in a greater expectation by regulatory agencies that a biochemical description be provided in addition to the purely chemical description of DM of an NCE.

1.2.3 The "Genetics" Phase (1990–Present)

In this phase, we began to account for individual variations in the pharmacokinetic rates and molecular sites of metabolism by genotyping human test subjects with respect to an ever-growing list of genetically polymorphic drug-metabolizing enzymes. Equally important, regulation of DM enzymes was recognized to occur mainly at the gene expression level, whether resulting from heredity, disease processes, or environment. This pharmacogenetic characterization has become a routine expectation for the registration packages of NCEs and continues to expand. As before, the new genetic information did not replace any previous requirements for DM information but instead added an additional dimension to that information package.

1.2.4 The "Biology" Phase (2010 and Beyond)

We are beginning to view drug metabolism in terms of systems biology. This involves taking a holistic view of the simultaneous interaction of a xenobiotic ule with all the enzymes and receptors in the human body. Some of ɹe receptors are the pharmacological targets that lead to therapeutic benefits, ome are unintended targets that generate adverse events and toxicities, and some are the enzymes and nuclear receptors of DM. In our overall description of the disposition of the compound, interactions of the compound with these DM targets are especially complex to relate to safety and efficacy. When designing a practical clinical medicine, we need to establish a balance between too rapid metabolism, leading to reduced efficacy, and too sluggish metabolism, leading to accumulation and possible toxicity. In the clinic, we need to determine whether metabolism decreases or increases the desired pharmacological effect (i.e., active metabolites). And, finally, in this holistic biological view, we need to assess how the metabolites interact with all the off-target human enzymes and receptors, especially the phenomenon of reactive metabolites covalently binding to proteins and nucleic acids, leading to toxic sequelae (Baillie, 2009).

These four phases are graphically depicted in Figure 1.1. They are layered in the figure because we continue to do all the activities of each preceding phase as we proceed through the evolution of industrial DM. Thus, the total amount of DM characterization work for a new drug has increased dramatically over the

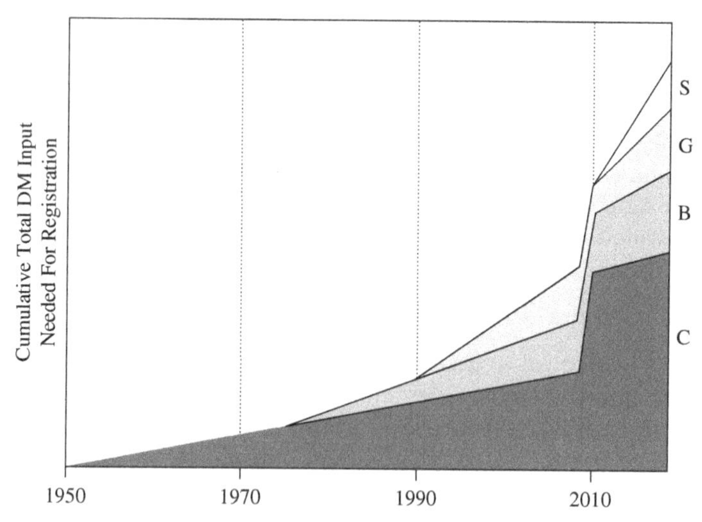

FIGURE 1.1 Progression of amounts of DM information required for regulatory filing of a new chemical entity. The vertical axis is the total amount of DM information in the registration application on a relative scale. The information classes are segregated as discussed in the text. The chemistry component (C, black) continually increases with time but abruptly increases about 2010 due to enhanced regulatory surveillance of metabolites. Starting around 1975, biochemistry (B, dark gray) begins to be included in the DM characterization and slowly increases with time. Genetics information (G, light gray) continues to increase, mainly in clinical trials. The apparent jump in B and G work around 2010 is due to the abrupt increase in C. Actual B and G work would not increase. Systems biology (S, white) is nearly zero in 2010 but is expected to increase subsequently.

years. The meaning of the step increase in work at around 2010 in the figure will be discussed in the next section.

1.3 NEXT STEP IN THE PROGRESSION OF DM

The first three of these historical phases of industrial DM serve to summarize and rationalize the scientific questions of yesteryear and today. The questions of tomorrow are described by the biology phase. However, now we can also discern the beginning of an additional new trend that could be called the "regulatory" phase. This phase is not primarily concerned with the physiological *process* of DM, as are the other phases. Instead, the regulatory phase is concerned with the human *safety* of the metabolites, once they are formed. But even though the focus of this phase is safety, it may well produce the greatest increment of additional DM work to be done in the future.

1.3.1 New Regulatory Expectation

Regulatory interest in metabolites has developed into a formal Guidance for Industry (Safety Testing of Drug Metabolites) issued by the U.S. Food and Drug Administration (FDA), which instructs sponsors on the qualitative and quantitative characterization of metabolites in both clinical and preclinical toxicological settings (U.S. FDA, 2008). A similar concern about metabolite safety is expressed in a Guidance from the International Conference on Harmonisation (ICH), currently in draft stage (ICH, 2009). We may succinctly state the requirement as follows: Human circulating metabolites that exceed 10% of the total exposure of all drug-related materials in circulation at pharmacokinetic steady state require safety assessment before large-scale clinical trials can proceed. These Guidances have implications for bioanalysis and safety assessment, but here we wish to focus on the implications for biotransformation studies. Development of these Guidances, though initiated by an industry-sponsored group (Baillie et al., 2002), has resulted in an inevitable regulatory emphasis on metabolite characterization much earlier than traditionally carried out. Importantly, the relative levels of parent drug and metabolites at pharmacokinetic steady state have been little studied previously. Consequently, we can expect surprises, possibly even new phenomena, as we watch the time evolution of drug metabolism during the approach to steady state. The traditional approach to definitive metabolite characterization (i.e., single-dose ^{14}C-labeled clinical ADME studies performed during Phase II or III) is inadequate for the new regulatory demands and either a new approach to ^{14}C studies or new nonradiolabel-based methodology is required.

1.3.2 New Challenges for Technology

Distressingly, both the qualitative and quantitative requirements of the new paradigm potentially push the existing technology past present limits.

Qualitatively, technology is now needed for metabolite detection in early clinical development. Up to this time, MS was only required to elucidate the chemical structures of metabolites, augmented as necessary by NMR and synthesis of authentic standards. However, *detection* of the presence of metabolites in a sample relied on the observation of nonparent radioactive peaks in the chromatogram. Now we are asking the mass spectroscopist to also detect the metabolites, without the benefit of radiolabel or prior knowledge of the structures. This requirement for MS-based detection as well as characterization has necessitated the development and validation of new algorithms of data acquisition and processing. In fact, as will be seen in later chapters, reliable nonradiometric MS-based methods of metabolite detection are now a reality.

Quantitatively, technology limits are pushed in two ways. First, the mass spectroscopist is asked to estimate the concentration of each metabolite in a plasma profile and categorize it as more or less than 10% of the total. This is problematic because the structures may not be completely known and authentic standards may not be available. Second, for those metabolites at levels more than 10%, a validated assay will be required, pushing the sensitivity limits that may be required in some cases. As the typical drug candidate becomes ever more potent, with ever smaller administered doses, the accurate estimation of analyte exposures that are as much as an order of magnitude *less* than that of the parent could become a challenge. Experts will discuss technological advances in MS related to the problems of detection and estimation of amounts of metabolites in Part II of this book and the experimental application of this technology to these problems in Part III. However, we can put some perspective here on the magnitude of the challenge.

1.4 PERSPECTIVE ON THE MAGNITUDE OF THE CHALLENGE

1.4.1 Ultimate Limits on Metabolite Quantitation

Let us start by considering whether there is any amount of metabolite that is too little to be meaningful. Since the present regulatory criterion for the amount of metabolite requiring assay is expressed as a *percentage* of a variable quantity (dose), then as new, more potent drugs are introduced, there is no regulatory lower limit on the absolute quantity of a metabolite that might need to be detected and assayed. So a literal reading of the Guidances is that no amount of metabolite is too little to be of regulatory interest. This approach to metabolite safety assessment has been questioned on the basis of the likelihood of metabolite-driven toxicity from minute body burdens (Smith and Obach, 2009), and it seems unlikely that regulatory authorities would apply the 10% rule to very low dose drugs (ICH, 2009). However, for this examination of limits, let us assume the worst case, that the 10% rule was always applied. Is there any other limitation of how much metabolite might need to be detected and quantitated? Actually, a moment's practical consideration of the question of lower limit reminds us that there is a fundamental limit imposed by nature (i.e., one molecule). Obviously, in this hypothetical limiting case, one molecule in a whole human body could only be assayed by exsanguinating the subject, so the stipulation that *the subject must survive the assay* clearly requires many more molecules than one in the body for detectability. To estimate *how* many more, in the next section we will use the "one-molecule" concept in the opposite direction (i.e., from the detector's point of view).

1.4.2 Practical Limits on Metabolite Quantitation

At very low concentrations, the discrete nature of molecules means that a detector signal no longer appears to be continuous. At the ultimate limit, either zero or one molecule will enter the inlet of the mass spectrometer; one cannot analyze half a molecule. Thus, one molecule entering the mass spectrometer inlet during a single duty cycle is the natural ultimate limit of sensitivity. Working backward from this limit and assuming that about 5% of the injected mass is actually sampled in a single duty cycle, we see that at least 20 molecules would have to be injected, on average, to get one into the inlet during a single duty cycle. Given a 10-μL injection from a 100-μL reconstituted extract of a 100-μL plasma aliquot of an original 1-mL plasma sample, we can estimate that plasma from a human subject would have to contain at least 2000 molecules per milliliter to be measurable in a practical sense by current liquid chromatography (LC)/MS laboratory methods. Although one could propose taking a larger plasma sample from a subject, reconstituting the extract in a smaller volume, and injecting a larger fraction onto the instrument, there are natural limitations for these volumes as well. For instance, a full pharmacokinetic profile for a subject that required 10 mL of plasma (i.e., about 20 mL of blood) for each of 10 time points would surely be close to the limit that physicians would ever accept under any circumstances, and even larger samples would clearly be out of the question. Thus, for this thought experiment, let us accept that a realistically and routinely measurable concentration could not be much less than 1000 molecules per milliliter of plasma based on the natural limit of one molecule interacting with a mass spectral detector. Application of Avogadro's number to this ensemble of 1000 molecules allows us to state that plasma concentrations less than about 1 attomolar (1 aM, 10^{-18} molar) will never be accessible in a practical sense.

1.4.3 Natural Limit Due to Dose Size

Now let us translate the natural-limit concept back to the real-world quantity of dose. If a midrange volume of distribution of 50 L is assumed for a typical drug, then a meaningful detection limit of 1 aM for a metabolite implies a concentration of 10 aM for the parent drug (i.e., metabolite is 10% of parent), which is equivalent to a total dose of 500 attomol. If the drug has a molecular weight (MW) of 400, then, the dose would be 0.2 pg. For comparison, what is the lowest dose of drug for which we are ever likely to be expected to characterize circulating metabolites? The most potent substances administered therapeutically are hormones, and their doses are exceedingly low. For instance, the labor-inducing hormone oxytocin is effective at a vanishingly low dose of about 40 ng, while the calcitriol dose is only about 4 μg. These examples may represent the lowest human doses that will ever be used for any drug. Amazingly, some actual drugs already approach these dose levels. For instance, the inhaled β-agonist formoterol is given at 12 μg, and the inhaled glucocorticoid beclomethasone is delivered at 40 μg. Circulating levels of these two drugs and their metabolites are almost unmeasurable (C_{max} in each case is ca. 10 pg/mL). The most potent oral drugs are also in the same range, with the dose of clonidine being about 100 μg. If, as suggested above, future drug candidates will not be more potent than the most potent drugs in use today, then we can say that levels of metabolites requiring detection and characterization will likely never be less than about 1 pg/mL. This is good news because, although they are not routine, MS-based assays with picogram/milliliter sensitivity are already in use. Thus, we can see that there is a natural limit to the ultimate sensitivity required to detect and characterize circulating metabolites, and it is not far from what our current MS technology already allows us to do. Most new drugs in development are dosed in the 1 to 1000-mg range, and metabolite levels for these drugs are well within contemporary MS sensitivity.

1.5 ARE THERE MORE SENSITIVE ALTERNATIVES TO MS?

An interesting extension of the concept of natural limitation is to remove the assumption that one molecule gives one quantum of signal (i.e., destructive analysis). Spectroscopic methods such as NMR and fluorescence can time integrate numerous signal pulses from a single molecule, resulting in no theoretical limit to how much signal could be accumulated with unlimited acquisition time. In fact, quantitative NMR spectroscopy has been proposed as a readily available solution to the main problems of implementing the Guidances, namely recognition and quantitation of metabolites without radiolabel at steady state (Espina et al., 2009; Vishwanathan et al., 2009). However, both NMR and fluorescence still face the indivisibility of individual molecules when applied to ex vivo samples such as plasma, so that a practical lower limit of concentration would still exist. Moreover, given the insensitivity of NMR relative to MS, it is unlikely that NMR could ever access concentrations that MS could not, even with the advantage of signal accumulation. Conversely, fluorescence might conceivably achieve the requisite sensitivity, but unlike MS and NMR, fluorescence is not generally applicable to all drug candidates. Clearly, then, the only currently available technology for low-level metabolite characterization is MS, which combines sensitivity, structural information, and general applicability.

1.6 SUMMARY

In summary, DM is a traditional yet dynamic discipline, comprising a constant core overlaid with evolving successive layers of related activities. The core activity of detecting and determining the chemical structure of human metabolites has changed little in decades and, in fact, is the starting point for all the other activities in areas such as enzymology, regulation, and genetics. For example, it is meaningless to inquire which cytochrome P450 (CYP) enzyme is principally responsible for the metabolism of a new drug until the chemical structure of the major metabolite shows that it was likely formed by a CYP enzyme. Today, structural elucidation of a metabolite almost always begins with MS, followed by complementary methods such as NMR, as necessary. Conversely, *detection* of metabolites has traditionally been accomplished by radiochromatography. However, in response to evolving regulatory expectations, it is likely that detection will become the job of MS also. Thus, MS is the most important single technique in DM and is likely to remain so going forward into the future. This conclusion explains the need for continued advancement of DM applications of MS technology, as described in the remainder of this book.

REFERENCES

Anderson S, Luffer-Atlas D, Knadler MP. Predicting circulating human metabolites: How good are we? Chem Res Toxicol 2009;22:243–256.

Baillie TA. Approaches to the assessment of stable and chemically reactive drug metabolites in early clinical trials. Chem Res Toxicol 2009;22:263–266.

Baillie TA, Cayen MN, Fouda H, Gerson RJ, Green JD, Grossman SJ, Klunk LJ, LeBlanc B, Perkins DG, Shipley LA. Drug metabolites in safety testing. Toxicol Appl Pharmacol 2002;182:188–196.

Espina R, Yu L, Wang J, Tong Z, Vashishtha S, Talaat R, Scatina J, Mutlib A. Nuclear magnetic resonance spectroscopy as a quantitative tool to determine the concentrations of biologically produced metabolites: Implications in metabolites in safety testing. Chem Res Toxicol 2009;22:299–310.

ICH (International Conference on Harmonisation of Technical Requirements for Registration of Pharmaceuticals for Human Use). Guidance on Nonclinical Safety Studies for the Conduct of Human Clinical Trials and Marketing Authorization for Pharmaceuticals, M3(R2); Step 4 version, 11, 2009, June available: www.ich.org/cache/compo/276-254-1.html.

Murphy PJ. The development of drug metabolism research as expressed in the publications of ASPET: Part 1, 1909–1958. Drug Metab Dispos 2008;36:1–5.

Smith DA, Obach RS. Metabolites in safety testing (MIST): Considerations of mechanisms of toxicity with dose, abundance, and duration of treatment. Chem Res Toxicol 2009;22:267–279.

U.S. Food and Drug Administration. Guidance for Industry: Safety Testing of Drug Metabolites. Center for Drug Evaluation and Research (CDER), Rockville, MD, 2008, available: www.fda.gov/cder/guidance/6897fnl.pdf.

Vishwanathan K, Babalola K, Wang J, Espina R, Yu L, Adedoyin A, Talaat R, Mutlib A, Scatina J. Obtaining exposures of metabolites in preclinical species through plasma pooling and quantitative NMR: Addressing metabolites in safety testing (MIST) guidance without using radiolabeled compounds and chemically synthesized metabolite standards. Chem Res Toxicol 2009;22:311–322.

White RE. Short and long-term projections about the involvement of drug metabolism in drug discovery and development. Drug Metab and Dispos 1998;26:1213–1216.

Wu CY, Benet LZ. Predicting Drug Disposition via Application of BCS: Transport/Absorption/Elimination Interplay and Development of a Biopharmaceutics Drug Disposition Classification System. Pharm Res 2005;22:11–23.

2 Common Biotransformation Reactions

BO WEN

Department of Drug Metabolism and Pharmacokinetics, Hoffmann-La Roche, Nutley, New Jersey

SIDNEY D. NELSON

Department of Medicinal Chemistry, University of Washington, Seattle, Washington

Mass Spectrometry in Drug Metabolism and Disposition: Basic Principles and Applications, First Edition. Edited by Mike S. Lee and Mingshe Zhu.

2.1 INTRODUCTION

Most chemicals, including drugs, are transformed in the human body to a wide variety of products by a host of enzymes present mostly intracellularly, though bacteria present in our gastrointestinal tract can metabolize some structures. The result of these reactions depends on the structures formed. The transformed product is often a metabolite with increased water solubility that either itself or as a sequential metabolite is excreted from the body as a "detoxication" product, which limits the time the drug is active in our system. Alternatively, the metabolite may have therapeutic activity, and some drugs are designed as prodrugs that lack a desired therapeutic activity until they are transformed either chemically or enzymatically in the human body into the active moiety. In a few cases, the metabolite or its sequential products may cause adverse reactions that can lead to toxic effects. The biotransformation reactions that lead to the various products or metabolites are governed by basic physicochemical principles, and most can be described with standard one- or two-electron chemical reactions. This chapter will survey the most common biotransformation reactions and is not intended to provide mechanistic details. For a more complete description of the chemical and enzymatic mechanistic aspects of drug metabolism, written for nonchemists and chemists, the reader is referred to a recent textbook on this topic (Uetrecht and Trager, 2007). For additional information on drug metabolism in general, the reader is referred to a recent handbook (Pearson and Wienkers, 2008).

2.2 OXIDATIVE REACTIONS

Oxidations are the most common biotransformation reactions that occur with most drugs. There are several classes of enzymes that carry out these reactions: cytochrome P450s, flavin monooxygenases, monoamine oxidases, xanthine oxidase, aldehyde oxidases, aldehyde dehydrogenases, and peroxidases. Typical reactions and substrate substructures for each of these classes of enzymes will be described.

2.2.1 Cytochrome P450 Oxidative Reactions

Cytochrome P450s are a superfamily of hemoproteins that exhibit a visible absorption band at approximately 450 nm when carbon monoxide is bound to the reduced (ferrous) protein (Ortiz de Montellano, 2005; Guengerich, 2008; P450 Homepage: http://drnelson.utmem.edu/CytochromeP450.html). Cytochrome P450s are ubiquitous in nature with over 8000 genes found as of 2008 and utilize reduced nicotinamide adenine dinucleotide phosphate (NADPH) as the cofactor. Although 115 cytochrome P450 genes have been identified in humans, only 57 are known to be functional, of which about half are known to metabolize drugs. Some of these enzymes, particularly those that oxidize physiological substrates, have high substrate selectivities, whereas many that metabolize drugs have broad and overlapping substrate selectivities. These enzymes catalyze a broad range of oxidative reactions that is usually driven by the reaction of an electron-deficient hypervalent iron-oxo species. Thus, for the most part, the reactions feature a one-electron radical abstraction/recombination that, depending on the particular drug substrate substructure, yields several different kinds of products as categorized below. In a few cases, the regiochemistry of the products may be dictated by electron-rich and radical stabilizing elements of particular substructures. However, in most cases, interactions of the substrate with specific active site residues are favored and dictate both the regio and stereochemistry of the products.

2.2.1.1 Aliphatic Hydrocarbon Hydroxylation These oxidations occur on allylic and benzylic sites (or similar aliphatic groups on heteroaromatic structures), as well as the ω and ω-1 positions on alkyl chains:

a $$R{-}CH_2{-}CH{=}CH_2 \longrightarrow R{-}CH(OH){-}CH{=}CH_2$$

b $$R{-}CH_2{-}C_6H_5 \longrightarrow R{-}CH(OH){-}C_6H_5$$

c $R-CH_2-CH_2-CH_3 \rightarrow R-CH_2-CH_2-CH_2-O-H$; $R-CH_2-CH(OH)-CH_3$

2.2.1.2 Aromatic Hydroxylation Phenols and phenol-like compounds (or their tautomers) are major metabolites of most benzenoid substructures, including heteroaromatic substructures. The hydroxylation may or may not proceed through an arene oxide, but the physicochemical/enzymic parameters that dictate which pathway will prevail are still largely unknown. Only a few aromatic epoxides have been stable enough to characterize:

OH ; O OH H O O H

R X R O X [FeO]$^{3+}$ R + O−Fe 2+ X − Fe^{3+} X H R O R X(H) OH

An uncommon form of aromatic hydroxylation, *ipso attack* at a site of substitution, can occur with some halogenated and other aromatic structures:

F Cl Cl Cl Cl O H $[FeO]^{3+}$ − H^+ F O−Fe 2+ Cl Cl Cl Cl O − Fe^{3+} − F^- O Cl Cl Cl Cl O

2.2.1.3 Unsaturated Carbon–Carbon Epoxidation Alkenes, other than those present in aromatic or heteroaromatic structures, usually form stable, isolable epoxides. Alkynes are thought to form unstable epoxides that rearrange to ketenes, which hydrolyze to carboxylic acids:

a $R-CH=CH-R' \longrightarrow R-CH-CH-R'$ (epoxide, O bridging)

For example:

Carbamazepine → Carbamazepine Epoxide

b $R-C\equiv C-H \longrightarrow R-C\overset{O}{=}C-H \longrightarrow R-\underset{H}{C}=C=O \xrightarrow{H_2O} R-CH_2-C(=O)-OH$

2.2.1.4 *Oxidative N, O, and S Dealkylation* Oxidative dealkylation is one of the most common biotransformation reactions observed with drugs. Essentially, all drugs that have an alkyl amine or amide, alkyl ether or ester, or alkyl thioether or thioester substructure that contains an α-carbon atom with at least one hydrogen atom, will be oxidized by cytochrome P450s at that carbon atom to form an intermediate carbinol (carbinolamine in the case of amines). Most of these semistable intermediates will spontaneously dealkylate to yield the corresponding heteroatom-containing product that has lost the alkyl group that was hydroxylated, and which forms an aldehyde or ketone in the process. However, several stable hydroxyamides, or similar structures that contain nonbasic nitrogens, do not spontaneously dealkylate. (Note that oxidative deamination is just another case of oxidative N-dealkylation in which the amine is a primary amine; thus, ammonia is lost.)

a $R_1-N(R_2)-C(R_3)(H)-R_4 \longrightarrow R_1-N(R_2)-C(R_3)(O-H)-R_4 \longrightarrow R_1-N(H)-R_2 + R_3-C(=O)-R_4$

b $R_1-O-C(R_2)(R_3)-H \longrightarrow R_1-O-C(R_2)(R_3)-O-H \longrightarrow R_1-O-H + R_2-C(=O)-R_3$ (S)

2.2.1.5 *Oxidative Dehalogenation* Drugs and other chemicals that contain halogen atoms attached to a carbon atom with at least one hydrogen atom undergo a similar reaction to that described for oxidative dealkylation, with intermediate formation of a halohydrin that spontaneously dehalogenates. The best leaving groups are iodide > bromide > chloride >>> fluoride:

$$R_1{-}C(R_2)(X){-}H \longrightarrow R_1{-}C(R_2)(X){-}O{-}H \longrightarrow R_1{-}C(=O){-}R_2 + H^+ + X^-$$

where X = halogen.

2.2.1.6 ***Heteroatom Oxidation*** Drugs and other chemicals that contain heteroatoms (mostly N and S) with hydrogen attached form the corresponding hydroxylamines and sulfenic acids. This oxidation is most commonly observed when the heteroatom is connected to an aromatic ring. Tertiary amines or heteroaromatic amines and sulfur ethers form N-oxides and sulfoxides, respectively, whereas imines can form oximes or nitrones.

a $$Ar{-}N(R){-}H \longrightarrow Ar{-}N(R){-}O{-}H$$

where Ar = aromatic or heteroaromatic ring.

b $$Ar{-}S{-}H \longrightarrow Ar{-}S{-}O{-}H$$

c $$R_1R_2R_3N{:} \longrightarrow R_1R_2R_3\overset{+}{N}{-}\overset{-}{O}$$

d $$\text{(ring)}{=}N{:} \longrightarrow \text{(ring)}{=}\overset{+}{N}{-}\overset{-}{O}$$

e $$R_1{-}S{-}R_2 \longrightarrow R_1{-}\overset{+}{S}(\overset{-}{O}){-}R_2$$

f $$R_1R_2C{=}\ddot{N}{-}H \longrightarrow R_1R_2C{=}N{-}OH$$

g $$R_1R_2C{=}\ddot{N}{-}R_3 \longrightarrow R_1R_2C{=}N^{+}(R_3){-}O^{-}$$

2.2.1.7 Alcohol and Aldehyde Oxidations Primary alcohols are oxidized to their respective aldehydes, which can be further oxidized to carboxylic acids. Secondary alcohols can be oxidized to ketones. Some diols and ketones can undergo oxidative C–C bond cleavage:

a $$R{-}CH_2{-}O{-}H \longrightarrow R{-}C({=}O){-}H \longrightarrow R{-}C({=}O){-}O{-}H$$

b $$R_1{-}CH(R_2){-}O{-}H \longrightarrow R_1{-}C({=}O){-}R_2$$

c $$R_1{-}C(OH)(R_2){-}C(OH)(H){-}R_3 \longrightarrow R_1{-}C({=}O){-}R_2 + R_3{-}C({=}O){-}H$$

2.2.1.8 Dehydrogenations Some alkanes can undergo dehydrogenation to alkenes. Catechols, *p*-hydroquinones, and *o*- and *p*-aminophenols or amidophenols can undergo dehydrogenation to reactive quinone or quinone-like structures. Some substructures that contain methyl groups on benzenoid or heteroaromatic structures with relatively low redox potentials because of other attached groups that stabilize benzylic radical formation can dehydrogenate to form very reactive methides:

a $$R{-}CH_2{-}CH_2{-}R_2 \longrightarrow R{-}CH{=}CH{-}R_2$$

b $$\text{catechol (1,2-dihydroxybenzene: OH, OH)} \longrightarrow \text{o-benzoquinone (O, O)}$$

c $$HO{-}C_6H_4{-}OH \longrightarrow O{=}C_6H_4{=}O$$

d HO–C₆H₄–NH–R ⟶ O=C₆H₄=N–R

where R = H or $C(=O)-R'$.

e 3-methylindole (CH_3, NH) ⟶ 3-methyleneindolenine (CH_2, N)

2.2.2 Oxidations by Flavin Monooxygenases

Flavin monooxygenases (FMOs) are a family of enzymes that catalyze the monooxygenation of soft nucleophilic groups (N, S, P, Se) through the formation of an enzyme-bound hydroperoxyflavin that is a stable, but relatively weak, oxidant (Krueger and Williams, 2005; Testa and Kramer, 2007; Strolin-Benedetti et al., 2006). Primary substructures are tertiary amines that are oxidized to N-oxides. FMOs also will metabolize primary alkylamines sequentially to hydroxylamines and oximes and secondary amines to N-hydroxy and nitrone products. Aromatic amines and amides are not substrates. Thioethers can be oxidized to sulfoxides; and thiols, thioamides, and thiocarbamates can be oxidized by both flavin monooxygenases and cytochrome P450s to reactive sulfenic acids, sulfines, and sulfenes:

1 $R_1R_2R_3N: \longrightarrow R_1R_2R_3N^{+}-O^{-}$

2 $R-CH_2-NH_2 \longrightarrow R-CH_2-NH-OH \longrightarrow R-CH=N-OH$

3 $R_1-CH_2-NH-R_2 \longrightarrow R_1-CH_2-N(R_2)-OH \longrightarrow R_1-CH=N^{+}(R_2)-O^{-}$

4 $R_1-S-R_2 \longrightarrow R_1-S^{+}(O^{-})-R_2$

5 $R-S-H \longrightarrow R-S-OH$

6 $R-C(=S)-NH_2 \longrightarrow R-C(=S^{+}-O^{-})-NH_2 \longrightarrow R-C(=S=O)-NH_2$

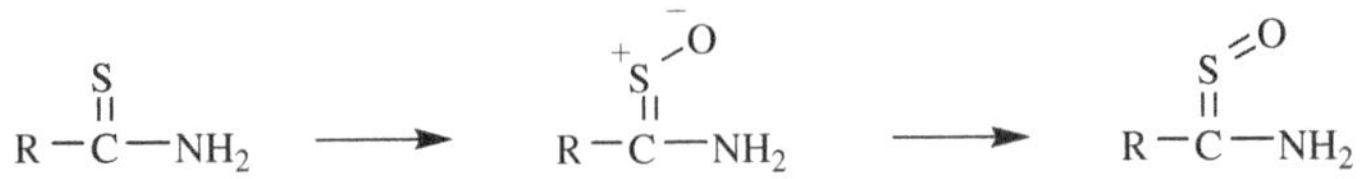

2.2.3 Oxidations by Monoamine Oxidases

Whereas cytochrome P450s and flavin monooxygenases are mostly microsomal enzymes, monoamine oxidases (MAOs) are located in mitochondria and are present in particularly high concentrations in nerve terminals (Testa and Kramer, 2007; Strolin-Benedetti et al., 2006; Youdim et al., 2006). Only two forms (MAO-A and MAO-B) have been characterized. Their substructure substrates are most commonly primary amines, though MAOs can oxidize some secondary and tertiary amines, such as the drug sumatriptan and the neurotoxin MPTP (1-methyl-4-phenyl-1,2,3,6-tetrahydropyridine). MAOs are flavin-containing enzymes, like FMOs, but they do not form a peroxyflavin oxidant. The MAOs apparently react via a radical abstraction mechanism that forms imines that are hydrolyzed to the amines and aldehydes with retention of oxygen from water rather than oxygen in the aldehyde product:

1 $R-CH_2-NH_2 \longrightarrow R-C(=O)-H + NH_3$

2

Sumatriptan

H₃C–NH–SO₂–CH₂–(indole)–CH₂–CH₂–N(CH₃)₂ → H₃C–NH–SO₂–CH₂–(indole)–CH₂–CHO + H–N(CH₃)₂

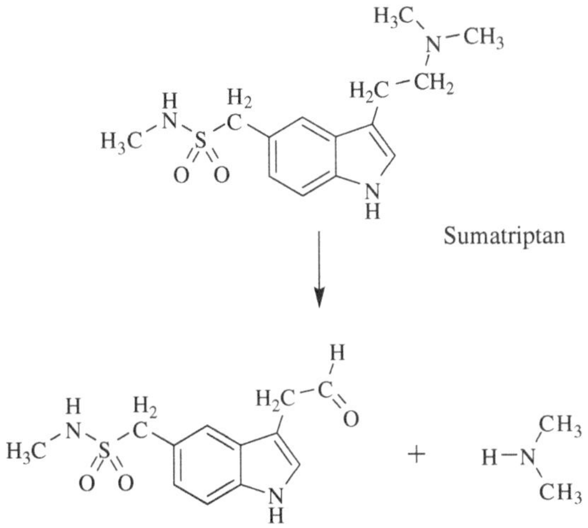

3

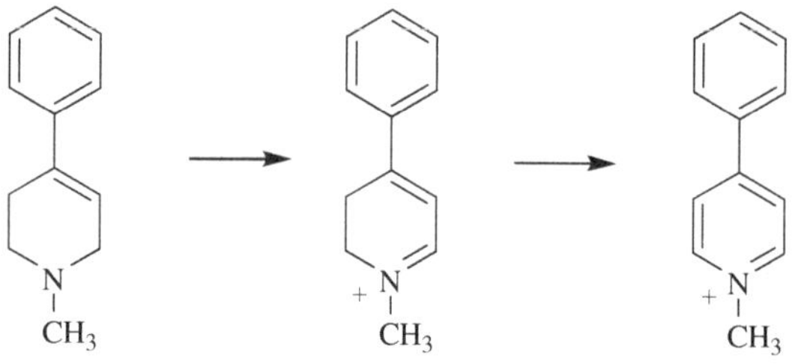

MPTP → MPDP+ → MPP+

2.2.4 Oxidations by Molybdenum Hydroxylases

Xanthine oxidase (XO) and aldehyde oxidase (AO) are molybdenum-containing cytosolic enzymes whose normal substrate substructures are iminelike *sp*^{2}-hybridized carbon atoms (Strolin-Benedetti et al., 2006; Garattini et al., 2008; Kitamura et al., 2006). The product amides contain oxygen that comes from water. These enzymes, particularly AO, which is present in relatively high concentrations in human liver, appear to play a greater role in the metabolism of new drugs that often contain nitrogen heterocycles:

General reactions:

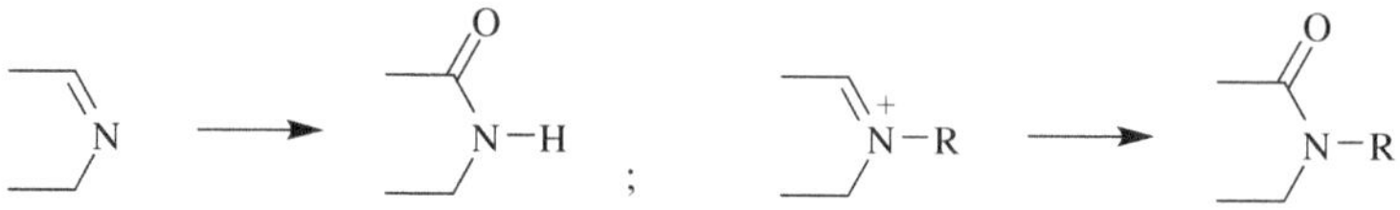

Specific drug:

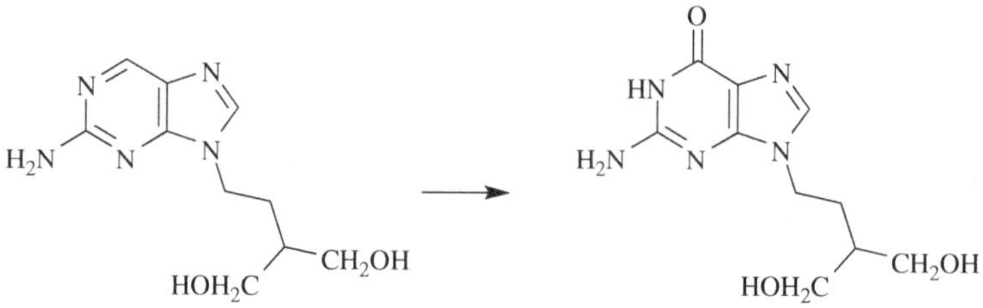

6-Deoxy penciclovir
(Prodrug)

Penciclovir
(Active drug)

2.2.5 Oxidations by Alcohol and Aldehyde Dehydrogenases

There are several classes of alcohol dehydrogenases that catalyze the reversible oxidation/reduction reaction of alcohols to aldehydes (Testa and Kramer, 2007;

Strolin-Benedetti et al., 2006). Most of these are zinc-containing cytosolic enzymes that use nicotinamide adenine dinucleotide/reduced nicotinamide adenine dinucleotide (NAD+/NADH) as the cofactor. In contrast, aldehyde dehydrogenases utilize NAD+ and catalyze the irreversible oxidation of aldehydes to carboxylic acids (Testa and Kramer, 2007; Strolin-Benedetti et al., 2006; Marchitti et al., 2008). Some forms of aldehyde dehydrogenases are cytosolic and others are mitochondrial:

$$1 \quad R{-}CH_2{-}OH \underset{NADH}{\overset{NAD^+}{\rightleftharpoons}} R{-}C({=}O){-}H$$

$$2 \quad R{-}C({=}O){-}H \xrightarrow{NAD^+} R{-}C({=}O){-}OH$$

2.2.6 Oxidations by Peroxidases

The two major peroxidases that have been found to oxidize drugs in humans are myeloperoxidase and the cyclooxygenase/prostaglandin H synthases, although lactoperoxidase and thyroid peroxidase have been found to oxidize some drugs (Uetrecht and Trager, 2007; Tafazoli and O'Brien, 2005). Myeloperoxidase is mostly localized in white blood cells and bone marrow, whereas the cyclooxygenases have a wide tissue distribution. Both classes of enzymes form a ferryl oxo protoporphyrin radical (referred to as Compound I) that can oxidize phenols to phenolic radicals and can also oxidize arylamines to radical cations. In addition, myeloperoxidase oxidizes chloride to hypochlorous acid that can oxidize (sometimes via chlorination) some drugs and physiological substrates. Depending on the remaining structure, the formed radical products of drugs can yield quinones, quinone imines, diimines, dimers (and other polymeric products), and hydroxylamines and/or nitroso compounds:

$$1 \quad Ph{-}O{-}H \longrightarrow Ph{-}O^{\bullet} \longrightarrow \text{quinones, dimers}$$

$$2 \quad Ph{-}N(R_1)(R_2) \longrightarrow Ph{-}N^{\bullet+}(R_1)(R_2) \longrightarrow Ph{-}N(H){-}OH \longrightarrow Ph{-}N{=}O$$

if R_1 and R_2 = H.

Example of a drug, amodiaquine, that is oxidized by HOCl formed by myeloperoxidase, to a reactive quinone imine.

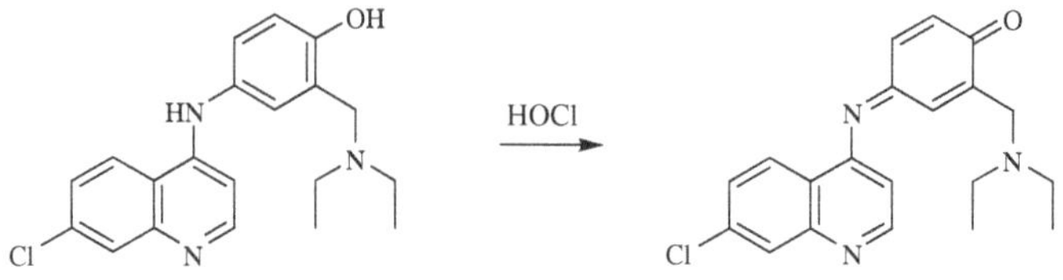

$$\text{Amodiaquine} \xrightarrow{HOCl} \text{quinone imine}$$

Amodiaquine

2.3 REDUCTIVE REACTIONS

The reductive biotransformation of drugs has been one of the least studied reactions, and many of the enzymes that are involved have not been well characterized. Some of the enzymes that catalyze reductive reactions of drugs are the cytochrome P450s, molybdenum reductases, alcohol dehydrogenases, carbonyl reductases, NADPH–cytochrome P450 reductase, NAD(P)H–quinone oxidoreductases, and enzymes of the intestinal microflora (Matsunaga et al., 2006; Rosemond and Walsh, 2004).

2.3.1 Reductions by Cytochrome P450s

2.3.1.1 Reductive Dehalogenation Halogenated compounds with redox potentials lower than that of the oxygen–superoxide couple, such as carbon tetrachloride (Mico et al., 1983) and halothane (Van Dyke and Gandolfi, 1976), can undergo reductive dehalogenation:

$$CCl_4 \longrightarrow CCl_3^{\bullet} + Cl^-$$

$$F_3C\text{–}CHBrCl \longrightarrow F_3C\text{–}\dot{C}HCl + Br^-$$

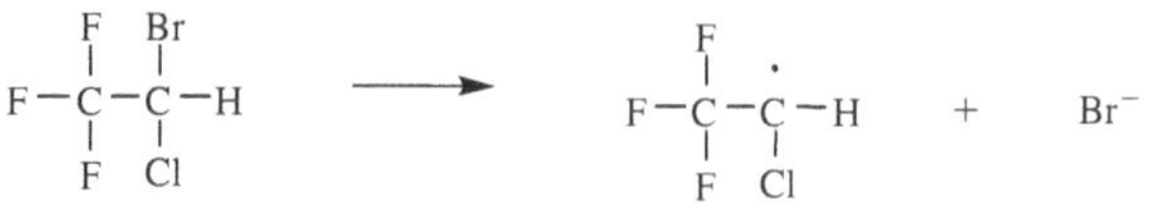

Halothane

2.3.1.2 Reduction of N-oxides Some N-oxides, such as nitric oxide, can be reduced by cytochrome P450s (Sugiura et al., 1976):

$$R_1R_2R_3N^+\text{–}O^- \longrightarrow R_1R_2R_3N\!: + {}^-OH$$

2.3.1.3 Reduction of Epoxides Some epoxides, including the arene oxide, benzene epoxide, can be reduced back to their alkenes (Kato et al., 1976; Yamazoe et al., 1978):

$$R_1(R_2)C(-O-)C(H)H \longrightarrow R_1(R_2)C=CH_2$$

2.3.1.4 Reduction of Peroxides Some peroxides can be reduced to alcohols, and depending on their structures, others can undergo reductive β-scission (Vaz and Coon, 1987; Vaz et al., 1990):

a $$R-O-OH \longrightarrow R-OH + H_2O$$

b $$R_1-C(R_2)(R_3)-O-OH \longrightarrow R_1-C(R_2)=O + R_3-H + H_2O$$

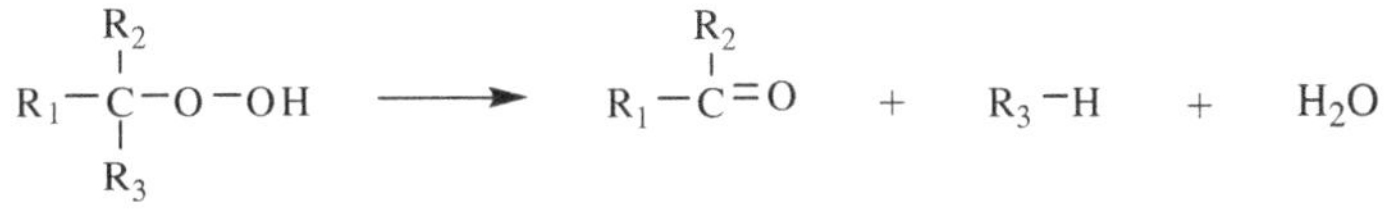

2.3.2 Reductions by Molybdenum-Containing Enzymes

Both XO and AO have been shown to reduce some heterocyclic rings, such as the isoxazole and thiazole structures, with resultant ring scission of the heteroatom–heteratom bonds (Kitamura et al., 2006). A new mammaliam molybdenum-containing enzyme, the mitochondrial amidoxime reducing component (mARC), has been shown to be a component, along with cytochrome b5 and cytochrome b5 reductase, in the reduction of amidoximes and N-hydroxyguanidines to their respective amidines and guanidines (Gruenewald et al., 2008):

1 (ring with N and X) $\longrightarrow$ (ring-opened, X−H, N−H)

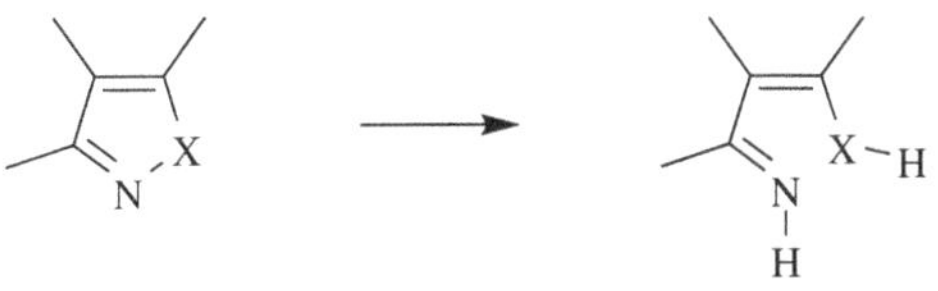

where X = O, S.

2 $$R-C(=N-OH)-NH_2 \longrightarrow R-C(=N-H)-NH_2 + H_2O$$

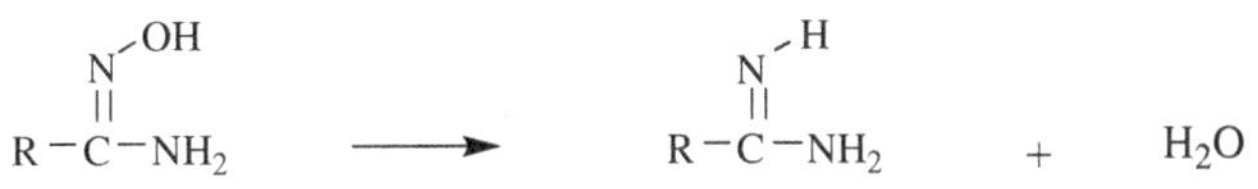

2.3.3 Reductions by Alcohol Dehydrogenases and Carbonyl Reductases

Both alcohol dehydrogenases and carbonyl reductases are cytosolic enzymes that catalyze the reduction of aldehydes to primary alcohols; however, carbonyl reductases also catalyze the reduction of ketones to alcohols. Whereas alcohol dehydrogenases utilize NADH as a cofactor, the carbonyl reductases utilize NADPH (Rosemond and Walsh, 2004):

1 $$R-C(=O)-H \longrightarrow R-CH_2-OH$$

2 $$R_1-C(=O)-R_2 \longrightarrow R_1-CH(OH)-R_2$$

2.3.4 Reductions by Cytochrome P450 Reductase and Quinone Oxidoreductase

Both NADPH–cytochrome P450 reductase (P450 reductase) and NAD(P)H-quinone oxidoreductase (NQO) are flavin adenine nucleotide-containing enzymes that catalyze the reduction of quinones and quinone-like structures. However, P450 reductase is a microsomal enzyme that catalyzes a one-electron reduction to yield semiquinone radicals that can redox cycle to produce superoxide anion radicals, whereas NQO is a cytosolic enzyme that catalyzes a two-electron reduction to yield hydroquinones (Matsunaga et al., 2006). P450 reductase also can catalyze the one-electron reduction of nitroaromatics to the nitro anion radical, which can redox cycle:

1

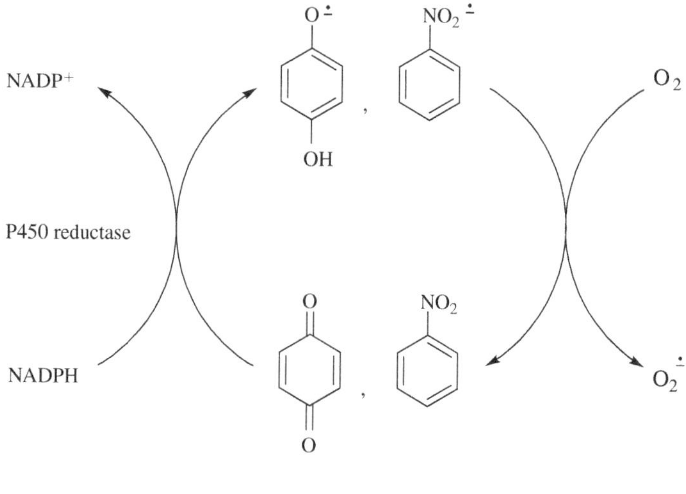

NADP+

P450 reductase

NADPH

$O^{\cdot-}$, OH, $NO_2^{\cdot-}$

O, O, NO_2

O_2

$O_2^{\cdot-}$

2

$$\text{1,4-benzoquinone} \xrightarrow[\text{NADH or NADPH}]{\text{NQO}} \text{hydroquinone (1,4-dihydroxybenzene)}$$

2.3.5 Reductions by Intestinal Microflora

Little is known about the enzymes of the intestinal microflora that reduce drugs and other chemicals (Matsunaga et al., 2006). However, these reductions can play an important role in the metabolism of some drugs. In particular, reductions of azo bonds and nitroaromatics are catalyzed by reductases in the intestinal microflora:

1 $$R_1{-}N{=}N{-}R_2 \longrightarrow R_1{-}\overset{H}{N}{-}\overset{H}{N}{-}R_2 \longrightarrow R_1{-}NH_2 + R_2{-}NH_2$$

2 $$Ar{-}NO_2 \longrightarrow Ar{-}NO_2^{\dot{-}} \longrightarrow Ar{-}N{=}O \longrightarrow Ar{-}\overset{H}{N}{-}OH \longrightarrow Ar{-}NH_2$$

where Ar = phenyl group or heteroaromatic ring.

2.4 HYDROLYTIC REACTIONS

Hydrolysis of epoxides, esters, amides, and related structures is an important biotransformation reaction that limits the therapeutic activity of many drugs and generates therapeutically active drugs from prodrug structures. In a few cases, hydrolytic reactions can generate a toxic structure. Epoxide hydrolases and esterases are members of the α/β hydrolase-fold family of enzymes (Morisseau and Hammock, 2005; Satoh and Hosokawa, 2006). Although their substrate specificities are radically different (e.g., lipids, peptides, epoxides, esters, amides, haloalkanes), their catalytic mechanisms are similar. All of these enzymes have an active site catalytic triad composed of a nucleophilic serine or cysteine residue (esterases/amidases), or aspartate residue (epoxide hydrolases) to activate the substrate, and histidine residue and glutamate or aspartate residues that act cooperatively in an acid–base reaction to activate a water molecule for the hydrolytic step.

2.4.1 Hydrolysis by Epoxide Hydrolases

There are at least five distinct epoxide hydrolases, some of which are cytosolic and others microsomal, that catalyze the hydrolysis of a variety of epoxides (Morisseau and Hammock, 2005). Some of these enzymes play a critical role in the biotransformation of endogenous substrates, such as leukotrienes, and others hydrolyze potentially reactive epoxide metabolites of drugs to diols that are significantly less reactive:

1

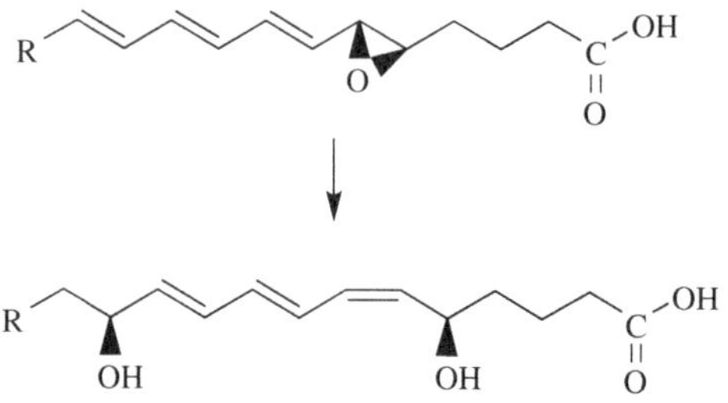

2

2.4.2 Hydrolysis of Esters, Amides, and Related Structures

Plasma butylcholinesterase hydrolyzes several simple esters and amides, while plasma paraoxonase hydrolyzes lactone substructures in drugs (Satoh and Hosokawa, 2006). Intestinal and liver microsomal carboxylesterases hydrolyze a variety of esters, amides, and related structures:

1 $$R_1-C(=O)-O-R_2 \xrightarrow[CES]{BuChE} R_1-C(=O)-OH + R_2-OH$$

2 $$R_1-C(=O)-N(R_2)-R_3 \xrightarrow[CES]{BuChE} R_1-C(=O)-OH + R_2-NH-R_3$$

3 PON1 / CES

4 $$R_1-O-P(=O)(OR_3)-OR_2 \xrightarrow[CES]{PON1} R_1-O-P(=O)(OH)-OR_2 + R_3-OH$$

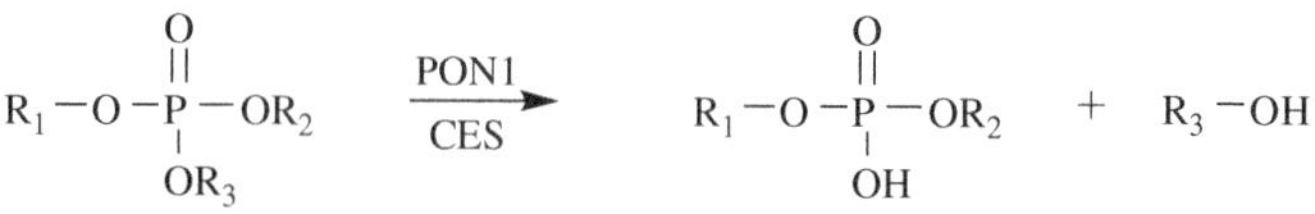

2.5 GLUCURONIDATION REACTIONS

Glucuronidation is the major conjugative biotransformation reaction for drugs in humans and most other mammalian species. The reactions are catalyzed by a large family of microsomal and nuclear membrane bound UDP-glucuronosyl transferases (UGTs) that utilize uridine diphosphoglucuronic acid (UDPGA) as a cofactor in a concerted displacement reaction on the anomeric carbon of glucuronic acid by nucleophilic groups on drug structures (O, N, S and sometimes carbon atoms) to yield β-glucuronides (Guillemette, 2003; Wells et al., 2004). Substrate substructures that form glucuronides include alcohols and other hydroxylated structures such as hydroxylamines, carboxylic acids and carbamic acids, amines of all types including tertiary amines and heterocyclic amines, some amides, thiols and thiocarbonyl groups, and imide carbon atoms that have acidic hydrogens. Glucuronidation is considered a high-capacity, low-affinity system in humans.

2.5.1 Glucuronidation of Hydroxy Groups

The hydroxy groups can be primary, secondary, or tertiary alcohols (with phenols and primary alcohols being most common), the hydroxy group on hydroxylamines, and the hydroxy group on carboxylic acids and carbamic acids:

1 $R_1-C(R_2)(R_3)-O-H \longrightarrow R_1-C(R_2)(R_3)-O-Gluc$

2 $R-C_6H_4-O-H \longrightarrow R-C_6H_4-O-Gluc$

3 $(R_1)(R_2)N-O-H \longrightarrow (R_1)(R_2)N-O-Gluc$

4 $R-C(=O)-O-H \longrightarrow R-C(=O)-O-Gluc$

5 $R-N(H)-C(=O)-O-H \longrightarrow R-N(H)-C(=O)-O-Gluc$

where Gluc = (glucuronic acid ring bearing C(=O)-OH, OH, OH, HO)

Note: Acyl glucuronides can undergo acyl migration reactions that may yield more reactive ring-opened products (Stachulski et al., 2006):

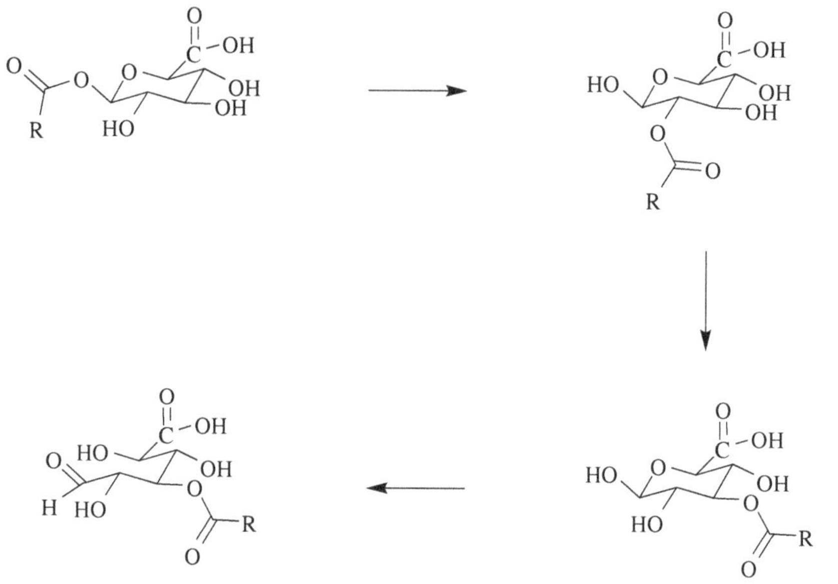

2.5.2 Glucuronidation of Amines and Amides

Essentially all classes of amines, except quaternary amines, can form glucuronides, as well as amides that contain a hydrogen atom, although this latter reaction is relatively rare:

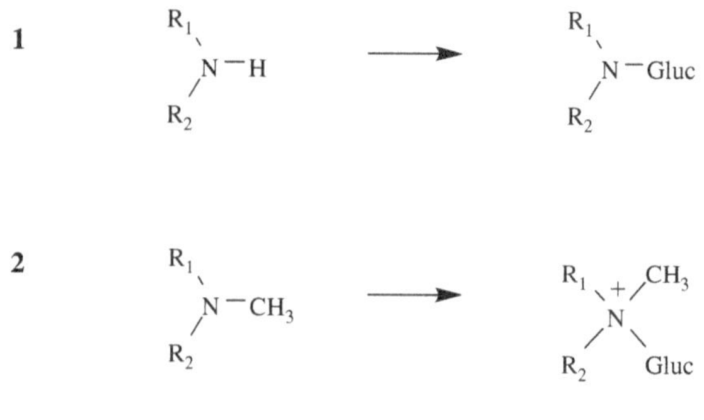

1 $R_1R_2N{-}H \longrightarrow R_1R_2N{-}Gluc$

2 $R_1R_2N{-}CH_3 \longrightarrow R_1R_2N^{+}(CH_3){-}Gluc$

3 [heterocyclic ring]=N: ⟶ [heterocyclic ring]=N⁺−Gluc

Heterocyclic amines

4 $R_1-C(=O)-N(R_2)(H) \longrightarrow R_1-C(=O)-N(R_2)(Gluc)$

2.5.3 Glucuronidation of Thiols and Thiocarbonyl Compounds

Since few drugs contain thiols and thiocarbonyl structures, only a few examples have been reported:

1 $R-S-H \longrightarrow R-S-Gluc$

2 $R_1-C(=S)-R_2 \longrightarrow R_1-C(=S^{+}-Gluc)-R_2$

2.5.4 Glucuronidation of Relatively Acidic Carbon Atoms

Examples of these substructures are imides:

$$R_1-C(=O)-CH_2-C(=O)-R_2 \longrightarrow R_1-C(=O)-CH(Gluc)-C(=O)-R_2$$

2.6 SULFATION REACTIONS

Sulfation is the second most common conjugative biotransformation for drugs in humans. The reactions are catalyzed by a family of sulfotransferases (SULTs), some of which are membrane bound and others that are cytosolic (Gamage et al., 2006). The reactions involve a nuclear substitution reaction of substrate nucleophilic groups with the reactive anhydride sulfate group of the cofactor 3′-phosphoadenosine-5′-phosphosulfate (PAPS). Substrates for the membrane bound forms of the enzymes are hydroxy groups or amino groups on proteins (e.g., tyrosine phenolic group), glycoproteins, and glycolipids. In contrast, most drugs are sulfated by cytosolic forms of SULT. Substrate substructures for these cytosolic SULTs are hydroxy groups on alcohols and on hydroxylamines and hydroxyamides, and nitrogens of amines and some amides. Sulfation is considered a low-capacity, high-affinity system in humans where sulfation of drugs may affect the sulfation of physiological substrates and visa versa. At physiological pH, the sulfates formed are essentially totally ionized.

2.6.1 Sulfation of Alcohols

The sulfation of alcohols is a very common biotransformation reaction, particularly for phenols, such as phenols of physiological substrates (Coughtrie et al., 1998):

1 $$R-\overset{H_2}{C}-O-H \longrightarrow R-\overset{H_2}{C}-O-\overset{\overset{O}{\|}}{\underset{\underset{O}{\|}}{S}}-O^-$$

2 $$R\ \ldots\ O-H \longrightarrow R\ \ldots\ O-\overset{\overset{O}{\|}}{\underset{\underset{O}{\|}}{S}}-O^-$$

2.6.2 Sulfation of Hydroxylamines and Hydroxyamides

Sulfation of hydroxylamines and hydroxyamides often leads to reactive electrophilic metabolites (Abu-Zeid et al., 1992; Banoglu, 2000), but in the case of the prodrug minoxidil, sulfation of the N-oxide yields the active drug (Anderson et al., 1998):

1 $$R-\overset{H}{N}-O-H \longrightarrow R-\overset{H}{N}-O-\overset{\overset{O}{\|}}{\underset{\underset{O}{\|}}{S}}-O^-$$

2 $$R_1-\underset{\underset{OH}{|}}{N}-\overset{\overset{O}{\|}}{C}-R_2 \longrightarrow R_1-\underset{\underset{O-SO_3^-}{|}}{N}-\overset{\overset{O}{\|}}{C}-R_2$$

3

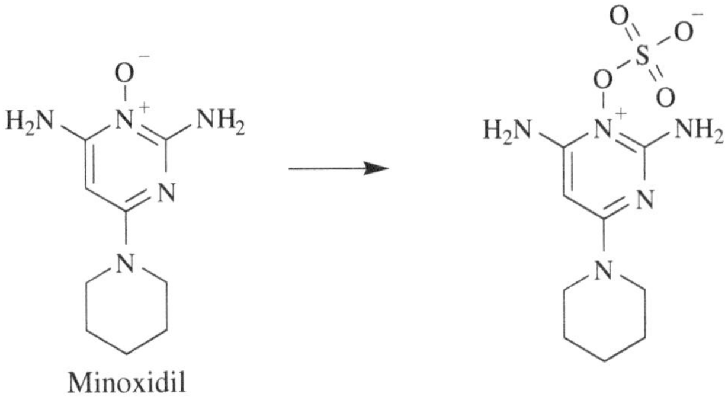

Minoxidil

2.6.3 Sulfation of Amines and Amides

Sulfation of primary amines is more common than sulfation of secondary amines. Sulfation of tertiary amines and amides is rare (Gamage et al., 2006):

1 $$R{-}NH_2 \longrightarrow R{-}NH{-}SO_2{-}O^-$$

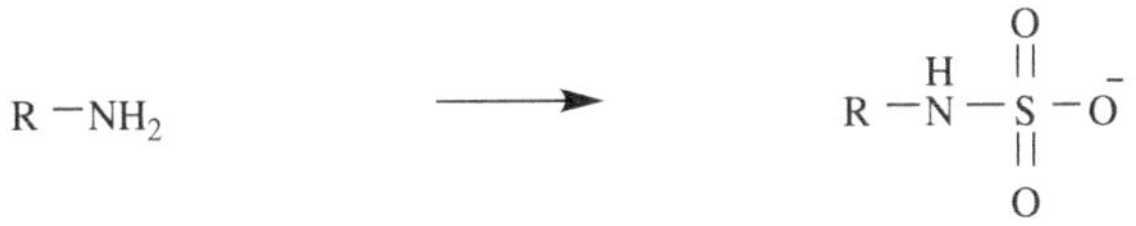

2 $$R{-}C(=O){-}NH_2 \longrightarrow R{-}C(=O){-}NH{-}SO_2{-}O^-$$

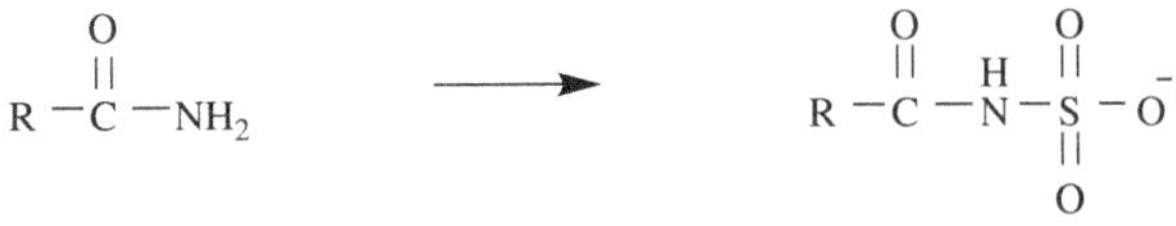

2.7 ACYLATION REACTIONS

Primary amines and hydrazines are acetylated by cytosolic polymorphic *N*-acetyltransferases (NATs) that utilize acetyl-CoA as a cofactor (Sim et al., 2008; Makarova, 2008). Some carboxylic acid drugs acylate amino acids through the formation of adenosine 5′-monophosphate (AMP) intermediates that subsequently form acyl CoA intermediates that react with amino acid *N*-acyltransferases to yield amides (Pearson and Wienkers, 2008; Testa and Kramer, 2008). A few chemical acylations can occur.

2.7.1 Acetylation of Primary Amines and Hydrazines

Anilines are the most common substructure is acetylated and only rarely are other amines. Drugs that contain hydrazine or hydrazide moieties with a primary amino group are almost always acetylated, as well. The extent of acetylation is dependent on both the overall drug structure and competing pathways of metabolism, and on the *N*-acetylation genotype of an individual taking the drugs. Phenotypically, slow acetylators are often more at risk of toxicities from aniline- and hydrazine-containing drugs because acetylation usually prevents the formation of reactive *N*-hydroxy drug metabolites:

1 $R{-}C_6H_4{-}NH_2 \longrightarrow R{-}C_6H_4{-}NH{-}C(=O){-}CH_3$

2 $R{-}NH{-}NH_2 \longrightarrow R{-}NH{-}NH{-}C(=O){-}CH_3$

2.7.2 Amino Acid Conjugation of Carboxylic Acids

In humans the most common conjugation reaction of drugs that contain carboxylic acids is the formation of glycine conjugates with benzoic acid substructures. Some drugs that contain an arylacetic acid substructure form glutamine conjugates, and some lipophilic steroid structures that contain carboxylic acid substructures (similar to bile acids) can form taurine conjugates:

1 $R{-}C_6H_4{-}C(=O){-}OH \longrightarrow R{-}C_6H_4{-}C(=O){-}NH{-}CH_2{-}C(=O){-}OH$

2 $R{-}C_6H_4{-}CH_2{-}C(=O){-}OH \longrightarrow R{-}C_6H_4{-}CH_2{-}C(=O){-}NH{-}CH(C(=O){-}OH){-}CH_2{-}CH_2{-}C(=O){-}NH_2$

3 Steroid∿$C(=O){-}OH \longrightarrow$ Steroid∿$C(=O){-}NH{-}CH_2{-}CH_2{-}SO_3^{-}$

2.7.3 Chemical Acylations

The fixation of carbon dioxide and trans-acetylation reactions have been observed with some amine-containing drugs:

1 (Cl, NH benzazepine) $\rightleftharpoons$ (Cl, $N{-}C(=O){-}OH$ benzazepine) $\longrightarrow$ Detected as its Gluc

2

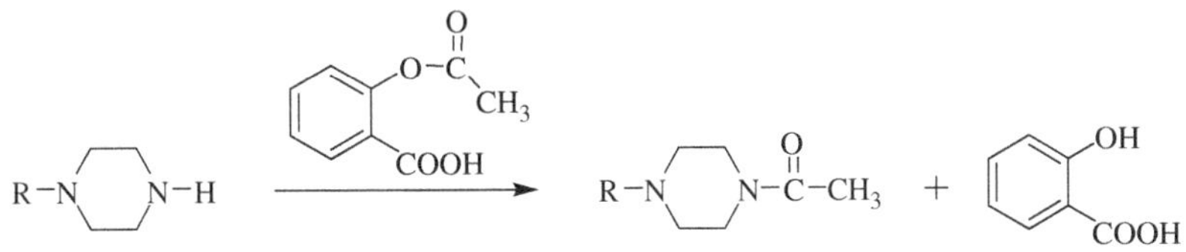

$$R-N\langle\rangle N-H \xrightarrow{\text{(2-acetoxybenzoic acid)}} R-N\langle\rangle N-C(=O)-CH_3 + \text{salicylic acid (OH, COOH)}$$

2.8 METHYLATION REACTIONS

Although not a common reaction for most drugs, methylation reactions are important because of genetic polymorphisms in the enzymes that catalyze methylation reactions, which can markedly affect therapy with a few drugs. All of the methyltransferase enzymes use *S*-adenosylmethionine (SAM) as the methyl donor cofactor wherein a nucleophilic group (O, S, N) on the substrate carries out a nucleophilic displacement reaction on the methyl sulfonium group of SAM. Drugs that contain catechol substructures are methylated by a cytosolic catechol-*O*-methyltransferase (COMT) that requires magnesium for activity (Testa and Kramer, 2008; Weinshilboum, 2006). Drugs that contain thiol moieties are methylated by either thiol methyltransferase (TMT), localized in the endoplasmic reticulum, or thiopurine methylransferase (TPMT) a cytosolic enzyme (Testa and Kramer, 2008; Weinshilboum, 2006). Only a few drugs are *N*-methylated by *N*-methyltransferase cytosolic enzymes.

2.8.1 Methylation of Catechols

Drugs that either contain catechol structures, such as α-methyldopa or L-dopa, are methylated to products that are therapeutically inactive. Several other drugs are metabolized to catechols that are subsequently methylated by COMT, usually to therapeutically inactive, detoxication products:

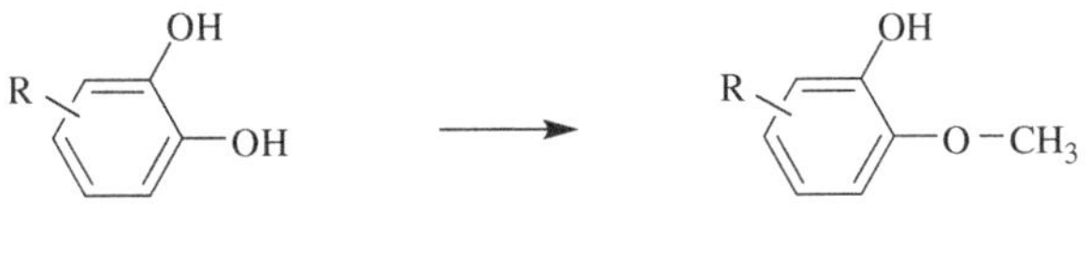

$$R-C_6H_3(OH)(OH) \longrightarrow R-C_6H_3(OH)(O-CH_3)$$

2.8.2 Methylation of Thiols

Drugs that contain aliphatic thiol groups are methylated by TMT, whereas drugs that contain aromatic or heteroaromatic thiols are methylated by TPMT:

1

$$R-CH_2-S-H \xrightarrow{TMT} R-CH_2-S-CH_3$$

2 R–C₆H₄–SH —TPMT→ R–C₆H₄–S–CH_3

3 (6-mercaptopurine: SH) —TPMT→ (S–CH_3)

2.8.3 Methylation of Amines

There are only a few examples of amine-containing drugs that are methylated:

1 Norephedrine (OH, CH_3, NH_2) → (OH, CH_3, HN–CH_3)

2 Nicotine (N–CH_3) → ($^+$N–CH_3, N–CH_3)

2.9 GLUTATHIONE CONJUGATION REACTIONS

Glutathione (GSH) is an unusual tripeptide, γ-glutamyl-cysteinyl-glycine, that is not hydrolyzed by normal peptidases because of its γ-glutamyl linkage. Its nucleophilic cysteinyl thiol group is involved in a number of reactions with electrophiles. GSH is present in high concentrations in most cells, and it can react nonenzymatically with highly reactive electrophiles. However, the reactivity of its thiol group is enhanced significantly by glutathione-*S*-transferases (GSTs) that catalyze most reactions of GSH with drug and/or drug metabolite substructures that are electrophilic (Testa and Kramer, 2008; Frova, 2006; Hayes et al., 2005; Anders, 2004). There are two unrelated families of GSTs. The membrane-associated family of GSTs are involved in eicosanoid metabolism, and some microsomal forms do react with electrophilic substructures on drugs or their metabolites. Most GSH conjugation reactions with xenobiotic electrophiles are catalyzed by members of a cytosolic family of GSTs with overlapping substrate selectivities. Drug substrate substructures for these biotransformation reactions include epoxides, enones/enals, and similarly conjugated systems, at saturated and unsaturated carbon atoms that have strong electron-withdrawing groups attached, and at heteroatoms that have good leaving groups attached.

2.9.1 GSH Conjugation of Epoxides

Both aliphatic and arene oxides form conjugates with GSH that decrease the reactivity and potential toxicity of these metabolites:

1 $R-CH(O)CH_2$ (epoxide) → $R-CH(OH)-CH_2-SG$

2 R-arene oxide → R-dihydrodiol conjugate (OH, SG)

2.9.2 GSH Conjugation of Conjugated Enone/Enal and Similar Systems

Reactions of GSH with α, β-unsaturated carbonyl systems in Michael-type addition reactions have been observed with drugs and their metabolites as exemplified below:

1 $H_3C-CH_2-C(=CH_2)-C(=O)-$[2,3-dichlorophenyl]$-O-CH_2-C(=O)-OH$ Etacrynic Acid (R = dichlorophenoxyacetic acid moiety)

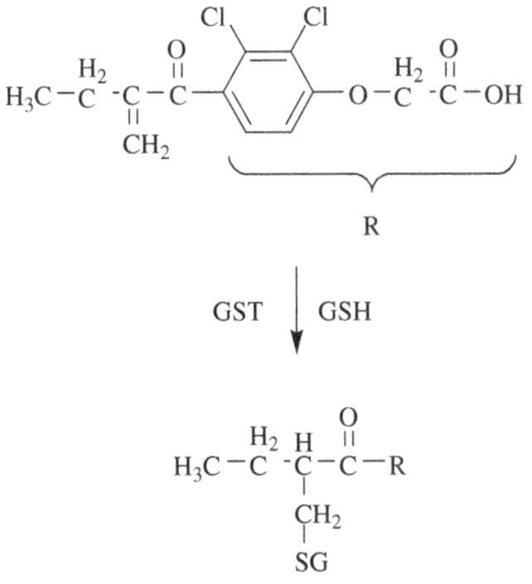

GST ↓ GSH

$H_3C-CH_2-CH(CH_2SG)-C(=O)-R$

2

$$\text{Acetaminophen} \xrightarrow{\text{P450}} \text{N-acetyl-}p\text{-benzoquinone imine} \xrightarrow[\text{(GST)}]{\text{GSH}} \text{3-(glutathion-S-yl)acetaminophen}$$

Acetaminophen

2.9.3 GSH Conjugations at Saturated and Unsaturated Carbon Atoms

Carbon atoms with good leaving groups such as halogens, sulfate, phosphate, and nitro groups may form conjugates with GSH with loss of the leaving groups:

$$R-\overset{H_2}{C}-X \xrightarrow[GST]{GSH} R-\overset{H_2}{C}-SG + H^+X^-$$

2.9.4 GSH Conjugation at Heteroatoms

Heteroatoms (N, O, S) with good leaving groups attached (refer to Section 2.9.3 above for a list of these groups) may form conjugates with GSH:

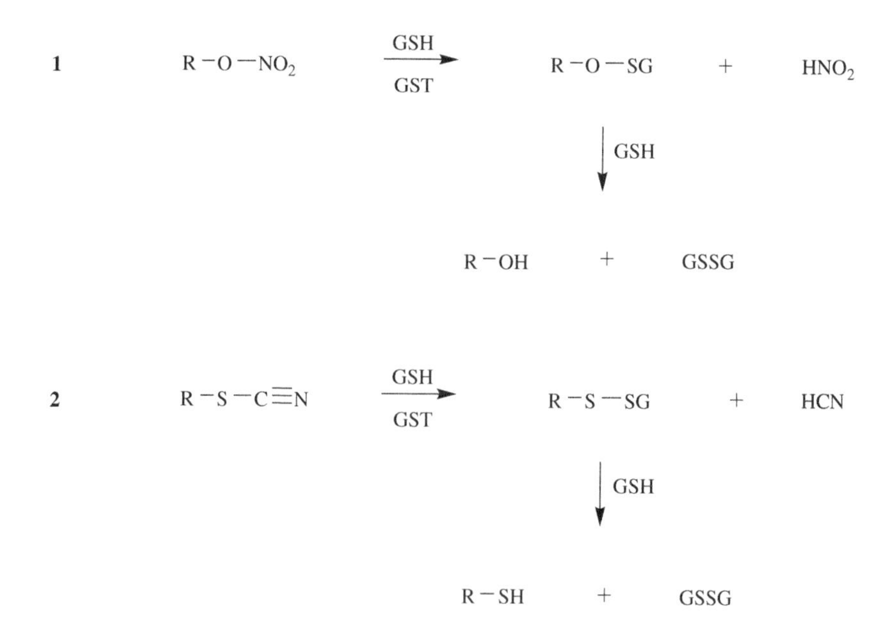

2.10 CONCLUSIONS

The wide variety of biotransformation reactions that drugs may undergo is a function of both the many enzymes that can catalyze various reactions and the many structures found in drugs that serve as substrates for the enzymes. Since these enzymes are distributed in different concentrations throughout the body, they can differentially affect the absorption and disposition and the pharmacology and toxicology of a drug. Furthermore, the efficacy and/or toxicity of a drug may vary widely in the patient population as a result of both genetic and environmental effects on drug metabolizing enzymes. Such factors are playing an increasingly important role in drug discovery and development as new drug structures are created to either avoid a particular biotransformation pathway (e.g., to avoid the formation of reactive, potentially toxic, metabolites) or to build "soft spots" into a structure to better control absorption, distribution, and elimination of a prodrug or drug. Finally, with the advent of methods to identify certain biotransformation genotypes and to characterize the phenotype of individual patients, drug therapy will become more individualized in order to decrease exposure to toxic metabolites in susceptible individuals. The purpose of this chapter was to survey the landscape of biotransformation reactions that can affect interindividual variation in drug efficacy and toxicity. Additional chapters in this book will provide a more complete picture of the significance of biotransformation reactions and transporters in the design and development of drugs.

REFERENCES

Abu-Zeid M, Yamazoe Y, Kato R. Sulfotransferase-mediated DNA binding of *N*–hydroxyarylamines(amide) in liver cytosols from human and experimental animals. Carcinogenesis 1992;13:1307–1314.

Anders MW. Glutathione-dependent bioactivation of haloalkanes and alkenes. Drug Metab Rev 2004;36:583–589.

Anderson RJ, Kudlacek PE, Clemens DL. Sulfation of minoxidil by multiple human cytosolic sulfotransferases. Chem Biol Interact 1998;109:53–67.

Banoglu E. Current status of the cytosolic sulfotransferases in the metabolic activation of promutagens and procarcinogens. Curr Drug Metab 2000;1:1–30.

Coughtrie MW, Sharp S, Maxwell K, Innes NP. Biology and function of the reversible sulfation pathway catalysed by human sulfotransferases and sulfatases. Chem Biol Interact 1998;109:3–27.

Frova C. Glutathione transferases in the genomics era. Biomol Eng 2006; 23:149–169.

Gamage N, Barnett A, Hempel N, Duggleby RG, Windmill KF, Martin JL, McManus ME. Human sulfotransferases and their role in chemical metabolism. Toxicol Sci 2006;90:5–22.

Garattini E, Fratelli M, Terao M. Mammalian aldehyde oxidases: Genetics, evolution and biochemistry. Cell Mol Life Sci 2008;65:1019–1048.

Gruenewald S, Wahl B, Bittner F, Hungeling H, Kanzow S, Kotthaus J, Schwering U, Mendel RR, Clement B. The fourth molybdenum containing enzyme mARC: Cloning and involvement in the activation of *N*-hydroxylated prodrugs. J Med Chem 2008;51:8173–8177.

Guengerich FP. Cytochrome P450 and chemical toxicology. Chem Res Toxicol 2008;21:70–83.

Guillemette C. Pharmacogenomics of human UDP glucuronosyltransferase enzymes. Pharmacogenomics J 2003;3:136–158.

Hayes JD, Flanagan JU, Jowsey IR. Glutathione transferases. Annu Rev Pharmacol Toxicol 2005;45:51–88.

Kato R, Iwasaki K, Shiraga T, Noguchi N. Evidence for the involvement of cytochrome P-450 in reduction of benzo(*a*)pyrene 4,5-oxide by rat liver microsomes. Biochem Biophys Res Commun 1976;70:681–687.

Kitamura S, Sugihara K, Ohta S. Drug metabolizing ability of molybdenum hydroxylases. Drug Metab Pharmacokinet 2006;21:83–98.

Krueger S, Williams DE. Mammalian flavin-containing monooxygenases: Structure, genetic polymorphisms and role in drug metabolism. Pharmacol Ther 2005; 106:357–387.

Makarova SI. Human *N*-acetyltransferases and drug-induced hepatotoxicity. Curr Drug Metab 2008;9:538–545.

Marchitti SA, Brocker C, Stagos D, Vasiliou V. Non-P450 aldehyde oxidizing enzymes: The aldehyde dehydrogenase superfamily. Expert Opin Drug Metab Toxicol 2008;4:697–720.

Matsunaga T, Shintani S, Hara A. Multiplicity of mammalian reductases for xenobiotic carbonyl compounds. Drug Metab Pharmacokinet 2006;21:1–18.

Mico BA, Branchflower RV, Pohl LR. Formation of electrophilic chlorine from carbon tetrachloride involvement of cytochrome P-450. Biochem Pharmacol 1983;32:2357–2359.

Morisseau C, Hammock BD. Epoxide hydrolases: Mechanisms, inhibitor design and biological roles. Ann Rev Pharmacol Toxicol 2005;45:311–333.

Ortiz de Montellano P. Cytochrome P450: Structure, Mechanism and Biochemistry, 3rd Ed. Plenum New York, 2005.

Pearson PG, Wienkers LC. Handbook of Drug Metabolism, 2nd ed. Informa Healthcare, New York, 2008.

P450 Homepage, http://drnelson.utmem.edu/CytochromeP450.html.

Rosemond MJC, Walsh JS. Human carbonyl reduction pathways: A strategy for their study in vitro. Drug Metab Rev 2004;36:335–361.

Satoh T, Hosokawa M. Structure, function and regulation of carboxylesterases. Chem-Biol Interact 2006;162:195–211.

Sim E, Lack N, Wang C-J, Long H, Westwood I, Fullam E, Kawamura A. Arylamine *N*-acetyltransferases: Structural and functional implications of polymorphisms. Toxicology 2008;254:170–183.

Stachulski AV, Harding JR, Lindon JC, Maggs JL, Park BK, Wilson ID. Acyl glucuronides: Biological activity, chemical reactivity and chemical synthesis. J Med Chem 2006;49:6931–6995.

Strolin-Benedetti M, Whomsley R, Baltes E. Involvement of enzymes other than CYPs in the oxidative metabolism of xenobiotics. Expert Opin Drug Metab Toxicol 2006;2:895–921.

Sugiura M, Iwasaki K, Kato R. Reduction of tertiary amine *N*-oxides by liver microsomal cytochrome P-450. Mol Pharmacol 1976;12:322–334.

Tafazoli S, O'Brien PJ. Peroxidases: A role in the metabolism and side effects of drugs. Drug Discov Today 2005;10:617–625.

Testa B, Kramer SD. The biochemistry of drug metabolism—An introduction: Part 2. Redox reactions and their enzymes. Chem Biodivers 2007;4:257–405.

Testa B, Kramer SD. The biochemistry of drug metabolism—An introduction: Part 4. Reactions of conjugation and their enzymes. Chem. Biodiversity 2008; 5:2171–2336.

Uetrecht JP, Trager WF. Drug Metabolism: Chemical and Enzymatic Aspects. Informa Healthcare, New York, 2007.

Van Dyke RA, Gandolfi AJ. Anaerobic release of fluoride from halothane. Relationship to the binding of halothane to hepatic cellular constituents. Drug Metab Dispos 1976;4:40–44.

Vaz AD, Coon MJ. Hydrocarbon formation in the reductive cleavage of hydroperoxides by cytochrome P-450. Proc Natl Acad Sci USA 1987;84:1172–1176.

Vaz AD, Roberts ES, Coon MJ. Reductive beta-scission of the hydroperoxides of fatty acids and xenobiotics: Role of alcohol-inducible cytochrome P-450. Proc Natl Acad Sci USA 1990;87:5499–5503.

Weinshilboum RM. Pharmacogenomics: Catechol *O*-methyltransferase to thiopurine methyltransferase. Cell Mol Neurobiol 2006;26:539–561.

Wells PG, Mackenzie PI, Chowdhury JR, Guillemette C, Gregory PA, Ishii Y, Hansen AJ, Kessler FK, Kim PM, Chowdhury NR, Ritter JK. Glucuronidation and the UDP-glucuronosyltransferases in health and disease. Drug Metab Dispos 2004;32:281–290.

Yamazoe Y, Sugiura M, Kamataki T, Kato R. Reconstitution of benzo[*a*]pyrene 4,5-oxide reductase activity by purified cytochrome p-450. FEBS Lett 1978; 88:337–340.

Youdim MB, Edmondson DE, Tipton KE. The therapeutic potential of monoamine oxidase inhibitors. Nature Rev Neurosci 2006;7:295–301.

3 Metabolic Activation of Organic Functional Groups Utilized in Medicinal Chemistry

AMIT S. KALGUTKAR
Pharmacokinetics, Dynamics, and Metabolism Department, Pfizer Global Research and Development, Eastern Point Road, Groton, Conneticut

3.1 INTRODUCTION

Safety-related attrition continues to be a major concern in the pharmaceutical industry (Kramer et al., 2007). Of a total of 548 drugs approved in the period

Mass Spectrometry in Drug Metabolism and Disposition: Basic Principles and Applications, First Edition. Edited by Mike S. Lee and Mingshe Zhu.

from 1975 to 1999, 45 drugs (8.2%) acquired 1 or more black-box warnings and 16 (2.9%) were withdrawn from the market owing to idiosyncratic adverse drug reactions (IADRs) *that were not predicted from animal testing and/or clinical trials* (Lasser et al., 2002). IADRs (also known as type B ADRs) are unrelated to known drug pharmacology and are generally dose independent. Because the frequency of occurrence of IADRs is very low (1 in 10,000 to 1 in 100,000), these reactions are often not detected until the drug has gained broad exposure in a large patient population. *Importantly, standard regulatory animal toxicity studies have traditionally shown a poor concordance with occurrence of IADRs in humans* (Olson et al., 2000). Life-threatening IADRs noted for drugs include hepatotoxicity, severe cutaneous reactions, aplastic anemia, and blood dyscrasias. Many pharmaceutical companies have recognized this issue and have increased their efforts to implement predictive in vitro tools and identify potential safety liabilities earlier in the drug discovery process. In this way, drug candidates can be eliminated via chemical intervention or these compounds can be suspended from further development. One component of such assays is aimed at understanding a drug candidate's propensity to undergo reactive metabolite formation.

3.2 BIOACTIVATION OF DRUGS

Drugs are metabolized via oxidative, reductive, and hydrolytic pathways known as phase I reactions. These reactions lead to a modest increase in aqueous solubility. Phase II reactions, also known as conjugation reactions, modify the newly introduced functionality to form *O*- and *N*-glucuronides, sulfate, and acetate esters, all with increased hydrophilicity relative to the unconjugated metabolite. In most cases metabolism results in the loss of biological activity of the parent drug, and such metabolic reactions are therefore regarded as detoxication pathways. However, depending on the structural features present in some compounds, the same metabolic events on occasion can generate electrophilic, reactive metabolites. Reactive metabolites can be formed by most, if not all, of the drug-metabolizing enzymes. Common phase I oxidative and phase II conjugation enzymes involved in reactive metabolite formation include the cytochrome P450 (CYP) family of hemoproteins and uridine glucuronosyl transferases (UGTs), respectively. In some cases a single enzymatic and/or chemical reaction is involved, and in other cases several enzymatic and/or chemical reactions are involved in the formation of reactive intermediates. The biotransformation of inert chemicals to electrophilic, reactive metabolites is commonly referred to as metabolic activation (bioactivation) and is now recognized to be the rate-limiting step in certain kinds of chemical-induced toxicities. Inadequate detoxication of reactive metabolites is thought to represent a pathogenic mechanism for tissue necrosis, carcinogenicity, teratogenicity, and/or certain immune-mediated idiosyncratic toxicities.

The consequences of covalent binding of reactive drug metabolites to proteins as it relates to IADRs remain poorly understood, even after some 40 years of research. In the case of acetaminophen, the dose-dependent hepatotoxicity observed in humans can be replicated in animals. For most other drugs, this is not the case; ADRs observed in humans cannot be reproduced in animals, which imply that there are no preclinical models to predict IADR potential of drug candidates. In addition, the downstream in vivo consequences of reactive metabolite formation and protein covalent modification as it relates to IADRs are poorly understood. Several hypotheses, however, have been proposed to explain these phenomena. The basic hypothesis that links the formation of reactive metabolites with IADRs (especially those with a possible immune component) is the process of haptenization wherein low-molecular-weight (< 1000 Da) reactive metabolites are converted to immunogens via binding to high-molecular-weight proteins as is the case with penicillin-induced anaphylactic reactions (Zhao et al., 2002). Examples of drugs associated with haptenization include halothane, tienilic acid, and dihydralazine, all of which are bioactivated to reactive metabolites and display mechanism-based inactivation of CYP isozymes responsible for their metabolism. Consistent with these observations, antibodies detected in sera of patients exposed to these drugs specifically recognize CYP isozymes and are responsible for their metabolism (Bourdi et al., 1992, 1996; Lecoeur et al., 1996).

Drug-metabolizing enzymes have evolved to biotransform a plethora of structurally diverse compounds encountered by the organism. These enzymes, however, cannot distinguish between molecules that are converted to reactive metabolites and those that are not. Furthermore, the likelihood of occurrence of bioactivation with a given compound will depend on several factors such as (1) the presence of functional group(s) (referred to as structural alerts or toxicophores) and/or chemical architecture, which is susceptible to bioactivation, (2) the presence of alternate metabolic soft spots within the molecule that compete with bioactivation, and (3) the detoxication of reactive metabolite and/or its precursor by metabolizing enzymes. The presence of structural alerts in a compound can be examined by visually inspecting the chemical structure or via the use of the DEREK software. DEREK for Windows is a knowledge-based expert system that is often used to identify structural alerts in a chemical (Dearden, 2003). Predictions are based largely on common occurrences of toxicophores in xenobiotics associated with some form of toxicity for which reactive metabolite formation has been observed in standard in vitro and in vivo systems.

3.3 EXPERIMENTAL STRATEGIES TO DETECT REACTIVE METABOLITES

With the possible exception of acyl glucuronides and cyclic iminium ions, most reactive metabolites are short-lived and are not detectable in circulation. Their formation can be inferred from the characterization of stable conjugates formed via reaction with the endogenous antioxidant glutathione (GSH). The presence of the soft nucleophilic thiol group in GSH ensures efficient conjugation to soft electrophilic centers on some reactive metabolites [e.g., Michael acceptors (quinonoids, α,β-unsaturated carbonyl compounds), epoxides (including arene oxides), alkyl halides, etc.] and yield stable sulfydryl conjugates. Qualitative in vitro assessment of reactive metabolite formation usually involves "trapping" studies conducted with reduced nicotinamide adenine dinucleotide phosphate (NADPH)-supplemented human liver microsomes and exogenously added GSH. Considering that nonmicrosomal enzymes can also participate in bioactivation, due consideration has to be given to the use of alternate metabolism vectors (e.g., liver cytosol, liver S-9 fractions, hepatocytes, neutrophils, etc.), which support the activity of non-CYP enzymes. It is noteworthy to point out that not all reactive metabolites can be trapped with GSH. "Hard" electrophiles including deoxyribonucleic acid (DNA)-reactive metabolites (e.g., aldehydes) will preferentially react with hard nucleophiles such as amines (e.g., semicarbazide and methoxylamine), amino acids (e.g., lysine), and DNA bases (e.g., guanine and cytosine) to afford the corresponding Schiff base (Sahali-Sahly et al., 1996; Olsen et al., 2005). Likewise, electrophilic iminium species, which are generated via metabolism of acyclic and cyclic tertiary amines, can be trapped using cyanide anion, which is a hard nucleophile (Gorrod et al., 1991; Argoti et al., 2005). Liquid chromatography tandem mass spectrometry is the preferred tool to detect reactive metabolite conjugates of nucleophiles (including exogenously added trapping agents and/or protein amino acids). Overall, characterization of the reactive metabolite conjugate structures provides an indirect insight into the structure of the reactive metabolite and the bioactivation pathway(s) leading to its formation.

Assessment of the amount of in vitro metabolism-dependent covalent binding to biological tissue is possible if radiolabeled drug is available (Evans et al., 2004). The assay provides quantitative estimates of radioactivity irreversibly bound to tissue but does not provide information about the nature of covalently modified proteins. Covalent binding studies can be performed in vivo as well. Either tissue or blood/plasma can be examined for the degree of covalent binding. However, covalent binding may require multiple dosing to establish the true impact of the compound. Reactive metabolites formed after the first dose may be efficiently trapped by GSH and eliminated from the body. Once GSH is depleted, the extent of covalent binding with cellular macromolecules may increase rapidly, resulting in toxicity.

3.4 FUNCTIONAL GROUP METABOLISM TO REACTIVE INTERMEDIATES

3.4.1 Two-Electron Oxidations on Electron-Rich Aromatic Ring Systems

By far, one on the most common pathways of reactive metabolite formation is via an enzymatic two-electron oxidation process on aromatic rings containing electron-rich functionalities in an *ortho* and/or *para* framework. The reactive intermediates derived from such bioactivation reactions can be generally categorized as quinone derivatives, which can react with GSH and/or protein nucleophiles in a typical 1,4-Michael fashion. Since 1,4-Michael addition to quinones represents a formal two-electron reduction process (regenerating the catechol/hydroquinone–nucleophile conjugate), their oxidant and electrophilic properties are intimately related. Many such enzymatic bioactivation reactions, leading to quinonelike species derived from, for example, electron-rich functional groups such as phenols, catechols, and aminophenols, have been efficiently mimicked via electrochemical techniques (Jurva et al., 2008; Madsen et al., 2008a, 2008b, 2007; Smith et al., 2003). Theoretical quantum chemical calculations have also been used to estimate the ease with which electron-rich aromatic systems undergo enzyme-catalyzed two-electron oxidations to reactive metabolites. The impact of adjacent aromatic substituents on the relative rates of oxidation can also be taken into account to rationalize differences in oxidation profiles. This approach has seen some success as demonstrated in the retrospective Ab initio analysis of acetaminophen oxidation (Koymans et al., 1989, 1990) and, more recently, with the atypical neuroleptic drug remoxipride (Erve et al., 2004). While not explored in great detail, there may be some additional scope for ab initio calculations in early discovery toward predicting oxidative instability of electron-rich aromatic ring systems as demonstrated in our studies on the hepatotoxic and nonhepatotoxic drugs, nefazodone and buspirone, respectively (Kalgutkar et al., 2005).

3.4.1.1 Quinones and Quinone-Methides A classic way of generating *ortho*-benzoquinones is via the CYP-mediated biotransformation sequence benzene → phenol → catechol → *ortho*-benzoquinone. *para*-Hydroxylation of phenols would generate *para*-benzoquinones via the two-electron oxidation of the corresponding *para*-hydroquinone intermediates. Mono- and dialkoxyphenols where substituents are located in an *ortho*- or *para*-arrangement can undergo CYP-mediated *O*-dealkylation reactions to catechols/hydroquinones followed by oxidation to quinones. Presence of electron-withdrawing substituents confers stronger oxidant properties to the quinone, and the corresponding hydroquinone or catechol metabolites are less readily oxidized to the quinone. In contrast, electron-donating groups confer weaker oxidant properties on the quinone, and the corresponding hydroquinone or catechol metabolites are more readily oxidized. These properties are particularly important since the catechol/hydroquinone conjugates with nucleophiles such as GSH can undergo reoxidation to quinones followed by conjugation with a second molecule of GSH (Zhao et al., 2007). In the redox environment of biological systems, quinones may also cause toxicity through the formation of reactive oxygen species (Bolton et al., 2000; Monks and Jones, 2002).

There are numerous examples of endogenous molecules, xenobiotics, and drugs that form electrophilic quinones as metabolites, which in turn appear to be responsible for their toxicity as illustrated with the antipsychotic remoxipride, which has been withdrawn from commercial use due to cases of aplastic anemia associated with its use. CYP-mediated hydroxylation followed by *O*-demethylation in remoxipride generates the catechol and hydroquinone derivatives that are oxidized to quinones, amenable to trapping with GSH (Erve et al., 2004) (Scheme 1). An additional example, which will be discussed later is that of the selective estrogen receptor modulator raloxifene, wherein the electron-rich phenol architecture is oxidized to quinonoid species (Scheme 24) (Chen et al., 2002).

CYP-catalyzed demethylenation of the methylenedioxyphenyl (1,3-benzdioxole) group in natural products and/or medicinal agents also results in quinone formation via the intermediate catechol intermediate. The mechanism (see Scheme 1) involves an initial hydroxylation at the methylene carbon followed by partitioning between demethylenation yielding a catechol intermediate and formaldehyde/formate or dehydration to a carbene (Murray, 2000). Further oxidation of the catechol generates the *ortho*-benzoquinone species. The selective serotonin reuptake inhibitor paroxetine is a classic example of a drug that undergoes this pathway (Zhao et al., 2007). As such, the mechanistic details of quinone formation with paroxetine will be discussed later (see Scheme 25).

In an analogous fashion to catechols and hydroquinones, *ortho*- and *para*-alkylphenols undergo two-electron π oxidation to form quinone-methides. Quinone-methides possess a significantly reduced propensity for redox chemistry than corresponding quinones and are therefore much more reactive

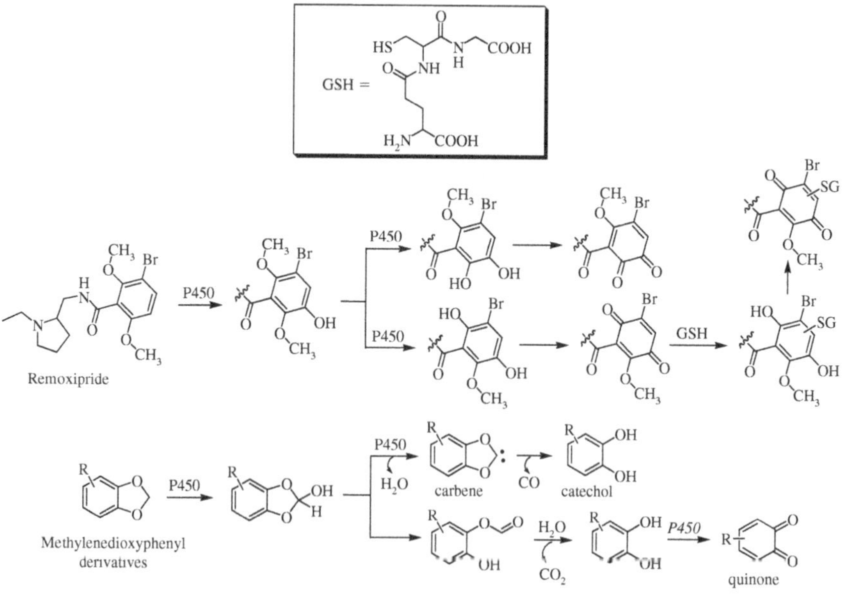

GSH =
Remoxipride
P450
GSH
Methylenedioxyphenyl derivatives
carbene
catechol
quinone

SCHEME 1 Bioactivation pathways leading to quinone formation.

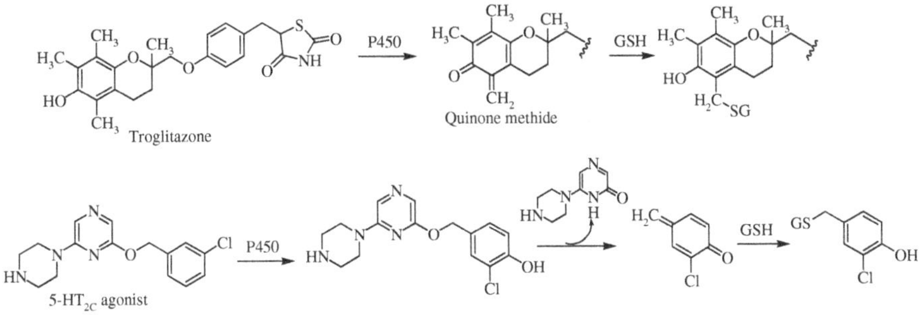

SCHEME 2 Bioactivation pathways leading to quinone-methide formation.

electrophiles (Thompson et al., 1993). Quinone-methides react with nucleophiles in a typical Michael fashion at the exocyclic methylene carbon. As such the exocyclic methylene carbon in a quinone-methide can be considered as a hard electrophile, which can react with both hard (DNA base) and soft (GSH) nucleophiles. A basic requirement for quinone-methide formation includes the presence of an aromatic system where an alkyl and hydroxyl group are separated in either *ortho* or *para* arrangement. The alkyl group must contain an abstractable hydrogen atom for the two-electron oxidation to occur as illustrated with troglitazone (Scheme 2) (Kassahun et al., 2001). *It is noteworthy to point out that oxidation of the alkylphenol motif in troglitazone to the quinone-methide intermediate can also occur nonenzymatically* (Mingshe Zhu, personal communication). The presence of a suitable leaving group on the alkyl carbon can also render quinone-methide formation via an elimination reaction as evident in metabolism studies on the 5-hydroxytryptamine $(5\text{-HT})_{2C}$ agonist and potential antiobesity agent 2-(3-chlorobenzyloxy)-6-(piperazin-1-yl)pyrazine (Scheme 2) (Kalgutkar et al., 2007a). Elucidation of the structure of the GSH conjugate discerned in S9/NADPH and GSH-supplemented incubations of the compound is consistent with a bioactivation pathway that involves initial aromatic ring hydroxylation on the 3-chlorobenzyl motif followed by β3 elimination to a quinone-methide, which is trapped by GSH.

Alkylbenzenes can also participate in this chemistry following aromatic hydroxylation by CYP enzymes. Likewise, alkoxyphenols can also form quinone-methides following oxidative dealkylation to the alkylphenol analogs as evident with tamoxifen (Scheme 3) (Fan et al., 2000; Fan and Bolton, 2001; Crewe et al., 2002). In one scenario, *O*-dealkylation occurs on the *N*, *N*-dimethylethylamine group to afford the phenol metabolite, which is oxidized to the quinone-methide. In another pathway, hydroxylation occurs on the aromatic ring geminal to the phenyl ether to the phenol metabolite, which upon further oxidation affords the quinone-methide. An interesting example of quinone-methide formation is highlighted in studies on the potassium channel opener maxipost (BMS-204352). *O*-Dealkylation of the methoxy group results in the phenol metabolite, which liberates the electrophilic *o*-quinone-methide

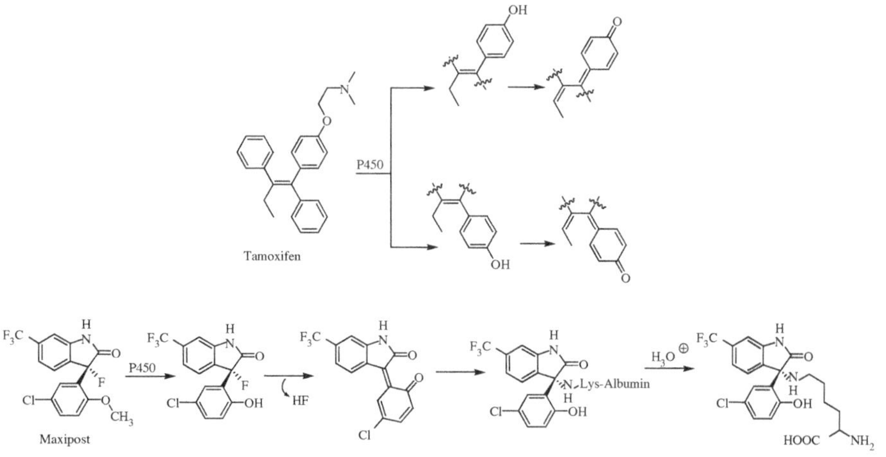

SCHEME 3 Bioactivation of tamoxifen and maxipost to quinone-methide species.

intermediate following the elimination of hydrogen fluoride (Scheme 3). The quinone-methide species is thought to be responsible for covalent binding of maxipost to albumin in vivo in animals and humans (Zhang et al., 2003, 2005). Acidic hydrolysis of plasma collected after intravenous administration of [^{14}C]-maxipost to rats and humans led to the characterization of a novel lysine conjugate of the des-fluoro des-*O*-methyl metabolite. The addition of lysine to the quinone-methide exocyclic carbon is consistent with the hard electrophilic character of the reactive species.

3.4.1.2 Quinone-imines, imine-methides, and diiminoquinones *Ortho*- and *para*-quinone-imines are analogous to catechol and hydroquinone derivatives in that one of the carbonyl oxygen atoms is replaced by NH functionality. Quinone-imines are derived from the CYP-catalyzed two-electron oxidation of *ortho*- and *para*-aminophenols as evident in the CYP and peroxidase-catalyzed oxidation of the antimalarial agent amodiaquine (Scheme 4) (Maggs et al., 1988). Quinone-imine formation is also discerned with *ortho*- and *para*-acetamidophenols as illustrated with acetaminophen, a dose-dependent hepatotoxin in animals and humans. Two-electron oxidation of the *para*-acetamidophenol architecture in acetaminophen by CYP (Dahlin et al., 1984), peroxidases including myeloperoxidase (Corbett et al., 1992), and cyclooxygenase (Potter and Hinson, 1987) generates the electrophilic *N*-acetyl-*para*-benzoquinone imine (NAPQI) species that reacts with GSH (Scheme 4).

The metabolism of nitrobenzenes to quinone-imines arises from a six-electron reduction of the nitro group to the corresponding aniline metabolite via the intermediate nitroso and hydroxylamine analogs. Aromatic ring hydroxylation by CYP *ortho* or *para* to the aniline nitrogen then generates the aminophenol derivative. The conversion of a nitrobenzene derivative to a quinone-imine is illustrated with the catechol-*O*-methyltransferase inhibitor tolcapone, an

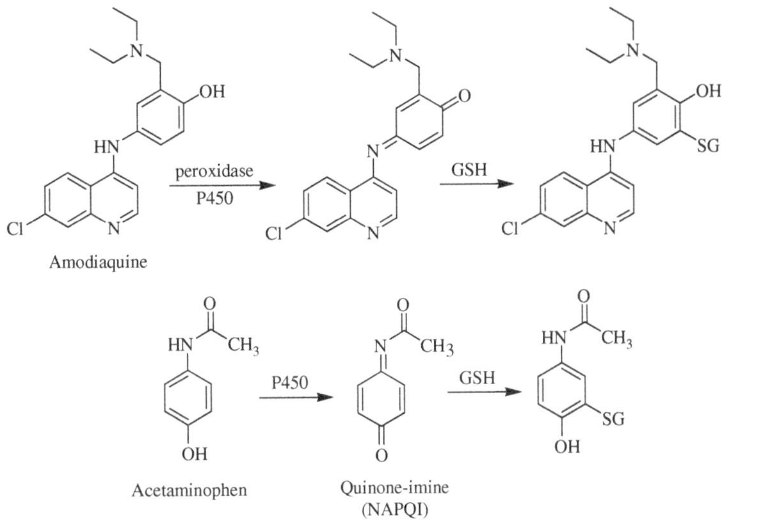

Amodiaquine → peroxidase, P450 → GSH →

Acetaminophen → P450 → Quinone-imine (NAPQI) → GSH →

SCHEME 4 Quinone-imine formation with amodiaquine and acetaminophen.

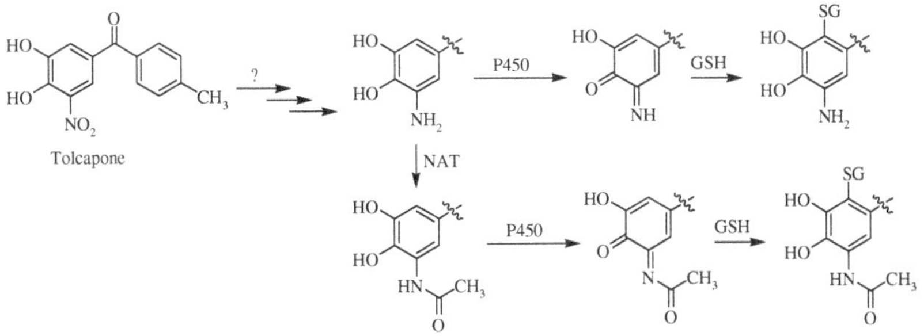

Tolcapone → ? → P450 → GSH; NAT; P450 → GSH

SCHEME 5 Metabolism of *ortho*-hydroxynitrobenzene derivative tolcapone to a quinone-imine species.

ortho-hydroxynitrobenzene derivative. The nitro group in tolcapone is first reduced to the aniline metabolite followed by an *N*-acetyltransferase-catalyzed acetylation of the aniline nitrogen. Both the *ortho*-hydroxyaniline and acetanilide metabolites are oxidized to the corresponding quinone-imine species by CYP and peroxidase enzymes (Scheme 5) (Smith et al., 2003).

In some cases, quinone-imine formation requires an initial hydroxylation step as illustrated with the nonsteroidal anti-inflammatory drug and idiosyncratic hepatotoxin diclofenac. Thus, CYP-catalyzed aromatic hydroxylation *para* to the aniline nitrogen affords the *para*-hydroxydiclofenac isomers, which can then undergo CYP or peroxidase-mediated oxidation to the quinone-imines amenable to trapping with GSH (Scheme 6) (Tang et al., 1999; Miyamoto et al., 1997). The aniline nitrogen in quinone-imine derivatives can also be part of a heterocyclic

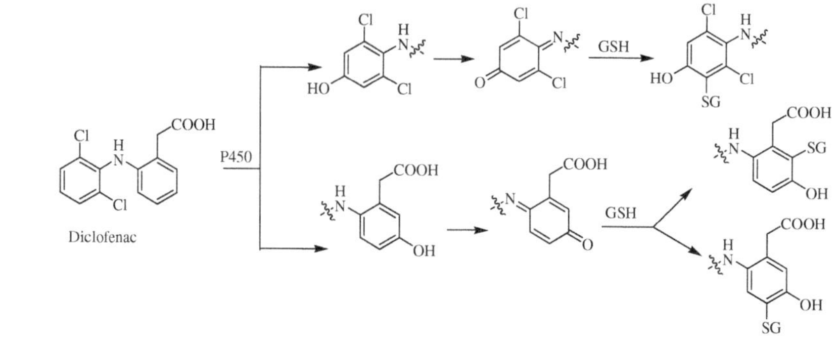

SCHEME 6 Bioactivation of diclofenac to electrophilic quinone-imine species.

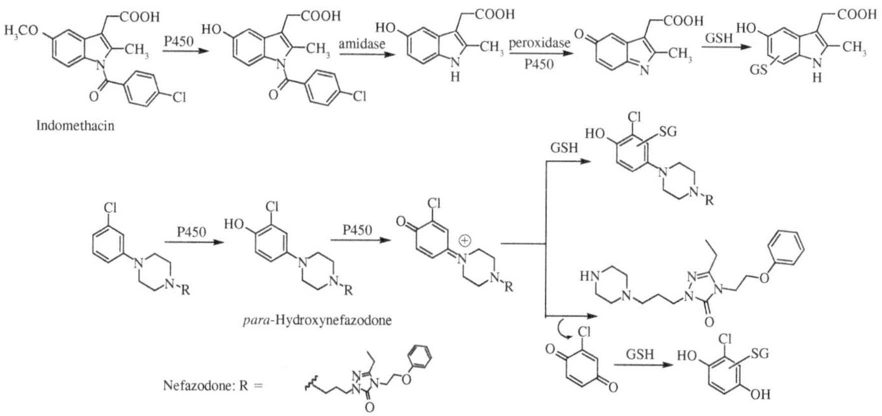

SCHEME 7 Quinone-imine formation with indomethacin and nefazodone.

ring system as is evident with 5-hydroxyindole and *para*-hydroxyphenylpiperazine derivatives. The former situation is illustrated with the nonsteroidal anti-inflammatory drug (NSAID) and idiosyncratic hepatotoxin indomethacin. Indomethacin bioactivation is a multistep process that involves: (1) CYP-catalyzed *O*-demethylation, (2) hydrolysis of the *N*-acylindole motif, and (3) CYP- or peroxidase-mediated two-electron oxidation of the 5-hydroxyindole metabolite to the quinone-imine, which can be trapped with GSH (Scheme 7) (Ju and Uetrecht, 1998). An example of quinone-imine formation with a cyclic tertiary aniline-containing molecule is evident with the antidepressant and hepatotoxin nefazodone. The bioactivation is initiated by a CYP-catalyzed aromatic hydroxylation *para* to the piperazine nitrogen to generate *para*-hydroxynefazodone as a metabolite, two-electron oxidation, which generates the quinone-imine (Scheme 7) (Kalgutkar et al., 2005). In the case of nefazodone, the quinone-imine intermediate has been shown to undergo a hydrolytic

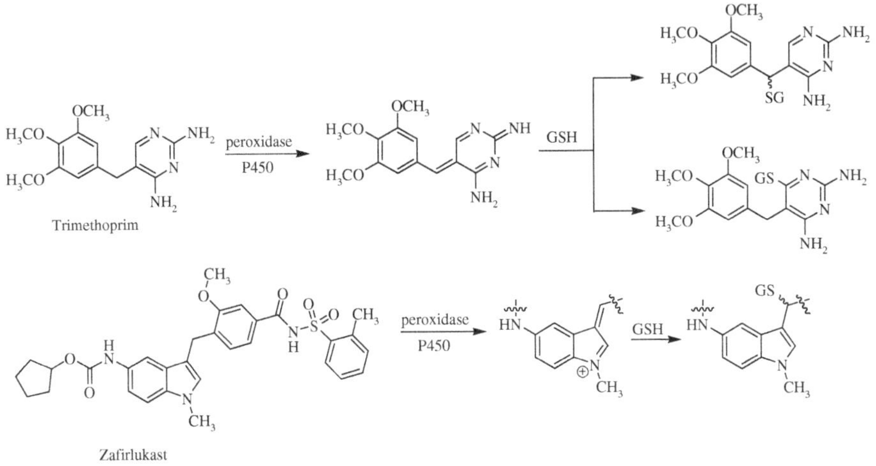

SCHEME 8 Bioactivation pathways leading to imine-methide formation with zafirlukast and trimethoprim.

cleavage liberating the corresponding *meta*-chloro-*para*-benzoquinone derivative and the *N*-dephenylated metabolite (Scheme 7) (Kalgutkar et al., 2005).

Imine-methides are analogous to quinone-methides in that the phenolic OH group is replaced by an NH group. The underlying chemistry that accompanies two-electron π oxidation of *ortho*- and *para*-alkylanilines to imine-methides is identical to that discerned with *ortho*- and *para*-alkylphenols, which yields quinone-methides. A classic example of a xenobiotic that undergoes imine-methide formation is evident with the pneumotoxin 3-methylindole. The pulmonary toxicity of 3-methylindole has been attributed to the action of the electrophilic imine-methide species 3-methyleneindolenine obtained by CYP-catalyzed dehydrogenation of the parent compound (Nocerini et al., 1985; Yan et al., 2007). Adducts of the imine-methide have been observed with thiol nucleophiles and DNA bases in microsomal incubations of 3-methylindole and exogenously added trapping agents (Nocerini et al., 1985; Yan et al., 2007), nucleosides, and DNA (Regal et al., 2001). The antiasthmatic drug zafirlukast is another example of a 3-methyleneindole derivative that is metabolized by CYP and peroxidases to an electrophilic imine-methide species, which can be trapped with GSH (Scheme 8) (Kassahun et al., 2005). An additional example of imine-methide formation is illustrated with the antibacterial agent trimethoprim. Unlike most imine-methides in which nucleophilic attack by GSH occurs almost exclusively on the exocyclic methylene carbon, in the case of trimethoprim, GSH preferentially adds to the pyrimidine ring (Scheme 8) (Lai et al., 1999).

Diiminoquinones are analogous to quinones in that both phenolic OH groups are replaced by NH substituents. A classic example of a drug, which forms a diiminoquinone metabolite is the antipsychotic agent clozapine wherein the *ortho-bis*-aniline framework undergoes a two-electron oxidation

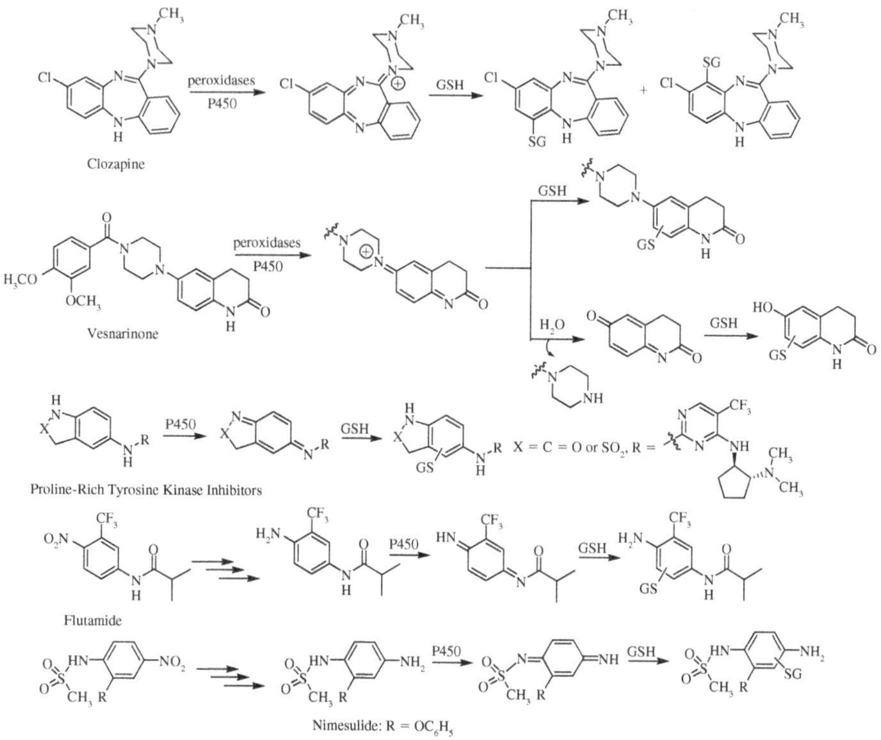

SCHEME 9 Bioactivation pathways leading to diiminoquinones.

process by CYP and/or peroxidases to form the reactive intermediate that can be conveniently trapped with GSH (Scheme 9) (Uetrecht, 1992; Liu and Uetrecht, 1995; Gardner et al., 1998). *Ortho*- and *para*-acetamidoanilines can also participate in this chemistry as discerned with the cardiotonic agent vesnarinone (Uetrecht et al., 1994) and proline-rich tyrosine kinase (PYK 2) inhibitors (Walker et al., 2008) (Scheme 9). It is interesting to note that the diiminoquinone derivative in vesnarinone has been shown to undergo hydrolytic cleavage in a manner similar to the pathway elucidated with nefazodone (Scheme 9). Nitrobenzenes can also participate in the bioactivation pathway after reduction to the corresponding aniline metabolites as highlighted with flutamide (Wen et al., 2008) and nimesulide (Li et al., 2009) (Scheme 9).

3.4.2 *N*-Hydroxylation of Anilines

Besides participating in quinone-imine formation, primary anilines can also undergo an alternate CYP- or peroxidase-catalyzed bioactivation pathway that involves an initial *N*-hydroxylation on the aniline nitrogen to afford the *N*-hydroxylamine metabolite followed by further oxidation to a reactive nitroso species, which, in some cases, can be trapped with GSH (Scheme 10)

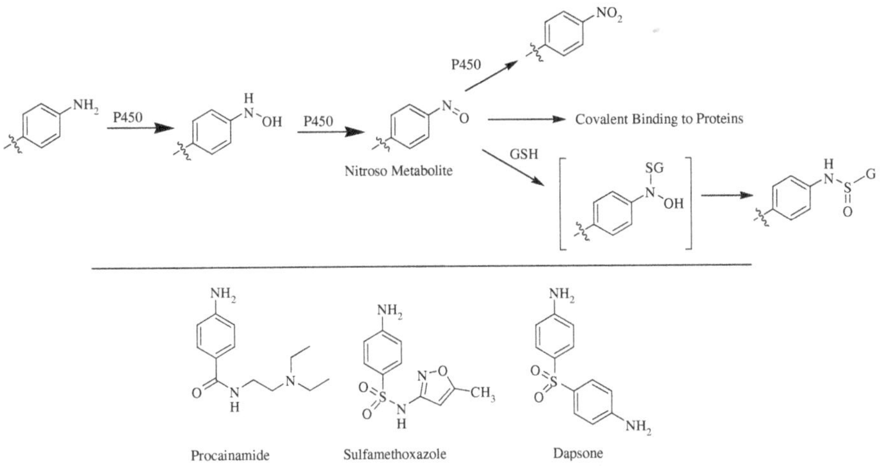

SCHEME 10 *N*-Hydroxylation of anilines to reactive nitroso species.

(Uetrecht, 1985). Procainamide, sulfamethoxazole, and dapsone are noteworthy examples of aniline-containing drugs that exhibit this bioactivation pathway (Israili et al., 1973; Rieder et al., 1988). Masked anilines (e.g., anilides) can also generate electrophilic nitroso intermediates provided they first undergo enzymatic hydrolysis to form the primary aniline metabolite as observed with the β-adrenoceptor antagonist practolol (see Scheme 18) (Amos et al., 1978; Orton and Lowery, 1981).

3.4.3 Hydrazines

Drugs that contain a pendant hydrazine and/or hydrazide motif are notorious as CYP, peroxidase, and monoamine oxidase inactivators. Investigations on the mechanisms of toxicity associated with drugs that contain the hydrazine motif also implicate bioactivation of this functionality as a potential cause for tissue injury. For example, the relationship between hepatotoxicity and metabolism of the antituberculosis drug, isoniazid and its metabolites, *N*-acetylisoniazid and *N*-acetylhydrazine, in rats has been investigated and toxic doses of the radiolabeled metabolites were shown to bind covalently to liver protein in vivo. Pretreatment of the rats with the acylamidase/esterase inhibitor, *bis*-para-nitrophenyl phosphate, prevented the hydrolysis of *N*-acetyisoniazid to *N*-acetylhydrazine and isonicotinic acid and concomitantly decreased covalent binding (Timbrell et al., 1980). These results suggest that *N*-acetylhydrazine is the ultimate hepatotoxin, whose bioactivation via the intermediate *N*-acetyldiazine leads to the formation of acetylating species that reacts covalently with hepatic tissue (Scheme 11). Electron spin resonance studies have confirmed the formation of the acetyl free radical (see Scheme 11) in incubations of isoniazid in perfused rat livers.

A second example involves the antihypertensive drug hydralazine. Like other monosubstituted hydrazines, hydralazine is polymorphically acetylated

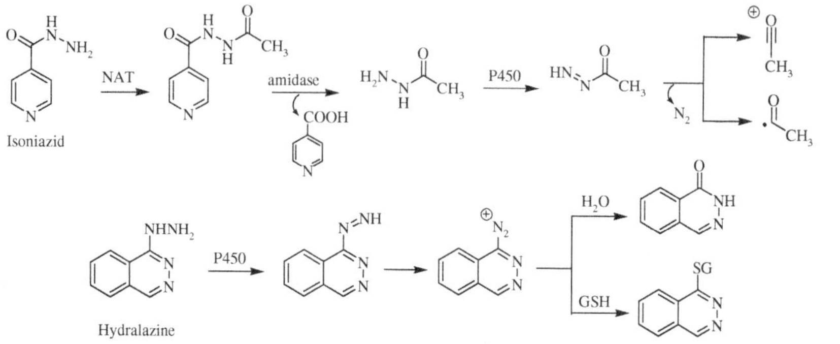

SCHEME 11 Bioactivation of the hydrazine/hydrazide motif in drugs to reactive metabolites.

by *N*-acetyltransferase (NAT) in humans, and the manifestations of toxicity (lupus syndrome) occurs almost exclusively in slow acetylators who produce lesser amounts of acetylated metabolites and larger amounts of oxidized metabolites such as phthalazinone (Scheme 11). Evidence for the role of CYP in hydralazine bioactivation was obtained following oxidation of [^{14}C]-hydralazine in rat liver microsomes in the presence of NADPH, which led to reactive species (diazine and diazonium ion) that covalently bound to microsomal protein (Streeter and Timbrell, 1985). A competing detoxification pathway involves the spontaneous reaction of water or GSH with the diazonium intermediate leading to phthalazinone or the mercapturic acid conjugate, respectively.

3.4.4 Bioactivation of Reduced Thiols

Free thiol groups react with cysteinyl-disulfide residues in proteins to form a new disulfide link with the concomitant release of a free cysteine residue within the protein (Scheme 12, pathway A). No metabolic activation is required for the covalent binding process to occur. Thiols are also readily oxidized to sulfenic acids, which react with cysteinyl residues and reduced GSH to form mixed disulfides (Scheme 12, pathway B) (Migdalof et al., 1984; Coleman et al., 1988). Methimazole, penicillamine, and captopril are examples of free thiol-containing drugs wherein evidence has been presented that traces toxicity of these compounds with oxidation of the thiol group.

3.4.5 Epoxidation of sp^2 and sp Centers

The isolation of a stable arene oxide metabolite of naphthalene in liver microsomal incubations provided evidence for the CYP- catalyzed epoxidation on the sp^2 carbons on a simple phenyl ring (Scheme 13). Carcinogenesis following exposure to polycyclic aromatic hydrocarbons (PAHs) such as benzo[*a*]pyrene is thought to arise from arene oxide formation. Arene oxides

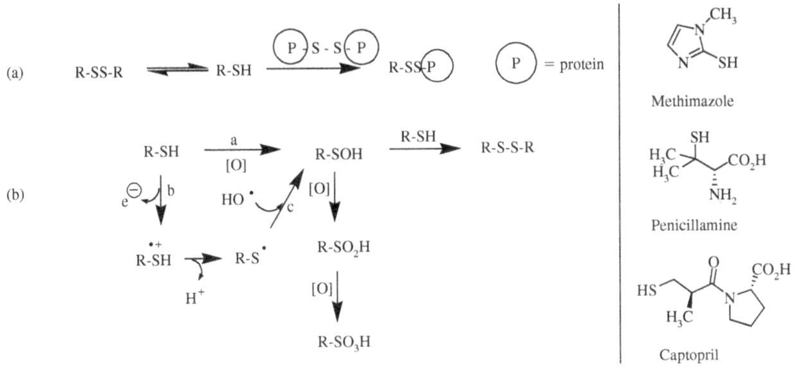

SCHEME 12 Oxidation of thiols to disulfides.

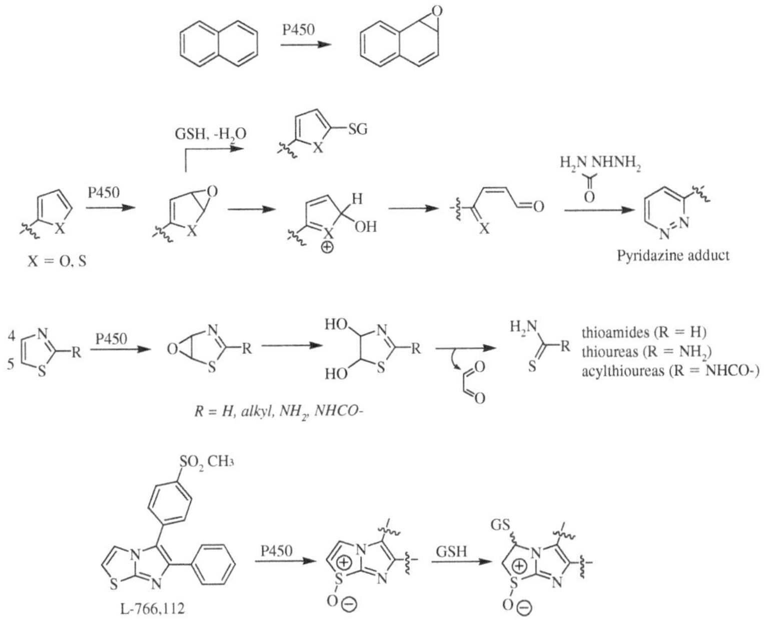

SCHEME 13 CYP-catalyzed epoxidation of aromatic and heteroaromatic rings.

are usually unstable and undergo ring opening by mechanism of general acid catalysis, ultimately leading to phenols. In many instances, hydrolysis of the arene oxide by epoxide hydrolase or reaction with GSH results in the formation of dihydrodiol metabolites or dihydrohydroxy–GSH conjugates. Dehydration of the dihydroxy–GSH conjugate provides a stable rearomatized GSH adduct. Carbamazepine and alpidem represent noteworthy examples of drugs where a circumstantial link between toxicity and arene oxide formation exists based upon characterization of stable sulfydryl conjugates both in vitro and in vivo in humans.

Heterocyclic rings (e.g., furans, thiophenes, and thiazoles) can also undergo epoxidation to yield electrophilic species. The mechanism involves an initial CYP-catalyzed 2,3- or 4,5-epoxidation on the furan/thiophene and thiazole rings, respectively, followed by ring opening and leads to the formation of electrophilic α,β-unsaturated dicarbonyl metabolites (Scheme 13) (Zhang et al., 1996; Dansette et al., 2005; Peterson, 2006). Evidence for furan ring epoxidation has been obtained from metabolism studies on *R*-(+)-pulegone and *R*-(+)-menthofuran by human liver CYPs using $^{18}O2$ and $H2^{18}O$ and C-2 deuterium-labeled menthofuran (Khojasteh-Bakht et al., 1999). The electrophilic ,-unsaturated dicarbonyl metabolites react with soft nucleophiles via a 1,4-Michael addition across the α,β-unsaturated dicarbonyl moiety or with hard nucleophiles via 1,2-addition to the aldehyde (Scheme 13) (Zhang et al., 1996; Khojasteh-Bakht et al., 1999; Erve et al., 2007).

Evidence for thiophene epoxidation has been derived from stable isotope studies on the diuretic tienilic acid and the NSAID suprofen (Scheme 13) (Koenigs et al., 1999; Dalvie et al., 2002; O'Donnell et al., 2003; Dansette et al., 2005). Incorporation of ^{18}O in the 5-hydroxythiophene metabolite of these compounds in microsomal incubations conducted with $^{18}O_2$ gas suggests that the source of oxygen in the 5-hydroxythiophene metabolite is exclusively derived from molecular oxygen and not from water. This is an important finding considering that previous studies have suggested thiophene bioactivation occurs exclusively via *S*-oxygenation to a Michael acceptor (Valadon et al., 1996). Additional evidence for epoxidation includes the trapping of the pyridazine derivative in microsomal incubations of suprofen (O'Donnell et al., 2003). The formation of the pyridazine conjugate of suprofen can be rationalized via the epoxidation and ring opening sequence to the γ-thioketo-α,β-unsaturated aldehyde intermediate in a manner analogous to that discerned with furans.

As observed with furans and thiophenes, the C4-C5 epoxide in thiazole analogs can be hydrolyzed by epoxide hydrolase to the corresponding dihydrodiol intermediate. Ring scission of the dihydrodiol then occurs, resulting in the liberation of glyoxal and the corresponding thioamides, thiourea, and/or acylthiourea as metabolites depending on the C-2 substituent (Mizutani et al., 1994, 1996; Dalvie et al., 2002). Once formed, thiourea and acylthiourea metabolites can undergo *S*-oxidation to electrophilic sulfenic and/or sulfinic acid intermediates that can covalently modify or oxidize critical proteins leading to toxicity (Scheme 13) (Obach et al., 2008a). For many thiourea derivatives that covalently bind to microsomes, addition of GSH abolishes covalent binding via reduction of the sulfenic acid metabolites and the concomitant formation of oxidized GSH (i.e., GSSG). Thus, GSH can be involved in the detoxication of thioureas as a cellular reductant for their reactive intermediates. It is noteworthy to point out that thiazole ring oxidation can also result in the formation of an electrophilic *S*-oxide, which can covalently bind to proteins or exogenously added thiol nucleophiles as demonstrated with L-766,112, a potent and selective cyclooxygenase-2 inhibitor (Scheme 13) (Trimble et al., 1997). Oxidative thiazole ring opening to reactive intermediates has also been proposed as the rate-limiting step in the mechanism-based inactivation of CYP3A4 by the human immunodeficiency virus (HIV) protease inhibitor ritonavir. This proposal is consistent with the observations that the HIV protease inhibitors indinavir and saquinavir, which do not possess the thiazole ring, are devoid of CYP3A4 inactivation properties (Kalgutkar et al., 2007b).

Olefins can also undergo CYP-catalyzed epoxidation; the NSAID alclofenac, which has been withdrawn from the clinic due to many cases of skin rash and hepatotoxicity, serves as a noteworthy example of this phenomenon. The epoxide metabolite of alclofenac (Scheme 14) has been detected in human urine as a stable metabolite (Slack and Ford-Hutchinson, 1980). The synthetic standard of the epoxide metabolite is a mutagen in the *Salmonella* Ames test and forms conjugates with sulfydryl nucleophiles (Slack et al., 1981; Mercier et al., 1983). The observation that the epoxide metabolite exhibits a NADPH-independent but

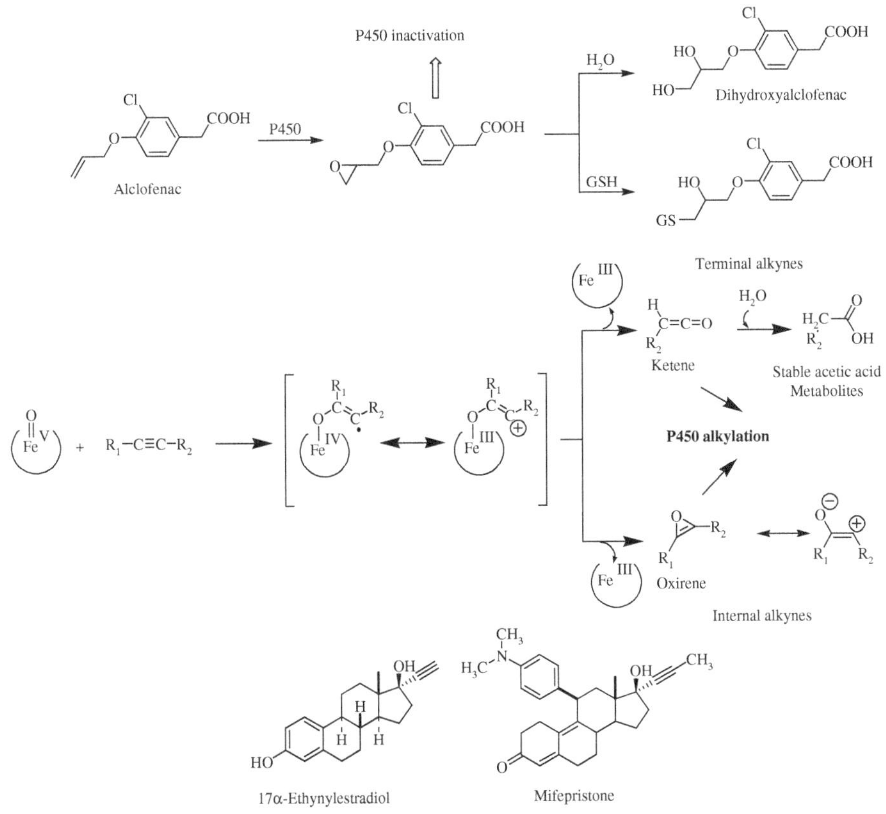

SCHEME 14 CYP-catalyzed epoxidation of olefins and alkynes.

time-dependent loss in CYP activity suggests that this intermediate covalently modifies CYP directly and does not require further metabolic processing to reactive species consistent with the characteristics of an affinity-labeling agent (Brown and Ford-Hutchinson, 1982). The sp-hybridized carbon atoms on both terminal and internal alkynes are susceptible to epoxidation by CYP enzymes, which leads to the formation of reactive metabolites capable of alkylating CYP isozymes (Scheme 14) (Kalgutkar et al., 2007b). With terminal alkynes, the bioactivation mechanism involves oxygenation on the terminal carbon followed by a 1,2-hydrogen shift of the terminal hydrogen to the vicinal carbon to form a reactive ketene intermediate. The ketene can acylate nucleophilic residues within the CYP active site resulting in enzyme inactivation or undergo hydrolysis to the acetic acid metabolite detectable as the stable downstream product of terminal alkyne metabolism. With internal alkynes, oxidation on the internal carbon is followed by rearrangement to the oxirene, which then leads to enzyme alkylation. Synthetic estrogens 17α-ethynylestradiol (a terminal alkyne) and mifepristone (an internal alkyne) are examples of drugs that are epoxidized on the alkyne by CYP3A4 to afford reactive metabolites, which covalently adduct to the CYP isozyme (see Scheme 14).

3.4.6 Thiazolidinedione Ring Bioactivation

Enzymatic conversion of the thiazolidinedione ring system to reactive metabolites was first documented with the antidiabetic agent and idiosyncratic hepatotoxin troglitazone (Kassahun et al., 2001). The mechanism of thiazolidinone ring opening (Scheme 15) involves CYP-mediated *S*-oxygenation to yield a sulfoxide intermediate, which spontaneously ring opens to the highly electrophilic α-ketoisocyanate derivative. Isocyanate hydrolysis followed by decarboxylation to the amide, accompanied by attack of GSH on the reactive sulfenic acid, would afford one GSH conjugate. Alternatively, conjugation of the isocyanate with GSH, followed by oxidation to the corresponding sulfinic acid derivative, provides a route for a second GSH conjugate. Dehydration of the sulfinic acid intermediate is thought to lead to the formation of a third

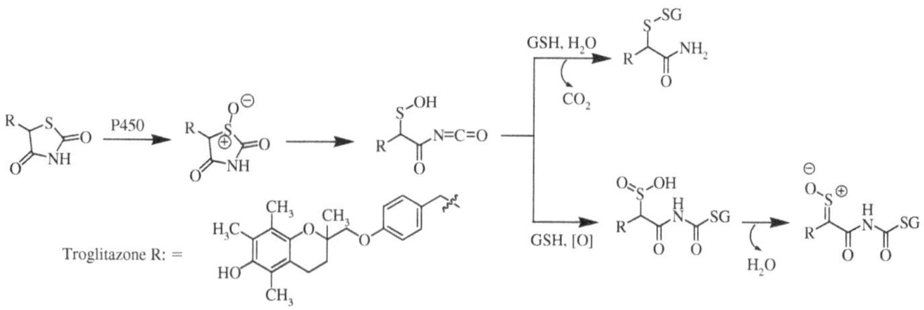

SCHEME 15 CYP-catalyzed thiazolidinedione ring scission.

sulfine-containing GSH conjugate. In separate studies, comparison of the hepatotoxic and nephrotoxic potential of imidazolinedione, oxazolidinedione, and thiazolidinedione analogs of the nephrotoxic fungicide, *N*-(3,5-dichlorophenyl)succinimide in rats after in vivo administration indicated that only *N*-(3,5-dichlorophenyl)thiazolidinedione caused hepatic damage (Kennedy et al., 2003). It is conceivable that bioactivation of the thiazolidinedione ring to electrophilic intermediates is responsible for the observed hepatotoxicity in rats.

3.4.7 α,β-Unsaturated Carbonyl Compounds

α,β-Unsaturated carbonyl compounds are usually avoided in medicinal chemistry because of their intrinsic electrophilic nature. However, there are several instances where in the process of metabolism latent drugs are converted into α,β-unsaturated carbonyl intermediate(s). In most cases these reactive intermediates are believed to be responsible for the toxicological consequences associated with the parent drug. Noteworthy examples include cyclophosphamide, terbinafine, and felbamate. With cyclophosphamide, β-elimination reaction following an initial CYP-mediated hydroxylation affords the corresponding α,β-unsaturated acrolein metabolite (Scheme 16) (Ramu et al., 1995). The characterization of a hydroxypropyl mercapturic acid conjugate (see Scheme 16) in human urine is consistent with the notion that this bioactivation

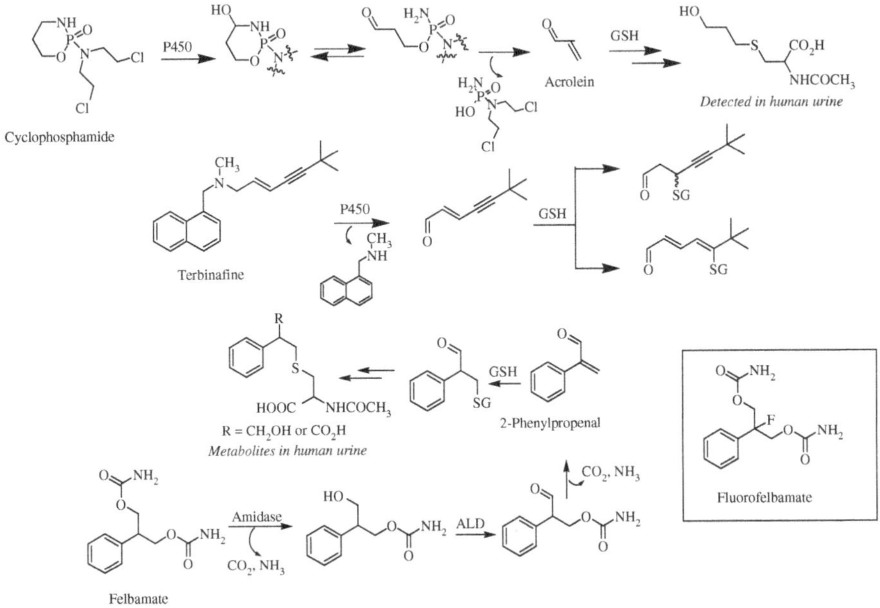

SCHEME 16 Generation of reactive, α,β-unsaturated carbonyl metabolites in the course of drug metabolism.

pathway operates in vivo. In the case of the antifungal agent and idiosyncratic hepatotoxin terbinafine, a simple CYP-catalyzed *N*-dealkylation reaction results in the formation of the α,β,γ,δ-unsaturated aldehyde metabolite, which can be trapped in a 1,4- and 1,6-Michael fashion with GSH to yield the corresponding sulfydryl adducts (Scheme 16) (Iverson and Uetrecht, 2001). Circumstantial evidence linking felbamate bioactivation to its toxicity (aplastic anemia and hepatotoxicity) has been presented by means of the in vivo characterization of mercapturic acid (downstream metabolites of GSH) conjugates of a highly reactive α,β-unsaturated carbonyl derivative 2-phenylpropenal following felbamate administration to rats and human (Scheme 16) (Dieckhaus et al., 2002). The mechanism is believed to involve esterase-mediated hydrolysis of one of the carbamate groups to afford the primary alcohol metabolite; oxidation of which by alcohol dehydrogenase to the intermediate aldehyde derivative followed by spontaneous β-elimination of the remaining carbamoyl group generates 2-phenylpropenal. Consistent with this finding, fluorofelbamate (see Scheme 16) does not succumb to bioactivation because the fluorine atom prevents the β-elimination process, which leads to the reactive 2-phenylpropenal (Parker et al., 2005).

3.4.8 Haloalkanes

Alkyl halides are rarely utilized as pharmacophores in drug design because of the susceptibility of halogens toward enzymatic and nonenzymatic nucleophilic displacement reactions as illustrated with chloramphenicol and inhaled anesthetics (e.g., halothane, isofluorane, and desfluorane). In the case of chloramphenicol, CYP inactivation results from the formation of the reactive acylating agent oxamyl chloride during oxidative dechlorination by CYP (Scheme 17). The characterization of oxalic and oxamic acid metabolites of chloramphenicol analogs upon alkaline hydrolysis of inactivated protein samples and the finding that the difluoromethyl derivative of chloramphenicol does not inactivate CYP

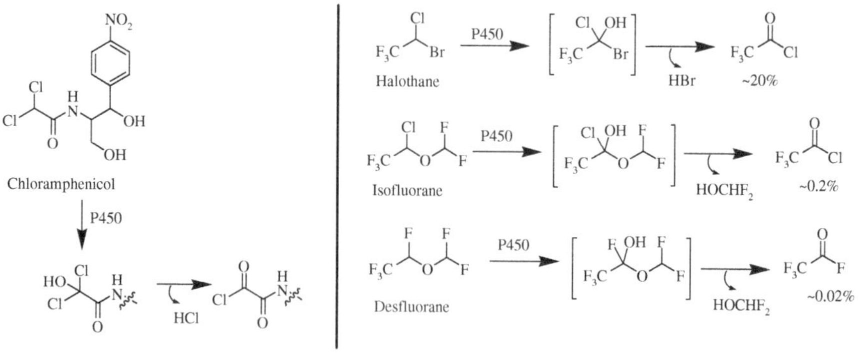

SCHEME 17 Bioactivation of haloalkanes to reactive metabolites.

lends further credence to the proposed pathway of inactivation (Miller and Halpert, 1986). Bioactivation of haloalkane substituents in inhaled anesthetics to reactive acylating agents is usually due to the availability of an extractable hydrogen atom on the halogenated alkyl carbon. The relative incidence of idiosyncratic hepatotoxicity due to these agents appears to directly correlate with the extent of their conversion to acyl halides by CYP, which in turn may be governed by the leaving group ability of the respective substituents within these drugs (Njoko et al., 1997). As seen in Scheme 17, halothane, which exhibits the greatest incidence of hepatotoxicity in the clinic, undergoes the most conversion to reactive acyl chloride, a feature that can be attributed to the presence of bromide substituent, which is a good leaving group. In contrast, isofluorane and desfluorane also undergo oxidative metabolism resulting in the formation of reactive acyl halides, but the degree to which these anesthetics are bioactivated is significantly lower than halothane due to the relatively poor leaving group ability of the difluoromethoxy group compared to the bromide.

3.4.9 Carboxylic Acids

UGT-catalyzed glucuronidation of the carboxylic acid group in drugs results in the formation of acyl glucuronides, which are intrinsically electrophilic in nature. Protein modification can occur via a simple transacylation reaction with a protein nucleophile(s) or by acyl migration within the β-*O*-glucuronide unit to a reactive aldehyde intermediate. Detailed mechanistic discussion on this issue is provided with the NSAIDs ibufenac and ibuprofen in the following section.

3.5 STRUCTURAL ALERTS AND DRUG DESIGN

Anecdotal evidence obtained from visual analysis of structures of several closely related toxic and nontoxic drugs suggests that drugs, which lack toxicophores, have a superior safety record, especially with regards to IADRs (Kalgutkar and Soglia, 2005). The evidence becomes even more compelling when metabolism data supports the hypothesis as illustrated with the cardio-selective β-adrenoceptor antagonists practolol, atenolol, and metoprolol. The mechanism of severe skin rashes induced by practolol is uncertain; however, a role for antinuclear antibodies, elicited by protein adducts of a reactive nitroso metabolite obtained from practolol biotransformation has been suspected (Scheme 18) (Amos et al., 1978; Orton and Lowery, 1981) in keeping with the observation that *cutaneous IADRs are not observed with atenolol and metoprolol, which lack the anilide toxicophore*. Consistent with this hypothesis are the findings that atenolol and metoprolol are metabolized by completely different pathways and are also subject to extensive urinary excretion as parent drugs (Borchard, 1990).

A second example is provided with the dibenzodiazepine derivatives and antipsychotic agents clozapine and quetiapine (Seroquel). While clozapine use

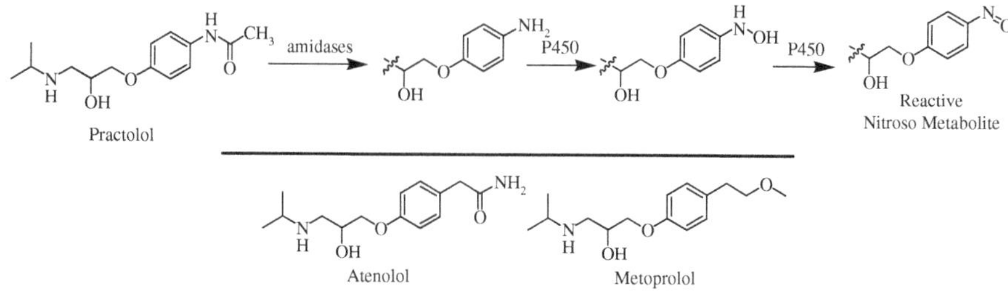

Practolol → amidases → P450 → P450 → Reactive Nitroso Metabolite

Atenolol Metoprolol

SCHEME 18 Structure–toxicity relationships with β-adrenoceptor antagonists practolol, atenolol, and metoprolol.

is limited by a high incidence of agranulocytosis and hepatotoxicty, quetiapine does not cause these toxic events. As demonstrated earlier (see Scheme 9), evidence has been presented that links clozapine toxicity to its propensity to form a reactive imine (Liu and Uetrecht, 1995; Uetrecht, 1992). Proteins covalently modified with clozapine were also observed in neutrophils of patients being treated with clozapine, which reaffirms the relevance of the in vitro studies (Gardner et al., 1998). In the case of quetiapine the bridging nitrogen in the benzodiazepine ring is replaced with a sulfur atom; consequently this drug is not bioactivated to the reactive iminium species as shown with close-in structural clozapine analogs (Uetrecht et al., 1997). Despite administration at doses comparable to clozapine, cases of agranulocytosis with quetiapine are extremely rare.

A third example becomes evident upon comparison of trovafloxacin with related fluoroquinolone antibiotics. The rare but serious idiosyncratic hepatotoxicity (~14 cases of acute liver failure; 4 patients required liver transplants, and an additional 5 died of liver-related injuries) led to the withdrawal of trovafloxacin from the market in many countries and a black-box warning with intensive monitoring requirements in the United States. Microarray analysis (Liguori et al., 2005) revealed substantial gene expression changes following treatment of human hepatocytes with trovafloxacin as compared to other marketed fluoroquinolones. The expression profile induced by trovafloxacin was markedly distinct from other fluoroquinolones in that genes involved in oxidative stress were regulated consistently by trovafloxacin. In HepG2 cells, trovafloxacin also induced oxidative stress and depleted intracellular glutathione levels to a greater extent than other fluoroquinolones (Liguori et al., 2008). A potential role of inflammatory mediators in trovafloxacin hepatotoxicity has also been established (Waring et al., 2006).

With the exception of trovafloxacin, none of the other drugs in the fluoroquinolone class of antibiotics have been associated with idiosyncratic hepatotoxicity. From a structure–toxicity standpoint, it is interesting to note the presence of the cyclopropylamine structural alert at the C7 position of the fluoroquinolone scaffold in trovafloxacin. Studies with a model compound have revealed cyclopropylamine ring bioactivation by CYP and myeloperoxidase to a reactive α,β-unsaturated aldehyde trapped as a sulfydryl conjugate

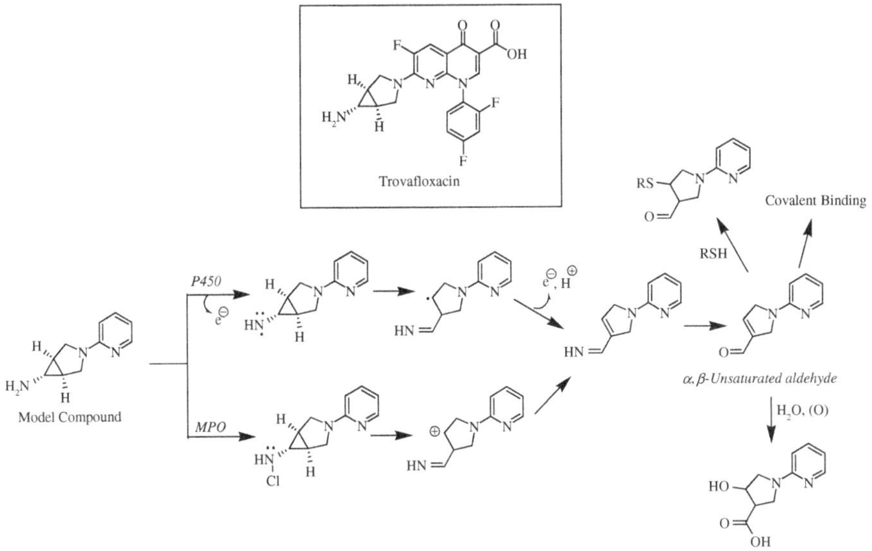

SCHEME 19 Insights into trovafloxacin bioactivation via the use of a model compound containing the cyclopropylamine functionality.

(Scheme 19) (Sun et al., 2008). The characterization of a hydroxycarboxylic acid metabolite of trovafloxacin in preclinical species (Dalvie et al., 1996) lends further support for the metabolism of the cyclopropylamine ring in trovafloxacin to a reactive intermediate. The formation of the hydroxycarboxylic acid can occur from the addition of water to the α,β-unsaturated aldehyde via Michael addition followed by oxidation as depicted for the model compound (Scheme 19). However, the proposal for reactive metabolite formation with trovafloxacin remains a speculation since the bioactivation studies did not involve the parent fluoroquinolone and no α,β-unsaturated aldehyde or the corresponding glutathione conjugate has been detected in trovafloxacin incubations in human liver microsomes (Sun et al., 2008). Furthermore, the primary pathways of trovafloxacin clearance in humans include phase II metabolism (*N*-acetylation, acyl glucuronidation, and *N*-sulfation) (Scheme 19) with very minor contributions from phase I oxidative pathways (Dalvie et al., 1997).

A final and perhaps an even more intriguing example of the influence that a subtle structural change can have on toxicity is highlighted with ibuprofen and ibufenac. While ibuprofen is considered to be one of the safest over-the-counter NSAID on the market, its close-in analog ibufenac was withdrawn due to severe hepatotoxicity. The daily doses of both NSAIDs were comparable and the only structural difference between the two drugs is the presence of the additional α-methyl substituent adjacent to the carboxylic acid moiety in ibuprofen (Scheme 20). Glucuronidation of the carboxylic acid moiety in most NSAIDs to the potentially electrophilic acyl glucuronide constitutes the

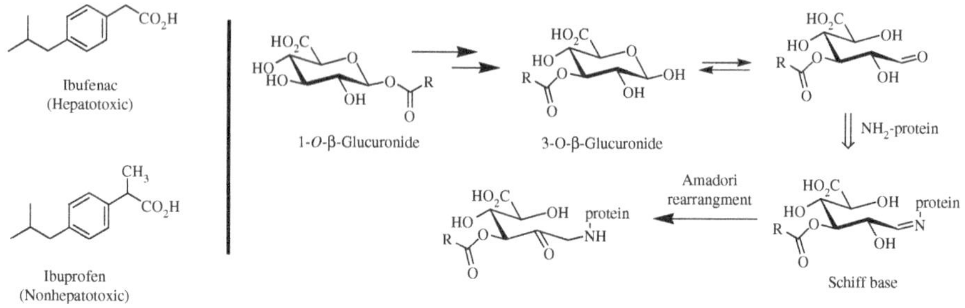

Ibufenac (Hepatotoxic)

Ibuprofen (Nonhepatotoxic)

1-*O*-β-Glucuronide

3-O-β-Glucuronide

NH_2-protein

Amadori rearrangment

Schiff base

SCHEME 20 Reaction of acylglucuronides with proteins.

principal elimination mechanism in vivo in humans (Johnson et al., 2007; Ding et al., 1993; Benet et al., 1993). As mentioned earlier, the proposed pathway of acyl glucuronide adduction with proteins involves condensation between the aldehyde group of a rearranged acyl glucuronide and a lysine residue or an amine group of the *N*-terminus, leading to the formation of a glycated protein. The formation of the iminium species is reversible but may be followed by an Amadori rearrangement of the imino sugar to the more stable 1-amino-2-keto product (see Scheme 20) (Walker et al., 2007; Wang et al., 2004; Ding et al., 1993). A structural relationship between acyl glucuronide degradation to the Schiff base and covalent binding has been established (Walker et al., 2007; Wang et al., 2004). Acyl glucuronides of ibufenac and other acetic-acid-based NSAIDs such as tolmetin and zomepirac, all of which have been withdrawn due to toxicity, exhibit the highest level of rearrangement and covalent binding, whereas mono-α-substituted acetic acids (2-substituted propionic acids) such as ibuprofen exhibit intermediate level of acyl glucuronide rearrangement and covalent binding. Overall, these observations imply that inherent electronic and steric properties must modulate the rate of acyl glucuronide rearrangement. Thus, in the case of ibuprofen, it is likely that the presence of the α-methyl substituent slows the rearrangement of the acyl glucuronide to the electrophilic carbonyl intermediate.

In retrospect, the examples discussed above imply that by avoiding toxicophores in drug design, one would lessen the odds that a drug candidate will lead to toxicity via a bioactivation mechanism. From a medicinal chemistry standpoint, this strategy seems to be an attractive option and a path forward toward the discovery of safer drugs, especially given the lack of methodology to predict IADRs. As noted earlier, an exhaustive listing of structural alerts is far too comprehensive and also includes a simple phenyl ring. Likewise, a strategy that avoids structural alerts altogether can lead to a missed opportunity to develop potentially important medicines. Atorvastatin (Lipitor) provides an example of such a scenario as the structure not only contains the acetanilide structural alert but metabolism by CYP3A4 results in the formation of acetaminophen-like metabolites (Scheme 21) (Jacobsen et al., 2000). Furthermore, glucuronidation of the carboxylic acid moiety results in the formation of

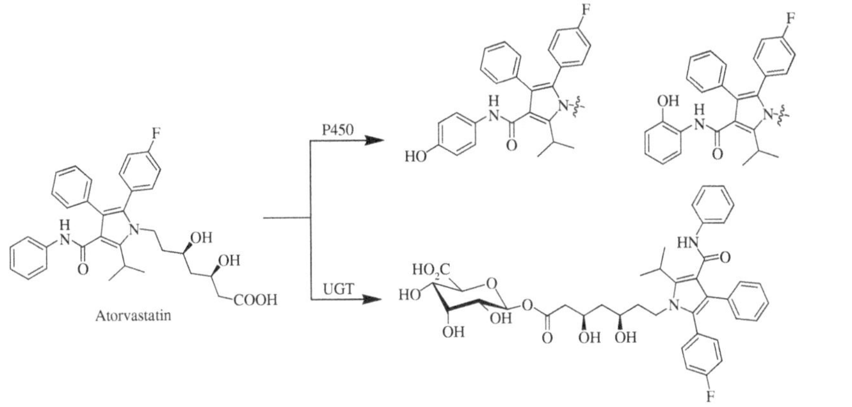

SCHEME 21 Chemical structures of atorvastatin and its metabolites derived from oxidative and conjugation pathways.

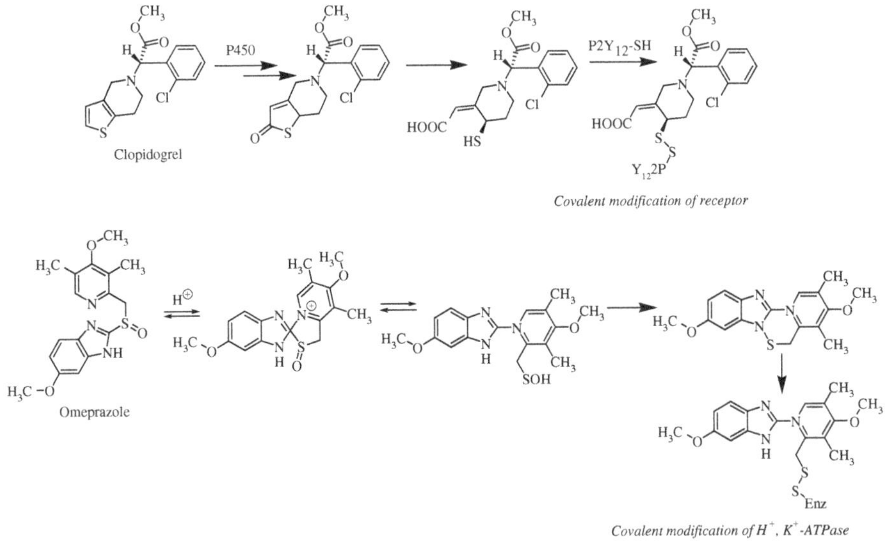

SCHEME 22 Examples of commercial blockbuster drugs that require reactive metabolite formation for their pharmacologic action.

the potentially electrophilic acyl glucuronide (Prueksaritanont et al., 2002) in a manner similar to that discerned with NSAIDs (see Scheme 21).

Finally, it is pivotal to point out that several blockbuster drugs contain toxicophores, which form reactive metabolites and covalently adduct to proteins, which in some cases is essential for pharmacological activity. For instance, the blockbuster cardiovascular drug and $P2Y_{12}$ antagonist clopidogrel (Plavix) by itself is inactive and requires P450-catalyzed bioactivation of its thiophene ring to form a reactive thiol metabolite, which forms a covalent disulfide bond with a cysteinyl residue on the $P2Y_{12}$ receptor in platelets (Scheme 22), a phenomenon that gives rise to its beneficial cardiovascular effects (Herbert and Savi, 2003; Savi et al., 2000, 2006). Likewise, the benzimidazole class of proton-pump inhibitors used to treat gastric disorders, exemplified by omeprazole by itself have no in vitro ability to inhibit the enzyme H^+, K^+-ATPase but are converted to a reactive sulfenamide intermediate in the acidic environment of the stomach. Covalent disulfide bond formation of this reactive species with an active site cytseine residue results in enzyme inactivation (Scheme 22) (Olbe et al., 2003). Irreversible enzyme inhibition on account of covalent binding (Fellenius et al., 1981) is one pharmacokinetic benefit that contributes to making omeprazole clinically superior to H_2-receptor antagonists initially used to treat gastric acid disorders.

3.6 REACTIVE METABOLITE TRAPPING AND COVALENT BINDING STUDIES AS PREDICTORS OF IDIOSYNCRATIC DRUG TOXICITY

The examples discussed above pose a significant challenge toward the reliability of structural alerts, reactive metabolite trapping, and covalent binding measurements as indicators of idiosyncratic drug toxicity. With regard to reactive metabolite formation potential with compounds possessing a structural alert, it is very important to consider mitigating factors such as reactive metabolite detoxication and alternate pathways of metabolism. This issue is highlighted with the benzodiazepine receptor ligands alpidem and zolpidem. While alpidem is hepatotoxic and has been withdrawn from the market, the commercial blockbuster zolpidem (Ambien) is devoid of the toxicity. A key structural difference in the two drugs is the replacement of the two chlorine atoms on the imidazopyridine nucleus in alpidem with two methyl groups in zolpidem. In alpidem, the imidazopyridine ring is bioactivated by CYP and leads to the formation of a reactive arene oxide that reacts with GSH to yield sulfydryl conjugates (Scheme 23), which have been detected in human excreta (Durand

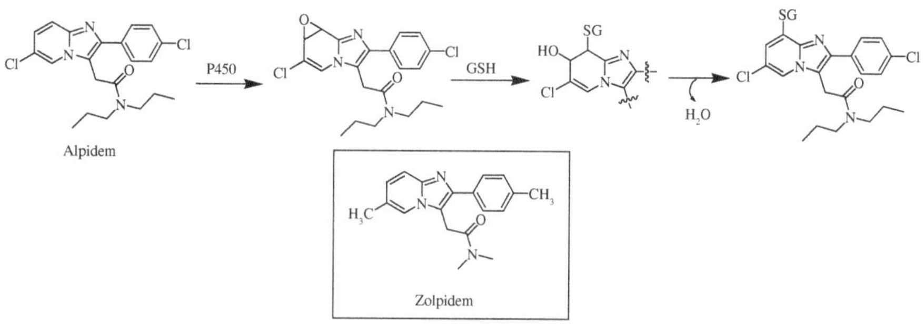

Alpidem — P450 → — GSH → — H_2O →

Zolpidem

SCHEME 23 Differential metabolism of the anxiolytic agents alpidem (hepatotoxin) and zolpidem (nonhepatotoxin).

et al., 1992). While bioactivation via epoxidation is also likely in zolpidem, the molecule does not undergo this metabolic fate; instead the two methyl groups function as metabolic soft spots and are oxidized to the corresponding alcohol and carboxylic acid metabolites (Scheme 23).

With regard to the importance of detoxication pathways, reactive metabolite formation may be discernible in standard in vitro systems; however, the principal clearance mechanism of the drug in vivo may involve a distinctly different and perhaps more facile metabolic fate that does not yield reactive intermediates as illustrated with the selective estrogen receptor modulator raloxifene. Raloxifene is known to undergo in vitro CYP3A4-catalyzed bioactivation on its phenolic groups to yield reactive quinonoid species (Scheme 24) (Chen et al., 2002); however, in vivo, glucuronidation of the same phenolic groups in the gut and liver constitute the principal elimination mechanism of raloxifene in humans (Scheme 24) (Kemp et al., 2002). Thus, the likelihood of raloxifene bioactivation in vivo is in question when compared with the phase II glucuronidation process, a phenomenon that may provide an explanation for the extremely rare occurrence of IADRs.

Although covalent binding data can provide a quantitative estimate of covalently bound radiolabeled drug to proteins and therefore an indirect measure of reactive metabolite formation, *there are no studies to date that have shown a correlation between amount of reactive metabolite formed and/or extent of covalent binding and the probability that a drug will be associated with toxicity* (Evans et al., 2004; Masubuchi et al., 2007). An example of this phenomenon is evident with the acetaminophen regioisomer, 3′-hydroxyacetanilide, which undergoes bioactivation to yield reactive metabolites that form covalently bound adducts to GSH and proteins (Rashed et al., 1989). However, despite dose normalization to provide comparable levels of covalent binding in vivo in mice, 3′-hydroxyacetanilide does not exhibit the hepatotoxicity observed with acetaminophen. To assess predictability of idiosyncratic drug toxicity, covalent binding measurements in human hepatic tissue has been examined for 18 drugs (9 hepatotoxins and 9 nonhepatotoxins) (Obach et al., 2008b; Bauman et al., 2009). This study also considers key factors such as

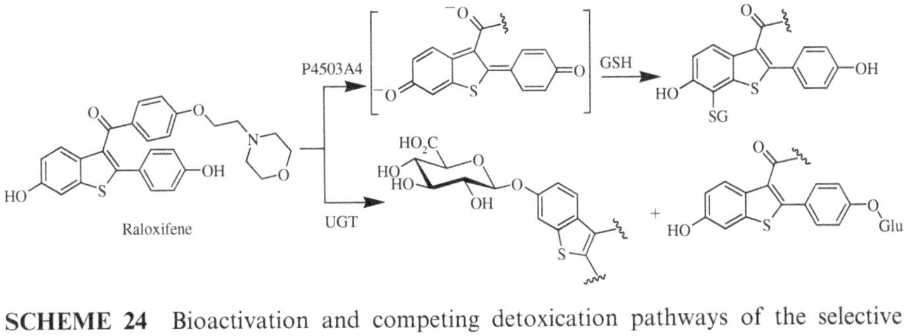

SCHEME 24 Bioactivation and competing detoxication pathways of the selective estrogen receptor modulator raloxifene.

reactive metabolite detoxication, relative importance of bioactivation that leads to covalent binding versus overall metabolism, and daily dose for each drug. While most of the hepatotoxic drugs (e.g., acetaminophen, nefazodone, tienilic acid, etc.) demonstrate covalent binding to some degree, it is interesting to note that several nonhepatotoxic and commercially successful drugs (e.g., buspirone, diphenhydramine, paroxetine, and simvastatin) also exhibit covalent binding. A quantitative comparison of covalent binding in vitro intrinsic clearance does not separate the two groups of compounds, and, in fact, paroxetine and diphenhydramine, both nonhepatotoxins, exhibit the greatest amount of covalent binding in microsomes. Including factors such as the fraction of total metabolism comprised by covalent binding and the total daily dose of each drug improves the discrimination between hepatotoxic and nonhepatotoxic drugs in liver microsomes, S-9, and hepatocytes; *however, the approach still would falsely identify some agents as potentially hepatotoxic.*

In the case of paroxetine, mechanistic studies further confirmed the importance of parallel metabolic and detoxication pathways in attenuating covalent binding to proteins (Zhao et al., 2007). As shown in Scheme 25, the catechol metabolite obtained via ring scission of the 1,3-benzdioxole group in paroxetine can partition between *O*-methylation by catechol-*O*-methyl transferase or undergo oxidation to the reactive quinone intermediate, which is efficiently detoxicated by GSH; both pathways lead to a significant reduction in covalent binding. In humans, the *O*-methylated catechol derivatives constitute the principal metabolic fate of the drug. When coupled with the fact that the daily dose of paroxetine is low (20 mg), some insight into the excellent safety record of this drug is obtained despite the bioactivation liability.

Finally, *a drug candidate that is devoid of reactive metabolite formation and/or covalent binding to proteins in "standard" systems is not a guarantee of its safety.* Despite the in vivo observations on felbamate bioactivation, evidence for the in

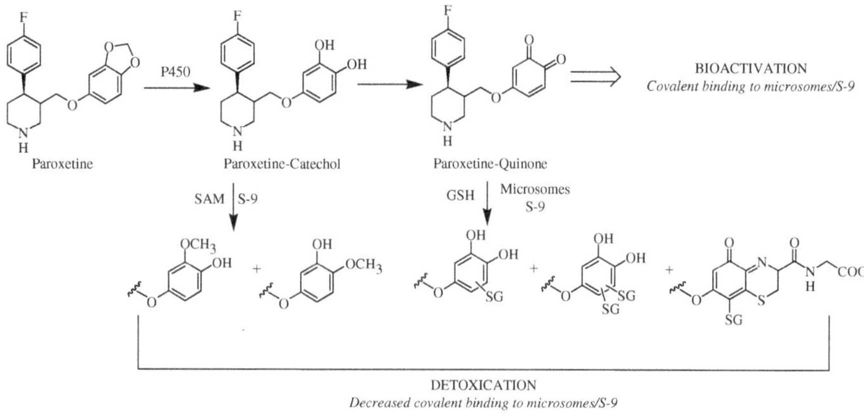

SCHEME 25 Parallel detoxication pathways that compete with the P450-catalyzed bioactivation pathway of the anti-depressant paroxetine as explanation for its wide safety margin.

vitro metabolism of felbamate to the reactive metabolite 2-phenylpropenal in human hepatic tissue is lacking. At therapeutically relevant concentrations of radiolabeled felbamate in in vitro incubations with human liver microsomes, S-9, and hepatocytes, no GSH adducts and/or covalent binding of felbamate has been discerned (Leone et al., 2007; Obach et al., 2008b; Bauman et al., 2009). While the reason(s) for this discrepancy remain unclear at the present time, *in a drug discovery paradigm relying solely on reactive metabolite trapping and liver microsomal covalent binding as means of predicting IADR potential of drug candidates, felbamate would have passed the hurdle with flying colors.*

3.7 DOSE AS AN IMPORTANT MITIGATING FACTOR FOR IADRs

A single most important factor in migrating IADR risks appears to be the daily dose of the drug. There are no examples of drugs that are dosed at < 20 mg/day that cause IADRs (whether or not these agents are prone to bioactivation). There are instances of two structurally related drugs that possess identical structural alerts susceptible to bioactivation; however, the drug administered at the lower dose is safer than the one given at a higher dose. It is likely that the improved safety of low-dose drugs arises from a marked reduction in the total body burden to reactive metabolite exposure, and therefore, unlikely to exceed the threshold needed for toxicity. For example, the dibenzodiazepine derivative olanzapine (Zyprexa) (Scheme 26) forms a reactive iminium metabolite very similar to the metabolite observed with clozapine, yet olanzapine is not associated with a significant incidence of agranulocytosis (Gardner et al., 1998). One difference between the two drugs is the daily dose; clozapine is given at a dose of >300 mg/day, while the maximum recommended daily dose of olanzapine is 10 mg/day. A second illustration of this concept is evident

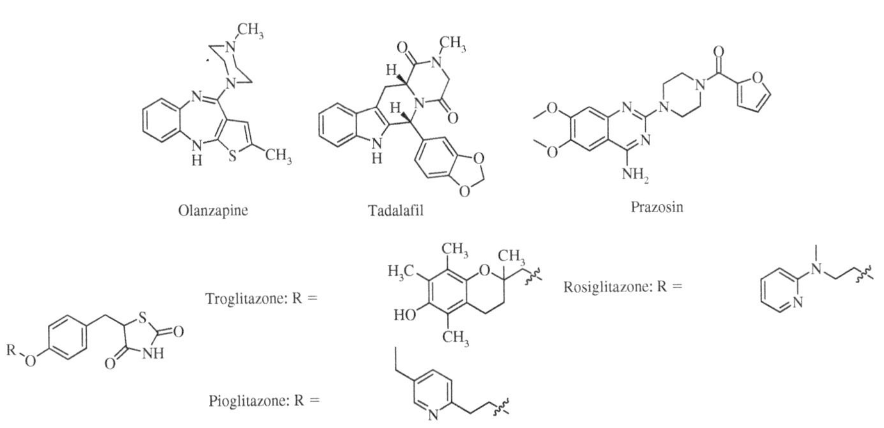

SCHEME 26 Examples of low daily dose drugs devoid of IADRs despite bioactivation liability.

upon comparison of the thiazolidinedione derivatives troglitazone, rosiglitazone, and pioglitazone. The idiosyncratic hepatotoxicity with troglitazone, which led to its withdrawal from the marketplace, has not been discerned with rosiglitazone and pioglitazone. As noted earlier, both the chromane and the thiazolidenedione ring systems in troglitazone succumb to CYP3A4-catalyzed bioactivation to reactive metabolites (Kassahun et al., 2001; He et al., 2004). While rosiglitazone and pioglitazone do not contain the chromane ring system found in troglitazone, they do contain the thiazolidinedione scaffold. Consistent with the findings with troglitazone, both rosiglitazone and pioglitazone have been shown to undergo thiazolidinedione ring scission mediated by P450 enzyme(s) in human microsomes resulting in reactive metabolites that are trapped by GSH (Baughman et al., 2005; Alvarez-Sanchez et al., 2006). While bioactivation of the thiazolidenedione ring is a common theme in these drugs, a key difference lies in their daily doses—troglitazone (200–400 mg/day) and rosiglitazone and pioglitazone (<10 mg/day).

Additional examples of this phenomenon are illustrated with tadalafil (Cialis) and the antihypertensive prazosin (Minipress) (see Scheme 26). The methylenedioxyphenyl group in tadalafil undergoes P4503A4-catalyzed bioactivation to an electrophilic catechol, a process that also leads to the suicide inactivation of P4503A4 activity in vitro (Ring et al., 2005). However, to date there are no reports of IADRs or P4503A4 drug–drug interactions associated with tadalafil use at the recommended dose of 10–20 mg/day. Likewise, there are no reports of IADRs with prazosin at the recommended daily dose of 1 mg/day, despite the bioactivation of the pendant furan ring to electrophilic intermediates, trapped with GSH and semicarbazide (Erve et al., 2007).

3.8 CONCLUDING REMARKS

The issue of reactive metabolites continues to receive widespread interest in the pharmaceutical industry. Should evidence for reactive metabolite formation cause abandonment of an otherwise attractive drug candidate or initiate the oftentimes challenging and time-consuming task of eliminating/minimizing their formation. The current evidence suggests that detection of reactive metabolites for a chemical series does not warrant an instant demise of the compounds per se, but this occurrence does trigger some additional due diligence, including the evaluation of competing detoxication pathways of the reactive metabolite and its precursor by phase I/phase II enzymes. Likewise, estimates of human dose can assist in decision making when considering advancement of drug candidates that form reactive metabolites. It is noteworthy to point out that the adduction of GSH with reactive metabolites is not necessarily a bad attribute; instead, it confirms the ability of the endogenous sulfydryl antioxidant to efficiently scavenge the electrophilic reactive intermediate. It is only in cases where the concentration of the reactive metabolite formed is so high that the endogenous antioxidant pool is depleted and leads to toxicity as has been demonstrated with acetaminophen. It is important to emphasize that bioactivation is only one aspect of the overall risk–benefit assessment for advancing a drug candidate into development. Consequently, data from reactive metabolite trapping and covalent binding studies need to be placed in proper and broader context with previously discussed factors such as the daily dose and alternate routes of metabolism/ detoxication. Likewise, appropriate consideration needs to be given for drug candidates for potential treatment options for unmet and urgent medical need.

The ability to predict the potential of a drug candidate to cause IADRs is dependent on a better understanding of the pathophysiological mechanisms of such reactions. IADRs are too complex to duplicate in a test tube, and their idiosyncratic nature precludes prospective clinical studies. Genetic factors also appear to have a crucial role in the induction of IADRs. A fruitful approach may therefore lie in focused and well-controlled phenotype/genotype studies of the rare patients who have survived this type of injury. For instance, results of a 500,000 single nucleotide polymorphism analysis in population exposed to the HIV agent abacavir-associated hypersensitivity reaction suggest that the known human leucocyte antigen B (HLA-B) gene region could be identified with as few as 15 cases and 200 population controls in a sequential analysis and as such has been instituted in practice to avoid the side effects (Hughes et al., 2009). An additional area of research includes studies on the identities of the protein targets of reactive metabolites discerned with toxic versus nontoxic drugs and on the combined application of covalent binding measurements with transcriptomic, metabonomic, and proteomic technologies in an effort to discern (and thereby predict) the characteristics of a toxic response. Until a better understanding of the risk of toxicity arising from the formation of reactive metabolites is developed, the advancement of potent (low-dose) drug candidates with only a limited propensity to form reactive intermediates would appear to be the most favored strategy in an ideal world.

REFERENCES

Alvarez-Sanchez R, Montavon F, Hartung T, Pähler A. Thiazolidinedione bioactivation: A comparison of the bioactivation potentials of troglitazone, rosiglitazone, and pioglitazone using stable isotope-labeled analogues and liquid chromatography tandem mass spectrometry. Chem Res Toxicol 2006;19(8):1106–1116.

Amos HE, Lake BG, Artis J. Possible role of antibody specific for a practolol metabolite in the pathogenesis of oculomucocutaneous syndrome. Br Med J 1978; 1(6110):402–404.

Argoti D, Liang L, Conteh A, Chen L, Bershas D, Yu CP, Vouros P, Yang E. Cyanide trapping of iminium ion reactive intermediates followed by detection and structure identification using liquid chromatography-tandem mass spectrometry (LC-MS/ MS). Chem Res Toxicol 2005;18(10):1537–1544.

Baughman TM, Graham RA, Wells-Knecht K, Silver IS, Tyler LO, Wells-Knecht M, Zhao Z. Metabolic activation of pioglitazone identified from rat and human liver microsomes and freshly isolated hepatocytes. Drug Metab Dispos 2005;33 (6):733–738.

Bauman JN, Kelly JM, Tripathy S, Zhao SX, Lam WW, Kalgutkar AS, Obach RS. Can in vitro metabolism-dependent covalent binding data distinguish hepatotoxic from nonhepatotoxic drugs? An analysis using human hepatocytes and liver S-9 fraction. Chem Res Toxicol 2009;22(2):332–340.

Benet LZ, Spahn-Langguth H, Iwakawa S, Volland C, Mizuma T, Mayer S, Mutschler E, Lin ET. Predictability of the covalent binding of acidic drugs in man. Life Sci 1993;53(8):PL141–146.

Bolton JL, Trush MA, Penning TM, Dryhurst G, Monks TJ. Role of quinones in toxicology. Chem Res Toxicol 2000;13(3):135–160.

Borchard U. Pharmacokinetics of beta-adrenoceptor blocking agents: Clinical significance of hepatic and/or renal clearance. Clin Physiol Biochem 1990;8 (suppl 2):28–34.

Bourdi M, Chen W, Peter RM, Martin JL, Buters JT, Nelson SD, Pohl LR. Human cytochrome P450 2E1 is a major autoantigen associated with halothane hepatitis. Chem Res Toxicol 1996;9(7):1159–1166.

Bourdi M, Gautier, JC, Mircheva J, Larrey D, Guillouzo C, Andre C, Belloc C, Beaune PH. Antiliver microsomes autoantibodies and dihydralazine-induced hepatitis: Specificity of autoantibodies and inductive capacity of the drug. Mol Pharmacol 1992;42(2):280–285.

Brown LM, Ford-Hutchinson AW. The destruction of cytochrome P-450 by alclofenac: Possible involvement of an epoxide metabolite. Biochem Pharmacol 1982;31 (2):195–199.

Chen Q, Ngui JS, Doss GA, et al. Cytochrome P450 3A4-mediated bioactivation of raloxifene: Irreversible enzyme inhibition and thiol adduct formation. Chem Res Toxicol 2002;15(7):907–914.

Coleman JW, Foster AL, Yeung JH, Park BK. Drug-protein conjugates—XV. A study of the disposition of D-penicillamine in the rat and its relationship to immunogenicity. Biochem Pharmacol 1988;37(4):737–742.

Corbett MD, Corbett BR, Hannothiaux MH, Quintana SJ. The covalent binding of acetaminophen to cellular nucleic acids as the result of the respiratory burst of neutrophils derived from the HL-60 cell line. Toxicol Appl Pharmacol 1992;113 (1):80–86.

Crewe HK, Notley LM, Wunsch RM, Lennard MS, Gillam EM. Metabolism of tamoxifen by recombinant human cytochrome P450 enzymes: Formation of the 4-hydroxy, 4′-hydroxy and *N*-desmethyl metabolites and isomerization of *trans*-4-hydroxytamoxifen. Drug Metab Dispos 2002;30(8):869–874.

Dahlin DC, Miwa GT, Lu AY, Nelson SD. *N*-Acetyl-*p*-benzoquinone imine: A cytochrome P-450-mediated oxidation product of acetaminophen. Proc Natl Acad Sci USA 1984;81(5):1327–1331.

Dalvie DK, Kalgutkar AS, Khojasteh-Bakht SC, Obach RS, O'Donnell JP. Biotransformation reactions of five-membered aromatic heterocyclic rings. Chem Res Toxicol 2002;15(3):269–299.

Dalvie DK, Khosla NB, Navetta KA, Brighty KE. Metabolism and excretion of trovafloxacin, a new quinolone antibiotic, in Sprague-Dawley rats and beagle dogs. Effect of bile duct cannulation on excretion pathways. Drug Metab Dispos 1996; 24(11):1231–1240.

Dalvie DK, Khosla N, Vincent J. Excretion and metabolism of trovafloxacin in humans. Drug Metab Dispos 1997;25(4):423–427.

Dansette PM, Bertho G, Mansuy D. First evidence that cytochrome P450 may catalyze both S-oxidation and epoxidation of thiophene derivatives. Biochem Biophys Res Commun 2005;338(1):450–455.

Dearden JC. In silico prediction of drug toxicity. J Comput Aided Mol Des 2003;17 (2–4):119–127.

Dieckhaus CM, Thompson CD, Roller SG, MacDonald TL. Mechanisms of idiosyncratic drug reactions: The case of felbamate. Chem Biol Interact 2002;142(1–2):99–117.

Ding A, Ojingwa JC, McDonagh AF, Burlingame AL, Benet LZ. Evidence for covalent binding of acyl glucuronides to serum albumin via an imine mechanism as revealed by tandem mass spectrometry. Proc Natl Acad Sci USA 1993;90(9): 3797–3801.

Durand A, Thenot JP, Bianchetti G, Morselli PL. Comparative pharmacokinetic profile of two imidazopyridine drugs: Zolpidem and alpidem. Drug Metab Rev 1992;24 (2):239–266.

Erve JC, Svensson MA, von Euler-Chelpin H, Klasson-Wehler E. Characterization of glutathione conjugates of the remoxipride hydroquinone metabolite NCQ-344 formed in vitro and detection following oxidation by human neutrophils. Chem Res Toxicol 2004;17(4):564–571.

Erve JC, Vashishtha SC, DeMaio W, Talaat RE. Metabolism of prazosin in rat, dog, and human liver microsomes and cryopreserved rat and human hepatocytes and characterization of metabolites by liquid chromatography/tandem mass spectrometry. Drug Metab Dispos 2007;35(6):908–916.

Evans DC, Watt AP, Nicoll-Griffith DA, Baillie TA. Drug-protein adducts: An industry perspective on minimizing the potential for drug bioactivation in drug discovery and development. Chem Res Toxicol 2004;17(1):3–16.

Fan PW, Bolton JL. Bioactivation of tamoxifen to metabolite E quinone methide: Reaction with glutathione and DNA. Drug Metab Dispos 2001;29(6):891–896.

Fan PW, Zhang F, Bolton JL. 4-Hydroxylated metabolites of the antiestrogens tamoxifen and toremifene are metabolized to unusually stable quinone methides. Chem Res Toxicol 2000;13(1):45–52.

Fellenius E, Berglindh T, Sachs G, Olbe L, Elander B, Sjöstrand SE, Wallmark B. Substituted benzimidazoles inhibit gastric acid secretion by blocking (H+ + K+) ATPase. Nature 1981;290(5802):159–161.

Gardner I, Leeder JS, Chin T, Zahid N, Uetrecht JP. A comparison of the covalent binding of clozapine and olanzapine to human neutrophils in vitro and in vivo. Mol Pharmacol 1998;53(6):999–1008.

Gorrod JW, Whittlesea CM, Lam SP. Trapping of reactive intermediates by incorporation of 14C-sodium cyanide during microsomal oxidation. Adv Exp Med Biol 1991;283:657–664.

He K, Talaat RE, Pool WF, Reily MD, Reed JE, Bridges AJ, Woolf TF. Metabolic activation of troglitazone: Identification of a reactive metabolite and mechanisms involved. Drug Metab Dispos 2004;32(6):639–646.

Herbert JM, Savi P. P2Y12, a new platelet ADP receptor, target of clopidogrel. Semin Vasc Med 2003;3(2):113–122.

Hughes AR, Brothers CH, Mosteller M, Spreen WR, Burns DK. Genetic association studies to detect adverse drug reactions: Abacavir hypersensitivity as an example. Pharmacogenomics 2009;10(2):225–233.

Israili ZH, Cucinell SA, Vaught J, Davis E, Lesser JM, Dayton PG. Studies of the metabolism of dapsone in man and experimental animals: Formation of *N*-hydroxy metabolites. J Pharmacol Exp Ther 1973;187(1):138–151.

Iverson SL, Uetrecht JP. Identification of a reactive metabolite of terbinafine: Insights into terbinafine-induced hepatotoxicity. Chem Res Toxicol 2001;14(2):175–181.

Jacobsen W, Kuhn B, Soldner A, Kirchner G, Sewing KF, Kollman PA, Benet LZ, Christians U. Lactonization is the critical first step in the disposition of the 3-hydroxy-3-methylglutaryl-CoA reductase inhibitor atorvastatin. Drug Metab Dispos 2000;28(11):1369–1378.

Johnson CH, Wilson ID, Harding JR, Stachulski AV, Iddon L, Nicholson JK, Lindon JC. NMR spectroscopic studies on the in vitro acyl glucuronide migration kinetics of Ibuprofen ((+/−)-(*R*,*S*)-2-(4-isobutylphenyl) propanoic acid), its metabolites, and analogues. Anal Chem 2007;79(22):8720–8727.

Ju C, Uetrecht JP. Oxidation of a metabolite of indomethacin (desmethyldeschlorobenzoylindomethacin) to reactive intermediates by activated neutrophils, hypochlorous acid, and the myeloperoxidase system. Drug Metab Dispos 1998;26(7):676–680.

Jurva U, Holmen A, Gronberg G, Masimirembwa C, Weidolf L. Electrochemical generation of electrophilic drug metabolites: Characterization of amodiaquine quinoneimine and cysteinyl conjugates by MS, IR, and NMR. Chem Res Toxicol 2008;21(4):928–935.

Kalgutkar AS, Dalvie DK, Aubrecht J, Smith EB, Coffing SL, Cheung JR, Vage C, Lame ME, Chiang P, McClure KF, Maurer TS, Coelho RV Jr, Soliman VF, Schildknegt K. Genotoxicity of 2-(3-chlorobenzyloxy)-6-(piperazinyl)pyrazine, a novel 5-hydroxytryptamine2c receptor agonist for the treatment of obesity: Role of metabolic activation. Drug Metab Dispos 2007a;35(6):848–858.

Kalgutkar AS, Obach RS, Maurer TS. Mechanism-based inactivation of cytochrome P450 enzymes: Chemical mechanisms, structure-activity relationships and relationship to clinical drug-drug interactions and idiosyncratic adverse drug reactions. Curr Drug Metab 2007b;8(5):407–447.

Kalgutkar AS, Soglia JR. Minimising the potential for metabolic activation in drug discovery. Expert Opin Drug Metab Toxicol 2005;1(1):91–142.

Kalgutkar AS, Vaz AD, Lame ME, Henne KR, Soglia J, Zhao SX, Abramov YA, Lombardo F, Collin C, Hendsch ZS, Hop CE. Bioactivation of the nontricyclic antidepressant nefazodone to a reactive quinone-imine species in human liver microsomes and recombinant cytochrome P450 3A4. Drug Metab Dispos 2005;33(2):243–253.

Kassahun K, Pearson PG, Tang W, McIntosh I, Leung K, Elmore C, Dean D, Wang R, Doss G, Baillie TA. Studies on the metabolism of troglitazone to reactive intermediates in vitro and in vivo. Evidence for novel biotransformation pathways involving quinone methide formation and thiazolidinedione ring scission. Chem Res Toxicol 2001;14(1):62–70.

Kassahun K, Skordos K, McIntosh I, Slaughter D, Doss GA, Baillie TA, Yost GS. Zafirlukast metabolism by cytochrome P450 3A4 produces an electrophilic alpha, beta-unsaturated iminium species that results in the selective mechanism-based inactivation of the enzyme. Chem Res Toxicol 2005;18(9):1427–1437.

Kemp DC, Fan PW, Stevens JC. Characterization of raloxifene glucuronidation in vitro: Contribution of intestinal metabolism to presystemic clearance. Drug Metab Dispos 2002;30(6):694–700.

Kennedy EL, Tchao R, Harvison PJ. Nephrotoxic and hepatotoxic potential of imidazolidinedione-, oxazolidinedione- and thiazolidinedione-containing analogues of *N*-(3,5-dichlorophenyl)succinimide (NDPS) in Fischer 344 rats. Toxicology 2003;186(1–2):79–91.

Khojasteh-Bakht SC, Chen W, Koenigs LL, Peter RM, Nelson SD. Metabolism of (R)-(+)-pulegone and (R)-(+)-menthofuran by human liver cytochrome P-450s: Evidence for formation of a furan epoxide. Drug Metab Dispos 1999;27(5):574–580.

Koenigs LL, Peter RM, Hunter AP, Haining RL, Rettie AE, Friedberg T, Pritchard MP, Shou M, Rushmore TH, Trager WF. Electrospray ionization mass spectrometric analysis of intact cytochrome P450: Identification of tienilic acid adducts to P450 2C9. Biochemistry 1999;38(8):2312–2319.

Koymans L, Van Lenthe JH, Donné-op Den Kelder GM, Vermeulen NP. Mechanisms of activation of phenacetin to reactive metabolites by cytochrome P-450: A theoretical study involving radical intermediates. Mol Pharmacol 1990;37(3):452–460.

Koymans L, van Lenthe JH, van de Straat R, Donné-Op den Kelder GM, Vermeulen NP. A theoretical study on the metabolic activation of paracetamol by cytochrome P-450: Indications for a uniform oxidation mechanism. Chem Res Toxicol 1989;2(1): 60–66.

Kramer JA, Sagartz JE, Morris DL. The application of discovery toxicology and pathology towards the design of safer pharmaceutical lead candidates. Nat Rev Drug Discov 2007;6(8):636–649.

Lai WG, Zahid N, Uetrecht JP. Metabolism of trimethoprim to a reactive iminoquinone methide by activated human neutrophils and hepatic microsomes. J Pharmacol Exp Ther 1999;291(1):292–299.

Lasser KE, Allen PD, Woolhandler SJ, Himmelstein DU, Wolfe SM, Bor DH. Timing of new black box warnings and withdrawals for prescription medications. JAMA 2002;287(17):2215–2220.

Lecoeur S, Andre C, Beaune PH. Tienilic acid-induced autoimmune hepatitis: Anti-liver and-kidney microsomal type 2 autoantibodies recognize a three-site conformational epitope on cytochrome P4502C9. Mol Pharmacol 1996;50(2):326–333.

Leone AM, Kao LM, McMillian MK, Nie AY, Parker JB, Kelley MF, Usuki E, Parkinson A, Lord PG, Johnson MD. Evaluation of felbamate and other antiepileptic drug toxicity potential based on hepatic protein covalent binding and gene expression. Chem Res Toxicol 2007;20(4):600–608.

Li F, Chordia MD, Huang T, MacDonald TL. In vitro nimesulide studies toward understanding idiosyncratic hepatotoxicity: Diiminoquinone formation and conjugation. Chem Res Toxicol 2009;22(1):72–80.

Liguori MJ, Anderson MG, Bukofzer S, McKim J, Pregenzer JF, Retief J, Spear BB, Waring JF. Microarray analysis in human hepatocytes suggests a mechanism for hepatotoxicity induced by trovafloxacin. Hepatology 2005;41(1):177–186.

Liguori MJ, Blomme EA, Waring JF. Trovafloxacin-induced gene expression changes in liver-derived in vitro systems: Comparison of primary human hepatocytes to HepG2 cells. Drug Metab Dispos 2008;36(2):223–233.

Liu ZC, Uetrecht JP. Clozapine is oxidized by activated human neutrophils to a reactive nitrenium ion that irreversibly binds to the cells. J Pharmacol Exp Ther 1995;275 (3):1476–1483.

Madsen KG, Olsen J, Skonberg C, Hansen SH, Jurva U. Development and evaluation of an electrochemical method for studying reactive phase-I metabolites: Correlation to in vitro drug metabolism. Chem Res Toxicol 2007;20(5):821–831.

Madsen KG, Skonberg C, Jurva U, Cornett C, Hansen SH, Johansen TN, Olsen J. Bioactivation of diclofenac in vitro and in vivo: Correlation to electrochemical studies. Chem Res Toxicol 2008a;21(5):1107–1119.

Madsen KG, Gronberg G, Skonberg C, Jurva U, Hansen SH, Olsen J. Electrochemical oxidation of troglitazone: Identification and characterization of the major reactive metabolite in liver microsomes. Chem Res Toxicol 2008b;21(10):2035–2041.

Maggs JL, Tingle MD, Kitteringham NR, Park BK. Drug-protein conjugates—XIV. Mechanisms of formation of protein-arylating intermediates from amodiaquine, a myelotoxin and hepatotoxin in man. Biochem Pharmacol 1988;37(2):303–311.

Masubuchi N, Makino C, Murayama N. Prediction of in vivo potential for metabolic activation of drugs into chemically reactive intermediate: Correlation of in vitro and in vivo generation of reactive intermediates and in vitro glutathione conjugate formation in rats and humans. Chem Res Toxicol 2007;20(3):455–464.

Mercier M, Poncelet F, de Meester C, McGregor DB, Willins MJ, Léonard A, Fabry L. In vitro and in vivo studies on the potential mutagenicity of alclofenac, dihydroxyalclofenac and alclofenac epoxide. J Appl Toxicol 1983;3(5):230–236.

Migdalof BH, Igdalof BH, Antonaccio MJ, McKinstry DN, Singhvi SM, Lan SJ, Egli P, Kriplani KJ. Captopril: Pharmacology, metabolism and disposition. Drug Metab Rev 1984;15(4):841–869.

Miller NE, Halpert J. Analogues of chloramphenicol as mechanism-based inactivators of rat liver cytochrome P-450: Modifications of the propanediol side chain, the *p*-nitro group, and the dichloromethyl moiety. Mol Pharmacol 1986;29(4):391–398.

Miyamoto G, Zahid N, Uetrecht JP. Oxidation of diclofenac to reactive intermediates by neutrophils, myeloperoxidase, and hypochlorous acid. Chem Res Toxicol 1997;10 (4):414–419.

Mizutani T, Suzuki K, Murakami M, Yoshida K, Nakanishi K. Nephrotoxicity of thioformamide, a proximate toxicant of nephrotoxic thiazoles, in mice depleted of glutathione. Res Commun Mol Pathol Pharmacol 1996;94(1):89–101.

Mizutani T, Yoshida K, Kawazoe S. Formation of toxic metabolites from thiabendazole and other thiazoles in mice. Identification of thioamides as ring cleavage products. Drug Metab Dispos 1994;22(5):750–755.

Monks TJ, Jones DC. The metabolism and toxicity of quinones, quinonimines, quinone methides, and quinone-thioethers. Curr Drug Metab 2002;3(4):425–438.

Murray M. Mechanisms of inhibitory and regulatory effects of methylenedioxyphenyl compounds on cytochrome P450-dependent drug oxidation. Curr Drug Metab 2000;1:67–84.

Njoko D, Laster MJ, Gong DH, Eger EI II, Reed GF, Martin JL. Biotransformation of halothane, enflurane, isoflurane, and desflurane to trifluoroacetylated liver proteins: Association between protein acylation and hepatic injury. Anesth Analg 1997;84:173–178.

Nocerini MR, Yost GS, Carlson JR, Liberato DJ, Breeze RG. Structure of the glutathione adduct of activated 3-methylindole indicates that an imine methide is the electrophilic intermediate. Drug Metab Dispos 1985;13(6):690–694.

Obach RS, Kalgutkar AS, Ryder TF, Walker GS. In vitro metabolism and covalent binding of enol-carboxamide derivatives and anti-inflammatory agents sudoxicam and meloxicam: Insights into the hepatotoxicity of sudoxicam. Chem Res Toxicol 2008a;21(9):1890–1899.

Obach RS, Kalgutkar AS, Soglia JR, Zhao SX. Can in vitro metabolism-dependent covalent binding data in liver microsomes distinguish hepatotoxic from nonhepatotoxic drugs? An analysis of 18 drugs with consideration of intrinsic clearance and daily dose. Chem Res Toxicol 2008b;21(9):1814–1822.

O'Donnell JP, Dalvie DK, Kalgutkar AS, Obach RS. Mechanism-based inactivation of human recombinant P450 2C9 by the nonsteroidal anti-inflammatory drug suprofen. Drug Metab Dispos 2003;31(11):1369–1377.

Olbe L, Carlsson E, Lindberg P. A proton-pump inhibitor expedition: The case histories of omeprazole and esomeprazole. Nat Rev Drug Discov 2003;2 (2):132–139.

Olsen R, Molander P, Øvrebø S, et al. Reaction of glyoxal with 2′-deoxyguanosine, 2′-deoxyadenosine, 2′-deoxycytidine, cytidine, thymidine, and calf thymus DNA: Identification of DNA adducts. Chem Res Toxicol 2005;18(4):730–739.

Olson H, Betton G, Robinson D, et al. Concordance of the toxicity of pharmaceuticals in humans and in animals. Regul Toxicol Pharmacol 2000;32(1):56–67.

Orton C, Lowery C. Practolol metabolism. III. Irreversible binding of [14C]practolol metabolite(s) to mammalian liver microsomes. J Pharmacol Exp Ther 1981;219 (1):207–212.

Parker RJ, Hartman NR, Roecklein BA, Mortko H, Kupferberg HJ, Stables J, Strong JM. Stability and comparative metabolism of selected felbamate metabolites and postulated fluorofelbamate metabolites by postmitochondrial suspensions. Chem Res Toxicol 2005;18(12):1842–1848.

Peterson LA. Electrophilic intermediates produced by bioactivation of furan. Drug Metab Rev 2006;38(4):615–626.

Potter DW, Hinson JA. The 1- and 2-electron oxidation of acetaminophen catalyzed by prostaglandin H synthase. J Biol Chem 1987;262(3):974–980.

Prueksaritanont T, Subramanian R, Fang X, Ma B, Qiu Y, Lin JH, Pearson PG, Baillie TA. Glucuronidation of statins in animals and humans: A novel mechanism of statin lactonization. Drug Metab Dispos 2002;30(5):505–512.

Ramu K, Fraiser LH, Mamiya B, Ahmed T, Kehrerh JP. Acrolein mercapturates: Synthesis, characterization, and assessment of their role in the bladder toxicity of cyclophosphamide. Chem Res Toxicol 1995;8(4):515–524.

Rashed MS, Streeter AJ, Nelson SD. Investigations of the *N*-hydroxylation of 3′-hydroxyacetanilide, a non-hepatotoxic potential isomer of acetaminophen. Drug Metab Dispos 1989;17(4):355–359.

Regal KA, Laws GM, Yuan C, Yost GS, Skiles GL. Detection and characterization of DNA adducts of 3-methylindole. Chem Res Toxicol 2001;14(8):1014–1024.

Rieder MJ, Uetrecht J, Shear NH, Spielberg SP. Synthesis and in vitro toxicity of hydroxylamine metabolites of sulfonamides. J Pharmacol Exp Ther 1988;244(2): 724–728.

Ring BJ, Patterson BE, Mitchell MI, et al. Effect of tadalafil on cytochrome P450 3A4-mediated clearance: Studies in vitro and in vivo. Clin Pharmacol Ther 2005; 77(1):63–75.

Sahali-Sahly Y, Balani SK, Lin JH, Baillie TA. In vitro studies on the metabolic activation of the furanopyridine L-754,394, a highly potent and selective mechanism-based inhibitor of cytochrome P450 3A4. Chem Res Toxicol 1996;9(6):1007–1012.

Savi P, Pereillo JM, Uzabiaga MF, Combalbert J, Picard C, Maffrand JP, Pascal M, Herbert JM. Identification and biological activity of the active metabolite of clopidogrel. Thromb Haemost 2000;84(5):891–896.

Savi P, Zachayus JL, Delesque-Touchard N, et al. The active metabolite of Clopidogrel disrupts P2Y12 receptor oligomers and partitions them out of lipid rafts. Proc Natl Acad Sci USA 2006;103(29):11069–11074.

Slack JA, Ford-Hutchinson AW, Richold M, Choi BC. Determination of a urinary epoxide metabolite of alclofenac in man. Drug Metab Dispos 1980;8(2):84–86.

Slack JA, Ford-Hutchinson AW, Richold M, Choi BC. Some biochemical and pharmacological properties of an epoxide metabolite of alclofenac. Chem Biol Interact 1981;34(1):95–107.

Smith KS, Smith PL, Heady TN, Trugman JM, Harman WD, Macdonald TL. In vitro metabolism of tolcapone to reactive intermediates: Relevance to tolcapone liver toxicity. Chem Res Toxicol 2003;16(2):123–128.

Streeter AJ, Timbrell JA. The in vitro metabolism of hydralazine. Drug Metab Dispos 1985;13(2):255–259.

Sun Q, Zhu R, Foss FW Jr, Macdonald TL. In vitro metabolism of a model cyclopropylamine to reactive intermediate: Insights into trovafloxacin-induced hepatotoxicity. Chem Res Toxicol 2008;21(3):711–719.

Tang W, Stearns RA, Bandiera SM, Zhang Y, Raab C, Braun MP, Dean DC, Pang J, Leung KH, Doss GA, Strauss JR, Kwei GY, Rushmore TH, Chiu SH, Baillie TA. Studies on cytochrome P-450-mediated bioactivation of diclofenac in rats and in human hepatocytes: Identification of glutathione conjugated metabolites. Drug Metab Dispos 1999;27(3):365–372.

Thompson DC, Thompson JA, Sugumaran M, Moldéus P. Biological and toxicological consequences of quinone methide formation. Chem Biol Interact 1993;86 (2):129–162.

Timbrell JA, Mitchell JR, Snodgrass WR, Nelson SD. Isoniazid hepatoxicity: The relationship between covalent binding and metabolism in vivo. J Pharmacol Exp Ther 1980;213(2):364–369.

Trimble LA, Chauret N, Silva JM, Nicoll-Griffith DA, Li C-S, Yergey JA. Characterization of the in vitro oxidative metabolites of the COX-2 selective inhibitor L-766,112. Bioorg Med Chem Lett 1997;7(1):53–56.

Uetrecht JP. Reactivity and possible significance of hydroxylamine and nitroso metabolites of procainamide. J Pharmacol Exp Ther 1985;232(2):420–425.

Uetrecht JP. Metabolism of clozapine by neutrophils. Possible implications for clozapine-induced agranulocytosis. Drug Safety 1992;7(suppl 1):51–56.

Uetrecht J, Zahid N, Tehim A, Fu JM, Rakhit S. Structural features associated with reactive metabolite formation in clozapine analogues. Chem Biol Interact 1997;104 (2–3)117–129.

Uetrecht JP, Zahid N, Whitfield D. Metabolism of vesnarinone by activated neutrophils: Implications for vesnarinone-induced agranulocytosis. J Pharmacol Exp Ther 1994;270(3):865–872.

Valadon P, Dansette PM, Girault JP, Amar C, Mansuy D. Thiophene sulfoxides as reactive metabolites: Formation upon microsomal oxidation of a 3-aroylthiophene and fate in the presence of nucleophiles in vitro and in vivo. Chem Res Toxicol 1996;9 (8):1403–1413.

Walker DP, Bi FC, Kalgutkar AS, Bauman JN, Zhao SX, Soglia JR, Aspnes GE, Kung DW, Klug-McLeod J, Zawistoski MP, McGlynn MA, Oliver R, Dunn M, Li JC, Richter DT, Cooper BA, Kath JC, Hulford CA, Autry CL, Luzzio MJ, Ung EJ, Roberts WG, Bonnette PC, Buckbinder L, Mistry A, Griffor MC, Han S, Guzman-Perez A. Bioorg Med Chem Lett 2008;18(23):6071–6077.

Walker GS, Atherton J, Bauman J, Kohl C, Lam W, Reily M, Lou Z, Mutlib A. Determination of degradation pathways and kinetics of acyl glucuronides by NMR spectroscopy. Chem Res Toxicol 2007;20(6):876–886.

Wang J, Davis M, Li F, Azam F, Scatina J, Talaat R. A novel approach for predicting acyl glucuronide reactivity via Schiff base formation: Development of rapidly formed peptide adducts for LC/MS/MS measurements. Chem Res Toxicol 2004;17 (9):1206–1216.

Waring JF, Liguori MJ, Luyendyk JP, Maddox JF, Ganey PE, Stachlewitz RF, North C, Blomme EA, Roth RA. Microarray analysis of lipopolysaccharide potentiation of trovafloxacin-induced liver injury in rats suggests a role for proinflammatory chemokines and neutrophils. J Pharmacol Exp Ther 2006;316(3):1080–1087.

Wen B, Coe KJ, Rademacher P, Fitch WL, Monshouwer M, Nelson SD. Comparison of in vitro bioactivation of flutamide and its cyano analogue: Evidence for reductive activation by human NADPH:cytochrome P450 reductase. Chem Res Toxicol 2008; Epub. 21(12):2393–2406.

Yan Z, Easterwood LM, Maher N, Torres R, Huebert N, Yost GS. Metabolism and bioactivation of 3-methylindole by human liver microsomes. Chem Res Toxicol 2007;20(1):140–148.

Zhang D, Krishna R, Wang L, Zeng J, Mitroka J, Dai R, Narasimhan N, Reeves RA, Srinivas NR, Klunk LJ. Metabolism, pharmacokinetics, and protein covalent binding of radiolabeled MaxiPost (BMS-204352) in humans. Drug Metab Dispos 2005;33(1):83–93.

Zhang KE, Naue JA, Arison B, Vyas KP. Microsomal metabolism of the 5-lipoxygenase inhibitor L-739,010: Evidence for furan bioactivation. Chem Res Toxicol 1996;9 (2):547–554.

Zhang D, Ogan M, Gedamke R, Roongta V, Dai R, Zhu M, Rinehart JK, Klunk L, Mitroka J. Protein covalent binding of maxipost through a cytochrome P450-mediated ortho-quinone methide intermediate in rats. Drug Metab Dispos 2003;31 (7):837–845.

Zhao Z, Baldo BA, Rimmer J. Beta-lactam allergenic determinants: Fine structural recognition of a cross-reacting determinant on benzylpenicillin and cephalothin. Clin Exp Allergy 2002;32(11):1644–1650.

Zhao SX, Dalvie DK, Kelly JM, Soglia JR, Frederick KS, Smith EB, Obach RS, Kalgutkar AS. NADPH-dependent covalent binding of [3*H*]paroxetine to human liver microsomes and S-9 fractions: Identification of an electrophilic quinone metabolite of paroxetine. Chem Res Toxicol 2007;20(11):1649–1657.

4 Drug-Metabolizing Enzymes, Transporters, and Drug–Drug Interactions

STEVEN W. LOUIE, and MAGANG SHOU
Department of Pharmacokinetics and Drug Metabolism, Amgen, Inc., Thousand Oaks, California

Mass Spectrometry in Drug Metabolism and Disposition: Basic Principles and Applications,
First Edition. Edited by Mike S. Lee and Mingshe Zhu.

4.1 INTRODUCTION

With the advent of polypharmacy, a competitive marketplace, and greater attention paid to public safety (i.e., risk versus benefit), pharmaceutical companies have increased their focus on new chemical entities (NCEs) with optimal pharmacokinetic absorption–distribution–metabolism–excretion (PK/ADME) properties that preclude costly therapeutic monitoring and genotyping of patients (Rodrigues and Rushmore, 2002; Lin and Lu, 1997). It is also advantageous to develop NCEs with minimal drug–drug interaction (DDI) liabilities, so that package insert (black-box) warnings can be avoided and patient safety maximized as much as possible. A drug metabolism- or transport-based drug interaction implies that one drug causes a change in the clearance of another drug and in turn either decreases or increases the concentration of the drug in plasma and presumably also causes a change at the site of action, which has led and can lead to either pharmacodynamic (PD) inhibition or enhancement of the clinical effects or serious toxicological consequences of another drug in human (Backman et al., 1994, 2002; Floren et al., 1997; Gomez et al., 1995; Greenblatt et al., 1998, 1999; Honig et al., 1993; Jaakkola et al., 2005; Niemi et al., 2003; Varhe et al., 1994; Williams et al., 2004). For example, the highly publicized market withdrawals of two DDI victim drugs, terfenadine and cerivastatin, were precipitated partially because of a drug interaction involving the inhibition of cytochrome P450 3A4 (CYP3A4) and CYP2C8, respectively (Honig et al., 1993; Backman et al. 1994, 2002; Emoto et al., 2006; Greenblatt et al., 1998; Azie et al., 1998; Gorski et al., 1998; Gomez et al., 1995; Floren et al., 1997; Varhe et al., 1994). More recently, drug interactions that involve the inhibition of tizanidine (CYP1A2-dependent), repaglinide (CYP2C8/CYP3A4-dependent), and fluticasone (CYP3A4-dependent) clearance have also received some attention (Niemi et al., 2003; Arrington-Sanders et al., 2006; Backman et al., 2006). In addition, there are examples of successfully marketed backup drugs that have improved DDI profiles and less dependency on polymorphic cytochromes P450s (CYPs) for clearance (e.g., fexofenadine versus terfenadine; desloratadine versus loratadine; esomeprazole versus omeprazole).

Most DDIs involve two or more drugs, with the perpetrator drug(s) causing changes in the PK profile of the victim drug(s). Consequently, most pharmaceutical companies now screen their NCEs in vitro for induction and inhibition of drug-metabolizing enzymes (DMEs) and transporters (Lin and Lu, 1997; Rodrigues and Lin, 2001; Mizuno et al., 2003). NCEs are also screened for metabolic stability and the enzyme(s) responsible for metabolic turnover. Because DMEs are responsible for the clearance of approximately three-quarters of approved drugs, identification and characterization of the enzyme(s) or transporter(s) responsible for the metabolism/disposition of a NCE (reaction phenotyping) and that is inhibited or induced by a NCE are important tasks during the drug discovery and development processes (Rodrigues and Rushmore, 2002; Rodrigues, 1999; Lu et al., 2003; Emoto et al., 2006; Williams et al., 2003). Major DMEs and transporters which play a large role in drug ADME are listed in Table 4.1.

4.2 DRUG-METABOLIZING ENZYMES

Many DMEs in phase I and II systems exist in the liver and in extrahepatic tissues such as intestine and kidney, which can contribute to the clearance of drugs from the systemic circulation. The liver serves as a port of entry for all absorbed drugs, some of which directly undergo the first-pass metabolism and then are excreted into bile before entering the systemic circulation. Therefore, multiple drugs can compete for the same clearance pathways and lead to potentially serious DDIs. The likelihood of such interactions is high when concomitant drugs are targeted by the same CYP. Also, genetic polymorphisms that reduce CYP function can cause serious gene–drug interactions. For both reasons, the distribution of clearance pathways among multiple CYPs or reliance on non-CYP pathways can be evaluated as a desirable feature during drug development.

4.2.1 CYPs

CYP1A Three genes, *CYP1A1*, *CYP1A2*, and *CYP1B1*, are members of the CYP1 family. There are no pseudogenes in this family. All three genes are similarly regulated; they are all transcriptionally controlled by the AhR-ARNT (aryl hydrocarbon receptor–aryl hydrocarbon receptor nuclear translocator) pathway (Schmidt and Bradfield, 1996) and induced by polycyclic aromatic hydrocarbons (PAHs), 2,3,7,8-tetrachlorodibenzo-*p*-dioxin (TCDD), and cigarette smoking. Importantly, these enzymes are capable of converting PAHs into reactive intermediates which bind to DNA and cause mutagenesis and carcinogenesis (Shimada et al., 1996). Constitutive expression of CYP1A1 is very low

TABLE 4.1 Major Drug-Metabolizing Enzymes and Transporters

	Phase I	Phase II	Phase III
Reaction	Oxidation, reduction, and hydrolysis	Conjugation • Glucuronidation • Sulfation	Transport • Hepatobiliary excretion • Renal secretion • Gut absorption • Blood–brain penetration
Enzyme and transporters	CYPs Molybdenum hydroxylases • Aldehyde oxidase (AO) • Xanthin oxidase (XO) Flavin-containing monooxygenase (FMO) Monoamine oxidase (MO) Aldehyde dehydrogenase CYP reductase Carbonyl reductase Epoxide hydrolase Esterases	UDP glucuronosyltransferase Sulfotransferase Glutathione *S*-transferase *N*-Acetyltransferase (NAT) Catechol-*O*-methyltransferase (COMT)	ABC family • MDR1(ABCB1) • BSEP (ABCB11) • MRP1 (ABCC1) • MRP2 (ABCC2) • MRP3 (ABCC3) • MRP4 (ABCC4) • MRP5 (ABCC5) • BCRP (ABCG2) SLC family • NTCP (SLC10A1) • PEPT1 (SLC15A1) • PEPT2 (SLC15A2) • OATP1B3 (SLC21A3) • OATP1B1 (SLC21A6) • OATP2B1 (SLC21A9) • OCT1 (SLC22A1) • OCT2 (SLC22A2) • OCT3 (SLC22A3) • OCTN1 (SLC22A4) • OCTN2 (SLC22A5) • OAT1 (SLC22A6) • OAT2 (SLC22A7) • OAT3 (SLC22A8)

in both hepatic and extrahepatic tissues. However, CYP1A1 is inducible by the inducers via the AhR transactivation in many tissues, including lung, lymphocytes, mammary gland, and placenta. In human primary hepatocytes, CYP1A1 is induced by 3-methylcholanthrene (3MC) and omeprazole (Rodriguez-Antona et al., 2000), which are commonly used as the positive controls for evaluation of NCEs as inducers. Seven variant alleles have been described, but none of them have been unequivocally shown to have altered catalytic activity of the CYP1A1 protein, such as T3801C (*MspI*) and I462V.

CYP1A2 is a liver-specific enzyme and constitutes about 13% of the total hepatic CYP content (Shimada et al., 1994). It metabolizes PAHs, nitrosamines, aflatoxin B_1, and aryl amines into the ultimate carcinogens. CYP1A2 is regulated by AhR and induced in vivo by cigarette smoking and foods that contain indole-3-carbinol, phenytoin, rifampicin, and omeprazole. Clozapine, phenacetin, caffeine, theophylline, imipramine, amitriptyline, and tacrine are known substrates for CYP1A2. The *CYP1A2* gene has six variant alleles, which were shown to have various enzyme activities (Kalow and Tang, 1991). Omeprazole reduced exposure of caffeine in a dose-dependent manner.

CYP1B1 is not expressed in the liver but in kidney, prostate, mammary gland, and ovary (Sutter et al., 1994; Shimada et al., 1996). The induction of CYP1B1 is regulated by the AhR-ARNT pathway (Sutter et al., 1994) 4) and induced by PAHs such as 3MC. CYP1B1 converts proximate carcinogen PAHs and aryl amines to the ultimate carcinogens. The *CYP1B1* gene is polymorphic with several alleles, which are functionally impaired and linked with human primary congenital glaucoma (Stoilov et al., 1997). In addition, these variants had varying enzyme activities toward the conversion of estradiol into carcinogenic 4-hydroxyestradiol.

CYP2 CYP2A6 expressed in liver and nasal mucosa accounts for 4% of the total CYP amount. CYP2A6 metabolizes mainly nicotine and halothane and bioactivates tobacco-specific nitrosamine NNK [4-(methylnitrosamino)-1-(3-pyridyl)-1-butanone] (Hecht, 1998). The genetic polymorphisms of CYP2A6 have been associated with inter individual differences in smoking behavior (Pianezza et al., 1998). Coumarin has been used as a probe drug to assess the activity of CYP2A6 in vivo. Very little is known about the mechanisms of its regulation, but it is induced by phenobarbital (PB), rifampicin, and other antiepileptic drugs (Sotaniemi et al., 1995). The CYP2A7 protein is nonfunctional due to its inability to incorporate heme. Heterologously expressed CYP2A13 is highly active in the activation of NNK.

CYP2B6 is detected in human liver and is about 1–2% of total hepatic CYP. The substrates for CYP2B6 include 6-aminochrysene, methoxychlor, NNK, and cyclophosphamide. CYP2B6 in human primary hepatocytes is typically induced by PB and rifampicin through the nuclear receptor CAR (constitutively active receptor) (Sueyoshi et al., 1999) and probably also by PXR (pregnane X receptor) (Pascussi et al., 2000). The human CYP2B subfamily also contains the *CYP2B7P* pseudogene.

The human CYP2C subfamily has four highly homologous genes: *2C8*, *2C9*, *2C18*, and *2C19*, which are located in a cluster on chromosome 10. Twenty percent of human hepatic CYPs is due to CYP2C (CYP2C9 > CYP2C8 > CYP2C19). CYP2C mRNA and protein are induced in primary hepatocytes by PB and rifampicin, which are mediated by CAR and PXR (Pascussi et al., 2000). The spectrum of CYP2C substrates is large, including commonly clinically used diazepam, omeprazole, mephenytoin, tolbutamide, and warfarin as well as many nonsteroidal anti-inflammatory drugs. Selective substrates are taxol for CYP2C8, tolbutamide for CYP2C9, and mephenytoin for CYP2C19. There are three allelic variants of *CYP2C9*, termed *CYP2C9*1* (wild type), *CYP2C9*2* (Arg144Cys), and *CYP2C9*3* (Ile359Leu). The latter two mutations show reduced activities for the substrates. CYP2C19-poor metabolizer is detected in 2–4% of Caucasians and in about 20% of Asians (Ingelman-Sundberg et al., 1999a). The common true null mutations that lead to poor metabolizer status are due to splicing defects (*CYP2C19*2*) and loss of stop colons (*CYP19*3*).

CYP2D6 constitutes about 2% of total hepatic CYP, and the protein is also expressed in duodenum and brain. CYP2D6 has one gene and four pseudogenes. The genetic polymorphism of CYP2D6 was observed and associated with altered expression of CYP enzyme (Gonzalez et al., 1988). About 6% of Caucasians and 1% of Asians are absent the gene (Ingelman-Sundberg et al., 1999) and are identified as poor metabolizers (PM) of many CYP2D6 substrates, tricyclic antidepressants, haloperidol, metoprolol, propranolol, codeine, and dextromethorphan. There are at least 30 different defective *CYP2D6* alleles, six of which contribute to 95–99% of PM phenotypes. Duplications of the *CYP2D6* gene up to 13 gene copies have been reported (Johansson et al., 1993) to lead to the rapid metabolism and decreased drug effects of CYP2D6 substrates in the population (4% ultrarapid metabolizer) (Dalen et al., 1998). CYP2D6 is not inducible.

CYP2E1 enzyme has been studied extensively due to its role in the metabolism of ethanol. Most CYP2E1 substrates are small and hydrophobic compounds, such as paracetamol, chlorzoxazone, enflurane, and halothane. Disulfiram is a clinically used inhibitor of CYP2E1. About 7% of the hepatic CYP consists of CYP2E1. It is also expressed in lung and brain. CYP2E1 is regulated at levels of the gene transcription, translation, and posttranslation and induced by acetone, ethanol, pyridine, pyrazole, and isoniazid. Several allelic variants of the *CYP2E1* gene have been detected. One of these alleles (*CYP2E1*2*) produces a protein with reduced stability (Itoga et al., 1998; Hu et al., 1997), while another (*CYP2E1*1D*) is associated with increased activity after alcohol exposure and in obese subjects (McCarver et al., 1998).

CYP2F1 has two full-length copies in addition to one pseudogene, but it is not known whether both of them produce functional transcripts (Hoffman et al., 1995). CYP2F1 mRNA has been identified in human lung and placenta, but not in liver (Carr et al., 2003). Recombinant CYP2F1 enzyme is shown to bioactivate two prototypical pneumotoxicants, naphthalene and 3-methylindole, to its highly pneumotoxic intermediates naphthalene-1,2-oxide and 3-methyleneindolenine, respectively.

CYP2J2 is expressed mainly extrahepatically in heart, kidney, lung, pancreas, and gastrointestinal tract. CYP2J2 is thought to be the predominant enzyme in cardiac tissue responsible for the metabolism of arachidonic acid into epoxyeicosatrienoic acids (EETs), which have physiological functions (Capdevila et al., 2000).

CYP3 CYP3A4 is expressed mainly in human liver with 30–40% of total hepatic CYPs. CYP3A4 is also present in small intestine and other tissues. More than 50% of the structural and therapeutic diverse drugs are metabolized by CYP3A4, such as quinidine, nifedipine, diltiazem, lidocaine, lovastatin, erythromycin, cyclosporin, triazolam, and midazolam, and endogenous substances such as testosterone, progesterone, and androstenedione. CYP3A4 also activates procarcinogens, aflatoxin B_1, PAHs, NNK, and 6-aminochrysene. Ketoconazole is a potent inhibitor of CYP3A4 and is often used in vitro and in vivo as a diagnostic inhibitor. CYP3A4 is induced in human hepatocytes and in vivo by rifampicin, dexamethasone, and PB. The induction of CYP3A4 is mainly regulated by the novel orphan receptor PXR (Lehmann et al., 1998; Moore et al., 2000), but also other receptors, including CAR. Some variant alleles have been detected for the *CYP3A4* gene. However, the functional significance of this alteration is uncertain (Westlind et al., 1999; Rodriguez-Antona et al., 2005).

CYP3A5 is expressed polymorphically in human liver (Wrighton et al., 1989), but consistently in lung, colon, kidney, oesophagus, and anterior pituitary gland. A few variant alleles of the *CYP3A5* gene were found (Jounaidi et al., 1996; Burckart, 2007; Dai et al., 2006; Hustert et al., 2001). CYP3A5 has a similar substrate specificity pattern to CYP3A4, but the turnover is usually lower. The CYP3A5 protein is not generally inducible in human liver and primary hepatocytes. However, its mRNA was induced in human primary hepatocytes by rifampicin and PB (Rodriguez-Antona et al., 2000). The promoter region of the *CYP3A5* gene has been shown to contain a functional glucocorticoid-responsive element that mediates induction by dexamethasone in the HepG2 cell line (Schuetz et al., 1996). Unlike the other CYP3A enzymes, CYP3A5 is probably not regulated by PXR.

CYP3A7 is mainly expressed in human fetal liver, where it is the major CYP form (Kitada and Kamataki, 1994). CYP3A7 has similar catalytic properties compared with other CYP3A enzymes, including testosterone 6β-hydroxylation (Kitada and Kamataki, 1994). It is induced in adult human primary hepatocytes by rifampicin, which is mediated by the PXR pathway (Pascussi et al., 1999).

CYP4 CYP4B1 mRNA was found in human lung, colon, placenta, and bladder but not in liver (Nhamburo et al., 1989). CYP4B1 bioactivates several protoxicants and procarcinogens in animal models such as 2-aminoanthracene, 2-aminofluorene, and valproic acid. Interindividual differences in CYP4B1 expression have been associated with bladder cancer susceptibility (Imaoka et al., 2000), and elevated CYP4B1 mRNA was demonstrated in lung carcinoma versus normal lung tissue (Czerwinski et al., 1994). Induction of CYP4B1 is unclear; however, sephadex protein 1 (Sp1) has been reported to be critical for CYP4B1 lung-selective regulation through both proximal element and distal, lung-selective enhancers (Poch et al., 2005).

4.2.2 UDP–Glucuronosyltransferases

The uridine diphosphate (UDP)–glycosyltransferases (UGTs) are a group of enzymes that catalyze the transfer of sugars (i.e., glucuronic acid) to a variety of molecules as acceptors. Glucuronidation is an important process in the elimination of many endogenous substances from the body (bilirubin, bile acids, steroids, and thyroids). Insufficient conjugation of bilirubin (UGT1A1) impairs biliary excretion and leads to hyperbilirubinemia (Crigler–Najjar syndromes and Gilbert's disease) (Monaghan et al., 1996). UGTs also catalyze a broad array of xenobiotics with diverse structures (amines, phenols, carboxylic acid–containing drugs, aliphatic and aromatic alcohols, etc.). UGT has two gene families in which about 18 human isoforms are identified (Mackenzie et al., 1997). Family 1 isoforms share a common C-terminus but are differentiated by N-terminal sequences. For example, the UGT1A locus yields different isoforms via differential splicing of a single variable N-terminal exon, with four common exons (exons 2–5). Thus, UGT1A isoforms have an identical C-terminus, approximately 245 amino acids. Family 2 isoforms have differences throughout the sequence. UGTs are prominent in hepatic, renal, gut, lung, and olfactory tissues. These isoforms are located in endoplasmic reticulum (ER) and nuclear membranes and found on the luminal side of the ER, in contrast to CYPs. Genetic polymorphisms have been identified in all major UGTs. *UGT1A1*28* (a TA insertion into a TATA box) is about 15% frequency of the variant in Caucasians and results in hyperbilirubinemia (Gilbert's syndrome) and high rate of neutropenia from irinotecan treatment due to lower activity of the enzyme. Like the CYP gene families, the UGT genes can be regulated by certain inducers. The regulatory elements of CYP genes appear to be present for the UGT genes. For example, 3MC, TCDD, and BNF (beta-naphthoflavone) induce metabolism of polycyclic aromatic hydrocarbon (PAH) phenols by UGT1A6 and UGT1A7. The xenobiotic response elements

(XREs) have been identified in the human UGT1A6. PB and phenytoin are known to induce glucuronidation of bilirubin, phenols, morphine, and steroids catalyzed by UGT1A1, UGT1A9, and UGT2B7. PB-responsive elements (PBREMs) have been identified in these genes. RIF is a PXR activator and is shown to increase the glucuronidation of a number of drugs, zidovudine (AZT) and morphine (GUT2B7), acetaminophen (UGT1A6), and lamotrigine (UGT1A4), in humans (Gardner-Stephen et al., 2004).

4.2.3 Sulfotransferases

Sulfotransferases (SULTs) catalyze the sulfation of many substrates in a similar preference for functional groups that glucuronidation has, such as phenols, alcohols, and arylamines. However, sulfation is less extensive than glucuronidation. Many endogenous substrates, steroids, bile acids and phenols, neurotransmitters, proteins, and carbohydrates are catalyzed by SULTs. Sulfation of chemicals involves the conjugation of the substrate with a sulfonyl group and the PAPS (3′-phospho-adenosine-5′phosphosulfate) serves as the sulfonyl donor, yielding either *O*- or *N*-sulfates. Cytosolic SULTs metabolize small molecules (endogenous and xenobiotic) and Golgi SULTs metabolize peptides, proteins, lipids, and glycosaminoglycans. About 11 human SULTs have been known and they are structurally related and share a common fold around a conserved active site. SULT1A1 and SULT1A2 are highly expressed in liver. However, SULT1A3 is highly present in gut, with negligible expression in hepatic tissue. SULT1B1 is expressed in liver and colon. Very little is known about SULT gene regulation. Much research has been done for SULT genetic polymorphism. Genetic polymorphism includes the *SULT1A1*, *SULT1A2*, *SULT2A1*, *SULT1C1*, and *SULT1E1* genes. The *SULT1A1* genotype has been reported to associate with prostate, breast, or colorectal cancer (Lilla et al., 2007; Pachouri et al., 2006; Langsenlehner et al., 2004).

4.2.4 Glutathione-*S*-Transferases

Glutathone-*S*-transferases (GSTs) are important in xenobiotic deactivation, regulation of the redox state of cells, maintenance of reduced cysteines, and general oxidative stress responses. GSTs catalyze the reaction of glutathione with electrophiles, for example, epoxides, quinones, quinonemethides, to form glutathione conjugates (Hayes et al., 2005). The isoforms are designated on the basis of sequence homology, but unlike CYPs, SULTs, and UGTs, the classes are defined by letter (family) and two numbers of subunit composition. In mammals, A (alpha), P (pi), M (mu), K (kappa) and T (theta), and S (sigma) classes are cytosolic. The cytosolic GSTs are dimeric enzymes, usually homodimers. Expression of GSTs is tissue specific. GSTA1 is expressed mainly in liver, and release of the enzyme to plasma is a highly sensitive marker of liver toxicity. GSTM2-2 in brain and skeletal muscle, GSTM3 in brain, lung, testes, and GSTM4 in lymphoblasts are found primarily.The *GSTM1-1* gene as a null phenotype is of high frequency in Caucasians (70%), Asians (70%), and Africans

(50%), which have possible increased risk of bladder and lung cancers. *GSTM3-3* has lower frequency in Caucasians and may be associated with increased risk of skin cell carcinomas. *GSTP1-1* with several allelic single-nucleotide polymorphism (PNP) variants exhibits altered catalytic functions and substrate specificity.

4.2.5 Regulation of Human CYPs

Aryl Hydrocarbon Receptor Aryl hydrocarbon receptor (AhR) is the Per-Arnt-Sim (PAS) family of transcription factors. It transcriptionally induces expression of human CYP1A1, CYP1A2, and CYP1B1 (Quattrochi et al., 1994; Whitlock, 1999) as well as some phase II metabolizing enzymes (Rodriguez-Antona et al., 2000; Schmidt and Bradfield, 1996). PAHs and TCDD are the known AhR ligands. AhR exists in cytoplasm as a complex comprising of a dimer of HSP90 (heat-shock protein 90), ARA9 (also called AIP1 and XAP2), and p23 for stabilization. Ligand binding to AhR dissociates the chaperone proteins and translocates to the nucleus, where it forms a heterodimer with AhR nuclear translocator (ARNT) (Hoffman et al., 1991). This heterodimer binds to the XREs of the CYP genes to activate transcription. A novel PAS protein called AhR repressor inhibits AhR signal transduction by competing with AhR for ARNT and also by binding to XRE.

Pregnane X Receptor Pregnane X receptor (PXR) is the orphan nuclear receptor. It mediates the induction of CYP3A4 (Bertilsson et al., 1998) and CYP3A7 (Pascussi et al., 1999). CYP2C8 and CYP2C9 are also regulated by PXR (Pascussi et al., 2000). PXR, similar to its principal target gene *CYP3A4*, is mainly expressed in liver, small intestine, and colon. A variety of structurally diverse, low-affinity exogenous and endogenous chemicals (steroid hormones, rifampicin, PB, nifedipine, clotrimazole, mifepristone, and metyrapone) were found to activate PXR. Many of the PXR ligands are also shared by CAR (Moore et al., 2000). The ligand binds to PXR and forms a heterodimer with the retinoid X receptor-α (RXRα). RXR serves as a common heterodimerization partner for many orphan nuclear receptors. The binding of PXR/RXR to the XRE results in transcriptional activation of the respective gene. PXR and RXRα are induced by glucocorticoid receptor (GR) (Pascussi et al., 2000). Expression of PXR and also CAR is down-regulated by the inflammatory cytokine interleukin-6 (Pascussi et al., 2000).

Constitutively Active Receptor Constitutive androstane receptor (CAR) is an orphan nuclear receptor. CAR is located in the cytoplasm and is translocated to the nucleus after inducer treatment (Kawamoto et al., 1999). CAR is predominantly expressed in liver, and it mediates the induction of CYP2B6 and, to a lesser extent, CYP3A4. CYP2C8 and CYP2C9 are also regulated by CAR (Pascussi et al., 2000e). CAR is down-regulated by the inflammatory cytokine interleukin-6, which could explain the repression of CYPs by inflammatory mediators (Abdel-Razzak et al., 1993). Importantly, CAR was shown to mediate the induction of the *CYP2B* genes by PB. However, PB is not a CAR ligand (Moore et al., 2000). Only activators (5β-pregnane) and inverse agonists (androstanol and clotrimazole) can bind to human CAR. Similar to PXR, CAR requires heterodimerization partner RXR to enable binding to DNA. CAR/RXR heterodimers bind to a conserved 51-base pair element called PBREM (PB-responsive enhancer module) in the 5′-flanking region of the *CYP2B* genes (Honkakoski et al., 1998). PBREM has been shown to mediate the induction by PB and PB-like inducers. PB not only induces CYP2B but also CYP2A, CYP2C, and CYP3A. The mechanism of PB induction is still unclear, but studies suggest that PB facilitates the translocation of CAR to the nucleus and activates CAR in the nucleus via phosphorylation, since translocation and activation are inhibited by protein phosphatase (PP) and CaM kinase (CK) inhibitors, respectively.

Glucocorticoid Receptor (GR) Glucocorticoids, such as dexamethasone, are related to CYP induction. However, their effects on CYP induction are not only dependent on GR binding to CYP genes but also on complex protein–protein interplays between GR and various other receptors (Honkakoski and Negishi, 2000). For example, dexamethasone has been shown to potentiate CYP1A1 induction by TCDD. Dexamethasone induces PXR and RXR expression and leads to an increase in CYP3A4 induction by PXR agonists (Pascussi et al., 2000). This increase explains the results on the dexamethasone-elicited induction of CYP3A4 in human hepatocytes (Pichard et al., 1992). The only human CYP gene induced directly by GR is *CYP3A5*. GR binds to the glucocorticoid responsive element in the 5′-flanking region of *CYP3A5* (Schuetz et al., 1996).

Other Regulatory Mechanisms CYP2A6 is induced by PB and rifampicin (et-Beluche et al., 1992) and suggests the involvement of CAR and/or PXR. CYP2A5 in mice is induced by cAMP-elevating agents and several hepatotoxic compounds and indicates that mRNA stabilization is involved in the regulation (Tilloy-Ellul et al., 1999). CYP2E1 is regulated in a complex manner, since it is regulated at levels of transcription and translation. The most important steps are probably the stabilization of mRNA and protein. However, starvation and chronic ethanol intake are thought to increase transcription as well as CYP2E1 protein stability. Unlike other cytokines, interleukin-4 induces human CYP2E1 in primary hepatocytes (Abdel-Razzak et al., 1993). Several other cytokines (interleukin-1β, interleukin-6, tumor necrosis factor-α, and interferon-γ) down-regulate CYP1A2, CYP2C, CYP2E1, and CYP3A.

4.3 METABOLISM-BASED DDIs

4.3.1 Reaction Phenotyping

Reaction phenotyping is the study that measures quantitative contribution of a DME to total clearance of a given drug. The CYP reaction-phenotyping process requires the integration of data obtained from various in vitro systems as well as integration of clinical data (e.g., human radiolabel study and drug interaction studies of selective inhibitors with probe drugs) and allows estimatation of the fraction of dose cleared via all CYPs (f_m) and the contribution of each CYP to total CYP-dependent clearance ($f_{m,\mathrm{CYP}}$). As such information becomes available, more effective clinical studies can be designed and conducted to avoid therapeutic failure (due to the induction of metabolic clearance) and unexpected toxicity (due to overdose that results from impaired metabolic clearance via polymorphic enzymes or enzyme inhibition). Likewise, in vitro drug metabolism studies can be valuable to help medicinal chemists effectively assess and address metabolic liabilities during the discovery stage.

Several polymorphisms in CYP enzyme have been reported as a result of SNPs, gene deletions, and gene duplications (Rodrigues and Rushmore, 2002). CYP2D6 and CYP2C19 are polymorphically expressed in different populations such as Asian and Caucasian (Nakamura et al., 1985; Kimura et al., 1998). If the in vivo clearance of a drug is largely mediated by a polymorphically expressed allelic variant of a CYP, then poor metabolizers will be anticipated to produce elevated exposure [area under the curve (AUC)] and/or an increased half-life that adversely affect drug safety and efficacy. In addition, the catalytic activity of CYP3A4-dependent midazolam hydroxylation has been reported to range from 95 to 651 pmol $\mathrm{min}^{-1}\,\mathrm{mg}^{-1}$ among 20 individual HLM (human liver microsomes) (Huang et al., 2004). This wide range of activity might cause interindividual variability for drug exposure. Therefore, it is important to determine the CYP contribution to the metabolism of drug candidates in the discovery process.

4.3.1.1 In Vitro Approaches for Reaction Phenotyping Given the importance of in vitro $f_{m,\mathrm{CYP}}$ data, CYP reaction-phenotyping studies are conducted at multiple stages during drug discovery and development. Possible steps to generate an integrated CYP reaction phenotype in vitro are needed before the data are integrated with clinical information (e.g., human radiolabel ADME and DDI studies).

cDNA-Expressed Human CYPs One of the approaches to study reaction phenotyping is the use of recombinant reagents. The specific activity of each CYP toward a given compound can be characterized with recombinant reagents. It is very important when attempting to study a particular biotransformation in the absence of competing enzymes and reactions. CYP reaction-phenotyping studies using a panel of recombinant CYPs (rCYPs) are regularly conducted by monitoring either substrate depletion or metabolite formation. Specific activity (pmol/min/pmol) of each rCYP can be determined. Reactions catalyzed by individual rCYPs can be optimized and definitive kinetic parameters such as K_m and $V_{\max}$ (or k_{cat}) can be determined. When multiple CYP enzymes are involved, the relative contribution of each CYP can be estimated using a CYP abundance-normalized approach [Eq. (4.1)] (Rodrigues,

1999), relative activity factors [Eqs. (4.2) and (4.3): v_i = velocity of rCYP isoform i, A_i = relative abundance of rCYP isoform i in human liver microsomes, $CL_{int,HLM}$ and $CL_{int,CYP}$ = intrinsic clearance for HLM and rCYP] (Venkatakrishnan et al., 2000) or intersystem extrapolation factors (Proctor et al., 2004):

$$f_{m,CYP_n} = \frac{\text{pmol/min/pmol}(\text{rCY}P_n) \cdot \text{pmol}(\text{mCY}P_n)/\text{mg(LM)}}{\sum \text{pmol/min/pmol}(\text{rCY}P_n) \cdot \text{pmol}(\text{mCY}P_n)/\text{mg(LM)}} \quad (4.1)$$

$$f_{m,CYP_n} = \frac{A_i \cdot v_i}{\sum_{i=1}^{n} A_i \cdot v_i} \quad (4.2)$$

$$\text{RAF} = \frac{CL_{int,HLM}(\mu L/\text{min/mg})}{CL_{int,rCYP}(\mu L/\text{min/pmol})} = \frac{V_{max,HLM} \cdot K_{m,rCYP}}{V_{max,rCYP} \cdot K_{m,HLM}} \quad (4.3)$$

Chemical Inhibition Chemical inhibitors have proven to be extremely valuable and in some cases have been used specifically for individual CYPs. These compounds have a number of inherent advantages in that they are readily available from either chemical synthesis or chemical or pharmaceutical companies. In addition, chemical inhibitors can be used with intact cells and in vivo, even with human subjects; thus a particular CYP may be linked with a specific toxicological or pharmacological response. The fractional inhibition of a reaction in human liver microsomes reflects the extent to which a particular CYP (or class of CYP isoforms) is responsible for a reaction. Table 4.2 contains a list of

TABLE 4.2 IC_{50} Values of Selective Chemical Inhibitors for Individual CYPs in Pooled Human Liver Microsomes

CYP Involved	Reaction	Inhibitor	IC_{50} (μM)
CYP1A2	Phenacetin *O*-deethylation	Fluvoxamine	0.4
		a-Naphthoflavone	0.014
CYP2A6	Coumarin 7-hydroxylation	Tranylcypromine	0.15
CYP2B6	Bupropion hydroxylation	MBA[a]	0.05
CYP2C8	Taxol 6*a*-hydroxylation	Montelukast	0.27
CYP2C9	Diclofenac 4′-hydroxylation	Sulfaphenazole	0.7
CYP2C19	(*S*)-Mephenytoin 4′-hydroxylation	(*R*)-*N*-3-Benzyl-phenobarbital	0.4
CYP2D6	Bufuralol 1′-hydroxylation	Quinidine	0.1
	Dextromethorphan	Quinidine	0.13
CYP2E1	Chlorzoxazone 6-hydroxylation	4-Methylpyrazole	0.74
CYP3A4	Testosterone 6β-hydroxylation	Ketoconazole	0.03
	Midazolam 1′-hydroxylation	Ketoconazole	0.03

[a]MBA = methylbenzotriazole.

chemical inhibitors used commonly for reaction phenotyping. Table 4.3 lists K_m values of individual marker substrates for the CYPs. However, many available chemical inhibitors used to target individual CYPs are poorly or incompletely characterized or exhibit considerable overlapping inhibition of CYP isoforms. Ketoconazole has been widely used as a specific inhibitor for CYP3A4 and exhibited very low half maximal inhibitory concentration (IC_{50}) values for inhibition of the metabolism of drugs by CYP3A4. However, ketoconazole also inhibited activities of other CYPs in a concentration-dependent manner (Sai et al., 2000; Otton et al., 1984; Back et al., 1988; Ward and Back, 1993)). Therefore, ketoconazole is thought to be a good inhibitor of a broad spectrum of CYP isoforms depending on the concentration used. When used to define a role of CYP3A4 in the metabolism of a given drug, ketoconazole must be used carefully and other approaches may be used to confirm the conclusion made.

Inhibitory Antibodies The antibodies, particularly monoclonal antibodies (MAbs), that are inhibitory to enzyme function possess a large and additional dimension of usefulness in that they serve to determine the quantitative contribution of individual epitope-specific CYPs to any reaction in a tissue preparation that contains multiple CYPs. Thus, the amount of inhibition by a MAb defines the quantitative role of a single CYP in the overall metabolism of a given substrate (Gelboin, 1993; Gelboin et al., 1995, 1996, 1997; Mei et al., 1999; Shou and Lu, 2009). Inhibitory MAb has been employed to identify the phenotype of the responsible CYP and in combination with other approaches can offer a more precise assessment of the quantitative contribution of its target CYP to the metabolism, which in turn aids in the elucidation of the role of that form toward the total metabolism of the substrate in question. Although MAbs have some advantages in CYP identification and quantitation studies, there are a number of concerns in the application of MAbs. First, a library of inhibitory MAbs for all human individual CYPs is not commercially available. Second, some MAbs may not achieve complete inhibition. The reason for this could be that MAb–enzyme–substrate complex is productive (Mei et al., 2002). For example, only 60% inhibition was observed for the *p*-nitroanisole *O*-demethylation by cDNA-expressed CYP3A4. Third, many present MAbs cannot distinguish between closely related CYP subfamily members.

Correlation Analysis This approach is performed with a bank of HLM ($n > 10$) that has been analyzed to determine the activity of several CYPs (namely, CYP1A2, CYP2A6, CYP2B6, CYP2C8, CYP2C9, CYP2C19, CYP2D6, CYP2E1, and CYP3A4/5). Such banks of HLM are commercially available as kits in which the content and specific activity of each individual CYPs with marker substrates are provided. A list of marker substrates used to probe specific activities of individual CYPs in human liver microsomes is shown in Table 4.3. Differences in the rates of metabolite formation in the panel of HLM are compared with the activities of CYP enzymes. Correlation analysis is then done by simple regression analysis (r^2 = regression coefficient), where the

TABLE 4.3 K_m Values for Individual CYPs in Pooled Human Liver Microsomes

Enzyme	Substrate	Metabolite	K_m (μM)
CYP1A2	Phenacetin	Acetaminophen	87.7
CYP2A6	Coumarin	7-OH-Coumarin	97.8
CYP2B6	Bupropion	OH-Bupropion	98.0
CYP2C8	Taxol	6α-OH-Taxol	14.4
CYP2C9	Diclofenac	4′-OH-Diclofenac	11.2
CYP2C19	(*S*)-Mephenytoin	4′-OH-Mephenytoin	77.9
CYP2D6	Bufuralol	1′-OH-Bufuralol	14.0
	Dextromethorphan	Dextrorphan	11.7
CYP2E1	Chlorzoxazone	6-OH-Chlorzoxazone	274
CYP3A4	Testosterone	6β-OH-Testosterone	47.2
	Nifedipine	Oxidized nifedipine	13.0
	Midazolam	1′-OH-Midazolam	2.98

[a]CYP-selective marker assays.
[b]*N*-(α-Methylbenzyl-)-1-aminobenzotriazole.

marker CYP enzyme activity is the independent variable and the rate of formation of drug metabolite is the dependent variable. The latter determination also provides the statistical significance of the relationship. If two kinetically distinct enzymes are involved in a given reaction, then the correlation analysis may be performed at multiple concentrations to identify the enzyme (affinity and capacity) that is more relevant at a given substrate concentration. The impact of concentrations used on the correlation is attributed to low or high K_m or V_{max} of the enzymes involved.

4.3.1.2 In Vivo Approaches for Reaction Phenotyping The degree to which CYP contributes to the clearance of a drug, that is, the fraction of the drug metabolized by this isozyme ($f_{m,CYP3A4}$), can be obtained from in vivo data (e.g.,human radiolabeled compound disposition studies or drug interaction studies with CYP isoform selective inhibitors). The $f_{m,CYP}$ values can be determined from clinical DDIs associated with selective CYP inhibition. The extent to which systemic or oral clearance of a drug is altered in the presence of an isozyme-specific inhibitor reveals the fraction of that drug cleared by the CYP. A number of CYP3A4-selective inhibitors have been widely used for coadministration with drugs suspected to be CYP3A4 substrates. These include the competitive inhibitors ketoconazole, fluconazole, itraconazole, and voriconazole and the mechanism-based inhibitors mibefradil, ritonavir, diltiazem, clarithromcyin, and saquinavir. The $f_{m,CYP3A4}$ values were calculated by Eq. (4.4) and are shown in Table 4.4 (Shou et al., 2008). As seen in Table 4.4, the $f_{m,CYP3A4}$ values from many drug interaction studies ranged markedly from 0.05 (ropivacaine) to 0.95 (buspirone). High fractions indicate that the drugs are cleared mainly by CYP3A4 and low fractions imply that the drugs are cleared not only by CYP3A4 but also possibly by other routes that

TABLE 4.4 Fractions of Drugs Cleared by CYP3A4 (f_m, $_{CYP3A4}$, CYP3A4 Reaction Phenotyping) Calculated by Decreased Systemic Clearance or Increased AUC in Presence of Individual CYP3A4 Inhibitors Reported in the Literature

Substrate	f_m	Substrate	f_m
Alfentanil	87(1)	Methadone	25(4)
Alprazolam	69	Midazolam	89
Amprenavir	22	Mirtazapine	35(10)
Antipyrine	25	Nefazodone	31
Atorvastatin	60(5)	Nifedipine	63(4)
Budesonide	76(5)	Omeprazole	37
Buspirone	93–95(5)	Pravastatin	52(6)
Codeine	32(9)	Prednisolone	27
Cyclophosphamide	43(4)	Quetiapine	84
Cyclosporine	79	Quinidine	61(5)
Diazepam	24(5)	Quinine	46(1)
Digoxin	23(3)	Repaglinide	30(6)
Ethinyl estradiol	19(4)	Ritonavir	23
Etizolam	35(5)	Ropivacaine	5(5)
Everolimus	77(8)	Saquinavir	66
Gefitinib	44(5)	Sildenafil	66(7)
Glyburide	25(6)	Simvastatin	91(5)
Imatinib	29	Tacrolimus	58
Indinavir	67–90(2)	Tirilazad	41
Lopinavir	43	Triazolam	91
losartan	21	Verapamil	39
Mefloquine	17	Zolpidem	26(3)

Note: $f_{m,\text{CYP3A4}}$ values were calculated [Eq.(4.18)] by the decreased oral clearance (or systemic clearance) or increased AUC of the drugs in the presence of ketoconazole or other individual CYP3A4 inhibitors as indicated by numbers in parentheses: (1) troleandomycin; (2) ritonavir; (3) mibefradil; (4) fluoconazole; (5) itraconazole; (6) clarithromycin; (7) Saquinavir; (8) erythromycin; (9) quinidine; (10) cemetidine.

Source: From Shou et al., 2008.

include conjugative enzyme- or transporter-mediated pathways, renal or hepatic excretion:

$$f_{m,\text{CYP3A4}} = 1 - \frac{\text{CL}'_{[\text{I}]}}{\text{CL}_{[\text{ctr}]}} = 1 - \frac{\text{AUC}_{[\text{ctr}]}}{\text{AUC}'_{[\text{I}]}} \tag{4.4}$$

4.3.2 Reversible CYP Inhibition

The exposure of a victim drug in human subjects may be altered by a coadministered perpetrator drug as a result of inhibition of a DME(s). Such phenomena are termed pharmacokinetics-related clinical DDIs that can sometimes severely compromise the safety and efficacy of the victim drug. Mechanisms of CYP inhibition can be divided broadly into two types: reversible inhibition and time-dependent inactivation.

Kinetic Model Reversible CYP inhibition is dependent on the mode of interaction between CYP enzymes and inhibitors and is further characterized as competitive, noncompetitive, uncompetitive, and mixed. Evaluation of reversible inhibition of CYP reactions is often conducted under conditions where Michaelis–Menten (MM) kinetics is obeyed. Based on Scheme 1 below, various types of reversible inhibition are described from the scheme during catalysis which can lead to enzyme inhibition:

$$\begin{array}{ccccc} E + S & \overset{K_S}{\rightleftharpoons} & ES & \xrightarrow{k_p} & E + P \\ + & & + & & \\ I & & I & & \\ K_i \updownarrow & & \alpha K_i \updownarrow & & \\ EI + S & \overset{\alpha K_S}{\rightleftharpoons} & SEI & & \end{array}$$

1

where E, S, I, and P represent enzyme, substrate, inhibitor, and product, respectively, EI the enzyme–inhibitor complex, ES the enzyme–substrate complex, and SEI the substrate–enzyme–inhibitor complex; K_i is the inhibition constant, K_s the dissociation constant of the ES complex, and the factor by which K_i and K_s values change following the formation of the SEI complex. The K_s value would approximate K_m if $k_p << k_{-1}$. The plots and rate equations for competitive, noncompetitive, uncompetitive, and mixed-type inhibition are indicated in Figure 4.1 and in Eqs. (4.5)–(4.8), respectively:

$$v = \frac{V_{\max}[\mathrm{S}]}{K_m(1 + [\mathrm{I}]/K_i) + [\mathrm{S}]} \tag{4.5}$$

$$v = \frac{V_{\max}[\mathrm{S}]}{K_m(1 + [\mathrm{I}]/K_i) + [\mathrm{S}](1 + [\mathrm{I}]/K_i)} \tag{4.6}$$

$$v = \frac{V_{\max}[\mathrm{S}]}{K_m + [\mathrm{S}](1 + [\mathrm{I}]/K_i)} \tag{4.7}$$

$$v = \frac{V_{\max}[\mathrm{S}]}{K_m(1 + [\mathrm{I}]/K_i) + [\mathrm{S}](1 + [\mathrm{I}]/\alpha K_i)} \tag{4.8}$$

Determination of IC_{50} and K_i Values In the pharmaceutical industry, NCEs are screened in vitro for their ability to inhibit P450 isoforms and attempts are made to predict the potential of clinical DDIs. The IC_{50} value, the concentration of the inhibitor that gives 50% of maximal inhibition of an individual P450 activity, usually is employed as an estimate of the P450 inhibitory potency, since the measurement of IC_{50} is simple and rapid. If the mechanism of inhibition is known, then the IC_{50} value also can be used to estimate K_i. When [S] approaches K_m ([S] = K_m) and velocity is half of $V_{\max}$ ($v = 0.5V_{\max}$ and [I] = IC_{50}), the above eqations

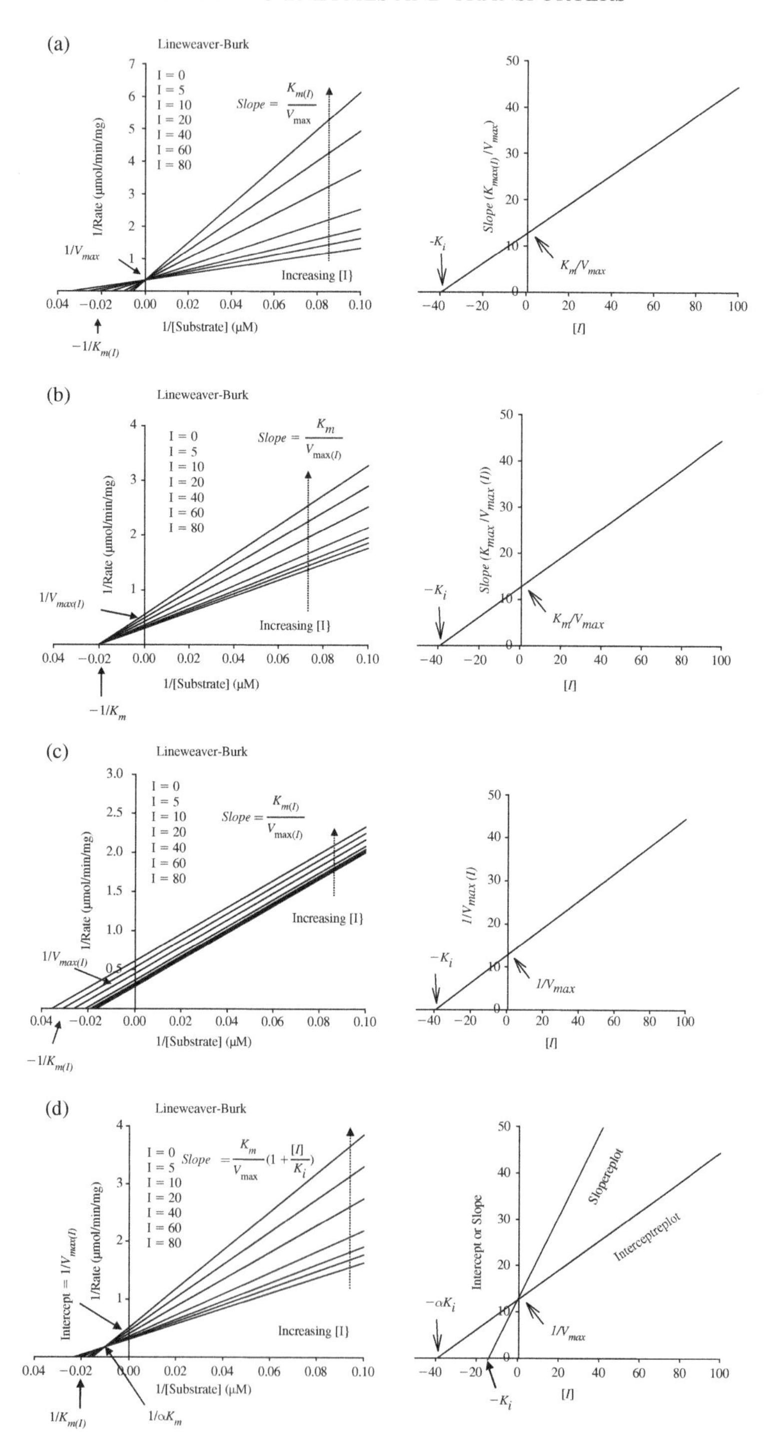

FIGURE 4.1 Methods to measure K_i values of (a) competitive, (b) noncompetitive, (c) uncompetitive, and (Shou, 2005), (d) mixed-type inhibition.

can be simplified greatly, and the IC_{50} are essentially equal to $2K_i$ for both competitive and uncompetitive inhibition and to K_i for noncompetitive inhibition.

Prediction of Clinical DDI from Reversible P450 Inhibition Prediction of the probable in vivo DDI potential is crucial at the discovery stage. In vitro enzyme inhibition kinetics (IC_{50} and K_i for reversible enzyme inhibition and K_I and k_{inact} for time-dependent inhibition) and reaction phenotyping ($f_{m,,CYP}$) are typically required to assess the degree of inhibition. Ratios of systemic exposure in the presence and absence of inhibitor ($AUC_{[I]}/AUC_{ctr}$) can be determined according to appropriate in vitro approaches (Bjornsson et al., 2003; Obach, 2003; Obach, et al., 2005, 2006; Shou, 2005; Ito et al., 2004, 1998; Mayhew et al., 2000; Kanamitsu et al., 2000; Azie et al., 1998; von Moltke et al., 1998).

If enzyme inhibition is competitive, then $CL_{int,\ [I]}$ may vary with regard to the type and concentration of inhibitor. The concentrations of an inhibitor (or drug) that are relevant to clinical application can be approached for the prediction in the in vivo situation. If a drug is eliminated due to both hepatic metabolism ($f_{m,hept} = 1 - f_{renal}$) and renal excretion ($f_{renal}$), then the fraction of the drug metabolized by the inhibited enzyme ($f_{m,hept} \cdot f_{m,CYP}$) can be introduced to the prediction. With inclusion of $f_{m,CYP}$, the ratio change in AUC in the presence and absence of an inhibitor can be expressed as

$$\frac{AUC_{[I]}}{AUC_{ctr}} = \frac{1}{f_{m,hept} \cdot f_{m,CYP}/(1 + [I]/K_i) + (1 - f_{m,hept} \cdot f_{m,CYP})} \tag{4.9}$$

where $f_{m,hept}$ and $f_{m,CYP}$ represent the fractions of total hepatic metabolism and CYP-dependent metabolism catalyzed by the inhibited CYP, respectively; K_i the inhibitor dissociation constant; $f_{u,p}$ the unbound fraction of inhibitor in plasma and [I] the projected plasma concentration of inhibitor (usually $C_{max,ss}$). If the hepatic clearance is due largely to a CYP ($f_{m,hept} = 1$) or drug is cleared by the CYP-mediated metabolism ($f_{m,hept} = 1$ and $f_{renal} = 1 - f_{m,hept} = 0$), then Eq. (4.9) can be simplified to Eqs. (4.10) and (4.11), respectively. The clinical implication of CYP inhibition by inhibitor is dependent on the in vivo concentration of the inhibitor and the role of that CYP in the metabolism of the coadministered drug (f_m):

$$\frac{AUC_{[I]}}{AUC_{ctr}} = \frac{1}{f_{m,CYP}/(1 + [I]/K_i) + (1 - f_{m,CYP})} \tag{4.10}$$

$$\frac{AUC_{[I]}}{AUC_{ctr}} = 1 + \frac{[I]}{K_i} \tag{4.11}$$

When intestinal metabolism is considered, the oral bioavailability in gut (F_g) can be affected and the change in AUC of a victim drug is described by the equations

$$\frac{\mathrm{AUC}_{\mathrm{po},i}}{\mathrm{AUC}_{\mathrm{po}}} = \frac{F_{g,i}}{F_g} \cdot \frac{1}{f_m/(1 + [\mathrm{I}]/K_i) + (1 - f_m)} \tag{4.12}$$

$$\frac{F_{g,[\mathrm{I}]}}{F_{g,\mathrm{ctr}}} = \frac{1}{F_{g,\mathrm{ctr}} + (1 - F_{g,\mathrm{ctr}}) \cdot (CL_{\mathrm{int},g,[\mathrm{I}]}/CL_{\mathrm{int},g})} \tag{4.13}$$

However, the intracellular inhibitor concentration available to the target enzyme cannot be measured. The concentration can be approximated using that in hepatic inlet circulation ($[\mathrm{I}]_{\mathrm{hept\ inlet}}$) and can be calculated as

$$[\mathrm{I}]_{\mathrm{hep.inlet}} = [\mathrm{I}]_{\max} + \frac{K_a \cdot F_a \cdot D}{Q_h} \tag{4.14}$$

where K_a, F_a, D, Q_h, and $[\mathrm{I}]_{\max}$ ($C_{\max} = [\mathrm{I}]_{\max}$) are the absorption rate constant, fraction of dose absorbed, dose of inhibitor, hepatic blood flow rate, and plasma inhibitor concentration, respectively. The concentration of inhibitor corrected with unbound fraction should also be considered when predicting DDIs. For example, montelukast was shown in vitro to be a potent CYP2C8 inhibitor with a K_i value of 0.11 μ*M* for inhibition of rosiglitazone *N*-demethylase activity. Using the total $C_{\max}$ value of 1 μ*M* following once-daily dosing of 10 mg of montelukast, the projected increase in AUC of rosiglitazone would be about eight-fold. However, montelukast is highly protein bound with a free fraction of 0.4% in human plasma. Using the unbound $C_{\max}$ of montelukast (around 4 n*M*), no interaction would be expected between montelukast and rosiglitazone. The later projection has been confirmed by clinical study which showed no DDIs between montelukast and rosiglitazone (Kim et al., 2007). Similar cases were also observed in the development of rolofylline.

4.3.3 Time-Dependent Inhibition

Time-dependent inhibition defined mainly by mechanism-based inhibition (MBI), which includes CYP suicide inactivation (irreversible inhibition, the more widely studied process) and metabolite–intermediate (MI) complex formation (quasi-irreversible inhibition), is responsible for most clinically significant DDIs (Silverman, 1995; Waley, 1980; Zhou et al., 2005). Suicide inactivation involves the formation of a reactive intermediate that irreversibly inactivates the CYP in the process of catalytic turnover. Quasi-irreversible inhibition occurs when the CYP produces a metabolite (e.g., nitroso intermediate) with the capacity to bind tightly to the CYP heme. TDI (time-dependent inhibition) can be characterized (1) to be dose dependent, (2) to be preincubation time dependent, (3) to have bioactivation of the inhibitor that is required for inactivation of the target enzyme, (4) to have de novo protein synthesis that is required to recover metabolic capacity, and (5) to have potentially slow onset of the effects but be more profound than reversible inhibition. If present, then TDI is the major component of overall enzyme inhibition and frequently leads to clinically relevant DDIs. Table 4.5 contains a list of inhibitors of TDI observed in vitro and in vivo.

Kinetic Model The kinetic model for enzyme inactivation can be expressed as

$$\mathrm{E} + \mathrm{I} \underset{k_{-1}}{\overset{k_1}{\rightleftharpoons}} \mathrm{EI} \xrightarrow{k_2} \mathrm{EI'} \xrightarrow{k_4} \mathrm{E_{inact}} \qquad \mathrm{EI'} \xrightarrow{k_3} \mathrm{E} + \mathrm{P}$$

2

where E, I, and P stand for enzyme, inhibitor, and product, respectively, EI is the initial binding of inhibitor to enzyme (the enzyme–inhibitor complex) and EI′ is the active form of the complex in which the inhibitor is catalyzed to the reactive intermediate, and E_{inact} is the enzyme inactivated by the reactive metabolite formed. Inactivation of the enzyme is an irreversible process over the time scale of the experiment. At the given concentrations of inhibitor and enzyme, the reactions indicated in Scheme 1 are governed by the first-order rate constants k_1, k_{-1}, k_2, k_3, and k_4, respectively. The inactivation of a given enzyme (percentage of remaining enzyme activity, $[E]_{remaining}/[E]_{total}$) can be evaluated by the equation (Kitz and Wilson, 1962; Jung and Metcalf, 1975)

$$\ln \frac{[E]}{[E]_{tot}} (\text{Fraction activity remaining}) = \frac{-k_{inact} \cdot [I] \cdot t}{K_I + [I]} \tag{4.15}$$

where [E] is the active enzyme concentration at time t; [I] the inhibitor concentration, k_{inact} the maximum inactivation rate constant [Eq. (4.16)], and K_I the inhibitor concentration that produces the half-maximal rate of inactivation [Eq. (4.17)]. The functions K_I and k_{inact} are functions of the rate constants in Scheme 2 (Waley, 1980; Tatsunami et al., 1981):

$$k_{inact} = \frac{k_2 \cdot k_4}{k_2 + k_3 + k_4} \tag{4.16}$$

$$K_I = \frac{k_{-1} + k_2}{k_1} \cdot \frac{k_3 + k_4}{k_2 + k_3 + k_4} \tag{4.17}$$

Determination of k_{inact} and K_I Values The experiments to evaluate time-dependent P450 inhibition are usually performed via incubation of the enzyme and cofactors with an inhibitor for different durations followed by at least a 10-fold dilution of the reaction mixture and further incubation of the diluted enzyme/inhibitor mixture with a probe substrate to assess the remaining enzyme activity. The dilution step is designed to minimize the contributions from potential competitive inhibition by the inhibitor. The apparent first-order inactivation rate constants (k_{obs}) at each inhibitor concentration are estimated from the initial slopes of a natural logarithm plot of the remaining enzyme activity versus the preincubation time, whereas the k_{inact} and K_I values are estimated by nonlinear regression using the equation

$$k_{obs} = \frac{k_{inact} \cdot [I]}{K_I + [I]} \tag{4.18}$$

Alternatively, the inhibitory parameters can be obtained using the double-reciprocal Lineweaver–Burk plot shown below:

$$\frac{1}{k_{obs}} = \frac{K_I}{k_{inact}} \times \frac{1}{I} + \frac{1}{k_{inact}} \tag{4.19}$$

The observed inactivation rates (k_{obs}) at individual inhibitor concentrations are calculated from the initial (negative) slopes of the linear regression lines of semilogarithmic plots (natural logarithm of remaining activity versus preincubation time) (Fig. 4.2a). Values of k_{inact} and K_I can be obtained by nonlinear regression using Eq. (4.18) (Fig. 4.2b). In many cases, the parameters can also be generated by the double-reciprocal Lineweaver–Burk plot ($1/k_{obs}$ vs. $1/[I]$) as depicted in Figure 4.2c [Eq. (4.19)], which shows the k_{inact} estimate at the reciprocal of the y intercept and K_I at the negative reciprocal of the x intercept. The K_I and k_{inact} values of some inhibitors are listed in Table 4.6.

Prediction of Clinical DDI from TDI The prediction of MBI-mediated DDI is more difficult than that for reversible CYP inhibition. Equation (4.20) is a commonly used model for the quantitative prediction of human DDIs (Mayhew et al., 2000; Brown et al., 2005; Galetin et al., 2006; Wang et al., 2004):

$$\frac{AUC_{[I]}}{AUC_{[contr]}} = \left(\frac{f_m}{1 + \dfrac{k_{inact} \cdot f_{u,p} \cdot [I]}{(K_I + f_{u,p} \cdot [I])k_E}} + (1 - f_m) \right)^{-1} \tag{4.20}$$

where f_m represents the fraction of CYP-dependent metabolism catalyzed by the targeted CYP; k_E the rate constant for endogenous enzyme degradation; k_{inact} the maximum rate constant for enzyme inactivation; K_I the inhibitor dissociation constant; $f_{u,p}$ the fraction of plasma-unbound inhibitor; and [I] the plasma concentration of inhibitor (usually $C_{max,ss}$). If the drug elimination is due largely to CYP3A4 ($f_m = 1$), Eq. (4.20) can be simplified to

$$\frac{AUC_{[I]}}{AUC_{ctr}} = 1 + \frac{k_{inact} \cdot f_{u,p} \cdot [I]}{(K_I + f_{u,p} \cdot [I])k_E} \tag{4.21}$$

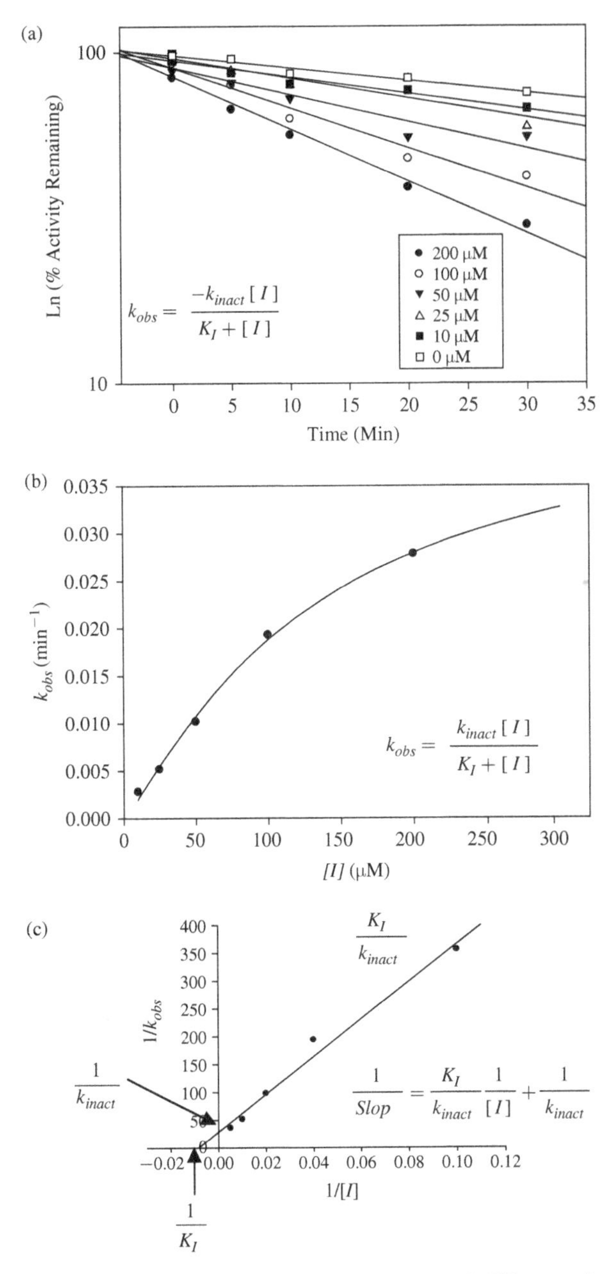

FIGURE 4.2 Method to determine KI and kinact of Zileuton in human liver microsomes (Lu et al., 2003). (a) aliquots were removed from the primary reaction mixture at the indicated time points and were assayed for residual marker activity (phenacetin *O*-deethylation). Percent activity remaining (related to time zero in the presence of solvent alone) was plotted in the logarithmic scale determined from a single experiment. The slopes in linear ranges at various inhibitor concentrations are defined as kobs. (b) Determination of kinact (0.045 min^{-1}) and K_I (131.7 μM) using nonlinear regression [Eq. (4.8)]. (c) Determination of kinact (0.035 min^{-1}) and K_I (117 μM) using Lineweaver-Burk plot [Eq. (4.9)].

TABLE 4.5 Known Mechanism-Based CYP Inhibitors

CYP	Mechanism-Based Inhibitor
1A	1-Aminobenzotriazole, dihydralazine, furafylline, isoniazid, *N*-desethylamiodarone, oltipraz, *trans*-resveratrol, zileuton, *N*-desethylamiodarone, rhapontigenin
2B6	Bergamottin (GF-I-2), clopidogrel, ethinyl estradiol, glabridin, *N*-desethylamiodarone, phencyclidine, tamoxifen, thiotepa, ticlopidine
2C8	Amiodarone, fluoxetine, gemfibrozil glucuronide, isoniazid, nortriptyline, phenelzine, verapamil
2C9	Silybin (milk thistle derivative), suprofen, ticrynafen (tienilic acid)
2C19	Isoniazid, ticlopidine, BPB
2D6	(−)-Chloroephedrine, cimetidine, ecstasy, EMTPP, L-754,394, metoclopramide, N-desethylamiodarone, Paroxetine
3A4	(−)-Clusin, (−)-dihydroclusin, (−)-dihydrocubebin, (−)-hinokinin, (−)-hydrastine, (−)-yatein, 1-aminobenzotriazole, 4-ipomeanol, 6,7-dihydroxybergamottin (DHB), bergamottin (GF-I-2), amiodarone, amprenavir, azithromycin, clarithromycin, delavirdine, diclofenac, dihydralazine, diltiazem, *N*-desmethyldiltiazem, DPC 681, erythromycin, ethinyl estradiol, fluoxetine, furanocoumarin dimer (GF-I-1), furanocoumarin dimer (GF-I-4), gestodene, glabridin, gomisin C, irinotecan, irinotecan metabolite SN-38, isoniazid, L-754,394, limonin, lopinavir, mibefradil, midazolam, mifepristone (RU 486), nelfinavir, nicardipine, oleuropein, raloxifene, resveratrol, ritonavir, rutaecarpine, saquinavir, silybin (milk thistle derivative), tadalafil, tamoxifen, *N*-desmethyltamoxifen, troleandomycin, verapamil, norverapamil, *N*-dealkylverapamil, zafirlukast

Source: Reported in the database WWW.druginteractioninfo.org.

TABLE 4.6 K_I and k_{inact} Values from in vitro Mechanism-Based CYP3A4 Inhibition

Inhibitor	Substrate	K_I (μ*M*)	k_{inact} (min^{-1})
Clarithromycin	Triazolam	5.5	0.07
	Cisapride	5.5	0.07
	Midazolam	5.5	0.07
Erythromycin	Simvastatin	10.9	0.05
	Buspirone	10.9	0.05
	Midazolam	10.9	0.05
Mibefradil	Midazolam	2.3	0.4
Nelfinavir	Simvastatin	1	0.22
Ritonavir	Triazolam	0.17	0.4
Saquinavir	Midazolam	0.65	0.26
Verapamil	Simvastatin	4.2	0.09
	Buspirone	4.2	0.09
	Midazolam	4.2	0.09

If the CYP in the gut contributes to the metabolism of a drug and is inhibited by a time-dependent inhibitor, then a change of bioavailability of the drug due to decreased gut metabolism should also be taken into account. The ratio of $F_{g,[I]}$ to $F_{g,\mathrm{ctr}}$ (the intestinal wall bioavailability in the presence and absence of inhibitor, respectively) can be incorporated into the model [Eq. (4.22)]. The ratio of $F_{g,[I]}$ to $F_{g,\mathrm{ctr}}$ can be estimated from the relative change in $\mathrm{CL}_{\mathrm{int},g}$ caused by the inhibitor [Eq. (4.13); see DDI prediction from reversible inhibition]:

$$\frac{\mathrm{AUC}_{[I]}}{\mathrm{AUC}_{ctr}} = \frac{F_{g,[I]}}{F_{g,\mathrm{ctr}}}\left(\frac{f_m}{1+\dfrac{k_{\mathrm{inact}}\cdot f_{u,p}\cdot[\mathrm{I}]}{(K_I+f_{u,p}\cdot[I])k_E}}+(1-f_m)\right)^{-1} \tag{4.22}$$

4.3.4 Prediction of Clinical DDIs from CYP Induction

CYP induction is not of concern for safety. However, CYP induction can lead to serious therapeutic failure due to increased clearance of a given drug (victim) by an inducer. Prediction of in vivo DDIs from in vitro data has been attempted in drug discovery. Unfortunately, little progress has been made in this regard. Development of predictive models of DDIs from CYP induction should be based on fundamental principles and assumptions of in vitro-in vivo extrapolation (IVIVE). The prediction can aid in (1) selecting compounds for further development; (2) developing structure–activity relationships (SARs) to avoid the potential for DDIs; and (3) planning of clinical DDI studies for compounds that are advanced into further drug development. Mathematical models for CYP3A4 induction have been proposed (Shou et al., 2008; Fahmi et al., 2008). Based on pharmacokinetic principles, the AUC ratio of a substrate in the presence and absence of an inducer is expressed as

$$\frac{\mathrm{AUC}'_{[\mathrm{Ind}]}}{\mathrm{AUC}_{\mathrm{Ctr}}} = \left[f_{m,\mathrm{CYP3A4}}\cdot\left(1+\frac{E_{\max}\cdot[\mathrm{Ind}]^n}{(\mathrm{EC}_{50}^n+[\mathrm{Ind}]^n)}\right)+(1-f_{m,\mathrm{CYP3A4}})\right]^{-1} \tag{4.23}$$

where $E_{\max}$ and EC_{50} are the maximum response (net maximum fold-increase) of induction and inducer concentration at 50% $E_{\max}$, respectively, and [Ind] is the maximum inducer concentration at steady state. In many cases, $E_{\max}$ and EC_{50} are not readily obtained from concentration–response curves measured by in vitro induction assays because of limitations imposed by drug solubility, cell permeability, or toxicity. In these instances the slope of the induction response curve (equivalent to $E_{\max}/\mathrm{EC}_{50}$) at a more experimentally feasible low concentration range of the inducer can be used for the prediction [Eq. (4.24)]. This equation is, however, only applicable if in vivo concentrations of an

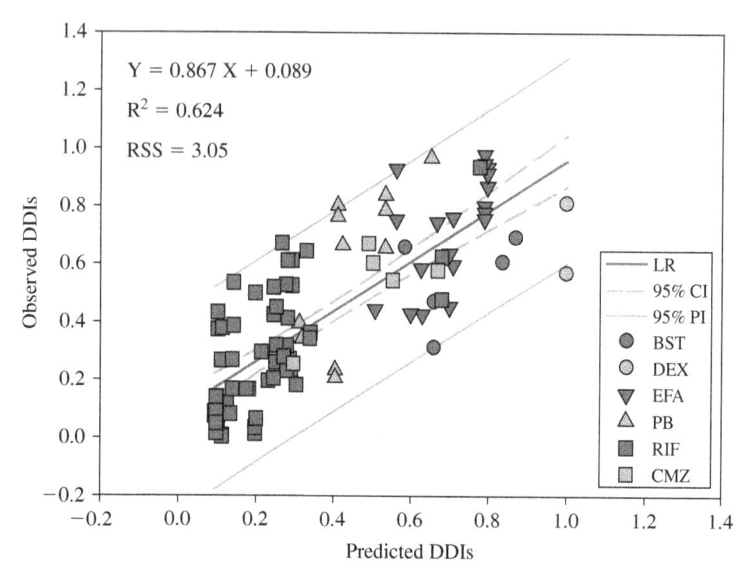

FIGURE 4.3 Prediction of clinical DDIs from CYP3A4 induction in human primary hepatocytes using Eq. (4.25) (Shou et al., 2008).

inducer are low ([Ind] << EC_{50}). Equation (4.25) shows the predictive model with inclusion of fractions unbound in plasma ($f_{u,p}$) and hepatocytes ($f_{u,\text{hept}}$), based on the hypothesis that only unbound drug (inducer) can access the intracellular nuclear receptor responsible for induction. A correlation analysis between predicted and observed DDIs mediated by CYP3A4 induction is given in Figure 4.3 (Shou et al., 2008):

$$\frac{\text{AUC}'}{\text{AUC}} = \frac{1}{f_{m,\text{CYP3A4}} \cdot (1 + \text{Slope} \cdot [\text{Ind}]^n) + (1 - f_{m,\text{CYP3A4}})} \tag{4.24}$$

$$\frac{\text{AUC}'}{\text{AUC}} = \left[f_{m,\text{CYP3A4}} \cdot \left(1 + \frac{E_{\max} \cdot [f_{u,p} \cdot \text{Ind}]^n}{[\text{EC}_{50} \cdot f_{u,\text{hept}}]^n + [f_{u,p} \cdot \text{Ind}]^n} \right) + (1 - f_{m,\text{CYP3A4}}) \right]^{-1} \tag{4.25}$$

Simcyp software (Simcyp Limited, Sheffield, United Kingdom) has been used as a population-based simulator to predict in vivo DDIs from in vitro DM data. Simcyp integrates information on genetic, physiological, and demographic variability relevant to the appropriate clinical studies. All the in vivo demographic data, such as the number of subjects, the minimum and maximum age, the proportion of males and females, and the ethnic mix of the groups, were incorporated into these simulations. The prediction in AUC with interindividual variability in the populations can be provided. The prediction of DDI from CYP inhibition and induction can be accomplished with the program.

4.3.5 Factors Affecting DDI Prediction

Prediction of drug interactions associated with CYP inhibition from in vitro kinetic data can be complicated by the accuracy of assumptions and many other unknown factors. These include:

1. The choice of steady-state plasma concentrations (C_{max}, C_{ave}, and C_{trough}) or the concentration in hepatic circulation ($C_{hept\ inlet,max}$) of the inhibitor as an estimate of the concentration at the active site of the enzyme (hepatic intracellular concentration). The $C_{hept\ inlet,max}$ is estimated using C_{max} and the rate of absorption after oral administration [Eq. (4.10)]. The DDI may be under- or overpredicted if the drugs are substrates of hepatic transporters (efflux or uptake), show poor membrane permeability, and/or are prone to rapid metabolism.
2. Incorporation of the unbound fraction of some drugs (f_u) in plasma and liver microsomes has been shown to improve the correlation, particularly for the drugs that are highly bound (Austin et al., 2002; Giuliano et al., 2005).
3. Accurate in vitro assessment of CYP contribution to metabolism (f_m). If a drug is metabolized by multiple enzymes which are inhibited by the inhibitor, then the f_m value of the drug is contributed by multiple enzymes ($f_m = f_{m,CYP3A4} + f_{m,CYP2C8} + \cdots$) (Rodrigues, 2005).
4. Involvement of both reversible and mechanism-based inhibitions by a drug.
5. Choice of K_E value. For CYP3A4, its rate constant of protein degradation k_{deg} has been reported to range from 0.00008 to 0.0006 min^{-1} (Shou et al., 2008; Yang et al., 2008).
6. Interindividual variability of enzymes in humans.
7. Enzyme inhibition by metabolite(s) formed.
8. Simultaneous induction (autoinduction) and inhibition (Greenblatt et al., 1999).
9. Potential complication due to renal or hepatic impairment.
10. Interactions with dietary ingredients.

4.4 CYP CONCLUSION

Evaluation of NCEs as substrates or inhibitors for human CYPs is routinely performed during the drug development process in industry. The data generated in vitro are well extrapolated to predict in vivo pharmacokinetics and DDIs. In addition, the role of drug transporters in drug disposition and drug interactions has been explored and this area has gained more attention. With progress in these areas, it will be possible to fully integrate in vitro ADME information not only for CYPs but also for transporters and phase II and non-CYP oxidative enzymes. These will enhance the ability to move drug candidates forward during the drug development process with adequate confidence and to minimize clinical adverse effects due to changed pharmacokinetics in either significant polymorphic enzymes or metabolism-based DDI.

4.5 DRUG TRANSPORTERS

Transporters are membrane-bound proteins that are expressed all over the human body. These transporters may play a critical role in small-molecule drug disposition (Beringer and Slaughter, 2005) or DDIs (Endres et al., 2006; Kindla et al., 2009) by taking up, extruding, or inhibiting excretion of drugs. The demand to understand the contributions of transporters in drug discovery (Ito et al., 2005; Tsuji, 2005; Terasaki and Ohtsuki, 2005; Niemi, 2007) by pharmaceutical companies has recently exploded in part due to the need to decrease lead optimization time (Hochman et al., 2007), because of the fiercely competitive market and short patent lifespan, the availability of new in vitro tools, and the 2006 Food and Drug Administration (FDA) Draft Guidance for Industry Drug Interaction Studies on transporters (http://www.fda.gov/cder/guidance/6695dft.pdf). There are many challenges for scientists who study transporters. These challenges range from the determination of which in vitro and in vivo assays may be appropriate to address the transporter issue in question to how to interpret those data generated from these assays. In this section, some of the key transporters that may require more extensive in vitro and in vivo studies are discussed. Numerous reviews are available that provide great detail on the transporters discussed in this chapter and their respective function. Some common transporter tools, from cultured cells to artificial membrane models, as well as advantages and limitations of these tools that are used in the pharmaceutical industry to address early- and late-stage ADME drug discovery studies are also discussed.

4.5.1 Key ADME Transporters

Currently over 800 transporters have been identified by the Human Genome Project (Venter et al., 2001) to transport ions, nutrients, sugars, vitamins, toxins, and so on, in and out of various parts of the human body. Some of these transporters play a role that contributes or limits drugs (Chu et al., 2007) from reaching their target sites in various organs. In addition, these transporters may contribute to adverse pharmacokinetic distribution (Zaher et al. 2008; Vaiyanathan et al., 2008) and pharmocodynamic effects (Elkiweri et al., 2009) that may lead to undesirable shifts in plasma concentration or DDIs of a therapeutic drug (Endres et al., 2006; Aszalos, 2007; Kindla et al., 2009; Yuan et al., 2009; Unakadt et al., 2010). In 2008, the FDA met with a panel of transporter experts from all around the world including scientists from academia and the pharmaceutical industry formed the International Transporter Consortium (ITC), to identify a "short list" of ADME transporters that may play a key role in clinical DDIs (http://www.fda.gov/downloads/

Drugs/DevelopmentApprovalProces/DevelopmentResources/DrugInteractions Labeling/ucm091581.pdf). Based on the current literature and clinical data, the ITC voted on which transporters remained on the "island of transporters" to be studied for potential contributions to clinical DDIs with the caveat that this so-called island is dynamic and can change at a later date to include other transporters if more clinically relevant data are available (Giacomini et al., 2010). The key transporters that were identified by the ITC are listed below with a brief review of their expression, function, and some DDI examples to warrant in vitro and in vivo investigations for potential contributions to clinical DDIs.

P-gp (ABCB1) P-glycoprotein, or P-gp, is the 170-kD protein product of the multiple-drug resistance (*MDR1*) gene. P-gp is the most studied mammalian efflux transporter to date (Krishna and Mayer, 2000; Ambudkar et al., 2006; Aszalos, 2007; Aller et al., 2009) Juliano and Ling (1976) demonstrated P-gp function in Chinese hamster ovary (CHO) cells that acquired drug resistance to chemotherapeutic agents. P-gp is expressed all over the body, including the gastrointestinal tract, liver, kidneys, adrenals, testes, ovaries, placenta, and brain (Schinkel, 1997; Ambudkar et al., 1999; Krishna and Mayer, 2000). P-gp is located in key organ sites that help eliminate compounds or xenobiotics from the body, for example, the luminal membrane of enterocytes, the biliary canicular membrane of hepatocytes, the luminal membrane of renal proximal tubules, and the apical membrane of endothelial cells of brain capillaries (Ambudkar et al., 1999). P-gp has a broad range of substrate specificity from beta-lactams to steroids and has been demonstrated to cause DDIs with a broad range of marketed compounds (Acharya et al., 2008; Elkiweri et al., 2009) (Table 4.7). The classic example of P-gp DDI is observed with digoxin and rifampin (Greiner et al., 1999; Ooi and Colucci, 2001). Because of its narrow therapeutic window, digoxin remains a concern for any coadministered drug(s) that may cause pharmacokinetic shifts in digoxin AUC. Numerous reviews have reported clinical implications to coadministered drugs that are P-gp substrates (Lin and Yamazaki, 2003; Ho and Kim, 2005).

BCRP (ABCG2) The breast cancer resistance protein (BCRP) transporter is also known as mitoxantrone resistance (MXR) protein (Miyake et al., 1999; Krishnamurthy and Schuetz, 2006). BCRP is expressed in plasma membrane of placental syncytiotrophoblasts of the chorionic villus (Mao, 2008; Robey et al., 2009), mammary glands, testis, canicular membrane of hepatocytes, cortical and proximal tubules in the kidney, villous tip of intestinal cells in the small intestine, hematopoietic stem cells, endothelial cells of venules and capillaries, and the luminal surface of the endothelial cells at the blood–brain barrier. The substrate specificity for BCRP overlaps broadly with P-gp; therefore, the clinical DDI implications for BCRP are as yet not well defined. The current limited data suggest that BCRP may have an impact on chemoresistance in cancer (Noguchi et al., 2009). Rosuvastatin and topotecan have been reported to increase plasma AUC for antivirals and cyclosporins (Allen et al., 2002; Maliepaard et al., 2001, Table 4.8).

TABLE 4.7 P-gp-Mediated Drug–Drug Interactions with ≥20% Change in AUC

Object	Precipitant	% Change in AUC	Object	Precipitant	% Change in AUC
(*R*)-Fexofenadine	Verapamil	119	Fexofenadine	Itraconazole	153.3
	Itraconazole	165.5		Ritonavir	168.4
(S)-Fexofenadine	Verapamil	186.6		Itraconazole	173.1–200.9
	Itraconazole	228.9		Lopinavir and ritonavir	248
Aliskiren	Atorvastatin	46.8		Verapamil	312.4
Atorvastatin	Erythromycin	32.5		Lopinavir and ritonavir	314.4
	Istradefylline	54.7		Ritonavir and indinavir	384.3
	Nelfinavir	74		Verapamil	151.2
	Cyclosporine	644.7	Ginkgo	Talinolol	21.6
Atorvastatin acid	Cyclosporine	1431.3	Glyburide (glibenclamide)	Clarithromycin	33.1
Azithromycin	Nelfinavir	106.7	Idarubicin	Cyclosporine	66.5–77
Buprenorphine	Ritonavir	57.4	Itraconazole	Cyclosporine	70.5
	Atazanavir and ritonavir	66.7	Lapatinib	Ketoconazole	266.8
	Atazanavir	93.2	Loperamide	Quinidine	80.2–148.1
Celiprolol	Itraconazole	80.6		Ritonavir	223
Cerivastatin	Cyclosporine	279.9–375.3		Itraconazole	279.6
Cetirizine	Pilsicainide	40.7		Gemfibrozil and itraconazole	1156.6
Cimetidine	Itraconazole	22.7	Methylprednisolone	Diltiazem	49.1
CP-481,715	Cyclosporine	428.3	Morphine	Quinidine	59.6
Cyclosporine	Posaconazole	33.2	Nelfinavir	Cyclosporine	28
	Carvedilol	35.5	Ornidazole	Ketoconazole	31.2

	Grapefruit juice	38.1	Paclitaxel	OC144-093 (ONT-093)	44.5
	Docetaxel	82.4		Cyclosporine	750
Dexamethasone	Valspodar (PSC 833)	23.5	Paroxetine	Itraconazole	55.2
Dicloxacillin	Cyclosporine	59.9	Pilsicainide	Cetirizine	36.1
Digoxin	Carvedilol	20	Prednisolone	Cyclosporine	45
	Cremophor RH40	21.4	Prednisone	Ritonavir	28.5–37
	Ritonavir	21.4	Quinidine	Itraconazole	142.2
	Talinolol	23.4	Ranitidine	Ketoconazole	74.3
	Carvedilol	24.1–56.5	Ranolazine	Diltiazem	66.1–66.3
	Clarithromycin	35.1–56.7		Verapamil	116.6
	Ranolazine	38.7–59.5	Risperidone	Verapamil	59.4
	Itraconazole	67.7	Ritonavir	Ketoconazole	27.8
	Valspodar (PSC 833)	73.9–204.5		Atazanavir	41.2
	Lopinavir and ritonavir	80.7	Saquinavir	Tenofovir	28.5
	Ritonavir	86.4		Cremophor EL	36.8–43.4
	Ranolazine	88.4		Atazanavir	44.6
	Quinidine	227.7		Ketoconazole	48.4
Docetaxel	Laniquidar (R101933)	26.9		Atazanavir	60–63
	MS209	38.5–96.1		Omeprazole	82.5
	Cyclosporine	632.4		Cyclosporine	329.8
Doxorubicin	Cremophor EL	23.3		Cremophor EL	401.2
	Zosuquidar (LY335979)	25		Ritonavir	1621.8–2968
	Paclitaxel	30.4	Simvastatin	Ranolazine	88.9

(*Continued*)

TABLE 4.7 (Continued)

Object	Precipitant	% Change in AUC	Object	Precipitant	% Change in AUC
	Elacridar (GF120918)	31.3–45.8		Verapamil	365
	Valspodar (PSC 833)	49.3–169.5		Nelfinavir	507.1
	Cyclosporine	55.2		Erythromycin	521.5
	Cyclosporine	81.4	Sirolimus	Cyclosporine	87.4–231
	Verapamil	104.4	Sitagliptin	Cyclosporine	28.6
	Valspodar (PSC 833)	119.4–169.5	Tacrolimus	Ketoconazole	139.3
Erythromycin	Grapefruit juice	48.6		Clotrimazole	147.5
Etoposide	Cyclosporine	47–58.5		Posaconazole	322.7
	Valspodar (PSC 833)	76.1–89.6	Talinolol	Surfactant TPGS	20
Everolimus	Cyclosporine	90.5		Ginkgo	24.7
Fentanyl	Quinidine	160		Milk thistle (silymarin; *Silybum marianum*)	30.2
Fexofenadine	St. Johns Wort extract (*Hypericum perforatum*)	30.7		Erythromycin	51.5
	Verapamil	38.3–46.4		Schisandra chinensis extract	52
	Quercetin	55.5	Teniposide	Cyclosporine	41.5
	Erythromycin	60.6	Verapamil	Atorvastatin	42.8
	Verapamil	62.1	Vinorelbine	Zosuquidar (LY335979)	29.7
	Azithromycin	71.5	Ximelagatran	Azithromycin	36
	Itraconazole	108.9–141.4		Erythromycin	81.6

Source: Reported in the database www.druginteractioninfo.org.

TABLE 4.8 BCRP-Mediated Drug–Drug Interactions

Object	Precipitant	AUC (%)
Rosuvastatin	Atazanavir and ritonavir	212.9
	Cyclosporine	608.2
	Lopinavir and ritonavir	107.6
Topotecan	Elacridar	142.9

Source: Reported in the database www.druginteractioninfo.org.

OAT1 (SLC22A6) and OAT3 (SLC22A8) The organic anion transporters (OATs) 1 and 3 are expressed in brain, smooth muscle, and kidney; OAT1 has been reported to be expressed in placenta (Sweet et. al., 2001; Eraly et al., 2004; Zhou and You, 2006). OAT1 and OAT3 are expressed in the basolateral membrane of the proximal tubules in the kidney. Organic anions are transported from the basolateral membrane through a tertiary mechanism. Organic anions are transported via an exchange through an antiporter for decarboxylates (i.e., ketoglutarate) which is generated by Na^+/K^+ ATPase, then transported in cytoplasmic vesicles and extruded to the apical surface and into urine (Sweet et al. 2003; Dantzler, 2002; Eraly et al, 2004; Nozaki et al., 2007). The details of the latter extrusion mechanism are still unclear. Many clinical DDIs have been reported for OAT-mediated transport, in particular with probenecid (Inotsume et al., 1990), cephalosporins, and antivirals (Eraly et al., 2004; Yuan et al., 2009; Ahn and Bhatnagar, 2008; Nigam et al., 2009; Table 4.9).

OCT2 (SLC22A2) Approximately 40% of all drugs on the market are organic cations (Kim and Shim, 2006). Organic cation transporters (OCTs) are expressed in liver, kidney, muscle, placenta, thymus, adrenal, neurons, choroid plexus, heart, intestine, spleen, and lung (Hediger et al., 2004). However, OCT2 is more abundantly expressed in kidney than OCT1 and OCT3 and is believed to be the critical transporter in the renal elimination of cations (Motohashi et al., 2002; Koepsell et al., 2007). Organic cation transport can occur in either directions and is electrogenic, requiring no Na^+ or proton gradient (Koepsell et al., 2007). Many investigators have studied physicochemical descriptors of compounds transported by OCT2. In the late 90's, studies demonstrated that the molecular polar surface area correlated well with intestinal absorption of drugs or their blood–brain barrier penetration (Clark, 1999; Ertl et al., 2000; Palm et al., 1997). Suhre et al. suggested that compound hydrophobicity, as calculated by the log P value, was the primary determinant in OCT2 interactions. Zolk et al. (2009) recently reported that the topical polar surface area (TPSA) correlated better to IC_{50} values generated by inhibition of MPP^+ uptake in a training set with 29 diverse compounds than cLog P values. Most compounds transported by OCT2 have a relative molecular mass below 500 (Zolk et al., 2009). Cimetidine appears to be the most reported perpetrator of clinical DDIs against procainamide (Somogyi et al., 1983), metformin, and others (Table 4.10).

TABLE 4.9 OAT-Mediated Drug–Drug Interactions

Drugs (Victims)	% AUC Increase	Drugs (Victims)	% AUC Increase
Fexofenadine	54.2	Famotidine	81.1
	74.6	Furosemide	69.1
(*R*)-Carprofen	156.4	Ganciclovir	167.9
(*S*)-Carprofen	176.2	Indomethacin	54.9
Acetaminophen	14.4	Isofezolac	63.8
Acyclovir	40		67.6
Cefaclor	114		124.8
Cefmetazole	58	Morphine 6-glucuronide	25
Cefonicid	109.2	Nafcillin	102.6
Cefoxitin	45.3	Nalidixic acid	25.6
	136.5	Naproxen	66.8
Ceftriaxone	27.2		147
	34	Olanzapine	26.3
Cephradine	105.8	Oseltamivir carboxylate	151.7
	120	Tapentadol	49.5
	138.8	Valacyclovir	21.7
	141.6	Zalcitabine	54.2
	192	Zidovudine	56.6
	192		106.2
	263.2	Zomepirac	329.1
Dicloxacillin	63.9		
	66.8		
Doripenem	74.6		

Source: Reported in the database www.druginteractioninfo.org.

TABLE 4.10 OCT-Mediated Drug–Drug Interactions

Drugs (Victims)	Drugs (Perpetrators)	AUC Increase (%)
Apricitabine	Cotrimoxazole	55.6
		65.8
Cephalexin	Cimetidine	9.2
Dofetilide	Cimetidine	47.9
Metformin	Cimetidine	46.2
		54.2
Procainamide	Cimetidine	24.2
		38.4
		39.7
		43.9
Ranitidine	Cimetidine	26.6
Metformin	Cephalexin	23.9
Procainamide	Levofloxacin	20.9

Source: Reported in the database (www.druginteractioninfo.org).

TABLE 4.11 OATP1B1-Mediated Drug–Drug Interactions

Drugs (Victims)	Drugs (Perpetrators)	Change in AUC (%)
Atorvastatin	Cyclosporine	644
Atorvastatin acid	Cyclosporine	769
		1431
Bosentan	Cyclosporine	97
Fluvastatin	Cyclosporine	89
		94
		210
		254
Pravastatin	Cyclosporine	1077
		2183
Repaglinide | CYP2C8 in vivo probe	Cyclosporine	143
Rosuvastatin	Cyclosporine	608
Simvastatin | CYP3A in vivo probe	Cyclosporine	155
		696
Atorvastatin acid	Rifampin	625
Caspofungin	Rifampin	61
Bosentan	Sildenafil	33
Rosuvastatin	Eltrombopag	55
	Gemfibrozil	88
	Ursodeoxycholic acid	59

Source: Reported in the database www.druginteractioninfo.org.

TABLE 4.12 OATP1B3-Mediated Drug–Drug Interactions

Substrate	Inhibitor	AUC Increase (%)
Bosentan	Cyclosporine	97.3
Atorvastatin acid	Rifampin	625.3
Bosentan	Sildenafil	33.1

Source: Reported in the database www.druginteractioninfo.org.

OATP1B1 (SLCO1B1), OATP1B3 (SLCO1B3), and OATP2B1 (SLCO2B1) Eleven human organic anion transporting polypeptide (OATP) family members have been identified to date that are expressed all over the body, including liver, kidney, brain, intestinal tract, lung, and placenta (Hagenbuch and Meier, 2004; Mikkaichi et al., 2004; König et al., 2000a, 2000b, 2006; Kindla et al., 2009). The OATP family of transporters is comprised of Na^+ ion–independent solute carriers. These uptake transporters have a broad range of substrate specificities that can mediate transport of cationic, neutral, zwitterionic, and anionic compounds (Kim, 2003). OATPs have been reported to transport hydrophilic (rousvastatin, pravastatin), neutral (digoxin, ouabain), and lipophilic (estrone sulfate) compounds (Mikkaichi et al., 2004). OATP transport does not require

ATP or membrane potential nor does it require ion gradients from potassium, sodium, or chloride (Leuthold et al., 2009; Mahagita et al., 2007). Of the OATP family members, OATP1B1 and OATP1B3 are the most studied with respect to drug transport. Both OATP1B1 and OATP1B3 are localized in the basolateral membrane hepatocytes and demonstrate 80% amino acid homology (Konig et al., 2000a, 2000b). Drugs transported by both transporters include rifampicin, fexofenadine, BQ123, pravastatin, fluvastatin, and pitavastatin (Hirano et al., 2004; Kindla et al., 2009). Drugs transported by OATP1B1 include atorvastatin (Hsiang et al., 1999; Lau et al., 2006) and capsofungin (Sandhu et al. 2005). OATP1B3 has been reported to transport digoxin (Kullak-Ublick et al., 2001; Noe et al., 2007), docetaxel (Baker et al., 2009), and paclitaxel (Smith et al., 2007). OATP2B1 has been reported to be responsible for gemfibrozil and fluvastatin DDI (Noe et al., 2007). Cyclosporine is the most reported perpetrator of clinical DDIs of statins that are transported by OATPs (Kim 2003; Konig et al, 2006; Simonson et al., 2004; Yamazaki et al., 2005; Niemi, 2007; Treiber et al., 2007) (Tables 4.11 and 4.12).

4.6 TOOLS OF THE TRANSPORTER TRADE

4.6.1 Absorption and Permeability

One of the key components for the development of small-molecule drugs from an industrial drug discovery perspective is for the drug to have good oral absorption. One of the methods to predict how well a compound may be absorbed in vivo is by measurement of the in vitro permeability. The higher the permeability, the more likely the drug will transit the gut and into the circulating blood stream to reach its target site. Doan et al. (2002) measured the passive permeability and P-gp-mediated efflux of 48 central nervous system (CNS) and 45 non-CNS marketed compounds to identify components of these molecules that allow transit into the blood–brain barrier (BBB). These researchers reported that for CNS delivery a drug should have ideally an in vitro passive permeability measurement of greater than 150 nm/s and P-gp efflux ratio of less than 2.5. Historically, there are many poorly permeable drugs on the market, such as digoxin, that demonstrated good oral bioavailability (Ooi and Colucci, 2001). The high bioavailability of digoxin had been inadvertently contributed by uptake transporters (Kullak-Ublick et al., 2001). This finding was not well understood at the time digoxin was being developed. However, new tools are available today to address permeability and transporter contributions for SARs. In addition to being transported by uptake transporters, digoxin is a substrate for the efflux transporter P-gp. And due to its narrow therapeutic window, the FDA is recommending close monitoring and evaluation of the P-gp inhibition potential (IC_{50}) of NCEs to increased digoxin exposure due to transporter-mediated DDIs. Digoxin is discussed in detail later in this chapter.

4.6.2 Caco-2 Permeability

The human colon carcinoma cell line (Caco-2) has historically been used to assay both permeability and P-gp efflux activity in a transwell system (Balimane et al., 2004; Marino et al., 2005). Numerous studies have reported good in vitro and in vivo correlations using the Caco-2 model (Artursson and Karlsson, 1991; Artursson et al., 2001; Shah et al,. 2006; Hilgendorf et al., 2007; Del Amo et al., 2009). However, the expression levels and function of endogenous transporters in Caco-2 vary significantly, potentially due to clonal variations worldwide from different culture conditions, passage numbers, and so on (Hayeshi et al., 2009). Therefore, the ability to generate comparable permeability and efflux data from laboratory to laboratory has been a challenge.

In the transwell system, Caco-2 cells are traditionally cultured into monolayers on porous membrane filter inserts for an average of 21 days. The monolayers represent the lipid bilayer barrier that drugs must transit, that is, from the luminal side of the gut to capillaries, to enter the blood circulation. The drug of interest (1–10 μM) is dosed in replicates to apical or basolateral chambers in the buffer. Most compounds at first screenings are assayed at physiological pH 7.4. More detailed screening may require adjusting pH for compound phenotyping studies (Thwaite et al., 1995; Neuhoff et al., 2005; Young et al., 2006). Dosed monolayers are incubated at 37°C, 5% CO_2, for 1–4 h. Samples are taken from both apical and basolateral compartments and processed for injection into liquid chromatography/tandon mass spectrometry (LC/MS/MS) or mixed with scintillation cocktail if radiolabeled drug was used. Analyzed concentrations are used to calculate the rates of compound transit through the lipid bilayers in the apical-to-basolateral (A-to-B) and basolateral-to-apical (B-to-A) directions. This measured rate is the calculated apparent permeability of the compound. Some companies use only the A-to-B rate as the apparent permeability (P_{app}). Other companies use the average of A-to-B plus B-to-A permeability rates as the P_{app}. The rationale for the former measurement is that they are only concerned with how much compound is absorbed, that is, transits from gut to blood. The latter rationale is concerned with how much compound is recycled back into the gut by either active efflux or passive permeability.

Direct interpretations of Caco-2 permeability data can be quite a challenge. If the compound of interest is transported by endogenous uptake or efflux transporters in Caco-2 cells, then the apparent or passive permeability may be skewed in the absorptive (A-to-B) or effluxed (B-to-A) directions. Efflux ratios (ER) are calculated from the permeability rates of vectorially transported drugs. For example, P-gp is expressed on the apical membrane of cultured Caco-2 cells. When compounds are dosed apically and are substrates for P-gp, they enter the cells and get recycled back or effluxed to the apical chamber in the transwells. When P-gp substrates are dosed basolaterally, the compounds enter the cells and then are effluxed apically at a higher rate than the opposite direction. The ER is calculated using the permeability rates with B-to-A divided by A-to-B. According to the FDA, if ER $\geq$ 2, then the NCE is considered a P-gp substrate; therefore, further in vitro and/or in vivo studies may be warranted. If the calculated ER is unity, where the A-to-B rate is equal to the B-to-A rate, then it is determined that the compound is not transporter mediated. If, however, the ER $<<<$ 1, then uptake transporters may play a role in transporting the NCE. To calculate the passive permeability in the Caco-2 system, a P-gp inhibitor, that is, GF120918 (Hyafil et al., 1993; Walstab et al., 1999) or cyclosporin A (Massart et al., 1995), will be necessary to be codosed with the NCE if it has already been determined to be a P-gp substrate.

The Caco-2 model has proven to be a reliable tool to address permeability and P-gp efflux for several decades (Balimane et al., 2004; Sun and Pang 2008). However, transporter scientists face many challenges in using these cells in a higher throughput manner. Although Balimane et al. (2004) have been able to optimize the Caco-2 cells to be assayed in a 96-well format, these cells still require a long culture period, up to 21 days. To circumvent this long culture period, many laboratories have implemented continuous transwell cultures of Caco-2 cells plated for weekly studies. The challenge, however, was to accurately predict the number of compounds that would be requested from program teams three weeks in advance. Sometimes, the number of compounds were overpredicted, and excess plates that were not used had to be discarded. Sometimes, the number of compounds were underpredicted, and program teams had to wait for data. For our applications, we found the Caco-2 system inefficient and it did not suit our needs. Because of these challenges in using the Caco-2 model, some companies have developed more efficient cell culture models to measure permeability and targeted transporter efflux activity, such as the LLC-PK1 or MDCK cell lines.

The LLC-PK1 and MDCK cell lines are pig and dog kidney epithelial cells, respectively, that require shorter culture (4–6 days) periods for monolayer formation in the transwell system. These two cell lines present their own set of challenges for data interpretation. LLC-PK1 cells have endogenous BCRP activity (Jonker et al., 2000). MDCKI cells have high endogenous P-gp activity (Raggers et al., 2002; Goh et al., 2002).

4.6.3 PAMPA

Another assay that has gained popularity and acceptance for the evaluation of permeability is the parallel artificial membrane permeability assay (PAMPA). Earlier versions of this system coated polyvinylidene fluoride (PVDF) filter plates with artificial membrane using dioleoyl-sn-glycerol-3-phosphocholine (Chen et al., 2008). Several companies attempted to develop this technology in the late 1990s with limited success (Kansy et al., 1998). These earlier versions suffered poor correlation to cell models, poor correlation to human absorption, and poor reproducibility. More modern systems have been developed with different lipid formulations and solvation techniques that seem to correlate better with Caco-2 and human data and are more reproducible (Chan et al., 2005). Some companies use the PAMPA as a tier 1 prescreen discovery compound before going into a cell system (Von Richter et al., 2009). The advantages of this assay is that it can be automated using a robotic liquid-handling system with minimal cost and compounds can be measured at gradient pHs. Either PAMPA plates can be made in-house or precoated PAMPA plates can be purchased and stored for up to six months at -20°C according to the manufacturer. One manufacturer indicated that over one hundred pharmaceutical companies worldwide are using its PAMPA system to measure permeability (personal communication).

The assay requires receiver buffer to be added to the transwell inserts. Compounds of interest are added to the basolateral wells in buffer. Plates with inserts and wells are coupled and incubated at room temperature or 37°C for 4–24 hours. At the end of the incubation period the plates are separated. Samples are taken from both inserts and wells and analyzed with either a UV-plate reader or LC-MS/MS. Concentrations of compound are calculated and the permeability rate is calculated over time. However, there are limitations with this system that may affect calculations and interpretations of permeability. Both solubility and nonspecific binding of the compound of interest may underestimate the permeability rate. Mass balance calculations may be needed to determine solubility and binding issues.

4.6.4 Immobilized Artificial Membrane

Another technology that is gaining momentum in evaluating drug permeability is immobilized artificial membrane (IAM) column chromatography. This technology was developed in the late 1980s (Pidgeon et al., 1995; Cohen and Leonard, 1995) and has been evaluated for Caco-2 (Yazdanian et al., 1998; Chan et al., 2005) and BBB (Yoon et al., 2006) drug correlations. This system requires very little sample or compound, and impurities do not interfere with permeability calculations. Phosphatidylcholine (PC) analogues are covalently bonded to silica in high-performance liquid chromatography (HPLC) columns. The phospholipids mimic the ionic charges of cell surface membranes. The principle is that when small molecules pass through the column they interact with the ionic charges of these phospholipids. Permeability is calculated by retention time: the longer the interaction, the greater the retention time. Valko et al. (2000) developed a rapid gradient HPLC method to measure drugs interacting with IAM. El-Gendya and Adejare (2004) evaluated a sulfonamide series of gamma-secretase inhibitors for physicochemical properties and found that the partitioning coefficient (c Log P) correlated well with the IAM partitioning data.

4.7 UPTAKE AND EFFLUX TRANSPORTER TOOLS

4.7.1 Transfected Cell Lines

New molecular biology techniques have become integral in manufacturing transporter tools. The commercialization of these tools has also exploded in recent years. Many transporter cDNA sequences are commercially available. Some transporter-transfected cell lines may also be commercially available for purchase or licensing. This commercialization has made generating transporter tools for pharmaceutical companies "easier." One of the newer and easier technologies that is used to generate transporter cell lines is the Flp-In™ system. The Flp-In™ system features a Flp recombination target (FRT) site and allows integration and expression of the gene of interest in mammalian cells at a specific genomic location. An expression vector that contains the gene of interest is then integrated or Flp-In into the genome by Flp recombinase-mediated DNA recombination at the FRT site (O'Gorman et al., 1991). Advantages of this system include a rapid transfection system that is capable of generating a stable monoclonal cell population (Gao et al., 2004; Liu et al., 2006; Pham et al., 2008). Cropp et al. (2008) recently generated a Flp-In rat Oat2-HEK cell line that transports cGMP more effectively than Oat1 or Oat3.

4.7.2 Uptake Assays

Although many drugs on the market have poor permeability but good oral bioavailability, transporters may play a critical role in the absorption of a particular drug, such as digoxin. Not only is digoxin a good substrate for P-gp, digoxin demonstrates poor passive permeability and poor solubility in vitro; however, this drug has 60–80% oral absorption in humans (Ooi and Colucci, 2001; Greiner, 1999). In vitro studies have demonstrated that digoxin uptake is transporter-mediated with OATP1B3 (SLCO1B3) and OATP4C1 (SLCO4C1) transfected cell lines (Kullak-Ublick et al., 2001; Noe et al., 2007; König et al., 2000b). Therefore, the development of uptake transporter assays is critical to any pharmaceutical company developing small molecules. These tools not only will identify potential mechanisms for limiting or enhancing oral absorption but also can help identify mechanisms of clearance and SAR.

The evaluation of the contributions of uptake transporters to discover ADME profiling remains a challenge for several reasons. First, unlike efflux transporters, in vitro tools to evaluate uptake transporters are not as readily available or well defined. Some of the more popular uptake transporter tools will be discussed below. Second, many of the limited substrates have overlap with other uptake transporters. Specific marker compounds for uptake transporters are not as well known compared to efflux transporters. Third, only a limited number of known inhibitors can provide specificity and a dynamic range that will allow sufficient kinetic profiling of a particular drug of interest with a specific uptake transporter.

Transfected uptake transporter cell lines and empty vector controls are seeded into multiwell plates for 1–5 days. Media are aspirated and cells are rinsed with buffer prior to dosing with the NCE in the presence and absence of selected inhibitors. Depending on the transporter, the dosing buffer may need to be adjusted, that is, pH 6 for PepT1/2 (Terada et al., 1997; Knutter et al., 2008),

OATP2B1 (Leuthold et al., 2009; Sai et al., 2006), Na^+ concentration for OCTN2 (Elimrani et al., 2003; Diao et al., 2009), and supplements with α-ketoglutarate (Nozaki et al., 2007). Uptake of substrates may be rapid, under 5 min, or over 30 min. Therefore, validation of the transporter cell line is critical prior to testing NCEs. Dose solutions are aspirated after selected time points. Cells are washed two to three times with chilled buffer. Cells are then extracted with a detergent or base and neutralized with acid before analyzing samples. If protein determination is desired, then detergent lysis and extraction are preferred. Uptake is calculated by picomoles per microgram of protein or rate by picomoles per minute per microgram of protein. The FDA is conservatively considering any compound that demonstrates twofold uptake over vector control a substrate for that uptake transporter.

4.7.3 Transwell Efflux Assays

P-gp (ABCB1) and BCRP (ABCG2) belong in the ATP-binding cassette family that requires energy from phosphate cleavage for transporter function. P-gp remains at the forefront of the most studied and characterized transporter. Polli et al. (2001) identified the rationale for screening NCEs for P-gp efflux. P-gp or BCRP efflux of NCEs can be measured by several assays. Direct efflux can be measured with transfected cell lines in the transwell system (Boothe-Genthe et al., 2005). Inhibition or indirect transporter interaction studies using a probe substrate is sometimes used when direct measurement of the NCE is not feasible (e.g., the NCE demonstrates poor analytical sensitivity or poor solubility or significant nonspecific binding to plastic or cells). However, poorly permeable compounds may require inverted membrane vesicles to assess efflux transporter substrate activity. This assay system will be discussed in the next section.

4.7.4 Membrane Vesicles

Another assay that is commonly used to assess transporter efflux activity is the inverted membrane vesicle system (Volk and Schneider 2003; Glavinas et al., 2008). Transporter membrane vesicles are commonly prepared using a baculovirus expression system to infect Sf9 insect cells (Pozza et al., 2009) or other stably transfected cell lines (Garrigues et al., 2002; Volk and Schneider 2003; Glavinas et al., 2008). Upon efficient transfection, insect cells are centrifuged to cell pellets prior to sonication and homogenization. Dounce homogenization or syringe-and-needle aspiration inverts the membrane vesicles, exposing the transporter binding sites externally. Membrane fractions are collected by high-speed centrifugation. Compounds can be tested directly or indirectly for transporter activity with these membrane vesicles. For direct substrate measurements, membrane vesicles are dosed with compound in the presence and absence of ATP. Radiolabeled compound is preferred for this assay to avoid the need for analytical preparations of samples. Compounds can be measured indirectly for transporter activity in two ways. The first approach features the use of a radiolabeled probe with the compound of interest as an inhibitor to the probe transport activity. The second approach measures the phosphate turnover of membrane vesicle ATPase activity (Glavinas et al., 2008; Von Richter et al., 2009). This assay is best used with radiolabeled compounds with poor permeability. Measuring membrane uptake with highly permeable compounds may require a short incubation time (e.g. < 1 min) and may not provide a significant dynamic range to determine clear substrate activity.

4.7.5 Hepatocyte Sandwich Cultures

The hepatocyte sandwich culture model measures uptake and efflux transporter-mediated transport of compounds. These measurements have been extrapolated to in vitro biliary excretion data of taurocholate (Chandra et al., 2001; Kemp et al., 2005), 7-ethoxycoumarin, and warfarin (Treijtel et al., 2005). Liu et al. (1999a) demonstrated good in vitro biliary clearance correlation with rat hepatocyte sandwich cultures to intrinsic in vivo biliary clearance in bile duct-cannulated rats using inulin, salicylate, methotrexate, enkephalin, and taurocholate model compounds. Ghibellini et al. (2007) generated good in vitro metabolism and clearance correlations with piperacillin to human in vivo data. Isolated hepatocytes are cultured in between gel matrices, hence the term *sandwich*. Over the course of four days, these hepatocytes form a three-dimensional architecture including tight junctions, transporter expression, and caniculi (LeCluyse et al., 1994). Calcium is required in the culture medium to maintain tight junctions and caniculi (Liu et al., 1999b). Transporter studies are performed in parallel with and without Ca^{2+}. When these hepatocytes are dosed in the presence of Ca^{2+}, compounds are taken up and extruded into the caniculi. The hepatocytes are then washed with buffer and lysed with detergent. The biliary excretion index is calculated by the amount accumulated in cells and caniculi minus the amount in cells alone divided by total accumulated in cells and caniculi (Liu et al., 1999b). The in vitro biliary clearance is calculated by the uptake difference in the presence and absence of Ca^{2+} divided by incubation time and medium concentration. This model requires the isolation of high-quality viable hepatocytes and may require some optimization before useful data are generated. Plated hepatocyte sandwich cultures are commercially available. However, they are expensive for the limited studies one is able to perform with these precultured plates.

4.7.6 Transgenic Mice

Numerous transporter knockout animals are available to evaluate pharmacokinetic distribution of NCEs. Schinkel et al. (1994) generated the knockout *mdr1a* (P-gp) mouse model that became 100-fold more sensitive to the neurotoxic pesticide ivermectin. This classic model has demonstrated how critical P-gp is to preventing xenobiotics from crossing the BBB (Löscher and Potschka, 2005; Kusuhara and Sugiyama, 2005; Terasaki and Ohtsuki, 2005;

Tsuji, 2005). Although no in vitro BBB model has been developed for high-throughput screening in the pharmaceutical industry, several models have demonstrated some advancement in this field, including transporter expression and tighter junction formation (Förster et al., 2008, Poller et al., 2008; Cohen-Kashi Malina et al., 2009). Numerous studies have been performed with P-gp knockout mice (Wijnholds et al., 2000; Kusuhara and Sugiyama, 2005; Zhao et al., 2009), including the recent demonstration that oseltamivir (Tamiflu) is transported by P-gp (Morimoto et al., 2008; Ose et al., 2008) but not BCRP (Ose et al., 2008). P-gp double- and triple-transporter knockout animals have been reported to affect the pharmacokinetics of vincristine and etoposide (Johnson et al., 2001), methotrexate and doxorubicin (Vlaming et al., 2006), edaravone (Mizuno et al., 2007), and Gleevec (Oostendorp et al., 2009). Early et al. (2006) generated Oat1 knockout mice that demonstrated approximately 75% diminished PAH transport compared to wild type. Sweet et al. (2002) generated Oat3 knockout mice and demonstrated diminished uptake of PAH, estrone sulfate, and taurocholate in kidney and fluorescein-methotrexate in the choroid plexus. When VanWert et al. (2007) generated the Oat3 knockout mice, they demonstrated not only significant decrease in estrone-3-sulfate and penicillin G clearance compared to wild type but also gender differences in these knockout animals. Chen et al. (2008) recently generated Oatp1b2 knockout mice, the homologues to human OATP1B1 and OATP1B3. The investigators demonstrated that these knockout animals expressed comparable mRNA levels of a panel of CYP and transporter genes to wild type. When injected subcutaneously (3 mg/kg) with six model compounds, the uptake liver-to-plasma concentration ratios ($K_{p,liver}$) of rifampicin and lovastatin were significantly diminished in the Oatp1b2 knockout animals compared to wild type. Similarily, Zaher et al. (2008) demonstrated diminished liver uptake of rifampicin and pravastatin in their Oatp1b2 knockout mice when dosed either intravenously or subcutaneously compared to wild type. Lu et al. (2008) demonstrated decreased hepatic uptake of phallodin and the blue-green algae toxin microcystin-LR.

4.8 SAMPLE ANALYSIS

Bioanalytical sample analysis by LC/MS/MS remains a rate-limiting step in many ADME processes, including transporter evaluation. The number of injection samples and MS run time can escalate rapidly and depend on whether compounds are tested in duplicate or triplicate and the number of points generated for a standard curve. For example, if one compound is evaluated in parental and MDR1-transfected cell lines in triplicate in both directions, 12 samples are generated from both apical and basolateral compartments. Add in an eight-point standard curve (1–500 n*M*), dose solutions, wash injections in-between triplicates, and approximately 58 injection samples per compound would result! If each injection were approximately 3 min per run, then each compound would take approximately 3 h to run. To increase efficiency, some laboratories have pooled bioanalytical samples, anywhere from 4 to 10 compounds per set. The challenge with the higher number of compounds per set is assuring that none of the compounds overlap in their atomic mass units (AMUs). In addition to pooling sets, some analytical scientists have evaluated newer technology that quantifies samples at a faster rate. The RapidFire® technology features the use of microscale solid-phase extraction coupled with flow injection MS/MS to analyze bioanalytical samples (Quercia et al., 2007; Shiau et al., 2008; Soulard et al., 2008; Fung et al., 2003).

4.9 AUTOMATION

Many aspects of transporter assays are labor intensive, from culturing cells, to preparing membrane vesicles, to preparing dosing solutions, to sampling transwells. In order to increase throughput and reduce manual labor and potential repetitive hand injuries, several robotic systems have been developed to reduce these laborious tasks, from cell culture maintenance to liquid handling.

4.9.1 Cell Maintenance Systems

Several automated cell maintenance systems are available on the market that range from huge room-size units to bench-size footprints. All of these systems have the capability of culturing cells from changing media to subcloning or passaging cells. All these systems are programmable with maintenance schedulers that calculate needed buffer or media for upcoming media changes or cell seeding. Some have alarms that are capable of sending error messages to the computer or cell phones if, for example, the CO_2 level drops in the incubators. Some systems feature the use of cameras to help determine whether these errors can be corrected remotely. Some systems can culture from tissue culture flasks; others use flasks in a 96-well footprint. However, not all systems are capable of seeding and maintaining cells in transwells. These automated cell culture units are expensive but may be worthwhile if a long "walk-away" time can be accommodated. For example, cultures can be maintained and cells seeded over the weekend or while operators are on vacation. Earlier versions of these systems demonstrated some disappointing results. The most frequent complaints were the large volume of waste fluids and frequent contaminations. However, most of these issues have been resolved with shorter media feed lines and better air-handling systems.

4.9.2 Robotic Liquid-Handling Systems

Designs of robotic liquid-handling systems have great flexibility to fit the needs of the user, such as aspirating, dispensing, mixing, and moving culture plates from incubators or hotels. These systems come in many sizes, from small bench-top units to room-size units. These systems can be programmed to accommodate most transporter assays, from transwell to accumulation studies. Some systems have schedulers that can perform time point studies. Many of these systems can accommodate a wide range of accessory equipment, from plate washers to plate mixers. There are many trade shows available to view the latest automation technologies.

4.10 IN VITRO DDI ASSAYS

With the FDA draft transporter guidance document pending implementation, many drug companies are scrambling to develop in vitro DDI assays or identify contract research organizations (CROs) capable of handling their respective needs. Many scientists are concerned with whether there is sufficient agreement in generating equivalent data from the current available tools. For example, as discussed previously, Caco-2 cells used worldwide express variable levels of endogenous transporters and demonstrate variable functional activity (Hayeshi et al., 2008). The current argument is that generating P-gp IC_{50} data of a NCE in one laboratory using Caco-2 cells may differ significantly in another laboratory depending on interactions of the NCE to endogenous uptake and/or efflux transporters. The use of efflux inhibitors such as GG918 (Hyafil et al., 1993; Wallstab et al., 1999) or Ko143 (Breedveld et al., 2006) may help define contributions from P-gp and BCRP, but Caco-2 cells express other efflux transporters that may contribute to efflux such as MRPs. With the expression of multiple endogenous transporters, Wang et al. (2008) discussed the merits and challenges of using inhibitors for drug–transporter interaction studies in Caco-2 cells. Nevertheless, many investigators have proposed and discussed the impact of several in vitro P-gp transporter assays and probes to help identify potential clinical DDI candidates (Keogh and Kunta, 2006; Unadkat et al., 2010). Rautio et al. (2006) proposed many companies are using digoxin as an in vitro and in vivo probe substrate for P-gp because of its low propensity for metabolism and clinical relevance for DDI. Fenner et al. (2009) used digoxin as a P-gp probe to correlate in vitro to clinical in vivo data. The MDR1-transfected LLC-PK1 model is used in our laboratory. Comparable P-gp IC_{50} data to most of those inhibitors indicated in the FDA transporter guidance document, such as with cyclosporin A (Fig. 4.4a) and verapamil (Fig. 4.4b) were generated with digoxin as a probe.

4.11 IN VITRO–IN VIVO CORRELATIONS

The bridge that divides in vitro and in vivo transporter correlation remains wide. Unlike direct enzyme kinetic correlations where a clear product is generated when substrate is added to enzyme, transporter outcomes are affected by many parameters. The physicochemical properties of compounds, such as lipophilicity, pK_a, permeability, solubility, and partitioning coefficient,

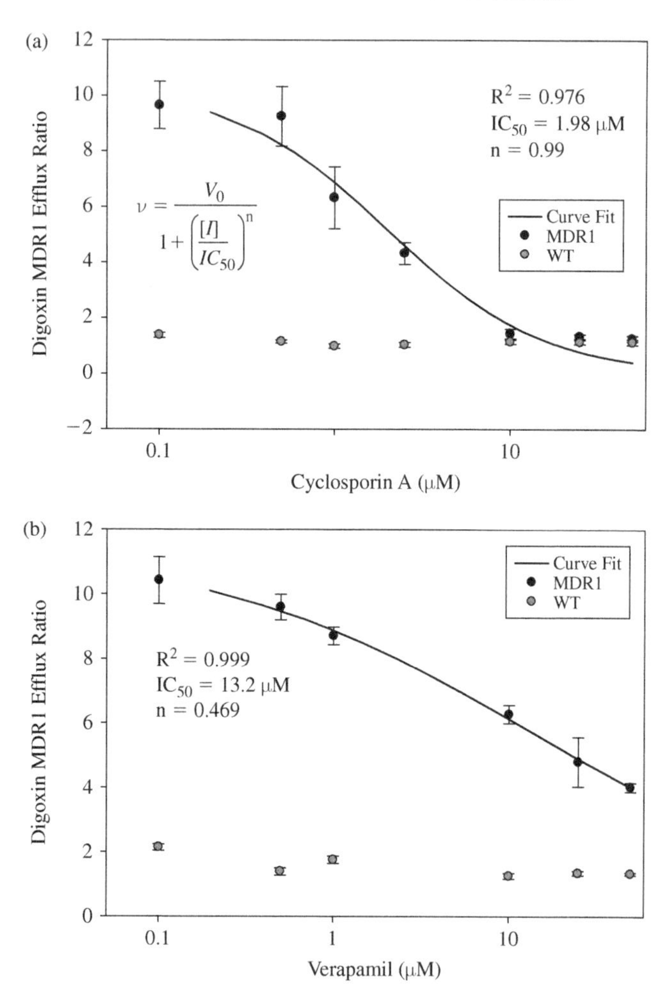

FIGURE 4.4 Inhibition (IC_{50}) of MDR1-mediated efflux of digoxin at K_m (3 μM) with increasing concentrations of cyclosporine (a, IC_{50} = 2 μM) and verapamil (b, IC_{50} = 13.2 μM).

all affect the function of transporters (Feng et al., 2001) and the micromilieu of where they are expressed. Transporter-mediated effects include contributions to absorption, metabolism, and efflux of compounds and many more parameters that make in vitro–in vivo correlations a challenge. Transporter polymorphism (Konig et al., 2006; Zhou and You, 2006; Smith et al., 2007), species differences (Dresser et al., 2000; Yamazaki et al., 2001; Booth-Genthe et al., 2005), and gender differences (Buist et al., 2002; Bebawy and Chetty, 2009) have recently emerged to contribute to the complications of in vitro–in vivo correlations. However, many investigators are attempting to narrow this divide. Since digoxin DDI remains a clinical concern, P-gp remains the most well studied transporter in the in vitro–in vivo transporter field (Fenner et al., 2009). Some investigators have studied compound permeability and gut absorption with respect to P-gp limiting bioavailability (Shirasaka et al., 2008; Del Amo et al., 2009). Others have used digoxin as an in vitro probe to predict adverse clinical DDIs (Fenner et al., 2009). As new uptake transporter tools are being developed, more in vitro studies will be performed to correlate in vivo outcomes. Feng et al. (2008) measured the pharmacokinetic effects of the smoking cessation drug varenicline on human OCT. Chu et al. (2007) evaluated the transporter susceptibility of the DPP-4 inhibitor Sitagpliptin for human OAT3, OATP4C1, and P-gp.

4.12 KINETIC MODELS

The identification and subsequent use of an appropriate kinetic model to define P-gp potential remain significant challenges among transporter scientists (Balimane et al., 2008). Because many parameters affect the inhibition potential (IC_{50}) of a NCE against a probe substrate, no model has been developed to address these parameters. Parameters such as permeability, molecular polar surface area, solubility, nonspecific binding, multiple binding sites (Aller et al., 2009), metabolism, contributions of uptake transporters (Eraly, 2008), tight junctions, species differences (Dresser et al., 2000; Yamazaki et al., 2001), and gender differences (Buist et al., 2002, Bebawy and Chetty, 2009) have been discussed but not agreed upon (Sun and Pang, 2008; Acharya et al. 2008). It is not clear whether a two-compartmental model (Poirer et al., 2008) sufficiently predicts equivalent P-gp potential in all cell models. However, it is clear that for transporter studies NCE input does necessarily translate to direct output.

4.13 TRANSPORTER CONCLUSION

Development of in vitro and in vivo tools to evaluate transporter-mediated SARs, pharmacokinetics, and pharmocodynamic effects has accelerated in recent years in the preclinical drug discovery setting. Investigators use these transporter tools to help predict and avoid adverse clinical DDIs. Transporter studies to correlate in vitro–in vivo data have not yet fully matured as those with CYP studies. However, the momentum to gain a better understanding of transporter contributions to clinical outcomes is advancing quickly.

ACKNOWLEDGMENT

We thank Drs. Gary L. Skiles (Amgen), Yihong Wang, and Wei Tang (Merck) for their valuable comments.

REFERENCES

Acharya P, O'Connor MP, Polli JW, Ayrton A, Ellens H, and Bentz J (2008) Kinetic identification of membrane transporters that assist P-glycoprotein-mediated transport of digoxin and loperamide through a confluent monolayer of MDCKII-hMDR1 cells. Drug Metab Dispos 36:452–460.

Ahn S-Y and Bhatnagar V (2008) Update on the molecular physiology of organic anion transporters. Curr Opin Nephrol Hypertens 17:499–505.

Allen JD, van Loevezijn A, Lakhai JM, van der Valk M, van Tellingen O, Reid G, Schellens JHM, Koomen G-J, Schinkel AH (2002) Potent and specific inhibition of the breast cancer resistance protein multidrug transporter in vitro and in mouse intestine by a novel analogue of Fumitremorgin C. Mol Cancer Therapeu 1:417–425.

Aller SG, Yu J, Ward A, Weng Y, Chittaboina S, Zhuo R, Harrell PM, Trinh YT, Zhang Q, Urbatsch IL, and Chang G (2009) Structure of P-glycoprotein reveals a molecular basis for poly-specific drug binding. Science 323:1718–1722.

Ambudkar SV, Dey S, Hrycyna CA, Ramachandra M, Pastan I, Gottesman MM (1999) Biochemical, cellular and pharmacological aspects of the multidrug transporter. Annu Rev Pharmacol Toxicol 39:361–398.

Ambudkar SV, Kim I-W and Sauna ZE (2006) The power of the pump: mechanisms of action of P-glycoprotein (ABCB1) Eur J Pharm Sci 27:392–400.

Arrington-Sanders R, Hutton N and Siberry GK (2006) Ritonavir-fluticasone interaction causing Cushing syndrome in HIV-infected children and adolescents. Pediatr Infect Dis J 25:1044–1048.

Artursson P and Karrlsson J (1991) Correlation between oral drug absorption in humans and apparent drug permeability coefficients in human intestinal epithelia (Caco-2) cells. Biochem Biophys Res Commun 175:880–890.

Artursson P, Palm K, Luthman K (2001) Caco-2 monolayers in experimental and theoretical predictions of drug transport. Adv Drug Deliv Rev 46:27–43.

Aszalos A (2007) Drug–drug interactions affected by the transporter protein, P-glycoprotein (ABCB1, MDR1): I. Preclinical aspects. Drug Discovery Today 12:833–837.

Austin RP, Barton P, Cockroft SL, Wenlock MC and Riley RJ (2002) The influence of nonspecific microsomal binding on apparent intrinsic clearance, and its prediction from physicochemical properties. Drug Metab Dispos 30:1497–1503.

Azie NE, Brater DC, Becker PA, Jones DR and Hall SD (1998) The interaction of diltiazem with lovastatin and pravastatin. Clinical Pharmacology & Therapeutics 64:369–377.

Back DJ, Tjia JF, Karbwang J and Colbert J (1988) In vitro inhibition studies of tolbutamide hydroxylase activity of human liver microsomes by azoles, sulphonamides and quinolines. British Journal of Clinical Pharmacology 26:23–29.

Backman JT, Karjalainen MJ, Neuvonen M, Laitila J and Neuvonen PJ (2006) Rofecoxib is a potent inhibitor of cytochrome P450 1A2: studies with tizanidine and caffeine in healthy subjects. British Journal of Clinical Pharmacology 62: 345–357.

Backman JT, Kyrklund C, Neuvonen M and Neuvonen PJ (2002) Gemfibrozil greatly increases plasma concentrations of cerivastatin. Clinical Pharmacology & Therapeutics 72:685–691.

Backman JT, Olkkola KT, Aranko K, Himberg JJ and Neuvonen PJ (1994) Dose of midazolam should be reduced during diltiazem and verapamil treatments. British Journal of Clinical Pharmacology 37:221–225.

Baker SD, Verweij J, Cusatis GA, van Schaik RH, Marsh S, Orwick SJ, Franke RM, Hu S, Schuetz EG, Lamba V, Messersmith WA, Wolff AC, Carducci MA and Sparreboom A (2009) Pharmacogenetic pathway analysis of docetaxel elimination. Clin Pharmacol Ther 85:155–163.

Balimane PV, Marino A, and Chong S (2008) P-gp inhibition potential in cell-based models: which "calculation" method is the most accurate? AAPS J 10:577–586.

Balimane PV, Patel K, Marino A, Chong S (2004) Utility of 96 well Caco-2 cell system for increased throughput of P-gp screening in drug discovery. Eur J Pharm Biopharm 58:99–105.

bdel-Razzak Z, Loyer P, Fautrel A, Gautier J-C, Corcos L, Turlin B, Beaune P and Guillouzo A (1993) Cytokines down-regulate expression of major cytochrome P-450 enzymes in adult human hepatocytes in primary culture. Mol Pharmacol 44(4):707–715.

Bebawy M and Chetty M (2009) Gender differences in P-glycoprotein expression and function: effects on drug disposition and outcome. Curr Drug Metab 10:322–328.

Beringer PM and Slaughter RL (2005) Transporters and their impact on drug disposition. Ann Pharmacother 39:1097–1108.

Bertilsson G, Heidrich J, Svensson K, Asman M, Jendeberg L, Sydow-Backman M, Ohlsson R, Postlind H, Blomquist P and Berkenstam A (1998) Identification of a human nuclear receptor defines a new signaling pathway for CYP3A induction. Proc Nat Acad Sci USA 95(21):12208–12213.

Bjornsson TD, Callaghan JT, Einolf HJ, Fischer V, Gan L, Grimm S, Kao J, King SP, Miwa G, Ni L, Kumar G, McLeod J, Obach RS, Roberts S, Roe A, Shah A, Snikeris F, Sullivan JT, Tweedie D, Vega JM, Walsh J, Wrighton SA, Pharmaceutical Research and Manufacturers of America (PhRMA) Drug Metabolism/Clinical Pharmacology Technical Working Group and FDA Center for Drug Evaluation and Research (CDER) (2003) The conduct of in vitro and in vivo drug-drug interaction studies: a Pharmaceutical Research and Manufacturers of America (PhRMA) perspective. Drug Metab Dispos 31:815–832.

Booth-Genthe CL, Louie SW, Carlini EJ, Li B, Leake BF, Eisenhandler R, Hochman JH, Mei Q, Kim RB, Rushmore TH and Yamazaki M (2005) Development and characterization of LLC-PK1 cells containing Sprague–Dawley rat Abcb1a (Mdr1a): Comparison of rat P-glycoprotein transport to human and mouse. J Pharmacol Tox Meth 54:78–89.

Breedveld P, Beijnen JH, and Schellens JHM (2006) Use of P-glycoprotein and BCRP inhibitors to improve oral bioavailability and CNS penetration of anticancer drugs. Trends Pharmacol Sci 27:17–24.

Brown HS, Ito K, Galetin A and Houston JB (2005) Prediction of in vivo drug-drug interactions from in vitro data: impact of incorporating parallel pathways of drug elimination and inhibitor absorption rate constant. Br J Clin Pharmacol 60:508–518.

Buist SCN, Cherrington NJ, Choudhuri S, Hartley DP and Klaassen CD (2002) Gender-specific and developmental influences on the expression of rat organic anion transporters. J Pharmacol Exp Ther 301:145–151.

Burckart GJ (2007) Looking beneath the surface of the CYP3A5 polymorphism. Pediatric Transplantation 11(3):239–240.

Capdevila JH, Falck JR and Harris RC (2000) Cytochrome P450 and arachidonic acid bioactivation: Molecular and functional properties of the arachidonate monooxygenase. J Lipid Res 41(2):163–181.

Carr BA, Wan J, Hines RN and Yost GS (2003) Characterization of the human lung CYP2F1 gene and identification of a novel lung-specific binding motif. J Bio Chem 278:15473–15483.

Chan ECY, Tan WL, Ho PC, Fang LJ (2005) Modeling Caco-2 permeability of drugs using immobilized artificial membrane chromatography and physicochemical descriptors J Chromatogr A 1072:159–168.

Chandra P, Lecluyse EL, and Brouwer KLR (2001) Optimization of culture conditions for determining hepatobiliary disposition of taurocholate in sandwich-cultured rat hepatocytes. In Vitro Cell Dev Biol – Animal 37:380–385.

Chen C, Stock JL, Liu X, Shi J, Van Deusen JW, DiMattia, DA, Dullea RG, de Morais SM (2008) Utility of a novel Oatp1b2 knockout mouse model for evaluating the role of Oatp1b2 in the hepatic uptake of model compounds. Drug Metab Dispos 36:1840–1845.

Chen X, Murawski A, Patel K, Crespi CL, Balimane PV (2008) A novel design of artificial membrane for improving the PAMPA model. Pharm Res 25:1511–1520.

Chu X-Y, Bleasby K, Yabut J, Cai X, Chan GH, Hafey MJ, Xu S, Bergman AJ, Braun MP, Dean DC, and Evers R (2007) Transport of the dipeptidyl peptidase-4 inhibitor Sitagliptin by human organic anion transporter 3, organic anion transporting polypeptide 4C1, and multidrug resistance P-glycoprotein. J Pharmacol Exp Ther 321:673–683.

Clark DE (1999) Rapid calculation of polar molecular surface area and its application to the prediction of transport phenomena. 2. Prediction of blood-brain barrier penetration. J Pharm Sci 88:815–821.

Cohen DE and Leonard MR (1995) Immobilized artificial membrane chromatography: a rapid and accurate HPLC method for predicting bile salt-membrane interaction. J Lipid Res 36:2251–2260.

Cohen-Kashi Malina K, Cooper I, Techberg VI (2009) Closing the gap between the in-vivo and in-vitro blood-brain barrier tightness. Brain Res 1284:12–21.

Cropp CD, Komori T, Shima JE, Urban TJ, Yee SW, More SS, and Giacomini KM (2008) Organic anion transporter 2 (SLC22A7) is a facilitative transporter of cGMP. Mol Pharmacol 73:1151–1158.

Czerwinski M, McLemore TL, Gelboin HV and Gonzalez FJ (1994) Quantification of CYP2B7, CYP4B1, and CYPOR messenger RNAs in normal human lung and lung tumors. Caner Res 54(4):1085–1091.

Dai Y, Hebert MF, Isoherranen N, Davis CL, Marsh C, Shen DD and Thummel KE (2006) Effect of CYP3A5 polymorphism on tacrolimus metabolic clearance in vitro. Drug Metab Dispos 34(5):836–847.

Dalen P, Dahl ML, Bernal Ruiz ML, Nordin J and Bertilsson L (1998) 10-Hydroxylation of nortriptyline in white persons with 0, 1, 2, 3, and 13 functional CYP2D6 genes. Clin Pharmacol Ther 63:444–452.

Danzler WH (2002) Renal organic anion transport: a comparative and cellular perspective. Biochim Biophys Acta 1566:169–181.

Del Amo EM, Heikkinen AT, Monkkonen J (2009) In vitro-in vivo correlation in p-glycoprotein mediated transport in intestinal absorption. Eur J Pharm Sci 36:200–211.

Diao L, Ekins S, Polli JE (2009) Novel inhibitors of human organic cation/carnitine transporter (hOCTN2) via computational modeling and in vitro testing. Pharm Res 26:1890–1900.

Doan, KMM, Humphreys JE, Webster LO, Wring SA, Shampine LJ, Serabjit-Singh CJ, Adkison KK, Polli JW (2002) Passive permeability and P-Glycoprotein-mediated efflux differentiate central nervous system (CNS) and non-CNS marketed drugs. J Pharmacol Exp Ther 303:1029–1037.

Dresser MJ, Gray AT and Giacomini KM (2000) Kinetic and selectivity differences between rodent, rabbit, and human organic cation transporters (OCT1) J Pharmacol Exp Ther 292:1146–1152.

El-Gendya AM and Adejare A (2004) Membrane permeability related physicochemical properties of a novel γ-secretase inhibitor. Int J Pharmaceutics 280:47–55.

Elimrani I, Lahjouji K, Seidman E, Roy M-J, Mitchell GA, Qureshi I (2003) Expression and localization of organic cation/carnitine transporter OCTN2 in Caco-2 cells. Am J Phyisiol Gastrointest Liver Physiol 284:G863–G871.

Elkiweri IA, Zhang YL, Christians Uwe, Ng K-Y, van Patot MCT, and Henthorn TK (2009) Competitive substrates for P-Glycoprotein and organic anion protein transporters differentially reduce blood organ transport of fentanyl and loperamide: Pharmacokinetics and pharmacodynamics in Sprague-Dawley rats. Anesth Analg 108:146–159.

Emoto C, Murase S and Iwasaki K (2006) Approach to the prediction of the contribution of major cytochrome P450 enzymes to drug metabolism in the early drug-discovery stage. Xenobiotica 36:671–683.

Endres CJ, Hsiao P, Chung FS, Unadkat JD (2006) The role of transporters in drug interactions. Eur J Pharm Sci 27:501–517.

Eraly SA (2008) Implications of the alternating access model for organic anion transporter kinetics. J Membrane Biol 226:35–42.

Eraly SA, Bush KT, Sampogna RV, Bhatnagar V, and Nigam SK (2004) The Molecular pharmacology of organic anion transporters: from DNA to FDA? Mol Pharmacol 65:479–487.

Eraly SA, Vallon V, Vaughn DA, Gangoiti JA, Richter K, Nagle M, Monte JC, Rieg T, Truong DM, Long JM, Barshop BA, Kaler G, and Nigam SK (2006) Decreased renal organic anion secretion and plasma accumulation of endogenous organic anions in OAT1 knock-out mice. J Biol Chem 281:5072–5083.

Ertl P, Rohde B, and Selzer P (2000) Fast calculation of molecular polar surface area as a sum of fragment-based contributions and its application to the prediction of drug transport properties. J Med Chem 43:3714–3717.

et-Beluche I, Boulenc X, Fabre G, Maurel P and Bonfils C (1992) Purification of two cytochrome P450 isozymes related to CYP2A and CYP3A gene families from monkey (baboon, Papio papio) liver microsomes – Cross reactivity with human forms. Euro J Biochem 204(2):641–648.

Fahmi OA, Maurer TS, Kish M, Cardenas E, Boldt S and Nettleton D (2008) A combined model for predicting CYP3A4 clinical net drug-drug interaction based on CYP3A4 inhibition, inactivation, and induction determined in vitro (Drug Metab Dispos (2008) 36: (1698–1708). Drug Metabolism and Disposition 36(9):1975.

Feng B, Dresser MJ, Shu Y, Johns SJ, and Giacomini KM (2001) Arginine 454 and lysine 370 are essential for the anion specificity of the organic anion transporter, rOAT3. Biochemistry 40:5511–5520.

Feng B, Obach RS, Burstein AH, Clark DJ, de Morais SM, and Faessel HM (2008) Effect of human renal cationic transporter inhibition on the pharmcokinetics of varenicline, a new therapy for smoking cessation: an in vitro-in vivo study. Clin Pharmacol Ther 83:567–576.

Fenner KS, Troutman MD, Kempshall S, Cook JA, Ware JA, Smith DA, and Lee CA (2009) Drug-drug interactions mediated through P-glycoprotein: clinical relevance and in vitro-in vivo correlation using digoxin as a probe drug. Clin Pharmacol Ther 85:173–181.

Floren LC, Bekersky I, Benet LZ, Mekki Q, Dressler D, Lee JW, Roberts JP and Hebert MF (1997b) Tacrolimus oral bioavailability doubles with coadministration of ketoconazole. Clin Pharmacol Ther 62:41–49.

Förster C, Burek M, Romero IA, Weksler B, Couraud P-O, and Drenckhahn D (2008) Differential effects of hydrocortisone and TNFα on tight junction proteins in an in vitro model of the human blood–brain barrier. J Physiol 586:1937–1949.

Fung E, Chu I, Li C, Liu T, Soares A, Morrison R, Nomier A (2003) Higher-throughput screening for Caco-2 permeability utilizing multiple sprayer liquid chromatography/tandem mass spectrometry system. Rapid Commun Mass Spectrom 17:2147–2152.

Galetin A, Burt H, Gibbons L and Houston JB (2006) Prediction of time-dependent CYP3A4 drug-drug interactions: impact of enzyme degradation, parallel elimination pathways, and intestinal inhibition. Drug Metab Dispos 34:166–175.

Gao J, Cha S, Jonsson R, Opalko J, Peck AB (2004) Detection of anti-type 3 muscarinic acetylcholine receptor autoantibodies in the sera of Sjögren's syndrome patients by use of a transfected cell line assay. Arth Rheum 50:2615–2621.

Gardner-Stephen D, Heydel J-M, Goyal A, Lu Y, Xie W, Lindblom T, Mackenzie P and Radominska-Pandya A (2004) Human PXR variants and their differential effects on the regulation of human UDP-glucuronosyltransferase gene expression. Drug Metab Dispos 32(3):340–347.

Garrigues A, Loiseau N, Delaforge M, Ferte J, Garrigos M, Andre F, Orlowski S (2002) Characterization of two pharmacophores on the multidrug transporter P-Glycoprotein Mol Pharmacol 62:1288–1298.

Gelboin HV (1993) Cytochrome P450 and monoclonal antibodies. Pharmacol Rev 45:413–453.

Gelboin HV, Goldfarb I, Krausz KW, Grogan J, Korzekwa KR, Gonzalez FJ and Shou M (1996) Inhibitory and noninhibitory monoclonal antibodies to human cytochrome P450 2E1. Chem Res Toxicol 9:1023–1030.

Gelboin HV, Krausz KW, Goldfarb I, Buters JT, Yang SK, Gonzalez FJ, Korzekwa KR and Shou M (1995) Inhibitory and non-inhibitory monoclonal antibodies to human cytochrome P450 3A3/4. Biochem Pharmacol 50:1841–1850.

Gelboin HV, Krausz KW, Shou M, Gonzalez FJ and Yang TJ (1997) A monoclonal antibody inhibitory to human P450 2D6: a paradigm for use in combinatorial determination of individual P450 role in specific drug tissue metabolism. Pharmacogenetics 7:469–477.

Ghibellini G, Bridges AS, Generaux CN, Brouwer, KLR (2007) In vitro and in vivo determination of piperacillin metabolism in humans. Drug Metab Dispo 35:345–349.

Giacomini KM, Huang S-M, Tweedie DJ, et al. (2010) Membrane transporters in drug development. Nat Rev 9:215–236.

Giuliano C, Jairaj M, Zafiu CM and Laufer R (2005) Direct determination of unbound intrinsic drug clearance in the microsomal stability assay. Drug Metab Dispos 33:1319–1324.

Glavinas H, Méhn D, Jani M, Oosterhuis B, Herédi-Szabó K, Krajcsi P (2008) Utilization of membrane vesicle preparations to study drug-ABC transporter interactions. Expert Opin Drug Metab Toxicol 4:721–732.

Goh L-B, Spears KJ, Yao D, Ayrton A, Morgan P, Wolf CR, and Friedberg T (2002) Endogenous drug transporters in in vitro and in vivo models for the prediction of drug disposition in man. Biochem Pharmacol 64:1569–1578.

Gomez DY, Wacher VJ, Tomlanovich SJ, Hebert MF and Benet LZ (1995) The effects of ketoconazole on the intestinal metabolism and bioavailability of cyclosporine. Clin Pharmacol Ther 58:15–19.

Gonzalez FJ, Skoda RC, Kimura S, Umeno M, Zanger UM, Nebert DW, Gelboin HV, Hardwick JP and Meyer UA (1988) Characterization of the common genetic defect in humans deficient in debrisoquine metabolism. Nature 331:442–446.

Gorski JC, Jones DR, Haehner-Daniels BD, Hamman MA, O'Mara EM, Jr. and Hall SD (1998) The contribution of intestinal and hepatic CYP3A to the interaction between midazolam and clarithromycin. Clin Pharmacol Ther 64:133–143.

Greenblatt DJ, von Moltke LL, Daily JP, Harmatz JS and Shader RI (1999) Extensive impairment of triazolam and alprazolam clearance by short-term low-dose ritonavir: the clinical dilemma of concurrent inhibition and induction. J Clin Psychopharmacol 19:293–296.

Greenblatt DJ, Wright CE, von Moltke LL, Harmatz JS, Ehrenberg BL, Harrel LM, Corbett K, Counihan M, Tobias S and Shader RI (1998) Ketoconazole inhibition of triazolam and alprazolam clearance: differential kinetic and dynamic consequences. Clin Pharmacol Ther 64:237–247.

Greiner B, Eichelbaum M, Fritz P, Kreichgauer H-P, von Richter O, Zundler J, and Kroemer HK (1999) The role of intestinal P-glycoprotein in the interaction of digoxin and rifampin. J Clin Invest 104:147–153.

Hagenbuch B and Meier PJ (2004) Organic anion transporting polypeptides of the OATP/ SLC21 family: phylogenetic classification as OATP/ SLCO superfamily, new nomenclature and molecular/functional properties. Pflügers Archiv 447:653–665.

Hayes JD, Flanagan JU and Jowsey IR (2005) Glutathione transferases. Annu Rev Pharmacol Toxicol 45:51–88.

Hayeshi R, Hilgendorf C, Artursson P, Augustijns P, Brodin B, Dehertogh P, Fisher K, Fossati L, Hovenkamp E, Korjamo T, Masungi C, Maubon N, Mols R, Müllertz A, Mönkkönen J, O'Driscoll C, Oppers-Tiemissen HM, Ragnarsson EGE, Rooseboom M,

Ungell A-L (2008) Comparison of drug transporter gene expression and functionality in Caco-2 cells from 10 different laboratories Eur J Pharm Sci 35:383–396.

Hecht SS (1998) Biochemistry, biology, and carcinogenicity of tobacco-specific N-nitrosamines. Chem Res in Toxicol 11:559–603.

Hediger MA, Romero MF, Peng J-B, Rolfs A, Takanaga H and Bruford EA (2004) The ABCs of solute carriers: physiological, pathological and therapeutic implications of human membrane transport proteins. Pflügers Archiv 447:465–468.

Hilgendorf C, Ahlin G, Seithel A, Artursson P, Ungell A-L, and Karlsson J (2007) Expression of thirty-six drug transporter genes in human intestine, liver, kidney, and organotypic cell lines Drug Metab Dispos 35:1333–1340.

Hirano M, Maeda K, Shitara Y, and Sugiyama Y (2004) Contribution of OATP2 (OATP1B1) and OATP8 (OATP1B3) to the hepatic uptake of pitavastatin in humans. J Pharmacol Exp Ther 311:139–146.

Ho RH and Kim RB (2005) Transporters and drug therapy: Implications for drug disposition and disease. Clin Pharmacol Ther 78:260–277.

Hochman J, Mei Q, Yamazaki M, Tang C, Prueksaritanont T, Bock M, Ha S and Lin J (2007) Role of mechanistic transport studies in lead optimization, editors Borchardt RT, Kerns EH, Hageman MJ, Thakker DR and Stevens JL, In: Optimizing the "Drug-Like" Properties of Leads in Drug Discovery, Springer, New York.

Hoffman EC, Reyes H, Chu F-F, Sander F, Conley LH, Brooks BA and Hankinson O (1991) Cloning of a factor required for activity of the Ah (dioxin) receptor. Science 252(5008):954–958.

Hoffman SM, Fernandez-Salguero P, Gonzalez FJ and Mohrenweiser HW (1995) Organization and evolution of the cytochrome P450 CYP2A-2B-2F subfamily gene cluster on human chromosome 19. J Mol Evol 41:894–900.

Honig PK, Wortham DC, Zamani K, Conner DP, Mullin JC and Cantilena LR (1993) Terfenadine-ketoconazole interaction. Pharmacokinetic and electrocardiographic consequences. JAMA 269:1513–1518.

Honkakoski P and Negishi M (2000) Regulation of cytochrome P450 (CYP) genes by nuclear receptors. Biochem J 347(2):321–337.

Hsiang B, Zhu Y, Wang Z, Wu Y, Sasseville V, Yang W-P and Kirchgessner TG (1999) A novel human hepatic organic anion transporting polypeptide (OATP2). Identification of a liver-specific human organic anion transporting polypeptide and identification of rat and human hydroxymethylglutaryl-CoA reductase inhibitor transporters. J Biol Chem 274:37161–37168.

Hu Y, Oscarson M, Johansson I, Yue QY, Dahl ML, Tabone M, Arinco S, Albano E and Ingelman-Sundberg M (1997) Genetic polymorphism of human CYP2E1: characterization of two variant alleles. Mol Pharmacol 51:370–376.

Huang W, Lin YS, McConn DJ, Calamia JC, Totah RA, Isoherranen N, Glodowski M and Thummel KE (2004) Evidence of significant contribution from CYP3A5 to hepatic drug metabolism. Drug Metab Dispos 32:1434–1445.

Hustert E, Haberl M, Burk O, Wolbold R, He Y-Q, Klein K, Nuessler AC, Neuhaus P, Klattig J, Eiselt R, Koch I, Zibat A, Brockmoller J, Halpert JR, Zanger UM and Wojnowski L (2001) The genetic determinants of the CYP3A5 polymorphism. Pharmacogenetics 11(9):773–779.

Hyafil F, Vergely C, Du Vignaud P, Thierry G-P (1993) In vitro and in vivo reversal of multidrug resistance by GF120918, an acridonecarboxamide derivative. Cancer Res 53:4595–4602.

Imaoka S, Yoneda Y, Sugimoto T, Hiroi T, Yamamoto K, Nakatani T and Funae Y (2000) CYP4B1 is a possible risk factor for bladder cancer in humans. Biochem Biophys Res Commun 277(3):776–780.

Ingelman-Sundberg M, Oscarson M and McLellan RA (1999) Polymorphic human cytochrome P450 enzymes: an opportunity for individualized drug treatment. Trends Pharmacol Sci 20:342–349.

Inotsume N, Nishimura M, Nakano M, Fujiyama S, and Sato T (1990) The inhibitory effect of probenecid on renal excretion of famotidine in young, healthy volunteers. J Clin Pharmacol 30:50–56.

Ito K, Brown HS and Houston JB (2004) Database analyses for the prediction of in vivo drug-drug interactions from in vitro data. Br J Clin Pharmacol 57:473–486.

Ito K, Iwatsubo T, Kanamitsu S, Ueda K, Suzuki H and Sugiyama Y (1998) Prediction of pharmacokinetic alterations caused by drug-drug interactions: metabolic interaction in the liver. Pharmacol Rev 50:387–412.

Ito K, Suzuki H, Horie T, Sugiyama Y (2005) Apical/basolateral surface expression of drug transporters and it role in vectorial drug transport. Pharm Res 22:1559–1577.

Itoga S, Harada S, Nomura F and Nakai T (1998) [Genetic polymorphism of human CYP2E1: new allels detected in exons and exon-intron junctions]. [Japanese]. Nihon Arukoru Yakubutsu Igakkai Zasshi 33:56–64.

Jaakkola T, Backman JT, Neuvonen M and Neuvonen PJ (2005) Effects of gemfibrozil, itraconazole, and their combination on the pharmacokinetics of pioglitazone. Clin Pharmacol Ther 77:404–414.

Johansson I, Lundqvist E, Bertilsson L, Dahl ML, Sjoqvist F and Ingelman-Sundberg M (1993) Inherited amplification of an active gene in the cytochrome P450 CYP2D locus as a cause of ultrarapid metabolism of debrisoquine. Proc Nat Acad Sci USA 90:11825–11829.

Johnson DR, Finch RA, Lin ZP, Zeiss CJ, Sartorelli AC (2001) The pharmacological phenotype of combined multidrug-resistance mdr1a/1b- and mrp1-deficient mice Cancer Res 61:1469–1476.

Jonker JW, Smit JW, Brinkhuis RF, Maliepaard M, Beijnen JH, Schellens JH, Schinkel AH (2000) Role of breast cancer resistance protein in the bioavailability and fetal penetration of topotecan. J Natl Cancer Inst 92:1651–1656.

Jounaidi Y, Hyrailles V, Gervot L and Maurel P (1996) Detection of a CYP3A5 allelic variant: A candidate for the polymorphic expression of the protein? Biochem Biophys Res Commun 221(2):466–470.

Juliano RL and Ling V (1976) A surface glycoprotein modulating drug permeability in Chinese hamster ovary cell mutants. Biochim Biophys Acta 455:152–162.

Jung MJ and Metcalf BW (1975) Catalytic inhibition of gamma-aminobutyric acid – alpha-ketoglutarate transaminase of bacterial origin by 4-aminohex-5-ynoic acid, a substrate analog. Biochem Biophys Res Commun 67:301–306.

Kalow W and Tang BK (1991) Use of caffeine metabolite ratios to explore CYP1A2 and xanthine oxidase activities. Clin Pharmacol Ther 50:508–519.

Kanamitsu S, Ito K and Sugiyama Y (2000) Quantitative prediction of in vivo drug-drug interactions from in vitro data based on physiological pharmacokinetics: use of maximum unbound concentration of inhibitor at the inlet to the liver. Pharm Res 17:336–343.

Kansy M, Senner F, and Guernator K (1998) Physicochemical high throughput screening: parallel artificial membrane permeation assay in the description of passive absorption processes. J Med Chem 41:1007–1010.

Kawamoto T, Sueyoshi T, Zelko I, Moore R, Washburn K and Negishi M (1999) Phenobarbital-responsive nuclear translocation of the receptor CAR in induction of the CYP2B gene. Mol Cell Bio 19(9):6318–6322.

Kemp DC, Zamek-Gliszczynski MJ, and Brouwer KLR (2005) Xenobiotics inhibit hepatic uptake and biliary excretion of taurocholate in rat hepatocytes. Toxicol Sci 83:207–214.

Keogh JP and R. Kunta JR (2006) Development, validation and utility of an in vitro technique for assessment of potential clinical drug–drug interactions involving P-glycoprotein. Eur J Pharm Sci 27:543–554.

Kim K-A, Park P-W, Kim KR and Park J-Y (2007) Effect of multiple doses of montelukast on the pharmacokinetics of rosiglitazone, a CYP2C8 substrate, in humans. Br J Clin Pharmacol 63(3):339–345.

Kim MK and Shim C-K (2006) The transport of organic cations in the small intestine: Current knowledge and emerging concepts. Arch Pharm Res 29:605–616.

Kim RB (2003) Organic anion-transporting polypeptide (OATP) transporter family and drug disposition. Eur J Clin Invest 33:1–5.

Kimura M, Ieiri I, Mamiya K, Urae A and Higuchi S (1998) Genetic polymorphism of cytochrome P450s, CYP2C19, and CYP2C9 in a Japanese population. Therapeutic Drug Monit 20:243–247.

Kindla J, Fromm M, and König J (2009) In vitro evidence for the role of OATP and OCT uptake transporters in drug-drug interactions. Expert Opin Drug Metab Toxicol 5:489–500.

Kitada M and Kamataki T (1994) Cytochrome P450 in human fetal liver: Significance and fetal-specific expression. Drug Metab Rev 26(1–2):305–323.

Kitz R and Wilson IB (1962) Esters of methanesulfonic acid as irreversible inhibitors of acetylcholinesterase. J Biol Chem 237:3245–3249.

Knutter I, Wollesky C, Kottra G, Hahn MG, Fischer W, Zebisch K, Neubert RHH, Daniel H, Brandsch M (2008) Transport of angiotensin-converting enzyme inhibitors by H^+/peptide transporters revisited. J Pharmacol Exp Ther 327:432–441.

Koepsell H, Lips K and Volk C (2007) Polyspecific organic cation transporters: Structure, function, physiological roles, and biopharmaceutical implications. Pharm Res 24:1227–1251.

König J, Cui Y, Nies AT, and Keppler D (2000a) A novel human organic anion transporting polypeptide localized to the basolateral hepatocyte membrane. Am J Physiol Gastrointest Liver Physiol 278:G156–G164.

König J, Cui Y, Nies AT, and Keppler D (2000b) Localization and genomic organization of a new hepatocellular organic anion transporting polypeptide. J Biol Chem 275:23161–23168.

König J, Seithel A, Gradhand U, and Fromm MF (2006) Pharmacogenomics of human OATP transporters. Naunyn-Schmiede Arch Pharmacol 372:432–443.

Krishna R, Mayer LD (2000) Multidrug resistance (MDR) in cancer. Mechanisms, reversal using modulators of MDR and the role of MDR modulators in influencing the pharmacokinetics of anticancer drugs. Eur J Pharm Sci 11:265–83.

Krishnamurthy P and Schuetz JD (2006) Role of ABCG2/BCRP in biology and medicine. Annu Rev Pharmacol Toxicol 46:381–410.

Kullak-Ublick GA, Ismair MG, Stieger B, Landmann L, Huber R, Pizzagalli F, Fattinger K, Meier PJ, Hagenbuch B (2001) Organic anion-transporting polypeptide B (OATP-B) and its functional comparison with three other OATPs of human liver. Gastroenterology 120:525–533.

Kusuhara H and Sugiyama Y (2005) Active Efflux across the Blood-Brain Barrier: Role of the Solute Carrier Family. NeuroRx 2:73–85.

Langsenlehner U, Krippl P, Renner W, Yazdani-Biuki B, Eder T, Wolf G, Wascher TC, Paulweber B, Weitzer W and Samonigg H (2004) Genetic variants of the sulfotransferase 1A1 and breast cancer risk. Breast Cancer Res Treat 87(1):19–22.

Lau YY, Okochi H, Huang Y, Benet LZ (2006) Multiple transporters affect the disposition of atorvastatin and its two active hydroxy metabolites: application of in vitro and ex situ systems. J Pharmacol Exp Ther 316:762–771.

LeCluyse EL, Audus KL and Hochman JH (1994) Formation of extensive canalicular networks by rat hepatocytes cultured in collagen-sandwich configuration. Am J Physiol Cell Physiol 266:C1764–C1774.

Lehmann JM, McKee DD, Watson MA, Willson TM, Moore JT and Kliewer SA (1998) The human orphan nuclear receptor PXR is activated by compounds that regulate CYP3A4 gene expression and cause drug interactions. J Clinl Invest 102(5):1016–1023.

Leuthold S, Hagenbuch B, Mohebbi N, Wagner CA, Meier PJ, and Stieger B (2009) Mechanisms of pH-gradient driven transport mediated by organic anion polypeptide transporters. Am J Physiol Cell Physiol 296:570–582.

Lilla C, Risch A, Verla-Tebit E, Hoffmeister M, Brenner H and Chang-Claude J (2007) SULT1A1 genotype and susceptibility to colorectal cancer. Int J Cancer 120 (1):201–206.

Lin JH and Lu AY (1997) Role of pharmacokinetics and metabolism in drug discovery and development. Pharmacol Rev 49:403–449.

Lin JH and Yamazaki M (2003) Role of P-Glycoprotein in Pharmacokinetics Clinical Implications. Clin Pharmacokinet 42:59–98.

Liu W, Xiong Y and Gossen M (2006) Stability and homogeneity of transgene expression in isogenic cells. J Mol Med 84:57–64.

Liu X, Chism JP, LeCluyse EL, Brouwer KR, Brouwer KLR (1999a) Correlation of biliary excretion in sandwich-cultured rat hepatocytes and in vivo in rats. Drug Metab Dispos 27:637–644.

Liu X, LeCluyse EL, Brouwer KR, Lightfoot RM, Lee JI, and Brouwer KLR (1999b) Use of Ca^{2+} modulation to evaluate biliary excretion in sandwich-cultured rat hepatocytes. J Pharmacol Exp Ther 289:1592–1599.

Löscher W and Potschka H (2005) Role of drug efflux transporters in the brain for drug disposition and treatment of brain diseases. Prog Neurobiol 76:22–76.

Lu AY, Wang RW and Lin JH (2003) Cytochrome P450 in vitro reaction phenotyping: a re-evaluation of approaches used for P450 isoform identification. Drug Metab Dispos 31:345–350.

Lu H, Choudhuri S, Ogura K, Csanaky IL, Lei X, Cheng X, Song P-z, and Klaassen CD (2008) Characterization of organic anion transporting polypeptide 1b2-null mice: Essential role in hepatic uptake/toxicity of phalloidin and microcystin-LR. Toxicol Sci 103:35–45.

Lu P, Schrag ML, Slaughter DE, Raab CE, Shou M and Rodrigues AD (2003) Mechanism-based inhibition of human liver microsomal cytochrome P450 1A2 by zileuton, a 5-lipoxygenase inhibitor. Drug Metab Dispos 31:1352–1360.

Mackenzie PI, Owens IS, Burchell B, Bock KW, Bairoch A, Belanger A, Fournel-Gigleux S, Green M, Hum DW, Iyanagi T, Lancet D, Louisot P, Magdalou J, Chowdhury JR, Ritter JK, Schachter H, Tephly TR, Tipton KF and Nebert DW (1997) The UDP glycosyltransferase gene superfamily: Rcommended nomenclature update based on evolutionary divergence. Pharmacogenetics 7(4):255–269.

Mahagita C, Grassl SM, Piyachaturawat P, and Ballatori N (2007) Human organic anion transporter 1B1 and 1B3 function as bidirectional carriers and do not mediate GSH-bile acid cotransport. Am J Physiol Gastrointest Liver Physiol 293:G271–G278.

Maliepaard M, Scheffer GL, Faneyte IF, van Gastelen MA, Pijnenborg ACLM, Schinkel AH, van de Vijver MJ, Scheper RJ, Schellens JHM (2001) Subcellular localization and distribution of the breast cancer resistance protein transporter in normal human tissues. Cancer Res 61:3458–3464.

Mao Q (2008) BCRP/ABCG2 in the placenta: expression, function, and regulation. Pharm Res 25:1244–1255.

Marino AM, Yarde M, Patel H, Chong S, Balimane PV (2005) Validation of the 96 well Caco-2 cell culture model for high throughput permeability assessment of discovery compounds. Int J Pharm 297:235–241.

Massart C, Gibassier J, Raoul M, Pourquier P, Leclech G, Robert J, Lucas C (1995) Cyclosporin A, verapamil and S9788 reverse doxorubicin resistance in a human medullary thyroid carcinoma cell line. Anticancer Drugs 6:135–46.

Mayhew BS, Jones DR and Hall SD (2000) An in vitro model for predicting in vivo inhibition of cytochrome P450 3A4 by metabolic intermediate complex formation. Drug Metab Dispos 28:1031–1037.

McCarver DG, Byun R, Hines RN, Hichme M and Wegenek W (1998) A genetic polymorphism in the regulatory sequences of human CYP2E1: association with increased chlorzoxazone hydroxylation in the presence of obesity and ethanol intake. Toxicol Appl Pharmacol 152:276–281.

Mei Q, Tang C, Assang C, Lin Y, Slaughter D, Rodrigues AD, Baillie TA, Rushmore TH and Shou M (1999) Role of a potent inhibitory monoclonal antibody to cytochrome P-450 3A4 in assessment of human drug metabolism. J Pharmacol Exp Ther 291:749–759.

Mei Q, Tang C, Lin Y, Rushmore TH and Shou M (2002) Inhibition kinetics of monoclonal antibodies against cytochromes P450. Drug Metab Dispos 30:701–708.

Mikkaichi T, Suzuki T, Tanemoto M, Ito S, and Abe T (2004) The organic anion transporter (OATP) family. Drug Metab Pharmacokin 19:171–179.

Miyake K, Mickley L, Litman T, Zhan Z, Robey R, Cristensen B, Brangi M, Greenberger L, Dean M, Fojo T, and Bates SE (1999) Molecular cloning of cDNAs which are highly overexpressed in mitoxantrone-resistant cells: demonstration of homology to ABC transport genes. Cancer Res 59:8–13.

Mizuno N, Niwa T, Yotsumoto Y and Sugiyama Y (2003) Impact of drug transporter studies on drug discovery and development. Pharmacol Rev 55:425–461.

Mizuno N, Takahashi T, Kusuhara H, Schuetz JD, Niwa T, and Sugiyama Y (2007) Evaluation of the role of breast cancer resistance protein (BCRP/ABCG2) and multidrug resistance-associated protein 4 (MRP4/ABCC4) in the urinary excretion of sulfate and glucuronide metabolites of Edaravone (MCI-186; 3-Methyl-1-phenyl-2-pyrazolin-5-one) Drug Metab Dispos 35:2045–2052.

Monaghan G, Ryan M, Seddon R, Hume R and Burchell B (1996) Genetic variation in bilirubin UDP-glucuronosyltransferase gene promoter and Gilbert's syndrome. Lancet 347(9001):578–581.

Moore LB, Parks DJ, Jones SA, Bledsoe RK, Consler TG, Stimmel JB, Goodwin B, Liddle C, Blanchard SG, Willson TM, Collins JL and Kliewer SA (2000) Orphan nuclear receptors constitutive androstane receptor and pregnane X receptor share xenobiotic and steroid ligands. J Biol Chem 275(20):15122–15127.

Morimoto K, Nakakariya M, Shirasaka Y, Kakinuma C, Fujita T, Tamai I, Ogihara T (2008) Oseltamvir (Tamiflu) efflux transport at the blood-brain barrier via P-glycoprotein. Drug Metab Dispos 36:6–9.

Motohashi H, Sakurai Y, Saito H, Masuda S, Urakami Y, Goto M, Fukatsu A, Ogawa O, Inui K (2002) Gene expression levels and immunolocalization of organic ion transporters in the human kidney. J Am Soc Nephrol 13:866–874.

Nakamura K, Goto F, Ray WA, McAllister CB, Jacqz E, Wilkinson GR and Branch RA (1985) Interethnic differences in genetic polymorphism of debrisoquin and mephenytoin hydroxylation between Japanese and Caucasian populations. Clin Pharmacol Ther 38:402–408.

Neuhoff S, Ungell AL, Zamora I, Artursson P (2005) pH-Dependent passive and active transport of acidic drugs across Caco-2 cell monolayers. Eur J Pharm Sci 3:211–220.

Nhamburo PT, Gonzalez FJ, McBride OW, Gelboin HV and Kimura S (1989) Identification of a new P450 expressed in human lung: Complete cDNA sequence, cDNA-directed expression, and chromosome mapping. Biochemistry 28 (20):8060–8066.

Niemi M (2007) Role of OATP transporters in the disposition of drugs. Pharmacogen 8:787–802.

Niemi M, Backman JT, Neuvonen M and Neuvonen PJ (2003) Effects of gemfibrozil, itraconazole, and their combination on the pharmacokinetics and pharmacodynamics of repaglinide: potentially hazardous interaction between gemfibrozil and repaglinide. Diabetologia 46:347–351.

Nigam SK, Bush KT and Bhatnagar V (2009) Drug and toxicant handling by the OAT organic anion transporters in the kidney and other tissues. Nature Clin Practice Nephrol 3:443–448.

Noé J, Portmann R, Brun M-E, and Funk C (2007) Substrate-dependent drug-drug interactions between gemfibrozil, fluvastatin and other organic anion-transporting peptide (OATP) substrates on OATP1B1, OATP2B1, and OATP1B3. Drug Metab Dispos 35:1308–1314.

Noguchi K, Katayama K, Mitsuhashi J, Sugimoto Y (2009) Functions of the breast cancer resistance protein (BCRP/ABCG2) in chemotherapy. Adv Drug Deliv Rev 61:26–33.

Nozaki Y, Kusuhara H, Kondo T, Hasegawa M, Shiroyanagi Y, Nakazawa H, Okano T, and Sugiyama Y (2007) Characterization of the uptake of organic anion transporter (OAT) 1 and OAT3 substrates by human kidney slices J Pharmacol Exp Ther 321:362–369.

Obach RS (2003) Drug-drug interactions: an important negative attribute in drugs. Drugs Today 39:301–338.

Obach RS, Walsky RL, Venkatakrishnan K, Gaman EA, Houston JB and Tremaine LM (2006) The utility of in vitro cytochrome P450 inhibition data in the prediction of drug-drug interactions. J Pharm Exp Ther 316:336–348.

Obach RS, Walsky RL, Venkatakrishnan K, Houston JB and Tremaine LM (2005) In vitro cytochrome P450 inhibition data and the prediction of drug-drug interactions: qualitative relationships, quantitative predictions, and the rank-order approach. Clin Pharmacol Ther 78:582–592.

O'Gorman S, Fox DT, Wahl GM (1991) Recombinase-mediated gene activation and site-specific integration in mammalian cells. Science 251:1351–1355.

Ooi H and Colucci WS (2001) Pharmacological treatment of heart failure. In: J.G. Hardman and L.E. Limbird, Editors, Goodman & Gilman's The pharmacological basis of therapeutics, McGraw-Hill, New York, p. 918.

Oostendorp RL, Buckle T, Beijnen JH, van Tellingen O, Schellens JH (2009) The effect of P-gp (Mdr1a/1b), BCRP (Bcrp1) and P-gp/BCRP inhibitors on the in vivo absorption, distribution, metabolism and excretion of imatinib. Invest New Drugs 27:31–40.

Ose, A, Kusuhara H, Yamatsugu K, Kanai M, Shibasaki M, Fujita T, Yamamoto A, Sugiyama Y (2008) P-glycoprotein restricts the penetration of oseltamivir across the blood-brain barrier. Drug Metab Dispos 36:427–434.

Otton SV, Inaba T and Kalow W (1984) Competitive inhibition of sparteine oxidation in human liver by beta-adrenoceptor antagonists and other cardiovascular drugs. Life Sci 34:73–80.

Pachouri SS, Sobti RC, Kaur P, Singh J and Gupta SK (2006) Impact of polymorphism in sulfotransferase gene on the risk of lung cancer. Cancer Genet Cytogenet 171(1):39–43.

Palm K, Stenberg P, Luthman K, and Artursson P (1997) Polar molecular surface properties predict the intestinal absorption of drugs in humans. Pharm Res 14:568–571.

Pascussi JM, Drocourt L, Fabre JM, Maurel P and Vilarem MJ (2000) Dexamethasone induces pregnane X receptor and retinoid X receptor-alpha expression in human hepatocytes: synergistic increase of CYP3A4 induction by pregnane X receptor activators. Mol Pharmacol 58:361–372.

Pascussi J-M, Jounaidi Y, Drocourt L, Domergue J, Balabaud C, Maurel P and Vilarem M-J (1999) Evidence for the presence of a functional pregnane X receptor response element in the CYP3A7 promoter gene. Biochem Biophys Res Commun 260(2):377–381.

Pham DH, Moretti PAB, Goodall GJ, Pitson SM (2008) Attenuation of leakiness in doxycycline-inducible expression via incorporation of 3′ AU-rich mRNA destabilizing elements. BioTechniques 45:155–162.

Pianezza ML, Sellers EM and Tyndale RF (1998) Nicotine metabolism defect reduces smoking. Nature 393:750.

Pidgeon C, Ong S, Liu H, Qiu X, Pidgeon M, Dantzig AH, Munroe J, Hornback WJ, Kasher JS, Glunz L, and Szczerba T (1995) IAM Chromatography: an in vitro screen for predicting drug membrane permeability. J Med Chem 38:590–594.

Poch MT, Cutler NS, Yost GS and Hines RN (2005) Molecular mechanisms regulating human CYP4B1 lung-selective expression. Drug Metab Dispos 33(8):1174–1184.

Poirier A, Lavé T, Portmann R, Brun M-E, Senner F, Kansy M, Grimm H-P, and Funk C (2008) Design, data analysis, and simulation of in vitro drug transport kinetic experiments using a mechanistic in vitro model. Drug Metab Dispos 36:2434–2444.

Poller B, Gutmann H, Krähenbühl S, Weksler B, Romero I, Couraud P-O, Tuffin G, Drewe J, Huwyler J (2008) The human brain endothelial cell line hCMEC/D3 as a human blood-brain barrier model for drug transport studies. J Neurochem 107:1358–1368.

Polli JW, Wring SA, Humphreys JE, Huang L, Morgan JB, Webster LO, Serabjit-Singh CS (2001) Rational use of in vitro P-glycoprotein assays in drug discovery. J Pharmacol Exp Ther 299:620–628.

Pozza A, Prez-Victoria JM, Di Pietro A (2009) Overexpression of homogeneous and active ABCG2 in insect cells. Protein Express Purif 63:75–83.

Proctor NJ, Tucker GT and Rostami-Hodjegan A (2004) Predicting drug clearance from recombinantly expressed CYPs: intersystem extrapolation factors. Xenobiotica 34:151–178.

Quattrochi LC, Vu T and Tukey RH (1994) The human CYP1A2 gene and induction by 3-methylcholanthrene. A region of DNA that supports Ah-receptor binding and promoter-specific induction. J Biol Chem 269(9):6949–6954.

Quercia AK, LaMarr WA, Myung J, Ozbal CC, Landro JA, Lumb KJ (2007) High-throughput screening by mass spectrometry: comparison with the scintillation proximity assay with a focused-file screen of AKT1/PKB alpha. J Biomol Screen 12:473–80.

Rodrigues AD (1999) Integrated cytochrome P450 reaction phenotyping: attempting to bridge the gap between cDNA-expressed cytochromes P450 and native human liver microsomes. Biochem Pharmacol 57:465–480.

Rodrigues AD and Lin JH (2001) Screening of drug candidates for their drug—drug interaction potential. Curr Opin Chem Biol 5:396–401.

Rodrigues AD and Rushmore TH (2002) Cytochrome P450 pharmacogenetics in drug development: in vitro studies and clinical consequences. Curr Drug Metab 3:289–309.

Rodrigues AD (2005) Impact of CYP2C9 genotype on pharmacokinetics: are all cyclooxygenase inhibitors the same? Drug Metab Dispos 33:1567–1575.

Rodriguez-Antona C, Jover R, Gomez-Lechon MJ and Castell JV (2000) Quantitative RT-PCR measurement of human cytochrome P-450s: application to drug induction studies. Arch Biochem Biophys 376:109–116.

Rodriguez-Antona C, Sayi JG, Gustafsson LL, Bertilsson L and Ingelman-Sundberg M (2005) Phenotype-genotype variability in the human CYP3A locus as assessed by the probe drug quinine and analyses of variant CYP3A4 alleles. Biochem and Biophys Res Commun 338(1):299–305.

Raggers RJ, Vogels I, van Meer G (2002) Upregulation of the expression of endogenous Mdr1 P-glycoprotein enhances lipid translocation in MDCK cells transfected with MRP2. Histochem Cell Biol 117:181–185.

Raviv Y, Pollard HB, Bruggemann EP, Pastan I, and Gottesman MM (1990) Photosensitized labeling of a functional multidrug transporter in living drug-resistant tumor cells. J Biol Chem 265:3975–3980.

Rautio J, Humphreys JE, Webster LO, Balakrishnan A, Keogh KP, Kunta JR, Serabjit-Singh CJ, and W. Polli JW (2006) In vitro P-glycoprotein inhibition assays for assessment of clinical drug interaction potential of new drug candidates: a recommendation for probe substrates Drug Metab Dispos 34:786–792.

Robey RW, To KKK, Polgar O, Dohse M, Fetsch P, Dean M, Bates SE (2009) ABCG2: a perspective. Adv Drug Deliv Rev 61:3–13.

Sai Y, Dai R, Yang TJ, Krausz KW, Gonzalez FJ, Gelboin HV and Shou M (2000) Assessment of specificity of eight chemical inhibitors using cDNA-expressed cytochromes P450. Xenobiotica 30:327–343.

Sai Y, Kaneko Y, Ito S, Mitsuoka K, Kato Y, Tamai I, Artursson P, Tsuji A (2006) Predominant contribution of organic anion transporting polypeptide OATP-B (OATP2B1) to apical uptake of Estrone-2-sulfate by human intestinal Caco-2 cells. Drug Metab Dispos 34:1423–1431.

Saier MH Jr, Beatty JT, Goffeau A, Harley KT, Heijne WH, Huang SC, Jack DL, Jahn PS, Lew K, Liu J, et al. (1999) The major facilitator superfamily. J Mol Microbiol Biotechnol 1:257–279.

Sandhu P, Lee W, Xu X, Leake BF, Yamazaki M, Stone JA, Lin JH, Pearson PG, Kim RB (2005) Hepatic uptake of the novel antifungal agent caspofungin. Drug Metab Dispos 33:676–682.

Schinkel AH (1997) The physiological function of drug-transporting P-glycoproteins. Sem Cancer Biol 8:161–170.

Schinkel AH, Smit JJM, van Tellingen O, Beijnen JH, Wagenaar E, van Deemter L, Mol CAAM, van der Valk MA, Robanus-Maandag EC, te Riele HPJ, Berns AJM, Borst P (1994) Disruption of the mouse mdr1a P-glycoprotein gene leads to a deficiency in the blood-brain barrier and to increased sensitivity to drugs. Cell 77:491–502.

Schmidt JV and Bradfield CA (1996) Ah receptor signaling pathways. Annu Rev Cell Dev Biol 12:55–89.

Schuetz EG, Beck WT and Schuetz JD (1996) Modulators and substrates of P-glycoprotein and cytochrome P4503A coordinately up-regulate these proteins in human colon carcinoma cells. Mol Pharmacol 49:311–318.

Shah P, Jogani V, Bagchi T, Misra A (2006) Role of Caco-2 cell monolayers in prediction of intestinal drug absorption. Biotechnol Prog 22:186–198.

Shiau AK, Massari ME, Ozbal CC (2008) Back to basics: label-free technologies for small molecule screening. Comb Chem High Throughput Screen 11:231–7.

Shimada T, Hayes CL, Yamazaki H, Amin S, Hecht SS, Guengerich FP and Sutter TR (1996) Activation of chemically diverse procarcinogens by human cytochrome P-450 1B1. Cancer Res 56:2979–2984.

Shimada T, Yamazaki H, Mimura M, Inui Y and Guengerich FP (1994) Interindividual variations in human liver cytochrome P-450 enzymes involved in the oxidation of drugs, carcinogens and toxic chemicals: studies with liver microsomes of 30 Japanese and 30 Caucasians. J Pharmacol Exp Ther 270:414–423.

Shirasaka Y, Masaoka Y, Kataoka M, Sakuma S, Yamashita S (2008) Scaling of in vitro membrane permeability to predict P-glycoprotein-mediated drug absorption in vivo Drug Metab Dispos 36:916–922.

Shou M (2005) Prediction of pharmacokinetics and drug-drug interactions from in vitro metabolism data. Curr Opin Drug Discov Devel 8:66–77.

Shou M and Lu AY (2009) Antibodies as a probe in cytochrome P450 research. Drug Metab Dispos 37:925–931.

Shou M, Hayashi M, Pan Y, Xu Y, Morrissey K, Xu L and Skiles GL (2008) Modeling, prediction, and in vitro in vivo correlation of CYP3A4 induction. Drug Metab Dispos 36(11):2355–2370.

Silverman RB (1995) Mechanism-based enzyme inactivators. Methods Enzymol 249:240–283.

Simonson SG, Raza A, Martin PD, Mitchell PD, Jarcho JA, Brown CDA, Windass AS, Schneck DW (2004) Rosuvastatin pharmacokinetics in heart transplant recipients administered an antirejection regimen including cyclosporine. Clin Pharmacol Ther 76:167–177.

Smith NF, Marsh S, Scott-Horton TJ, Hamada A, Mielke S, Mross K, Figg WD, Verweij J, McLeod HL, and Sparreboom A (2007) Variants in the SLCO1B3 gene: Interethnic distribution and association with paclitaxel pharmacokinetics. Clin Pharmacol Ther 81:76–82.

Somogyi A, McLean A and Heinzow B (1983) Cimetidine-procainamide pharmacokinetic interaction in man: Evidence of competition for tubular secretion of basic drugs. Eur J Clin Pharmacol 25:339–345.

Sotaniemi EA, Rautio A, Backstrom M, Arvela P and Pelkonen O (1995) CYP3A4 and CYP2A6 activities marked by the metabolism of lignocaine and coumarin in patients with liver and kidney diseases and epileptic patients. Br J Clin Pharmacol 39:71–76.

Soulard P, McLaughlin M, Stevens J, Connolly B, Coli R, Wang L, Moore J, Kuo M-ST, LaMarr WA, Ozbal CC and Bhat BG (2008) Development of a high-throughput screening assay for stearoyl-CoA desaturase using rat liver microsomes, deuterium labeled stearoyl-CoA and mass spectrometry Anal Chimica Acta 627:105–111.

Stoilov I, Akarsu AN and Sarfarazi M (1997) Identification of three different truncating mutations in cytochrome P4501B1 (CYP1B1) as the principal cause of primary congenital glaucoma (Buphthalmos) in families linked to the GLC3A locus on chromosome 2p21. Hum Mol Genet 6:641–647.

Sueyoshi T, Kawamoto T, Zelko I, Honkakoski P and Negishi M (1999) The repressed nuclear receptor CAR responds to phenobarbital in activating the human CYP2B6 gene. J Biol Chem 274:6043–6046.

Suhre WM, Ekins S, Chang C, Swaan PW, and Wright SH (2005) Molecular determinants of substrate/inhibitor binding to the human and rabbit renal organic cation transporters hOCT2 and rbOCT2. Mol Pharm 67:10067–1077.

Sun H and Pang KS (2008) Permeability, transport, and metabolism of solutes in Caco-2 cell monolayers: A theoretical study. Drug Metab Dispos 36:102–123.

Sutter TR, Tang YM, Hayes CL, Wo YY, Jabs EW, Li X, Yin H, Cody CW and Greenlee WF (1994) Complete cDNA sequence of a human dioxin-inducible mRNA identifies a new gene subfamily of cytochrome P450 that maps to chromosome 2. J Biol Chem 269:13092–13099.

Sweet DH, Bush KT, and Nigam SK (2001) The organic anion transporter family: from physiology to ontogeny and the clinic. Am J Physiol 281:F197–F205.

Sweet DH, Chan LM, Walden R, Yang XP, Miller DS, and Pritchard JB (2003) Organic anion transporter 3 (Slc22a8) is a dicarboxylate exchanger indirectly coupled to the Na+ gradient. Am J Physiol 284:F763–F769.

Sweet DH, Miller DS, Pritchard JB, Fujiwara Y, Beier DR, and Nigam SK (2002) Impaired organic anion transport in kidney and choroid plexus of organic anion transporter 3 (Oat3 (Slc22a8)) knockout mice. J Biol Chem 277:26934–26943.

Tatsunami S, Yago N and Hosoe M (1981) Kinetics of suicide substrates. Steady-state treatments and computer-aided exact solutions. Biochim Biophys Acta 662:226–235.

Terada T, Saito H, Mukai M, Inui K-I (1997) Recognition of β-lactam antibiotics by rat peptide transporter PEPT1 and PEPT2, in LLC-PK1 cells. Am J Physiol Renal Physiol 273:F706–F711.

Terasaki T and Ohtsuki S (2005) Brain-to-blood transporters for endogenous substrates and xenobiotics at the blood-brain barrier: An overview of biology and methodology. NeuroRx 2:63–72.

Thwaites DT, Cavet M, Hirst BH, and Simmons NL (1995) Angiotensin-converting enzyme (ACE) inhibitor transport in human intestinal epithelial (Caco-2) cells. Br J Pharmacol 114:981–986.

Tilloy-Ellul A, Raffalli-Mathieu F and Lang MA (1999) Analysis of RNA-protein interactions of mouse liver cytochrome P4502A5 mRNA. Biochem J 339(3):695–703.

Treiber A, Schneiter R, Häusler S, and Stieger B (2007) Bosentan is a substrate of human OATP1B1 and OATP1B3: Inhibition of hepatic uptake as the common mechanism of its interactions with cyclosporin A, rifampicin, and sildenafil. Drug Metab Dispos 35:1400–1407.

Treijtel N, van Helvoort H, Barendregt A, Blaauboer BJ, van Eijkeren JC (2005) The use of sandwich-cultured rat hepatocytes to determine the intrinsic clearance of compounds with different extraction ratios: 7-ethoxycoumarin and warfarin. Drug Metab Dispos 33:1325–1332.

Tsuji A (2002) Transporter-mediated drug interactions. Drug Metabol Pharmacokin 17:253–274.

Tsuji A (2005) Small molecular drug transfer across the blood-brain barrier via carrier-mediated transport systems. NeuroRx 2:54–62.

Unadkat JD, Kirby BJ, Endres, CJ, Zolnerciks JK (2010) The impact of in vitro to in vivo prediction of transporter based drug-drug interactions in humans. In: S Pang, D Rodrigues, and R Peter, Editors, Springer's Enzymatic- and Transporter-Based Drug-Drug Interactions: Progress and Future Challenges, pp. 517–553.

Vaidyanathan S, Camenisch G, Schuetz H, Reynolds C, Yeh C-M, Bizot M-N, Dieterich HA, Howard D, Dole WP (2008) Pharmacokinetics of the oral direct renin inhibitor Aliskiren in combination with digoxin, atorvastatin, and ketoconazole in healthy subjects: The role of P-glycoprotein in the disposition of Aliskiren. J Clin Pharmacol 48:1323–1338.

Valko K, Du CM2, Bevan CD, Reynolds DP, Abraham MH (2000) Rapid-gradient HPLC method for measuring drug interactions with immobilized artificial membrane: Comparison with other lipophilicity measures. J Pharm Sci 89:1085–1096.

VanWert AL, Bailey RM, and Sweet DH (2007) Organic anion transporter 3 (Oat3/Slc22a8) knockout mice exhibit altered clearance and distribution of penicillin G Am J Physiol Renal Physiol 293:F1332–F1341.

Varhe A, Olkkola KT and Neuvonen PJ (1994) Oral triazolam is potentially hazardous to patients receiving systemic antimycotics ketoconazole or itraconazole. Clin Pharmacol Ther 56:601–607.

Venkatakrishnan K, von Moltke LL, Court MH, Harmatz JS, Crespi CL and Greenblatt DJ (2000) Comparison between cytochrome P450 (CYP) content and relative activity approaches to scaling from cDNA-expressed CYPs to human liver microsomes: ratios of accessory proteins as sources of discrepancies between the approaches. Drug Metab Dispos 28:1493–1504.

Venter JC, Adams MD, Myers EW, Li PW, Mural RJ, Sutton GG, Smith HO, Yandell M, Evans CA, Holt RA, Gocayne JD, Amanatides P, Ballew RM, Huson DH, Wortman JR, Zhang Q, Kodira CD, Zheng XH, Chen L, Skupski M, Subramanian G, Thomas PD, Zhang J, Gabor Miklos GL, Nelson C, Broder S, Clark AG, et al. (2001) The sequence of the human genome. Science 291:1304–51.

Vlaming MLH, Mohrmann K, Wagenaar E, de Waart DR, Elferink RPJ, Oude, Lagas JS, van Tellingen O, Vainchtein LD, Rosing H, Beijnen JH, Schellens JHM, Schinkel AH. (2006) Carcinogen and anticancer drug transport by Mrp2 in vivo: Studies using Mrp2 (Abcc2) knockout mice. J Pharmacol Exp Ther 318:319–327.

Volk EL and Schneider E (2003) Wild-type Breast Cancer Resistance Protein (BCRP/ABCG2) is a methotrexate polyglutamate transporter. Cancer Res 63:5538–5543.

von Moltke LL, Greenblatt DJ, Schmider J, Wright CE, Harmatz JS and Shader RI (1998) In vitro approaches to predicting drug interactions in vivo. Biochem Pharmacol 55:113–122.

Von Richter O, Glavinas H, Krajcsi P, Liehner S, Siewert B, Zech K (2009) A novel screening strategy to identify ABCB1 substrates and inhibitors. Naunyn-Schmiede Arch Pharmacol 279:11–26.

Waley SG (1980) Kinetics of suicide substrates. Biochem J 185:771–773.

Wallstab A, Koester M, Böhme M, and Keppler D (1999) Selective inhibition of MDR1 P-glycoprotein-mediated transport by the acridone carboxamide derivative GG918. Br J Cancer 79:1053–1060.

Wang YH, Jones DR and Hall SD (2004) Prediction of cytochrome P450 3A inhibition by verapamil enantiomers and their metabolites. Drug Metab Dispos 32:259–266.

Wang Q, Strab R, Kardos P, Ferguson C, Li J, Owen A, Hidalgo IJ (2008) Application and limitation of inhibitors in drug–transporter interactions studies. Int J Pharm 356:12–18.

Ward S and Back DJ (1993) Metabolism of gestodene in human liver cytosol and microsomes in vitro. J Steroid Biochem Mol Biol 46:235–243.

Westlind A, Lofberg L, Tindberg N, Andersson TB and Ingelman-Sundberg M (1999) Interindividual differences in hepatic expression of CYP3A4: Relationship to genetic polymorphism in the 5′-upstream regulatory region. Biochem Biophys Res Commun 259(1):201–205.

Whitlock J (1999) Induction of cytochrome P4501A1. Ann Rev Pharmacol Toxicol 39:103–125.

Wijnholds J, de Lange ECM, Scheffer GL, van den Berg D-J, Mol CAAM, van der Valk M, Schinkel AH, Scheper RJ, Breimer DD and Borst P (2000) Multidrug resistance protein 1 protects the choroid plexus epithelium and contributes to the blood-cerebrospinal fluid barrier. J Clin Invest 105:279–285.

Williams JA, Hurst SI, Bauman J, Jones BC, Hyland R, Gibbs JP, Obach RS and Ball SE (2003) Reaction phenotyping in drug discovery: moving forward with confidence? Curr Drug Metab 4:527–534.

Williams JA, Hyland R, Jones BC, Smith DA, Hurst S, Goosen TC, Peterkin V, Koup JR and Ball SE (2004) Drug-drug interactions for UDP-glucuronosyltransferase substrates: a pharmacokinetic explanation for typically observed low exposure (AUCi/AUC) ratios. Drug Metab Dispos 32:1201–1208.

Wrighton SA, Ring BJ, Watkins PB and Vandenbranden M (1989) Identification of a polymorphically expressed member of the human cytochrome P-450III family. Mol Pharmacol 36(1):97–105.

Yamazaki M, Li B, Louie SW, Pudvah NT, Stocco R, Wong W, Abramovitz M, Demartis A, Laufer R, Hochman JH, Prueksaritanont T, Lin JH (2005) Effects of fibrates on human organic anion-transporting polypeptide 1B1-, multidrug resistance protein 2- and P-glycoprotein-mediated transport. Xenobiotica 35:737–53.

Yamazaki M, Neway WE, Ohe T, Chen I-W, Rowe JF, Hochman JH, Chiba M, Lin JH (2001) In vitro substrate identification studies for P-glycoprotein-mediated transport: species difference and predictability of in vivo results. J Pharmacol Exp Ther 296:723–735.

Yang J, Liao M, Shou M, Jamei M, Yeo KR, Tucker GT and Rostami-Hodjegan A (2008) Cytochrome p450 turnover: regulation of synthesis and degradation, methods for determining rates, and implications for the prediction of drug interactions. Curr Drug Metab 9:384–394.

Yazdanian M, Glynn SL, Wright JL, and Hawi A (1998) Correlating partitioning and Caco-2 Cell permeability of structurally diverse small molecular weight compounds. Pharm Res 15:1490–1494.

Yoon CH, Kim SJ, Shin BS, Lee KC, and Yoo SD (2006) Rapid screening of blood-brain barrier penetration of drugs using immobilized artificial membrane phosphatidylcholine column chromatography J Biomol Screen 11:13–20.

Young AM, Audus KL, Proudfoot J, Yazdanian M (2006) Tetrazole compounds: The effect of structure and pH on Caco-2 cell permeability. J Pharm Sci 4:717–725.

Yuan H, Feng B, Yu Y, ChupkaJ, Zheng JY, Heath TG, and Bond BR (2009) Renal organic anion transporter-mediated drug-drug interaction between gemcabene and quinapril. J Pharmacol Exp Ther 330:191–197.

Zaher H, Meyer zu Schwabedissen HE, Tirona RG, Cox ML, Obert LA, Agrawal N, Palandra J, Stock JL, Kim RB, and Ware JA (2008) Targeted disruption of murine organic anion-transporting polypeptide 1b2 (oatp1b2/Slco1b2) significantly alters disposition of prototypical drug substrates pravastatin and rifampin. Mol Pharmacol 74:320–329.

Zhao R, Kalvass JC, Yanni SB, Bridges AS, and Pollack GM (2009) Fexofenadine brain exposure and the influence of blood-brain barrier P-glycoprotein after fexofenadine and terfenadine administration. Drug Metab Dispos 37:529–535.

Zhou F and You G (2006) Molecular insights into the structure–function relationship of organic anion transporters OATs. Pharm Res 24:28–36.

Zhou S, Yung Chan S, Cher Goh B, Chan E, Duan W, Huang M and McLeod HL (2005) Mechanism-based inhibition of cytochrome P450 3A4 by therapeutic drugs. Clin Pharmacokinet 44:279–304.

Zolk O, Solbach TF, König J, and Fromm MF (2009) Structural determinants of inhibitor interaction with the human organic cation transporter OCT2 (SLC22A2). Naunyn-Schmied Arch Pharmacol 379:337–348.

5 Experimental Models of Drug Metabolism and Disposition

GANG LUO

Drug Metabolism and Pharmacokinetics, Covance Laboratories, Madison, Wisconsin

CHUANG LU

Millennium Pharmaceuticals, Inc., Cambridge, Massachusetts

XINXIN DING

Wadsworth Center, New York State Department of Health, Albany, New York

DONGLU ZHANG

Department of Biotransformation, Bristol-Myer Squibb, Princeton, New Jersey

Mass Spectrometry in Drug Metabolism and Disposition: Basic Principles and Applications, First Edition. Edited by Mike S. Lee and Mingshe Zhu.

5.1 INTRODUCTION

Drug discovery and development involve the applications of in vitro and in vivo models to find therapeutically useful agents that intervene with human diseases and enhance human life. All preclinical (in vitro and animal) studies and even clinical studies can be considered model studies. Different models are used at different stages of drug discovery and development to determine efficacy and safety—from test tube experiments to cell cultures, animals, healthy human subjects, and even small numbers of patients that are involved in clinical trials. The selection and application of correct models and appropriate data interpretation are critically important in decision making to successfully advance compounds. All critical phases of drug discovery and development are completed through the design, execution, interpretation of results, and documentation of individual efficacy and safety studies. This work must be conducted in the most efficient manner possible to advance a drug precursor to a drug for patient concerns and meet or exceed worldwide regulatory standards.

The investigation of absorption, distribution, metabolism, and excretion (ADME) properties of a compound is conducted to support drug discovery, safety evaluation, and clinical development. ADME processes determine the exposure and duration of a drug and relevant metabolites and therefore affect the efficacy and safety of a drug. The ability to find a compound that is bioavailable and pharmacologically active and has minimal toxicity potential through lead selection and optimization is the ultimate goal and cornerstone of early drug discovery. At the preclinical stage of drug development, a compound will need to be extensively tested in the laboratory to ensure the safe administration to humans and to maximize the success potential of a compound in the clinical trials. Metabolism studies are needed during the clinical development stage to investigate the dispositional properties of a compound in humans and to address toxicological findings and safety concerns, including those from drug–drug interaction (DDI).

As shown in Figure 5.1, ADME studies are important for drug discovery and development. In fact, the ADME parameters obtained from in vitro and in vivo models are important in decision making to advance or stop a drug candidate to the next stage and help predict the drug behaviors in patients. Incomplete ADME studies or misinterpretation of ADME data may cause a failure in drug development. ADME studies are conducted with in vitro, in vivo, or in silico models. In vitro models should be able to generate many ADME parameters, including apparent permeability, metabolic stability, reaction phenotyping, protein binding, blood-to-plasma partitioning, drug–drug interaction potentials [such as cytochrome P450 (CYP) and transporter inhibition and induction], cell proliferation and cytotoxicity, and human ether à-go-go related gene (hERG) inhibition. In vivo models of animals and healthy subjects will provide information of drug oral bioavailability, exposures, distribution, clearance, and duration of a drug and its metabolites.

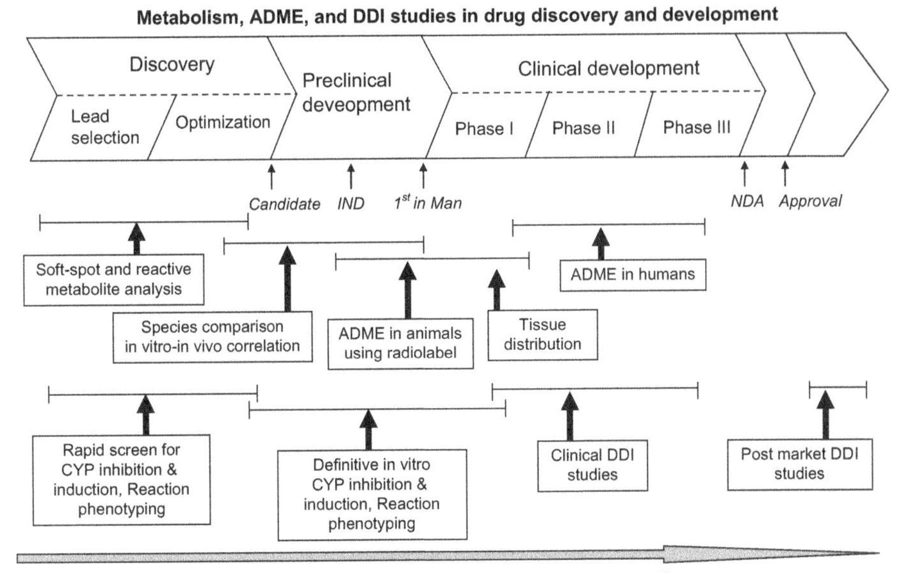

FIGURE 5.1 ADME and drug–drug interaction studies in drug discovery and development.

Model systems needed for particular studies are listed in Figure 5.2. In addition, in silico models are developed to predict drug behaviors based on physicochemical properties of drugs or drug candidates, crystal structures of a protein (an enzyme or a transporter), and database of ADME properties generated in laboratories. Therefore, experimental models are important for ADME studies. The objective of this chapter is to discuss strategy and applications of experimental models in drug metabolism and disposition.

In the last two decades, the pharmaceutical industry has steadily increased its investment in drug discovery and development. However, the return on investment has been declined as fewer drugs have been approved by the Food and Drug Administration (FDA). Several factors could be blamed for the lackluster performance, but an in-depth discussion will not be given here because it is out of the scope of this chapter. A review of annual failure rates of drug candidates indicates that drug metabolism and pharmacokinetics (DMPK) has improved compared to the others (Spalding et al., 2000; Di and Kerns, 2003, 2005; Kerns and Di, 2003), though there is still room for better performance (Alavijeh and Palmer, 2004). The improvement can be attributed to at least three reasons. First, DMPK has been incorporated into early drug discovery, which has become the major trend since about 10 years ago. Second, biomedical sciences have been advancing at a steady pace. DMPK scientists have a better understanding of drug metabolic enzymes, transporters, and differences among individuals and species. Scientists also have many research tools and advanced instrumentation to make relatively reliable predictions

Model Systems for Particular studies

Metabolic stability →	Liver preps, enzymes
Metabolite ID →	Liver preps, bioreactors, in vivo
Reaction phenotyping →	Microsome, hepatocyte, enzyme
CYP inhibition →	Microsome, hepatocyte, enzyme
CYP induction →	Hepatocyte, ex vivo
Transporters →	Caco-2 and cell lines
Plasma protein binding →	plasma
Mass balance →	Animals and human subjects
Metabolite profiling →	Animals and human subjects
Disposition →	Healthy subjects or patients
Species comparison →	Animals and humans
Tissue Distribution →	Rat

FIGURE 5.2 Typical model systems used in drug metabolism and dispositional studies.

from the in vitro to the in vivo data and from preclinical animals to humans. Third, strategies for the optimization of appropriate DMPK properties have been formed, improved, and applied to drug discovery and development.

The objective of this chapter is to discuss model applications in drug metabolism and disposition.

5.2 ADME STUDY STRATEGY IN DRUG DISCOVERY

The task involved with DMPK is not a passive process of simply measuring and reporting the in vitro and in vivo data. In fact, DMPK program leaders in a large pharmaceutical company are expected to provide proper DMPK data interpretation and, more importantly, strategy recommendations and guidance to the project team as well as lead the troubleshooting for DMPK-related issues. The common strategies may include: step-by-step, issue-driven, pharmacokinetic–pharmacodynamic (PK–PD), and PK–toxicokinetic (PK–TK) considerations.

5.2.1 Step-by-Step Strategy

For most drug discovery and development programs, particularly during the early stages, a step-by-step strategy is generally followed (Fig. 5.3). After a therapeutic target is validated, therapeutic molecules are synthesized and tested for the potency on the targeted enzyme or receptor. Structure–activity relationships (SARs) are expected to be gradually established. During the early and middle stages of drug discovery (also called lead identification and lead

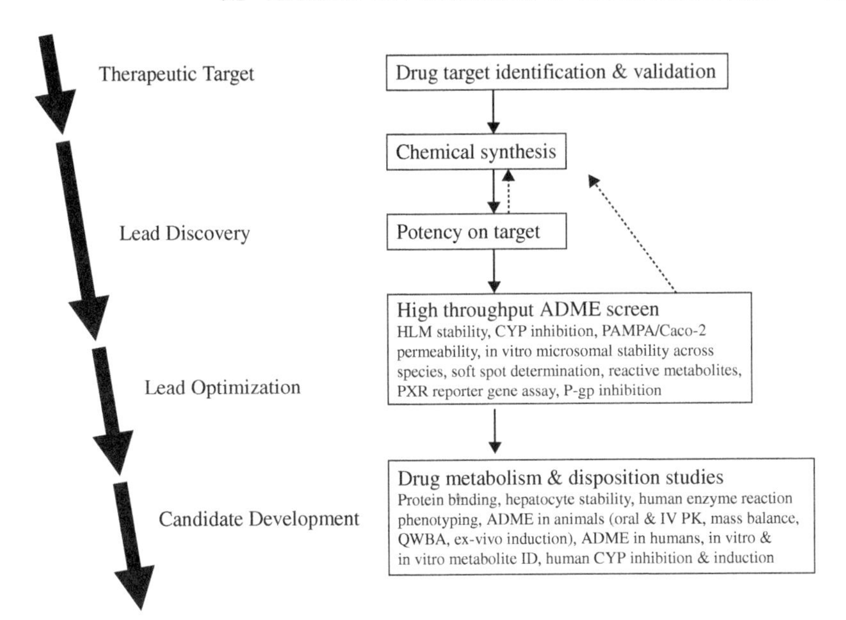

FIGURE 5.3 ADME study strategies in drug discovery and development.

optimization), ADME screening, which is usually in a high-throughput mode, is conducted. ADME screening usually includes, but is not limited to, human and animal liver microsomal stability, human CYP inhibition, apparent permeability using parallel artificial membrane permeation assay (PAMPA) and/or Caco-2 cells, cassette PK screening, metabolic soft-spot determination, reactive metabolite screening, and pregnane X receptor (PXR) reporter gene assay. In the late stage of drug discovery (also called clinical candidate characterization), more comprehensive ADME properties of drug candidates will be determined. DMPK studies at this stage include but are not limited to protein binding (as well as blood-plasma partitioning and plasma stability), hepatocyte stability, metabolic enzyme phenotyping, human CYP inhibition, mechanism-based inactivation, ADME studies in animals (including oral and IV PK studies, mass balance, drug distribution using quantitative whole-body autoradiography (QWBA), drug elimination pathways using bile duct cannulated animals, ex vivo induction, metabolite identification (Met ID), P-glycoprotein (P-gp) substrate and/or inhibitor evaluation using Caco-2 cells or P-gp-transfected cell lines, human CYP induction using human hepatocytes, drug–drug interaction evaluation, and DMPK modeling that includes scaling up from in vitro to in vivo, from preclinical animals to humans, and human dosing projection.

5.2.2 Issue-Driven Strategy

As aforementioned, DMPK work is neither simple nor routine. DMPK scientists should proactively identify DMPK issues (even potential issues) and provide the right strategy to address these issues. The list of DMPK issues may include poor (and variable) oral bioavailability; low PAMPA/Caco-2 permeability; significant first-pass effect; high clearance; low solubility; potent CYP as well as transporter inhibition; DDI; reactive metabolites; metabolism- or transporter-mediated toxicity; species, *in vitro*/*in vivo,* and individual differences; food effect; and human CYP induction. Some issues may indeed be related. Poor bioavailability, for example, can be caused by one or any combination of the following factors: poor intestinal permeability, low solubility, significant first-pass effect, and high clearance. Poor oral bioavailability may be species specific. Furthermore, poor oral bioavailability could in turn lead to high individual variations in drug exposure.

The working team should identify the major issues and take appropriate actions. Presently, many drugs are developed for oral administration, ideally at one dose a day. This strategy requires a reasonably good oral bioavailability and a relatively low clearance. Unfortunately, poor oral availability is a common hurdle in drug discovery. To address this issue, a DMPK scientist should examine the relevant physicochemical properties (application of Lipinski's "rule of five") (Lipinski et al., 2001), permeability from PAMPA/Caco-2 cells, solubility from pharmaceutical analysis, and blood/plasma clearance from a PK study with intravenous administration. Regarding the first-pass effect, a three-leg PK study is commonly employed to dissect the respective intestinal and hepatic contributions. Namely, the test article is given to three groups of animals via intravenous, intraportal vein, and gastrointestinal administration, and then the area under the curve (AUC) is compared. Regarding the intestinal first-pass effect, both poor absorption and active metabolism should be taken into consideration. PAMPA/Caco-2 permeability assays should help estimate the intestinal absorption.

To overcome poor absorption, DMPK scientists should understand its root causes before deciding on which approach to take. The common cause of poor absorption is that a drug candidate does not have the proper physicochemical properties regarding solubility, lipophilicity or hydrophilicity, intestinal permeation, ionization characteristics, and zwitterionicity (Aungst, 1993). In addition, transporter-mediated efflux of the parent or metabolite which is produced in the intestine should also be considered (Malingre et al., 2001; Kakumanu et al., 2006). With feedback from DMPK scientists, chemists can modify the chemical structure to improve physicochemical properties, which is called the "chemical approach." Prodrug strategy is a special chemical approach to improve oral bioavailability. The successful prodrug examples to enhance oral bioavailability include enalapril (to enalaprilat) (Vlasses et al., 1985) and valacyclovir (to acyclovir) (Purifoy et al., 1993). Although enzymes such as esterase and amidase are often involved in converting prodrug to active drug to enhance the bioavailability, it is worth noting that peptide transporter 1 (PEPT1), with its broad substrate spectrum and high capacity (Ganapathy and Leibach, 1996), also plays an important role in the intestinal absorption of valacyclovir. Experimental results demonstrate that valacyclovir, but not acyclovir, is a substrate of PEPT1 (Balimane et al., 1998; Han et al., 1998). A formulation strategy could also be applied to improve oral bioavailability.

5.2.3 PK–PD and PK–TK Considerations

Pharmacodynamics (PD), PK, and TK are three equally important but different properties of a drug candidate. The success of a lead drug candidate is highly dependent on PD, PK, and TK properties. Ideally, a lead drug candidate possesses favorable properties for all three categories. Often, an attractive lead drug candidate has good properties in one aspect but not in others. For example, chemists are able to synthesize very potent compounds that are not orally bioavailable (Stratford et al., 1999). Furthermore, modifications to chemical structures to improve PK properties will also likely alter PD and TK properties, but not necessarily in a more favorable direction.

PK and PD are linked at least in the following aspects: therapeutics, population of patients, potential alternations to metabolic enzymes and transporters in liver and kidney by targeted disease, target organ or tissue by a drug for therapeutics, mechanism of drug therapeutics, pharmacological activation by metabolism, length of clinical dosage (acute or chronic), concentration required to have effectiveness, and the period required to maintain a minimally effective concentration after each dose.

PK and TK are also linked in several aspects: potential impairments to liver and kidneys by drug candidate or its metabolites, toxification via drug metabolism, species-dependent metabolism and toxicity, and toxification via drug transporting. Toxification via drug metabolism includes special toxicity such as mutagenesis, genotoxic carcinogenesis, and teratogenesis (Gibson and Skett, 2001). Species-dependent metabolism and consequent toxicity have been observed in many cases. For example, a glutathione conjugate from efavirenz was formed and led to nephrotoxicity only in rats but not in cynomolgus monkeys or humans (Mutlib et al., 2000). Drug uptake and efflux could also lead to toxicity (Ho et al., 2000).

5.3 ADME EXPERIMENTAL MODELS

5.3.1 In Vitro Models

In vitro assays play a very important role in drug discovery. First, these assays provide a simple, convenient, and fast way to test the potency and drug properties of chemical entities to help advance them rapidly. The drug-like properties commonly refer to respectable absorption, adequate distribution, low metabolism, and complete elimination (ADME) from body and minimal toxicological risk. Second, the amount of compound available is often limited in the early drug discovery stage and it is not always feasible for preclinical animal studies. Thus, in vitro assays could be a more rapid alternative to screen compounds. Third, in vitro assays are also designed to answer specific questions, such as SAR for metabolic stability or DDI potential of the drug candidates. Therefore, these assays are useful tools to optimize DMPK and toxicity properties or SAR after an issue has been identified. In contrast, in vivo results are multifactorial. For example, in vivo results provide combined effect of permeability, distribution, metabolism, and elimination, but they may not be able to provide adequate direction if a researcher would like to pinpoint the effect on any one of those factors. For example, to evaluate bioavailability among a series of drug candidates, it is hard to rely only on the difference in permeability without considering their metabolism. Another advantage of an in vitro assay is that at the preclinical stage human-based in vitro assays could provide close estimation of human clinical outcomes, especially for some properties that are known to have species differences (across from preclinical animals to humans), but good in vitro–in vivo correlations have been established in humans. An example of such a property is the CYP enzyme induction. It is known that preclinical species cannot predict human CYP induction potential due to species difference, and it is also known that there is a good in vitro–in vivo correlation in humans (especially lack of false negative in human in vitro studies using primary hepatocytes cultures); therefore, the FDA can waive clinical DDI studies if the drug candidate is tested negative in a human in vitro CYP induction study.

Furthermore, prediction of human efficacy and DMPK/toxicity properties from preclinical studies is perhaps the ultimate goal of drug discovery. A common practice in the pharmaceutical industry is to establish in vitro–in vivo correlation for an ADME parameter in preclinical studies, then find the species that correlates to humans. If such animal-to-human in vitro correlation could be established, then the human in vitro information could be used to predict the clinical ADME outcome (Fig. 5.4). On the other hand, if the in vivo correlation for a parameter between animal and human is well understood, then animal in vivo data could be used to predict human outcomes. In vitro assays play an invaluable role in these processes. Examples of such predictions often include human efficacy and pharmacokinetic properties (e.g., target inhibition,

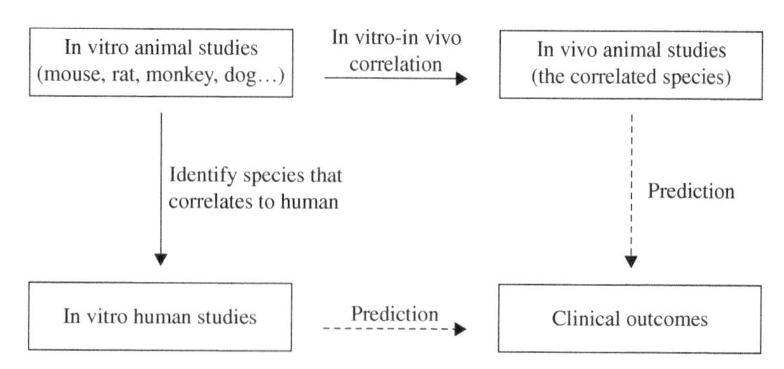

FIGURE 5.4 Model of using in vitro–in vivo correlation (IVIVC) in preclinical species and in human in vitro studies to predict human clinical outcomes.

absorption, clearance) and toxicological endpoints (e.g., QT prolongation). Once the in vitro–in vivo or animal-to-human correlations are established, estimations, for example, of human clearance from human in vitro clearance data, human efficacy from xerograph mouse models, or human QT prolongation risk from dog studies could be made. DMPK parameters can be extrapolated from (1) comparison of in vitro data across species and/or (2) in vitro–in vivo correlation in animals.

5.3.1.1 Expressed Enzymes Expressed enzymes were introduced to drug metabolism in the late 1980s (Dochmer et al., 1988) and they are now available from multiple commercial sources, including Supersomes® from BD Gentest (Woburn, MA). The set of cloned or expressed single-CYP isozymes could be used as an alternative to microsomal system in in vitro drug metabolism and DDI studies. One of the very first applications was to use this purified CYP enzyme to study CYP inhibition with several fluorescence substrates (Stresser et al., 2000). Unlike the CYP inhibition assays in human liver microsomes that require using specific substrate to monitor the inhibitory effect of a test drug on a specific CYP isozyme, the expressed enzymes could share a set of common substrates. Because each expressed enzyme only contains one isozyme, the substrates do not have to be specific to the certain isozyme. The fluorescent substrates offer high throughput by using a microplate reader for detection instead of depending on expensive instruments, such as liquid chromatography/tandem mass spectrometry (LC/MS/MS) systems, in the microsomal assays. However, there could be potential interference if the test compounds also have fluorometric properties or it could interfere with light (e.g., absorbance or luminescence). Another undesired factor is that all fluorescent substrates are not drug-like compounds; therefore the DDI prediction may be less relevant. The most significant disadvantage of using the expressed enzymes–fluorescent substrate approach is that it often overpredicts CYP inhibition potential compared to human liver microsomal assays. Cohen and colleagues (2003) compared CYP inhibition in expressed enzyme assays and human liver microsomal assays and found that expressed enzyme assays, in general, had three- to fivefold lower IC_{50} values across various CYP isozymes. This overprediction and variation among different expressed enzymes (CYPs) could be attributed to multiple factors, such as:

1. Expressed enzymes are purified enzymes and often used in small amount in an incubation; therefore the protein binding is different compared to that in microsomal incubation.
2. Not all the coenzymes, such as P450 NADPH oxidoreductase (OR) and cytochrome b_5, are expressed in each of the expressed enzymes.
3. The amount of coenzymes in different expressed enzymes, if they do express, are different and they are in turn different compared to liver microsomes.

4. The use of non-drug-like small-molecule substrates in the supersomal assay could be another factor contributing to the differences in DDI prediction.

In recent years, LC/MS/MS systems have become more sensitive, affordable, and automated. As a consequence, microsomal assays have become routine to accommodate the high-throughput needs within drug discovery. Most pharmaceutical companies are no longer using the expressed enzyme–fluorescent substrate approach for CYP inhibition evaluation. Recent regulatory or Pharmaceutical Research and Manufacturers of America (PhRMA) guidance also recommends the use of human liver microsomes or primary hepatocytes, as they are more close to human in vivo conditions, for evaluation of DDI potential (Food and Drug Administration, 2006; Grimm et al., 2009).

Another application of expressed enzymes is to study time-dependent inhibition (TDI). In the FDA (2006) draft DDI guidance, the only required CYP to be studied for TDI is CYP3A4, The underlying reason is that CYP3A4 is the most abundant CYP and involves metabolism of greater than 50% of marketed drugs. For compounds that generate reactive metabolites, the possibility to inactivate CYP3A4 could be high. However, there are reports suggesting that a metabolite generated by one CYP may be a time-dependent inhibitor of another CYP (Ohyama et al., 2000). Similar to Ohyama's study, Li et al. (2004) reported that an investigational compound metabolized by CYP3A caused a TDI of CYP2D6 in human liver microsomes without affecting CYP3A4 activity. They further studied the compound in the CYP2D6 and 3A4 expressed enzymes individually and found that neither system resulted in TDI. When the preincubation was performed in a mixture of expressed enzymes CYP2D6 and CYP3A4, the compounds were shown to be TDI for CYP2D6 but not CYP3A4. These data suggest that this investigational compound was metabolized by CYP3A4 to a metabolite(s), which in turn was further metabolized by CYP2D6 to a reactive species, which eventually inactivated CYP2D6 (Li et al., 2004). Thus, using expressed enzymes for TDI study may generate false-negative results. Therefore, the recent published PhRMA TDI white paper suggests using human liver microsomes for TDI studies (Grimm et al., 2009).

A main application of expressed enzymes is to study the reaction phenotyping of drug candidates. The FDA DDI guidance requires conducting a clinical DDI trial if a compound has greater than 25% of its clearance mediated by a particular pathway. CYP metabolism is the most common clearance pathways for elimination of drugs. Thus, identifying the relative contribution of each CYP isozyme to the total hepatic clearance is an important task. Together with the data on other nonmetabolic clearance pathways, such as the renal or biliary clearance, one could estimate the contribution of each isozyme to the total body clearance of the drug. Crespi (1995) first used the expressed enzymes along with microsomes to study the reaction phenotyping. Since different expressed enzymes carry different coexpressed enzymes and thus have different catalytic activities per unit base, the CYP activity of each expressed enzyme cannot be quantitatively linked to that in human liver microsomes. They proposed the relative activity factor (RAF) to normalize the activities in Supersomes to that in native liver microsomes (Crespi, 1995):

$$\mathrm{RAF} = \frac{V_{max} \text{ or } \mathrm{CL_{int}} \text{ for specific substrate for human liver microsomes}}{V_{max} \text{ or } \mathrm{CL_{int}} \text{ for specific substrate for expressed enzymes}} \quad (5.1)$$

From the equation above, the RAF is determined using the CYP probe substrate where the CYP-specific metabolic pathways are known. Values of V_{max} or intrinsic clearance (usually in V_{max}/K_m format) are commonly used to determine the relative activities of probe substrates in these two in vitro systems under an arbitrarily selected enzyme concentration, for example, 10 pmol expressed enzymes/mL and 0.5 mg microsomes/mL. Then, similar incubations are performed for the test compound to determine the relative metabolic activities in expressed enzymes and microsomes. Since the metabolic pathways of the test compound may not have been identified at the drug discovery stage, a total clearance is usually determined using a parent compound disappearance assay (Uttamsingh et al., 2005). The RAF determined for the probe substrate is then applied to the relative activities of test compound to calculate the percentage or relative contributions of each tested CYP to the total hepatic microsomal clearance:

$$\text{Percent contribution} = \mathrm{RAF} \times \frac{V_{max} \text{ or } \mathrm{CL_{int}} \text{ for test compound forexpressed enzymes}}{V_{max} \text{ or } \mathrm{CL_{int}} \text{ for test compound for human liver microsomes}} \quad (5.2)$$

This approach, as well as other approaches using chemical inhibitor or inhibitory monoclonal antibodies, assumes that the probe substrates behave like the test compound. A simplified approach is to study the test compound in expressed enzymes only and then assume hepatic metabolism is covered by the several expressed enzymes that have been assayed (Venkatakrishnan et al., 2001). In other words, this simplified approach uses the sum of various supersomal clearances instead of the clearance in liver microsomes as 100% of hepatic metabolic clearance. While in most cases several major CYPs may have covered the majority of hepatic metabolism, other enzymes could also contribute significantly to the overall metabolism. Examples of non-CYP microsomal enzymes often involved in drug metabolism include flavin-containing monooxygenase (FMO), UDP–glucuronosyltransferase (UGT), sulfotransferase (SULT), and many others (Parkinson and Ogilvie, 2008).

Expressed enzymes are often used to qualitatively identify the involvement of CYPs in metabolic pathways (often called CYP mapping). For example, after metabolites have been identified in microsomal incubation, a set of incubations of the test compound with expressed enzymes could be applied to identify which CYP isozymes are involved in the formation of a given metabolite (Zhang et al., 2007a; Wang et al., 2008). This assay provides important information on whether the drug candidate is metabolized by a single isozyme and furthermore whether polymorphic enzymes such as CYP2D6 and 2C19 are the major contributors to the metabolic clearance of this drug candidate. Expressed enzymes are also used in mechanistic studies. The previously described example (Li et al., 2004) featured the use of expressed enzymes incubated separately or in combination with each other, and the results indicated that CYP3A4 was involved in the generation of a metabolite that was responsible for the TDI of CYP2D6.

Since expressed CYP systems are concentrated enzymes with high activity toward a specific CYP isozyme, they, especially those with protein purification, are often used as "bioreactors" to generate metabolites of interest for further testing. For example, to identify whether a metabolite of interest is pharmacologically active or to quantify metabolites, metabolite standards often need to be characterized and then synthesized. For some compounds, the chemical synthesis of the metabolites could be technically challenging and both time and resource consuming. The use of large quantities of expressed enzymes is an alternate way to generate metabolites. Furthermore, metabolites collected as chromatographic fractions could be tested for pharmacological activity without the need to identify the structures. In this scenario, the high activity and selectivity of expressed enzymes provide a superior choice over liver microsomes. In addition, HLM could generate more metabolites, which can make LC separation difficult; however, expressed CYP is much more expensive.

5.3.1.2 Subcellular Fractions (Microsomes, Cytosol, S9) Along with the expressed enzyme systems, subcellular fractions from drug-metabolizing tissues such as microsomes, cytosol, and S9 fractions from liver and gut are widely used in drug metabolism studies. Liver tissues that are not suitable for transplant in patients are often donated for research purposes. Usually, liver tissue is minced in 3–4 volumes of buffer on ice and then homogenized. A 20-min centrifugation at 9000*g* is applied to the homogenate to remove mitochondria, nuclei, fat, and membrane proteins. The resulting supernatant, the S9 fraction (supernatant at 9000*g*), contains most of the hepatic enzymes except monoamine oxidase (MAO), which resides in mitochondrial membrane. The S9 fraction could then be further centrifuged at 105,000g for 1 h to pellet down the microsomes, containing major metabolism enzymes, such as the CYP, UGT, certain glutathione *S*-transferases (GST) and FMO. The supernatant from the centrifugation, called cytosol, contains soluble enzymes, such as SULT, *N*-acetyl transferase (NAT), certain GSTs, aldehyde oxidase (AO), and xanthine oxidase (XO).

Cytosol is the least used subcellular fraction compared to the S9 and microsomes. Cytosol is often used to conduct mechanistic studies and identify if any of the above enzymes could be involved in certain metabolic pathways. For example, regioselective sulfation of the phytoestrogens daidzein (DZ) and genistein (GS) was investigated using human liver cytosol and purified recombinant human sulfotransferase (SULT) isoforms (Nakano et al., 2004). The study found that the average ratios for 7- to 4′-sulfate formation were 4.5 : 1 from DZ and 8.4 : 1 from GS in human liver cytosol. In addition, SULT1A1 and SULT1E1 exhibited much higher catalytic activities than the other two isozymes, SULT1A3 and SULT2A1. Cytosol is also used to complement the microsomal studies to assess drug metabolism pathways (Zhang et al., 2008). The preference to use S9 fractions in drug metabolism studies varies among different laboratories. Some laboratories prefer to use microsomes to study the CYP- and UGT-mediated metabolism and use cytosol as a complementary tool to check the possible involvement of other enzymes not present in microsomes, whereas other laboratories use S9 or primary hepatocytes to capture the whole spectrum of drug metabolism pathways. In the authors' laboratories, S9 fractions are often used to study intrinsic clearance of drug candidates, especially those that have structural alert of undergoing potential conjugations or metabolism through enzymes other than CYP, such as SULT, NAT, or FMO. One issue regarding the use of S9 fractions is that the enzymes in S9 are not as concentrated as those in microsomes. A large amount of S9 protein (usually 5 times) is used compared to that in microsomal incubation to achieve a similar activity. This condition could result in higher protein binding of the test compound. The advantage of using the S9 fractions over microsomes is that additional non-CYP enzyme-mediated metabolism could be captured, such as sulfation and acetylation. Compared to the primary hepatocyte incubation, S9 is cheaper and easier to handle. The scaleup factor from S9 incubation to intrinsic clearance is not available in the literature. Our laboratories have determined this value as 165 mg S9 per gram of rat liver. This value is used to calculate intrinsic clearance across all species (including humans) to extrapolate this calculation to whole-body clearance. In the isolation of subcellular fractions, the cofactors that either mediate the oxidative metabolism (e.g., NADPH) or supplement the conjugations [e.g., uridine diphosphate glucuronic acid (UDPGA), 3′-phosphoadenosine-5′-phosphosulfate (PAPS), and *S*-adenosylmethionine (SAM)] are lost. Therefore, supplementing the subcellular incubations with these cofactors could pinpoint the different metabolic pathways and their relative contributions. This application has been well demonstrated in the 17α-ethinylestradiol metabolism study (Li et al., 1999a).

Microsomes are the most widely used subcellular fractions for drug metabolism studies. They provide the advantages of being inexpensive and easy to handle and cover the major drug metabolism enzymes (e.g., CYPs and UGTs). At the drug discovery stage to study "what the drug metabolism enzyme does to the drug," microsomal assays are the default assays that are used to determine intrinsic clearance to screen for stable compounds and establish in vitro correlation between animal and humans. Primary hepatocyte assays have become popular in recent years for this application. Microsomes from different tissues could also be applied to study extrahepatic metabolism and further strengthen the in vitro prediction to the total body clearance by considering multiorgan metabolism together if applicable for the drug of interest.

TABLE 5.1 Prediction of Clinical Relevance of Reversible CYP Inhibition

$[I] / k_i$ ratio	Prediction
$[I] / k_i > 1$	Likely
$1 > [I]/k_i > 0.1$	Possible
$0.1 > [I]/k_i$	Remote

On the other hand, to study "what the drug does to the drug metabolism enzyme," microsomal assays are also the preferred choice for the study of DDI potential. Both reversible inhibition and time-dependent inhibition studies could be conducted to investigate if the drug candidate might be a perpetrator of DDI for a co-administered drug. CYP reaction phenotyping studies could also be conducted to study whether the drug candidate might be a victim, if co-administered with a CYP inhibitor. Microsomes contain relatively concentrated CYP enzymes; thus, they are often used to conduct enzyme kinetic studies, such as to determine the V_{max} and K_m of a metabolic pathway, k_i or IC_{50} for reversible inhibition, and k_{inact} and K_I for mechanism-based inhibition. The current DDI guidance suggests that if the ratio of total plasma C_{max} [I] over the inhibition constant k_i is greater than 0.1 (Table 5.1), a clinical trial to evaluate reversible inhibition is warranted. When in vitro assays indicate a compound is a mechanism-based inhibitor, clinical evaluation should be followed (Food and Drug Administration, 2006; Grimm et al., 2009).

5.3.1.3 Whole-Cell Systems (Hepatocytes, Other Cells, and Expressed Cell Lines) Hepatocytes are the primary liver cells where drug metabolism occurs. Hepatocytes have the cell membrane which contains various uptake or efflux transporters. Drugs could passively diffuse through the membrane or be taken up by transporters such organic anion transporting polypeptide (OATP) and sodium taurocholate cotransporting polypeptide (NTCP) to enhance the intracellular concentration. Once the drug enters into the hepatocytes, the efflux pumps, such as the MRP or MDR transporters, could pump the drug out to reduce intracellular drug concentration. The primary hepatocytes also carry enzymes and cofactors at physiological concentrations and provide a drug metabolism environment that closely mimics in vivo conditions. Freshly prepared hepatocytes are a good model for drug metabolism and transporter studies, however, hepatocytes have some limitations, such as (1) they are not readily available from higher species and (2) a preparation could only be used once, so it is difficult to repeat studies or compare studies between laboratories. The development and commercialization of cryopreserved human hepatocytes have provided a big advantage for the use of this in vitro system in drug discovery (Ruegg et al., 1997; Li et al., 1999b). The precharacterized cryopreserved hepatocytes could serve as a "reagent" and allow researchers to perform experiments at their own schedule and test drug candidates over a period of time using the same lot(s) of hepatocytes. Therefore, the same study can be compared between different laboratories. In addition, researchers could select specific lots (donors) of hepatocytes to meet their research needs, such as a study that involves drug metabolism in special populations (e.g., smokers or alcohol users) or a population with poor metabolism in some polymorphic CYPs, such as CYP2D6 or 2C19.

Hepatocyte incubations require experienced researchers to either prepare fresh hepatocytes from liver tissue or thaw and use cryopreserved hepatocyte working suspension. There are several formats for conducting hepatocyte incubations that are generally being used in drug discovery studies, suspension, and sandwich cultures. The suspension is used to study the metabolic stability or transporter-mediated uptake of drug candidates. This format has the limitation that hepatocytes are viable for only acute dose studies, generally up to 4 h (Li et al., 1999b; Shitara et al., 2003). Researchers have used hepatocytes suspended in plasma incubations to improve the in vitro–in vivo correlations in drug clearance and DDI prediction (Shibata et al., 2002; Bachmann et al., 2003; Blanchard et al., 2006; Skaggs et al., 2006; Lu et al., 2007). Cultured hepatocytes on collagen-coated plastic surfaces could extend the period of hepatocyte viability to about 7–10 days. Besides the disadvantage of CYP activities decreasing over the culturing period, the hepatocyte cultures could still be used for metabolism study, providing longer exposure for the test compound and increasing secondary metabolite formation similar to in vivo situations. This assay condition is also reported to better identify some time-dependent inhibitors, which would otherwise be missed in microsomal incubation due to shorter exposure time (Zhao et al., 2005; McGinnity et al., 2006). A major application of cultured hepatocytes is in CYP induction study. Because enzyme induction involves gene transcription and translation, requiring time to take effect, primary cultured human hepatocytes have become the gold standard for conducting CYP induction studies (Food and Drug Administration, 2006; Luo et al., 2007). There are several ways to perform CYP induction studies; for example, after 2–3 days of treatment of the hepatocytes with the test drug, one could measure the effect on activity directly in the hepatocytes by incubating with the probe substrate or one could collect the hepatocytes, make microsomes from the hepatocytes, and then measure the CYP activity and protein expression changes. The main advantage of the latter approach is that any inhibitory effect of the drug residue would be removed, which would otherwise mask the induction readout. The microsomes prepared could also be stored for assays later at a convenient time. Some cryopreserved hepatocytes are platable while others are not. The reason for this is not very clear yet. Since the first report of using platable cryopreserved hepatocytes to conduct a CYP induction study (Ruegg et al., 1997), the technology has matured and using platable cryopreserved hepatocytes is now an alternative to using freshly isolated hepatocytes to study CYP induction. The two systems generate comparable fold of induction even though cryopreserved cultures may have lower absolute activity and protein amount/well due to fewer hepatocytes surviving until the end of the study, compared to the fresh hepatocyte cultures.

The current guidance from the regulatory agency and PhRMA suggest that, if a compound causes greater than 40% induction compared to the positive control compound, it would have a potential to cause clinical risk, and therefore clinical DDI trials should be followed (Food and Drug Administration, 2006; Chu et al., 2009). Some cell lines, such as Fa2N-4 and HepaGR, are also used for CYP induction evaluations (Hariparsad et al., 2008; McGinnity et al., 2009); however, some reports suggest that some CYP enzymes, such as CYP2B6, may be underrepresented in these cell lines (McGinnity et al., 2009).

Liu and colleagues (1999a; 1999b) have modified the sandwich culture condition by adding a thick layer of Matrigel™ on top of the hepatocytes to study biliary excretion of compounds. Under these conditions, healthy bile canaliculi between hepatocytes were developed. When the test compound is incubated with hepatocytes, part of the test compound is excreted into bile canaliculi and the rest remains in hepatocytes or medium. The tight junctions between the hepatocytes prevent the compound from escaping the bile pockets. Therefore, upon treatment of the hepatocytes with calcium-free medium [or ethylenediaminetetraacetic acid (EDTA)–containing medium], the tight junctions are opened up, releasing the compound from bile pockets into the media (Fig. 5.5). The extent of bile excretion is measured as the biliary excretion index (BEI), which is calculated using Eq. (5.3). Furthermore, the clearance, a kinetic parameter, of biliary excretion could be determined in a well-designed study [Eq. (5.4)]. As illustrated earlier in Figure 5.4, this application provides a major advantage that one could identify species similarities to find a preclinical species that could predict humans in vitro. One could also use the human hepatocyte results to predict human clinical outcomes once the in vitro–in vivo correlation is established in animals:

$$\mathrm{BEI} = \frac{\mathrm{Accumulation}_{\mathrm{cells+bile}} - \mathrm{Accumulation}_{\mathrm{cells}}}{\mathrm{Accumulation}_{\mathrm{cells+bile}}} \times 100\% \tag{5.3}$$

$$\mathrm{CL}_{\mathrm{bile}} = \frac{\mathrm{Accumulation}_{\mathrm{cells+bile}} - \mathrm{Accumulation}_{\mathrm{cells}}}{\mathrm{AUC}_{\mathrm{media}}} \tag{5.4}$$

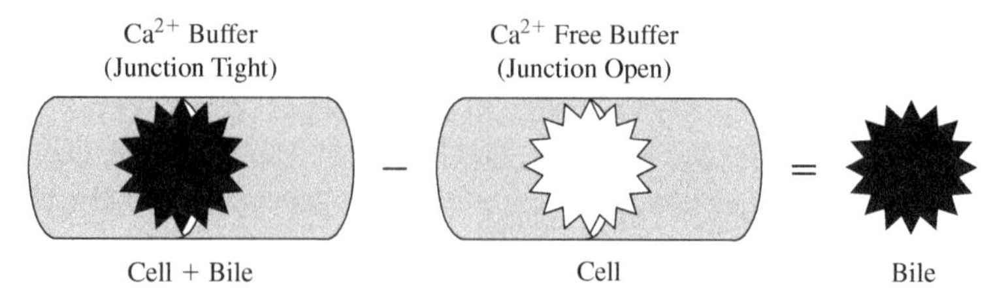

FIGURE 5.5 Experimental model for determination of biliary uptake of test compounds in cultured hepatocytes. Incubation with calcium-free medium would release the compound from bile ducts into medium and allow determination of the amount of compound excreted in bile.

Caco-2, an immortalized human colorectal carcinoma-derived cell line (Artusson and Karlsson, 1991), is the most commonly used model for studying drug permeability and identifying transporter involvement in drug uptake or efflux. Caco-2 cells are usually cultured in Transwells® for 21 days to allow full expression of P-gp and other transporters. Adding drug to the apical side and monitoring the appearance of the drug at the basolateral side over a period of time or vice versa allow one to study the kinetics (the apparent permeability coefficient P_{app}) of the absorption of the test compound. The P_{app} values are usually expressed in centimeters per second times 10^{-6} and calculated using the equation

$$P_{app} = \frac{dQ}{dt} \times \frac{1}{AC_0} \tag{5.5}$$

where dQ/dt is the total amount of drug present in the receiver chamber per unit time (e.g., nanomoles per second), A is surface area of the Transwells® (0.33 cm^2 for a 24-well plate), and C_0 is the initial drug concentration in the donor chamber (e.g., micromolar).

In a study, if the P_{app} value from the apical side to the basolateral side is close to that from the basolateral side to the apical side, then the compound likely enters the cells by passive diffusion. A greater P_{app} value from the basolateral side to the apical side indicates involvement of efflux transporters, such as P-gp, MRP2, and/or BCRP. Incubation of the test compound with different transporter inhibitors or substrates could help to differentiate whether the test compound is a substrate or inhibitor of certain transporters. For example, diminished P_{app} values from the basolateral side to the apical side in the presence of LY335979, indomethacin, or Ko143 suggest that the test compound may be a substrate of P-gp, MRP2, or BCRP, respectively. On the other hand, a decrease in the P_{app} values from the basolateral side to the apical side of the transporter substrates paclitaxel (P-gp), estradiol-17b-glucuronide (MRP2), or estradiol-3-sulfate (BCRP) in the incubations with test compound indicates that the test compound is an inhibitor of these transporters.

Caco-2 cells are heterogeneous cell lines. Cultures from different laboratories may generate different P_{app} values. These differences may be partially due to different techniques, media, or passages of the cells used in different laboratories. Thus, individual laboratories should routinely calibrate their results with a set of reference compounds (Food and Drug Administration, 2000).

5.3.1.4 Artificial Systems (PAMPA) Some artificial systems are widely used as an alternative to Caco-2 cell culture to evaluate drug permeability. An example of these is the parallel artificial membrane permeation assay (PAMPA). PAMPA is usually performed in a 96-well microtiter plate format that consists of test wells fixed with an artificial hydrophobic membrane and reference wells (Kansy et al., 1998). Test compounds are applied at one side of the membrane, and the permeability rates are monitored at the receiving side. The properties of the membrane (lipophilicity and thickness) control the rate of permeability. Different membranes could be designed to mimic the passive perfusions in different tissues, such as the gastrointestinal tract (GI) or the blood–brain barrier (BBB). PAMPA offers a high-throughput way to screen compounds. Beside the throughput, PAMPA also allows researchers to test compound permeability at different pHs to compare the permeability at different ionization stages whereas in Caco-2 cells the study is usually done at close to neutral pH values (e.g., 6.7–7.4). PAMPA is also insensitive to a high percentage of organic solvents, so one could test the intrinsic permeability of the compound under solution conditions. PAMPA is widely used in early drug discovery, and it is found to have good correlation with GI permeability (Kansy et al., 1998) as well as BBB permeability (Di et al., 2009).

5.3.1.5 ***Microbial Systems*** Besides using a large quantity of enzymes, such as microsomes or expressed CYP, to generate metabolites, microbial fermentation is another way to prepare drug metabolites or even drugs themselves from precursor compound (Zhang et al., 1995; 2006). Many steroid hormones, antibiotics, anticancer drugs, and human insulin could be effectively made through fermentation. For example, the immunosuppressant tacrolimus is often prepared from microbial fermentation. Fermentation is also viewed as a "green" process as the microbes can regenerate. Chemical waste is not generated, and thus, there are no health and environmental concerns. From a DMPK perspective, this process is often applied to synthesize drug metabolites of interest or some building blocks on which metabolites could be further synthesized. These "biosyntheses" could overcome some difficult chemical reactions due to the high selectivity of the bacterial enzymes (Glueck et al., 2005). A well-designed fermentation process could be very effective in handling reactions involving chiral centers. For example, researchers have developed an immobilized microbial esterase modified from pig liver esterase to prepare enantiopure (−)-2′,3′-dideoxy-5-fluoro-3′-thiacytidine [(-)-FTC], the active ingredient in the anti-HIV drug Emtriva. This process avoided costly chemical separation of the racemic drug mixture [(±)-FTC] and additional processes to degrade the inactive by-product of [(+)-FTC] (Osborne et al., 2006). Hu and colleagues (2006a; 2006b) have successfully applied a chemoenzymatic reaction in the synthesis of the antiangiogenesis agent pelitrexol and a chiral intermediate for an HIV protease inhibitor. In both cases, the chemoenzymatic approaches provided optimal reaction conditions and resulted in high yields. In such cases, the cost of drug synthesis was estimated to be reduced by as much as 95% (Hu et al., 2006a; 2006b). Another example involves the preparation of vitamin C—a mutant strain of *Gluconobacter oxydans* selectively converts the D-sorbitol to L-sorbose, which is the precursor for vitamin C synthesis. A chemical reaction would produce unwanted D-sorbose along with L-sorbose (Demain and Sanchez, 2006).

5.3.2 In Situ and Ex Vivo Models

5.3.2.1 ***In Situ* Models (Perfusion)** The organ perfusion model most closely mimics in vivo drug absorption, metabolism, and elimination. Among various organ perfusion models, the liver perfusion model is the most studied. This model has the advantage over hepatocyte and other subcellular systems and maintains the liver structure and architecture, preserves all the transporters and cell populations other than hepatocytes (e.g., the Kupffer cells, which are important in the regulation of proinflammatory cytokines), and preserves hepatocyte cell-to-cell interaction and zonal differentiation. As an endpoint, the perfusate, bile, and liver tissue itself can be analyzed for the parent compound and its metabolites. This approach could provide information on the extent of the hepatic first-pass effect, effect of protein binding, parent uptake from the perfusate, metabolism, parent and metabolite elimination via canalicular transporters, and the potential for toxicity from reactive metabolites. The liver perfusion technique is also a useful tool for toxicological and pharmacological studies (Compagnon et al., 2001).

The liver perfusion technique has been standardized regarding the experimental setup (Mehendale, 1976; Meijer et al., 1994) and the constituents of the perfusates usually vary depending on the purpose of the study. Liver perfusion can be performed in situ or in a setting where the liver is isolated. The in situ perfusion model requires minimal organ preparation, and therefore, the potential of organ damage is minimized. However, with the isolated liver setting, liver-specific mechanistic studies can be conducted to answer specific questions with less interference from other organs and therefore is generally a more widely used model.

Isolated liver perfusion systems are commercially available (e.g., Harvard Apparatus, Holliston, MA). Commonly used perfusates contain various amounts of albumin and red blood cells (RBCs) that affect the protein binding of the compound as well as other components that may affect the metabolism of the test compound and hence the bioavailability of the compound. Heparinized whole blood from the same species is perhaps the best perfusate. However, heparinized whole blood is not always readily available and potential issues such as the formation of clots need to be taken into consideration. Bovine blood, bovine serum albumin, and bovine or human erythrocytes have also been used to mimic in vivo conditions because they provide hemoglobin as the oxygen carrier and ideal protein binding conditions.

***5.3.2.2 Ex Vivo* Models for Induction and Toxicity Issues** Ex vivo studies refer to studies where animals are dosed with a drug and then the organ tissues are removed and processed (e.g., making liver into microsomes) to investigate the changes in enzyme or transporter level (or any biomarkers) upon drug treatment. This information could then be linked back to the toxicology or pathology findings in the animal in-life study. One of the most useful practices is to perform the ex vivo study in parallel with the TK study where drug is dosed at a high level (close to the maximum tolerable dose) and PK information is collected and used to link or explain the PD or TK observations. Usually, at the end of the subchronic or chronic dosing study, organs are collected. Liver, the organ which is exposed to high drug concentrations, is often studied for the effect of drug or enzyme levels. This change could in turn explain the changes in PK or could be linked to the toxicity findings. Among various observations of enzyme level change, PK, and toxicity, some are directly linked to each other, whereas others may show delayed responses. PK/PD modeling could then help dissect these response relationships and aid in further prediction. In general, in an animal repeat-dosing study, if there is a decrease in drug exposure ($AUC_{(0-24),\text{day } x} < AUC_{(0-\infty),\text{day1}}$), it is often considered to be due to an autoinduction by the drug to the enzyme(s) responsible for its own metabolism. By testing with CYP probe substrates in an ex vivo study, the induction of CYPs by a drug can be determined. On the other hand, if there is an accumulation of AUC ($AUC_{(0-24),\text{day } x} > AUC_{(0-\infty),\text{day 1}}$), the drug could be a mechanism-based inhibitor that inactivates the enzyme(s) responsible for its own metabolism. One advantage of the ex vivo study is that the parent compound has already been removed during the preparation of the subcellular fractions, so the inhibitory effect from reversible inhibition via the parent compound is unlikely. Inducible CYP enzymes are often the enzymes to be studied. Enzymes responsible for xenobiotic metabolism, such as the CYP1–3 families, are routinely studied to investigate whether enzymes are induced to enhance reactive metabolites, which in turn could cause toxicity. Enzymes responsible for detoxification of reactive species are also studied to see if these detoxification pathways are inhibited or down-regulated upon drug treatment. Other CYPs which involve the metabolism of endogenous substrates such as CYP4A11 and CYP7A1 are often studied. CYP4A11 is responsible for fatty acid metabolism and its induction is believed to attribute to liver enlargement via peroxisomal proliferation, resulting in either hypertrophy (enlargement of hepatocytes) or hyperplasia (increase in the number of hepatocytes). CYP4A11 is also believed to participate in the defense mechanism against UV-induced oxidative damage (Gonzalez et al., 2001).

CYP7A1 is critical in bile salts synthesis and cholesterol homeostasis. Imbalances in the level of these two enzymes can cause abnormal liver function.

5.3.3 In Vivo Models

The drug-like ADME properties contain an integrated set of parameters commonly referring to good absorption, adequate distribution, low metabolism, and complete elimination from the body and minimal toxicology risk. In contrast to in vitro models that only generate or reflect one particular aspect of the whole picture, in vivo results are multifactorial and provide the combined effect of permeability, distribution, metabolism, and elimination, leading to a measurable set of pharmacokinetic parameters and toxicology endpoints. Regardless of how thorough and complete in vitro tests had been performed prior to human dose, animals will eventually be dosed to measure drug exposures and test toxicities. There is now a trend that rat is the first animal species to be tested for the drug exposure. In addition, a drug candidate should be dosed in human subjects as early as possible since that is the ultimate target for required drug exposures. It should be pointed out that human-based in vitro assays could provide closer estimation of human clinical outcomes, especially for some properties that are known to have species differences, such as P450 enzyme induction. However, good in vitro–in vivo correlations need to be established in humans. Furthermore, the prediction of human efficacy and DMPK/toxicity properties from preclinical studies is the ultimate goal of drug discovery.

5.3.3.1 Pharmacokinetic Studies Understanding the pharmacokinetics of a compound in preclinical species with various dose routes (oral, intravenous, subcutaneous, transdermal, intraperitoneal, continuous infusion, intratracheal, as well as by access ports of the portal vein) is an essential component for lead selection and optimization and clinical candidate nomination and development. Early rapid PK screening using various dose regimens and administration routes in rodents or nonrodents can assist in lead candidate selection from multiple compounds by eliminating compounds with unfavorable PK profiles and optimization. Early TK studies with limited histopathology evaluation can help lead optimization to produce the candidate that is more likely to succeed in preclinical testing and in the clinic. More extensive PK studies are needed to evaluate the linearity, dose proportionality, bioavailability, or food effects in single- or multiple-dose administration. To define the PK profiles of a compound, compartmental and noncompartmental methods are used to analyze multiple parameters, including maximum concentration (C_{max}), time of maximum concentration (T_{max}), AUC, volume distribution, clearance (CL), terminal elimination half life ($t_{1/2}$), and bioavailability. Dispositional studies also provide data on mass balance, biliary excretion, and tissue distribution. Surgical animal models for dosing and sampling can be used to provide insight on biliary excretion from bile-duct cannulation and to investigate first-pass metabolism from vascular and portal vein cannulation. The animal models that are used in these experiments have been extensively reviewed by Salyers (2009).

Table 5.2 lists typical PK studies in drug discovery, preclinical development and safety, and clinical and biopharmacology studies.

5.3.3.2 Preclinical ADME Models During the discovery stage, ADME studies are conducted with nonlabeled compound and limited quantitation data can be obtained for parent and metabolites. In drug development, ADME studies are performed with either C-14 or tritium-labeled material to provide detailed quantitative information on the circulating metabolites and the extent of metabolism and routes of excretion for drug and its metabolites. Zhang and Comezoglu (2007) have provided a detailed review on ADME study design and data presentation.

Tissue distribution studies in pigmented Long-Evans rats to provide dosimetry to various tissues and organs are required to support human ADME with radiolabeled compound. Typically, these studies are limited to a single dose by the intended route of administration (PO, IV, etc.). The radioactivity levels in various tissues at different time points can be measured by quantitative whole-body autoradiography (QWBA) where sections of the

TABLE 5.2 Role of Pharmacokinetic Studies in Drug Discovery and Development

Stages	PK Study Types	Objectives	PK or TK Parameters
Discovery	Fast exposure and PK screening in rat	Finding bioavailable compound	AUC, C_{max}
	Single/multiple/IV dose in mouse, rat, dog, or monkey	Species PK comparison	AUC, C_{max}, T_{max}, $T_{1/2}$, V_d, F
	Allometric scaling from in vivo and in vitro data to predict human dose and PK	Human dose and PK prediction	CL, AUC, C_{max}, $T_{1/2}$, and efficacious dose projection
Preclinical develop/ safety	Formulation in animal species	Finding best formulation for drug exposures	AUC, C_{max}
	Single ascending-dose TK in rat, dog, or monkey	Finding NOAEL	AUC, C_{max}, exposure multiples
	Dose range finding in rat, dog, or monkey	Finding safe doses	Same as above
	One-month IND toxicology in rat	Finding NOAEL	Same as above
	One-month IND toxicology in dog or monkey	Finding NOAEL	Same as above
	Long-term (0.25–1-year) tox and two-year carcinogen in mouse and rat	Ensure safety in rat and mouse, including carcinogenicity	Same as above
	Long-term (0.5–1-year) toxicology in dog or monkey	Ensure safety in animal species	Same as above
Clinical/ biopharma	Single/multiple ascending dose	Yes	AUC, C_{max}, T_{max}, $T_{1/2}$, (V_d, F)
	Radiolabeled ADME	Finding parent metabolite profile	Parent and metabolite exposures
	DDI with inhibitors (e.g., ketoconazole)	Test DDI potential	AUC, C_{max}, T_{max}, $T_{1/2}$, (V_d, F)
	DDI with an inducer (e.g., rifampin)	Test DDI potential	Same as above
	DDI with other drugs (e.g,. comeds, pH modifiers)	Test DDI potential	Same as above
	Special populations (renal/ hepatic impair/age/gender, weight)	Test if dose needs be adjusted	Same as above
	Special studies (e.g., food effect, bioavailability)	Yes	Same as above
Late clinical	Population kinetics	Yes	Same as above

[a]NOAEL: No observed adverse effect level.

whole animals are exposed to a phosphorimaging screen which is then scanned with a phosphor imager. Based on body surface area and body weight, exposure of radioactivity to the tissues in rat is extrapolated to human tissues [World Health Organization (WHO), 1977]. For most compounds, administration of a 100-Ci radioactive dose typically exposes the human subjects to an effective dose equivalent of <1 mSiv, well below the radiation limit set by the Nuclear Radiation Committee (Valentin, 2002). In addition to the tissue distribution for human dosimetry, other tissue distribution studies such as maternal–fetal transfer studies are also conducted during the drug development process. These studies are required for the Investigation New Drug (IND) and New Drug Application (NDA) filings and are typically conducted in rat species used in the toxicological evaluation. Data from pregnant rat studies provide information about the drug's potential to cross the placental barrier and expose the fetus.

The nonclinical species used in the ADME studies are based on toxicology species used in the long-term safety evaluation of a compound. These are typically rodent (rat and mouse) and nonrodent (dog or monkey) species. The study is designed to closely resemble the toxicology studies where the dose selected is close to the toxicology dose, and the route of administration and the dosing vehicle are the same as in the toxicology studies. The amount of radioactive dose in animal species is determined by the pharmacokinetic properties of the parent. The typical administered radioactivity range is 1.5–100 μCi/kg. This dose level should be sufficient to generate metabolic profiles in plasma at multiple time points with adequate radioactive sensitivity. The duration of the study is determined by the terminal half-life of the parent (in that species) and is set to at least five plasma half-lives. For compounds with long half-lives, the duration of the study can be based on the criteria where 1% of the total dose is excreted in a given 24-h interval in both urine and feces. Plasma, urine, and fecal samples are collected during the duration of the study to analyze for radioactivity and for metabolite profiling. If needed, additional tissue samples are collected to determine the concentration and accumulation of drug and drug-related component in a particular tissue.

The first set of information generated from the ADME studies are the overall plasma profiles of total radioactivity (TRA) versus time which is compared to the plasma profile of parent versus time measured by a validated LC/MS/MS assay. For those drugs where the parent is the major component at all time points in plasma, the total radioactivity profile usually parallels the profile of the parent. Metabolic profiles of plasma samples generated at different time points by high-performance liquid chromatography (HPLC) analysis followed by radioactivity and mass spectrometric detection provide exposure-related information for parent and metabolites in humans and animal species. In addition, this profile provides information about metabolites that humans are exposed to and how this compares to exposures in animal species.

Mass balance and metabolic profiles of urine and fecal samples generated in ADME studies provide information about extent of metabolism and routes of excretion for parent and metabolites. The excretion of radioactivity after administration of [^{14}C]-apixaban to healthy human volunteers shows that the compound is excreted in both urine and feces (Raghavan et al., 2009; Zhang et al., 2009). Based on urinary excretion and metabolites in feces, approximately at least 50% of the dose is absorbed when administered orally. These types of representation help understand the quantitative and qualitative differences in metabolism across species. All the metabolites that were generated in humans are also observed in animals and suggest that the primary pathways of metabolism are similar across species. The combined metabolic profiles in plasma, urine, and fecal profiles help understand the complete distribution and disposition of drug. For drugs that tend to be in the 300–700-Da range and are highly lipophilic in nature, metabolism and excretion through the bile play a major role in their disposition (Luo et al., 2010). For these types of molecules, ADME studies conducted in bile-duct cannulated (BDC) animals, where bile is collected during the duration of the study, are extremely useful. BDC studies are particularly helpful if conjugative pathways such as glucuronidation or sulfation are involved in the metabolic clearance of the drug. Glucuronides and sulfate conjugates can hydrolyze in the large intestine when excreted through the bile into the GI tract (Parker et al., 1980; Wang et al., 2006). Therefore, in the absence of a bile profile, the role of conjugation in the overall metabolic clearance of the drug would be missed. BDC studies are usually run for a shorter duration, 0–24 h in rat and 0–72 h in dog or monkey. A reasonable mass balance for compounds that have short half-lives can be achieved from BDC animal studies. Where nonclinical data show that most of the drug is excreted in the feces through bile as conjugative metabolites, it is useful to include a panel in the human ADME where bile can be collected for a short duration with incubation. This is exemplified for ADME study with [^{14}C]-muraglitazar where bile was collected for a short duration (3–8 h) after dosing. Comparing the bile profile with fecal profile, the dose in the bile is excreted as conjugates that are hydrolyzed during their passage through the GI tract (Wang et al., 2006; Zhang et al., 2007b).

Data from the human ADME study provide information about the primary pathways of metabolism for the compound. The metabolic pathways are based on the identification of metabolites in plasma, urine, and feces/bile, which in turn can lead to detailed reaction phenotyping studies that are performed to identify the enzymes that generate the primary metabolites. Furthermore, metabolism data obtained from human ADME studies in conjunction with reaction phenotyping can drive decisions that involve the conduct of clinical DDI studies. Reaction phenotyping studies showed that the O-demethylation and hydroxylation of apixaban were catalyzed mainly by CYP3A4/5 with minor contribution from CYP1A2, 2C8, 2C9, 2C19, and 2J2 (Wang et al., 2010). Moreover, metabolism through these pathways, based on the metabolites identified in urine and feces, accounted for approximately 25% of the parent dose. Therefore, a decision was made to conduct a ketoconazole DDI study to assess the effect of inhibiting CYP3A4/5 on the exposure of the parent.

Some dispositional properties of a compound can only be investigated adequately through using an animal model. For example, there is no technology that can be used to assess the intestinal excretion of a compound in humans. Bile-duct cannulated rats after intravenous administration of [^{14}C]-apixaban showed that approximately 22% of dose was recovered in 0–24 h in feces, indicating that the intestinal excretion represented approximately 22% of apixaban clearance (Zhang et al., 2009).

5.3.3.3 MIST Considerations Recent publications on the metabolite in safety testing (MIST) have provided guidance for preclinical and clinical studies to support the safety testing of human metabolites in toxicology studies (Baillie et al., 2002; Zhang et al., 2007b; Food and Drug Administration, 2008; Leclercq et al., 2009; Pang, 2009; Zhu et al., 2009). Measurements of metabolite exposures in early clinical development provide metabolite exposure multiples, the indexes in safety assessment, in toxicology species for human metabolites. Although qualitative and quantitative comparisons of metabolite profiles in vitro are important when comparing metabolites in the nonclinical species relative to humans, in vivo quantitation data for metabolites are most relevant. Qualitative differences in metabolism between humans and animals are rare, but a quantitative difference is common, where a metabolite is produced to a significantly greater or lesser extent in humans than animal species. In addition to quantities, understanding the types of metabolites, including pharmacologically active metabolites, toxic or reactive metabolites, or conjugate metabolites, is also a critically important consideration.

There are multiple approaches to assess the metabolite exposures in animal species and humans after single and multiple doses. Radiolabeled ADME studies provide information on metabolite exposure in animal species and humans through radioactivity profiling of plasma samples at multiple time points (Zhang and Comezoglu, 2007). The metabolite exposures obtained this way are normally derived from single-dose studies. Major human plasma metabolites can then be isolated for structural characterization and subsequently synthesized and measured in single- or repeat-dose clinical and nonclinical studies using validated bioanalytical methods. Because of the lengthy process and significant resource requirement, the ADME studies are often conducted in the late stage of drug development to support drug registrations.

Because of the recent MIST guidance, the pharmaceutical industry began to develop alternative strategies with advanced analytical technologies to assess metabolite exposure in humans earlier than the human ADME study.

Practical considerations of various strategies for the assessment of metabolite exposures in humans and toxicology species with or without radiolabeled materials have been discussed in detail by Zhu et al. (2009). The methodologies include ADME study-based approaches and early analysis of human plasma metabolites in single ascending dose (SAD) and multiple ascending dose (MAD) studies using either nonradiolabeled drugs or microtrace-radiolabeled drugs. In addition, utility, limitations, and recommendation of recently developed analytical technologies and methodologies for executing these strategies were described. It is important to point out that the need for independent toxicity testing for major human metabolites is not frequent. In addition, the disposition of a metabolite administered directly might be different from that of a metabolite formed from the parent drug in vivo.

5.3.4 Engineered Mouse Models

Engineered mouse models are increasingly used for the determination of the roles of cytochrome P450 (CYP) enzymes in drug metabolism and toxicity. Mouse *Cyp* gene(s) can be removed from, while human *CYP* gene(s) can be added to, the mouse genome in order to study the specific involvement of the given mouse and/or human CYP(s) in various aspects of ADME. The combined functions of an entire *Cyp* gene subfamily can be studied by simultaneously removing a cluster of multiple *Cyp* genes. The composite functions of all microsomal CYP enzymes can be studied by removal or modification of the NADPH–CYP reductase (*Cpr*) gene. All of these genetic changes can be programmed in a cell-type-specific or tissue-selective fashion so that the specific contributions of an organ (e.g., the liver or the intestine) to the metabolism and/or toxicity of a drug can be directly determined in vivo. The potential of these genetic mouse models for ADME applications is considerable. In this section, selected mouse models will be introduced, and examples of their applications in drug metabolism, pharmacokinetics, and toxicology studies will be briefly described, with an emphasis on practical advice regarding how to avoid potential confounding factors.

5.3.4.1 Available Mouse Models A partial list of available *Cyp*-knockout mouse models and human *CYP*-transgenic mouse models is shown in Table 5.3.

TABLE 5.3 List of Selected *Cyp*-Knockout and Human *CYP*-Transgenic Mouse Models

Cyp-Knockout Mouse	Reference(s)	Human *CYP*-Transgenic Mouse	Reference(s)
Cyp1a1	Dalton et al., 2000	*CYP1A1/2*	Jiang et al., 2005
Cyp1a2	Liang et al., 1996	*CYP2A6*	Q-Y. Zhang et al., 2005
Cyp1b1	Buters et al., 1999	*CYP2C18/2C19*	Löfgren et al., 2008
Cyp2a5	Zhou et al., 2010	*CYP2D6*	Corchero et al., 2001
Cyp2e1	Lee et al., 1996	*CYP2E1*	Cheung et al., 2005
Cyp3a (gene cluster)	van Herwaarden et al., 2007	*CYP3A4*	Cheung et al., 2006; Granvil et al., 2003; van Herwaarden et al., 2005; 2007;

Notably, two knockout mouse lines (e.g., *Cyp1a1* null and *Cyp1b1* null) can be intercrossed to produce a double-knockout mouse (Uno et al., 2006), except in cases where the two targeted genes are located close to each other (e.g., *Cyp1a1* and *Cyp1a2*). In the latter case, a double-knockout model can be generated through genetic engineering in the embryonic stem (ES) cells. An even more challenging case is when multiple mouse *Cyp* genes occur in a gene cluster, such as the mouse *Cyp3a* gene cluster. Other mouse *Cyp* gene clusters include the *Cyp2j* cluster, the *Cyp2d* cluster, the *Cyp2c* cluster, and the *Cyp2a-b-f-g-s-t* cluster. The presence of multiple copies of structurally similar *Cyp* genes makes it difficult (and sometimes irrelevant) to identify the specific CYP isoform that is normally active in the metabolism of a given drug in vivo. Instead, it makes sense to knock out all members of the subfamily and determine the combined functions of the CYP enzymes impacted by the genetic manipulation. Indeed, this strategy was successfully used for the *Cyp3a* gene cluster (van Herwaarden et al., 2007).

The human *CYP*-transgenic mouse models are most often prepared using a large gene fragment contained in a bacterial artificial chromosome (BAC) clone, although cDNA-based transgene constructs have also been used successfully (e.g., for CYP2A6; Q-Y. Zhang et al., 2005). The tendency for two *CYP* genes in a given *CYP* gene subfamily to be located in close proximity in the genome and potentially sharing common regulatory sequences for gene expression has made it difficult to produce single-gene transgenic models for some CYPs, such as *CYP1A1/2* (Jiang et al., 2005) and *CYP2C18/19* (Löfgren et al., 2008). As a result, mouse models expressing two closely associated *CYP* genes were produced. On the other hand, single-gene *CYP*-transgenic mice can be intercrossed, in order to produce double transgenic mouse models (e.g., *CYP2D6/CYP3A4* transgenic; Felmlee et al., 2008).

A *Cyp*-knockout mouse and a human *CYP*-transgenic mouse can be crossbred in order to produce so-called CYP-humanized mouse models, in which the human *CYP* is expressed, but the orthologous mouse *Cyp* gene is inactivated. For example, a human *CYP2E1*-transgenic mouse was intercrossed with *Cyp2e1*-null mouse to produce *CYP2E1*-humanized mouse (Cheung et al., 2005). More recently, a *CYP3A4*-transgenic mouse with either hepatic or intestinal *CYP3A4* expression was intercrossed with a *Cyp3a*-null mouse, yielding *CYP3A4*-humanized mouse models (van Herwaarden et al., 2007).

A unique group of knockout mouse models, in which the *Cpr* gene is targeted, is also worth noting. CPR is the obligate redox partner for microsomal P450 enzymes; therefore, the deletion of the *Cpr* gene causes the inactivation of all microsomal P450 enzymes in targeted cells. Several mouse models, in which the *Cpr* gene is deleted in a tissue-specific fashion, are available, such as liver-specific *Cpr*-null mice (Gu et al., 2003; Henderson et al., 2003) or intestinal epithelium-specific *Cpr*-null mice (Q-Y. Zhang et al., 2009). A transgenic mouse (known as "Cpr-low" mouse) in which *Cpr* expression was globally down-regulated (Wu et al., 2005) is also available.

5.3.4.2 Utility of Engineered Mouse Models Drug metabolism can play an essential role in the extent of tissue exposure to either the parent compound or, in the case of a prodrug, the active metabolite(s). Numerous mechanistic questions related to drug metabolism or toxicity can be answered through the use of a proper engineered mouse model, thus facilitating preclinical drug safety and PK studies. Examples of the types of questions that can be addressed are given below:

1. Is in vivo metabolism mediated mainly by P450 or by other drug metabolism enzymes? Examples of this type of application are the use of the *Cpr*-low mouse for demonstration of the role of P450 enzymes in the clearance of nifedipine (Q-Y. Zhang et al., 2007) and the use of the liver *Cpr*-null mouse for demonstration of the lack of a significant contribution by P450 enzymes to systemic acetaminophen clearance (Gu et al., 2005).
2. Which P450 enzyme(s) are critical for in vivo metabolism? For example, the *Cyp2a5*-null mouse was used for demonstration of the role of CYP2A5 in the in vivo clearance of nicotine; the deletion of the *Cyp2a5* gene led to substantial increases in the bioavailability of nicotine (Zhou et al., 2010).
3. Is a given drug metabolism enzyme responsible for bioactivation and consequent toxicity? For example, the critical role of CYP2E1 in acetaminophen-induced hepatotoxicity (Lee et al., 1996) and the lack of an essential role by CYP1A2 in acetaminophen-induced olfactory toxicity (Genter et al., 1998) were demonstrated through the use of the respective *Cyp*-knockout mouse models.
4. Does the small intestine play a major role in first-pass clearance of a given oral drug? For example, the intestinal epithelium-*Cpr*-null mouse was used to show that intestinal P450 enzymes play a major role in the first-pass clearance of oral nifedipine (Q-Y. Zhang et al., 2009). The small intestinal CYP3A4-humanized mouse model (SI-selective expression of CYP3A4 and whole-body knockout of all *Cyp3a* genes) was used to demonstrate the capability of SI CYP3A4 to reduce the bioavailability of oral docetaxel (van Herwaarden et al., 2007).
5. For compounds metabolized through multiple pathways, which pathway leads to active (or reactive) metabolites in vivo? For example, CPR/P450 of the cardiomyocyte was found not to be essential for the cardiotoxicity of doxorubicin, an anticancer drug, which can be metabolized by multiple enzymes, including CPR and carbonyl reductase, in a study that used a cardiomyocyte-specific *Cpr*-null mouse (Fang et al., 2008). On the other hand, hepatic CPR/P450 enzymes were found to be essential for the hepatotoxicity and renal toxicity of acetaminophen, which can be metabolized by both phase I and phase II enzymes (Gu et al., 2005).
6. What is the capability of a human P450 enzyme to produce a given metabolite (including reactive intermediates) in vivo? For examples, the

in vivo capabilities of human CYP2A6 to generate 7-hydroxycoumarin from coumarin (Q-Y. Zhang et al., 2005), of human CYP2D6 to produce 4-hydroxydebrisoquine from debrisoquine (Corchero et al., 2001), and of human CYP3A4 to produce various metabolites from docetaxel (van Herwaarden et al., 2007) or to produce 1′-hydroxymidazolam from midazolam (Granvil et al., 2003) have been assessed using the respective CYP-transgenic mouse models.

7. What is the impact of genetic polymorphisms in drug metabolism genes on drug clearance? Few transgenic mouse models that express allelic variants of *P450* genes have been reported; however, a wild-type mouse model, as compared to a human *CYP*-transgenic mouse model, or a *Cyp*-knockout mouse model, as compared to a *CYP*-humanized mouse model, can represent a mouse model containing a null allele for the human *P450* gene. Furthermore, hemizygote and homozygote human *CYP*-transgenic mouse models can be compared for studying the effects of losing one *CYP* allele on systemic drug clearance.

5.3.4.3 How to Avoid Potential Confounding Factors The extent to which each of the available mouse models has been characterized varies considerably, but it is essential that the user is aware of the properties and limitations of each model. While it is beyond the scope of this chapter to provide detailed information about each of the available mouse models, a partial list of questions to consider when acquiring or utilizing a mouse model is given here.

For mouse *Cyp*-knockout models:

1. Is it single-gene deletion or multiple-gene deletion? While only a single *Cyp* gene is deleted in most existing *Cyp*-null mouse models, the entire mouse *Cyp3a* subfamily was deleted in the *Cyp3a*-null model (van Herwaarden et al., 2007). Additional gene cluster-null models are expected to be available in the near future.
2. Is the gene deletion germline or conditional (i.e., inducible and/or tissue specific)? Available *Cyp*-null models (e.g., *Cyp2e1* null or *Cyp2a5* null) are essentially all germline; however, the *Cpr* nulls are conditional, with the *Cpr* gene deleted in specific cell types in selected organ(s), such as hepatocytes in the liver (Gu et al., 2003; Henderson et al., 2003) or enterocytes in the SI (Q-Y. Zhang et al., 2009).
3. What is the genetic background of the embryonic stem cells used for gene targeting and has the mouse strain been backcrossed to a different strain? Most existing *Cyp*-null models were produced using ES cells derived from the 129/Sv strain, such as the *Cyp2e1* null (Lee et al., 1996) or the *Cyp2g1* null (Zhuo et al., 2004). Mice produced using ES cells derived from the C57BL/6 (B6) strain are also available, such as the *Cyp2a5* null (Zhou et al., 2010). Note that a mouse originally produced on a mixed 129/Sv and B6 genetic background can be made congenic on B6 by backcrossing

for 10 or more generations to the B6 strain. Nevertheless, genes located in close proximity with each other (e.g., *Cyp2g1* and *Cyp2a5*) will unlikely segregate during backcrossing; thus, when using such a congenic mouse colony for drug metabolism studies, attention should still be paid to potential confounding by strain differences.

4. How thorough was the analysis of potential compensatory gene expression? The extent to which available mouse models have been characterized for potential gene expression changes varies greatly, ranging from brief immunoblot analysis of a few P450 proteins to broad microarray analysis of genomewide expression changes and metabolic profiling with several probe substrates. It is important to confirm that the expression of enzymes known to be capable of metabolizing the drugs under study is not altered and that unintended changes in the bioavailability or tissue distribution of the drug under study do not occur in a knockout model. Notably, elevated levels of CYP2C enzymes were found in the *Cyp3a*-null mouse, with corresponding increases in rates of midazolam metabolism (van Waterschoot et al., 2008).
5. Did the genetic manipulations leave behind any exogenous promoter (e.g., one that was associated with an antibiotic selection marker) at the site of gene deletion? The presence of a strong exogenous promoter associated with the neomycin marker gene at the *Cyp2g1* locus led to the down-regulation of the expression of the neighboring *Cyp2a5* gene in a *Cyp2g1*-null mouse (Zhuo et al., 2004). Selection marker genes are also present in a number of other *Cyp*-null models; therefore, attention should be paid to the identity of the genes located near the targeted *Cyp* gene and any potential impact of these genes on drug metabolism. In newer models, the selection markers are usually removed through the use of the Cre/LoxP technology.

For human *CYP*-transgenic models:

1. Was the expression of the human CYP controlled by an exogenous promoter (in which case, the expression of the human CYP will not reflect regulation of the CYP promoter in either mice or humans)? For examples, the *CYP2A6* gene was controlled by a transthyretin promoter in the *CYP2A6*-transgenic model (Q-Y. Zhang et al., 2005). In the *CYP3A4*-humanized mouse models, the *CYP3A4* transgene was controlled by either the albumin promoter, for hepatocyte expression, or the villin promoter, for intestinal epithelial expression (van Herwaarden et al., 2007).
2. Alternatively, was the human CYP controlled by its own promoter (as is the case when the entire human *CYP* gene is inserted into the mouse genome)? Several transgenic mice have been prepared using a full-length gene sequence contained in a BAC clone, such as the *CYP2D6*-transgenic

mouse (Corchero et al., 2001). The use of the authentic human CYP promoter and the inclusion of a large amount of flanking DNA sequence help to achieve a humanlike tissue distribution of the transgene expression.

3. Does the transgene fragment contain a single *CYP* or multiple *CYPs*? The close proximity of neighboring *CYP* genes to each other in a human gene cluster sometimes makes it necessary for the transgene construct to contain more than one *CYP* genes, as was the case for the *CYP3A4-3A7*-transgenic mouse (Cheung et al., 2006), the *CYP1A1-1A2*-transgenic mouse (Jiang et al., 2005), and the *CYP2C18-2C19*-transgenic mouse (Löfgren et al., 2008). These mouse models are suitable for studying the combined functions of the transgene present (or those expressed), but it may be necessary to also study single-*CYP* transgenic mouse models, in order to identify specific contributions by the given CYP enzyme.
4. Is the transgene expressed at the expected tissue and cellular site(s)? Tissue specificity of transgene expression can be influenced by a number of factors, including the DNA sequence at the site of transgene integration and potential strain differences in transcription factor expression between mouse and humans. Attention should be paid to whether the transgene is expressed adequately in the tissue of interest and whether the tissue distribution profile of the transgene in the mouse model agrees with that in humans. For example, an early version of the *CYP3A4*-transgenic mouse model expresses CYP3A4 in the intestine but not in the liver (Granvil et al., 2003); this mouse model would not be suitable for studying drug metabolism by hepatic CYP3A4.
5. Is the transgene inducible by major classes of P450 inducers? Inducibility of the transgene by the drug under study can impact interpretation of the data obtained using a human *CYP*-transgenic mouse. For example, the *CYP3A4* gene in the *CYP3A4-3A7*-transgenic mouse model was found to be inducible by phenobarbital (Cheung et al., 2006), whereas the *CYP3A4* gene in the *CYP3A4*-humanized model (van Herwaarden et al., 2007) would not respond to phenobarbital or other activators of xenobiotic receptors. Such differences in transgene inducibility may impact the experimental outcome for drugs that are activators of xenobiotic receptors.
6. Which allele of the human *CYP* gene was used for preparing the transgenic mouse? Although the **1A* allele (the wild type) is usually used for the preparation of transgenic mouse, attention should be paid to whether such a distinction was made in the original paper reporting the generation of the mouse model.
7. What is the genetic background of the transgenic mouse? Transgenic models are usually prepared using either the B6 or the FVB mouse strain. Potential strain differences for either the expression of the human transgene or the expression and/or function of background mouse *Cyp* genes may occur. Thus, attention should be paid to potential strain differences among study groups.

For either transgenic or knockout models, the user should determine whether there is any biological phenotype associated with the gene deletion. Some phenotypes, such as embryonic lethality or infertility, can make it difficult to maintain the mouse colony, whereas others, such as alterations in body or organ weights or in lipid metabolism, can confound the interpretation of experimental observations. Notably, in a routinely prepared transgenic model, the transgene is randomly integrated into the mouse genome; thus, each transgenic line can be different in the potential occurrence of any phenotypes resulting from transgene integration. In contrast, random integration should not be present in a knockout model, given the screening procedures implemented during model preparation.

5.3.5 In Silico Modeling

In the field of drug metabolism and PK, laboratory experiments are critical to generate reliable in vitro and in vivo (in both animals and humans) data. However, this approach has certain limitations. Laboratory work is usually expensive and the throughput is relatively low. In addition, experiments have to work with the existing compounds but cannot be done on virtual drug candidates. Computational (in silico) modeling may partially compensate for these limitations, reduce laboratory work load, and provide valuable feedback to chemical synthesis for better drug candidates. In fact, in silico modeling has been applied to many aspects of DMPK, including CYP substrates, inhibitors and inducers, transporter substrates, and oral absorption and bioavailability.

There are many different approaches regarding in silico modeling for DMPK. They can be divided into structure-based approaches, data-based approaches, and integrated PK models (Yamashita and Hashida, 2004). A structure-based approach focuses on one of the two key components in drug metabolism: protein or ligand. Here the protein represents any active metabolic enzyme, transporter, receptor, or transcriptional factor for an enzyme or transporter. On the other hand, the ligand is a drug-related chemical, either parent or metabolite, representing for any substrates, inhibitors, or inducers of enzymes or transporters. Protein (target)–based approaches directly calculate the three-dimensional structures of the target protein. However, the crystal structure of a target protein may not be available. If that is the case, then the calculation will be based on the homology to the available crystal structures of genetically and functionally related protein. For example, the three-dimensional models of CYP1A2, CYP2D6, and CYP3A4 were first built using the X-ray crystallographic structures of bacterial CYPs (CAM, BM-3, TERP, ERYF) as templates (De Rienzo et al., 2000). Notably, in this homology modeling, the targets were mammalian CYPs while the templates were bacterial CYP, and furthermore, sequence identity percentages were only around 20% between the target and template CYPs. Rabbit CYP2C5 is the first mammalian CYP isozyme whose crystal structure was determined (Williams et al., 2000), followed by rabbit CYP2B4 (Scott et al., 2003; 2004). Since rabbit CYPs are much more closely related to human CYPs than bacterial CYPs in structure, the homology models of the human CYPs based on the rabbit CYP crystal structures achieved higher quality and better accuracy (Kirton et al., 2002). Fortunately, the X-ray crystal structures of several important human CYPs have now been determined, including CYP2C9 (Williams et al., 2003; Wester et al., 2004), CYP3A4 (Williams et al., 2004; Yano et al., 2004), CYP2C8 (Schoch et al., 2004), CYP2D6 (Rowland et al., 2006), and CYP1A2 (Rowland et al., 2006). Therefore, there may be no need to use homology modeling approaches for major human CYP enzymes.

The ligand-based approaches involve analyzing the observed activities for a set of ligands (substrates and inhibitors) and deriving protein–ligand interaction models without prior knowledge of protein structure. The most commonly used approaches are quantitative structure–activity relation (QSAR) and pharmacophore modeling. Ligand-based approaches are more widely used than protein-based approaches, likely because the three-dimensional structures are not known for many proteins.

The pharmacophore was first introduced by Paul Ehrlich in 1909 as "a molecular framework that carries the essential features responsible for a drug's biological activity." Typical pharmacophore features include whether a molecule is hydrophobic, aromatic, a hydrogen bond acceptor or donor, and a cation or an anion and has positive or negative charges. The most common pharmacophore features have been built into various in silico modeling programs.

Data-based approaches have been widely used for the prediction of ADME properties, including solubility, intestinal and BBB permeability, intestine bioavailability, active transport, and metabolic stability (Yamashita and Hashida, 2004). Essentially, data-based modeling proceeds in a stepwise fashion. Reliable DMPK data are first collected from an established experimental model with many drug candidates whose chemical structures are diversified. Molecular descriptors are then obtained from the tested drug candidates. These various kinds of quantitative descriptors include fragment descriptors, topological descriptors, and global physicochemical descriptors based on two- or three-dimensional molecular structures. Observed DMPK properties are correlated to the descriptors using multivariate statistical analysis employing linear methods such as multiple linear regression and partial least-square and nonlinear methods. Ideally, the modeling established from this training mode is tested and confirmed with data obtained from molecules which are not included in the training mode. A good example is in silico prediction of biliary excretion in rats based on physiochemical properties published recently (Luo et al., 2010). Authors collected rat bilairy excretion data for 50 structurally diverse compounds and then established an in silico model using a multiple-correlation method based on three physicochemical properties (polar surface area, free energy of aqueous salvation, and presence or absence of a carboxylic acid moiety) to predict rat biliary excretion.

The integrated PK model is used to predict PK properties of a drug candidate in the whole body by assessing all the kinetics with a global model. Obviously there is a long way to go before this ideal model can be achieved. In fact, in silico modeling is facing a great challenge. This challenge exists because the target proteins (metabolic enzymes, transporters, and transcriptional factors) involved in drug disposition are promiscuous (Ma and Lu, 2008). The target proteins are flexible and can form different conformations when interacting with different substrates and ligands and have large substrate/ligand binding sites to accommodate a large variety of molecules with varying sizes and shapes, and a single metabolic enzyme is able to biotranform a substrate to multiple metabolites. Moreover, metabolic enzymes, transporters, and transcriptional factors form a tangled network at different layers. In addition, substrates and ligands are mobile in their binding site, leading to unusual kinetic properties. More than one substrate can simultaneously bind to or compete for one active site, resulting in drug interactions. On the other hand, current knowledge of the target proteins is not detailed enough to meet the needs of in silico modeling.

5.4 DATA INTERPRETATION

Proper interpretation of data plays a critical role at any stage of drug discovery and development, from deciding whether to advance a compound to choosing the next level of experimental models. The aspects that should be recognized when interpreting drug metabolism data include species difference, in vitro–in vivo discrepancy, enzyme–transporter interplay, individual difference, DDI, and factors affecting metabolic pathways.

5.4.1 Species Difference

There are fundamental and mechanistic similarities between animal species and humans that serve as the basis to using animal species as models for drug pharmacological activity, drug toxicity, and drug ADME properties. There are also differences between animal species and between animals and humans. For example, P450 induction in animals does not always correlate with induction in humans. Expression levels and binding affinities could be different for the enzymes, transporters, and receptors between different species although they could have similar functions.

5.4.2 In vitro–In Vivo Discrepancy

It is not uncommon to see a high clearance in animals for a slow metabolism drug or a significant fraction of dose as a glucuronide metabolite yet the metabolite was only a minor one in the in vitro incubations. There are still limitations for quantitative extrapolation of in vitro results to in vivo outcomes.

5.4.3 Enzyme–Transporter Interplay

Multiple uptake and efflux transporters are expressed on important organs for metabolism and disposition, such as the kidney and liver. The efficiency of drug metabolism enzyme in the liver could be dependent on the available substrate that is transported in by an uptake hepatic transporter. Recently developed double-knockout mouse model demonstrate the importance of enzyme–transporter interplay in drug metabolism and disposition (Benet, 2009).

5.4.4 Interindividual Differences

The target receptor or enzyme could have polymorphism in addition to polymorphic drug metabolism enzymes and transporters such as CYP2C9/2C19, UGT1A1, and P-gp that determine the drug exposure and duration. Furthermore, there could be huge differences in expression levels of receptors or metabolism enzymes which could lead to variable levels of response to a drug. In addition, animal species are more homogeneous than humans and the healthy subjects do not always represent the patients. It is important to take these individual differences into consideration and apply the average data collected from clinical trials to large patient populations.

5.4.5 Drug–Drug Interaction

The inhibition or induction of drug metabolism enzymes or transporters by one drug could affect the distribution and disposition of itself or another drug. It is always important to consider DDI potential of any drug because of multiple-drug therapy.

5.4.6 Multiple Other Factors Affecting Metabolic Pathways

Identification of species differences of metabolic pathways qualitatively and quantitatively will help address safety concerns such as MIST in Parkinson's diseases due to any active metabolite. Conjugation enzyme activities could be missed or inactive (due to lacking cofactors) in liver microsomes while the uptake of a drug candidate into hepatocytes can be a limiting factor for certain drug candidates in metabolism by hepatocytes. In addition, liver microsomes are usually prepared from pooled donors in order to properly represent the population regarding metabolism. Fresh hepatocytes from humans and large animals (rabbits, dogs, and monkeys) are often prepared from a single donor, whereas cryopreserved hepatocytes are generally obtained from pooled donors but have lower activities compared to the fresh hepatocytes. An in vitro study is a test system that reflects the short metabolism period of the liver, while an in vivo study is a relatively open test system that reflects the sum of the metabolism in the whole body, the distribution of the parent and metabolites to tissues and organs, and elimination of the parent and metabolites into excreta (urine, bile, and feces) for a longer period. The impact of metabolic enzymes on drug metabolites (amounts and kinds of metabolites produced) with regard to metabolites from in vivo samples should not be ignored. If a drug candidate is a metabolic enzyme inducer and the dosing period is long (>7 days), then autoinduction may occur and will affect drug metabolism (Svensson et al., 2003). Alternatively, when a drug candidate is concomitantly administered with an inducer, the metabolic pathways may shift, in particular for an noninducible enzyme such as CYP2D6 (Dilger et al., 1999). Dosages in in vivo studies and initial incubation concentrations in in vitro studies may also affect the metabolic pathways because of concentration-dependent metabolism. High dose/concentration may result in decreased percentages of overall metabolism and even shift metabolic pathways. For example, CYP2D6 contributes primarily to the formation of dextrorphan from dextromethorphan (via O-demethylation) at $\leq 10 \mu M$, while CYP3A4 will biotransform dextromethorphan to 3-methoxymorphinan (via N-demethylation) when the dextromethorphan concentration is $\geq 25 \mu M$ (von Moltke et al., 1998). Therefore, a therapeutic dose or an observed plasma concentration in clinical trials as well as liver/plasma concentration ratios should be used as references for the design of DMPK studies that involve the evaluation of metabolic pathways.

5.5 SUMMARY

Drug discovery and development remain as complicated models-based experimental scientific exploration. These models provide a variety of data that range from in vitro systems to in vivo animal species and healthy human subjects that should predict the behavior of a drug in patients. These data either deal with a particular aspect of drug metabolism such as permeability and transporter properties as derived from Caco-2 models or provide the entire distributional and dispositional properties of a drug as obtained from a radiocarbon human ADME study. Figure 5.6 is a summary of the mechanistic input and

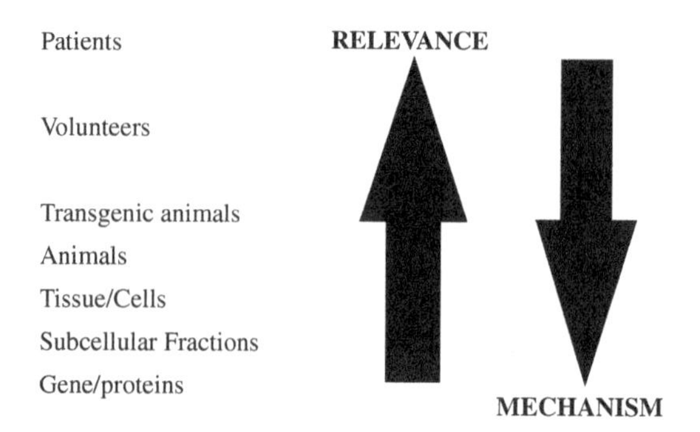

FIGURE 5.6 Applications of models in metabolism and disposition.

relevance of major experimental models that are used in drug discovery and development. The selection of an appropriate model followed by the application of the specific model(s) with a specific strategy and correct data interpretation is critical. This chapter has described some important experimental models used in drug metabolism and disposition. These experimental models predict the clinical behavior in patients such as that the experimental data for CYP inhibition from human liver microsomal incubations together with clinical drug concentrations can predict the clinical metabolic DDI fairly well. Most modern drugs are discovered and developed with the timely application of appropriate experimental models. Certainly, there will be new and better modified models, such as humanized mouse models, to be discovered and developed to support more efficient drug discovery and development processes.

ACKNOWLEDGMENTS

Xinxin Ding was supported in part by Public Health Service grants CA092596 and ES007462 from the National Institutes of Health.

REFERENCES

Alavijeh MS, Palmer AM. The pivotal role of drug metabolism and pharmacokinetics in the discovery and development of new medicines. IDrugs 2004;7(8):755–763.

Alexander JH, Becker RC, Bhatt DL, Cools F, Crea F, Dellborg M, Fox KA, Goodman SG, Harrington RA, Huber K, Husted S, Lewis BS, Lopez-Sendon J, Mohan P, Montalescot G, Ruda M, Ruzyllo W, Verheugt F, Wallentin L. Apixaban, an oral, direct, selective factor Xa inhibitor, in combination with antiplatelet therapy after acute coronary syndrome: Results of the Apixaban for Prevention of Acute Ischemic and Safety Events (APPRAISE) trial. Circulation 2009;119(22): 2877–2885.

Artursson P, Karlsson J. Correlation between oral drug absorption in humans and apparent drug permeability coefficients in human intestinal epithelial (CACO-2) cells. Biochem Biophys Res Commun 1991;175:880–885.

Aungst BJ. Novel formulation strategies for improving oral bioavailability of drugs with poor membrane permeation or presystemic metabolism. J Pharm Sci 1993; 82(10):979–987.

Bachmann K, Byers J, Ghosh R. Prediction of in vivo hepatic clearance from in vitro data using cryopreserved human hepatocytes. Xenobiotica 2003;33:475–483.

Baillie, TA, Cayen MN, Fouda H, Gerson RJ, Green JD, Grossman SJ, Klunk LJ, LeBlanc B, Perkins DG, Shipley LA. Drug metabolites in safety testing. Toxicol Appl Pharmacol 2002;182:188–196.

Balimane PV, Tamai I, Guo A, Nakanishi T, Kitada H, Leibach FH, Tsuji A, Sinko PJ. Direct evidence for peptide transporter (PepT1)-mediated uptake of a nonpeptide prodrug, valacyclovir. Biochem Biophys Res Commun 1998;250(2):246–251.

Benet LZ. The drug transporter-metabolism alliance: uncovering and defining the interplay. Mol Pharm. 2009;6:1631–1643.

Blanchard N, Hewitt NJ, Silber P, Jones H, Coassolo P, Lave T. Prediction of hepatic clearance using cryopreserved human hepatocytes: A comparison of serum and serum-free incubations. J Pharm Pharmacol 2006;58:633–641.

Buters JTM, Sakai S, Richter T, Pineau T, Alexander DL, Savas U, Doehmer J, Ward JM, Jefcoate CR, Gonzalez FJ. Cytochrome P450 CYP1B1 determines susceptibility to 7,12-dimethylbenz[*a*]anthracene-induced lymphomas. Proc Ntil Acad Sci USA 1999;96:1977–1982.

Cheung C, Yu AM, Chen CS, Krausz KW, Byrd LG, Feigenbaum L, Edwards RJ, Waxman DJ, Gonzalez FJ. Growth hormone determines sexual dimorphism of hepatic cytochrome P450 3A4 expression in transgenic mice. J Pharmacol Exp Ther 2006;316:1328–1334.

Cheung C, Yu AM, Ward JM, Krausz KW, Akiyama TE, Feigenbaum L, Gonzalez FJ. The CYP2E1-humanized transgenic mouse: Role of CYP2E1 in acetaminophen hepatotoxicity. Drug Metab Dispos 2005;33:449–457.

Chu V, Einolf HJ, Evers R, Kumar G, Moore D, Ripp S, Silva J, Sinha V, Sinz M, Skerjanec A. In vitro and in vivo induction of cytochrome P450: A survey of the current practices and recommendations: A Pharmaceutical Research and Manufacturers of America Perspective. Drug Metab Dispos 2009;37:1339–1354.

Cohen LH, Remley MJ, Raunig D, Vaz AND. In vitro drug interactions of Cytochrome P450: An evaluation of fluorogenic to conventional substrates. Drug Metab Dispos 2003;31:1005–1015.

Compagnon P, Clement B, Campion JP, Boudjema K. Effects of hypothermic machine perfusion on rat liver function depending on the route of perfusion. Transplantation 2001;72:606–614.

Corchero J, Granvil CP, Akiyama TE, Hayhurst GP, Pimprale S, Feigenbaum L, Idle JR, Gonzalez FJ. The CYP2D6 humanized mouse: Effect of the human CYP2D6 transgene and HNF4alpha on the disposition of debrisoquine in the mouse. Mol Pharmacol 2001;60:1260–1267.

Crespi CL. Xenobiotic-metabolizing human cell as tools for pharmacological and toxicological research. Adv Drug Res 1995;26:180–233.

Dalton TP, Dieter MZ, Matlib RS, Childs NL, Shertzer HG, Genter MB, Nebert DW. Targeted knockout of *Cyp1a1* gene does not alter hepatic constitutive expression of other genes in the mouse [Ah] battery. Biochem Biophys Res Commun 2000;267:184–189.

DeRienzo F, Fanelli F, Menziani MC, De Benedetti PG. Theoretical investigation of substrate specificity for cytochromes P450 IA2, P450 IID6 and P450 IIIA4. J Comput Aided Mol Des 2000;14(1): 93–116.

Demain AL, Scanchez S. Microbial synthesis of primary metabolites: Current advances and future prospects. In Fermentation Microbiology and Biotechnology, El-Mansi EMT, Bryce CFA, Demain AL, Allman AR, Eds. CRC, New York, pp. 99–130.

Di L, Kerns EH. Profiling drug-like properties in discovery research. Curr Opin Chem Biol 2003;7(3): 402–408.

Di L, Kerns EH. Application of pharmaceutical profiling assays for optimization of drug-like properties. Curr Opin Drug Discov Devel 2005;8(4): 495–504.

Di L, Kerns EH, Bezar IF, Petusky SL, Huang Y. Comparison of blood-brain barrier permeability assays: in situ brain perfusion, MDR1-MDCKII and PAMPA-BBB. J Pharm Sci 2009;98:1980–1991.

Dilger K, Greiner B, Fromm MF, Hofmann U, Kroemer HK, Eichelbaum M. Consequences of rifampicin treatment on propafenone disposition in extensive and poor metabolizers of CYP2D6. Pharmacogenetics 1999;9(5):551–559.

Dochmer J, Dogra S, Friedberg T, Moniter S, Adesnik M, Glatt H, Oesch F. Stable expression of rat cytochrome P450IIB1 cDNA in Chinese hamster cells (V79) and metabolic activation of aflatoxin B1, Proc Natl Acad Sci USA 1988;85: 5769–5773.

Fang C, Gu J, Xie F, Behr M, Yang W, Abel ED, Ding X. Deletion of the NADPH-cytochrome P450 reductase gene in cardiomyocytes does not protect mice against doxorubicin-mediated cardiac toxicity. Drug Metab Dispos 36:2008;1722–1828.

Felmlee MA, Lon HK, Gonzalez FJ, Yu AM. Cytochrome P450 expression and regulation in CYP3A4/CYP2D6 double transgenic humanized mice. Drug Metab Dispos 2008;36:435–441.

Food and Drug Administration (FDA). Waiver of in vivo bioavailability and bioequivalence studies for immediate-release solid oral dosage foems based on a biopharmaceutics classification system, 2000, available: http://www.fda.gov/downloads/Drugs/GuidanceComplianceRegulatoryInfomaion/Guidances/ucm070246.pdf.

Food and Drug Administration (FDA). Drug interaction studies—Study design, data analysis, and implications for dosing and labeling, 2006, available: http://WWW.FDA.gov/cder/guidance/6695dft.htm.

Food and Drug Administration (FDA). Guidance for Industry: Safety Testing of Drug Metabolites, 2008, available: www.fda.gov/CDER/GUIDANCE/6897fnl.pdf.

Ganapathy V, Leibach FH. Peptide transporters. Curr Opin Nephrol Hypertens 1996; 5(5): 395–400.

Gibson CG, Skett P. Pharmacological and toxicological aspects of drug metabolism. *In Introduction to Drug Metabolism*, Gibson CG, Skett P, Eds. Nelson Thrones Publisher, Cheltenham, 2001, pp. 171–202.

Genter MB, Liang HC, Gu J, Ding X, Negishi M, McKinnon RA, Nebert DW. Role of CYP2A5 and 2G1 in acetaminophen metabolism and toxicity in the olfactory mucosa of *cyp1A2(-/-)* mouse. Biochem Pharmacol 1998;55:1819–1826.

Glueck SM, Pirker M, Nest BM, Ueberbacher BT, Larissegger-Schnell B, Csar K, Hauer B, Stuermer R. Biocatalytic reaction of aliphatic, arylaliphatic, and aromatic a-hydroxycarboxylic acids. J Org Chem 2005;70:4028–4232.

Gonzalez MC, Marteau C, Franchi J, Migliore-Samour D. Cytochrome P450 4A11 expression in human keratinocytes: Effects of ultraviolet irradiation. Br J Dermatol 2001;145:749–757.

Granvil CP, Yu AM, Elizondo G, Akiyama TE, Cheung C, Feigenbaum L, Krausz KW, Gonzalez FJ. Expression of the human CYP3A4 gene in the small intestine of transgenic mice: In vitro metabolism and pharmacokinetics of midazolam. Drug Metab Dispos 2003;31:548–558.

Grimm SW, Einolf HJ, Hall SD, He K, Lim HK, Ling J, Lu C, Nomeir AA, Seibert E, Skordos KW, Tonn G, Van Horn R, Wang RW, Wong YN, Yang T, Obach RS. The conduct of in vitro studies to address time-dependent inhibition of drug metabolizing enzymes: A perspective of the pharmaceutical research and manufactures of America. Drug Metab Dispos 2009;37:1355–1370.

Gu J, Cui H, Behr M, Zhang L, Zhang Q-Y, Yang W, Hinson JA, Ding X. In vivo mechanisms of tissue-selective drug toxicity: Effects of liver-specific knockout of the NADPH-cytochrome P450 reductase gene on acetaminophen toxicity in kidney, lung, and nasal mucosa. Mol Pharmacol 2005;67:623–630.

Gu J, Weng Y, Zhang Q-Y, Cui H, Behr M, Wu L, Yang W, Zhang L, Ding X. Liver-specific deletion of the NADPH-cytochrome P450 reductase gene: Impact on plasma cholesterol homeostasis and the function and regulation of microsomal cytochrome P450 and heme oxygenase. J Biol Chem 2003;278:25895–25901.

Han H, de Vrueh RL, Rhie JK, Covitz KM, Smith PL, Lee CP, Oh DM, Sadee W, Amidon GL. 5′-Amino acid esters of antiviral nucleosides, acyclovir, and AZT are absorbed by the intestinal PEPT1 peptide transporter. 1998;Pharm Res 15(8):1154–1159.

Hariparsad N, Carr BA, Evers R, Chu X. Comparison of immortalized Fa2N-4 cells and human hepatocytes as in vitro models for Cytochrome P450 induction. Drug Metab Dispos 2008;36:1046–1055.

Henderson CJ, Otto DME, Carrie D, Magnuson MA, McLaren AW, Rosewell I, Wolf CR. Inactivation of the hepatic cytochrome P450 system by conditional deletion of hepatic cytochrome P450 reductase. J Biol Chem 2003;278:13480–13486.

Ho ES, Lin DC, Mendel DB, Cihlar T. Cytotoxicity of antiviral nucleotides adefovir and cidofovir is induced by the expression of human renal organic anion transporter 1. J Am Soc Nephrol 2000;11(3):383–93.

Hu S, Kelly S, Lee S, Tao J, Flahive E. Efficient chemoenzymatic synthesis of pelitrexol via enzymic differentiation of a remote stereocenter. Org Lett 2006a;8:1653–1655.

Hu S, Martinez CA, Kline B, Yazbeck D, Tao J, Kucera DJ. Efficient enzymatic process for the production of (2*S*)-4,4-difluoro-3,3-dimethyl-*N*-Boc-proline, a key intermediate in the synthesis of HIC protease inhibitors. Org Process Res Devel 2006b;10:650–654.

Iyer R, Zhang D. Design metabolism in drug development. *In Drug Metabolism in Drug Design and Development: Principle and Applications*, Zhang D, Zhu M, Humphreys WG, Eds. Wiley, Hoboken, NJ, pp. 261–286.

Jiang ZW, Dalton TR, Jin L, Wang B, Tsuneoka Y, Shertzer HG, Deka R, Nebert DW. Toward the evaluation of function in genetic variability: Characterizing human SNP frequencies and establishing BAC-transgenic mice carrying the human CYP1A1_CYP1A2 locus. Human Mutat 2005;25:196–206.

Kakumanu VK, Arora V, Bansal AK. Investigation of factors responsible for low oral bioavailability of cefpodoxime proxetil. Int J Pharm 2006;317(2):155–160.

Kansy M, Senner F, Gubernator K. Physicochemical high throughput screening: Parallel Artificial Membrane Permeability Assay in the description of passive absorption processes. J Med Chem 1998;41:1007–1010.

Kerns EH, Di L. Pharmaceutical profiling in drug discovery. Drug Discov Today 2003; 8(7):316–323.

Kirton SB, Kemp CA, Tomkinson NP, St-Gallay S, Sutcliffe MJ. Impact of incorporating the 2C5 crystal structure into comparative models of cytochrome P450 2D6. Proteins 2002;49(2):216–231.

Leclercq L, Cuyckens F, Mannens GS, de Vries R, Timmerman P, Evans DC. Which human metabolites have we MIST? Retrospective analysis, practical aspects, and perspectives for metabolite identification and quantification in pharmaceutical development. Chem Res Toxicol 2009;22:280–293.

Lee ST, Buters JM, Pineau T, Fernandezsalguero P, Gonzalez FJ. Role of CYP2E1 in the hepatotoxicity of acetaminophen. J Biol Chem 1996;271:12063–12067.

Li AP, Hartman NR, Lu C, Collins JM and Strong JM. Effects of cytochrome P450 inducers on 17a-ethynylestradiol (EE2) conjugation in primary human hepatocytes. Br J Clin Pharmacol 1999a;48:733–742.

Li AP, Lu C, Brent JA, Pham C, Fackett A, Ruegg CE, Silber PM. Cryopreserved human hepatocytes: Characterization of drug metabolizing activities and applications in higher throughput screening assays for hepatotoxicity, metabolic stability, and drug-drug interaction potential, Chem-Biol Interact 1999b;121:17–35.

Li P, Lu C, Daniels JS, Miwa G, Gan LS. CYP3A4 mediated time- and concentration-dependent CYP2D6 inhibition by a melanocortin receptor antagonist. Drug Metab Rev 2004;36(S1):572.

Liang HL, Li H, Mckinnon RA, Duffy JJ, Potter SS, Puga A, Nebert DW. Cyp1a2(/) null mutant mice develop normally but show deficient drug metabolism. Proc Natl Acad Sci USA. 1996;93:1671–1676.

Lipinski CA, Lombardo F, Dominy BW, Feeney PJ. Experimental and computational approaches to estimate solubility and permeability in drug discovery and development settings. Adv Drug Deliv Rev 2001;46:3–26.

Liu X, LeCluyse EL, Brouwer KR, Gan L-SL, Lemasters JJ, Stieger B, Meier PJ, Brouwer KLR. Biliary excretion in primary rat hepatocytes cultured in a collagen-sandwich configuration. Am J Physiol 1999a;277:G12–G21.

Liu X, Lecluyse EL, Brouwer KR, Lightfoot RM, Lee JI, Brouwer KLR. Use of Ca^{2+} modulation to evaluate biliary excretion in sandwich-cultured rat hepatocytes. J Pharmacol Exper Ther 1999b;289:1592–1599.

Löfgren S, Baldwin RM, Hiratsuka M, Lindqvist A, Carlberg A, Sim SC, Schulke M, Snait M, Edenro A, Fransson-Steen R, Terelius Y, Ingelman-Sundberg M. Generation of mice transgenic for human CYP2C18 and CYP2C19: Characterization of the sexually dimorphic gene and enzyme expression. Drug Metab Dispos 2008;36:955–962.

Lu C, Miwa GT, Prakash SR, Gan LS, Balani SK. A novel model for the prediction of drug-drug interactions in humans based on in vitro CYP phenotyping. Drug Metab Dispos 2007;35:79–85.

Luo G, Gan L, Guenthner T. Testing drug candidates for CYP3A4 induction. *In Drug Metabolism in Drug Design and Development*, Zhang D, Zhu M, Humphreys WG, Eds. Hoboken, NJ, 2007, pp. 545–571.

Luo G, Johnson S, Hsueh MM, Zheng J, Cai H, Xin B, Chong S, He K, Harper T. In silico prediction of biliary excretion of drugs in rats based on physicochemical properties. Drug Metab Dispos. 2010;38:422–430.

Ma Q, Lu AY. The challenges of dealing with promiscuous drug-metabolizing enzymes, receptors and transporters. Curr Drug Metab 2008;9(5):374–383.

Malingre MM, Beijnen JH, Rosing H, Koopman FJ, Jewell RC, Paul EM, Ten Bokkel Huinink WW, Schellens JH. Co-administration of GF120918 significantly increases the systemic exposure to oral paclitaxel in cancer patients. Br J Cancer 2001; 84(1):42–47.

McGinnity DF, Berry A, Kenny JR, Grime K, Riley RJ. Evaluation of time-dependent cytochrome P450 inhibition using cultured human hepatocytes. Drug Metab Dispos 2006;34:1291–1300.

McGinnity DF, Zhang G, Kenny JR, Hamilton GA, Otmani S, Stams KR, Haney S, Brassil P, Stresser DM, Riley RJ. Evaluation of multiple in vitro systems for assessment of CYP3A4 induction in drug discovery: Human hepatocytes, pregnane X receptor reporter gene, and Fa2N-4 and HepaRG Cells. Drug Metab Dispos 2009;37:1259–1268.

Mehendale HM. Uptake and disposition of chlorinated biphenyls by isolated perfused rat liver. Drug Metab Dispos 1976;4:124–132.

Meijer DKF, Groothuis GMM, Mulder GJ, Swart PJ. The isolated perfused rat liver as a tool to study drug transport, drug metabolism and cell specific drug delivery. *In In Vitro and Ex Vivo Test Systems to Rationalize Drug Design and Delivery*, Crommelin D, Couvreur P, Duchene D, Eds. Edition de Sante, Paris, 1994, pp. 17–31.

Mutlib AE, Gerson RJ, Meunier PC, Haley PJ, Chen H, Gan LS, Davies MH, Gemzik B, Christ DD, Krahn DF, Markwalder JA, Seitz SP, Robertson RT, Miwa GT. The species-dependent metabolism of efavirenz produces a nephrotoxic glutathione conjugate in rats. Toxicol Appl Pharmacol 2000;169(1):102–113.

Nakano H, Ogura K, Takahashi E, Harada T, Nishiyama T, Muro K, Hiratsuka A, Kadota S, Watabe T. Regioselective monosulfation and disulfation of the phytoestrogens daidzein and genistein by human liver sulfotransferases. Drug Metab Pharmacokinet 2004;19:216–226.

Ohyama K, Nakajima M, Suzuki M, Shimada N, Yamazaki H, Yokoi T. Inhibitory effects of amiodarone and its N-deethylated metabolite on human cytochrome P450 activities: Prediction of in vivo drug interactions. Br J Clin Pharmacol 2000;49: 244–253.

Osborne AP, Brick D, Ruecroft G, Taylor IN. Immobilization of cholesterol esterase for use in multiple batch biotransformations to prepare (-)-FTC (Emtricitabine). Org Process Res Dev 2006;10:670–672.

Pang, KS. Safety testing of metabolites: Expectations and outcomes. Chem-Biol Interact 2009;179:45–59.

Parker RJ, Hirom PC, Millburn P. Enterohepatic recycling of phenolphthalein, morphine, lysergic acid diethylamide (LSD) and diphenylacetic acid in the rat. Hydrolysis of glucuronic acid conjugates in the gut lumen. Xenobiotica. 1980; 10(9):689–703.

Parkinson A, Ogilvie BW. Biotransformation of xenobiotics. in Casarett and Doull's Toxicology, Klaassen CD, Ed. McGraw-Hill, New York, 2008, pp. 161–304.

Pinto DJ, Orwat MJ, Wang S, Fevig JM, Quan ML, Amparo E, Cacciola J, Rossi KA, Alexander RS, Smallwood AM, Luettgen JM, Liang L, Aungst BJ, Wright MR, Knabb RM, Wong PC, Wexler RR, Lam PY. Discovery of 1-[3-(aminomethyl) phenyl]-*N*-3-fluoro-2′-(methylsulfonyl)-[1,1′-biphenyl]-4-yl]-3-(trifluoromethyl)-1*H*-pyrazole-5-carboxamide (DPC423), a highly potent, selective, and orally bioavailable inhibitor of blood coagulation factor Xa. J Med Chem 2001;44(4):566–578.

Purifoy DJ, Beauchamp LM, de Miranda P, Ertl P, Lacey S, Roberts G, Rahim SG, Darby G, Krenitsky TA, Powell KL. Review of research leading to new anti-herpesvirus agents in clinical development: Valaciclovir hydrochloride (256U, the L-valyl ester of acyclovir) and 882C, a specific agent for varicella zoster virus. J Med Virol 1993;Suppl 1:139–145.

Quan ML, Lam PY, Han Q, Pinto DJ, He MY, Li R, Ellis CD, Clark CG, Teleha CA, Sun JH, Alexander RS, Bai S, Luettgen JM, Knabb RM, Wong PC, Wexler RR. Discovery of 1-(3′-aminobenzisoxazol-5′-yl)-3-trifluoromethyl-*N*-[2-fluoro-4-[(2'-dimethylaminomethyl)imidazol-1-yl]phenyl]-1*H*-pyrazole-5-carboxyamide hydrochloride (razaxaban), a highly potent, selective, and orally bioavailable factor Xa inhibitor. J Med Chem 2005;48(6):1729–1744.

Raghavan N, Frost C, Yu Z, He K, Zhang H, Humphreys W, Pinto D, Chen S, Bonacorsi S, Wong P, Zhang D. Apixaban metabolism and pharmacokinetics after oral administration to humans. Drug Metab Dispos 2009;37:74–81.

Rowland P, Blaney FE, Smyth MG, Jones JJ, Leydon VR, Oxbrow AK, Lewis CJ, Tennant MG, Modi S, Eggleston DS, Chenery RJ, Bridges AM. Crystal structure of human cytochrome P450 2D6. J Biol Chem 2006;281(11):7614–7622.

Ruegg CE, Silber PM, Mughal RA, Ismail J, Lu C, Bode DC, Li AP. Cytochrome-P450 induction and conjugated metabolism in primary human hepatocytes after cryopreservation. In Vitro Toxicol 1997;10:217–222.

Salyers KL. Preclinical pharmacokinetic models for drug discovery and development. In Handbook of Drug Metabolism, 2nd ed., Pearson P, Wienkers L, Eds. Informa Healthcare, New York, 2009, pp. 659–673.

Schimmel RJ, Knobil E. Role of free fatty acids in stimulation of gluconeogenesis during fasting. Am J Physiol 1969;217:1803–1808.

Schoch GA, Yano JK, Wester MR, Griffin KJ, Stout CD, Johnson EF. Structure of human microsomal cytochrome P450 2C8. Evidence for a peripheral fatty acid binding site. J Biol Chem 2004;279(10):9497–9503.

Scott EE, He YA, Wester MR, White MA, Chin CC, Halpert JR, Johnson EF, Stout CD. An open conformation of mammalian cytochrome P450 2B4 at 1.6-A resolution. Proc Natl Acad Sci USA 2003;100(23):13196–13201.

Scott EE, White MA, He YA, Johnson EF, Stout CD, Halpert JR. Structure of mammalian cytochrome P450 2B4 complexed with 4-(4-chlorophenyl)imidazole at 1.9-A resolution: Insight into the range of P450 conformations and the coordination of redox partner binding. J Biol Chem 2004;279(26):27294–27301.

Shantsila E, Lip GY. Apixaban, an oral, direct inhibitor of activated Factor Xa. Curr Opin Investig Drugs 2008;9(9):1020–1033.

Shibata Y, Takahashi H, Chiba M, Ishii Y. Prediction of hepatic clearance and availability by cryopreserved human hepatocytes: An application of serum incubation method. Drug Metab Dispos 2002;30:892–896.

Shitara Y, Li AP, Kato Y, Lu C, Ito K, Itoh B, Sugiyama Y. Effect of cryopreservation on the uptake of taurocholate and estradial 17beta-D-glucuronide in isolated human hepatocytes. Drug Metab Pharmacokinet 2003;18:33–41.

Skaggs A, Foti RS, Fisher MB. A streamlined method to predict hepatic clearance using human liver microsomes in the presence of human plasma. J Pharmacol Toxicol Methods 2006;55:284–290.

Spalding DJ, Harker AJ, Bayliss MK. Combining high-throughput pharmacokinetic screens at the hits-to-leads stage of drug discovery. Drug Discov Today 2000; 5(12 Suppl 1):70–76.

Stratford RE , Jr, Clay MP, Heinz BA, Kuhfeld MT, Osborne SJ, Phillips DL, Sweetana SA, Tebbe MJ, Vasudevan V, Zornes LL, Lindstrom TD. Application of oral bioavailability surrogates in the design of orally active inhibitors of rhinovirus replication. J Pharm Sci 1999;88(8):747–757.

Stresser DM, Blanchard AP, Turner SD, Erve JC, Dandeneau AA, Miller VP, Crespi CL. Substrate-dependent modulation of CYP3A4 catalytic activity: Analysis of 27 tst compounds with four fluorometric substrates. Drug Metab Dispos 2000;28: 1440–1448.

Svensson US, Maki-Jouppila M, Hoffmann KJ, Ashton M. Characterisation of the human liver in vitro metabolic pattern of artemisinin and auto-induction in the rat by use of nonlinear mixed effects modelling. Biopharm Drug Dispos 2003;24(2):71–85.

Uno S, Dalton TP, Dragin N, Curran CP, Derkenne S, Miller ML, Shertzer HG, Gonzalez FJ, Nebert DW. Oral benzo[*a*] pyrene in Cyp1 knockout mouse lines: CYP1A1 important in detoxication, CYP1B1 metabolism required for immune damage independent of total-body burden and clearance rate. Mol Pharmacol 2006;69:1103–1114.

Uttamsingh V, Lu C, Miwa G, Gan LS. Relative contributions of the five major human Cytochromes P450, 1A2, 2C9, 2C19, 2D6, and 3A4, to the hepatic metabolism of the proteasome inhibitor Bortezomib. Drug Metab Dispos 2005;33:1723–1728.

Valentin J. Basic anatomical and physiological data for use in radiological protection: Reference values: ICRP Publication 89. Ann ICRP 2002;32(3–4):1–277.

Van Herwaarden AE, Smit JW, Sparidans RW, Wagenaar E, Van der Kruijssen CMM, Schellens JHM, Beijnen JH, Schinkel AH. Midazolam and cyclosporin A metabolism in transgenic mice with liver-specific expression of human CYP3A4. Drug Metab Dispos 2005;33:892–895.

van Herwaarden AE, Wagenaar E, van der Kruijssen CMM, van Waterschoot RAB, Smit JW, Song JY, van der Valk MA, van Tellingen O, van der Hoorn JWA, Rosing H, Beijnen JH, Schinkel AH. Knockout of cytochrome P450 3A yields new mouse models for understanding xenobiotic metabolism. J Clin Investig 2007;117:3583–3592.

van Waterschoot RAB, van Herwaarden AE, Lagas JS, Sparidans RW, Wagenaar E, van der Kruijssen CMM, Goldstein JA, Zeldin DC, Beijnen JH, Schinkel AH. Midazolam metabolism in cytochrome P450 3A knockout mice can be attributed to up-regulated CYP2C enzymes. Mol Pharmacol 2008;73:1029–1036.

Venkatakrishnan K, von Moltke LL, Greenblatt DJ. Application of the relative activity factor approach in scaling from heterologously expressed cytochromes P450 to human liver microsomes: Studies on amitriptylline as a model substrate. J Pharmacol Exp Ther 2001;297:326–337.

Vlasses PH, Larijani GE, Conner DP, Ferguson RK. Enalapril, a nonsulfhydryl angiotensin-converting enzyme inhibitor. Clin Pharm 1985;4(1):27–40.

von Moltke LL, Greenblatt DJ, Grassi JM, Granda BW, Venkatakrishnan K, Schmider J, Harmatz JS, Shader RI. Multiple human cytochromes contribute to biotransformation of dextromethorphan in-vitro: Role of CYP2C9, CYP2C19, CYP2D6, and CYP3A. J Pharm Pharmacol 1998;50(9):997–1004.

Wang L, Chistopher L, Cui D, Li W, Iyer R, Humpheys W, Zhang D. Reaction-phenotyping of oxidative metabolism of dasatinib. An effective approach for determining metabolite formation kinetics. Drug Metab Dispos 2008;36:1828–1939.

Wang L, Zhang D, Raghavan N, He K, Frost C, Humphreys W, Grossman SJ. In vitro assessment of metabolic drug-drug interaction potential of apixaban through cytochrome P450 enzyme phenotyping, inhibition and induction studies. Drug Metab Dispos 2010;38:448–458.

Wang L, Zhang D, Swaminathan A, Xue Y, Cheng PT, Wu S, Mosqueda-Garcia R, Aurang C, Everett DW, Humphreys WG. Glucuronidation as a major clearance pathway of muraglitazar in humans: Different metabolic profiles in subjects with and without bile collection. Drug Metab Dispos 2006;34:427–439.

Wester MR, Yano JK, Schoch GA, Yang C, Griffin KJ, Stout CD, Johnson EF. The structure of human cytochrome P450 2C9 complexed with flurbiprofen at 2.0-A resolution. J Biol Chem 2004;279(34):35630–35637.

Williams PA, Cosme J, Sridhar V, Johnson EF, McRee DE. Microsomal cytochrome P450 2C5: Comparison to microbial P450s and unique features. J Inorg Biochem 2000;81(3):183–190.

Williams PA, Cosme J, Vinkovic DM, Ward A, Angove HC, Day PJ, Vonrhein C, Tickle IJ, Jhoti H. Crystal structures of human cytochrome P450 3A4 bound to metyrapone and progesterone. Science 2004;305(5684):683–686.

Williams PA, Cosme J, Ward A, Angove HC, Matak Vinkovic D, Jhoti H. Crystal structure of human cytochrome P450 2C9 with bound warfarin. Nature 2003; 424(6947):464–468.

Wong PC, Crain EJ, Watson CA, Xin B. Favorable therapeutic index of the direct factor Xa inhibitors, apixaban and rivaroxaban, compared with the thrombin inhibitor dabigatran in rabbits. J Thromb Haemost 2009;7(8):1313–1320.

World Health Organisation (WHO). Use of ionizing radiation and radionuclides on human beings for medical research, training and non-medical purposes. WHO Publication 611 WHO, Geneva, 1977.

Wu L, Gu J, Cui H, Zhang Q-Y, Behr M, Fang C, Weng Y, Kluetzman K, Swiatek P, Yang W, Kaminsky LS, Ding X. Transgenic mice with a hypomorphic NADPH-cytochrome P450 reductase gene: Effects on development, reproduction, and microsomal cytochrome P450. J Pharmacol Expt Ther 2005;312:35–43.

Yamashita F, Hashida M. In silico approaches for predicting ADME properties of drugs. Drug Metab Pharmacokinet 2004;19(5):327–38.

Yano JK, Wester MR, Schoch GA, Griffin KJ, Stout CD, Johnson EF. The structure of human microsomal cytochrome P450 3A4 determined by X-ray crystallography to 2.05-A resolution. J Biol Chem 2004;279(37):38091–38094.

Yao M, Chang S, Zhu M, Zhang D, Rodrigues D. Applications of recombinant and purified human drug metabolizing enzymes: An industrial perspective. In Handbook of Drug Metabolism, 2nd ed., Pearson P, Wienkers L, Eds. Informa Healthcare, New York, 2009, pp. 393–444.

Zhang D, Comezoglu N. ADME Studies in Animals and Humans: Experimental Design, Metabolite Profiling and Identification, and Data Presentation. In Drug Metabolism in Drug Design and Development: Principle and Applications, Zhang D, Zhu M, Humphreys WG, Eds. Wiley, Hoboken, NJ, 2007, pp. 573–604.

Zhang D, Evans FE, Freeman JP, Duhart B, Cerniglia CE. Biotransformation of amitriptyline by *Cunninghamella elegans*. Drug Metab Dispos 1995;23:1417–1425.

Zhang D, Hanson R, Roongta V, Dischino D, Gao Q, Sloan CP, Polson C, Keavy D, Zheng M, Mitroka J, Yeola S. *In vitro* and *in vivo* metabolism of a gamma-secretase inhibitor BMS-299897 and generation of active metabolites in milligram quantities with a microbial bioreactor. Curr Drug Metabol 2006;7:1–14.

Zhang D, Wang L, Chandrasena G, Ma L, Zhu M, Zhang H, Davis CD, Humphreys WG. Involvement of multiple cytochrome P450 and UDP-glucuronosyltransferase enzymes in the in vitro metabolism of muraglitazar. Drug Metab Dispos 2007a;35:139–149.

Zhang D, Raghavan N, Chen S, Zhang H, Quan M, Lecureux L, Patrone LM, Lam P, Bonacorsi S, Knabb R, Skiles G, He K. Reductive isoxazole ring-opening of the anticoagulant razaxaban as the major metabolic clearance pathway in rats and dogs. Drug Metab Dispos 2008;36:303–315.

Zhang D, Raghavan N, Wang L, He K, Pinto D, Chen S, Knabb R, Humphreys W, Grossman SJ. Comparative metabolism of 14C-labeled apixaban in mice, rats, rabbits, dogs, and humans. Drug Metab Dispos 2009;37:1738–1748.

Zhang D, Wang L, Raghavan N, Zhang H, Li W, Cheng PT, Yao M, Zhang L, Zhu M, Bonacorsi S, Mitroka J, Hariharan N, Hosagrahara V, Chandrasena, Shyu W, Humphreys GW. Comparative metabolism of radiolabeled muraglitazar in animals and humans by quantitative and qualitative metabolite profiling. Drug Metab Dispos 2007b;35:150–167.

Zhang Q-Y, Fang C, Dunbar D, Zhang J, Kaminsky LS, Ding X. An intestinal epithelium-specific cytochrome P450 reductase-knockout mouse model: Direct evidence for a role of intestinal cytochromes P450 in first-pass clearance of oral nifedipine. Drug Metab Dispos 2009;37:651–657.

Zhang Q-Y, Gu J, Su T, Cui H, Zhang X, D'Agostino J, Zhuo X, Yang W, Swiatek P, Ding X. Generation and characterization of a transgenic mouse model with hepatic expression of human CYP2A6. Biochem Biophys Res Commun 2005;338:318–324.

Zhang Q-Y, Kaminsky LS, Dunbar D, Ding X. Role of small intestinal cytochrome P450 in the bioavailability of oral nifedipine. Drug Metab Dispos 2007;35: 1617–1623.

Zhao P, Kunze KL, Lee CA. Evaluation of time-dependent inactivation of CYP3A in cryopreserved human hepatocytes. Drug Metab Dispos 2005;32:335–346.

Zhou X, Zhuo X, Xie F, Kluetzman K, Shu Y-Z, Humphreys WG, Ding X. Role of CYP2A5 in the clearance of nicotine and cotinine: Insights from studies on a Cyp2a5-null mouse model. J Pharmacol Expt Ther 2010;332:578–587.

Zhu M, Zhang D, Zhang H, Shyu W. Integrated strategies for assessment of metabolite exposure in humans during drug development: Analytical challenges and clinical development considerations. Biopharm Drug Dispos 2009;30:163–184.

Zhuo X, Gu J, Behr M, Swiatek P, Cui H, Zhang Q-Y, Xie Y, Collins DN, Ding X. Targeted deletion of the olfactory mucosa-specific *CYP2G1* gene: Impact on acetaminophen toxicity in the lateral nasal gland and neighboring effects on *CYP2A5* expression in the liver and kidney. J Pharmacol Expt Ther 2004;308:719–728.

6 Principles of Pharmacokinetics: Predicting Human Pharmacokinetics in Drug Discovery

TAKEHITO YAMAMOTO, HIROSHI SUZUKI

Department of Pharmacy, The University of Tokyo Hospital, Faculty of Medicine, The University of Tokyo, Tokyo, Japan

AKIHIRO HISAKA

Pharmacology and Pharmacokinetics, 22nd Medical Center, The University of Tokyo Hospital, Faculty of Medicine, The University of Tokyo, Tokyo, Japan

Mass Spectrometry in Drug Metabolism and Disposition: Basic Principles and Applications, First Edition. Edited by Mike S. Lee and Mingshe Zhu.

6.1 INTRODUCTION

6.1.1 General Introduction

In current drug development procedures, pharmacokinetic (PK) properties of new drug candidates are extensively evaluated at the exploratory stage. Through a considerable number of failures, the pharmaceutical industry has become aware of the necessity to appropriately evaluate PK properties to develop superior drugs (Frank and Hargreaves, 2003). However, even in current drug development procedures, many drug candidates drop out at the clinical trial stage because of inadequate PK properties in humans (Frank and Hargreaves, 2003; Singh, 2006). Moreover, as in the cases of terfenadine, astemizole and sorivudine, these approved drugs had to be withdrawn from the market because of severe adverse effects attributed to the unexpected elevation of drug concentrations by drug–drug interactions (DDIs) (Ito et al., 1998; Wienkers and Heath, 2005). Therefore, to minimize such failures, more systematic research that focuses on the precise prediction of PK properties for new drug candidates in humans is needed in the early stages of drug development (Singh, 2006).

In this chapter, basic concepts and mathematical models of PK and general methods and approaches used to predict PK properties in humans from preclinical data will be reviewed.

6.1.2 Relationship between Drug Efficacy and Concentration

Drugs are delivered to target organs via the bloodstream after oral or systemic administration, and exhibit pharmacological actions there, except in cases of local application. Since drug concentrations in the blood and target organs are closely correlated, the blood concentrations of a drug are generally considered as useful indicator that reflects the drug efficacy (Buxton, 2006). Indeed, this relationship has been proved in many cases, and the prediction of the PK properties of compounds is very important for successful drug development. In the following sections, the general methods used to predict the PK properties of new drug candidates in humans will be summarized.

6.1.3 Prediction of Pharmacokinetics by Extrapolation from Animal Data

The prediction methods for PK properties in humans are classified into two categories; an empirical estimation from animal data such as allometric scaling and a theoretical approach such as physiological PK modeling (Buxton, 2006). In allometric scaling, many PK parameters are determined from correlations between observed data in animals assuming no causal relationships or physiological structure (Adolph, 1949; Sawada et al., 1985; Kita et al., 1986). In modern drug exploratory research, however, allometric scaling is used less for estimations of metabolic clearance across species because of poor predictability; in many cases, species differences in drug metabolism are evident. In contrast, it has become relatively easy to reliably evaluate metabolizing activities in humans from the results of in vitro experiments with human enzymes, due to the development of physiological pharmacokinetics (Chiba et al., 2009).

However, extrapolations from animal data are useful for predicting the volume of distribution (V_d); the V_d in humans can often be extrapolated directly from observations in animals (Fagerholm, 2007). In addition, organ clearance (CL_{org}) is often estimated from in vitro data using human specimens with a correction factor that is calculated from a discrepancy between in vitro and in vivo experiments in animals (Naritomi et al., 2001, 2003; Kitamura et al., 2008). In this approach, it is implicitly assumed that the discrepancy arises from mechanisms common to humans and animals, although it is difficult to specify these discrepancies precisely with current knowledge.

6.2 PHYSIOLOGICAL PHARMACOKINETICS

6.2.1 Why Is a Physiological Pharmacokinetic Model Necessary?

In general, a physiological PK model is used to predict the human PK properties of new drug candidates in a deductive manner. With the physiological PK model, we can explain the drug disposition and total body clearance from events in local tissues or organs based on the physiological structure (Fig. 6.1a). In contrast, although simpler models such as compartment models are easier to use, they are not applicable to many situations since they do not consider the physiological structure. The physiological PK model describes the behavior of drugs in local organs and tissues with two key parameters: the V_d and CL_{org} (Buxton, 2006). Using this model, in vivo drug disposition can be predicted by connecting each organ and tissue with the blood flow.

6.2.2 Clearance

Clearance represents the ability of drug elimination and is defined by dividing the velocity of elimination by the blood or plasma concentration of a drug (Buxton, 2006). Based on the physiological PK model theory, the $CL_{org,blood}$ is described by Eq. (6.1):

$$CL_{org,blood} = \frac{v}{C_{B,in}} = Q_B \times \frac{C_{B,in} - C_{B,out}}{C_{B,in}} = Q_B \times E \tag{6.1}$$

where $CL_{org,blood}$ represents the organ clearance based on the blood concentration; v represents the velocity of elimination of the drug; $C_{B,\,in}$ and $C_{B,\,out}$ represent the concentrations of the drug in the inflow and outflow of blood of the organ, respectively; Q_B represents the blood flow rate of organs; and E represents the extraction ratio. In relation to this equation, it would be useful to describe the definition of availability (F):

$$F = 1 - E = \frac{C_{B,\,out}}{C_{B,\,in}} \tag{6.2}$$

As indicated by Eq. (6.1), $CL_{org,\,blood}$ is given as a function of Q_B and cannot exceed the Q_B value. Although the blood-concentration-based clearances are straightforward theoretically, plasma- (or serum)-concentration-based clearances are more frequently used because drug concentrations in plasma or serum are usually quantified. Therefore, all organ clearances are presented as plasma concentration-based in the following in this chapter. The relationship between these clearances is given by Eq. (6.3):

$$CL_{org} = R_B \times CL_{org,blood} \tag{6.3}$$

where CL_{org} represents the organ clearance based on the plasma concentration; R_B represents the blood-to-plasma concentration ratio. The total body clearance (CL_{tot}) is given by the sum of the CL_{org} of each organ and tissue. The major organs responsible for drug elimination are the liver and kidney, and consequently CL_{tot} is described as the sum of hepatic and renal clearances (CL_h and CL_r, respectively).

Generally, the efficacy and the safety of a drug depend on the area under the plasma concentration–time curve (AUC), and CL_{tot} is the determining factor of the AUC as shown by Eq. (6.4):

$$CL_{tot} = \frac{D_{iv}}{AUC_{iv}} \tag{6.4}$$

where AUC_{iv} and D_{iv} represent the AUC after intravenous dose and the dose of drug administrated intravenously, respectively. Consequently, CL_{tot} is an important parameter for predicting drug efficacy and safety.

It is noteworthy that blood (C_B) and plasma (alternatively serum) concentrations (C_P) of a drug must be clearly differentiated when applying the physiological model. Although drug concentrations are measured as the plasma or serum concentration in most cases, drug molecules can be distributed to erythrocytes. The relationship between C_B and C_P is given by Eq. (6.5):

$$C_B = C_{\mathrm{RBC}} \times \mathrm{Hct} + C_P \times (1 - \mathrm{Hct}) \tag{6.5}$$

where C_{RBC} represents the drug concentration in red blood cells and Hct represents the hematocrit value. The mass balance equation in each organ and tissue should be described based on blood concentrations, although unbound concentrations of a drug are equal in plasma and blood by their definition as given by Eq. (6.6):

$$C_B \times f_B = C_P \times f_P \Leftrightarrow \frac{C_B}{C_P} = R_B = \frac{f_P}{f_B} \tag{6.6}$$

where f_B and f_P represent the unbound fraction of drugs in blood and plasma, respectively.

6.2.3 Volume of Distribution

The volume of distribution (V_d) of a drug in an organ or tissue is generally defined as

$$V_d = \frac{X}{C_P} \tag{6.7}$$

where X represents the amount of drug in an organ or tissue (Buxton, 2006). X is the sum of the amount of drug in blood and tissues as represented by Eq. (6.8):

$$X = C_B \times V_B + C_T \times V_T = \frac{R_B \times C_P \times V_P}{1 - \mathrm{Hct}} + C_T \times V_T \tag{6.8}$$

where V_P and V_B represent the volume of plasma and blood, respectively; C_T represents the tissue concentration of a drug; and V_T represents the tissue volume.

Then, Eq. (6.8) is converted to the following equation:

$$V_d = \frac{R_B \times V_P}{1 - \mathrm{Hct}} + K_P \times V_T \tag{6.9}$$

where K_p, the tissue-to-plasma drug concentration ratio, is given by

$$K_P = \frac{C_T}{C_P} \tag{6.10}$$

Assuming that there are no active uptake mechanisms, the unbound concentrations of a drug in plasma and tissue are considered to be equal. Thus, Eq. (6.10) can be converted to Eq. (6.11):

$$f_P \times C_P = f_T \times C_T \Leftrightarrow \frac{C_T}{C_P} = K_P = \frac{f_P}{f_T} \tag{6.11}$$

where f_T represents the unbound fraction of a drug in tissue.

6.2.4 Relationship between Intrinsic Clearance and Organ Clearance

By definition, the intrinsic clearance (CL_{int}) is calculated by dividing the elimination velocity of a drug (v) by the drug concentration at the target site (Jusko, 1992). The activity of metabolizing enzymes and drug transporters can also be determined as the intrinsic clearance by in vitro experiments. As discussed later, we can extrapolate CL_{int} values determined in vitro to in vivo CL_{int} values.

In order to calculate the CL_{org} from the CL_{int}, we need to consider the Q_B. In addition, it is necessary to assume mathematical models to describe the behavior of drug molecules in organs. For mathematical models, the "well-stirred model" (Rowland et al., 1973; and Rowland, 1977a), the "parallel tube model" (Pang and Rowland, 1977b), and the "dispersion model" (Roberts and Rowland, 1986) are frequently used in PK analysis (Fig. 6.1a, c). The well-stirred model assumes that the drug concentration in the blood is discontinued at the entrance and that in the organ vessels is equal to the drug concentration in the distal venous blood due to complete stirring within the organ. In contrast, the parallel tube model assumes no stirring along the blood flow and thus, a concentration gradient along the blood flow is formed. The dispersion model assumes the certain degree of stirring along the blood flow.

Here, we discuss the advantages and disadvantages of these three mathematical models used to describe the behavior of drug molecules in organs, taking the liver as an example (Fuse et al., 1995; Yamamoto et al., 2005; Liu and Pang, 2006). Equations (6.12)–(6.14) are the calculation formulas of hepatic availability (F_h) and CL_h derived from each model.

Well-stirred model:

$$R_B \times CL_h = \frac{Q_h \times f_B \times CL_{int,h}}{Q_h + f_B \times CL_{int,h}} \tag{6.12a}$$

$$F_h = \frac{Q_h}{Q_h + f_B \times CL_{int,h}} \tag{6.12b}$$

Parallel tube model:

$$R_B \times CL_h = Q_h \times \left[1 - \exp\left(-\frac{f_B \times CL_{int,h}}{Q_h}\right)\right] \tag{6.13a}$$

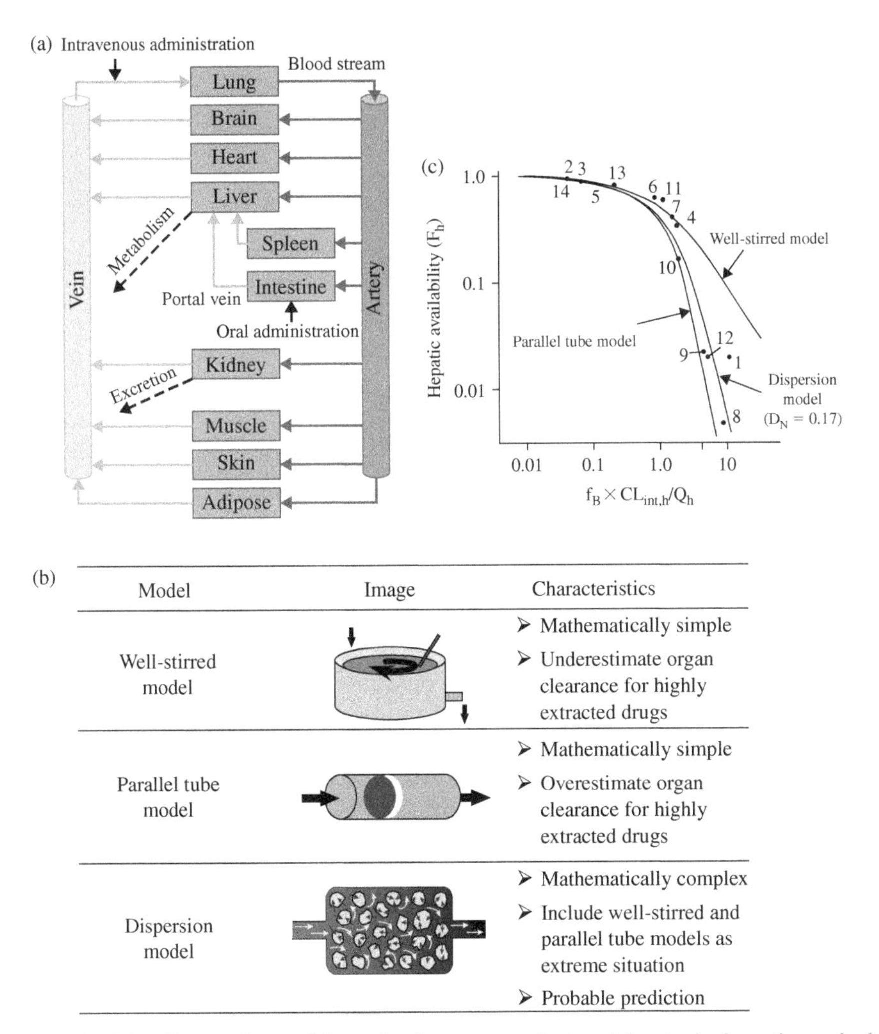

FIGURE 6.1 Comparison of hepatic clearances calculated by typical mathematical models for description of pharmacokinetics in organs and tissues. (*a*) Schematic diagram illustrating the physiological pharmacokinetic model. Arrows indicate the blood stream. (*b*) Comparison of characteristics of three pharmacokinetic models. The well-stirred model assumes that the drug concentration in the blood within the organ vessels is equal to the drug concentration in the venous blood. The parallel tube model assumes no stirring along the blood flow, and, thus, a concentration gradient along the blood flow is formed. The dispersion model assumes the certain degree of stirring along the blood flow. (*c*) Prediction of in vivo F_h from $CL_{int,\,h}$ determined in vitro based on three pharmacokinetic models. Observed F_h values of 14 drugs in in vivo experiments were plotted against the predicted $f_P \times CL_{int,\,h}/Q_h$ values. $CL_{int,\,h}$ was determined in in vitro experiments. The lines indicate the F_h values predicted from in vitro $CL_{int,\,h}$ values by using the well-stirred, parallel tube, and dispersion models. 1: alprenolol; 2: antipyrine; 3: carbamazepine; 4: diazepam; 5: ethoxybenzamide; 6: hexobarbital; 7: 5-hydroxytryptamine; 8: lignocaine; 9: pethidine; 10: phenacetin; 11: phenytoin; 12 propranolol; 13: thiopental; 14: tolbutamide. (reproduced with permission from Fuse et al., 1995.)

$$F_h = \exp\left(-\frac{f_B \times \mathrm{CL}_{\mathrm{int},h}}{Q_h}\right) \tag{6.13b}$$

Dispersion model:

$$R_B \times \mathrm{CL}_h = Q_h \times (1 - F_h) \tag{6.14a}$$

$$F_h = \frac{4a}{(1+a)^2 \times \exp\left(\frac{a-1}{2D_N}\right) - (1-a)^2 \times \exp\left(-\frac{a+1}{2D_N}\right)} \tag{6.14b}$$

$$a = \sqrt{1 + 4R_N \times D_N} \tag{6.14c}$$

$$R_N = \frac{f_B \times \mathrm{CL}_{\mathrm{int},h}}{Q_h} \tag{6.14d}$$

where $\mathrm{CL}_{\mathrm{int},h}$ represents the intrinsic hepatic clearance, Q_h represents the hepatic blood flow rate, and D_N represents the normalized dispersion number.

The well-stirred model is widely used because of its mathematical simplicity. Indeed, there is compatibility between the observed $\mathrm{CL}_{\mathrm{org}}$ and the $\mathrm{CL}_{\mathrm{org}}$ calculated from in vitro experiments when using any of the three mathematical models for low-clearance drugs. The difference between $C_{B,\mathrm{in}}$ and $C_{B,\mathrm{out}}$ is minimal for low-clearance drugs, and consequently, the well-stirred model can be used to describe the physiological conditions for these drugs (Fuse et al., 1995; Yamamoto et al., 2005). However, for high-clearance drugs, the difference is significant. Therefore, the well-stirred model is not suitable for describing the physiological conditions for these drugs (Fuse et al., 1995; Yamamoto et al., 2005). For high-clearance drugs, that is, those with an E value greater than 0.5, the dispersion model gives the most reasonable prediction from in vitro experimental results, whereas the well-stirred model and the parallel tube model tend to underestimate or overestimate the organ clearance, respectively (Fig. 6.1b) (Fuse et al., 1995; Yamamoto et al., 2005).

The normalized dispersion model, D_N, indicates the extent of stirring along the blood flow, and Eq. (6.14b) is equivalent to Eq. (6.12b) (well-stirred model) or Eq. (6.13b) (parallel tube model) when D_N reaches infinity or zero, respectively. This means that the dispersion model is the inclusive model of the other two models (Liu and Pang, 2006). When calculating CL_h from in vitro data, D_N is empirically assumed to be 0.17 (Roberts and Rowland 1986), and Q_h is set at 0.95 mL/min/g liver. In addition, the normal Hct value is assumed to be 0.45 (Montandon et al., 1975). It is worth emphasizing that CL_h/R_B approaches $f_B \times \mathrm{CL}_{\mathrm{int},h}$ when the $f_B \times \mathrm{CL}_{\mathrm{int},h}$ is negligible compared with Q_h (clearance-limited kinetics). This is equivalent with that CL_h approaches $f_P \times \mathrm{CL}_{\mathrm{int},h}$. In contrast, the CL_h approaches $R_B \times Q_h$ when the Q_h is negligible compared with $f_B \times \mathrm{CL}_{\mathrm{int},h}$ (blood flow-limited kinetics).

Under the assumption that the extrahepatic clearance of a drug is negligible and intestinal absorption is complete, oral clearance can be simplified to Eq. (6.15), based on the well-stirred model:

$$CL_{oral} = \frac{CL_{tot}}{BA} \approx \frac{CL_h}{F_h} = f_P \times CL_{int,h} \tag{6.15}$$

$$AUC_{oral} = \frac{D_{oral}}{CL_{oral}} \approx \frac{D_{oral}}{f_P \times CL_{int,h}} \tag{6.16}$$

where CL_{oral} represents the total body clearance after oral administration and AUC_{oral} represents the AUC after oral administration. D_{oral} represents oral dose. BA is bioavailability which is explained in 6.3.

Equation (6.16) demonstrates that AUC_{oral} is independent of Q_h and inversely proportional to $CL_{int,\ h}$. This relationship is quite important in the screening of new drug candidates. However, it must be kept in mind that Eqs. (6.15) and (6.16) are derived using the well-stirred model. Thus, the prediction error may be significant for high-clearance drugs.

6.2.5 Estimation of Permeability-Limited Clearance

In Eqs. (6.12)–(6.16), hepatic intrinsic clearance ($CL_{int,\ h}$) is calculated directly from in vitro metabolism experiments when the drug is completely cleared by the metabolism and an equilibrium between the blood and the tissue is rapidly reached. However, if these conditions are not satisfied, it is necessary to consider the permeability of the drug across the basal and apical membranes.

Figure 6.2 depicts a model that describes membrane permeability and metabolism in hepatocytes. Under such situations, $CL_{int,\ h}$ is given by Eq. (6.17) (Shitara et al., 2006):

$$CL_{int,h} = \frac{PS_1 \times (PS_3 + CL_{int,m,h})}{PS_2 + PS_3 + CL_{int,m,h}} \tag{6.17}$$

where PS_1 and PS_2 represent the permeability surface area product for the influx and efflux of the drug across the sinusoidal membrane; PS_3 represents the permeability surface area product for excretion of the drug across the bile canalicular membrane; and $CL_{int,\ m,\ h}$ represents the intrinsic hepatic metabolic clearance.

If $PS_3 + CL_{int,\ m,\ h}$ is much higher than PS_2, then $CL_{int,\ h}$ is given by (6.18) and is independent from $CL_{int,m,h}$.

$$CL_{int,h} = PS_1 \tag{6.18}$$

In contrast, if PS_2 is much higher than $PS_3 + CL_{int,\ m,\ h}$, $CL_{int,\ h}$ is given by

$$CL_{int,\ h} = \frac{PS_1 \times (PS_3 + CL_{int,\ m,\ h})}{PS_2} \tag{6.19}$$

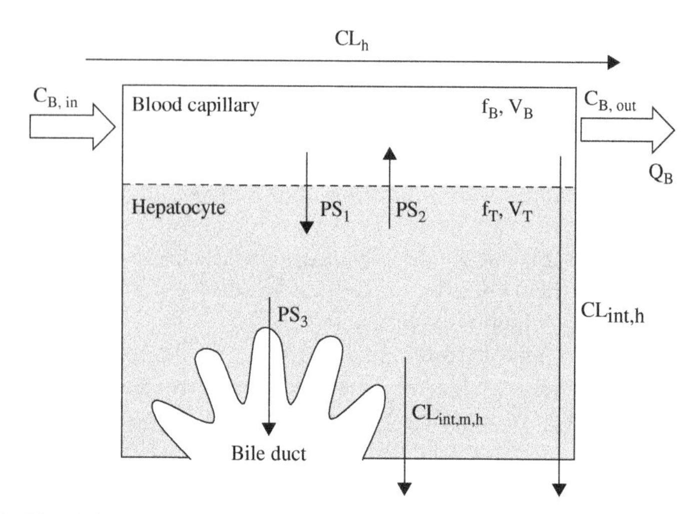

FIGURE 6.2 Schematic diagram illustrating the factors affecting on hepatic clearance of drugs. Drug molecules are taken up into hepatocytes from the blood capillary with the permeability surface area product of PS_1. The drug molecules in hepatocytes are metabolized with $CL_{int, m, h}$ and secreted or excreted into the blood and bile with PS_2 and PS_3, respectively. $CL_{int, m, h}$ is defined as the elimination velocity of drugs divided by hepatic unbound drug concentration. $CL_{int, h}$, which is defined as the elimination velocity of drugs divided by $C_{B, out}$, is given as a function of PS_1, PS_2, PS_3, and $CL_{int, m, h}$, as shown in Eq. (6.17) in the text.

6.3 PREDICTION OF ABSORPTION

6.3.1 Determinants of Bioavailability

In pharmacotherapy, oral formulations are most popular because they are easy to administer. After oral administration, drug molecules are absorbed in the gastrointestinal tract and pass through the liver via the portal vein before migrating to the systemic circulation (Fig. 6.3). Therefore, drugs may undergo elimination in the intestine and liver, where many metabolizing enzymes are highly expressed, before entering the systemic circulation.

The reduction in the amount of drug by this process is ascribed to the *first-pass effect*, which is the determinant of bioavailability (BA) (Buxton, 2006). BA is described by

$$\mathrm{BA} = F_a \times F_g \times F_h \tag{6.20}$$

where F_a and F_g represent absorption ratio and intestinal availability, respectively. BA is also given by Eq. (6.21) (Wagner, 1975):

$$\mathrm{BA} = \frac{D_{\mathrm{iv}} \times \mathrm{AUC}_{\mathrm{oral}}}{D_{\mathrm{oral}} \times \mathrm{AUC}_{\mathrm{iv}}} \tag{6.21}$$

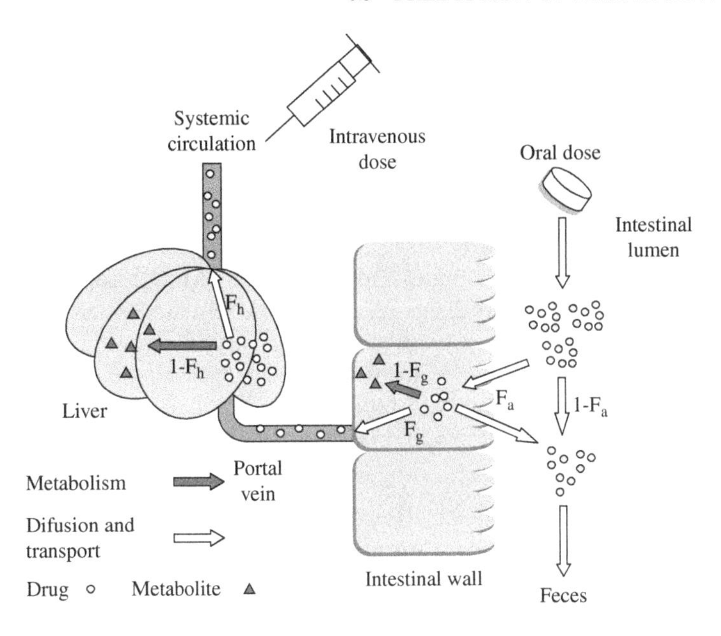

FIGURE 6.3 Schematic diagram illustrating the first-pass effect. After oral administration, drugs undergo intestinal and hepatic extraction before entering into the systemic circulation. Bioavailability (BA), which represents the fraction entering into the systemic circulation, is described as $F_a \times F_g \times F_h$.

The prediction of these availabilities will be described in the following sections.

6.3.2 Absorption Ratio

The small intestine is the most important organ for drug absorption in the gastrointestinal tract. The surface area of the intestine is approximately 200 m^2 (Caspary, 1992) because of the presence of well-developed villi. Collapsibility, drug solubility, and membrane permeability are the most influential factors in determining F_a. Solubility in aqueous solution is a critical factor in optimizing F_a (Ku, 2008).

6.3.3 Dosing Vehicle and Feeding State

The selection of dosing vehicle is also an important factor that affects the absorption of candidate drugs in animal experiments (Ku, 2008). Although methylcellulose suspension is usually preferred as a first-line vehicle, there may be a need to use detergents or organic solvents as vehicles to increase absorption. In some situations, the addition of cyclodextorin facilitates absorption. Furthermore, pulverization, forming amorphous, forming salts with appropriate counterions, or other particular optimizations may need to be considered in a drug formulation, even for animal experiments. However, it should be taken into consideration whether these particular means are absolutely necessary to exert pharmacological actions in animal experiments. Optimization of a formulation for clinical use would be a significant burden in the course of drug development.

It is well known that feeding conditions may have an effect on drug absorption (Lentz, 2008). Particularly for lipophilic compounds, F_a may significantly increase when administered under fed conditions (Mueller et al., 1994). In contrast, absorption of many drugs is delayed or attenuated when administered under fed conditions. For example, absorption of bisphosphonate, a hydrophilic compound that is assumed to be absorbed via the paracellular route, is almost entirely abolished when given with food (Bell and Johnson, 1997). Hence, the influence of food intake on F_a should be investigated during the earlier stage of drug development.

6.3.4 Evaluation Methods for Absorption

Methods used to determine drug absorption kinetics often involve in situ perfusion of the intestinal tract or in vitro everted tract experiments. In addition to these techniques, in vitro experiments using Caco-2 monolayer cells (Fig. 6.4*a*) are now most frequently used for this purpose in the exploratory stages of drug development (Hubatsch et al., 2007). However, there is a considerable difference between drug permeability measured in vitro and the absorption observed in vivo (Yamashita et al., 1997). This difference may be ascribed to the anatomical structure of the intestine. Under in vivo conditions, the blood capillaries are situated nearby to the intestinal tract and absorbed drugs are rapidly removed; however, such a mechanism does not exist in Caco-2 monolayer cells or the everted tract. In addition, although Caco-2 cells resemble intestinal epithelial cells in morphology, the expression patterns of drug transporters and metabolizing enzymes in Caco-2 cells are distinct from those in intestinal epithelial cells.

Nevertheless, in vitro screening using a Caco-2 monolayer is advantageous due to its convenience, and, therefore, it is routinely used as a qualitative evaluation method to screen the absorption of candidate compounds [Food and Drug Administration (FDA), 2008]. In fact, as shown in Figure 6.4*b*, there is a good relationship between the in vivo absorption and the apparent permeability rate constant (P_{app}) across the Caco-2 monolayer (Hubatsch et al., 2007). From a practical viewpoint, researchers should pay attention to the experimental conditions when conducting drug permeability studies using a Caco-2 monolayer, since the absolute values of P_{app} vary depending on the experimental conditions, such as passage number of cells (Shah et al., 2006). For this reason, the U.S. FDA provides a list of model drugs that are recommended for use as standard compounds to confirm the suitability and reliability of the methods in each permeation study (FDA, 2008).

When examining the role of active transporters located on the plasma membrane, such as P-glycoprotein in the intestinal absorption, transport from the apical to

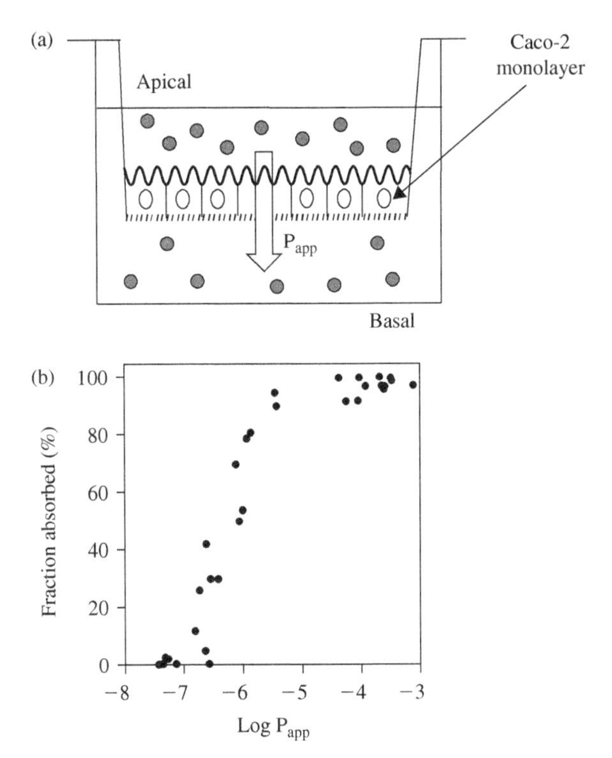

FIGURE 6.4 Prediction of in vivo absorption from in vitro experiments with Caco-2 monolayer. (*a*) Schematic diagram illustrating the transcellular transport experiments across Caco-2 monolayer. Solution of a test drug is applied to the apical chamber, and concentration of the test drug appeared in the basal chamber is measured as a function of time, in order to calculate the apparent permeability (P_{app}). (*b*) Relationship between absorption fraction observed in vivo and P_{app} determined in vitro using the Caco-2 monolayer. Because of the correlation between the two parameters, transport across the Caco-2 monolayer became a popular measure for predicting drug absorption in humans. (reproduced with permission from Hubatsch et al., 2007.)

basolateral surface and in the opposite direction is determined. Then, the extent of active transport across the cell monolayer is evaluated using the ratio of transport activity in both directions. This evaluation method is usually conducted using an exogenous expression system of membrane transporters of interest and is useful to clarify the transport mechanisms of drugs (Adachi et al., 2003).

Studying in vitro transport using an exogenous expression system is valuable when the F_a of a new drug candidate is unexpectedly high or low irrespective of its physical properties. For example, β-lactam antibiotics are well absorbed via the small intestine, although they are ionized at intestinal pH and predicted to exhibit low absorption. Using exogenous expression systems, it has been proven that β-lactam antibiotics are extensively taken up by dipeptide transporters (PEPT1/SLC15A1 and PEPT2/SLC15A2), which are responsible for the uptake of dipeptides as nutrients under physiological conditions (Brandsch, 2009).

6.3.5 Intestinal Availability

The existence of intestinal metabolism was first reported in humans in 1991 (Kolars et al., 1991). In this report, the metabolites of cyclosporin, a potent immunosuppressant, were detected in the portal venous blood of patients during the anhepatic phase of liver transplantation. Since then, the important role of intestinal metabolism in presystemic drug elimination has been reported for cyclosporin (Wu et al., 1995), midazolam (Thummel et al., 1996), simvastatin (Lilja et al., 1998), and some other drugs (Doherty and Charman, 2002). Among these reports, the simvastatin study was striking. The AUC_{oral} of simvastatin, a typical cytochrome P450 (CYP)3A4 substrate, was increased by about 16-fold, whereas the AUC_{iv} was unchanged by concomitant intake of grapefruit juice, which contains inhibitor(s) of CYP3A4, indicating that the major site of interaction is intestinal metabolism by CYP3A4. This series of studies suggest that intestinal metabolism significantly contributes to the BA of certain kinds of drugs. Furthermore, a recent report suggested that the expression of CYP3A4 in human intestine would be comparable with those in the liver (Galetin and Houston, 2006).

However, there remain two major problems in predicting the F_g from in vitro or animal data. The first problem is the species differences in intestinal

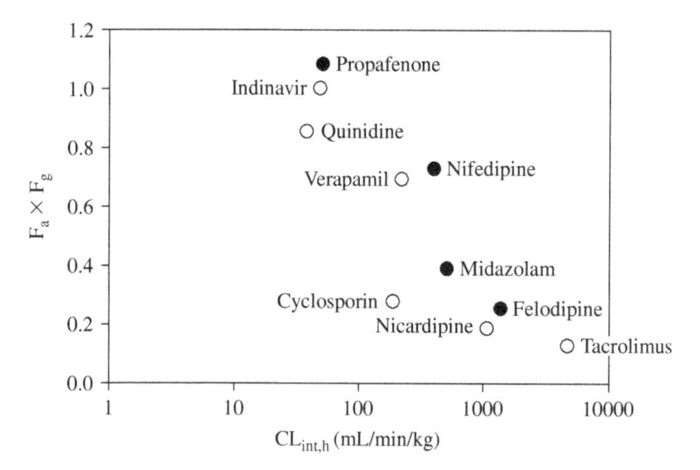

FIGURE 6.5 Relationship between susceptibility to metabolism ($CL_{int,\,h}$) and intestinal availability ($F_a \times F_g$). For 10 drugs, their $F_a \times F_g$ and $CL_{int,\,h}$ were compared. The values of $F_a \times F_g$ were calculated by dividing the bioavailability (BA) by F_h, and the values of F_h were calculated from CL_h and Q_h $(1\text{-}CL_h/Q_h)$. CL_h is equivalent to CL_{tot} when extrahepatic clearance is negligible, after intravenous dose and Q_h is 25.5 mL/min/kg. Open and closed circles represent substrate and nonsubstrate of P-glycoprotein, respectively. Drugs with high $CL_{int,\,h}$ tend to exhibit a low $F_a \times F_g$ value irrespective of the fact that the drug is a substrate of P-gp or not. It is assumed that $F_a \times F_g$ may be primarily determined by F_g since lipophilicity of the drugs presented in this figure is relatively high, and, therefore, excellent absorption ($F_a \approx 1$) is expected. $F_a \times F_g$ values tend to decrease with the increase in $CL_{int,\,h}$. A $CL_{int,\,h}$ of 100 mL/min/kg corresponds to approximately 78 μL/min/mg microsomal protein. (reproduced with permission from Kato et al., 2003.)

CYP expression. In humans, the major intestinal CYP is CYP3A4, whereas the expression level of human CYP3A4 homologs is relatively low in the intestine of rodents. Thus, researchers should be aware that F_g may be underestimated when predicted from rodent data (Komura and Iwaki, 2008). The second problem is the lack of a suitable pharmacokinetic model to describe the behavior of drugs in the intestine. Only limited reports are available concerning the construction of such a model (Galetin et al., 2008). It has been demonstrated that the extent of intestinal metabolism becomes significant only for drugs that suffer from extensive metabolism by CYP3A4 in humans (Fig. 6.5) (Kato et al., 2003).

In the human intestine, the activities of drug-metabolizing enzymes other than CYPs, such as esterase, peptidase, and conjugating enzymes, are very high, although species differences in these enzymes are also considerable. Therefore, when developing a prodrug that is activated by an esterase or other enzymes, species differences should be carefully investigated (Yoshigae et al., 1998).

6.4 DISTRIBUTION

6.4.1 Plasma Protein Binding

Plasma protein binding is also an important factor that affects drug disposition. Although plasma albumin is the major protein capable of binding many kinds of drugs, some basic drugs bind to α1-acidic glycoprotein (AGP), the plasma concentrations of which are affected by the extent of inflammation (Tozer, 1981). The disposition of some drugs is affected by AGP (O'Connor and Feely, 1987). In addition, although in rare case, there are striking species differences in protein binding (Fuse et al., 1999). Therefore, it is necessary to investigate the species differences in in vitro studies.

6.4.2 Relationship between Drug Efficacy and Protein Binding

For drugs with an efficacy (and safety) that depend on the unbound concentration in the blood, the drug efficacy (Ef) after intravenous administration is given by Eq. (6.22):

$$\mathrm{Ef} \propto f_P \times \mathrm{AUC_{iv}} = f_P \times \frac{D_{\mathrm{iv}}}{CL_{\mathrm{tot}}} \tag{6.22}$$

After intravenous administration of drugs that are extensively metabolized in the liver (blood-flow-limited elimination; see Section 6.2.4 for details), Eq. (6.22) can be simplified to

$$\mathrm{Ef} \propto f_B \times \frac{D_{\mathrm{iv}}}{Q_h} \tag{6.23}$$

Equation (6.23) demonstrates that the drug efficacy under such conditions is affected by the extent of plasma–protein binding (Buxton, 2006).

In contrast, after oral administration, Eq. (6.22) can be simplified to

$$\mathrm{Ef} \propto \frac{D_{\mathrm{oral}}}{\mathrm{CL}_{\mathrm{int},\,h}} \tag{6.24}$$

Equation (6.24) holds true for both blood-flow-limited and clearance-limited drugs after oral administration. Equation (6.24) also holds true after intravenous administration of clearance-limited drugs. Under such conditions, drug efficacy is not affected by changes in f_B values (Buxton, 2006).

6.5 METABOLISM AND EXCRETION

6.5.1 Estimation of Clearance

Generally, the liver and the kidney are responsible for drug elimination via hepatic metabolism and biliary excretion and renal excretion, respectively. Therefore, the $\mathrm{CL_{tot}}$ of drugs after intravenous administration is usually described by Eq. (6.25) (Buxton, 2006):

$$\mathrm{CL_{tot}} = \mathrm{CL}_h + \mathrm{CL}_r \tag{6.25}$$

In contrast, the clearance of a drug after oral administration is described by Eq. (6.26) since orally administrated drugs undergo the first-pass effect before entering the systemic circulation (Fig. 6.4) (Buxton, 2006):

$$\mathrm{CL_{oral}} = (\mathrm{CL}_h + \mathrm{CL}_r) \times F_a \times F_g \times F_h = (\mathrm{CL}_h + \mathrm{CL}_r) \times \mathrm{BA} \tag{6.26}$$

Strategies for prediction of F_a and F_g are explained in Section 6.3 and F_h is calculated from CL_h. In the following section, general strategies for predicting hepatic and renal clearance of drugs are described.

6.5.2 Estimation of Hepatic Intrinsic Clearance

The liver is the most important organ in drug metabolism. Many kinds of drug-metabolizing enzymes are expressed in the liver. Among them, CYPs are the most abundant and consequently have a great impact on the PK properties of drugs. Indeed, in vitro experiments using microsome prepared from human and insect cells that exogenously express human CYPs, cytosol and microsomes prepared from human liver and intestine, and cryopreserved human hepatocytes are extensively used to predict the disposition of new drug candidates (Chiba et al., 2009).

In the following section, strategies for estimation of $\mathrm{CL}_{\mathrm{int},\,m,\,h}$ are discussed assuming that the membrane penetration process is rapid and does not affect the metabolic clearance of a drug.

6.5.3 Determination of Hepatic Intrinsic Metabolic Clearance from in vitro Experimental Data

Generally, the relationship between the initial metabolic velocity and unbound drug concentration is described by the Michaelis–Menten equation [Eq. (6.27)] in in vitro experiments:

$$v_m = \frac{V_{\max} \times S}{K_M + S} \tag{6.27}$$

where v_m represents the initial metabolic velocity, $V_{\max}$ represents the maximum metabolic velocity, K_M represents the Michaelis–Menten constant, and S represents the unbound drug concentration (Jusko, 1992). Human liver microsomes are usually used as the source of CYPs, and consequently v_m is given in terms of velocity/milligram microsomal protein. When S is negligible compared with K_M, Eq. (6.27) is simplified to Eq. (6.28):

$$v_m = \frac{V_{\max}}{K_M} \times S \tag{6.28}$$

Under linear conditions, $CL_{\text{int}, m, h}$ is given by dividing v_m by S [Eq. (6.29)] (Jusko, 1992):

$$CL_{\text{int}, m, h} = \frac{V_{\max}}{K_M} \tag{6.29}$$

It is important to confirm the following three points to calculate the $CL_{\text{int}, m, h}$ by use of Eq. (6.29): (1) unbound drug concentration is used in the calculation, (2) initial velocity is measured correctly, and (3) data are obtained under linear conditions. Many drugs bind to some degree to microsomal proteins; therefore, it is necessary to determine the unbound drug concentrations in in vitro incubations to calculate the $CL_{\text{int}, m, h}$. In general, microsomal protein-bound drugs are separated by ultracentrifuge or ultrafiltration and the unbound drug concentration in the supernatant or ultrafiltrate is measured. However, for drugs extensively bound to microsomal protein, the unbound drug concentration is extremely low and difficult to measure.

In such cases, the use of liquid chromatography–tandem mass spectroscopy (LC–MS/MS) is beneficial, since LC–MS/MS provides a sensitive and reproducible measurement compared with conventional methods such as LC–ultraviolet detection or LC–fluorescence detection (Hermo et al., 2008). Furthermore, particularly in the early stages of drug development, v_m is frequently determined by measuring the reduction in the amount of substrate, not the formation of metabolites, because of the difficulty in synthesizing enough amounts of many kinds of metabolites. In such cases, it is necessary to carry out the in vitro reaction so that substrate reduction can be accurately determined.

TABLE 6.1 Parameters Commonly Used for Prediction of Human Hepatic Clearance from in vitro Experiments

	Reported Value	Reference
Microsomal protein contained in the liver	52.5 mg/g liver	Iwatsubo et al., 1996
Volume of the liver	19.5 mL/kg	Wynne et al., 1989
Hepatic blood flow	0.95 mL/min/g liver	Montandon et al., 1975

The intrinsic metabolic hepatic clearance ($CL_{int, m, h}$) calculated in the above experiments is usually given in terms of microliters/minute/miligrams of microsomal protein and, therefore, should be converted to per liver, per body weight, or other usable units using reported physiological data (Roberts and Rowland 1986; Wynne et al., 1989; Iwatsubo et al., 1996) (Table 6.1), although the physiological data may vary somewhat depending on the report.

The typical scheme to predict the in vivo PK from in vitro data has been illustrated in Figure 6.6*a*. Many reports are available in which in vivo drug clearance were predicted from in vitro data (Houston and Galetin, 2008; Chiba et al., 2009). Among them, studies with YM796, a compound with antidementia activity, will be described as a typical example (Iwatsubo et al., 1997a, 1997b, 1998). YM796 is a substrate of CYP3A4 (Iwatsubo et al., 1997b), and species differences in PK properties are considerable; the BA of YM796 in rats is dose dependent, whereas in dogs the BA of YM796 is approximately 20% (Fig. 6.6*b*). In contrast, high AUC_{oral} values for YM796 are obtained in humans (Fig. 6.6*b*) (Iwatsubo et al., 1998). As shown in Figure 6.6*b*, the AUC_{oral} and BA of YM796 in humans, dogs, and rats can be accurately predicted by hepatic metabolic clearance determined in in vitro experiments with microsomes using the nonlinear dispersion model (Iwatsubo et al., 1998). It has also been demonstrated that the AUC_{oral} in humans can be quantitatively predicted using recombinant CYP enzymes. For YM796, the metabolic clearance in rodents is extremely high compared with that in humans at low doses (Iwatsubo et al., 1997b). During the process of new drug development, species differences in PK properties often hinder a project; thus, it is worthwhile to consider the prediction of human in vivo clearance based on the physiological PK theory.

In addition, the in vitro–in vivo extrapolation of biliary excretion clearance, the other pathway of hepatic drug elimination, can be achieved by use of a double-transfected Madin–Darby canine kidney II (MDCK II) monolayer stably expressing uptake and efflux transporters. Figure 6.7 illustrates that the observed in vivo biliary excretion clearance of compounds in rats may be predicted from in vitro experiments with an MDCK II monolayer expressing the uptake (rat organic anion transporting polypeptide 4; Oatp4/SLC21a10) and efflux (multidrug resistance-associated protein 2; Mrp2/ABCC2) transporters by assuming the appropriate common scaling factor (α). In addition to this combination, several types of doubly transfected cells, including cells

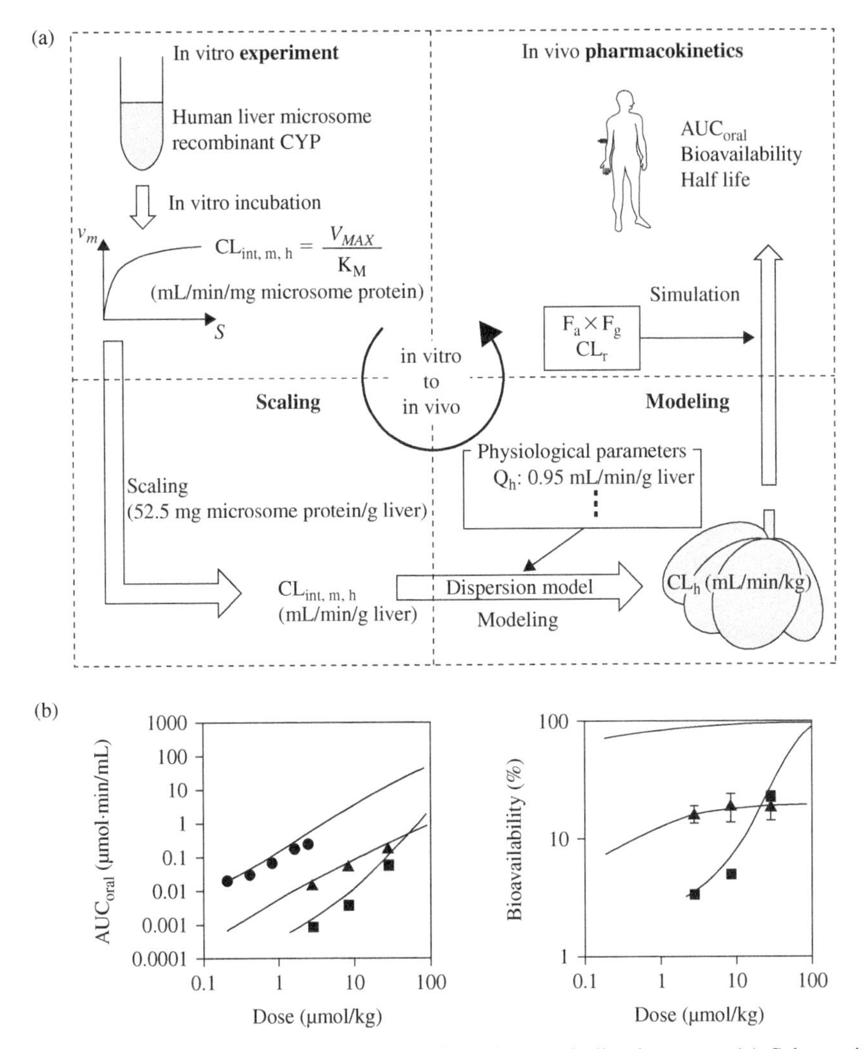

FIGURE 6.6 In vitro–in vivo scaling of hepatic metabolic clearance. (*a*) Schematic diagram illustrating the standard strategy of in vitro–in vivo scaling of hepatic metabolic clearance. Using the human microsomes, parameters for the metabolism of drugs are determined in vitro. The unit of these parameters are converted by considering the content of microsome in the liver. CL_h is calculated from $CL_{int,\ m,\ h}$ based on the dispersion model, using Eq. (6.14) in the text. Then, in vivo pharmacokinetic parameters can be predicted. (*b*) Comparison of the predicted and observed values for the oral AUC (left panel) and bioavailability (right panel) of YM796 in rats, dogs, and humans. The solid lines indicate the predicted AUC after oral administration (left panel) and bioavailability (right panel) of YM796. $CL_{int,\ h}$ was determined in in vitro experiments using liver microsomes from rats, dogs, and humans. Based on these in vitro values, the AUC after oral administration and bioavailability in these animal species and in humans were calculated as depicted in panel (*a*). Squares, triangles, and circles represent the observed in vivo data from rats, dogs, and humans, respectively, and bars indicate the mean ± SD. (reproduced with permission from Iwatsubo et al., 1998.)

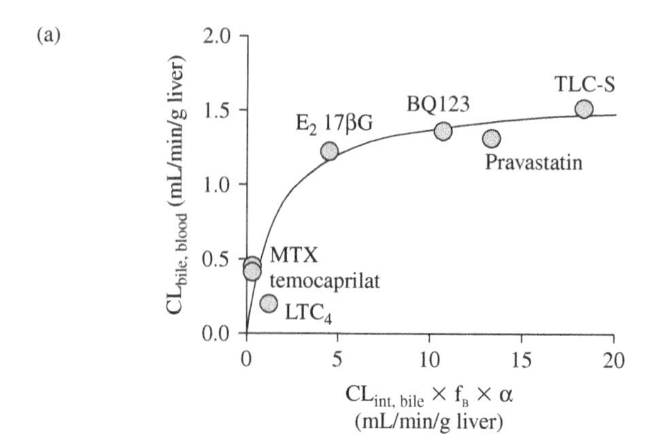

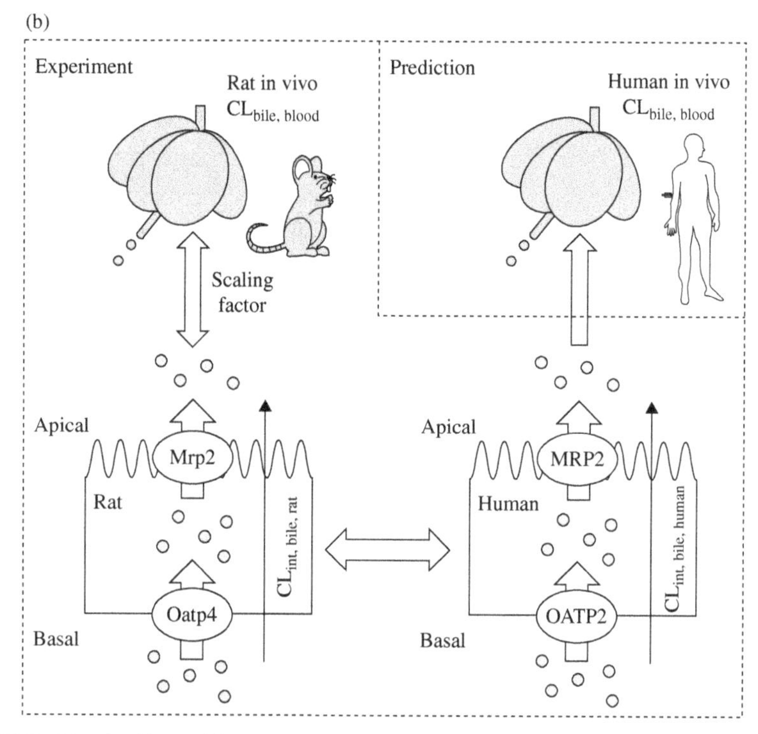

FIGURE 6.7 Scaling of in vitro transcellular transport clearance to in vivo biliary excretion clearance. (*a*) Relationships between in vivo biliary excretion clearance ($CL_{bile, blood}$, vertical axis) and in vitro transcellular transport clearance ($CL_{int, bile}$, horizontal axis) across the monolayer of double transfectant. The basal-to-apical and apical-to-basal flux of substrates was determined across the MDCK II monolayer expressing uptake (rat Oatp4/SLC21a10) and efflux (rat Mrp2/Abcc2) transporters. A scaling factor (α) of 17.9 was obtained by fitting predicted and observed values of rat biliary excretion clearance ($CL_{bile, blood}$) using the well-stirred model as presented in the figure. (reproduced with permission from Sasaki et al., 2004.) (*b*) Schematic demographics illustrating the strategy of in vitro–in vivo scaling of biliary excretion clearance. By considering the scaling factor obtained from the rat experiments, human in vivo biliary excretion clearances may be predicted in a quantitative manner from in vitro experiments using double transfectant of human transporters. Mrp2/Abcc2 and Oatp4/SLC21a10 are rat ortholog for human MRP2/ABCC2 and OATP2/SLCO1B1, respectively.

expressing human transporters, have been established and are used in predicting human biliary excretion clearance (Fig. 6.7*b*) (Sasaki et al., 2002). Although for some drugs it is necessary to take transport and metabolic processes into consideration simultaneously (Fig. 6.2), suitable in vitro models need to be established in future.

In addition, although many types of enzymes and transporters are responsible for drug elimination in the liver, it is important to consider the contribution of each enzyme and transporter in predicting human in vivo clearance. This aspect is discussed in Section 6.6.

6.5.4 Estimation of Renal Clearance

The human kidney is composed of approximately one million nephrons. There is a size barrier of approximately 30 kDa in the glomerulus, and compounds with a lower molecular weight than this undergo glomerular filtration. The glomerular filtration rate (GFR) is defined as the volume of plasma filtered and passed into the proximal tubule per unit time, which is approximately 100–120 mL/min in healthy adults. Tubular reabsorption or tubular secretion of compounds by transporters also takes place at the proximal and distal tubules (Kusuhara and Sugiyama, 2009). For example, glucose molecules are filtered by the glomerulus and reabsorbed in the tubule to avoid its urinary excretion.

By definition, CL_r is given by Eq. (6.30) (Jusko, 1992):

$$CL_r = \frac{v_E}{C_P} \tag{6.30}$$

where v_E represents the urinary excretion rate. By integration, Eq. (6.30) is converted to Eq. (6.31) (Jusko, 1992):

$$CL_r = \frac{X_E}{AUC} \tag{6.31}$$

where X_E represents the total amount of unchanged drug excreted in the urine. Although it is empirically known that the CL_r in humans is accurately predicted by allometric scaling (Kita et al., 1986), it is worth determining the CL_r in clinical studies. For the quantitative prediction of elimination processes, the clearance of drugs by glomerular filtration is given by $f_P \times GFR$ (Jusko, 1992).

In addition, a number of transporters expressed on the basal and apical sides of the proximal and distal tubules are responsible for tubular secretion (Fig. 6.8) (Kusuhara and Sugiyama, 2009); thus, it is necessary to determine the relative contribution of each transporter to elucidate the mechanism for the renal excretion of drugs. For this purpose, an uptake assay using kidney slices and cultured cells transfected with complementary deoxyribonucleic acid (cDNA) of transporters can be conducted (Hasegawa et al., 2003).

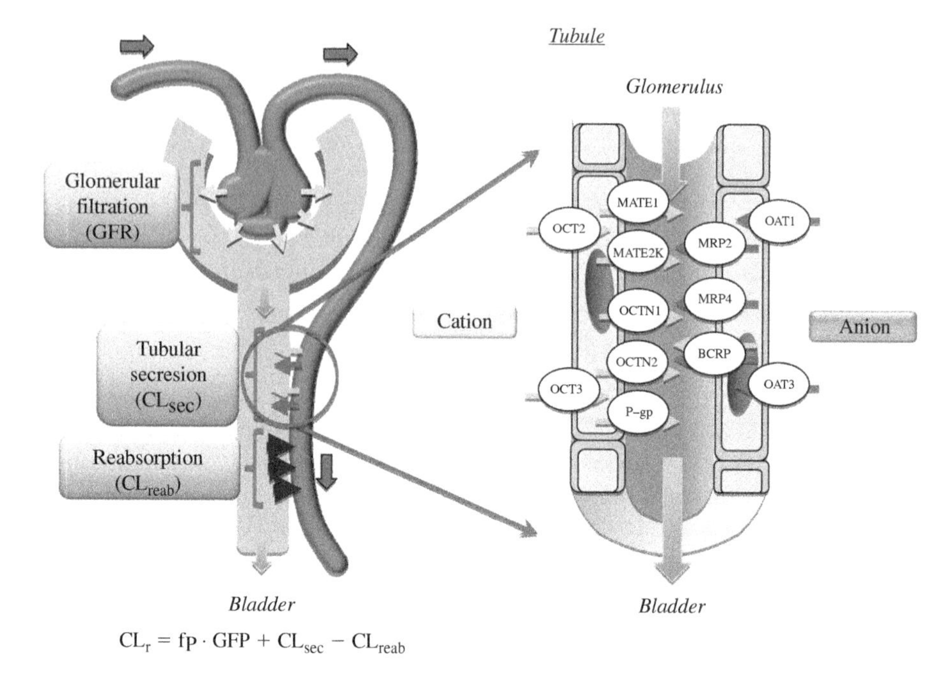

$CL_r = fp \cdot GFP + CL_{sec} - CL_{reab}$

FIGURE 6.8 Schematic diagram illustrating the mechanisms of urinary excretion. The left panel illustrates the structure of a nephron, and the right panel contains the enlarged illustration of the renal tubule. Drugs in plasma are excreted into the urine via glomerular filtration and tubular secretion. Some of the drugs are reabsorbed in the tubule. As shown in the right panel, many kinds of transporters are expressed in the tubule and mediate the urinary drug excretion. In general, lipophilic drugs filtered or secreted into the renal tubule are reabsorbed extensively into the bloodstream. OCT/SLC22A, organic cation transporter (OCT2/SLC22A2, OCT3/SLC22A3, OCTN1/SLC22A4, and OCTN2/SLC22A5); MATE/SLC47A, multidrug and toxin compound extrusion (MATE1/SLC47A1 and MATEK2/SLC47A2); OAT/SLC22A, organic anion transporter (OAT1/SLC22A6 and OAT3/SLC22A8); MRP/ABCC, multidrug-resistance-associated protein (MRP2/ABCC2 and MRP4/ABCC4); BCRP/ABCG2, breast-cancer-resistance-associated protein.

6.6 DRUG–DRUG INTERACTIONS

6.6.1 Importance of Determining the Contribution Ratio for Prediction of Drug–Drug Interactions

Among DDIs, the inhibition of drug-metabolizing enzymes is frequently observed, and in some cases this results in severe adverse effects due to the unexpected elevation of drug concentrations. Therefore, it is necessary to quantitatively predict the extent of DDIs of new drug candidates in drug development (Zhang et al., 2009). Because detailed PK theory and mathematical handling of enzyme inhibition is explained in the literature, the importance of the contribution ratio in DDIs is discussed in this chapter.

The extent of increase in the AUC_{oral} of a drug that is co-administered with an inhibitor of a drug-metabolizing enzyme (enzyme X) is described by Eq. (6.32) when extrahepatic clearance is minimum (Ito et al., 1998):

$$\frac{AUC_{oral,I}}{AUC_{oral,C}} = \frac{1}{f_{m,X} \times \frac{1}{1+[I]/K_i} + 1 - f_{m,X}} \quad (6.32)$$

where $AUC_{oral,\ I}$ and $AUC_{oral,\ C}$ represent AUC_{oral} after the same dose of a substrate drug with and without co-administration of an inhibitor, $f_{m,X}$ represents the proportion of intrinsic metabolic clearance of enzyme X ($CL_{int,\ m,\ X}$) to $CL_{int,\ h}$, [I] represents unbound concentration of inhibitor at the site of drug-metabolizing enzymes, and K_i represents the inhibition constant of the inhibitor. By definition, $f_{m,X}$ is given by

$$f_{m,\ X} = \frac{CL_{int,\ m,\ X}}{CL_{int,\ h}} \quad (6.33)$$

Figure 6.9 illustrates the relationship between an increase in the AUC ($AUC_{oral,\ I}/AUC_{oral,\ C}$) and $[I]/K_i$ with various $f_{m,X}$ (Ito et al., 2005a). Strategies for the estimation of $[I]/K_i$ are described in a previous report (Ito et al., 2004).

6.6.2 Methods for Determination of the Contribution Ratio

Obviously, $f_{m,X}$ has a significant impact on the intensity of DDIs, and the increase in the AUC is estimated to be not more than twofold when the $f_{m,X}$ is lower than 0.5 (Fig. 6.9). This means that if a drug is metabolized by multiple enzymes and the contribution of a particular CYP isozyme is not extensive (i.e., $f_{m,\ X}$ is not high enough), the AUC increase is moderate by selectively inhibiting the individual drug-metabolizing enzyme. Consequently, new drug candidates with a significant CL_r compared with their CL_h, or candidates that are metabolized by multiple hepatic drug-metabolizing enzymes, may be at low risk of DDIs due to inhibition of hepatic drug-metabolizing enzymes (Ito et al., 2005b).

As described above, $f_{m,X}$ is a crucial determinant of the extent of DDIs; therefore, it is important to estimate $f_{m,X}$ accurately. Although a number of reports discuss the methods for determining $f_{m,X}$, among them the relative activity factor (RAF) method seems to be promising. Figure 6.10 depicts the principles of the RAF method. First, the metabolic clearance of the specific/selective substrate (probe compound) for a CYP enzyme (e.g., testosterone or midazolam for CYP3A4) is determined using human liver microsomes and the recombinant CYP enzyme, and the ratio (i.e., RAF) of these values is calculated. Then, the metabolic clearance of test compounds is determined using the recombinant enzyme and extrapolated using the RAF value as a scaling factor. The RAF method has also been used to determine the

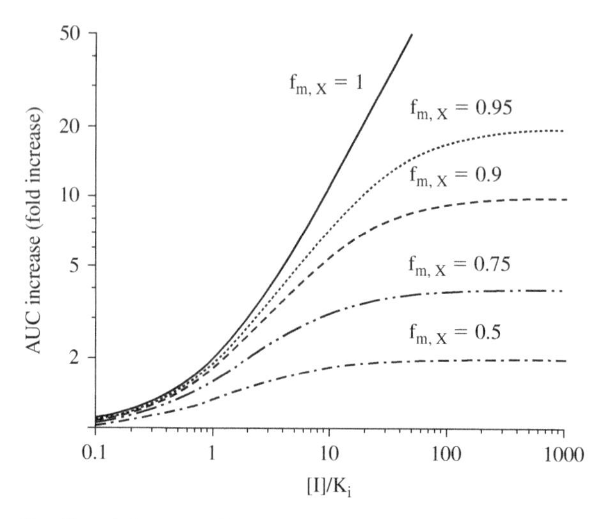

FIGURE 6.9 Predicted increase in AUC by co-administration of an inhibitor of drug-metabolizing enzymes. An increase in the AUC (fold increase) was simulated for five $f_{m,\ X}$ values (1, 0.95, 0.9, 0.75, and 0.5) as a function of $[I]/K_i$. It is indicated that the higher AUC increase is observed for substrates with higher $f_{m,\ X}$ values in the presence of inhibitors with higher $[I]/K_i$ values. (reproduced with permission from Ito et al., 2005a.)

contribution of each transporter in the hepatic and renal uptake of drugs (Hasegawa et al., 2003).

6.7 PRACTICAL ISSUES THAT NEED TO BE CONSIDERED

6.7.1 Evaluation of PK During the Exploratory Stage

One of the most important purposes of PK evaluation during the exploratory stage is to select new drug candidates with high BA. This is usually accomplished by investigating the metabolism stability of new drug candidates. However, pitfalls in selecting drug candidates with high BA need to be recognized. Candidates with high BA are often associated with high protein-binding potential to plasma protein and/or strong irreversible inhibitory activity for CYP enzymes. Currently, many pharmaceutical companies have become nervous about the irreversible binding of drug candidates to proteins including CYPs with regard to preventing hepatotoxicity (Evans et al., 2004; Masubuchi et al., 2007; Nakayama et al., 2009). In addition, selection of a drug candidate with an unusual PK profile, such as strong induction potential for drug-metabolizing enzymes, sometimes makes pharmacological or safety evaluation complicated and may result in delaying a project, even if the profile is only specific to experimental animals. In the course of selecting new drug candidates, such possibilities must also be considered during PK evaluations.

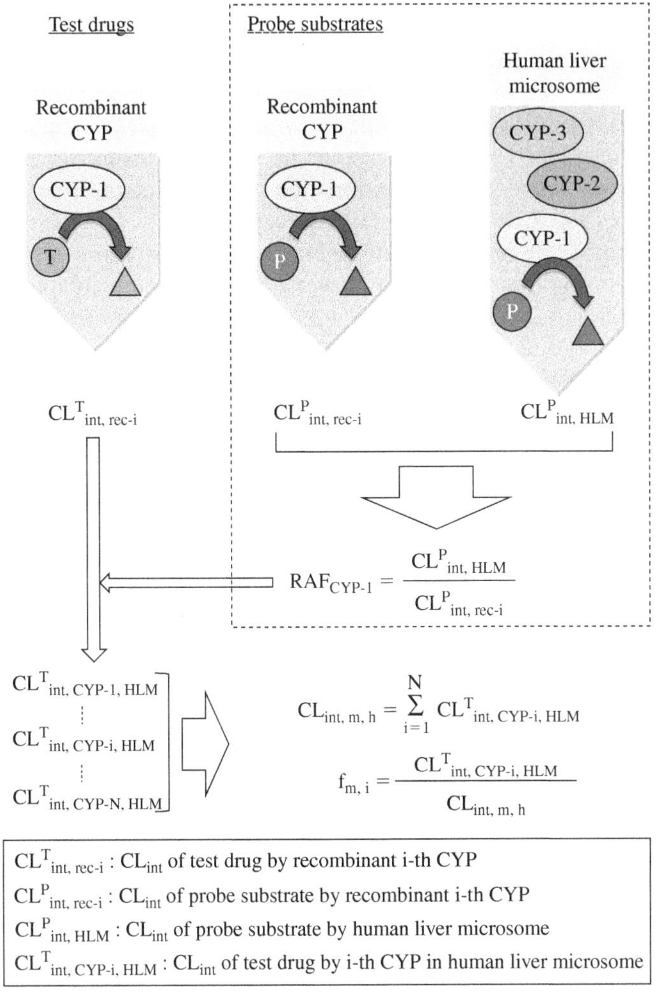

FIGURE 6.10 Schematic diagram illustrating the relative activity factor (RAF) methods. In general, drugs are metabolized by multiple enzymes. RAF method is one of the reliable methods to determine the contribution of *i*th enzyme to the hepatic metabolism of drugs ($f_{m,i}$). The procedure of caluculation of $f_{m,i}$ is described as follows: At first, in order to calculate the CL_{int} mediated by *i*th enzyme of test drugs in microsome ($CL^{T}_{int, CYP\text{-}i, HLM}$), intrinsic clearance of a test drug ($CL^{T}_{int, rec\text{-}i}$) is determined using recombinant CYP *i*th enzyme. $CL^{T}_{int, rec\text{-}i}$ is then multiplied by the RAF value for *i*th enzymes, which is calculated as the ratio between intrinsic clearances of a probe (selective) substrate determined using human liver microsomes ($CL^{P}_{int, HLM}$) and recombinant CYP *i*th eyzyme ($CL^{P}_{int, rec\text{-}i}$). The products are determined for each *i*th CYP enzyme and summed up to calculate the hepatic intrinsic metabolic clearance ($CL_{int, m, h}$).

From an aspect of PK/pharmacodynamic (PK/PD) relationships, the difference in time points to provide the maximum blood concentration of drugs and maximum efficacy or toxicity must be examined carefully. In cases where time gaps are considerable, exploration of appropriate biomarkers may be beneficial to clarify the underlying mechanisms. In such situations, it is also worth considering the irreversible inhibition of a target protein or markedly slow dissociation of the drug from the target receptors.

The evaluation of PK/PD relationships is an important subject in projects that are aimed to develop drugs with novel pharmacological activity. Careful consideration should be given as to whether the efficacy of a drug is dependent on blood or tissue concentration. For example, it is possible that unbound drug concentrations in tissues are not equivalent to those in blood because of the presence of a transport process across the plasma membrane. The tissue selectivity of efficacy may be accounted for by considering the selective uptake mediated by drug transporters and/or by considering the expression of subtypes of target proteins present in selective tissues. Furthermore, the possibility of formation of active metabolites should be considered concurrently.

6.7.2 Evaluation of PK During the Development Stage

The evaluation of PK during the drug development stage is significantly different in the exploratory stage. Consequently, along with the precise prediction of efficacy and safety of new drug candidate in humans, the evaluation procedure must be carried out to prove that the new drug application is reliable. In other words, precise evaluation of given endpoints is desired in the evaluation of PK during the development stage.

There are three trends that have recently become important considerations for the evaluation of new drug candidates promoted to the development stage. First, a strategy for the safety evaluation of metabolites has to be proposed in accordance with the guidance of the FDA (2008). As a general rule, the guidance requires the preclinical safety study of metabolites with the AUC value more than 10% of parent compounds in humans, when the metabolites are not detected in preclinical studies. Second, microdosing studies are becoming a feasible development strategy during drug development (Lappin and Garner, 2008). Although the use of radiolabeled compounds was considered to be required in microdosing studies, studies with unlabeled compounds have become possible in some cases because of the recent improvement in analytical methods. Third, as mentioned in the FDA White Paper (March 2004), the evaluation of validated biomarkers and application of modeling and simulation are emphasized to improve the success rate of drug development. To deal with these trends, researchers responsible for PK evaluation during the development stage are required to understand and apply techniques to strategically develop new drug candidates, their metabolites, and biomarkers. For these purposes, the use of LC–MS/MS may be essential.

ABBREVIATIONS AND NOTATIONS

Please note that the organ and body clearance values in this chapter (i.e., CL_h, CL_{oral}, CL_{org}, CL_r, and CL_{tot}) are defined for the plasma concentration, which are given by the velocity of elimination divided by the plasma drug concentration, except for $CL_{org,blood}$, which is defined for the blood drug concentration.

ABC	ATP binding cassette transporter super family
AGP	α1-acidic glycoprotein
AUC	Area under the plasma concentration–time curve
AUC_{iv}	AUC after intravenous administration
AUC_{oral}	AUC after oral administration
$AUC_{oral, C}$	AUC_{oral} without co-administration of inhibitor
$AUC_{oral, I}$	AUC_{oral} with co-administration of inhibitor
BA	Bioavailability
C_B	Drug concentration in blood
$C_{B, in}$	Concentration of drugs in inflow blood flow
$C_{B, out}$	Concentration of drugs in outflow blood flow
CL_h	Hepatic organ clearance
CL_{int}	Intrinsic clearance
$CL_{int, h}$	Hepatic intrinsic clearance
$CL_{int, m, h}$	Hepatic intrinsic metabolic clearance
$CL_{int, m, X}$	Hepatic intrinsic metabolic clearance of enzyme X
CL_{oral}	Total body clearance after oral administration
CL_{org}	Clearance of each organ
$CL_{org,blood}$	Blood concentration-based clearance of each organ
CL_r	Renal organ clearance
CL_{tot}	Total body clearance
C_P	Drug concentration in plasma
C_{RBC}	Drug concentrations in red blood cells
C_T	Drug concentration in tissue
CYP	Cytochrome P450
DDI	Drug–drug interaction
D_{iv}	Dose of drug administrated intravenously
D_N	Normalized dispersion number
D_{oral}	Dose of drug administrated orally
E	Extraction ratio
Ef	Efficacy of drug
F	Availability
F_a	Absorption ratio
f_B	Unbound fraction in blood
FDA	Foods and Drug Administration (United States)
F_g	Intestinal availability
F_h	Hepatic availability
$f_{m, X}$	Proportion of $CL_{int, m, X}$ to $CL_{int, h}$

f_P	Unbound fraction in plasma
f_T	Unbound fraction in tissue
GFR	Glomerular filtration rate
Hct	Hematocrit value
[I]	Unbound concentration of inhibitor nearby the drug-metabolizing enzyme
K_i	Inhibition constant of the inhibitor
K_M	Michaelis–Menten constant
K_P	Tissue-to-plasma drug concentration ratio
LC–MS/MS	Liquid chromatography–tandem mass spectroscopy
MDCK II	Madin–Darby canine kidney II cell
Mrp	Multidrug resistance associated protein
Oatp	Organic anion transporting polypeptide
P_{app}	Apparent permeability rate constant
PEPT	Peptide transporter
PK	Pharmacokinetics
PK/PD	Pharmacokinetics/pharmacodynamics
PS_1	Permeability surface area product for the influx across the sinusoidal membrane
PS_2	Permeability surface area product for the efflux across the sinusoidal membrane
PS_3	Permeability surface area product for the excretion across the canalicular membrane
Q_B	Blood flow rate of each organ
Q_h	Hepatic blood flow rate
RAF	Relative activity factor
R_B	Blood-to-plasma concentration ratio
S	Unbound concentration of substrate drug in in vitro experiment
SLC	Solute carrier family
v	Velocity of elimination of drug
V_B	Volume of blood
V_d	Volume of distribution
v_{E}	Urinary excretion rate
v_{m}	Initial metabolic velocity
V_{max}	Maximum metabolic velocity
V_P	Volume of plasma
V_{T}	Volume of tissue
X	Amount of drug in the organ or tissue
X_E	Total amount of drug excreted into urine as unchanged form

REFERENCES

Adachi Y, Suzuki H, Sugiyama Y. Quantitative evaluation of the function of small intestinal P-glycoprotein: Comparative studies between in situ and in vitro. Pharm Res 2003;20:1163–1169.

Adolph EF. Quantitative relations in the physiological constitutions of mammals. Science 1949;109:579–585.

Bell NH, Johnson RH. Bisphosphonates in the treatment of osteoporosis. Endocrine 1997;6:203–206.

Brandsch M. Transport of drugs by proton-coupled peptide transporters: Pearls and pitfalls. Expert Opin Drug Metab Toxicol 2009;5:887–905.

Buxton ILO. Pharmacokinetics and pharmacodynamics: The dynamics of drug absorption, distribution, action, and elimination. In Goodman & Gilman's The Pharmacological Basis of Therapeutics, Laurence L, Brunton JSLaKLP, Eds. McGraw-Hill, New York, 2006, pp. 1–40.

Caspary WF. Physiology and pathophysiology of intestinal absorption. Am J Clin Nutr 1992;55:299S–308S.

Chiba M, Ishii Y, Sugiyama Y. Prediction of hepatic clearance in human from in vitro data for successful drug development. AAPS J 2009;11:262–276.

Doherty MM, Charman WN. The mucosa of the small intestine: How clinically relevant as an organ of drug metabolism? Clin Pharmacokinet 2002;41:235–253.

Evans DC, Watt AP, Nicoll-Griffith DA, Baillie TA. Drug-protein adducts: An industry perspective on minimizing the potential for drug bioactivation in drug discovery and development. Chem Res Toxicol 2004;17:3–16.

Fagerholm U. Prediction of human pharmacokinetics—Evaluation of methods for prediction of volume of distribution. J Pharm Pharmacol 2007;59:1181–1190.

Food and Drug Administration (FDA), Guidance for Industry: Safety Testing of Drug Metabolites, FDA, Washington, DC, 2008.

Food and Drug Administration (FDA), White Paper: Innovation or Stagnation? Challenge and Opportunity on the Critical Path to New Medical Products. U.S. Department of Health and Human Services, FDA, Washington, DC, 2004.

Frank R, Hargreaves R. Clinical biomarkers in drug discovery and development. Nat Rev Drug Discov 2003;2:566–580.

Fuse E, Tanii H, Asanome K, Kurata N, Kobayashi H, Kuwabara T, Kobayashi S, Sugiyama Y. Altered pharmacokinetics of a novel anticancer drug, UCN-01, caused by specific high affinity binding to alpha 1-acid glycoprotein in humans. Cancer Res 1999;59:1054–1060.

Fuse E, Kobayashi T, Inaba M, Sugiyama Y. Prediction of the maximal tolerated dose (MTD) and therapeutic effect of anticancer drugs in humans: Integration of pharmacokinetics with pharmacodynamics and toxicodynamics. Cancer Treat Rev 1995;21:133–157.

Galetin A, Gertz M, Houston JB. Potential role of intestinal first-pass metabolism in the prediction of drug-drug interactions. Expert Opin Drug Metab Toxicol 2008;4:909–922.

Galetin A, Houston JB. Intestinal and hepatic metabolic activity of five cytochrome P450 enzymes: Impact on prediction of first-pass metabolism. J Pharmacol Exp Ther 2006;318:1220–1229.

Hasegawa M, Kusuhara H, Endou H, Sugiyama Y. Contribution of organic anion transporters to the renal uptake of anionic compounds and nucleoside derivatives in rat. J Pharmacol Exp Ther 2003;305:1087–1097.

Hermo MP, Nemutlu E, Kir S, Barrón D, Barbosa J. Improved determination of quinolones in milk at their MRL levels using LC-UV, LC-FD, LC-MS and LC-MS/MS and validation in line with regulation 2002/657/EC. Anal Chim Acta 2008;613:98–107.

Houston JB, Galetin A. Methods for predicting in vivo pharmacokinetics using data from in vitro assays. Curr Drug Metab 2008;9:940–951.

Hubatsch I, Ragnarsson EG, Artursson P. Determination of drug permeability and prediction of drug absorption in Caco-2 monolayers. Nat Protoc 2007;2:2111–2119.

Ito K, Brown HS, Houston JB. Database analyses for the prediction of in vivo drug-drug interactions from in vitro data. Br J Clin Pharmacol 2004;57:473–486.

Ito K, Hallifax D, Obach RS, Houston JB. Impact of parallel pathways of drug elimination and multiple cytochrome P450 involvement on drug-drug interactions: CYP2D6 paradigm. Drug Metab Dispos 2005a;33:837–844.

Ito K, Iwatsubo T, Kanamitsu S, Ueda K, Suzuki H, Sugiyama Y. Prediction of pharmacokinetic alterations caused by drug-drug interactions: Metabolic interaction in the liver. Pharmacol Rev 1998;50:387–412.

Ito K, Suzuki H, Horie T, Sugiyama Y. Apical/basolateral surface expression of drug transporters and its role in vectorial drug transport. Pharm Res 2005b;22: 1559–1577.

Iwatsubo T, Hirota N, Ooie T, Suzuki H, Sugiyama Y. Prediction of in vivo drug disposition from in vitro data based on physiological pharmacokinetics. Biopharm Drug Dispos 1996;17:273–310.

Iwatsubo T, Hisaka A, Suzuki H, Sugiyama Y. Prediction of in vivo nonlinear first-pass hepatic metabolism of YM796 from in vitro metabolic data. J Pharmacol Exp Ther 1998;286:122–127.

Iwatsubo T, Suzuki H, Shimada N, Chiba K, Ishizaki T, Green CE, Tyson CA, Yokoi T, Kamataki T, Sugiyama Y. Prediction of in vivo hepatic metabolic clearance of YM796 from in vitro data by use of human liver microsomes and recombinant P-450 isozymes. J Pharmacol Exp Ther 1997a;282:909–919.

Iwatsubo T, Suzuki H, Sugiyama Y. Prediction of species differences (rats, dogs, humans) in the in vivo metabolic clearance of YM796 by the liver from in vitro data. J Pharmacol Exp Ther 1997b;283:462–469.

Jusko WJ. Guidelines for collection and analysis of pharmacokinetic data. In Applied Pharmacokinetics, William E, Evans JJSaWJJ, Eds. Applied Therapeutics, Vancouver, WA, 1992, pp. 2-1–2-44.

Kato M, Chiba K, Hisaka A, Ishigami M, Kayama M, Mizuno N, Nagata Y, Takakuwa S, Tsukamoto Y, Ueda K, Kusuhara H, Ito K, Sugiyama Y. The intestinal first-pass metabolism of substrates of CYP3A4 and P-glycoprotein-quantitative analysis based on information from the literature. Drug Metab Pharmacokinet 2003;18:365–372.

Kita Y, Fugono T, Imada A. Comparative pharmacokinetics of carumonam and aztreonam in mice, rats, rabbits, dogs, and cynomolgus monkeys. Antimicrob Agents Chemother 1986;29:127–134.

Kitamura S, Maeda K, Sugiyama Y. Recent progresses in the experimental methods and evaluation strategies of transporter functions for the prediction of the pharmacokinetics in humans. Naunyn Schmiedebergs Arch Pharmacol 2008;377:617–628.

Kolars JC, Awni WM, Merion RM, Watkins PB. First-pass metabolism of cyclosporin by the gut. Lancet 1991;338:1488–1490.

Komura H, Iwaki M. Species differences in in vitro and in vivo small intestinal metabolism of CYP3A substrates. J Pharm Sci 2008;97:1775–1800.

Ku MS. Use of the Biopharmaceutical Classification System in early drug development. AAPS J 2008;10:208–212.

Kusuhara H, Sugiyama Y. In vitro–in vivo extrapolation of transporter-mediated clearance in the liver and kidney. Drug Metab Pharmacokinet 2009;24:37–52.

Lappin G, Garner RC. The utility of microdosing over the past 5 years. Expert Opin Drug Metab Toxicol 2008;4:1499–1506.

Lentz KA. Current methods for predicting human food effect. AAPS J 2008;10: 282–288.

Lilja JJ, Kivisto KT, Neuvonen PJ. Grapefruit juice–simvastatin interaction: Effect on serum concentrations of simvastatin, simvastatin acid, and HMG-CoA reductase inhibitors. Clin Pharmacol Ther 1998;64:477–483.

Liu L, Pang KS. An integrated approach to model hepatic drug clearance. Eur J Pharm Sci 2006;29:215–230.

Masubuchi N, Makino C, Murayama N. Prediction of in vivo potential for metabolic activation of drugs into chemically reactive intermediate: Correlation of in vitro and in vivo generation of reactive intermediates and in vitro glutathione conjugate formation in rats and humans. Chem Res Toxicol 2007;20:455–464.

Montandon B, Roberts RJ, Fischer LJ. Computer simulation of sulfobromophthalein kinetics in the rat using flow-limited models with extrapolation to man. J Pharmacokinet Biopharm 1975;3:277–290.

Mueller EA, Kovarik JM, van Bree JB, Grevel J, Lucker PW, Kutz K. Influence of a fat-rich meal on the pharmacokinetics of a new oral formulation of cyclosporine in a crossover comparison with the market formulation. Pharm Res 1994;11:151–155.

Nakayama S, Atsumi R, Takakusa H, Kobayashi Y, Kurihara A, Nagai Y, Nakai D, Okazaki O. A zone classification system for risk assessment of idiosyncratic drug toxicity using daily dose and covalent binding. Drug Metab Dispos 2009;37:1970–1977.

Naritomi Y, Terashita S, Kagayama A, Sugiyama Y. Utility of hepatocytes in predicting drug metabolism: Comparison of hepatic intrinsic clearance in rats and humans in vivo and in vitro. Drug Metab Dispos 2003;31:580–588.

Naritomi Y, Terashita S, Kimura S, Suzuki A, Kagayama A, Sugiyama Y. Prediction of human hepatic clearance from in vivo animal experiments and in vitro metabolic studies with liver microsomes from animals and humans. Drug Metab Dispos 2001;29:1316–1324.

O'Connor P, Feely J Clinical pharmacokinetics and endocrine disorders. Therapeutic implications. Clin Pharmacokinet 1987;13:345–364.

Pang KS, Rowland M. Hepatic clearance of drugs. I. Theoretical considerations of a "well-stirred" model and a "parallel tube" model. Influence of hepatic blood flow, plasma and blood cell binding, and the hepatocellular enzymatic activity on hepatic drug clearance. J Pharmacokinet Biopharm 1977a;5:625–653.

Pang KS, Rowland M. Hepatic clearance of drugs. II. Experimental evidence for acceptance of the "well-stirred" model over the "parallel tube" model using lidocaine in the perfused rat liver in situ preparation. J Pharmacokinet Biopharm 1977b;5:655–680.

Roberts MS, Rowland M. A dispersion model of hepatic elimination: 1. Formulation of the model and bolus considerations. J Pharmacokinet Biopharm 1986;14:227–260.

Rowland M, Benet LZ, Graham GG. Clearance concepts in pharmacokinetics. J Pharmacokinet Biopharm 1973;1:123–136.

Sasaki M, Suzuki H, Ito K, Abe T, Sugiyama Y. Transcellular transport of organic anions across a double-transfected Madin-Darby canine kidney II cell monolayer expressing both human organic anion-transporting polypeptide (OATP2/SLC21A6) and multidrug resistance-associated protein 2 (MRP2/ABCC2). J Biol Chem 2002;277:6497–6503.

Sasaki M, Suzuki H, Aoki J, Ito K, Meier PJ, Sugiyama Y. Prediction of in vivo biliary clearance from the in vitro transcellular transport of organic anions across a double-transfected Madin-Darby canine kidney II monolayer expressing both rat organic anion transporting polypeptide 4 and multidrug resistance associated protein 2. Mol Pharmacol 2004;66:450–459.

Sawada Y, Hanano M, Sugiyama Y, Iga T. Prediction of the disposition of nine weakly acidic and six weakly basic drugs in humans from pharmacokinetic parameters in rats. J Pharmacokinet Biopharm 1985;13:477–492.

Shah P, Jogani V, Bagchi T, Misra A. Role of Caco-2 cell monolayers in prediction of intestinal drug absorption. Biotechnol Prog 2006;22:186–198.

Shitara Y, Horie T, Sugiyama Y. Transporters as a determinant of drug clearance and tissue distribution. Eur J Pharm Sci 2006;27:425–446.

Singh SS. Preclinical pharmacokinetics: An approach towards safer and efficacious drugs. Curr Drug Metab 2006;7:165–182.

Thummel KE, O'Shea D, Paine MF, Shen DD, Kunze KL, Perkins JD, Wilkinson GR. Oral first-pass elimination of midazolam involves both gastrointestinal and hepatic CYP3A-mediated metabolism. Clin Pharmacol Ther 1996;59:491–502.

Tozer TN. Concepts basic to pharmacokinetics. Pharmacol Ther 1981;12:109–131.

Wagner JG. In Fundamentals of Clinical Pharmacokinetics, Wagner JG, Ed. Drug Intelligence Publication, Hamilton, IL, 1975.

Wienkers LC, Heath TG. Predicting in vivo drug interactions from in vitro drug discovery data. Nat Rev Drug Discov 2005;4:825–833.

Wu CY, Benet LZ, Hebert MF, Gupta SK, Rowland M, Gomez DY, Wacher VJ. Differentiation of absorption and first-pass gut and hepatic metabolism in humans: Studies with cyclosporine. Clin Pharmacol Ther 1995;58:492–497.

Wynne HA, Cope LH, Mutch E, Rawlins MD, Woodhouse KW, James OF. The effect of age upon liver volume and apparent liver blood flow in healthy man. Hepatology 1989;9:297–301.

Yamamoto T, Itoga H, Kohno Y, Nagata K, Yamazoe Y. Prediction of oral clearance from in vitro metabolic data using recombinant CYPs: Comparison among well-stirred, parallel-tube, distributed and dispersion models. Xenobiotica 2005;35:627–646.

Yamashita S, Tanaka Y, Endoh Y, Taki Y, Sakane T, Nadai T, Sezaki H. Analysis of drug permeation across Caco-2 monolayer: Implication for predicting in vivo drug absorption. Pharm Res 1997;14:486–491.

Yoshigae Y, Imai T, Aso T, Otagiri M. Species differences in the disposition of propranolol prodrugs derived from hydrolase activity in intestinal mucosa. Life Sci 1998;62:1231–1241.

Zhang L, Zhang YD, Zhao P, Huang SM. Predicting drug-drug interactions: an FDA perspective. AAPS J 2009;11:300–306.

7 Drug Metabolism Research as Integral Part of Drug Discovery and Development Processes

W. GRIFFITH HUMPHREYS
Department of Biotransformation, Bristol-Myers Squibb Research & Development, Princeton, New Jersey

7.1 INTRODUCTION

The understanding that drug metabolites often play a critical role in the efficacy and side effect profile of drugs has propelled drug metabolism research to being an integral part of the lead optimization and development phases of modern

---

Mass Spectrometry in Drug Metabolism and Disposition: Basic Principles and Applications, First Edition. Edited by Mike S. Lee and Mingshe Zhu.

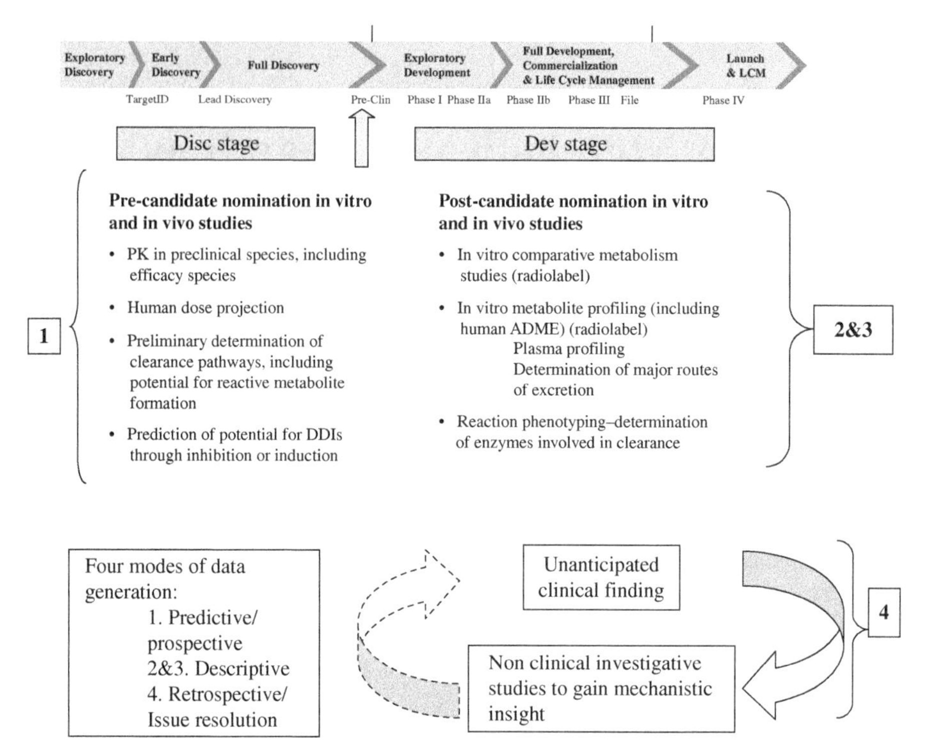

FIGURE 7.1 ADME data generation modes segregated by phase during drug discovery and development.

drug research. The importance of determination of metabolite profiles has been followed hand-in-hand by dramatic developments in the tools necessary to perform this research.

Work in the area can be divided into four phases that follow a drug's development path and have slightly different goals (Fig. 7.1). The phases can be described as follows: (1) design and optimization phase to aid in discovering the best molecule, (2) initial characterization and prediction phase to aid in compound selection and early development planning, (3) descriptive phase where full absorption, distribution, metabolism, and excretion (ADME) profile of the compound is determined and used to optimize clinical and safety plans, and (4) retrospective studies designed to help understand unexpected clinical or toxicological findings (Fig. 7.1).

The goal of the work in the discovery phase is to attempt to optimize the metabolism properties of clinical candidates in parallel with the optimization of the potency and efficacy. Drugs that have rapid metabolic clearance and low absolute oral bioavailability (%F) due to high preabsorptive and first-pass metabolism are likely to have a high degree of inter- and intrapatient variability (Hellriegel et al., 1996) and be more likely to suffer from drug–drug interactions.

Also, drugs with rapid clearance and low %F are likely to require suboptimal dosing regimes, that is, twice daily (bid) or three times per day (tid), and/or relatively high doses. The former leads to poor patient compliance, suboptimal efficacy, and marketing issues and the later can lead to unanticipated toxicities due to a large flux of drug and drug metabolites. Other benefits to understanding and optimizing the metabolism of new chemical entity (NCE)s are:

1. A decreased clearance may translate into a lower overall dose.
2. Lower rates of formation and overall amounts of reactive intermediates that may mediate acute or idiosyncratic toxicities.
3. Increased pharmacokinetic half-life ($t_{1/2}$) that will hopefully translate into a longer duration of action, less frequent dosing, and better patient compliance.
4. Better understanding of the extrapolation of animal data to humans, making human dose projections more reliable and reducing risk upon entry into clinical development.
5. Lower risk of drug–drug interactions, even if the compound is still dependent on metabolism for the bulk of its clearance, low-clearance drugs are less susceptible in drug–drug interaction caused by the co-administration of inhibitor.
6. Lower risk of drug–food interactions due to the reduced dose.
7. Decreased formation of metabolites that may have pharmacological activity against the target or may have significant off-target activity.

These considerations make it important to understand the metabolism characteristics of candidate drug molecules and to optimize these characteristics during the preclinical stage of drug development. Drug metabolism plays a central role in modern drug discovery, and candidate optimization and recent reviews have detailed how metabolism has impacted the discovery and design process and challenges that the field faces in the future (Baillie, 2006; Caldwell et al., 2009; Tang and Lu, 2009; Zhang et al., 2009b, Sun and Scott, 2010).

Work in the late discovery phase after candidate selection involves the further characterization of the selected drug to allow prediction of clinical pharmacokinetic and metabolism properties, which may include prediction of efficacious human dose (Huang et al., 2008). These properties are important in the design of regulatory toxicology and first-in-man studies and can affect things such as choice of toxicology species, dose selection, and important metabolites to monitor in early studies (Walker et al., 2009; Zhu et al., 2009; Baillie, 2009).

The development phases shifts from a predictive, prospective nature trying to influence drug design to a descriptive study of the complete metabolic profile of the candidate drug. Work done in this phase has significant impact on both the safety plan as well as the clinical pharmacology plan. Additional characterization of the metabolism of the drug after all the metabolic pathways have been determined can yield important information on potential for drug–drug interactions and variability in special populations.

This chapter will give an overview of metabolism-related topics that are often incorporated into discovery and development efforts in an effort to advance the optimal drug candidate into clinical investigations and then fully characterize it as part of development efforts. Many of the topics covered briefly in this chapter will also be covered in more depth in other chapters of this book.

7.2 METABOLIC CLEARANCE

7.2.1 General

The metabolic clearance of NCEs is most often studied with a combination of in vitro and in vivo approaches. There are several in vivo approaches that can be used to study metabolism in preclinical species, and these along with in vitro results can often shed mechanistic insight onto the problems associated with rapid metabolic clearance and incomplete oral bioavailability due to first-pass metabolism. Modern liquid chromatography–tandem mass spectrometry (LC–MS/MS) measurement of plasma drug concentrations provide a rapid tool to assess oral bioavailability of new candidate compounds and allows for early definition of bioavailability problems. When bioavailability concerns do arise, there are several avenues discussed below that can be followed to isolate the factor(s) limiting the oral delivery of a compound.

In vitro experiments often provide valuable insight into the clearance mechanisms for NCEs. The experiments that are most often employed in tandem to understand bioavailablilty are determinations of compound solubility, membrane permeability, and stability in subcellular fractions. The subcellular fractions most often employed are plasma (for ester-containing compounds) and liver subcellular fractions with the addition of either nicotinamide adenine dinucleotide phosphate (NADPH) or UDP–glucuronic acid (UDPGA) as cofactor. Hepatocytes are also a very useful in vitro tool that provides a more complete system for studying metabolism. All of the assays mentioned can be set up in an automated, medium-throughput system; however, all require a specific assay to measure compound concentration at the end of the assay, which places significant limitations on throughput. There have been recent attempts to solve the throughput problems inherent with this type of assay by developing generic endpoint assays for metabolic stability, but there has been no clear solution to date.

7.2.2 Prediction of Human Clearance

The rapid determination of pharmacokinetic parameters, solubility, permeability, and in vitro stability in plasma or liver tissue can often provide a reasonable explanation of the mechanisms limiting oral bioavailability. The in vitro rate of metabolism is often used to estimate the hepatic clearance using in vitro–in vivo correlation methodology (Rane et al., 1977; Lave et al., 1999; Obach et al., 1999; Pelkonen and Turpeinen, 2007; Barter et al., 2007; Houston and Galetin, 2008; Stringer et al., 2009; Klopf and Worboys, 2010; Hallifax and Houston, 2009). These methods use in vitro kinetic parameters, usually V_{max}/K_m or in vitro $t_{1/2}$, to determine an intrinsic clearance, which is then scaled to hepatic clearance using amount of tissue in the in vitro incubation, the weight of the liver, and the well-stirred model for hepatic clearance.

Generation of the data used for scaling to determine predicted clearance is often dependent on the stage of a program, with predictions on early-stage compounds being accomplished with in vitro $t_{1/2}$ data and later predictions done with more rigorous data such as Michaelis–Menten parameters (Kramer and Tracy, 2008). Care should be taken in the design of all experiments with reasonable kinetic assumptions made and full knowledge of the limitation of the predictions (Kramer and Tracy, 2008). While full Michaelis–Menten kinetic parameters along with corrections for microsomal protein binding and plasma protein binding (Grime and Riley, 2006) may be considered the gold standard, practical limitations often dictate the use of less rigorous data. An important step in the scaling exercise often involves the prediction of animal clearance from in vitro data and comparison to measured clearance in that species. This process can be used with all preclinical species and can add confidence to a prediction of human clearance.

This method has most often been applied in conjunction to rates of cytochrome P450 (CYP)-mediated oxidative metabolism in microsome incubation systems but can also be applied to conjugation (Fisher et al., 2000; Lin and Wong, 2002; Soars et al., 2002; Kilford et al., 2009; Miners et al., 2010), data from expressed CYP enzymes (Yamamoto et al., 2005; Stringer et al., 2009), flavin-containing monooxygenase (FMO)-catalyzed reactions (Fisher et al., 2002), data derived from hepatocytes (McGinnity et al., 2004; Soars et al., 2007b; Chao et al., 2009), or other in vitro systems (Ghibellini et al., 2007). Data generated with hepatocytes has also been used to predict clearance via hepatic uptake (Soars et al., 2007a).

7.2.3 In vivo Methods to Study Metabolism

The methods outlined above may not lead to a satisfactory determination of mechanism of incomplete bioavailability, and additional methods may be necessary in some cases to fully characterize the factors responsible. Bile duct cannulated animals provide a powerful model to examine incomplete bioavailability issues, and the rat provides the most flexibility because terminal studies can routinely be done. For compounds that are thought to have dissolution limitations or instability in the gastrointestinal tract, the GI tract can be removed at the end of the experiment and the contents assayed for drug and metabolites. The amount of parent in bile and urine can be quantitated by LC-MS/MS, and this method can also be used for any metabolites where authentic standards have been prepared. Alternatively, bile and urine metabolites can be estimated with high-pressure liquid chromatography–ultraviolet (HPLC– UV) spectroscopy using the extinction coefficient of the parent. These measurements will begin to define the total absorption of the compound and how that relates to the systemic bioavailabilty. The availability of radiolabeled compound at this point makes the bile duct cannulated rat experiment especially powerful.

A second in vivo model system that is very useful in sorting through problems of low oral bioavailability is portal vein cannulated animals. There are two ways this experiment can be conducted to determine hepatic extraction: (1) measure systemic plasma concentration after oral, portal vein, and systemic administration and (2) measure portal vein and hepatic vein concentrations after an oral dose. Both methods yield information on hepatic extraction and the percentage of dose reaching the portal circulation (the product of the fraction absorbed and the fraction metabolized by the gut wall).

Both the bile duct cannulated model and the portal vein cannulated model can be combined with a number of methods of modulating absorption or metabolism to ask specific mechanistic questions regarding stability, permeability, and metabolism. The following are several methods that can be used to modulate metabolism in vivo. The most commonly used method to modulate oxidative metabolism in vivo is to co-administer either ketoconazole or 1-aminobenzotriazole (Balani et al., 2002) to inhibit CYP enzymes. Both of these compounds will inhibit intestinal and liver CYP enzymes after oral administration. Care must be exercised when using ketoconazole for this purpose because the compound also has effects on transporters, and this effect may make interpretation of results ambiguous. Alternatively, a CYP inducer can be co-administered to determine the effect on the clearance of the compound of interest. This experiment can be accomplished with one of many known inducers of CYP enzymes. For inhibition of esterase enzyme activity, *bis*-[*p*-nitrophenyl] phosphate (BNPP) can be co-administered because this compound does not inhibit cholinesterase activity (Buch et al., 1969). Methods for inhibiting conjugation enzymes in vivo are not as well defined since in most cases good inhibitors of the enzymes have not been identified. The depletion of cofactor stores is possible by co-administering large amounts of substrate to ask mechanistic questions, although this method is most easily applied to sulfotransferase (SULT) enzymes (Kim et al., 1995). Many conjugation enzymes are subject to induction, which may be an avenue available for modulating enzyme activities in vivo.

7.2.4 Screening Strategies

The approach often taken in candidate optimization is to try to isolate bioavailability problems for a member of a chemotype of interest with a combination of in vitro and in vivo experiments and then to try to devise rapid techniques to screen for the liability. This approach relies on a good deal of up-front work to fully understand the bioavailability limitations and periodic checking of the property to ensure that the screen is providing reliable results. Systematic decision trees can be employed to allow clear pathways for evaluation of new compounds. Alternatively, screens can be run in parallel so that all information is generated for all compounds. This strategy makes work flow simpler and increases efficiency but can lead to information overload and complex decision making.

For metabolic stability screens to be most effective, the screens must be tightly linked to some means of gathering information on metabolite structure. Methods for rapidly determining metabolite molecular weight and limited structural information have improved dramatically and allow this approach to be routinely employed (Anari and Baillie, 2005; Wrona et al., 2005, Tiller et al., 2008a; Tolonen et al., 2009; Hsieh and Korfmacher, 2009). The goal of this type of approach is to allow the identification of metabolic "soft spots," which can then be altered to produce compounds with improved metabolic stability. Reviews are available on successes and challenges inherent in the application of a structural modification paradigm to increase stability (Thompson, 2001; Nassar et al., 2004; Fisher et al., 2006; Zhang et al., 2009b).

7.3 METABOLITE PROFILING/MASS BALANCE STUDIES

Although advances in LC-MS/MS technology have made the determination of metabolic profiles without radiolabeled material more sensitive and reliable, methods for obtaining reliably quantitative profiles using cold material are still lacking. The use of radiolabeled material for the rigorous determination of metabolic profiles remains the "gold standard." This simply sets up a situation similar to that found in bioanalytical analysis where there is a progression from early discovery bioanalytical assays to fully validated quantitative assay in which early information on metabolite profiles is qualitative to semiquantitative in nature and progresses to the detailed work with radiolabel and/or authentic metabolite standards. It is important to match the needs of the compound with the type of information provided to properly answer questions while doing cost-disciplined science.

Early in vitro metabolite profiling experiments are designed to evaluate interspecies differences in profiles and determine specific metabolite pathways to aid in (1) prediction of the validity of toxicology and/or efficacy models and (2) determine metabolite soft spots leading to unfavorable metabolism (usually manifest as overly rapid metabolic clearance, discussed above) or "hot spots" leading to reactive metabolite formation to aid in the development of improved analogs. A major recent focus of improvements in profiling methods has been in decreasing analysis times using fast HPLC and faster scanning mass spectrometers. Others have developed powerful integrated profiling methods that allow parent quantitation in the same run as metabolite profiling (Bateman et al., 2009; Zhang et al., 2009a). New highly sensitive triple quadrupole and ion trap mass spectrometers along with techniques such as neutral loss and product ion scanning allow routine detection of metabolites even from complex matrices (Yao et al., 2009). Accurate mass–mass spectrometers coupled with mass filters and other data–mining techniques have become an important new technology for metabolite profiling (Ruan et al., 2008; Zhang et al., 2009a, 2009c).

An important consideration for profiling experiments with non-radiolabeled drug is whether or not to generate semiquantitative information and, if so the method used to get the semiquantitative estimates of metabolite concentration. Simple generation of qualitative information can be done rapidly, but the information generated in this fashion complicates the translation of the information into meaningful actions in the medicinal chemistry program, that is, creation of structure–metabolism relationships. The use of response data from SIM profiles of metabolite relative to parent is simple to collect but may be misleading, even with improvements such as the use of nanoflow LC. This leaves UV detection as the generic method of choice for semiquantitative estimations, but care should be exercised in that extinction coefficients can be significantly altered through biotransformation reactions.

Profiling with radiolabeled drug is used to produce definitive information for in vitro and in vivo studies. These studies generally involve HPLC separation of metabolites followed by quantitation by flow, microplate, or standard scintillation counting. There continues to be advances in technology that improves the coupling of mass spectrometers in line with radioactive flow detectors possible so that metabolites can be identified in the same run that they are being quantified. Also, new counting technologies are available that allow high-throughput plate counting via high-throughput imaging.

The major goals of in vivo studies with radiolabeled drugs are to (1) determine quantitative interspecies metabolic profiles in plasma, (2) determine the routes of elimination of drug-related material in human urine and feces (and possibly bile), and (3) aid in the investigation of toxicological findings, either preclinical or clinical. Much recent emphasis has been placed on the importance of completely determining the metabolite profile in human plasma to ensure these metabolites are covered in toxicological species (U.S. FDA, 2008). An equally important goal of the human ADME study is to gain complete knowledge of clearance pathways as this information is critical for design of a reasonable clinical pharmacology plan that investigates the potential for polymorphic clearance, variable clearance in special populations (e.g., renally impaired patients), and drug–drug interactions. Metabolic profiling from excreta collected in the clinical study allows the determination of major routes of clearance for the drug and along with reaction phenotyping data (discussed below) allow for the complete determination of enzymes important for clearance.

Achievement of mass balance (i.e., complete recovery of administered dose) is an important part of the first goal listed above. Incomplete recovery can lead to questions about missing elimination pathways and subsequent incomplete understanding of the drug's disposition. The typical standard for good recovery, 90% of administered dose, leaves little in the way unaccounted for drug, but is often difficult to achieve due to a variety of factors. Recoveries in the range of 80–90% are often encountered in human studies and should be adequate to address the goals of the study. Recoveries under 80% do leave open questions regarding incomplete recovery, as >20% of dose is unaccounted for, and do occur in human studies, especially for lipophilic drugs eliminated predominately in feces (Roffey et al., 2007). However, there is very little evidence that incomplete recovery in the range of >20% is a sign of missing pathways, including pathways leading to tissue retention, but is simply the result of difficulties in sample collection and recovery of radioactivity (Smith and Obach, 2009).

7.4 SAFETY TESTING OF DRUG METABOLITES

The topic of safety testing of drug metabolites, the process through which sponsors ensure for the suitability of the species used for toxicological investigations, has received a great deal of recent attention. Much of this is because of the publication of the Safety Testing of Drug Metabolites Guidance by the U.S. Food and Drug Administration (FDA, 2008). This guidance, along with the even more recent International Conference on Harmonisation (ICH) guidance (ICH, 2009), provides a framework to ensure proper exposure of human metabolites in the species used for toxicological investigations and also a path forward in the event the toxicology species are judged to be not adequate to model human metabolite exposure.

The most significant toxicological concern related to metabolite formation is generally thought to be the generation of reactive metabolites. However, the monitoring of plasma has major limitations with regard to detection of reactive metabolite formation. The species detected in plasma are almost always downstream products formed from the reactive metabolites and not the metabolites themselves, but these species can still be a useful signal and possible biomarker of reactive metabolite formation. The absence of circulating reactive metabolites or downstream products of reactive metabolites does not rule out the formation of reactive species, which still could be formed at high levels but be undetectable in the plasma. The property that is much better covered with the monitoring of plasma is the potential for metabolites to mediate either on-target (active metabolites) or off-target pharmacology. The potential for metabolites to mediate on-target pharmacology is relative high as many drugs have metabolites with significant activity; however, the potential for a metabolite to mediate receptor-based pharmacology not already displayed by the parent is relatively low (Humphreys and Unger, 2006). The potential for this type of toxicity becomes even lower for drugs that are administered at low dose; this point has been made in several recent reviews (Humphreys and Unger, 2006; Smith and Obach, 2009). As well, it is the basis for the special treatment of drugs administered at less than 10 mg with regard to metabolite characterization in the ICH M3 guidance on nonclinical toxicity studies.

To allow for better early decision making and more complete early development, metabolite profiling has been moved forward in the development programs employed by many pharmaceutical companies. Traditionally, this activity was not done until well into a development program and was initiated through the conduct of a clinical ADME study with radiolabeled drug. Many companies now profile human plasma obtained in early clinical trials and detect metabolites through the use mass spectrometry, often with advanced tools such as accurate mass–mass spectrometery and data-mining techniques (Tiller et al., 2008a). There are various methods and strategies that allow comparisons to the metabolite levels found in animals. The information gathered in these studies can be used to trigger characterization activities for important human metabolites and can potentially be used to prioritize or deprioritize the conduct of the human ADME study. There have been recent suggestions that the incorporation of trace levels of radioactive material in first-in-man studies (done with a pharmacologically active dose level of the parent) along with accelerator mass spectrometry would allow very early quantitative plasma profiling and potentially full elimination profiling. This type of study could potentially be done with multiple doses of trace-labeled material during the multiple ascending dose protocol to allow metabolite profiling at or close to steady state. Overall, it is important to have a strategy in place that allows a thorough characterization of metabolites (Baillie, 2009; Zhu et al., 2009; Walker et al., 2009).

7.5 REACTION PHENOTYPING

An early understanding of the enzymes that are involved with the metabolic clearance of an NCE is important so that some level of prediction can be made as to the potential for drug–drug interactions and polymorphic clearance in humans. It should be noted that the true contribution of individual enzymes to the clearance of a compound will in most cases be difficult to predict until the human ADME study is complete and the contribution of metabolic clearance to overall clearance of the drug has been determined.

The major goal of phenotyping at the discovery stage is to make predictions regarding the potential for the dependence of a single enzyme for a major fraction (usually defined as >20%) of the clearance of a NCE. The type of information that can be reasonably generated at this stage is the contribution of individual CYP enzymes to the intrinsic clearance in human liver microsomes using well-established experimental techniques (Zhang et al., 2007; Harper and Brassil, 2008). These techniques consist of the use of (1) enzyme-specific chemical or antibody inhibitors (Shou and Lu, 2009), (2) expressed enzymes, and (3) correlation of an unknown activity with a characterized enzyme-specific probe activity across a panel of microsomes. The in vitro information can be augmented with in vivo information in animal models establishing the contribution of oxidative biotransformation to the overall clearance, but this preclinical information cannot be directly extrapolated to humans and does not add to the prediction of contribution of individual CYP enzymes. The estimation of fractional clearance by transferase enzymes producing direct conjugates, esterases, or other drug-metabolizing enzymes remains problematic as does the prediction of transporters to the direct clearance via biliary or urinary excretion. This information is certainly important for the design of early clinical drug–drug interaction studies but in some cases may be used in the discovery stage for decision-making purposes.

With radiolabeled drug in hand much more detailed reaction phenotyping studies can be preformed. This is especially the case with compounds that are poorly metabolized in vitro and where there is difficulty in accurately measuring parent disappearance. Radiolabeling allows quantitation of all metabolites formed even in small quantity. The same in vitro techniques are used as described above, but this time with accurate quantitation of all components. There are methods that allow more rapid phenotyping studies, either through the use of authentic metabolite standards and LC–MS or a combination of radiolabeling and LC–MS.

7.6 ASSESSMENT OF POTENTIAL TOXICOLOGY OF METABOLITES

7.6.1 Reactive Metabolite Studies—General Considerations

The metabolism of drugs to reactive intermediates followed by covalent binding to cellular components is generally considered to be the basis for the acute or idiosyncratic toxicities caused by some drugs (Kaplowitz, 2005; Liebler and Guengerich, 2005; Walgren et al., 2005; Antoine et al., 2008; Uetrecht, 2009; Srivastava et al., 2010). The testing of new drug candidates for their potential to form reactive metabolites and the challenges associated with data interpretation from those experiments have been reviewed (Uetrecht, 2003; Baillie, 2006; Caldwell et al., 2009; Kalgutkar and Didiuk, 2009). Many pharmaceutical companies examine new drug candidates for the potential to form reactive metabolites and if present make attempts to design the property out through targeted structural modification (Doss and Baillie, 2006; Kumar et al., 2010). Reactive intermediates are most commonly thought to arise through the generation of high-energy intermediates during the oxidation of drugs by CYP enzymes (Guengerich, 2003, 2006; Amacher, 2006). Examples of these intermediates are epoxides, oxirenes, arene oxides, and quinoid species. Myeloperoxidase is another human oxidative enzyme that is known to catalyze the formation of reactive intermediates. Reactive esters formed by the conjugation of carboxylic acids with glucuronic acid or acyl coenzyme A are also thought to be a source of reactive metabolites.

The study of the interaction of drugs with cellular components can be broken down into two types of experiments, either measurement of covalent binding after reaction with nucleophilic sites on cellular macromolecules or studies with small-molecule nucleophiles.

7.6.2 Reactive Metabolite Studies—In Vitro

Trapping experiments are most often done with unlabeled drug and glutathione with detection by mass spectrometry. Because unlabeled drug is used, these experiments can be done earlier in the discovery cycle than the covalent binding experiments (Shu et al., 2008; Kumar et al., 2010). The detection of glutathione adducts is aided by the characteristic fragmentation pattern of glutathione. Other analytical strategies have been developed to provide qualitative and/or quantitative determination of reactive metabolite formation at the screening stage (Argoti et al., 2005; Dieckhaus et al., 2005; Gan et al., 2005, 2009; Mutlib et al., 2005; Yan et al., 2005). These assays provide only qualitative information or utilize a nonphysiological trapping agent and are thus somewhat limited in utility. The approach most often employed to circumvent this problem is through the use of radiolabeled trapping agents, typically tritiated glutathione. The information generated from this approach can be utilized in a number of ways: (1) after determination of the formation of adduct and quantitation of adduct level, the adduct structure can be characterized. The adduct structure can then lead to medicinal chemistry approaches with the goal of limiting the amount of adduct formed; (2) as a trigger to perform more advanced covalent binding studies; and (3) as a trigger to do advanced toxicology studies in animals or in hepatocytes. The link between the absolute amount of glutathione adduct formed upon reaction of a compound with liver microsomes in the presence of glutathione and prediction of toxicological endpoints has not been made and is likely to remain elusive. For this reason, the information gained in trapping experiments is usually used to modify structure to minimize the problem or to trigger more definitive studies.

Protein covalent binding (PCB) studies require radiolabeled drug and are thus usually carried out late in the discovery phase or in early development. A typical experiment is done either with labeled drug in microsomes or by administering the labeled drug to rodents. Both types of studies involve isolation of cellular proteins through precipitation, followed by extensive washing steps to remove noncovalently bound drug. Residual radioactivity bound to proteins is then determined by scintillation counting. The use of this type of data has received much recent attention as a potential predictor of toxicity, especially of idiosyncratic toxicities, although this remains a controversial subject. Multiple recent publications have described the PCB properties of drugs known to cause clinical drug-induced toxicity (DIT) and have importantly included data on drugs that are not associated with clinical toxicity. This type of data on "safe drugs" has not been available previously and is essential in judging how to use information from reactive metabolite studies. The results from the three major reports from the Pfizer, Daiichi, and Dianippon groups were similar. They all conclude that: (1) some compounds not associated with DIT do form measurable amounts of reactive intermediate, (2) there is little correlation between reactive intermediates and whether a compound is considered generally safe or associated with DIT, (3) there are multiple factors that complicate the simple extrapolation of protein covalent binding/glutathione (GSH) adduct data to a prediction of toxicology liabilities, and (4) there is a large "gray zone" between clear positives and clear negatives (Masubuchi et al., 2007; Takakusa et al., 2008; Obach et al., 2008; Nakayama et al., 2009; Bauman et al., 2009; Usui et al., 2009). A study with a broad range of positive and negative compounds that measured thiol adduct levels after metabolic activation also reached similar conclusions (Gan et al., 2009) (Fig. 7.2; example of the data set from Gan et al., 2009). All studies do conclude that it is important to consider the total adduct burden rather than simply the level of adduct formed in the in vitro incubation, as the correlation of covalent binding measurement with DIT improved in all cases. The important parameters to consider for calculation of adduct burden are total dose and some measure of the fraction of drug predicted to proceed down the pathway to reactive metabolite formation (e.g., Total burden of adduct = Dose $\times F_a \times F_m \times F_{\text{adduct}}$, where F_a is fraction absorbed, F_m is fraction metabolized, and F_{adduct} is the ratio of covalent adduct/total metabolite).

Multiple recent publications have linked genetic polymorphisms in genes for immune system components to DIT (Daly and Day, 2009). These associations have been seen with abacavir (Hughes et al., 2008), flucloxacillin (Daly et al., 2009), ximelgatran (Kindmark et al., 2008), and even acetaminophen (Harrill

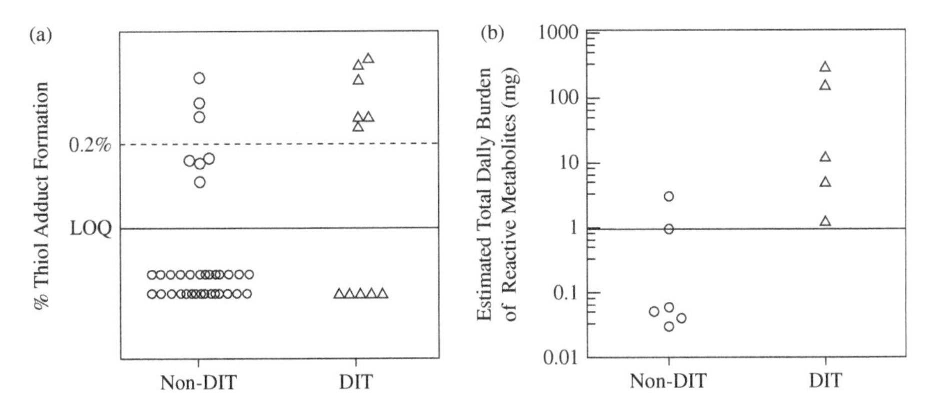

FIGURE 7.2 Scatter plot of (*a*) % dGSH adduct formation and (*b*) estimated total daily burden in the DIT and non-DIT groups. Open circles and triangles represent drugs not associated and associated with DIT, respectively. For illustrational purposes, a horizontal dotted line is plotted at 0.2% adduct level in (*a*) and another is plotted at the 1-mg level in (*b*). (Results reprinted from Gan et al., 2009, with permission)

et al., 2009) and underscore the fact that while covalent binding may serve as an initiating event for idiosyncratic DIT, the downstream responses are very complicated and difficult to predict based on early steps in the process.

7.6.3 Reactive Metabolite Studies—In Vivo

Questions regarding the formation of metabolites that covalently bind proteins can be addressed in vivo with experiments similar in nature to the experiments described in the in vitro section. Animal studies are often focused on the liver proteins as a target for drug binding, while human studies are usually focused on blood components due to obvious ethical limitations. Baillie and colleagues proposed the use of extent of covalent modification as a measure of risk of adverse events for new chemical entities. In this proposal, a level of greater than 50 pmol compound bound/milligram microsomal protein was put forth as a level of binding that would provide a safety margin (approximately 20-fold) over compounds that have shown liver toxicities in the clinic and are thought to produce toxicity through formation of reactive intermediate (Evans et al., 2004).

Even with the full consideration of all factors that may be involved in the prediction of toxicity as discussed above, it must be understood that there are large gaps in the basic understanding of the science involved that add a great deal of uncertainty to the prediction process. The degree with which this information is used in drug design and development must be carefully considered on a program-by-program and compound-by-compound basis (Kalgutkar and Didiuk, 2009).

7.6.4 Metabolite Contribution to Off-Target Toxicities

The biotransformation of parent molecules often produces metabolites with relatively minor structural alterations (addition of a hydroxyl group, demethylation, etc.). As these changes may not be dramatic enough to depart from the parent structure–activity relationship, these metabolites may have potent interactions with the same pharmacological target as the parent. Although it is rare for metabolites to have greater potency than the parent, a loss of less that one order of magnitude is fairly common, and the metabolites may be potent enough to contribute to the overall pharmacological response.

Although it is fairly common for metabolites to potently interact with the same pharmacological target as the parent, it is fairly uncommon for metabolites to have potent interactions with other "off-target" receptors that are not already affected by the parent. Of course, this relationship would likely break down as the dose and plasma concentration of drug and metabolites increases. This relationship between dose and consideration of metabolite toxicity forms the basis of arguments put forth that the metabolites of low-dose drugs do not need to be characterized as extensively as high-dose drugs (Smith and Obach, 2005, 2009). Duration of dosing administration is also an important consideration

(Smith et al., 2009). The chances for new pharmacology would be expected to increase for metabolites that result from major structural alterations, but even in these cases there are relatively few examples of this type of behavior. The chances for off-target pharmacology of metabolites would also be expected to increase when the target is a member of a receptor family with multiple related members (e.g., peroxisome proliferator activated receptors (PPARs), kinase families, etc.) (Humphreys and Unger, 2006).

7.7 ASSESSMENT OF POTENTIAL FOR ACTIVE METABOLITES

In most cases, the metabolism of drugs leads to pharmacological inactivation through biotransformation to therapeutically inactive molecules. However, drug metabolism can also result in pharmacological activation, where pharmacologically active metabolites are generated. Although formation of pharmacologically active metabolites can be mediated by oxidation and conjugation reactions, biotransformation resulting from oxidative metabolism mediated by CYP enzymes is the more common pathway leading to active metabolites.

Active metabolites may have superior pharmacological, pharmacokinetic, and safety profiles compared to their respective parent molecules (Fura et al., 2004; Fura, 2006). As a result, a number of active metabolites have been developed and marketed as drugs with improved profiles relative to their parent molecules. Examples of active metabolites of marketed drugs that have been developed as drugs include acetaminophen, oxyphenbutazone, oxazepam, cetirizine (Zyrtec), fexofenadine (Allegra), and desloratadine (Clarinex). Each of these drugs provides a specific benefit over the parent molecule and is superior in one or more properties important for the drug's action.

During lead optimization, drug candidates are routinely screened for metabolic stability or in vivo systemic exposure and rank ordered according to the rate and extent of metabolism or systemic exposure level (White, 2000; Cox et al., 2002). In the case of metabolic screening, this is usually performed in vitro after incubations of the drug candidates with subcellular fractions such as liver microsomes or intact cellular systems (e.g., hepatocytes) containing full complement of drug-metabolizing enzymes. Compounds with low metabolic stability are then excluded from further consideration because most therapeutic targets require compounds with an extended pharmacokinetic half-life. The same is true with in vivo exposure studies, where high clearance compounds are discarded. In these early screens, the concentration of the parent compound is typically the only measurement made. Consequently, there is no information on the number, identity, and pharmacological significance of metabolites that may have been formed. Even when metabolic profiling is completed and metabolites are identified, the information is typically used to direct synthesis of analogs with improved metabolic stability through the modification of metabolic soft/hot spots. Thus, the information is rarely used for the purpose of searching for pharmacologically active products as new analogs. However, rapid metabolism of parent compounds could lead to the formation of pharmacologically active metabolites that may have comparatively superior developability characteristics. As a result, metabolic instability, which otherwise may be considered a liability, can become advantageous as a method of drug design.

There are a number of advantages for screening drug candidates for active metabolites during drug discovery. The primary reason is that the process could lead to the discovery of a drug candidate with superior drug developability attributes such as:

1. Improved pharmacodynamics
2. Improved pharmacokinetics
3. Lower probability for drug–drug interactions
4. Less variable pharmacokinetics and/or pharmacodynamics
5. Improved overall safety profile
6. Improved physicochemical properties (e.g., solubility)

Other advantages of early screening for active metabolites include the potential for modifications of the entire chemical class (chemotype) to improve overall characteristics (Clader, 2004; Fura et al., 2004). Furthermore, early discovery of active metabolites will allow for more complete patent protection of the parent molecule. Additionally, tracking active metabolites at the drug discovery stage will allow for the correct interpretation of the pharmacological effect observed in preclinical species in relation to a predicted effect in humans. In other words, if an active metabolite is responsible for significant activity in a species used for preclinical efficacy determination, then there is a significant risk that the effect will be dramatically different in humans unless similar levels of metabolite can be expected in humans.

An active metabolite may have a low potential for off-target toxicity as it, in most cases, leads to the formation of a fewer number of metabolites compared to the parent compound. Moreover, most active metabolites are products of functionalization reactions, and as such are more susceptible to conjugation reactions. Conjugation reactions result in the formation of secondary metabolites that, in general, are safely cleared from the body. For example, phenacetin, which is no longer in use in humans, is metabolized to a number of metabolites. Of the many phenacetin metabolic pathways, the *O*-deethylation pathway leads to the formation of acetaminophen, a more analgesic agent, whereas *N*-hydroxylation of phenacetin leads to the formation of a toxic metabolite. On the other hand, the corresponding active metabolite, acetaminophen, is predominantly cleared via conjugation reactions (sulfation and glucuronidation) and has a greater margin of safety relative to phenacetin.

In general, drug metabolism reactions convert lipophilic compounds to more hydrophilic, more water-soluble products. An improvement in the solubility profile is an added advantage, particularly in the current drug discovery paradigm where many drug candidates generated during lead optimization have poor aqueous solubility.

The discovery of an active metabolite can serve as a modified lead compound around which new structure–activity relationships can be examined during the lead optimization stage of drug discovery. For example, this approach was used in the discovery of ezetimibe, a cholesterol absorption inhibitor (Van Heek et al., 1997; Clader, 2004). In these studies, a lead candidate (SCH48461) gave rise to a pharmacologically active biliary metabolite, which upon oral administration to rats was approximately 30-fold more potent than the parent molecule. Further optimization of the metabolite through structural modification led to the discovery of ezetimibe, a molecule that was approximately 400-fold more potent than the initial lead candidate.

In summary, tracking active metabolites at the drug discovery stage is not only important to correctly interpret the pharmacological effects in preclinical species but may also lead to the discovery of a lead candidate with superior drug developability characteristics.

7.7.1 Detection of Active Metabolites During Drug Discovery

The exploration of the potential for formation of active metabolites can be carried out with varying degrees of direction from information gathered through metabolism, pharmacokinetics, and biological/pharmacological assays. An example of undirected screening of active metabolites would be the modification of chemical libraries by subjecting them to metabolizing systems and subsequently using these modified libraries for high-throughput screens, either against the intended target or more broadly. This example is a way to generate increased molecular diversity from a given chemical library. However, this approach requires significant "deconvolution" efforts when activity is found in mixtures. To increase the success rate and decrease the number of compounds screened to a manageable size, the search for active metabolites could be limited to those compounds/chemotypes showing high clearance rates in in vitro metabolic stability or in vivo exposure screens.

However, activity assays may serve as a more rationale approach to the exploration of active metabolites. This is most often and most effectively done in the setting of an in vivo efficacy experiment that allows for both pharmacodynamic (PD) and pharmacokinetic (PK) information to be gathered. Analysis of the relationship between the PD endpoint and the PK profile will sometimes demonstrate an apparent disconnect between the two data sets and point to the possibility that an active metabolite is responsible for some of the activity. These disconnects can serve as clear trigger points for the initiation of active metabolite searches.

For example, Van Heek and co-workers observed a lead candidate that underwent extensive first-pass metabolism and yet elicited a significant level of pharmacological activity (Van Heek et al., 1997). To evaluate the biological activity of the in vivo biotransformation products, they collected samples of bile from rats dosed with a lead compound and directly administered the samples to bile duct cannulated rats via an intraduodenal cannula. As a control study, the parent compound prepared in blank bile was dosed in a similar fashion to the recipient rats. The results indicated that the in vivo activity elicited by the bile samples was higher than the parent control sample, clearly indicating the presence of an active metabolite(s) that was more potent than the parent compound. To identify the active component, the bile sample was then fractionated and each fraction tested for biological activity. The structure of the metabolite was then established following the detection of the active fraction. Further modification of the active metabolite led to the discovery of ezetimibe.

Although a lack of correlation between PK and PD data is the clearest trigger point for pursuing the possibility of metabolite contributions to the observed pharmacology, there are several other potential triggers that can be used that include (1) the observation of a greater pharmacological effect upon extravascular administration of a compound relative to parenteral administration, (2) a reduced pharmacological effect upon coadministration in vivo/in vitro with compounds that inhibit metabolism (e.g., aminobenzotriazole, ketoconazole, etc.), and (3) a prolonged PD effect despite rapid in vitro metabolism. Examples of the utilization of metabolite structural information in drug design can be found in several reviews (Fura et al., 2004; Fura, 2006).

7.8 SUMMARY

The last several decades have witnessed an explosion in our knowledge of drug metabolism and drug-metabolizing enzymes. It is now possible to fully characterize and predict the metabolic fate of new chemical entities in humans with reasonable certainty. The in vitro methods used to do this, that is, human liver microsomes, expressed enzymes, cryopreserved, or freshly isolated hepatocytes, can be adapted to be run in medium-throughput fashion and allow the metabolic properties of new compounds to be optimized during the discovery phase. Also, medium-throughput screens for bioavailability are now possible with rapid quantitation by LC–MS/MS techniques. These techniques will need to continue to evolve to allow metabolism scientists the ability to keep pace with the increasing speed of drug discovery. The increasing complexity of the process will place an ever greater dependence on information management, and the ability to turn that information into knowledge will be tantamount. This is especially true when one considers the advances in other fields of ADME-related research (e.g., transporters, interactions with nuclear hormone receptors) and the fact that more studies will likely need to be conducted during drug discovery. The relationship of drug metabolites to toxicity will remain a major challenge in the years to come. These considerations also make drug development an increasingly challenging endeavor as all of the characteristics of a candidate need to be synthesized in a complete picture of the molecule's disposition that will adequately capture its disposition in all patients prescribed the drug. This complexity ensures that the field of drug metabolism will continue to play an increasing role in the endeavor to provide high-quality drug candidates that will become safe, highly efficacious medicines.

REFERENCES

Amacher DE. Reactive intermediates and the pathogenesis of adverse drug reactions: The toxicology perspective. Curr Drug Metab 2006;7:219–229.

Anari MR, Baillie TA. Bridging cheminformatic metabolite prediction and tandem mass spectrometry. Drug Discov Today 2005;10:711–717.

Antoine DJ, Williams DP, Park BK. Understanding the role of reactive metabolites in drug-induced hepatotoxicity: State of the science. Expert Opin Drug Metab Toxicol 2008;4(11):1415–1427.

Argoti D, Liang L, Conteh A, Chen L, Bershas D, Yu CP, Vouros P, Yang E. Cyanide trapping of iminium ion reactive intermediates followed by detection and structure identification using liquid chromatography–tandem mass spectrometry (LC-MS/MS). Chem Res Toxicol 2005;18:1537–1544.

Baillie TA. Future of toxicology-metabolic activation and drug design: Challenges and opportunities in chemical toxicology. Chem Res Toxicol 2006;19:889–893.

Baillie TA. Approaches to the assessment of stable and chemically reactive drug metabolites in early clinical trials. Chem Res Toxicol 2009;22(2):263–266.

Balani SK, Zhu T, Yang TJ, Liu Z, He B, Lee FW. Effective dosing regimen of 1-aminobenzotriazole for inhibition of antipyrine clearance in rats, dogs, and monkeys. Drug Metab Dispos 2002;30:1059–1062.

Barter ZE, Bayliss MK, Beaune PH, Boobis AR, Carlile DJ, Edwards RJ, Houston JB, Lake BG, Lipscomb JC, Pelkonen OR, Tucker GT, Rostami-Hodjegan A. Scaling factors for the extrapolation of in vivo metabolic drug clearance from in vitro data: Reaching a consensus on values of human microsomal protein and hepatocellularity per gram of liver. Curr Drug Metab 2007;8(1):33–45.

Bateman KP, Kellmann M, Muenster H, Papp R, Taylor L. Quantitative-qualitative data acquisition using a benchtop Orbitrap mass spectrometer. J Am Soc Mass Spectrom 2009;20(8):1441–1450.

Bauman JN, Kelly JM, Tripathy S, Zhao SX, Lam WW, Kalgutkar AS, Obach RS. Can in vitro metabolism-dependent covalent binding data distinguish hepatotoxic from nonhepatotoxic drugs? An analysis using human hepatocytes and liver S-9 fraction. Chem Res Toxicol 2009;22:332–340

Buch H, Buzello W, Heymann E, Krisch K. Inhibition of phenacetin- and acetanilide-induced methemoglobinemia in the rat by the carboxylesterase inhibitor bis-[*p*-nitrophenyl] phosphate. Biochem Pharmacol 1969;18:801–811.

Caldwell GW, Yan Z, Tang W, Dasgupta M, Hasting B. ADME optimization and toxicity assessment in early- and late-phase drug discovery. Curr Top Med Chem 2009;9(11):965–980.

Chao P, Maguire T, Novik E, Cheng KC, Yarmush ML. Evaluation of a microfluidic based cell culture platform with primary human hepatocytes for the prediction of hepatic clearance in human. Biochem Pharmacol 2009;78(6):625–632.

Clader JW. The discovery of ezetimibe: A view from outside the receptor. J Med Chem 2004;47:1–9.

Cox KA, White RE, Korfmacher WA. Rapid determination of pharmacokinetic properties of new chemical entities: In vivo approaches. Comb Chem High Throughput Screen 2002;5:29–37.

Daly AK, Day CP. Genetic association studies in drug-induced liver injury. Semin Liver Dis 2009;29:400–411.

Daly AK, Donaldson PT, Bhatnagar P, Shen Y, Pe'er I, Floratos A, Daly MJ, Goldstein DB, John S, Nelson MR, Graham J, Park BK, Dillon JF, Bernal W, Cordell HJ, Pirmohamed M, Aithal GP, Day CP, DILIGEN Study; International SAE Consortium. HLA-B*5701 genotype is a major determinant of drug-induced liver injury due to flucloxacillin. Nat Genet 2009;41:816–819.

Dieckhaus CM, Fernandez-Metzler CL, King R, Krolikowski PH, Baillie TA. Negative ion tandem mass spectrometry for the detection of glutathione conjugates. Chem Res Toxicol 2005;18:630–638.

Doss GA, Baillie TA. Addressing metabolic activation as an integral component of drug design. Drug Metab Rev 2006;38(4):641–649.

Evans DC, Watt AP, Nicoll-Griffith DA, Baillie TA. Drug-protein adducts: An industry perspective on minimizing the potential for drug bioactivation in drug discovery and development. Chem Res Toxicol 2004;17:3–16.

Fisher MB, Campanale K, Ackermann BL, VandenBranden M, Wrighton SA. In vitro glucuronidation using human liver microsomes and the pore-forming peptide alamethicin. Drug Metab Dispos 2000;28:560–566.

Fisher MB, Henne KR, Boer J. The complexities inherent in attempts to decrease drug clearance by blocking sites of CYP-mediated metabolism. Curr Opin Drug Discov Devel 2006;9(1):101–109.

Fisher MB, Yoon K, Vaughn ML, Strelevitz TJ, Foti RS. Flavin-containing monooxygenase activity in hepatocytes and microsomes: In vitro characterization and in vivo scaling of benzydamine clearance. Drug Metab Dispos 2002;30:1087–1093.

Fura A. Role of pharmacologically active metabolites in drug discovery and development. Drug Discov Today 2006;11:133–142.

Fura A, Shu YZ, Zhu M, Hanson RL, Roongta V, Humphreys WG. Discovering drugs through biological transformation: Role of pharmacologically active metabolites in drug discovery. J Med Chem 2004;47:4339–4351.

Gan J, Harper TW, Hsueh MM, Qu Q, Humphreys WG. Dansyl glutathione as a trapping agent for the quantitative estimation and identification of reactive metabolites. Chem Res Toxicol 2005;18:896–903.

Gan J, Ruan Q, He B, Zhu M, Shyu WC, Humphreys WG. In vitro screening of 50 highly prescribed drugs for thiol adduct formation—Comparison of potential for drug-induced toxicity and extent of adduct formation. Chem Res Toxicol 2009;22:690–698.

Ghibellini G, Vasist LS, Leslie EM, Heizer WD, Kowalsky RJ, Calvo BF, Brouwer KL. In vitro–in vivo correlation of hepatobiliary drug clearance in humans. Clin Pharmacol Ther 2007;81(3):406–413.

Grime K, Riley RJ. The impact of in vitro binding on in vitro–in vivo extrapolations, projections of metabolic clearance and clinical drug-drug interactions. Curr Drug Metab 2006;7:251–264.

Guengerich FP. Common and uncommon cytochrome P450 reactions related to metabolism and chemical toxicity. Chem Res Toxicol 2001;14:611–650.

Guengerich FP. Cytochrome P450 oxidations in the generation of reactive electrophiles: Epoxidation and related reactions. Arch Biochem Biophys 2003;409:59–71.

Guengerich FP. Principles of covalent binding of reactive metabolites and examples of activation of bis-electrophiles by conjugation. Arch Biochem Biophys 2005;433:369–378.

Guengerich FP. Cytochrome P450S and other enzymes in drug metabolism and toxicity. AAPS J 2006;8:E101–111.

Hallifax D, Houston JB. Methodological uncertainty in quantitative prediction of human hepatic clearance from in vitro experimental systems. Curr Drug Metab 2009;10(3):307–321.

Harper TW, Brassil PJ. Reaction phenotyping: Current industry efforts to identify enzymes responsible for metabolizing drug candidates. AAPS J 2008;10(1):200–207.

Harrill AH, Watkins PB, Su S, Ross PK, Harbourt DE, Stylianou IM, Boorman GA, Russo MW, Sackler RS, Harris SC, Smith PC, Tennant R, Bogue M, Paigen K, Harris C, Contractor T, Wiltshire T, Rusyn I, Threadgill DW. Mouse population-guided resequencing reveals that variants in CD44 contribute to acetaminophen-induced liver injury in humans. Genome Res 2009;19:1507–1515.

Hawkins DR. Use and value of metabolism databases. Drug Discov Today 1999;4:466–471.

Hellriegel ET, Bjornsson TD, Hauck WW. Interpatient variability in bioavailability is related to the extent of absorption: Implications for bioavailability and bioequivalence studies. Clin Pharmacol Ther 1996;60:601–607.

Houston JB, Galetin A. Methods for predicting in vivo pharmacokinetics using data from in vitro assays. Curr Drug Metab 2008;9(9):940–951.

Hsieh Y, Korfmacher W. The role of hyphenated chromatography–mass spectrometry techniques in exploratory drug metabolism and pharmacokinetics. Curr Pharm Des 2009;15(19):2251–2261.

Huang C, Zheng M, Yang Z, Rodrigues AD, Marathe P. Projection of exposure and efficacious dose prior to first-in-human studies: How successful have we been? Pharm Res 2008;*25*(4):713–726.

Hughes AR, Spreen WR, Mosteller M, Warren LL, Lai EH, Brothers CH, Cox C, Nelsen AJ, Hughes S, Thorborn DE, Stancil B, Hetherington SV, Burns DK, Roses AD. Pharmacogenetics of hypersensitivity to abacavir: From PGx hypothesis to confirmation to clinical utility. Pharmacogenom J 2008;28:365–374.

Humphreys WG, Unger SE. Safety assessment of drug metabolites: Characterization of chemically stable metabolites. Chem Res Toxicol 2006;19(12):1564–1569.

International Conference on Harmonisation. Non-Clinical Safety Studies for the Conduct of Human Clinical Trials for Pharmaceuticals, ICH Guidance M3(R2), 2009, available: http://www.emea.europa.eu/pdfs/human/ich028695en.pdf.

Kalgutkar AS, Didiuk MT. Structural alerts, reactive metabolites, and protein covalent binding: How reliable are these attributes as predictors of drug toxicity? Chem Biodivers 2009;6(11):2115–2137.

Kaplowitz N. Idiosyncratic drug hepatotoxicity. Nat Rev Drug Discov 2005;4: 489–499.

Kilford PJ, Stringer R, Sohal B, Houston JB, Galetin A. Prediction of drug clearance by glucuronidation from in vitro data: Use of combined cytochrome P450 and UDP-glucuronosyltransferase cofactors in alamethicin-activated human liver microsomes. Drug Metab Dispos 2009;37(1):82–89.

Kim HJ, Cho JH, Klaassen CD. Depletion of hepatic 3'-phosphoadenosine 5'-phosphosulfate (PAPS) and sulfate in rats by xenobiotics that are sulfated. J Pharmacol Exp Ther 1995;275:654–658.

Kindmark A, Jawaid A, Harbron CG, Barratt BJ, Bengtsson OF, Andersson TB, Carlsson S, Cederbrant KE, Gibson NJ, Armstrong M, Lagerström-Fermér ME, Dellsén A, Brown EM, Thornton M, Dukes C, Jenkins SC, Firth MA, Harrod GO, Pinel TH, Billing-Clason SM, Cardon LR, March RE. Genome-wide pharmacogenetic investigation of a hepatic adverse event without clinical signs of immunopathology suggests an underlying immune pathogenesis. Pharmacogenomics J 2008;8:186–195.

Klopf W, Worboys P. Scaling in vivo pharmacokinetics from in vitro metabolic stability data in drug discovery. Comb Chem High Throughput Screen. 2010;13(2):159–69.

Kramer MA, Tracy TS. Studying cytochrome P450 kinetics in drug metabolism. Expert Opin Drug Metab Toxicol 2008;4(5):591–603.

Kumar S, Mitra K, Kassahun K, Baillie TA. Approaches for minimizing metabolic activation of new drug candidates in drug discovery. Handb Exp Pharmacol 2010;(196):511–544.

Lave T, Coassolo P, Reigner B. Prediction of hepatic metabolic clearance based on interspecies allometric scaling techniques and in vitro–in vivo correlations. Clin Pharmacokinet 1999;36:211–231.

Liebler DC, Guengerich FP. Elucidating mechanisms of drug-induced toxicity. Nat Rev Drug Discov 2005;4:410–420.

Lin JH, Wong BK. Complexities of glucuronidation affecting in vitro in vivo extrapolation. Curr Drug Metab 2002;3:623–646.

Masubuchi N, Makino C, Murayama N. Prediction of in vivo potential for metabolic activation of drugs into chemically reactive intermediate: Correlation of in vitro and in vivo generation of reactive intermediates and in vitro glutathione conjugate formation in rats and humans. Chem Res Toxicol 2007;20:455–464.

McGinnity DF, Soars MG, Urbanowicz RA, Riley RJ. Evaluation of fresh and cryopreserved hepatocytes as in vitro drug metabolism tools for the prediction of metabolic clearance. Drug Metab Dispos 2004;32(11):1247–1253.

Miners JO, Mackenzie PI, Knights KM. The prediction of drug-glucuronidation parameters in humans: UDP-glucuronosyltransferase enzyme-selective substrate and inhibitor probes for reaction phenotyping and in vitro–in vivo extrapolation of drug clearance and drug-drug interaction potential. Drug Metab Rev 2010;42(1):189–201.

Mutlib A, Lam W, Atherton J, Chen H, Galatsis P, Stolle W. Application of stable isotope labeled glutathione and rapid scanning mass spectrometers in detecting and characterizing reactive metabolites. Rapid Commun Mass Spectrom 2005;19:3482–3492.

Nakayama S, Atsumi R, Takakusa H, Kobayashi Y, Kurihara A, Nagai Y, Nakai D, Okazaki O. A zone classification system for risk assessment of idiosyncratic drug toxicity using daily dose and covalent binding. Drug Metab Dispos 2009;37:1956–1962.

Nassar AE, Kamel AM, Clarimont C. Improving the decision-making process in the structural modification of drug candidates: Enhancing metabolic stability. Drug Discov Today 2004;9(23):1020–1028.

Obach RS. Prediction of human clearance of twenty-nine drugs from hepatic microsomal intrinsic clearance data: An examination of in vitro half-life approach and nonspecific binding to microsomes. Drug Metab Dispos 1999;27:1350–1359.

Obach RS, Kalgutkar AS, Soglia JR, Zhao SX. Can in vitro metabolism-dependent covalent binding data in liver microsomes distinguish hepatotoxic from nonhepatotoxic drugs? An analysis of 18 drugs with consideration of intrinsic clearance and daily dose. Chem Res Toxicol 2008;21:1814–1822.

Pelkonen O, Turpeinen M. In vitro–in vivo extrapolation of hepatic clearance: Biological tools, scaling factors, model assumptions and correct concentrations. Xenobiotica 2007;37(10–11):1066–1089.

Rane A, Wilkinson GR, Shand DG. Prediction of hepatic extraction ratio from in vitro measurement of intrinsic clearance. J Pharmacol Exp Ther 1977;200:420–424.

Roffey SJ, Obach RS, Gedge JI, Smith DA. What is the objective of the mass balance study? A retrospective analysis of data in animal and human excretion studies employing radiolabeled drugs. Drug Metab Rev 2007;39(1):17–43.

Ruan Q, Peterman S, Szewc MA, Ma L, Cui D, Humphreys WG, Zhu M. An integrated method for metabolite detection and identification using a linear ion trap/Orbitrap mass spectrometer and multiple data processing techniques: Application to indinavir metabolite detection. J Mass Spectrom 2008;43(2):251–261.

Shou M, Lu AY. Antibodies as a probe in cytochrome P450 research. Drug Metab Dispos 2009;37(5):925–931.

Shu YZ, Johnson BM, Yang TJ. Role of biotransformation studies in minimizing metabolism-related liabilities in drug discovery. AAPS J 2008;10(1):178–192.

Smith DA, Obach RS. Seeing through the mist: Abundance versus percentage. Commentary on metabolites in safety testing. Drug Metab Dispos 2005;33: 1409–1417.

Smith DA, Obach RS. Metabolites in safety testing (MIST): Considerations of mechanisms of toxicity with dose, abundance, and duration of treatment. Chem Res Toxicol 2009;22:267–279.

Smith DA, Obach RS, Williams DP, Park BK. Clearing the MIST (metabolites in safety testing) of time: The impact of duration of administration on drug metabolite toxicity. Chem Biol Interact 2009;179(1):60–67.

Soars MG, Burchell B, Riley RJ. In vitro analysis of human drug glucuronidation and prediction of in vivo metabolic clearance. J Pharmacol Exp Ther 2002;301: 382–390.

Soars MG, Grime K, Sproston JL, Webborn PJ, Riley RJ. Use of hepatocytes to assess the contribution of hepatic uptake to clearance in vivo. Drug Metab Dispos 2007a;35 (6):859–865.

Soars MG, McGinnity DF, Grime K, Riley RJ. The pivotal role of hepatocytes in drug discovery. Chem Biol Interact 2007b;168(1):2–15.

Srivastava A, Maggs JL, Antoine DJ, Williams DP, Smith DA, Park BK. Role of reactive metabolites in drug-induced hepatotoxicity. Handb Exp Pharmacol 2010; (196):165–194.

Stringer RA, Strain-Damerell C, Nicklin P, Houston JB. Evaluation of recombinant cytochrome P450 enzymes as an in vitro system for metabolic clearance predictions. Drug Metab Dispos 2009;37(5):1025–1034.

Sun H, Scott DO. Structure-based drug metabolism predictions for drug design. Chem Biol Drug Des 2010;75(1):3–17.

Takakusa H, Masumoto H, Yukinaga H, Makino C, Nakayama S, Okazaki O, Sudo K. Covalent binding and tissue distribution/retention assessment of drugs associated with idiosyncratic drug toxicity. Drug Metab Dispos 2008;36:1770–1779.

Tang W, Lu AY. Drug metabolism and pharmacokinetics in support of drug design. Curr Pharm Des 2009;15(19):2170–2183.

Thompson TN. Optimization of metabolic stability as a goal of modern drug design. Med Res Rev 2001;21:412–449.

Tiller PR, Yu S, Bateman KP, Castro-Perez J, McIntosh IS, Kuo Y, Baillie TA. Fractional mass filtering as a means to assess circulating metabolites in early human clinical studies. Rapid Commun Mass Spectrom 2008a;22(22):3510–3516.

Tiller PR, Yu S, Castro-Perez J, Fillgrove KL, Baillie TA. High-throughput, accurate mass liquid chromatography/tandem mass spectrometry on a quadrupole time-of-flight system as a "first-line" approach for metabolite identification studies. Rapid Commun Mass Spectrom 2008b;22(7):1053–1061.

Tolonen A, Turpeinen M, Pelkonen O. Liquid chromatography–mass spectrometry in in vitro drug metabolite screening. Drug Discov Today 2009;14 (3–4):120–133.

Uetrecht J. Screening for the potential of a drug candidate to cause idiosyncratic drug reactions. Drug Discov Today 2003;8:832–837.

Uetrecht J. Immune-mediated adverse drug reactions. Chem Res Toxicol 2009; 22(1):24–34.

U.S. Food and Drug Administration (FDA). Guidance for Industry: Safety Testing of Drug Metabolites, 2008, available: http://www.fda.gov/downloads/Drugs/GuidanceComplianceRegulatoryInformation/Guidances/ucm79266.pdf.

Usui T, Mise M, Hashizume T, Yabuki M, Komuro S. Evaluation of the potential for drug-induced liver injury based on in vitro covalent binding to human liver proteins. Drug Metab Dispos 2009;37:2383–2392.

Van Heek M, France CF, Compton DS, McLeod RL, Yumibe NP, Alton KB, Sybertz EJ, Davis HR , Jr. In vivo metabolism-based discovery of a potent cholesterol absorption inhibitor, SCH58235, in the rat and rhesus monkey through the identification of the active metabolites of SCH48461. J Pharmacol Exp Ther 1997;283:157–163.

Walgren JL, Mitchell MD, Thompson DC. Role of metabolism in drug-induced idiosyncratic hepatotoxicity. Crit Rev Toxicol 2005;35:325–361.

Walker D, Brady J, Dalvie D, Davis J, Dowty M, Duncan JN, Nedderman A, Obach RS, Wright P. A holistic strategy for characterizing the safety of metabolites through drug discovery and development. Chem Res Toxicol 2009;22(10):1653–1662.

White RE. High-throughput screening in drug metabolism and pharmacokinetic support of drug discovery. Annu Rev Pharmacol Toxicol 2000;40:133–157.

Wrona M, Mauriala T, Bateman KP, Mortishire-Smith RJ, O'Connor D. "All-in-one" analysis for metabolite identification using liquid chromatography/hybrid quadrupole time-of-flight mass spectrometry with collision energy switching. Rapid Commun Mass Spectrom 2005;19(18):2597–2602.

Yamamoto T, Itoga H, Kohno Y, Nagata K, Yamazoe Y. Prediction of oral clearance from in vitro metabolic data using recombinant CYPs: Comparison among well-stirred, parallel-tube, distributed and dispersion models. Xenobiotica 2005;35:627–646.

Yan Z, Maher N, Torres R, Caldwell GW, Huebert N. Rapid detection and characterization of minor reactive metabolites using stable-isotope trapping in combination with tandem mass spectrometry. Rapid Commun Mass Spectrom 2005;19:3322–3330.

Yao M, Ma L, Duchoslav E, Zhu M. Rapid screening and characterization of drug metabolites using multiple ion monitoring dependent product ion scan and post-acquisition data mining on a hybrid triple quadrupole-linear ion trap mass spectrometer. Rapid Commun Mass Spectrom 2009;23(11):1683–1693.

Zhang H, Davis CD, Sinz MW, Rodrigues AD. Cytochrome P450 reaction-phenotyping: An industrial perspective. Expert Opin Drug Metab Toxicol 2007;3(5):667–687.

Zhang NR, Yu S, Tiller P, Yeh S, Mahan E, Emary WB. Quantitation of small molecules using high-resolution accurate mass spectrometers — A different approach for analysis of biological samples. Rapid Commun Mass Spectrom 2009a; 23(7):1085–1094.

Zhang Z, Zhu M, Tang W. Metabolite identification and profiling in drug design: Current practice and future directions. Curr Pharm Des 2009b;15(19):2220–2235.

Zhang H, Zhang D, Ray K, Zhu M. Mass defect filter technique and its applications to drug metabolite identification by high-resolution mass spectrometry. J Mass Spectrom 2009c 44(7):999–1016.

Zhu M, Zhang D, Zhang H, Shyu WC. Integrated strategies for assessment of metabolite exposure in humans during drug development: Analytical challenges and clinical development considerations. Biopharm Drug Dispos 2009;30 (4):163–184.

PART II
Mass Spectrometry in Drug Metabolism: Principles and Common Practice

8 Theory and Instrumentation of Mass Spectrometry

GÉRARD HOPFGARTNER

Life Sciences Mass Spectrometry, School of Pharmaceutical Sciences, University of Geneva, University of Lausanne, Geneva, Switzerland

8.1 BASIC CONCEPTS AND THEORY OF MASS SPECTROMETRY

8.1.1 Historical Perspective

Mass spectrometry is an analytical technique that analyzes gaseous ions by their mass-to-charge ratio (m/z). Prior separations of ions by their mass-to-charge ratio in a mass analyzer require that solid, liquid, or gases are ionized. Mass spectrometry started in the early nineteenth century with the work of

Mass Spectrometry in Drug Metabolism and Disposition: Basic Principles and Applications, First Edition. Edited by Mike S. Lee and Mingshe Zhu.

J. J. Thompson on electrons and "positive rays." Thompson constructed the first mass spectrometer, called the parabola spectrograph. Most of the early applications of the new technique focused on the analysis of the isotopes. J. W. Ashton from the Cavendish Laboratory reported the analysis of about 212 naturally occurring isotopes. One of the first analysis of volatile organic compounds was reported by J. J. Dempster (1918). For that purpose he developed an instrument based on a magnetic field where the gas molecules where ionized by a beam of electrons. Nowadays, this technique is still in use and known as electron impact (EI) ionization. In the 1940s most applications with mass spectrometry focused on the analysis of volatile organic molecules. Other types of mass analyzers such as the time-of-flight analyzer appeared. The benefit of the combination of mass spectrometry with a chromatographic separation technique became obvious for the analysis of complex organic mixtures. While the interfacing of gas chromatography with mass spectrometry (GC–MS) was already described in the late 1950s (Gohlke and McLafferty, 1993), the combination of liquid chromatography became available much later (Arpino et al., 1974). This was mainly due to the difficulties of introducing a liquid in a vacuum to perform EI or chemical ionization (CI). The quadrupole and the ion trap mass analyzers were invented by W. Paul in the 1950s and the commercialization of the quadrupole analyzer started in the 1960s. In 1989, the Nobel Prize in Physics was awarded to Hans Dehmelt and Wolfgang Paul for the development of the ion trap technique in the 1950s and 1960s. Yost and Enke (1978) described a triple quadrupole instrument composed by two mass-analyzing quadrupoles and a collision cell capable of performing selected ion fragmentation by collision-induced dissociation. Nowadays, triple quadrupole instruments play an essential role for the quantitative liquid chromatography–tandem mass spectrometry (LC–MS/MS) analysis in the selected reaction monitoring mode (SRM) (Hopfgartner and Bourgogne, 2003).

In took almost 50 years for alternative ionization methods to EI to became available, allowing the ionization of thermolabile and high-molecular-weight molecules, which is essential to analyses of pharmaceuticals and compounds involved in biological processes (Vestal, 2001). A successful approach to improving the utility of low-energy impact ionization for nonvolatile molecules was developed by Barber and co-workers. This technique, known as fast atom bombardment (FAB) or liquid secondary ion mass spectrometry (SIMS), depending of the charge on the incident beam, has been successfully applied to analyze peptides, glycoside antibiotics, organometallics, and vitamin B_{12} and its coenzyme (Barber et al., 1982). In the early 1980s a spray-based technique, thermospray, which was particularly suitable for the analysis of low-molecular-weight compounds, became available (Vestal, 1984). The success of thermospray was related to the possibility of using standard liquid chromatography. The real breakthrough for the analysis of macromolecules and the interfacing of liquid chromatography with mass spectrometry came with the development of the electrospray by J. Fenn and matrix-assisted laser desorption by F. Hillenkampf and M. Karas in the mid-1980s (Karas and Hillenkamp, 1988). In

2002, the Nobel Prize in Chemistry was awarded to J. B. Fenn for the development of the analysis of macromolecules by electrospray ionization (ESI) and K. Tanaka for the development of soft laser desorption (SLD) in 1987 (Tanaka et al., 1988). The development of new hybrid mass spectrometers has been particularly productive over the last 10 years. To be mentioned are the combination quadrupole time-of-flight (Morris et al., 1996), ion trap time-of-flight (Bilsborough et al., 2004), triple quadrupole linear ion trap (Hager, 2002), ion trap ion cyclotron resonance, and ion trap orbitrap spectrometers (Hu, et al., 2005). More recently, new ambient desorption techniques, such as direct analysis in real time (DART) (Cody et al., 2005) or desorption electrospray ionization (DESI) (Cooks et al., 2006), showed the potential of mass spectrometry to analyze solid samples such as tablets or paper surfaces.

8.1.2 Isotopes

Most organic molecules are formed by carbon, oxygen, nitrogen, or sulfur atoms. These atoms have natural occurring isotopes (see Table 8.1) at a certain isotopic ratio. An isotope is an atom that differentiates from another only by the number of atomic mass (number on neutrons). At unit mass resolution,

TABLE 8.1 Isotopic Abundance of Common Elements[a]

Element	Symbol	Atomic Mass	Isotopic Mass	Abundance (%)
Carbon	^{12}C	12.0110	12.000000	98.9
	^{13}C		13.003354	1.10
Hydrogen	H	1.0080	1.007825	99.985
	D		2.013999	0.015
Oxygen	^{16}O	15.993	15.994915	99.76
	^{17}O		16.999133	0.04
	^{18}O		17.999160	0.20
Nitrogen	^{14}N	14.0067	14.0030698	99.64
	^{15}N		15.00010	0.36
Sulfur	^{32}S	32.066	31.97207	95.02
	^{33}S		32.971456	0.75
	^{34}S		33.96787	4.21
	^{36}S		35.96708	0.02
Chlorine	^{35}Cl	35.4610	34.968849	75.77
	^{37}Cl		36.999988	24.23
Bromine	^{79}Br	79.9035	78.918348	50.50
	^{81}Br		80.916344	49.50
Fluorine	^{19}F	18.998404	18.998404	100

[a] Cl and Br, which are present in numerous organic molecules, have two naturally intense isotopes, while F is monoisotopic.

mass spectrometry is able to differentiate between isotopes, and, in general, the most abundant ion corresponds to the most intense isotope composing the molecules (^{12}C, ^{1}H, ^{14}N, ^{16}O, or ^{32}S).

The sum of the atomic weight of the atoms composing a molecule is defined as the molecular weight (MW). The common unit for the molecular weight is the dalton (Da) or the unified atomic mass unit (u). The relative molecular mass M_r , more commonly used, is the ratio of the mass of a molecule to the unified atomic mass unit. The exact mass is given with certain accuracy (three or four digits). The nominal mass of an ion or molecule is calculated using the mass of the most abundant isotope of each element rounded to the nearest integer (see Table 8.2).

At unit mass resolution or better, mass spectrometry is able to differentiate between isotopes, and, in general, the most abundant ion corresponds to the most intense isotope composing the molecules (^{12}C, ^{1}H, ^{14}N, ^{16}O, or ^{32}S). Most of the isotopic contributions observed for low-molecular-weight compounds (i.e, < 500 Da) is due to the atom carbon. Therefore, if the number of carbon atoms increases, the isotopic contribution increases. Figure 8.1 depicts a simulation of the increase of the isotopic contribution for a series of linear hydrocarbons for EI.

TABLE 8.2 Nominal and Exact Mass for Selected Analytes

Analyte	Structure	Formula	M_r	Nominal Mass	Exact Mass
Tolcapone	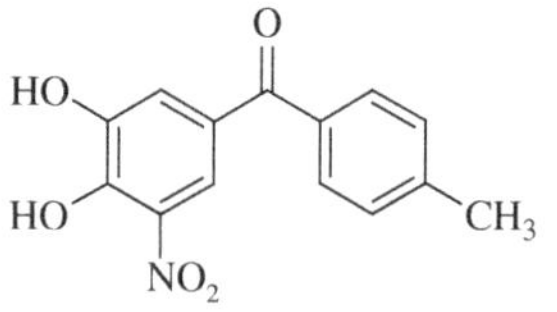	$C_{14}H_{11}NO_5$	273.25	273	273.0632
Moclobemide	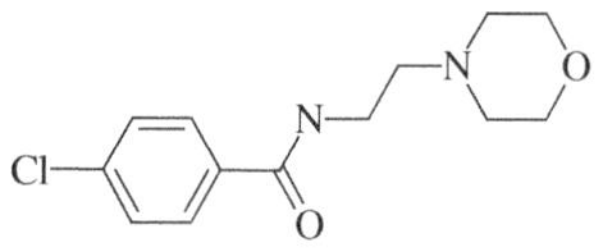	$C_{13}H_{17}ClN_2O_2$	268.75	269	268.0973
Bosentan	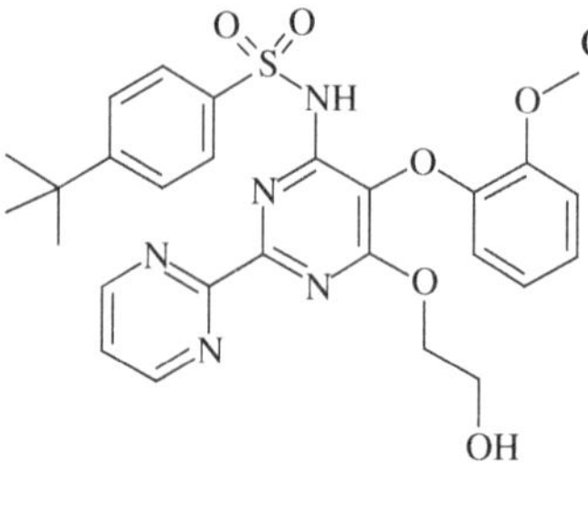	$C_{27}H_{29}N_5O_5S$	551.63	552	551.1833

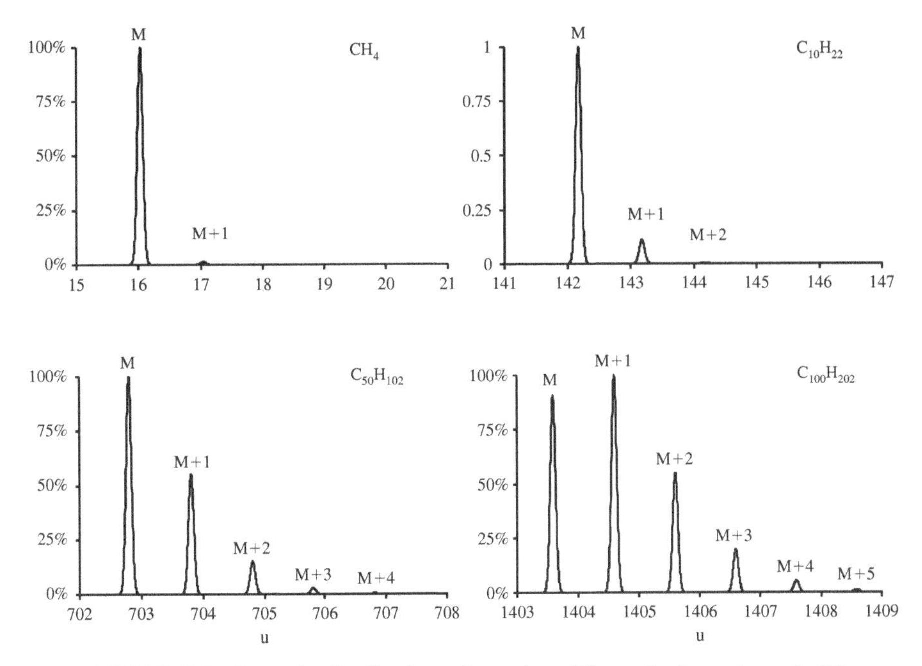

FIGURE 8.1 Isotopic distribution of a series of linear hydrocarbons in EI.

As can be seen in Figure 8.1 the isotopic contribution of carbon to M+1 correlates quite well with the number of carbon atoms. Another important aspect is that isotopic contributions are additive also between elements. Therefore, Cl and Br contribution at M+2 can be clearly visualized. In addition, the isotopic contribution can be correlated to an elemental formula. With ions the m/z difference between isotopic peaks is of one unit for a singly charge ion, 0.5 and 0.3 with a doubly or triply charged ion, respectively.

Mass spectrometry has the ability to differentiate isotopes and to measure the precise mass of an ion or the accurate mass (Webb et al., 2004). The accurate mass is the experimentally measured mass of an ion with at least three to four significant figures after the decimal point. Mass accuracy is the mass difference between the theoretical mass of an ion and its measured mass by a mass spectrometer. Mass accuracy is generally expressed in millimass unit (mmu) and in parts per million (ppm). Accurate mass measurements are typically performed on time-of-flight or Fourier transform instruments (ion cyclotron resonance, orbitrap), but accurate mass has also been demonstrated on a quadrupole mass analyzer (Tyler et al., 1996). The exact mass of the protonated remikiren ($C_{33}H_{51}N_4O_6S$) corresponds to m/z 631.35238. In the case of a positively charged ion, it is important that the mass of the electron is subtracted [$9.10938215(45) \times 10^{-31}$ kg]. In the case of remikiren not subtracting the electron would generate an error of 0.8 ppm. Based on accurate mass measurements, it is possible to calculate elemental formulas. Depending

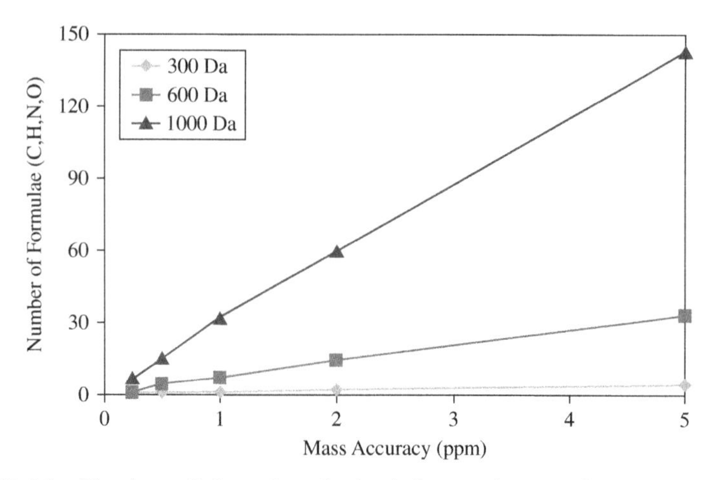

FIGURE 8.2 Number of formulas obtained for a given analyte mass and mass accuracy (Webb et al., 2004).

on the search constrain (elements), the mass of the analyte, and the accuracy of mass spectrometer, a different number of formulas are obtained. More mass accuracy is needed at higher masses, as illustrated in Figure 8.2.

With remikren 90, 17, and 3 possible formulas would be generated, considering a mass accuracy of 5, 1, and 0.1 ppm and restricting the search with the following elements: C, H, O, N, and S. An elegant way to significantly reduce the number of possible elemental formulas is to include as a constrain the isotopic distribution of the measured analyte versus the calculated one.

The mass resolution is defined as the smallest mass difference (Δm) between two equal magnitude peaks so that the valley between them is a specified fraction of the peak height. The mass resolving power (R) in a mass spectrum is the observed m/z of a singly charged ion (m) divided by the difference [between M and an adjacent ion that can be separated (Δm)] (Murray et al., 2006). While the mass resolution is not mass dependent, the resolving power can change with the mass, depending on the type of instrument. In practice, the full peak width at half-mass or half-height (FWHM, FWHH) is used to determine the mass resolution. For singly charged ions, the minimal mass resolution to differentiate isotopes would be in the range of 0.5–0.7 m/z, which is typical for a quadrupole instrument. It is also referred to as unit mass resolution. The mass resolving power is calculated by dividing the FWHM by the m/z of the compound. The mass resolving power (MRP) at m/z 500 is typically in the range of 8000–12,000 for time-of-flight instruments while the MPR can vary from 5000 to 100,000 for the orbitrap instruments depending on the duty cycle. On Ion cyclotron resonance instruments an MPR of 500,000 and above can be reached. The calculated isotopic distribution of the protonated remikiren is illustrated in Figure 8.3.

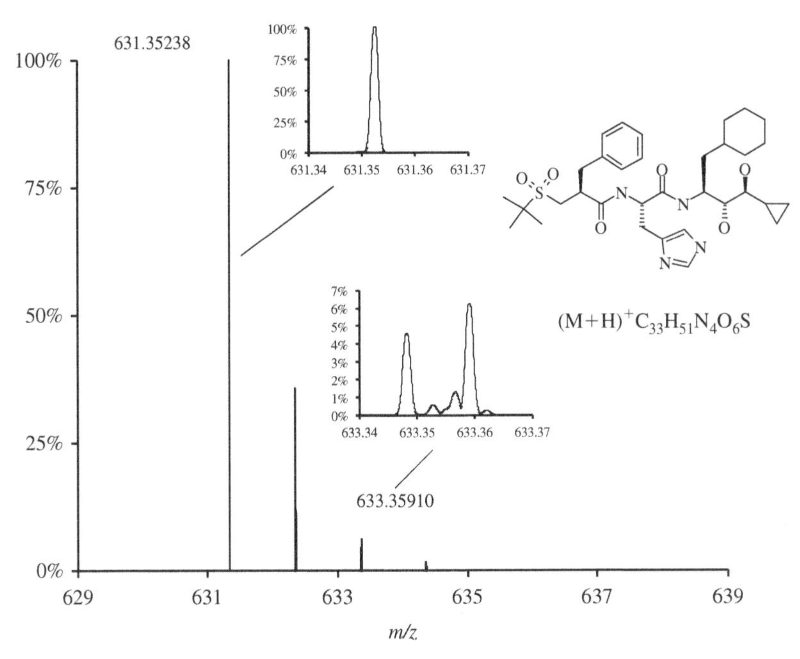

FIGURE 8.3 Isotopic distribution of remikiren at a theoretical mass resolving power of 500,000.

To avoid interferences resolution is important for accurate mass measurements, and a mass resolving power of 5000 is generally sufficient. In the case of the isotopes or overlapping ions, much higher resolutions are necessary, as illustrated in Figure 8.3 for the separation of the C, O, S, and N isotopes.

8.2 MAJOR COMPONENTS OF A MASS SPECTROMETER

Mass spectrometry is an analytical technique that can determine precisely the atomic or the molecular weight of atoms or molecules once they have been ionized. There are four key elements in mass spectrometer: the sample introduction, the source where ionization occurs, the mass analyzer, and the detector. Mass spectrometry can analyze many different types of samples that range from solid, liquid, or gases. First, the molecules have to be ionized either under vacuum or at atmospheric pressure. Depending on the ionization technique, either molecular ions ($M^{+\bullet}$) with an odd electron number or protonated ions ($[M + H]^+$ with an even electron number are formed in the positive mode and M^-, $M^{\cdot -}$ or ($[M - H]^-$ in the negative mode. Ionization techniques are often classified into soft ionization, where little or no fragmentation occurs, and hard ionization, where fragmentation is extensive. Electrospray ionization (ESI) and matrix-assisted laser desorption ionization (MALDI) are considered as soft ionization techniques while electron impact (EI) is considered as a hard ionization technique. Over the years, many different types of ion sources and mass analyzers have been developed.

Most mass analyzers operate under high-vacuum conditions to ensure that charged particles do not deviate from their trajectories due to collision with residual gas. Mass spectrometers can be grouped into different types of operation mode: continuous mode (magnetic sector, quadrupole), pulsed mode (time of flight), and ion trapping mode (quadrupole traps, Fourier transform ion cyclotron, orbitrap).

8.3 ION SOURCES

8.3.1 Electron Impact Ionization

Electron impact ionization is one of the most classical ionization techniques used in mass spectrometry, in particular in combination with gas chromatography. The sample (solids, liquids, and gases) is vaporized into the vacuum where the formed gas-phase molecules are bombarded with electrons generated by a glowing filament. One or more electrons are removed from the molecules to form odd-electron ions ($M^{+\bullet}$) or multiply charged ions. The mass range of the ionization technique is typically below 1000 u. The ionization energy of most organic compounds to form a radical cation is below 15–20 eV and any excess of energy transferred to the molecules causes fragmentation. The amount of $M^{+\bullet}$ and fragment ions is characteristic for each compound. A standard energy of 70 eV is mostly used because it provides the best sensitivity and causes reproducible fragmentation, allowing searches in large commercial libraries to rapidly identify compounds present in a sample (Ausloos et al., 1999). Although most analytical applications feature the use of EI in the positive ion mode, negative mode operation is also possible. EI is mostly combined with single MS mass analyzers such as quadrupole, ion trap, or time of flight. EI spectra are quite informative; however, often the molecular ion is too small to clearly identify the right compound between different structural analogs. In such a case on repeats the analysis with vacuum chemical ionization generates mostly a protonated ion of the analyte.

8.3.2 Electrospray Ionization

Electrospray (Fig. 8.4) is a process where a high potential is applied onto a liquid to generate a fine aerosol. Electrospray or electrostatic spraying has been applied to electrostatic painting, rocket propulsion, or fuel atomization. The phenomenon of electrospray was observed and investigated long before it was practical to the analysis of gas-phase ions transferred from solution into a mass spectrometer. One of the earliest reports of the effect of an electrical tension applied to a liquid was made in 1600 by William Gilbert. He observed that a

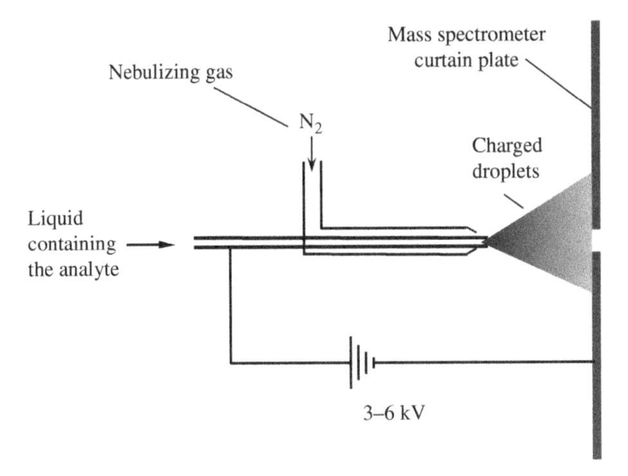

FIGURE 8.4 Schematic of electrospray.

spherical drop of water on a dry surface is drawn up into a cone when a piece of rubber amber is held at a suitable distance above it (Taylor, 1969).

Electrospray ionization for mass spectrometry, in which gas-phase ions are generated from electrolytes dissolved in a solution, was developed by J. Fenn and co-workers in an attempt to analyze large biomolecules by mass spectrometry (Fenn et al., 1989). Practically, a liquid is introduced into a small stainless steel or fused silica capillary (sprayer < 150 μm). A strong electric potential (3–6 kV) is set between the sprayer tip and a counterelectrode. An electric field gradient is generated, which induces the deformation of the liquid into a conical shape called the Taylor cone (Fig. 8.5). Then the solution forms a charged aerosol. After size reduction of the droplets by evaporation at atmospheric pressure, ions escape from the droplets and are sampled into the mass analyzer. Electrospray works best within a flow rate range from 1 to 50 μL/min and with high organic solutions (methanol or acetonitrile). In electrospray the liquid flow range can be increased by adding coaxially nitrogen as the nebulizing gas and is referred to as ion spray (Bruins et al., 1987) or as pneumatically assisted electrospray. During the electrospray process current flows around the conducting loop similar to an electrolysis process. In the positive ion mode oxidation reduction will occur at the sprayer tip. In some configurations, the potential is not set at the sprayer tip but at the interface of the mass spectrometer.

Electrospray ionization works best with preformed ions in solution and when preformed ions are separated from their counterions. In 1991 Blades et al. reported on the electrophoretic nature of electrospray, in which the charge balance requires the conversion of ions into electrons. Therefore, oxidation occurs at the needle (Fig. 8.5) and the interface of the mass spectrometer acts as a counterelectrode. Electrospray is particularly suitable for the analysis of inorganic ions and molecules, which have acidic or basic functional groups. Organic molecules are generally observed as protonated or deprotonated

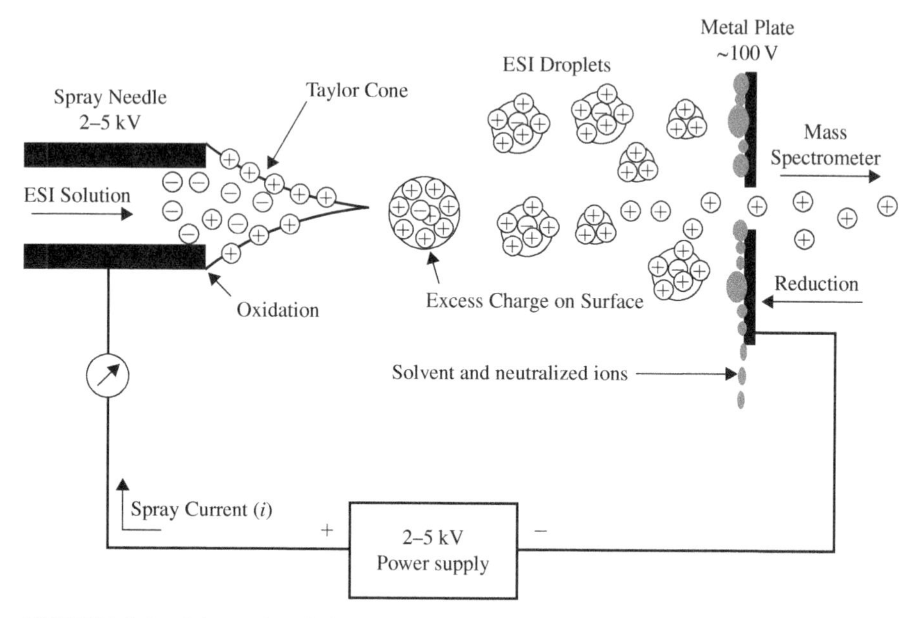

FIGURE 8.5 Schematic of electropheric nature of electrospray process (Cech et al., 2001).

molecules, depending on the polarity. Because signal intensities are dependent on the pK_a of the analytes and the organic content, the infusion or LC mobile-phase composing must be optimized. For peptides, it has been shown that intense signals can be observed either in positive or negative mode using strongly acidic or basic solutions, respectively. These observations are reported as "wrong way round" and have been discussed by Zhou and Cook (2000). Analytes can also be detected through the formation of adducts: sodium, ammonium, silver (Rentel et al., 1998), and the like in positive mode and fluorine, formate, and the like in negative mode.

The mechanism of formation of gas-phase ions from ions in solution is still under investigation (Bruins et al., 1987; Fenn et al., 1989; Kebarle and Peschke, 2000) but two major mechanisms have been proposed: (i) the charge residue model (CRM) described by Dole et al. (1968) and (ii) the ion evaporation model (IEM) proposed by Iribarne and Thomson (1976). Evaporation of charged droplets to produce free gas-phase ions from analyte species in solution was first proposed for mass spectrometry by Dole et al. in (1968). In 1966, when Dole first had the idea that lead to this study, there was no evidence of any possible solution to the "vaporization problem," which stated that large polyatomic molecules, including the complex and fragile species of biological interest that were proteins, could not be vaporized without extensive fragmentation and decomposition (Fenn, 2002). Dole's proposition was that evaporation of solvent would increase the surface charge density until it reached the Rayleigh limit at which the forces due to Coulombic repulsion and surface tension become comparable. Ultimately, droplets so small that they contain one single solute molecule are formed, after disruption of parent droplets into smaller droplets continue to evaporate by "Coulombic explosions." This molecule becomes an ion, thus a "charge residue," when it retains some of the droplet charge as the last of the solvent is vaporized.

In 1976 Iribarne and Thomson proposed the atmospheric pressure ion evaporation model (IEM). This model is consistent with the scenario described by Dole and co-workers where a sequence of evaporation and Coulombic explosions leads to droplets where charge densities are so high that the resulting electrostatic field at their surface is high enough ($> 10^9$ V/m) to lift (or "push") solute ions into the gas phase. In the experiments conducted by Iribarne and Thomson, charged droplets where generated by pneumatic nebulization, sometimes referred to as aerospray (Fenn, 1993). Most published work suggests that that the two models strongly depend on the nature of the analyte and that most molecules follow the IEM proposed by Iribarne and Thomson while large macromecules undergo mostly the charge residue mechanism. A recent study from Nguyen and Fenn (2007), where the authors demonstrated the benefit of the addition solvent vapor to the bath gas, showed that in electrospray most gas-phase ions are produced by the IEM rather than the CRM. Typical flow rates for electrospray and pneumatic electrospray ranges from microliters/milliliter to milliters/minute and low-flow electrospray where the samples are infused into the mass spectrometer at the nanoliter flow rate range is called nanoelectrospray (Wilm and Mann, 1996). The infusion of a few microliters will result in a stable signal for more than 30 min using pulled capillaries or chip-based emitter (Zhang and Van Pelt, 2004). With infusion signal, averaging allows to improve the limit of detection in tandem mass spectrometry. Nanoelectrospray is particularly important in combination with nanoflow liquid chromatography or chip-based infusion (Sikanen et al., 2009).

8.3.3 Chemical Ionization and Atmospheric Pressure Chemical Ionization

Chemical ionization is an ionization mechanism that allows the formation of protonated or deprotonated molecules via a gas-phase ion–molecule reaction. It exists under two different forms: one under vacuum (CI) and the second one at atmospheric pressure referenced as atmospheric pressure chemical ionization (APCI). The principal difference between CI and EI mode is the presence of reagent gas, which is typically methane, isobutene, or ammonia (Munson, 2000). The electrons ionize the gas to form the radical cations (in the case of methane, $CH_4 + e^- \rightarrow CH_4^{+\bullet} + 2e^-$). In positive chemical ionization (PCI), the radical cations undergo various ion–molecule reactions to form CH_5^+ and finally lead to the formation, after proton transfer ($CH_5^+ + M \rightarrow [M + H]^+$), of protonated molecules. Negative chemical ionization (NCI) (Budzikiewicz, 1986), after proton abstraction, leads to the formation of deprotonated molecules $[M - H]^-$. Negative ions can be produced by different processes such as by capture of low-energy electrons present in the chemical ionization plasma. The major advantages of negative CI over positive EI or CI are higher sensitivity, the occurrence of the molecular ion, and less fragmentation. Due to its high sensitivity NCI is mainly used in quantitative analysis after derivatization of the analyte (Maurer, 2002).

Atmospheric pressure chemical ionization is a gas-phase ionization process based on ion–molecule reactions between a neutral molecule and reactant ions (Carroll et al., 1975). The method is very similar to chemical ionization with the difference that ionization occurs at atmospheric pressure. APCI requires that the liquid sample is completely evaporated (Fig. 8.6). Typical flow rates are in the range of 200–1000 μL/min. First, an aerosol is formed with the help of a pneumatic nebulizer using nitrogen. The aerosol is directly formed in a heated quartz or ceramic tube (typical temperatures 200–500°C) where the mobile phase and the analytes are evaporated. The temperature of the aerosol itself remains in the range 120–150°C due to the evaporation enthalpy. In a second step, the evaporated liquid is bombarded with electrons formed by corona discharge. In positive-mode primary ions such as $N_2^{+\bullet}$ are formed by EI. These ions react further with water in several steps by charge transfer to form H_3O^+. Ionization of the analyte A occurs by proton transfer. In negative mode, ions are formed either by (i) resonance capture ($AB \rightarrow AB^-$), (ii) dissociative capture ($AB \rightarrow B^-$), or (iii) ion–molecule reaction ($BH \rightarrow B^-$). Generally, APCI is limited to low-molecular-weight compounds (MW < 2000 Da), which do not undergo thermal decomposition, and singly charged ions $[M + H]^+$ or $[M - H]^-$ are predominantly observed. While electrospray is a condensed-phase ionization process, APCI is a gas-phase ionization process where analyte ionization efficiency is dependent on its gas-phase proton affinity. APCI has become very popular for liquid chromatography coupled with mass spectrometry because it can handle very easily liquid flow rates from 200 μL/min to 1 mL/min. Contrary to electrospray, the application of heat may generate thermal decomposition of the analyte. At atmospheric pressure the ionization occurs with high collision frequency of the ambient gas, and rapid desolvation and vaporization limits the thermal decomposition of the analyte.

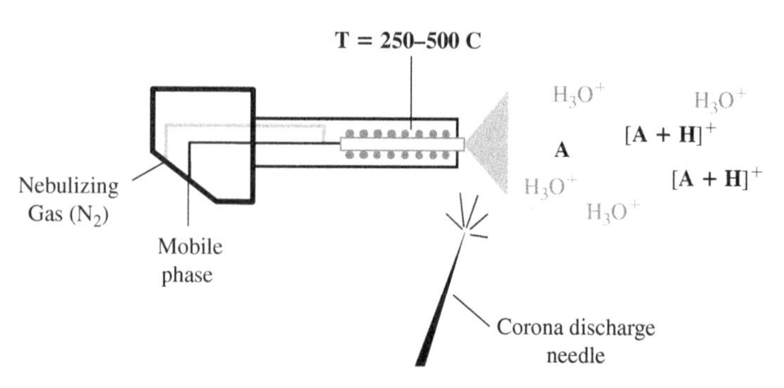

FIGURE 8.6 Schematic of the heated nebulizer probe.

8.3.4 Atmospheric Pressure Photoionization

The setup for atmospheric pressure photoionization (APPI) (Bos et al., 2006; Hanold et al., 2004; Raffaelli and Saba, 2003; Robb et al., 2000) is very similar to that of APCI (Fig. 8.7). Only the corona discharge is replaced by a gas discharge lamp (krypton,10.0 eV) that generates ultraviolet (UV) photons in vacuum. The liquid phase is also vaporized by a pneumatic nebulizer and different geometries are used. Most analytes have ionization potentials below 10 eV, while high-pressure liquid chromatography (HPLC) solvents have higher ionization potentials (water 12.6 eV, methanol 10.8 eV, and acetonitrile 12.2 eV).

The absorption of a photon by the molecule and the ejection of an electron forms a radical cation. Better sensitivities have been reported with the addition of dopants such as toluene or acetone. The mechanism of ionization is not fully understood, but two different mechanisms can occur: (i) dopant radical cations react with the analyte by charge transfer or (ii) the dopant radical cation can ionize the solvent molecules by proton transfer, which can then ionize the analyte. APPI can also be performed in the negative mode. As APCI, APPI can handle a large range of analytes. The performance of APPI is dependant on the flow rate, and better sensitivities, compared to APCI, have been reported at lower flow rates. Because APCI and APPI are gas ionization processes, it appears that compared to ESI they are less or differently affected by matrix effects (Robb and Blades, 2008). APPI is attractive for a large variety of neutral analytes such as steroids (Cai et al., 2005).

8.3.5 Matrix-Assisted Laser Desorption Ionization

Matrix-assisted laser desorption ionization (MALDI) has grown from the efforts to analyze nonvolatile macromolecules by mass spectrometry (Fig. 8.8). Two groups were able to obtain mass spectra of proteins in the late 1980s. The first group was led by Tanaka (Tanaka et al., 1988) and featured a methodology where the analyte was mixed in a matrix of glycerol and cobalt and ionized with a laser. Karas and Hillenkamp (1988) developed a MALDI methodology where

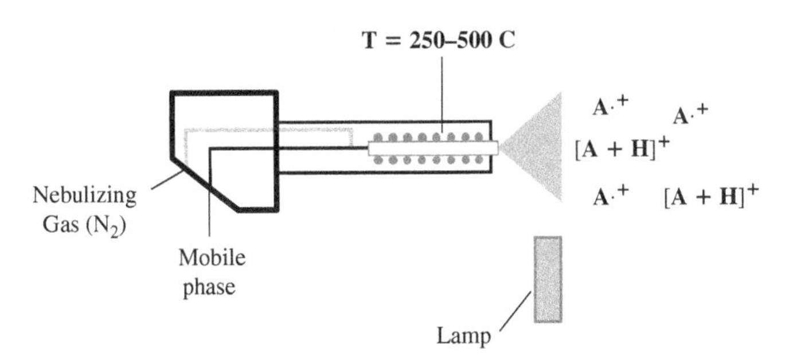

FIGURE 8.7 Schematic of the photoionization probe.

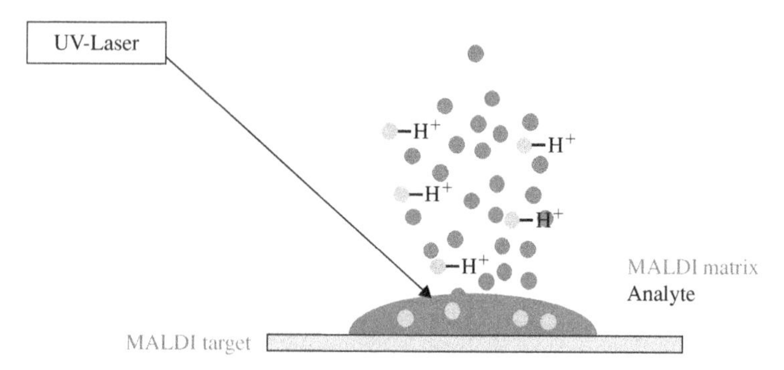

FIGURE 8.8 Matrix-assisted laser desorption ionization (MALDI).

the analyte was mixed with a matrix solution containing UV-absorbing molecules (Table 8.3). A few microliters of the solution are then spotted onto a MALDI target. After introduction of the target into the vacuum, an UV laser pulse is used to desorb and ionize the sample. A nitrogen laser emitting at 337 nm and Nd:YAG laser emitting at 355 nm are the most widely used for MALDI. Two parameters can be tuned: (i) the laser energy and (ii) the laser frequency (typically 20–200 Hz).

Matrix-assisted laser desorption ionization is a very powerful technique for the analysis of synthetics and biopolymers, and it has completely replaced former techniques such as fast atom bombardment (FAB). In most cases, singly charged ions are predominantly detected while very little fragmentation or multiply charged ions are observed. MALDI is commonly used for the analysis of high-molecular-weight compounds such as peptides and proteins (Rappsilber et al., 2003), synthetic polymers (Nielen, 1999), deoxyribonucleic acid (DNA) (Gut, 2004), and lipids (Schiller et al., 2004).

MALDI has the intrinsic high sample throughput advantage over electrospray ionization LC–MS. Sample preparation and separation can be decoupled from the mass spectrometric analysis. The MALDI target plate can be easily archived for the facile reanalysis of selected samples, if necessary. MALDI or ESI are suitable for the analysis of proteins as depicted in Figure 8.9. One of the key advantages of ESI over MALDI is the formation of multiply charged ions, which allows the analysis of proteins on almost any type of mass analyzer while MALDI requires a time-of-flight (TOF) mass analyzer in the linear mode to cover the high mass range.

The high-throughput capability of MALDI and the different ionization mechanisms make this technique an attractive alternative to ESI for the analysis of low-molecular-weight compounds (LMWC) (Cohen and Gusev, 2002). However, interferences of matrix ions and the ionization of the LHWCs remains significant with this technique (Donegan et al., 2004; McCombie and Knochenmuss, 2004). Desorption/ionization on porous silicon (DIOS) without any matrix has been described for the analysis of LMWC with no chemical background (Go et al., 2003; Lewis et al., 2003). The use of MALDI for the

TABLE 8.3 Commonly Used Matrices for Matrix-Assisted Laser Desorption Ionization

Matrix	Name	Comments
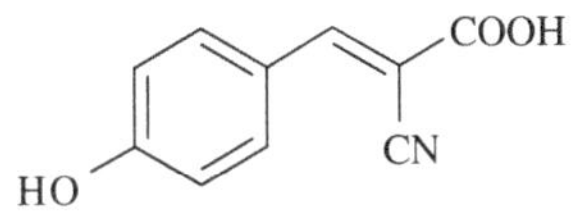 [structure: HO–C6H4–CH=C(CN)–COOH]	α-Cyano-4-hydroxycinnamic acid (4-HCCA)	Peptides, low-molecular-weight compounds
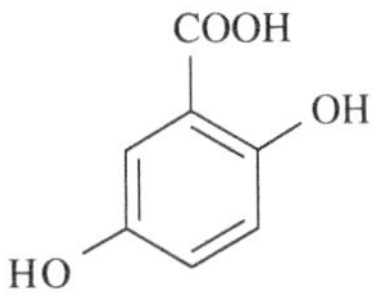 [structure: 2,5-dihydroxybenzoic acid; COOH, OH, HO]	2, 5-Dihydroxy benzoic acid (DHB)	Proteins
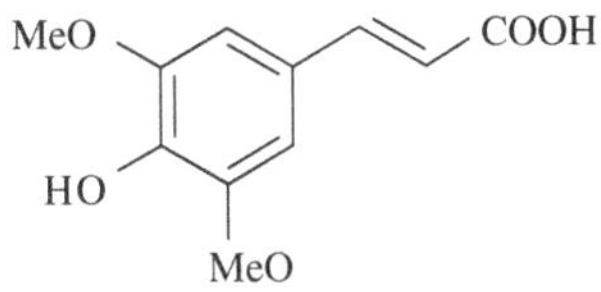 [structure: MeO, HO, MeO substituted cinnamic acid; COOH]	Sinnapinic acid (SA)	Proteins
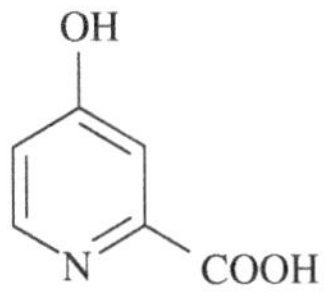 [structure: OH, N, COOH pyridine]	4-Hydroxypicolinic acid (HPA)	Oligonucleotides
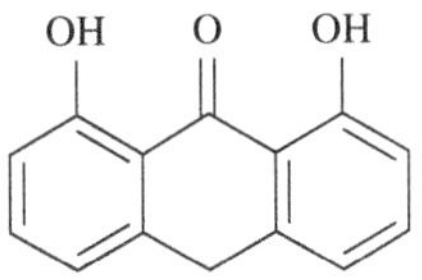 [structure: OH, O, OH anthrone]	1, 8, 9 Antrancetriol (dithranol)	Polar and apolar polymers

analysis of small molecules has been recently reported. Particularly attractive is the coupling of a MALDI source with a triple quadrupole mass analyzer for quantitative analysis in the selected reaction monitoring (SRM) mode due to very high analysis speed (Kovarik et al., 2007; Wagner et al., 2008).

MALDI ionization is mainly performed in the vacuum region of the mass spectrometer; however, it can also be performed at atmospheric pressure (Moyer and Cotter, 2002; Schneider et al., 2005). One of the major advantages of MALDI is that in can be adapted very easily, using the atmospheric pressure interface on virtually any type of mass spectrometer.

Surface-enhanced laser desorption/ionization (SELDI) is a distinctive form of laser desorption ionization where the target plays an active role in the sample preparation procedure and ionization process (Tang et al., 2003). Depending on the chemical or biochemical treatment, the SELDI surface acts as solid-phase extraction or an affinity probe. A chromatographic surface is used for

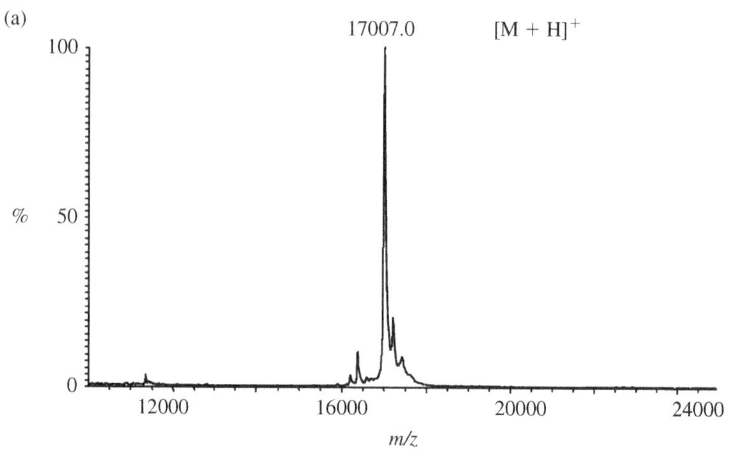

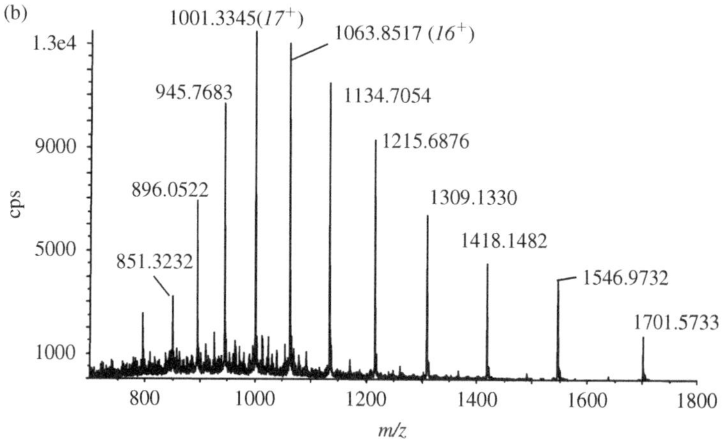

FIGURE 8.9 Mass spectra of recombinant protein obtained by (*a*) matrix-assisted laser desorption ionization–time-of-flight and (*b*) electrospray–quadrupole time of flight.

sample fractionation and purification of biological samples prior to direct analysis by laser desorption/ionization. SELDI is mainly used for protein profiling applications and in biomarker discovery and features, the direct comparison of protein profiles obtained from control and patient groups.

Because MALDI is a desorption technique, it is suited for the analysis of surfaces such as tissue slices such as brain or liver (Caprioli et al., 1997). In this application, the matrix is, generally, sprayed on the complete surface of the tissue. The laser resolution is about 50–100 μm and complete analyte distribution (low-molecular-weight compounds, peptides, proteins) images can be recorded (Reyzer et al., 2003; Rohner et al., 2005). The technique is attractive to monitor the distribution of pharmaceutical compounds or their metabolites in tissues (Hopfgartner et al., 2009; Signor et al., 2007)

8.3.6 Other Ionization Techniques

Direct analysis of solid samples or analytes present on solid or liquid surfaces without any sample preparation has recently gained much interest. Desorption electrospray ionization (DESI) is an atmospheric pressure desorption ionization method introduced by Cooks et al. (2006) where ions are produced directly from the surface to be analyzed. The DESI technique features the use of charged liquid droplets that are directed by a high velocity as jet (on the order of 300 m/s) to the surface to be analyzed. Analytes are desorbed from the surface and analyzed by the mass spectrometer.

Direct analysis in real time (DART) has been reported by Cody et al. (2005). This technique is based on the reactions of metastable helium atoms generated by corona discharge with oxygen/water (negative mode) or water clusters (positive mode). The reactant ions ionize the analytes either by cluster-assisted desorption or proton exchange. Both DESI and DART methods generate mostly protonated or deprotonated molecular ions. More recently, the analysis of liquid samples by extractive electrospray ionization (EESI) has been reported (Chen et al., 2006). EESI uses two separate sprayers. The first sprayer nebulizes the sample and the second sprayer is essentially identical to an electrospray source. The principle of analysis is based on liquid extraction of the colliding microdroplets. More recently, the same EESI concept was applied to analyze volatile compounds on surfaces after desorption of the analytes on a neutral gas jet such as nitrogen (Chen et al., 2007).

8.4 MASS ANALYZERS

Most mass analyzers operate under high vacuum or at low pressure to ensure that the charged particles do not significantly deviate from their trajectories due to collision with residual gas. Mass spectrometers can be grouped into different types of operation mode: continuous mode (i.e., magnetic sector, quadrupole), pulsed mode (i.e., time of flight), and ion trapping mode (i.e., quadrupole traps, ion cyclotron, orbitrap).

8.4.1 Triple Quadrupole

A quadrupole mass analyzer (Fig. 8.10) consists of four hyperbolic or circular rods placed in parallel with identical diagonal distances from each other. The rods are electrically connected in diagonal. In addition to an alternating current (ac) potential (V) at a fixed frequency, a positive direct current (dc) potential (U) is applied on one pair of rods while a negative potential is applied on the other pair. The ion trajectory is affected in the x and y directions by the total electric field composed by a quadrupolar alternating field and a constant field. When accelerated ions enter the quadrupole, they maintain their velocity along the z axis. Because the ac potential is in the radio-frequency range, it is generally named RF potential.

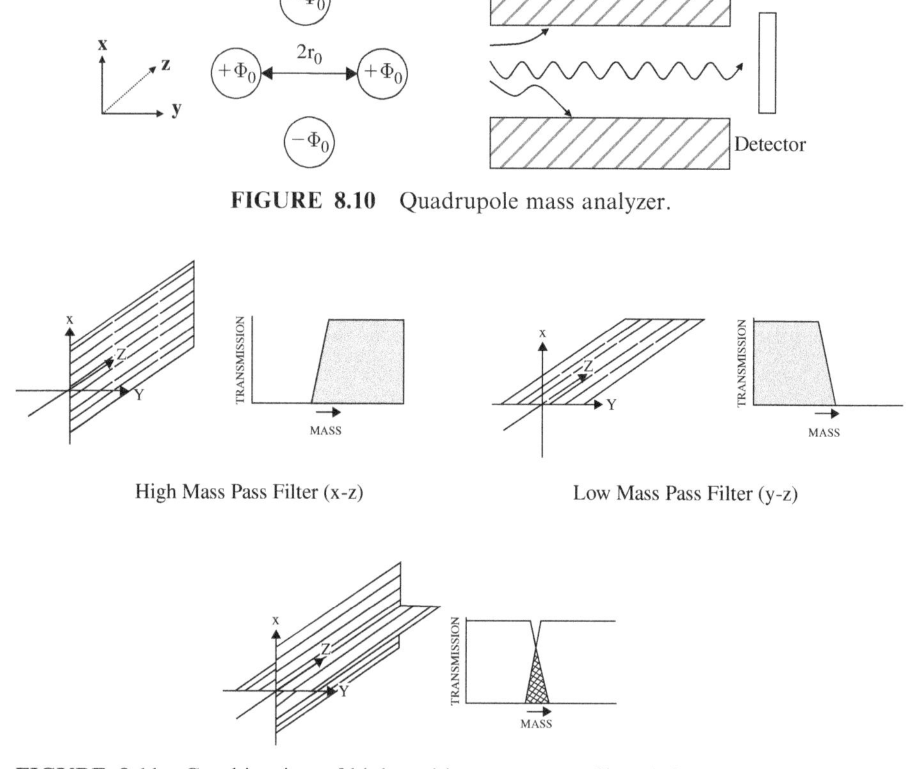

FIGURE 8.10 Quadrupole mass analyzer.

FIGURE 8.11 Combination of high and low mass pass filter defines the resolution of the mass analyzer.

The ion trajectories of ions are stable when they do not reach the rods of the quadrupole and are dependent from the ac and dc potentials. At constant ac frequency, m/z is proportional to the amplitude of U (dc) and V (ac) in a linear fashion. The operation of a quadrupole mass analyzer results from the overlap of a high mass filter of the x-z plane and a low mass filter of the y-z plane as illustrated in Figure 8.11 (Miller and Denton, 1986). This figure illustrates the strong relation between resolution and transmission.

A mass spectrum is obtained with a quadrupole mass analyzer by increasing U and V at a constant ratio. When the dc voltage of a quadrupole is set to zero and the RF voltage is maintained, ions remain focused with no mass selectivity. Therefore, RF quadrupoles serve as ideal ion guides or as a collision cell. Typically, quadrupole mass analyzers operate at unit mass resolution (FWHM 0.6–0.7 u), but resolution corresponding to a peak width of 0.2 u without significant loss in sensitivity has also been reported. The mass range of a quadrupole is generally from 5 to 4000 u and dependent on the frequency of the ac voltage, typically several hundred kilohertz. A quadrupole that operates at a higher frequency has a smaller mass range. A quadrupole can be operated at fixed U/V ratio (i.e., selected ion monitoring mode) or by scanning U and V at a constant ration (i.e., full scan mode). The time at which a quadrupole stays at a fixed U/V ratio is called the dwell time and ranges from 0.5 to 500 ms.

A triple quadrupole instrument (QqQ) is a combination of two quadrupole mass filters (tandem mass spectrometry) separated by a collision cell that is also a quadrupole operating in RF-only mode (Yost and Enke, 1978) (Fig. 8.12). In the past triple quadrupole mass spectrometers were built in a linear geometry. Today, modern instrumentation features the use of a curved collision cell to minimize space and pumping requirements. A common nomenclature is to use (Q) to describe a quadrupole that is operated is in RF/dc mode and (q) for a quadrupole that is operated in RF-only mode. Tandem mass spectrometry (MS/MS) is particularly attractive to obtain additional mass spectral information. In a first step, a specific m/z ion (precursor ion) is selected in the first mass analyzer (Q_1). Collision-induced dissociation (CID) between the accelerated precursion ions and a neutral gas (argon or nitrogen) occurs in the collision cell (q_2). The fragment ions (product ions) are then sorted according to their mass-to-charge ratio in the second mass analyzer (Q_3) and recorded by the ion detector. This method of performing MS/MS experiments is called *MS/MS in space* contrarily to quadrupole ion traps where MS/MS experiments are performed in time. The potentials used to carry out CID are in the range of 0–250 V. The collision energy is defined in electrons volts (eV) and is, therefore, dependent on the charge of the ions. For a potential difference of 20 V the collision energy of a singly charged precursor ion would be of 20 V and 40 eV for a doubly charged precursor ion. The intensities of the fragments of the product ion spectra varies from instrument to instrument, which makes it challenging to build up libraries. The nature of the collision gas (N_2 or Ar) does not affect the fragmentation process while the gas pressure in the collision cell mainly influences the sensitivity.

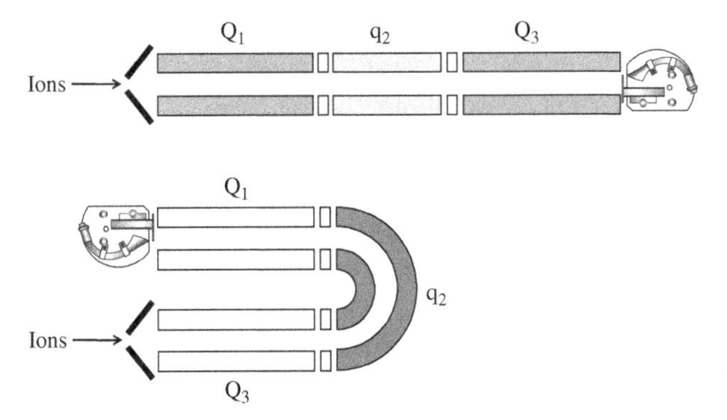

FIGURE 8.12 Schematic of triple quadrupole instrument, Q_1 and Q_3 mass analyzing quadrupoles, q_2 collision cell.

TABLE 8.4 Settings of Q_1 and Q_3 Quadrupoles for Various Scan Modes of Triple Quadrupole

Mode	Q_1	Q_3
Full-scan Q_1/single ion monitoring (SIM) Q_1	Scan/fixed	RF mode
Full-scan Q_3/single ion monitoring (SIM) Q_3	RF mode	Scan/fixed
Product ion scan (PIS)	Fixed	Scan
Precursor ion scan (PC)	Scan	Fixed
Neutral loss (NL)	Scan	Scan-neutral loss offset
Selected reaction monitoring (SRM)	Fixed	Fixed

Various types of MS and MS/MS experiments can be performed on a QqQ, which are summarized in Table 8.4. Symbols to describe various MS/MS or multistage MS^n experiments have also been proposed (de Hoffmann, 1996; Schwartz et al., 1990).

Precursor ion and neutral loss scans are performed on a QqQ to identify structurally related components in a mixture using a common fragment with the parent compound or the specific neutral loss of conjugated compounds such as phase II metabolites. These selective scan modes do not require any knowledge of the molecular weight or structure of the compounds under investigation.

In the selected reaction monitoring (SRM) mode, Q_1 is set at the mass of the precursor $[M+H]^+$ and Q_3 is set at the m/z of the most important fragment of the analyte. Because in SRM mode both quadrupoles are set at a fixed m/z, much better detection limits can be achieved compared to the scanning mode. Therefore, this mode has become the work horse for quantitative analysis. Typical dwell times are in the range of 5–250 ms allowing the analysis of several hundred analytes in a single LC–MS analysis. Quantitative analysis in single ion monitoring (SIM) mode can also be performed on a QqQ either using Q_1 and Q_3. Generally, with SIM in Q_3 mode, the collision cell is filled with collision gas (nitrogen or argon) and serves as a further declustering device to improve signal-to-noise (S/N) ratio.

8.4.2 Ion Trap and Linear Ion Trap

The quadrupole ion trap is a device that utilizes ion path stability of ions to separate them by their mass-to-charge ratio (March, 2000). The quadrupole ion trap and the related quadrupole mass filter were invented by Paul and Steinwedel (1953). Quadrupole ion trap (QIT) mass spectrometer operates with a three-dimensional (3D) quadrupole field. The QIT is formed by three electrodes: a ring electrode with a donut shape placed symmetrically between two end cap electrodes (Fig. 8.13).

The QIT can be described as a small ion storage device where the ions are focused toward the center of the trap by collision with helium gas. The x and y components of the field are combined to a single radial r component where

$r^2 = x^2 + y^2$ because of the cylindrical symmetry of the trap. The motion of ions in the trap is characterized by secular frequencies, one radial and one axial. As for quadrupoles, the motion of ions can be described by the solutions of Matthieu's equations (a and q). Ions can be stored in the trap with the condition that trajectories are stable in r and z directions (Fig. 8.13). Each ion of a specific m/z will be trapped at a specific q_z value. The lower m/z will be located at the higher q_z values.

The quadrupole ion trap can store only a limited number of ions before space charging occurs. To circumvent this effect, most instruments have an automatic gain control (AGC) procedure. This procedure determines the exact fill time of the trap to maximize sensitivity and minimize resolution losses due to space charge. A mass spectrum can be obtained by mass-selective ejection where the amplitude of the RF potential is continuously increased at a certain rate. The mass-selective axial instability mode requires that the ions are confined at the center of the trap and at a limited mass range. Resonant mass ejection is another procedure that allows for the generation of a mass spectrum with a higher mass range. Ion motion can be modified either by exciting the radial or the axial frequencies by applying a small oscillating potential at the end cap electrodes during the RF ramp. In both mass analyzing modes, the resolution of the spectrum is strongly dependent on the speed at which the RF amplitude is increased. Higher resolution can be achieved with slower scan speed. Compared to quadrupole instruments, higher sensitivity can be obtained in full-scan mode with the quadrupole ion trap due to the ability of ion accumulation in the trap before mass analysis. Rapid mass analysis with the mass instability scan allows one to scan at a speed of several thousand units per/second. There are several important components that affect the time necessary to obtain a mass spectrum (duty cycle): (i) the injection time from

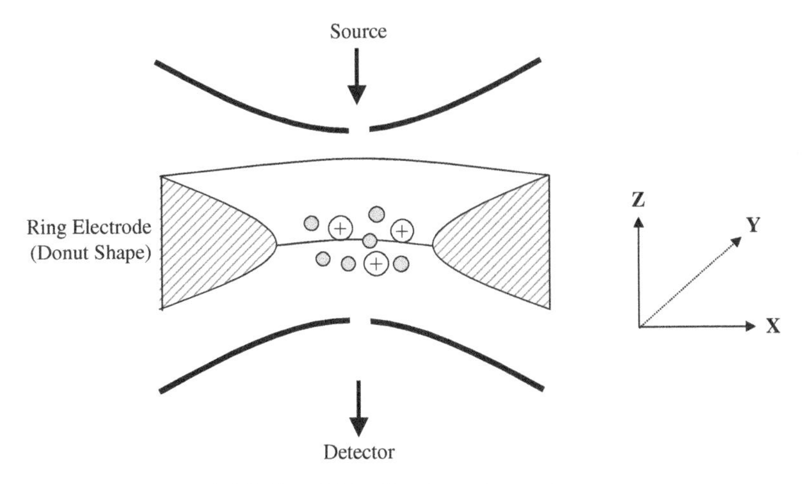

FIGURE 8.13 Quadrupole ion trap.

0.05 to 500 ms, (ii) the scan speed (in the range of 5000–20,000 u/s), and (iii) isolation of the precursor ion and fragmentation in tandem MS or MS^n. Contrary to the triple quadrupole, MS/MS is not performed in *space* but in *time*. Another significant difference is the use of helium as the collision gas. Because the ion trap is permanently filled with gas, the instrument can switch very rapidly from single MS to MS/MS mode. High sensitivity can be achieved in the QIT because of ion-selective accumulation of the precursor. Another advantage compared to the triple quadrupole is the short duty cycle for an MS/MS experiment. A typical MS^n sequence requires several steps. First, after injection, the precursor ion is isolated and then excited while fragments are trapped and then the fragments are ejected from the trap to obtain an MS^2 spectrum. Because MS/MS is performed in time in the same physical device, the operation can be repeated several times. The next step involves the isolation of a fragment ion to perform isolation and CID fragmentation to generate an MS^3 spectrum. Most commercial instruments can technically perform MS^n to the 10th or to the 11th power, but in practice most analytes do generate fragments after four or five MS/MS experiments. The challenge of MS/MS in ion traps is to excite the precursor ions efficiently and trap the product ions in the same device. Generally, solely the precursor is excited in a specific window corresponding to 1–4 u. The consequence is that fragment ions are not further excited and cannot produce second-generation fragments. In many cases MS^2 collision-induced dissociation generates similar spectra than quadrupole collision-induced dissociation, but there are cases where the spectra differ significantly. The consequence is that for molecules that can easily lose water or ammonia, the most abundant fragment observed in MS^2 will be M-18 or M-17, which is not very informative. To overcome this limitation, wide-band excitation (range 20 u) can be applied. Another difference compared to QqQ is that QIT has a low mass cutoff of about one-third of the mass of the precursor ion.

In a linear ion trap (LIT) the ions are confined radially by a two-dimensional (2D) RF field. To avoid the axial escape of ions, a dc potential is applied to the end electrodes. LIT has gained interest for various applications either as a stand-alone mass analyzer or coupled with a Fourier transform ion cyclotron, 3D ion trap (IT), TOF, or orbitrap mass analyzer (Douglas et al., 2005). Due to the small volume, 3D ion traps have a limited capacity for ion storage. Overfilling the IT results in a deterioration of the mass spectrum and the dynamic response range. The number of ions introduced in the trap can be controlled in different ways to avoid space charging. There are several advantages to trap ions in a 2D trap compared to 3D traps: (i) no quadrupole field along the *z* axis, (ii) enhanced trapping efficiencies, (iii) more ions can be stored before observing space charging effects, and (iv) strong focusing along the centerline instead of focusing ions to a point. Schwartz et al. (2002) have described a stand-alone linear ion trap where mass analysis is performed by ejecting the ions radially through slits of the rods using the mass instability mode (Fig. 8.14). To maximize sensitivity the detection is performed by two detectors placed axially on either side of the rods.

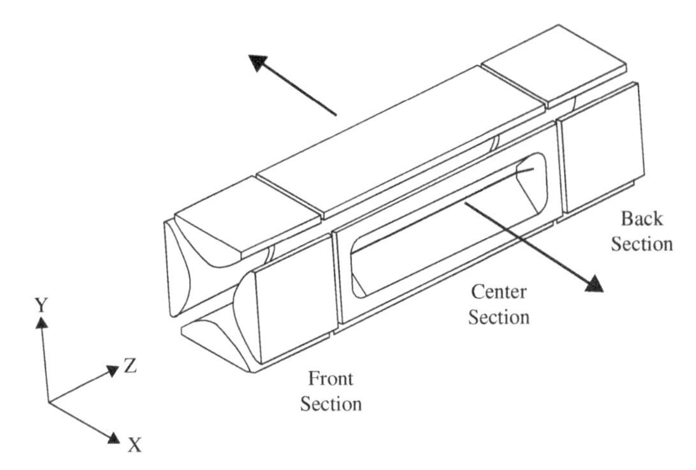

FIGURE 8.14 Stand-alone linear ion trap (Schwartz et al., 2002).

The single setting of the helium pressure in an ion trap compromises ion trapping, isolation, and fragmentation efficiencies and isolation time to fulfill all requirements of the instrument, affecting overall sensitivity and cycle time *r*. To improve the overall performance, a dual-pressure LIT was developed consisting of two trapping cells separated by a single aperture lens to allow differential pumping between the traps (Pekar Second et al., 2009). LITs have been successfully coupled to time-of-flight (LIT–TOFMS) (Collings et al., 2001) and Fourier Transform Ion Cyclotron Resonance (FTICR-MS) (Belov et al., 2001; Syka et al., 2004). The intention of such hybrid instruments is to combine ion accumulation and MS^n features with the superior mass analysis (accuracy and resolution) and high sensitivity of TOF–MS or FTICR–MS. The ions stored in the trap are then axially ejected in a non-mass-dependent fashion to the mass analyzer.

One of the most efficient ways to perform mass analysis with an LIT is to eject ions radially. However, this approach requires that holes be drilled into the quadrupoles and limits the possibility to operate the instrument in RF/dc mode. Hager (2002) has demonstrated the use of fringe filed effects so that ions can be mass selectively ejected in the axial direction. A hybrid mass spectrometer was developed based on a triple quadrupole platform where Q_3 can be operated either in normal RF/dc mode or in the LIT ion trap mode (Fig. 8.15).

In the present system, MS/MS is performed in space where the LIT serves only as a trapping and mass analyzing device. The commercial form of this instrument is named Q Trap and already four different generations have been developed using different quadrupole geometries (Q Trap, 3200 Qtrap, 4000 Qtrap, and 5500 Qtrap). While quadrupole CID spectra are obtained in MS^2, MS^3 are typical trap CID spectra.

MS^3 is performed in the following manner: The first stage of fragmentation is accomplished by accelerating the precursor ions chosen by Q_1 into the pressurized collision cell, q_2. The fragments and residual precursor ions are transmitted into the Q_3 linear ion trap mass spectrometer and are cooled for

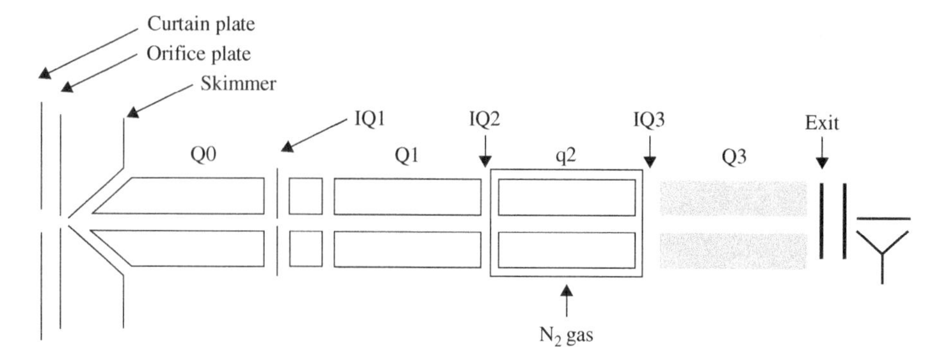

FIGURE 8.15 Schematic of triple quadrupole linear ion trap. Q_3 can be operated either with RF/dc as a standard quadrupole or with RF as linear ion trap with axial ejection.

approximately 10 ms. The next-generation precursor ion is isolated within the linear ion trap by application of resolving dc near the apex of the stability diagram. The ions are then excited by a single frequency of 85 kHz auxiliary signal and fragmented. The particularity of the QqQ_{LIT} is that the instrument can be operated in various ways as described in Table 8.5 (Hopfgartner et al., 2004a, 2004b).

Qualitative and quantitative analysis can be performed in the same LC–MS run. As an example in a data-dependent experiment, the SRM mode can be used as a survey scan and the enhanced product ion (EPI) as a dependent scan. The consequence is that for each quantified analyte a confirmatory MS/MS spectrum can be obtained.

8.4.3 Time-of-Flight Spectrometry

Time of flight (TOF) may be the simplest way to perform mass spectrometric analysis. TOF is the measure of the time that ions need to cross in a field-free tube of about 1 m in length. It is a pulsed technique and requires a starting time point. The motion of an ion is characterized by its initial kinetic energy ($E_c = \frac{1}{2} mv^2$). Because all ions are accelerated with the same potential, they have the same kinetic energy, and, therefore, the speed of ions or the time to fly through the tube is proportional to their $\sqrt{m/z}$ value. Low-mass ions reach the detector more rapidly than high-mass ions. Because the velocity of the ions after ionization is generally low, they need to be accelerated by electric fields (2–30 kV). Due to the short fly time (50–100 μs) a good-quality spectrum can be generated within 100 ms over a large mass range.

Because only time is measured, the mass range is almost unlimited in linear ionization, which is particularly attractive for MALDI ionization. On the other hand, with conventional multichannel plate detector the response decreases with the increase of m/z of the ions. The resolution is strongly dependent on the ability to produce a very focused ion beam and to avoid energy dispersion in

TABLE 8.5 Mode of Operation of Triple Quadrupole Linear Ion Trap (QqQ_{LIT})

Mode of Operation	Q_1	Q_2	Q_3
Q_1 scan	Resolving (scan)	RF only	RF only
Q_3 scan	RF only	RF only	Resolving (scan)
Product ion scan (PI)	Resolving (fixed)	Fragment	Resolving (scan)
Precursor ion scan (PC)	Resolving (scan)	Fragment	Resolving (fixed)
Neutral loss scan (NL)	Resolving (scan)	Fragment	Resolving (scan offset)
Selected reaction monitoring mode (SRM)	Resolving (fixed)	Fragment	Resolving (fixed)
Enhanced Q_3 single MS (EMS)	RF only	No fragment	Trap/scan
Enhanced product ion (EPI)	Resolving (fixed)	Fragment	Trap/scan
MS^3	Resolving (fixed)	Fragment	Isolation/fragment trap/scan
Time-delayed fragmentation (TDF)	Resolving (fixed)	Trap/No fragment	Fragment/trap/scan
Enhanced resolution Q_3 single MS (ER)	RF only	No fragment	Trap/scan
Enhanced multiply charged (EMC)	RF only	No fragment	Trap/scan

the ionization/acceleration region. One way to reduce the kinetic energy spread is to introduce a time delay between ion formation and acceleration, referred to as delayed pulsed extraction. After a certain time delay, ranging from nanoseconds to microseconds, a voltage pulse is applied to accelerate the ions out of the source.

By applying an electrostatic mirror (mass reflectron) placed (Fig. 8.16) in the drift region of ions, the resolution can be improved significantly. Shortly, ions with a given m/z but having a higher energy penetrate deeper into the ion mirror region than those with the same m/z at lower energy. Because of the different trajectories, all ions with the same m/z reach the detector at the same time. Thus, all ions of the same mass have a much lower energy dispersion. With the reflectron the fly path is increased without changing the physical size of the instrument. In reflectron mode 15,000 mass resolving power is common, but the mass range is limited compared to the linear mode. TOF instruments detect simultaneously a large mass range of ions resulting in an increased sensitivity compared to scanning instruments. Many factors (Vestal, 2009) can affect the resolving power of TOF instruments, and many efforts are undertaken to commercialize instruments with resolving power in the range of 30,000

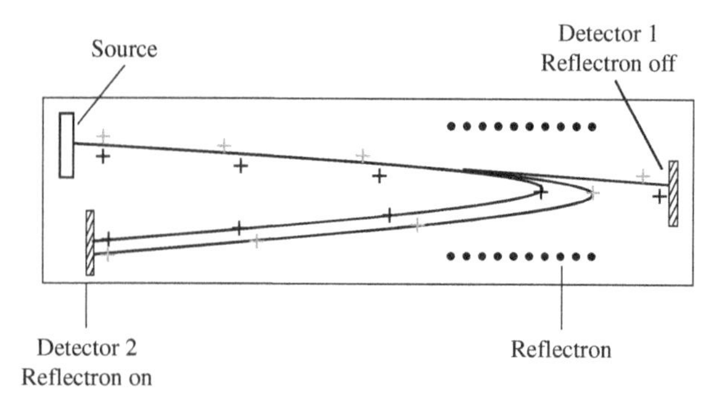

FIGURE 8.16 Schematic of simplest form of time-of-flight mass spectrometer: (1) reflectron off linear mode and (2) reflectron on reflectron mode.

or higher. To arrange continuous beam as generated by ESI, orthogonal acceleration was developed (Guilhaus et al., 2000).

The ion beam is introduced perpendicularly to the TOF, and packets are accelerated orthogonally (oa-TOF) at similar frequencies, improving the sensitivity. While a packet of ions is analyzed by the reflectron, a new beam is formed in the orthogonal acceleration.

MS/MS with QqQ suffers from relatively poor sensitivity and limited resolution. The replacement of the last quadrupole section (Q_3) of a triple quadrupole by a TOF analyzer to form a hybrid quadrupole–time-of-flight instrument (QqTOF) (Fig. 8.17) represents a powerful combination of mass range, resolution, and sensitivity (Chernushevich et al., 2001; Morris et al., 1996). In single MS mode the quadrupole serves as RF ion guide and the mass analysis is performed in the TOF.

With a large mass range the lighter ions may be discriminated, and it becomes necessary to scan or step the RF voltages of the quadrupole. In tandem MS mode the quadrupole can isolate the precursor ion at unit mass resolution. Accurate mass measurements of the elemental composition of product ions greatly facilitate spectra interpretation. The versatile precursor ion scan, another specific feature of QqQ, is maintained in the QqTOF instrument. However, in precursor scan mode the sensitivity is lower in QqTOF than in QqQ instruments. In general, QqTOF instruments are equipped with ESI sources. However, it is also possible to mount an orthogonal MALDI (o-MALDI) source with collisional cooling for analysis of low-molecular-weight compounds or peptides.

Because the TOF mass analyzer has a low duty cycle, the combination with an ion accumulation device such as an ion trap is very advantageous. This format also offers MS^n capabilities with accurate mass measurement. In all modes for mass analysis the ions are accelerated into the TOF. rQIT (Martin and Brancia, 2003) as well as linear ion trap have been combined with TOF either with MALDI or ESI sources.

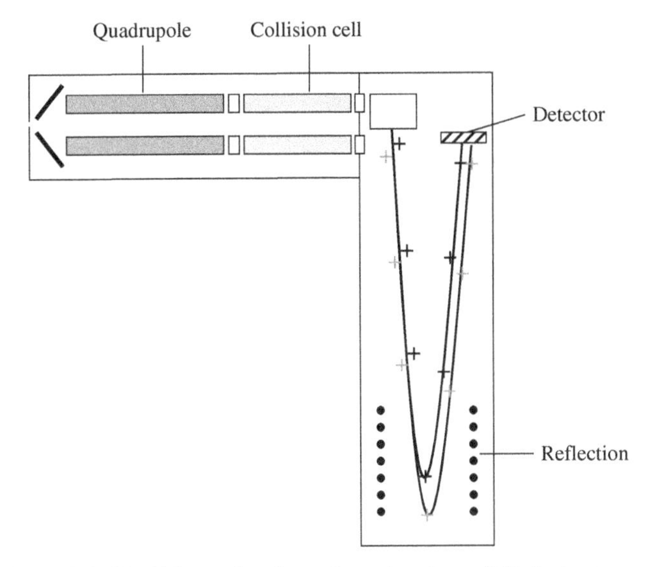

FIGURE 8.17 Schematic of quadrupole–time-of-flight instrument.

8.4.4 Fourier Transform Mass Spectrometry

While most mass analyzers such as quadrupole, ion trap, or TOF require destructive detectors such as electron multiplers or multichannel plate detectors, in Fourier transform mass spectrometry (FTMS) the detector uses a nondestructive detection mode. Ion cyclotron resonance (ICR) and the orbitrap uses Fourier transform detection.

The main components of an ion cyclotron resonance mass spectrometer (Fig. 8.18) are a superconducting magnet and a cubic or cylindrical cell. Typically, field strengths (B) are in the range of 4–12 tesla. Ions are stored in the cell according to their cyclotronic motion, which arises from the interaction of an ion with the unidirectional constant homogenous magnetic field. A static magnetic field applied in the z direction confines ions in the x and y directions according the cyclotronic motion. A low electrostatic potential is applied to the end cap electrodes to minimize the loss of ions along the z axis.

The trapping of ions generates a further fundamental motion of ions called magnetron motion. Magnetron frequencies are independent of m/z of the ions and are much lower frequencies (1–100 Hz) than cyclotron motion (5 kHz to 5 MHz). Cyclotron motion is characterized by its cyclotron frequency (f), which depends from (i) the magnetic field, (ii) charge of the ion, and (iii) the mass of the ion:

$$f = kB\frac{z}{m}$$

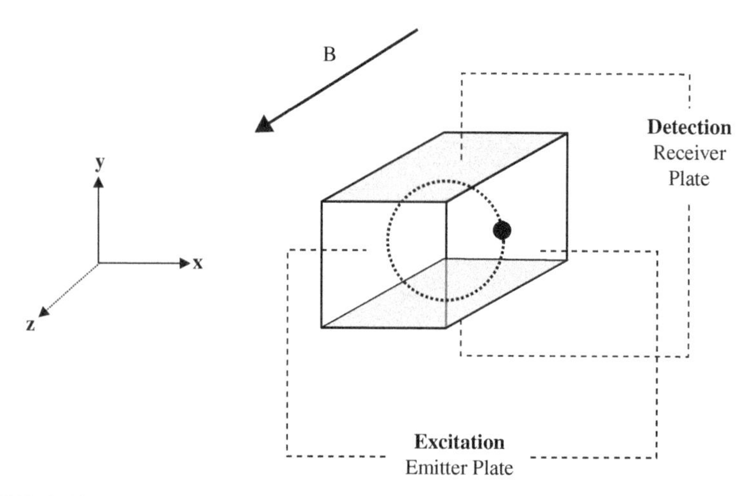

FIGURE 8.18 Diagram of ion cyclotron resonance instrument. The magnetic field is oriented along the z axis and ions (•) are trapped according to the same axis. Due to the cyclotronic motion the ions rotate around the z axis in the x-y plane.

where f is the cyclotron frequency, B is the magnetic field, z is the charge, m is the mass, and k is the constant.

Contrarily to other types of mass spectrometers, the detection is performed in a nondestructive way. The ions are detected by excitation applying a coherent broadband excitation on the excitation plates. The ions undergo cyclotron motion as a packet with a larger radius. When the ion packet approaches the detection plates, an alternating current is generated and referred to as an image current. Ions of any mass can be detected simultaneously with FTMS. The image current is a composite of different frequencies and amplitudes, which is converted by applying a Fourier transformation to frequency components and further to a mass spectrum.

Mass resolution is best with high field strength, and it decreases when the mass increases and is dependent on acquisition time. The resolution R is strongly dependent on the length of the transient time. Typical transient times are in the range of 0.1–2 s:

$$R = \frac{fT}{2}$$

where R is resolution, T is transient time, and f is cyclotron frequency.

The transient signal decreases with collision of ions and neutral gas molecules. It is therefore essential to operate at very high vacuum (10×10^{-10} torr). The dynamic range of FT-ICR is relatively poor because the instrument suffers from the fact that the number of ions in the trap must be in a specified range. Over- or underfilling of the trap results in mass shifts toward high and low values, respectively.

Markarov (Makarov, 2000; Perry et al., 2008) invented a novel type of mass spectrometer based on the orbital trapping of ions around a central electrode using electrostatic fields named orbitrap. Kingdon (1923) has already described the orbiting of ions around a central electrode using electrostatic fields in 1923, but the device had been only used for ion capturing and not as a mass analyzing device. The orbitrap is formed by a central spindlelike electrode surrounded by a barrel-like electrode. The m/z is reciprocal proportionate to the frequency of the ions oscillating along the z axis. Stability of trajectories in the orbitrap is achieved only if the ions have sufficient tangential velocity to collide with the inner electrode and short injection times are used:

$$\omega = \sqrt{(z/m)k}$$

where ω is ion frequency, z is charge, m is mass, and k is constant.

There is no collisional cooling inside the orbitrap, which operates at very high vacuum (2×10^{-10} mbar). Detection is performed by measuring the current image of the axial motion of the ions around the inner electrode. The mass spectrum is obtained after Fourier transformation of the image current. The mass resolving power depends on the time constant of the decay transient. The orbitrap provides resolution that exceeds 100,000 (FWHM) and a mass accuracy $\leq$ 3 ppm. To be operational as a mass spectrometer, the orbitrap requires external ion accumulation, cooling, and fragmentation. The commercial setup (LTQ XL orbitrap) is depicted in Figure 8.19. The instrument consists of an LIT with two detectors connected to the obitrap via a C-trap. Various MS or MS^n experiments can be performed with the LIT. When the orbitrap is used as a detector, the ions are transferred into the C-trap where

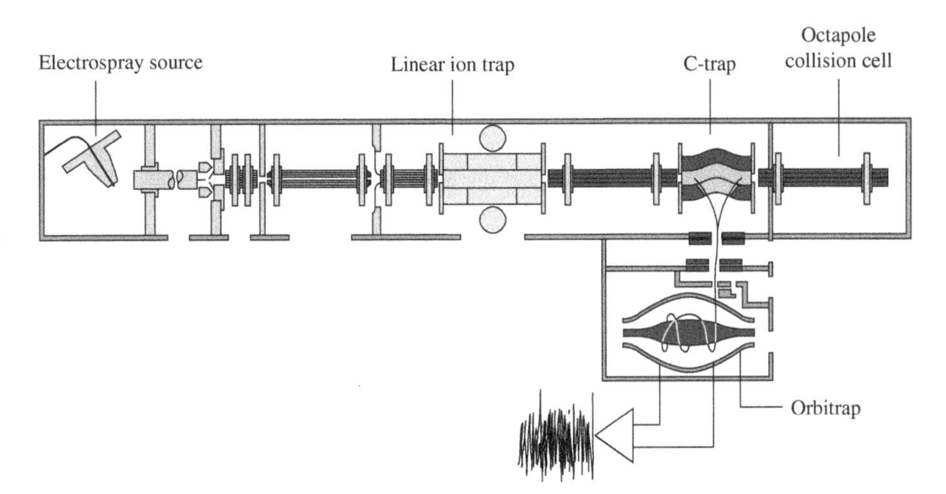

FIGURE 8.19 Schematic of LTQ orbitrap XL (Olsen et al., 2007).

they are collisionally damped by nitrogen at low pressure. Injection into the orbitrap is then performed with short pulses of high voltage.

The particularity of the LTQ orbitrap instrument is the possibility to operate the orbitrap and the LIT independently. Because high resolution requires longer transient time, further data can already be collected in the LIT at the same time. As an example accurate mass measurements of the precursor ion can be performed in the orbitrap while MS^2 and MS^3 spectra are recorded with the LIT. One other specificity of the system is that collision-induced dissociation can also be performed in the octapole collision cell.

REFERENCES

Arpino PJ, Dawkins BG, McLafferty FW. Liquid chromatography/mass spectrometry system providing continuous monitoring with nanogram sensitivity. J Chromatogr Sci 1974;12(10):574–578.

Ausloos P, Clifton CL, Lias SG, Mikaya AI, Stein SE, Tchekhovskoi DV, Sparkman OD, Zaikin V, Zhu D. The critical evaluation of a comprehensive mass spectral library. J Am Soc Mass Spectrom 1999;10(4):287–299.

Barber M, Bordoli RS, Elliott GJ, Sedgwick RD, Tyler AN. Fast atom bombardment mass spectrometry. Anal Chem 1982;54(4):645A–657A.

Belov ME, Nikolaev EN, Alving K, Smith RD. A new technique for unbiased external ion accumulation in a quadrupole two-dimensional ion trap for electrospray ionization Fourier transform ion cyclotron resonance mass spectrometry. Rapid Commun Mass Spectrom 2001;15(14):1172–1180.

Bilsborough S, Loftus N, Taniguchi J, Miseki K. A new hybrid electrospray–ion trap/time of flight mass spectrometer (ES-IT-TOF-MS) capable of high mass accuracy and resolution. Spectra Analyse 2004;33(239):35–38.

Blades AT, Ikonomou MG, Kebarle P. Mechanism of electrospray mass spectrometry. Electrospray as an electrolysis cell. Anal Chem 1991;63(19):2109–2114.

Bos SJ, van Leeuwen SM, Karst U. From fundamentals to applications: Recent developments in atmospheric pressure photoionization mass spectrometry. Anal Bioanal Chem 2006;384(1):85–99.

Bruins AP, Covey TR, Henion JD. Ion spray interface for combined liquid chromatography/atmospheric pressure ionization mass spectrometry. Anal Chem 1987;59 (22):2642–2646.

Budzikiewicz H. Studies in negative ion mass spectrometry. Part XI. Negative chemical ionization (NCI) of organic compounds. Mass Spectrom Rev 1986;5(4):345–380.

Cai Y, Kingery D, McConnell O, Bach AC , 2nd. Advantages of atmospheric pressure photoionization mass spectrometry in support of drug discovery. Rapid Commun Mass Spectrom 2005;19(12):1717–1724.

Caprioli RM, Farmer TB, Gile J. Molecular imaging of biological samples: Localization of peptides and proteins using MALDI-TOF MS. Anal Chem 1997;69 (23):4751–4760.

Carroll DI, Dzidic I, Stillwell RN, Haegele KD, Horning EC. Atmospheric pressure ionization mass spectrometry. Corona discharge ion source for use in a liquid chromatograph–mass spectrometer–computer analytical system. Anal Chem 1975;47(14):2369–2372.

Cech NB, Enke CG. Practical implications of some recent studies in electrospray ionization fundamentals. Mass Spectrom Rev 2001;20(6):362–87.)

Chen H, Venter A, Cooks RG. Extractive electrospray ionization for direct analysis of undiluted urine, milk and other complex mixtures without sample preparation. Chem Commun (Camb) 2006(19):2042–2044.

Chen H, Yang S, Wortmann A, Zenobi R. Neutral desorption sampling of living objects for rapid analysis by extractive electrospray ionization mass spectrometry. Angew Chem Int Ed Engl 2007;46(40):7591–7594.

Chernushevich IV, Loboda AV, Thomson BA. An introduction to quadrupole-time-of-flight mass spectrometry. J Mass Spectrom 2001;36(8):849–865.

Cody RB, Laramee JA, Durst HD. Versatile new ion source for the analysis of materials in open air under ambient conditions. Anal Chem 2005;77(8): 2297–2302.

Cohen LH, Gusev AI. Small molecule analysis by MALDI mass spectrometry. Anal Bioanal Chem 2002;373(7):571–586.

Collings BA, Campbell JM, Mao D, Douglas DJ. A combined linear ion trap time-of-flight system with improved performance and MSn capabilities. Rapid Commun Mass Spectrom 2001;15(19):1777–1795.

Cooks RG, Ouyang Z, Takats Z, Wiseman JM. Detection Technologies. Ambient mass spectrometry. Science 2006;311(5767):1566–1570.

de Hoffmann E. Tandem mass spectrometry: A primer. J Mass Spectrom 1996;31 (2):129–137.

Dempster AJ. A new method of positive ray analysis. Phys Rev 1918;11(4):316.

Dole M, Mack LL, Hines RL, Mobley RC, Ferguson LD, Alice MB. Molecular beams of macroions. J Chem Phys 1968;49(5):2240–2249.

Donegan M, Tomlinson AJ, Nair H, Juhasz P. Controlling matrix suppression for matrix-assisted laser desorption/ionization analysis of small molecules. Rapid Commun Mass Spectrom 2004;18(17):1885–1888.

Douglas DJ, Frank AJ, Mao D. Linear ion traps in mass spectrometry. Mass Spectrom Rev 2005;24(1):1–29.

Fenn J. Electrospray ionization mass spectrometry: How it all began. J Biomol Tech 2002;13(3):101–118.

Fenn JB. Ion formation from charged droplets: Roles of geometry, energy, and time. J Am Soc Mass Spectrom 1993;4(7):524–535.

Fenn JB, Mann M, Meng CK, Wong SF, Whitehouse CM. Electrospray ionization for mass spectrometry of large biomolecules. Science 1989;246(4926):64–71.

Go EP, Prenni JE, Wei J, Jones A, Hall SC, Witkowska HE, Shen Z, Siuzdak G. Desorption/ionization on silicon time-of-flight/time-of-flight mass spectrometry. Anal Chem 2003;75(10):2504–2506.

Gohlke RS, McLafferty FW. Early gas chromatography/mass spectrometry. J Am Soc Mass Spectrom 1993;4(5):367–371.

Guilhaus M, Selby D, Mlynski V. Orthogonal acceleration time-of-flight mass spectrometry. Mass Spectrom Rev 2000;19(2):65–107.

Gut IG. DNA analysis by MALDI-TOF mass spectrometry. Hum Mutat 2004;23 (5):437–441.

Hager JW. A new linear ion trap mass spectrometer. Rapid Commun Mass Spectrom 2002;16:512–526.

Hanold KA, Fischer SM, Cormia PH, Miller CE, Syage JA. Atmospheric pressure photoionization. 1. General properties for LC/MS. Anal Chem 2004;76 (10):2842–2851.

Hopfgartner G, Bourgogne E. Quantitative high-throughput analysis of drugs in biological matrices by mass spectrometry. Mass Spectrom Rev 2003;22(3):195–214.

Hopfgartner G, Varesio E, Stoeckli M. Matrix-assisted laser desorption/ionization mass spectrometric imaging of complete rat sections using a triple quadrupole linear ion trap. Rapid Commun Mass Spectrom 2009;23(6):733–736.

Hopfgartner G, Varesio E, Tschäppät V, Grivet C, Emmanuel Bourgogne E, Leuthold LA. Triple quadrupole linear ion trap mass spectrometer for the analysis of small molecules and macromolecules. J Mass Spectrom 2004a;39:845–855.

Hopfgartner G, Zell M. Q Trap MS: A new tool for metabolite identification. In Using Mass Spectrometry for Drug Metabolism studies, Korfmacher WA, Ed. CRC Press, Boca Raton, FL, 2004b.

Hu Q, Noll RJ, Li H, Makarov A, Hardman M, Graham Cooks R. The Orbitrap: A new mass spectrometer. J Mass Spectrom 2005;40(4):430–443.

Iribarne JV, Thomson BA. On the evaporation of small ions from charged droplets. J Chem Phys 1976;64(6):2287–2294.

Karas M, Hillenkamp F. Laser desorption ionization of proteins with molecular masses exceeding 10,000 daltons. Anal Chem 1988;60(20):2299–2301.

Kebarle P, Peschke M. On the mechanisms by which the charged droplets produced by electrospray lead to gas phase ions. Anal Chim Acta 2000;406(1):11–35.

Kingdon KH. A method for neutralizing the electron space charge by positive ionization at very low pressures. Phys Rev 1923;21:408–418.

Kovarik P, Grivet C, Bourgogne E, Hopfgartner G. Method development aspects for the quantitation of pharmaceutical compounds in human plasma with a matrix-assisted laser desorption/ionization source in the multiple reaction monitoring mode. Rapid Commun Mass Spectrom 2007;21(6):911–919.

Lewis WG, Shen Z, Finn MG, Siuzdak G. Desorption/ionization on silicon (DIOS) mass spectrometry: Background and applications. Int J Mass Spectrom 2003;226 (1):107–116.

Makarov A. Electrostatic axially harmonic orbital trapping: A high-performance technique of mass analysis. Anal Chem 2000;72(6):1156–1162.

March RE. Quadrupole ion trap mass spectrometer. In Encyclopedia of Analytical Chemistry, Wiley, Chichester, editor RAM. 2000, pp. 11848–11872.

Martin RL, Brancia FL. Analysis of high mass peptides using a novel matrix-assisted laser desorption/ionisation quadrupole ion trap time-of-flight mass spectrometer. Rapid Commun Mass Spectrom 2003;17(12):1358–1365.

Maurer HH. Role of gas chromatography–mass spectrometry with negative ion chemical ionization in clinical and forensic toxicology, doping control, and biomonitoring. Ther Drug Monit 2002;24(2):247–254.

McCombie G, Knochenmuss R. Small-molecule MALDI using the matrix suppression effect to reduce or eliminate matrix background interferences. Anal Chem 2004;76 (17):4990–4997.

Miller PE, Denton MB. The quadrupole mass filter: Basic operating concepts. J Chem Ed 1986;63(7):617–622.

Morris HR, Paxton T, Dell A, Langhorne J, Berg M, Bordoli RS, Hoyes J, Bateman RH. High sensitivity collisionally-activated decomposition tandem mass spectrometry on a novel quadrupole/orthogonal-acceleration time-of-flight mass spectrometer. Rapid Commun Mass Spectrom 1996;10(8):889–896.

Moyer SC, Cotter RJ. Atmospheric pressure MALDI. Anal Chem 2002;74 (17):468A–476A.

Munson B. Development of chemical ionization mass spectrometry. Int J Mass Spectrom 2000;200(1/3):243–251.

Murray KK, Boyd RK, Eberlin MN, Langley GJ, Li L, Naito Y. Standard definitions of terms relating to mass spectrometry. International Union of Pure and Applied Chemistry Analytical Chemistry Division, 2006. http://old.iupac.org/reports/provisional/abstract06/murray_310107.html (21 November 2010).

Nguyen S, Fenn JB. Gas-phase ions of solute species from charged droplets of solutions. Proc Natl Acad Sci USA 2007;104(4):1111–1117.

Nielen M. MALDI Time-of-flight mass spectrometry of synthetic polymers. Mass Spectrom Rev 1999;18:309–344.

Paul W, Steinwedel H. A new mass spectrometer without magnetic field. Zeitschr Naturforsch 1953;8a:448–450.

Second PT, Blethrow JD, Schwartz JC, Merrihew GE, Maccoss MJ, Swaney DL, Russell JD, Coon JJ, Zabrouskov V. Dual-Pressure Linear Ion Trap Mass Spectrometer Improving the Analysis of Complex Protein Mixtures. Anal Chem 2009;81:7757–7765. The citation should be corrected in the text page 269 line 6.

Perry RH, Cooks RG, Noll RJ. Orbitrap mass spectrometry: Instrumentation, ion motion and applications. Mass Spectrom Rev 2008;27(6):661–699.

Raffaelli A, Saba A. Atmospheric pressure photoionization mass spectrometry. Mass Spectrom Rev 2003;22(5):318–331.

Rappsilber J, Moniatte M, Nielsen ML, Podtelejnikov AV, Mann M. Experiences and perspectives of MALDI MS and MS/MS in proteomic research. Int J Mass Spectrom 2003;226(1):223–237.

Rentel C, Strohschein S, Albert K, Bayer E. Silver-plated vitamins: A method of detecting tocopherols and carotenoids in LC/ESI-MS coupling. Anal Chem 1998;70 (20):4394–4400.

Reyzer ML, Hsieh Y, Ng K, Korfmacher WA, Caprioli RM. Direct analysis of drug candidates in tissue by matrix-assisted laser desorption/ionization mass spectrometry. J Mass Spectrom 2003;38(10):1081–1092.

Robb DB, Blades MW. State-of-the-art in atmospheric pressure photoionization for LC/MS. Anal Chim Acta 2008;627(1):34–49.

Robb DB, Covey TR, Bruins AP. Atmospheric pressure photoionization: An ionization method for liquid chromatography–mass spectrometry. Anal Chem 2000;72 (15):3653–3659.

Rohner TC, Staab D, Stoeckli M. MALDI mass spectrometric imaging of biological tissue sections. Mech Ageing Dev 2005;126(1):177–185.

Schiller J, Suss R, Arnhold J, Fuchs B, Lessig J, Muller M, Petkovic M, Spalteholz H, Zschornig O, Arnold K. Matrix-assisted laser desorption and ionization time-of-flight (MALDI-TOF) mass spectrometry in lipid and phospholipid research. Prog Lipid Res 2004;43(5):449–488.

Schneider BB, Lock C, Covey TR. AP and vacuum MALDI on a QqLIT instrument. J Am Soc Mass Spectrom 2005;16(2):176–182.

Schwartz JC, Senko MW, Syka JEP. A two-dimensional quadrupole ion trap mass spectrometer. J Am Soc Mass Spectrom 2002;13:659–669.

Schwartz JC, Wade AP, Enke CG, Cooks RG. Systematic delineation of scan modes in multidimensional mass spectrometry. Anal Chem 1990;62(17):1809–1818.

Signor L, Varesio E, Staack RF, Starke V, Richter WF, Hopfgartner G. Analysis of erlotinib and its metabolites in rat tissue sections by MALDI quadrupole time-of-flight mass spectrometry. J Mass Spectrom 2007;42(7):900–909.

Sikanen T, Franssila S, Kauppila TJ, Kostiainen R, Kotiaho T, Ketola RA. Microchip technology in mass spectrometry. Mass Spectrom Rev 2010;29:351–391.

Syka JEP, Marto JA, Bai DL, Horning S, Senko MW, Schwartz JC, Ueberheide B, Garcia B, Busby S, Muratore T, Shabanowitz J, Hunt DF. Novel linear quadrupole ion trap/FT mass spectrometer: Performance characterization and use in the comparative analysis of histone H3 post-translational modifications. J Proteome Res 2004;3(3):621–626.

Tanaka K, Waki H, Ido Y, Akita S, Yoshida Y, T Y. Protein and polymer analysis up to m/z 100,000 by laser ionization time-of-flight mass spectrometry. Rapid Commun Mass Spectrom 1988;2:151.

Tang N, Tornatore P, Weinberger SR. Current developments in SELDI affinity technology. Mass Spectrom Rev 2003;23(1):34–44.

Taylor G. Electrically driven jets. Proc Roy Soc Lond A 1969;313:453–475.

Tyler AN, Clayton E, Green BN. Exact mass measurement of polar organic molecules at low resolution using electrospray ionization and a quadrupole mass spectrometer. Anal Chem 1996;68(20):3561–3569.

Vestal ML. High-performance liquid chromatography–mass spectrometry. Science 1984;226(4672):275–281.

Vestal ML. Methods of ion generation. Chem Rev 2001;101(2):361–375.

Vestal ML. Modern MALDI time-of-flight mass spectrometry. J Mass Spectrom 2009;44(3):303–317.

Wagner M, Varesio E, Hopfgartner G. Ultra-fast quantitation of saquinavir in human plasma by matrix-assisted laser desorption/ionization and selected reaction monitoring mode detection. J Chromatogr B Analyt Technol Biomed Life Sci 2008;872 (1–2):68–76.

Webb K, Bristow T, Sargent M, Stein B. Methodology for accurate mass measurement of small molecules: Best practise; LGC 2004, Teddington, UK, ISBN-0-948926-22-8.

Wilm M, Mann M. Analytical properties of the nanoelectrospray ion source. Anal Chem 1996;68(1):1–8.

Yost RA, Enke CG. Selected ion fragmentation with a tandem quadrupole mass spectrometer. J Am Chem Soc 1978;100(7):2274–2275.

Zhang S, Van Pelt CK. Chip-based nanoelectrospray mass spectrometry for protein characterization. Exp Rev Prot 2004;1(4):449–468.

Zhou S, Cook KD. Protonation in electrospray mass spectrometry: Wrong-way-round or right-way-round? J Am Soc Mass Spectrom 2000;11(11):961–966.

9 Common Liquid Chromatography–Mass Spectrometry (LC–MS) Methodology for Metabolite Identification

LIN XU, LEWIS J. KLUNK, AND CHANDRA PRAKASH

Department of Drug Metabolism and Pharmacokinetics, Biogen Idec, Cambridge, Massachusetts

9.1 INTRODUCTION

Drug metabolite identification is a challenging area of research with widespread interest. Information on the metabolic fate of the new chemical entities (NCEs) can direct medicinal chemists to synthesize metabolically more stable analogs by blocking sites of metabolism and potentially synthesizing NCEs with superior pharmacology and safety profiles (Kostiainen et al., 2003; Watt et al., 2003; Baranczewski et al., 2006). In the development stage, elucidation of biotransformation pathways of a drug candidate is important to understand

Mass Spectrometry in Drug Metabolism and Disposition: Basic Principles and Applications, First Edition. Edited by Mike S. Lee and Mingshe Zhu.

its physical and biological effects. Metabolic pathways of drug candidates in animals used for safety evaluation studies are required to ensure that the selected animal species are exposed to all major metabolites formed in humans (Baillie et al., 2002). Recently, the Food and Drug Administration (FDA, 2008) suggested that additional toxicological testing on metabolites that have higher exposure in humans than preclinical species may be warranted. In addition, pharmaceutical companies are mandated by the regulatory agencies to define the metabolic profiles of a drug candidate in laboratory animals and humans for its approval.

Drugs are metabolized primarily by two reaction classes: phase I and phase II. Phase I or oxidative reactions include hydroxylation, dealkylation, deamination, *N*- or *S*-oxidation, reduction, and hydrolysis. These reactions introduce a functional group within a molecule to enhance its hydrophilicity. Phase II or conjugation biotransformations include glucuronidation, sulfation, methylation, acetylation, and amino acid (glycine, glutamic acid, and taurine) and glutathione (GSH) conjugation. Phase II biotransformation reactions result in a large increase in drug hydrophilicity, thus greatly promoting the excretion of foreign chemicals via urine and/or bile. In general, metabolites are pharmacologically less active and less toxic than the corresponding parent compound. However, there are several examples where biotransformation reactions can also lead to the formation of pharmacologically active metabolites (Fura et al., 2004), drug–drug interactions via inhibition or induction of drug-metabolizing enzymes (Rock et al., 2008), and/or formation of reactive/toxic metabolites (Kalgutkar et al., 2005; Prakash et al., 2008b). Therefore, determination of an NCE's biotransformation pathways in animals and humans and pharmacological and toxicological consequences of its metabolites are critical to pharmaceutical research and compound progress.

Gas chromatography coupled with mass spectrometry (GC–MS) has been used extensively for the quantitation and identification of trace components in complex mixtures for a wide variety of compounds. However, due to the laborious process of analyte derivatization prior to GC–MS and chemical decomposition of labile phase II metabolites, this technique has limited utility for metabolite identification. During the last two decades, liquid chromatography–atmospheric ionization tandem mass spectrometry (LC–API MS/MS) has been probed to be a very powerful analytical tool for the structural characterization and quantification of drug metabolites. Triple quadrupole and ion trap mass spectrometers are routinely used for this purpose because of their MS/MS and MS^n capabilities (Kamel and Prakash, 2006; Prakash et al., 2007c). Recently, high-resolution mass spectrometers such as Orbitrap and time of flight (TOF) have become more popular due to their enhanced full-scan sensitivity, scan speed, improved resolution and ability to measure mass-to-charge (m/z) values for protonated molecules and fragment ions accurately (Prakash et al., 2008a; Ruan et al., 2008; Xu et al., 2008).

Identification of drug metabolites involves three steps: (1) detection, (2) structure elucidation, and (3) quantitation. This chapter focuses on the most common and useful application of various LC–MS/MS techniques for the detection and structural elucidation of drug metabolites from biological fluids.

9.2 STRATEGIES FOR METABOLITE IDENTIFICATION

Several MS acquisition and data processing strategies are used for detection and structure elucidation of metabolites. The common metabolite detection strategies are summarized in Section 9.2.1, which include full MS scan, constant neutral loss, parent ion scan, multiple reactions monitoring, and mass defect filtering. The structure elucidation strategies feature product ion scan, multistage scan, and accurate mass measurement, which are reviewed in Section 9.2.2.

9.2.1 Detection of Metabolites

9.2.1.1 Full MS Scanning Full MS scanning is used to measure the m/z ratio of parent compound and its metabolites. Most current mass analyzers, such as quadrupole mass filters (single and triple quadrupole mass spectrometers), linear ion trap, time of flight, Orbitrap, and Fourier transform mass spectrometers, can provide m/z ratios of all ions. This nonselective nature of full-scan MS data acquisition ensures that most ionizable metabolites are represented in the spectrum. However, the presence of interfering ions from non-drug-related components in the complex biological samples often leads to difficulty for the identification of the corresponding metabolite parent ions. To reduce or eliminate these interferences, high-pressure liquid chromotography (HPLC) is coupled with MS to separate complex mixtures. The use of UHPLC—ultra-high chromatographic resolution technique—allows further reduction of the coeluting ions significantly (Castro-Perez et al., 2005a). Thus, full MS scan becomes more useful as a survey scan in the data-dependent scanning approach for metabolite detection (Triolo et al., 2005).

Another common procedure for the metabolite detection involves the analysis of test and control samples in a full mass range scan (Sinz and Podoll, 2002; Mutlib and Shockor, 2003) with subsequent subtraction of control sample to reduce background. This approach is very effective for the analysis of in vitro incubation samples where background ions in control and test samples are similar (Tiller et al., 2008; Zhang and Yang, 2008). A typical background subtracted LC–MS chromatogram is shown in Figure 9.1. After the LC–MS data file of the control sample was subtracted from the test sample data file, peaks that correspond to metabolites became dominant in a subtracted LC–MS profile (Fig. 9.1b). The background subtraction approach is less successful for complex biological matrices from in vivo experiments, partially because true control samples are difficult or impossible to obtain.

Therefore, the full-scan approach alone is rarely sufficient for analysis of in vivo samples, such as urine, bile, and plasma. Additional data interrogations

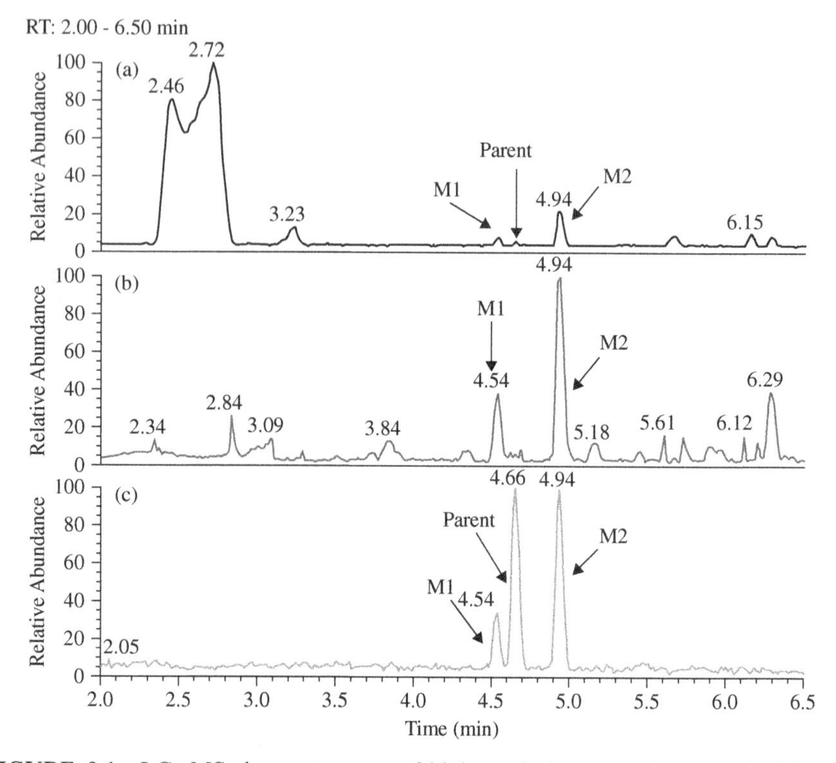

FIGURE 9.1 LC–MS chromatograms of high-resolution mass data acquired by the Orbitrap. (a) unprocessed full-scan MS data, (b) background subtracted full-scan MS data, and (c) MDF filtered MS data.

such as accurate mass measurement and mass defect filtering (MDF) techniques are necessary (Fig. 9.1c). An algorithm for background subtraction based on accurate mass (Zhang et al., 2008a) holds promise for full MS scanning approach for samples from in vivo studies Both accurate mass measurement and MDF methodology are described in more detail in Section 9.2.1.4.

9.2.1.2 Constant Neutral Loss and Precursor Ion Scanning Constant neutral loss (CNL) and precursor ion (PI) scanning are two common approaches used in tandem (MS/MS) spectrometers such as triple quadrupole (QqQ), hybrid quadrupole time of flight (QqTOF), and triple quadrupole linear ion trap (QqLIT) MS. These two scanning approaches require very little knowledge of metabolite structure. These approaches are often used as survey scans at the initial step of metabolite detection (Prakash et al., 2007c) in drug discovery and development.

The CNL approach is utilized to search for metabolites, which exhibit a characteristic loss of a neutral fragment such as a fragment of the parent drug or typical neutral loss of phase II metabolites. Characteristic PI and CNL of parent drug are determined from its MS/MS spectrum and then used in the survey scan to search for drug-related metabolites. For example, the acetyl or cyano group in the parent molecule typically yields 42 Da (Ketene) or 27 Da (HCN) CNL, respectively, which can be used in the survey scan to detect the metabolites that contain the acetyl or cyano groups (Zhang et al., 2005). Fluorine substitution is commonly used in medicinal chemistry to block the metabolically labile site of drug candidates during lead optimization. Metabolites of fluorine-containing drug candidates sometimes have typical CNL at 20/40 (HF/2HF) Da. The most widely used application of CNL is to detect phase II metabolites, notably the CNL of 176 Da for glucuronide conjugates and the loss of 80 Da for sulfates. Conjugated metabolites can be detected without prior knowledge of fragmentation of parent drug or corresponding phase I metabolites. This approach can also provide indirect evidence of formation of phase I metabolites when a newly formed hydroxyl group is further conjugated. Common CNL used for metabolite identification are listed at Table 9.1.

The CNL scanning is also commonly used for detecting the GSH conjugates (Castro-Perez et al., 2005b; Holcapek et al., 2008; Yan et al., 2008). Identification of *S*-linked GSH conjugates is a valuable indirect approach to determine the existence of the reactive metabolites of drug candidates. This information is very important in mechanistic studies of metabolic activation to support drug discovery in the pharmaceutical industry. Hence, a sensitive and specific MS acquisition method is preferred. A traditional and the widely used MS/MS acquisition method is based on the CNL scan of 129 Da, which corresponds to the loss of pyroglutamic acid moiety from the protonated molecular ion $[M+H]^+$ (Baillie and Davis, 1993). However, this CNL approach also has its limitations. The sensitivity is highly dependent on the fragmentation pattern of GSH conjugates and the abundance of a product ion with CNL of 129-Da (Yan et al., 2008). Some GSH adducts provide different CNL and may be missed in this 129 Da CNL approach. For example, diclofenac-*S*-acyl-glutathione thioester has a different CNL of 147 Da by loss of glutamic acid in tandem MS spectrum (Grillo et al., 2003). In addition, the CNL of 129 Da is less useful when a GSH conjugate is present in the doubly charged $[M+2H]^{2+}$ state under positive ionization conditions (Zhang and Yang, 2008).

Precursor ion experiments yield a spectrum of all parent ions, which have the specified product ion in their spectra. A PI scan is a powerful technique for identification of metabolites that contain a predefined fragment ion, which can be derived from the MS/MS spectrum of the parent drug or based on common biotransformation pathways, that is, +16 Da shift of the parent compound's fragment ion. For example, the typical fragment *m/z* 85 from carnitine is used in the PI scan mode for the analysis of acylcarnitine metabolites (Paglia et al., 2008). For the detection of the GSH conjugate, a recent study described the use of the PI scan with diagnostic fragment ions at *m/z* 272 and 254 in the negative ion mode and was determined to be more sensitive and selective than the CNL of 129 Da in the positive ion mode (Mahajan and Evans, 2008).

The utility of PI and CNL scanning techniques for metabolite identification of a compound A (5*S*, 6*S*)-5-(5-(5-isopropyl-2-methoxy-benzylamino-)-6-benzhydrylquinuclidine-3-carboxylic acid (Fig. 9.2) have been reported

TABLE 9.1 Common Neutral Losses from Conjugates Used in CNL Scan for Metabolite Identification

Conjugation Type	Mass Shift	Characteristic CNL	Polarity of Parent Ion	Mass of CNL	Reference
Glucuronidation	176	Glucuronic acid	Positive	176	Clayton et al., 2001; Lampinen-Salomonsson et al., 2006; Dai et al., 2008
Sulfation	80	SO_3	Positive	80	Clayton et al., 2001; Lampinen-Salomonsson et al., 2006; Dai et al., 2008
Glutathione	305	pyroglutamic acid	Positive	129	Baillie and Davis, 1993; Yan et al., 2008
Taurine	107	Taurine H_2SO_4	Positive	125 153	Li et al., 2006
N-acetylcysteine	161	*N*-acetyl-2-iminopropionic acid	Negative	129	Scholz et al., 2005
N-acetylglucosamine	203	*N*-acetylglucosamine	Positive	203	Johnson et al., 2008
Glycine	57	Glycine	Positive	75	Zhang et al., 2005
Carnitine	143	Carnitine	Positive	59	Paglia et al., 2008
Glucosidation		Anhydroglucose	Positive	162	Nakazawa et al., 2006

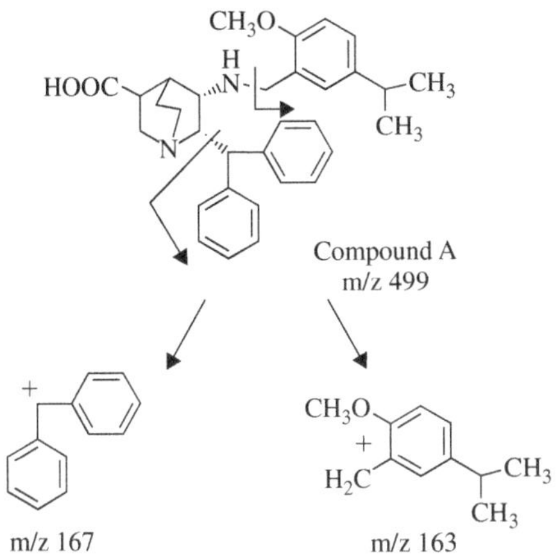

FIGURE 9.2 MS/MS fragmentation of compound A.

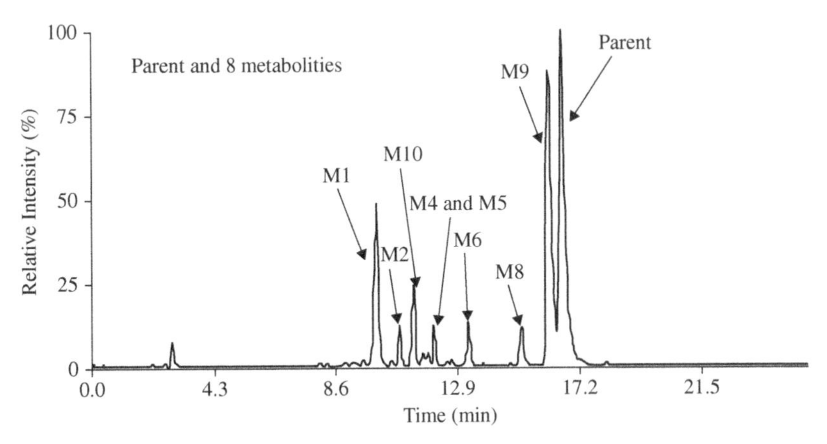

FIGURE 9.3 Total ion chromatogram from PI scanning (*m/z* 167) of metabolites of compound A in human urine.

(Prakash et al., 2007c). The MS/MS spectrum of its protonated molecular ion (*m/z* 499) showed prominent fragment ions *m/z* 167 and 163 formed from the benzhydryl and anisole parts of the molecule, respectively. The PI scanning of ion *m/z* 167 detected eight metabolites, which contained unchanged benzhydryl moiety (Fig. 9.3), whereas CNL scanning of 176 showed four glucuronide conjugates. Identified glucuronide metabolites M11 and M12 indirectly confirmed the existence of phase I oxidation (M5 and M6) at the isopropyl benzyl group (Fig. 9.4).

Both PI and CNL scanning methodologies are undoubtedly selective and specific approaches to detect metabolites, especially in complex biological matrices such as urine and bile. However, endogenous components could result

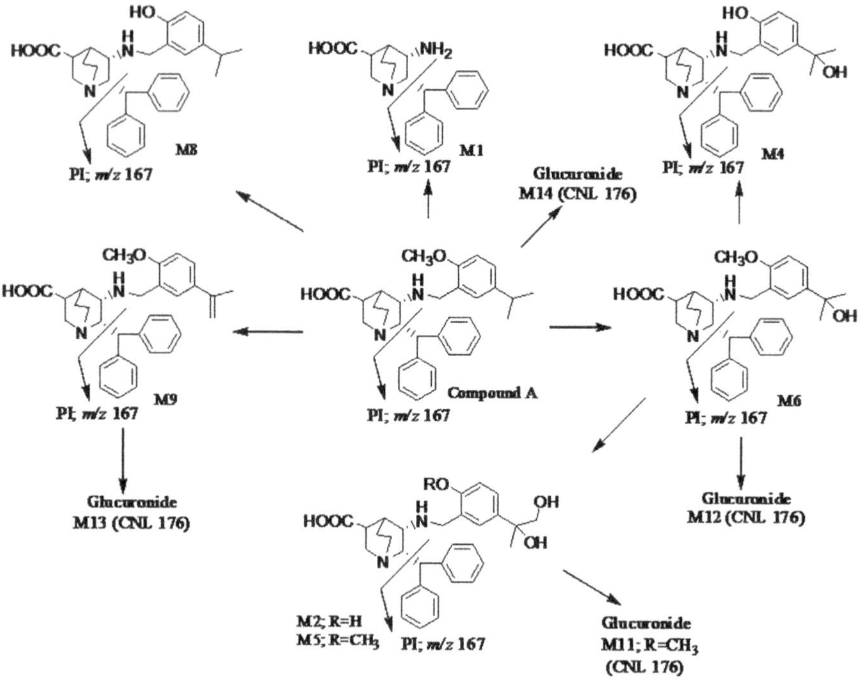

FIGURE 9.4 Proposed metabolic pathway of compound A in humans.

in false positives in general CNL scans. For example, taurine conjugates of the bile acids could be identified when a CNL of 125 Da (Table 9.1) is used to detect the taurine conjugates. The selectivity and specificity of PI and CNL could be further increased via high-resolution mass spectrometer (Jemal et al., 2003). For example, the sensitivity and selectivity of GSH adducts detection can be increased using an exact CNL detection of m/z 129.0426 at positive mode via high-resolution mass spectrometer (Castro-Perez et al., 2005b). In combination with full-scan and high-resolution MS, PI and CNL scanning are valuable tools for metabolite identification.

9.2.1.3 Multiple-Reaction Monitoring Although the use of PI and CNL data acquisition improves the selectivity of metabolite detection when compared to full scan, these approaches usually lack detection sensitivity. The recently introduced triple quadrupole/linear ion trap (API 4000 Qtrap and 5500 Qtrap) hybrid mass spectrometers provide superior sensitivity, selectivity, and fast multiple-reaction monitoring (MRM) approaches to detect metabolites (Hager and Le Blanc, 2003). Traditionally, MRM scans are used for drug quantitation in biological samples due to its high sensitivity. When the MRM transition of parent drug is determined, MRM ion pairs of most common phase I and II metabolites can be predicted with a software script. This MRM list

(Gao et al., 2007) is used as a survey scan, similar to PI or CNL, to detect metabolites. The MRM transition list can also be generated from metabolites identified from in vitro incubations (Mauriala et al., 2005). The peak, which reaches threshold in MRM transition, triggers the acquisition of a corresponding enhanced product ion (MS/MS) spectrum. This procedure, known as the MRM approach, is an alternative to rapid metabolite detection and structural elucidation with a triple quadrupole mass spectrometer.

The advantage of the MRM approach is the superior selectivity due to filtering out the complicated background ions. The low background interference in MRM allows for detection of the minor metabolites present at the trace levels in the biological matrices. Coeluting metabolites can also be distinguished and identified because of the specificity of their respective MRM transitions. The relative ion intensities of metabolites versus parent provides some insight into estimating dominant metabolites, which is especially important for the circulating metabolites (Mauriala et al., 2005). Another advantage of MRM approach is ultra-fast speed of a survey scan of 20,000 Da/s. The MRM survey scan can be set to more than 300 MRM transitions without a significant loss of sensitivity. This approach has been reported to successfully identify phase I and II metabolites (Gao et al., 2007), GSH conjugates (Zheng et al., 2007), *N*-acetylcysteine conjugates (Scholz et al., 2005), and to detect over 400 drugs commonly examined in forensic toxicology labs (Herrin et al., 2005).

Utilizing metabolism prediction and understanding of the MS/MS fragmentation of the parent compound, the MRM approach gives a significant increase in sensitivity over traditional PI or CNL scanning in a triple quadrupole MS and enables a wide range of potential transitions to be targeted as a result of the rapid cycle times. Although there is a potential to miss some of the metabolites, the approach is a powerful alternative for metabolite detection when sensitivity is an issue. In a recent study, the MRM approaches (Zheng et al., 2007) were shown to improve the detection of glutathione conjugates compared to the traditional CNL scan of 129 Da and thereby facilitate the identification of compounds with the potential to form reactive metabolites.

However, the success of comprehensive metabolite identification via the MRM approach heavily depends on the correct selection of putative metabolite ions and product ions predicted based on similarity of fragmentation patterns between the parent drug and its metabolites. The effectiveness of the MRM approach still relies to some extent on the ability of the analyst to predict the metabolic pathways of the compound. For example, certain types of metabolites arise from amide hydrolysis and heterodealkylation, which often have unpredictable fragmentation patterns. To overcome this limitation, Yao et al. (2008) developed a multiple ion monitoring (MIM) scanning approach in which the collision energy in the quadrupole (Q_2) was reduced to the minimum such that ions isolated in Q_1 pass through Q_2 without fragmentation. These ions are then monitored in the third quadrupole (Q_3) or linear ion trap. Therefore, the MIM scan can be used as a survey scan without prediction of fragmentation patterns. The total number of MIM transitions are significantly reduced compared to the MRM scan because one MIM transition can detect all putative metabolites that have the same molecular weight, regardless of the location of metabolic modification. However, the selectivity of MIM at unit resolution is less than that of MRM. The likelihood of false positives and background noise increases. The MIM approach is quite similar to the target-list-dependent metabolite detection with full scan and, therefore, shares the limitations of the latter. This approach is likely to be reliable for common biotransformation pathways and useful for increasing throughput of metabolite profiling; however, there is a possibility of missing the unusual or multiple-step metabolic reactions.

9.2.1.4 Accurate Mass Measurement and Mass Defect Filtering Accurate mass measurements are traditionally used to determine the possible empirical formulas of drug metabolites. Using the parent drug's molecular formula, elemental composition analysis of drug metabolites can be narrowed down to a limited number and types of atoms. In many cases, large portions of molecular structures of metabolites are known due to a similarity to the parent compound, and molecular formula analysis with high-resolution MS (HRHS) may not be necessary for structure identification. However, it has been reported (Xu et al., 2008; Hobby et al., 2009) that elemental composition analysis of molecular ion in accurate mass can provide unambiguous structure assignment, especially if the metabolites have a large mass shift compared to the parent. The recent mass spectral correlation method, called FuzzyFit, aids in empirical formula determination of metabolites by utilizing both mass accuracy and isotope patterns (Hobby et al., 2009). Overall, the accurate mass fragment ion data in the MS/MS spectrum allow for easy fragment ion assignment and, thereby, facilitates accurate spectral interpretation and structural determination.

Application of accurate mass measurement for unambiguous structure assignment of a GSH conjugate of colchicine (Xu et al., 2008) is illustrated in Figure 9.5. The identified GSH conjugate of colchicine M1 showed a protonated molecular ion at *m/z* 675.23332 in high-resolution MS. As *O*-demethylation was the major metabolic pathway of colchicine, the mechanism of M1 formation was proposed by first *O*-demethylation to dihydroquinone with further dehydrogenation to *ortho*-quinone followed by GSH addition. Based on this assumption, the molecular formula of M1 was calculated to be $C_{30}H_{34}N_4O_{12}S$, with a calculated exact mass of 674.18939. In low-resolution MS, the measured molecular ion of M1 seemed to match with the proposed structure. However, accurate mass measurement by HRMS of M1 showed 52.7 ppm mass difference from the proposed structure. Elemental composition analysis derived from HRMS was $C_{31}H_{38}N_4O_{11}S$, suggesting that M1 had a net loss of CH_3O from parent colchicine after subtraction of the GSH moiety instead of C_2H_7 as proposed originally.

High-resolution MS obviously has become a critical tool in the unambiguous structure assignment of metabolites in drug discovery and development. The power of HRMS in unambiguous structural elucidation of expected and

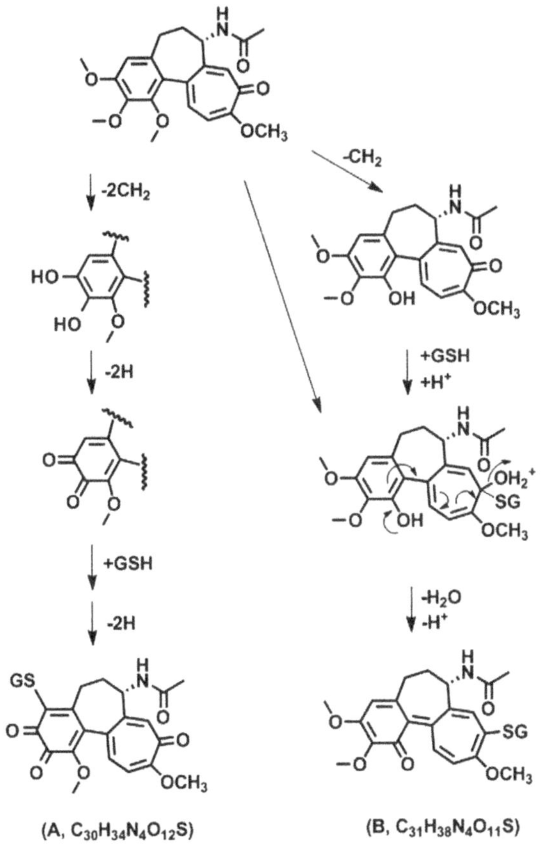
-CH2
-2CH2
HO
HO
-2H
+GSH
+H+
+GSH
-2H
-H2O
-H+
GS
SG
OCH3
OH
(A, C30H34N4O12S)
(B, C31H38N4O11S)

FIGURE 9.5 Proposed biotransformation pathways and structural variants of M1. A was excluded based on HRMS results. (With permission of American Society for Pharmacology and Experimental Therapeutics.)

unusual metabolites has been presented in many recent studies. Recently, Prakash et al. (2008a) reported the identification of several novel and unusual quinoline metabolites of torcetrapib (M3, M4, M5, M9, M17, M22) and a sulfate conjugate M28 with accurate mass measurements.

With modern QTOF and Orbitrap mass spectrometers, mass resolution of 10^4–10^5 with a mass accuracy within 5 ppm is routinely achieved. The advances in HRMS have not only provided accurate MS measurement but also significantly improved the specificity of LC–MS-based metabolite identification approaches in recent years (Tiller et al., 2008; Zhang et al., 2008a). For example, specificity of background subtraction in high-resolution full scanning is significantly improved. Current HRMS measures m/z values with precision at sub-ppm levels, which allows distinguishing the analyte ions from isobaric interfering ions from control sample (Zhang et al., 2008a). The occurrence of instrumental noise and spikes associated with unit resolution full-scan MS is less of an issue in HRMS. HRMS simplifies background subtraction without the need of the complicated the Biller–Biemann algorithm and allows for the routine application of background subtraction tools with data-processing software packages of commercial HRMS instruments such as Metworks from Thermo Fisher Scientific and Metabolynx from Waters.

Mass defect filtering is a unique function in HRMS to identify metabolites. The atom mass of C-12 is defined as exactly 12,0000 Da; therefore, masses of all other elements have a unique difference between the exact and nominal atomic mass. This difference is called the mass defect; therefore, each organic molecule has a unique mass defect based on its elemental compositions. Typical phase I and II biotransformations change the mass defect of the parent drug within 70 mDa; for example, a single oxidation will introduce the mass defect change of −5 mDa, demethylation and dehydrogenation of −16 mDa, and phase II glucuronidation of +32 mDa. Therefore, when coupled with LC, the HRMS method is able to differentiate drug-related material from endogenous components via the MDF process. The ability to detect unexpected metabolites is significantly improved. After Zhang and co-workers (2003) first reported the successful application of MDF to metabolic profiling in the dog bile samples, the effectiveness of the MDF approach for metabolite detection has been widely demonstrated. This methodology is now available as a data-processing tool in many software packages. The common biotransformation and the corresponding mass defect changes are listed in Table 9.2. Only GSH conjugate has a mass defect more than 80 mDa due to large element change. A typical MDF LC–MS chromatogram based on phase I biotransformation is shown in Fig. 9.1c. The metabolite peaks (M1 and M2) were clearly observed. The original full MS spectrum of M1 (Fig. 9.6a) contains many background

TABLE 9.2 Common Biotransformation and Corresponding Elemental Composition and Mass Defect Changes

Nominal Mass Shift (ΔDa)	Element Change	Biotransformation	Mass Defect Change (mDa)
−14	$-CH_2$	Demethylation (PhI)	−15.7
+2	$+H_2$	Reduction (PhI)	+15.7
+16	+O	Oxidation (PhI)	−5.1
+18	$+H_2O$	Hydration (PhI)	+10.6
+80	$+SO_3$	Sulfation (PhII)	−43.2
+119	$+C_3H_5NO_2S$	Cysteine conjugation (PhII)	+4.1
+176	$+C_6H_8O_6$	Glucuronidation (PhII)	+32.1
+307	$+C_{10}H_{17}N_3O_6S$	GSH conjugation (PhII)	+83.9

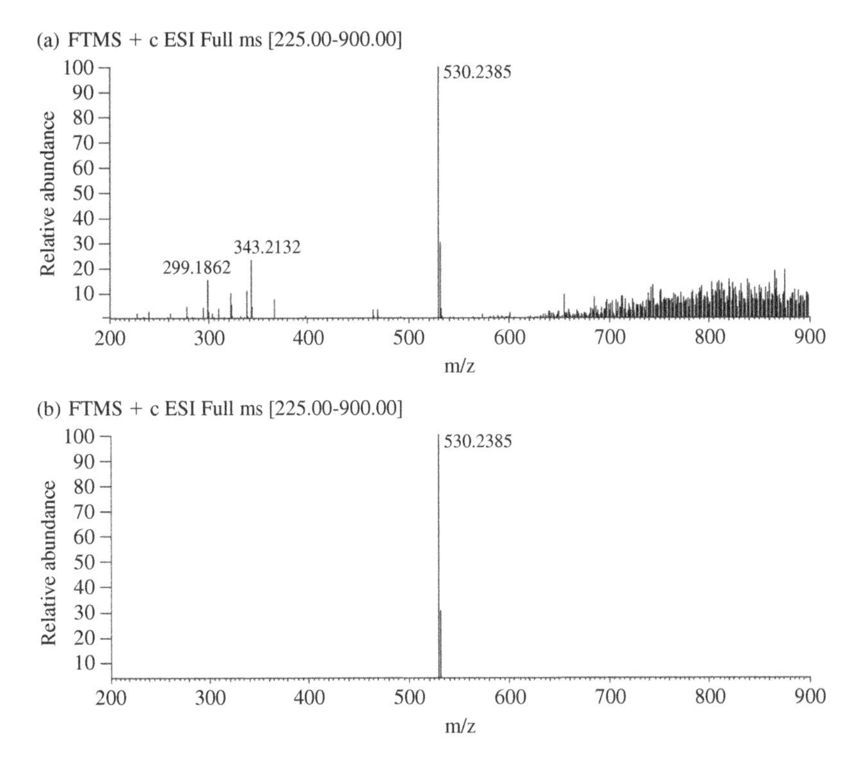

FIGURE 9.6 Full MS spectrum of M1: (*a*) original and (*b*) after MDF.

ions; while only the parent ion (*m*/*z* 530.2385) of M1 remained after MDF (Fig. 9.6b).

The exclusive use of MDF might not be sufficient to accomplish comprehensive metabolite profiling. The following limitations need to be taken into consideration. When biotransformation leads to significant changes arising from multiple phase I and phase II biotransformations or dealkylation (cleavage of a large part of molecule), metabolites could be missed during MDF data processing. A multiple MDF approach from MetWorks processing software (Huang et al., 2008) and the *N*-dealkylation tool (Mortishire-Smith et al., 2009) in Metabolynx eliminate this shortage. In addition, it is reported that MDF alone may not filter out all interference peaks, particularly for the detection of urinary metabolites (Zhang et al., 2008b). Mass defects of many endogenous urinary components fall into the range of most commercial drug molecules, and, therefore, MDF is less specific for urine sample analysis compared to other in vivo matrices such as plasma and bile. Therefore, the combination of MDF with other MS data acquisition and processing strategies is recommended. Several recent studies (Ruan et al., 2008; Tiller et al., 2008) demonstrated that integration of MDF into background subtraction, CNL, and product ion analysis is an effective and efficient approach for drug metabolite profiling.

9.2.2 Structure Elucidation of Metabolites

9.2.2.1 Data-Dependent Product Ion Scanning (MS^2) Typically, structural information is obtained via interpretation of the fragmentation patterns from the MS/MS data of the parent drug and metabolites. The advent of data-dependant acquisition enables the simultaneous acquisition of MS^2 data based on full MS or other types of survey scans.

The data-dependent acquisition method is based on the ion intensity of the full MS, which is widely used in ion trap MS. The most and second most abundant ions in the full MS, spectrum are selected via the acquisition method to trigger MS^2 data collection. Furthermore, the highly resolved chromatographic peaks obtained from UHPLC assists with the MS^2 data collection, and MS^2 spectra are generated from the most abundant to include all the metabolites of interest. Thus, both metabolite detection and structure elucidation information are acquired in the same analytical run, which allows maximizing instrumental throughput. The selectivity of this method is sufficient to identify even the metabolites of low abundance in the presence of abundant coeluting endogenous compounds (Sang et al., 2008). Meanwhile, data processing postacquisition allows looking for any desired CNL and PI ions for metabolite detection, which is more sensitive than traditional CNL and PI scans owing to the higher duty cycle and fragmentaion at the same location in the ion trap. The third advantage of this method is the possibility of acquiring MS^n spectra within the same run to facilitate structure elucidation (Sang et al., 2008).

The challenge of ion intensity-based data-dependent acquisition is the study of in vivo metabolites where most of the metabolites exist in relatively low concentrations accompanied by large amounts of endogenous matrices (Sun et al., 2007; Song et al., 2009). The data-dependent scan may not even be acquired for the metabolites of interest. Therefore, other survey scans mentioned above have been developed to trigger product ion scans on a variety of modern mass spectrometers. Knowledge-based full-scan lists (Lim et al., 2007), stable isotope patterns (Yan et al., 2008), MRM or MIM scan, and CNL/PI can be used as the survey scans. The full-scan (parent ion) list and MRM/MIM specific data-dependent analysis for potential metabolites are found to be effective in triggering MS^2/MS^3 data acquisition for the corresponding metabolites even though the signal intensity of these metabolites was low. The optimal use of data-dependent acquisition for in vivo sample effectively remains a subject of studies in many research groups.

Another approach to the simultaneous acquisition of both full MS and fragment ion spectra from a single run is an MS^E (where E represents collision energy) method used in hybrid quadrupole QTof MS (Wrona et al., 2005). Two staggered scans are used in this method: (1) the first scan acquires full MS with low collision energy and (2) the second scan acquires nonselective fragment ion data of the same full MS ions with high collision energy. As a result, the acquired MS^2 spectra do not arise from preselected precursor ions; instead, the MS^2 spectra correspond to the fragment ion spectra of all eluted ions. The entire data set is then processed to identify metabolites via parent ions, PI, CNL, and high-resolution MDF, facilitated by data-processing software. Thus, any MS^2 spectrum of a potential metabolite will not be missed in this approach. The success of this approach heavily depends on chromatographic resolution. Overall, this approach offers substantial time savings for drug metabolite identification in early drug discovery. The acquisition method is very generic and does not need any preliminary information about parent drug or predicted metabolites. The parent ion, product ion, precursor ion, and neutral loss chromatograms can be extracted and searched postacquisition via data processing software. The potential limitations include the possibility of coeluting metabolites, which may confound the observed product ion spectra. When the metabolite of interest has multiple ionization forms in full MS, such as $[M+H]^+$, $[M+Na]^+$, and doubly charged $[M+2H]^{2+}$ ions, the product ion spectra could attribute to all different ion adducts. In some cases, the observed product ion spectra may contain ions unrelated to parent compound or its metabolites when the sample matrices become complicated and coeluted. Therefore, this approach is often applied in conjunction with high-resolution MDF and background subtractions. Several successful cases in metabolite identification with this approach are reported in recent articles (Bateman et al., 2007; Tiller et al., 2008).

After detection and acquisition of MS^2 spectra of metabolites, the assignment of product ion spectra is followed to elucidate the structure information. Interpretation of the product ion spectrum for metabolite structural elucidation is illustrated below to differentiate regioisomeric metabolites of compound B, {1-[2-hydroxy-2-(4-hydroxy-phenyl)-1-methyl-ethyl]-4-phenyl-piperidin-4-ol} (Fig. 9.7). The full-scan LC-MS of metabolites B1 and B2 both displayed a

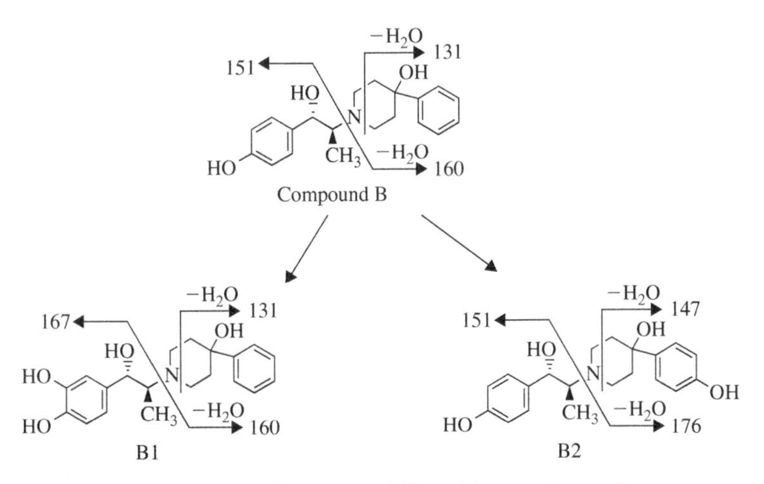

FIGURE 9.7 Fragmentation of compound B and its two isomeric monooxygenated metabolites, B1 and B2.

protonated molecular ion of *m/z* 344, 16 Da higher than that of compound B, suggesting that B1 and B2 were monooxygenated and regioisomers. The presence of characteristic fragment ions in the Collision induced dissociation (CID) product ion spectra of B1 and B2 allowed to assign the site of hydroxylation at the phenyl-piperidinol and phenol moieties, respectively. CID product ion spectrum of B1 (*m/z* 344) contained intense fragment ions at *m/z* 167, 160, and 131 (Fig. 9.7). The fragment ions at *m/z* 160 and 131, similar to the parent compound, suggested that the phenyl-piperidine moiety was unchanged. The fragment ion at *m/z* 167, 16 Da higher than that of the parent compound, indicated that the oxidation had occurred on the hydroxy-phenyl ring (Prakash et al., 2007a). On the other hand, the CID product ion spectrum of B2 contained fragment ions at *m/z* 176, 151, and 147. The fragment ions at *m/z* 176 and 147 suggested the addition of an oxygen atom on the phenyl piperidine portion of the molecule. The prominent fragment ions at *m/z* 151 indicated that the hydroxy phenyl ring was unsubstituted.

The structural interpretation of product ion spectra is a time-consuming step and traditionally relies on the knowledge and experience of analysts. The development of accurate mass techniques for acquisition of product ion spectra and improvement in chemical intelligence in processing software (Hobby et al., 2009; Leclercq et al., 2009) have led to increased accuracy and speed of data interpretation.

9.2.2.2 Multistage Scanning (MS^n) The MS^n scans are product ion scans that can be performed only in the ion trap MS and provide additional information about fragmentation pathways of molecules to further facilitate structural elucidation. The collision energy in ion trap MS is relatively soft compared to the triple quadrupole MS and few major product ions generated from labile bond breakdown. The major product ion can be further fragmented (MS^3) to generate more fragments that can be illustrated in a fragmentation tree ($MS \rightarrow MS^2 \rightarrow MS^3 \rightarrow MS^4$). This feature is very useful to determine the site of metabolite modification and to suggest which part of the molecule was changed. In most cases, generation of an MS^3 scan is a common approach sufficient to obtain this information. Occasionally, MS^4 and MS^5 may be required to assign the site of modification.

The usefulness of MS^n is demonstrated with the assignment of GSH conjugation site of compound C, {1-(2-(4-fluorophenylamino)thiazol-5-yl) ethanone} via high-resolution Orbitrap MS (Fig. 9.8). The presence of the GSH conjugate of compound C is determined from the ions at *m/z* 431.0838 and 413.0770 in the MS^2 spectrum and contained a typical CNL of 129 and 147 Da from the parent ion, respectively. The GSH conjugation site was suspected on either the 4-fluoroaniline or the thiazole moiety because both of them have bioactivation potential for formation of GSH adducts. Most major fragments in MS^2 data arise from amide bond breakage within GSH and do not contain structural information about the binding site. The ion at *m/z* 413.0770 was further fragmented in an MS^3 experiment (Fig. 9.8). The ion at *m/z* 269.0221

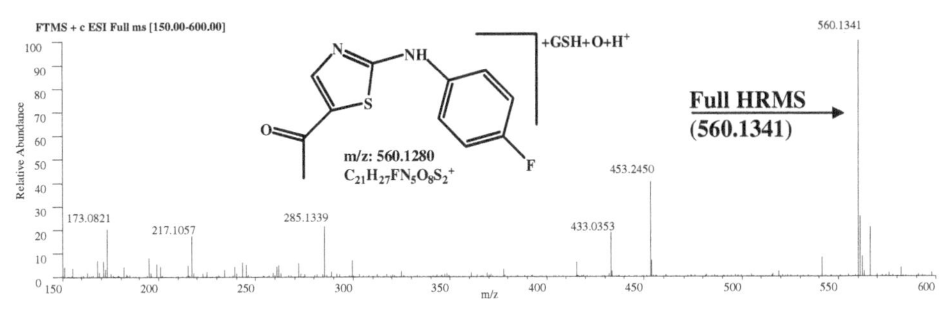

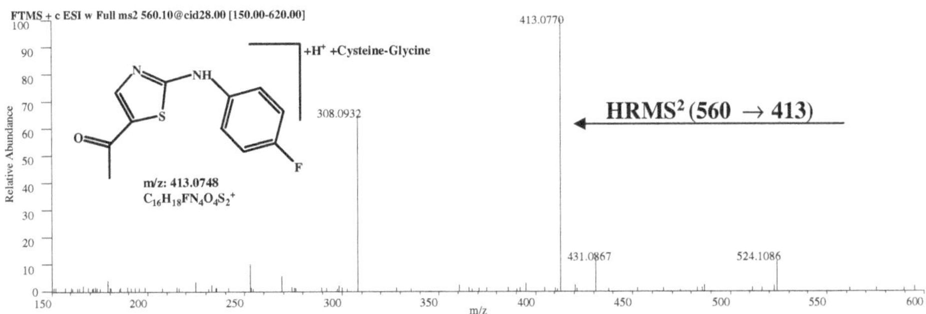

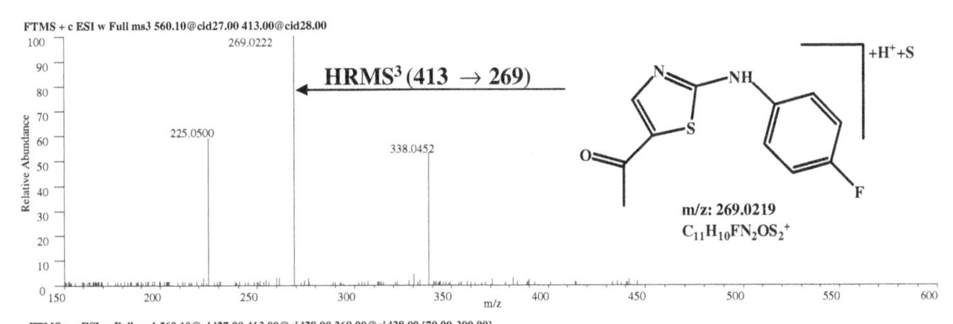

FIGURE 9.8 MS^n of glutathione conjugate of compound C.

has an elemental composition of $C_{11}H_{10}FN_2OS_2$ and indicates all amino acid moieties are lost and only the sulfur atom from GSH and parent molecule remained. Therefore, the ion at *m/z* 269.0221 was fragmented further in order to provide information on a location of the sulfur atom from GSH. The major fragment ion at *m/z* 210.0387 in MS^4 spectrum results from CHSN loss from its parent ion at *m/z* 269, which could only occur when sulfur is attached to the thiazole moiety. Thus, the fragmentation pathways of the GSH conjugate of compound C are shown in Figure 9.9. The data strongly suggested that the

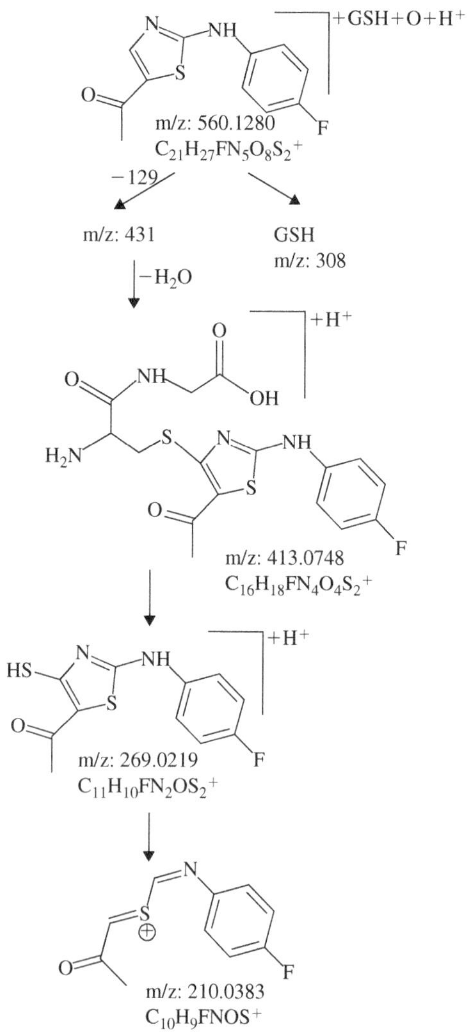

FIGURE 9.9 Proposed fragmentation pathway of glutathione conjugate of compound C via MS^n experiments.

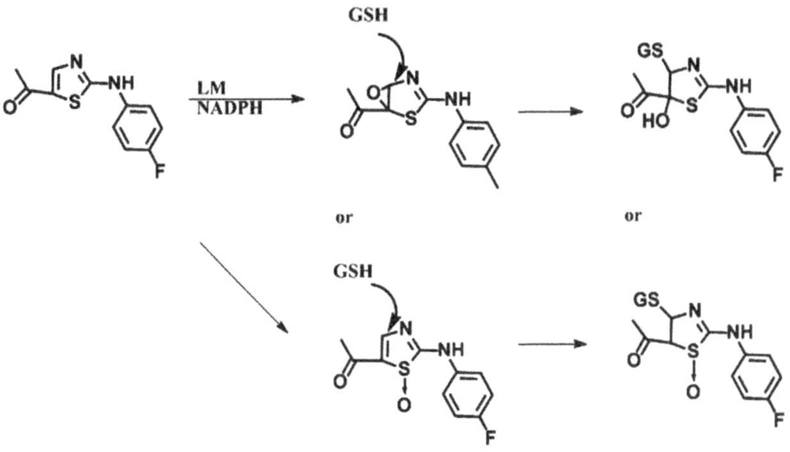

FIGURE 9.10 Proposed bioactivation pathway of compound C (LM: liver microsomes).

thiazole moiety was bioactivated via an epoxide or sulfoxide reactive intermediate that was further trapped by GSH (Fig. 9.10). The determination of the reactive moiety and its location is critical for medicinal chemists to block or mitigate liability of drug candidates during the drug discovery process (Prakash et al., 2008b). This information in most cases can be obtained via the MS^n approach with accurate mass measurement.

It is clearly demonstrated that MS^n can provide detailed structural information and narrow down modification site of metabolites. The use of accurate mass measurement in each MS^n spectrum, (i.e., in current Obitrap HRMS), elemental composition of each fragment can be calculated and greatly increases the reliability of the interpretation of fragmentation pathway and structural information. The fragmentation pathways determined for different structures and fragment ions can in turn be saved in the database to assist future interpretation and prediction of fragments. Commercial software, such as HighChem Mass Frontier and ACD/MS Fragmenter, offers automatic fragmentation prediction and interpretation to reduce this time-consuming and knowledge-based task for the analyst. However, the exact structure of the metabolites is still impossible to determine in most cases. Additionally, the applicability of this approach is limited by the intensity of a characteristic product ion, which may not be sufficient to obtain many MS^n steps.

9.2.2.3 Stable Isotope Cluster Techniques Stable isotopes to identify drug metabolites and elucidate metabolite structures and biotransformation mechanisms has been widely used in drug discovery and development for more than a half century. Application of stable isotope patterns in metabolite identification was extensively reviewed (Bjorge, 1997; Ma et al., 2006).

The advantage of stable isotope patterns for metabolite identification is the ease of identification of metabolites with the same specific isotope patterns as parent drug in a complex biological sample. For example, chlorine and bromine exhibit unique natural isotopic patterns. Chlorine or bromine-containing compounds will have similar isotopic ratio patterns arising from ^{35}Cl: ^{37}Cl (1:0.33) or ^{79}Br: ^{81}Br (1:1). Thus, potential metabolites of chlorine or bromine-containing drugs can be detected by searching molecular ions with similar isotope patterns in each MS spectrum (Tarning et al., 2006). Most current commercial MS data-processing software contain an isotopic pattern recognition function, which facilitates data mining and metabolite identification.

Since not all compounds contain diagnostic natural isotopes such as chlorine or bromine, stable isotope labeling of xenobiotics is often used to assist in metabolite identification by MS. By mixing of natural and synthetically enriched isotopes, preferably in a 1:1 ratio, a characteristic isotope ion cluster can be generated in the mass spectra of compounds. This isotope cluster filtered chromatogram significantly reduces background interference in MS and simplifies drug metabolite detection, especially when a radiolabeled parent compound is not available (Wienkers et al., 1996; Liu et al., 2008). Commonly used stable isotopes are ^{2}H, ^{13}C, and ^{15}N. For example, Wienkers et al. (1996) identified metabolites of tirilazad, a potent inhibitor of membrane lipid peroxidation and employed a mixture of unlabeled and [2,4,6,-^{13}C,1,3-$^{15}N_2$-pyrimidine]tirilazad. Metabolites were easily identified by the characteristic ions of equal abundance separated by 5 Da. Chowdhury et al. (2005) utilized 1:1 ratio of ribavirin and $^{13}C_3$-labeled ribavirin to determine rat metabolites in vivo. Ribavirin-related metabolites were identified by the detection of dual equivalent isotopic peaks with a 3 Da difference. Hydrolyzed metabolites of ribavirin were identified by the detection of dual equivalent isotopic peaks with a 2-Da difference due to loss of a ^{13}C-enriched carbon. Metabolites of penehyclidine in rats, a novel anticholinergic drug, were identified by administering equal amounts of penehyclidine and $^{2}H_5$-penehyclidine (Liu et al., 2008).

Another important application of the stable-isotope labeling technique involved the detection and identification of reactive metabolites with an isotopically labeled trapping agent. Yan and Caldwell (2004) applied an equimolar ratio of glutathione and $^{13}C_2$-^{15}N-labeled glutathione (γ-glutamyl-cysteinyl-[$^{13}C_2$-^{15}N]-glycine) to trap the reactive metabolites from liver microsomal incubations. Unambiguous identification of GSH conjugates was realized by the search for a unique pair of ions with equally abundant isotopic peaks separated by 3 Da in the MS spectra. High confidence of identification further enhanced the acquisition of characteristic MS/MS spectra, which exhibited neutral loss of 75 Da (glycine) and 129 Da (pyroglutamic acid) from GSH conjugates, and 78 Da ([$^{13}C_2$-^{15}N]-glycine) and 129 Da from isotopic GSH conjugates. This isotopic cluster method (Mutlib et al., 2005; Yan et al., 2008) demonstrated high selectivity and reliability for high-throughput screening for reactive metabolites in comparison to traditional neutral loss or precursor ion scanning methods on triple quadrupole mass spectrometers. A similar concept to facilitate the detection and characterization of reactive metabolites was utilized with other stable isotope trapping agents such as potassium cyanide ($K^{13}C^{15}N$) and ^{13}C-$^{15}N_2$-semicarbazide (Rousu et al., 2009).

Similar to radioactive isotope labeling, stable isotopes and labeling sites should be carefully chosen. Deuterium is widely used in metabolic studies because of its availability and low cost. However, limitations of deuterium labeling have been observed in many metabolic studies. When biotransformation involves a $C{-}^2H$ bond, a kinetic isotope effect may occur in addition to the loss of the deuterium label (Dybala-Defratyka et al., 2008). Occasionally, the reduced rate of reaction caused by the isotopic effect can lead to a change in the metabolic pathway of a drug (Ling and Hanzlik, 1989). Thus, the sites for deuterium labeling should be carefully chosen in order to avoid loss of label through proton exchange during keto-enol tautomerism or metabolic NIH shift in aromatic rings. Stable isotopes of ^{13}C and ^{15}N are usually preferred due to their stability and lack of deuterium isotope effects. Dual ^{13}C and ^{15}N labeling has been used for many trapping agents mentioned above. Stable isotope ^{34}S was also reported in the metabolite studies of omeprazole (Jensen Berit et al., 2004). ^{18}O is not commonly used for metabolite identification studies due to oxygen exchange and kinetic isotope effect (Mirica et al., 2008), but it is widely used in mechanistic studies of drug metabolism.

9.2.2.4 Online H/D Exchange H/D exchange followed by mass spectrometry has been widely recognized as a valuable tool to identify the metabolites of drugs as well as to differentiate the isomeric structures. H/D exchange techniques help to determine the presence, number, and position of functional groups subjected to H/D exchange and serve to aid in the structure elucidation of metabolites. Isomeric mono- and dihydroxylated metabolites were differentiated from *N*-oxide and *S*-oxide, and sulfone metabolites, respectively, with this approach (Liu et al., 2001). Miao et al. (2005) recently reported the characterization of an unusual metabolite of ziprasidone with online H/D exchange and mass spectrometry.

9.2.2.5 Chemical Derivatization Strategies Derivatization techniques combined with LC-MS/MS have proven to be very useful for the characterization of novel and unusual metabolites (Prakash and Cui, 1997; Miao et al., 2005; Prakash et al., 2007c). Derivatization is useful when: (1) a metabolite is unstable, (2) a metabolite is very polar, (3) a metabolite is volatile, (4) to characterize the functional group, (5) to prove MS fragmentation, and (6) to improve sensitivity when metabolite is present in trace amounts.

Derivatization with phenyl isothiocynate and methylation have been used for the structural characterization of regioisomeric glucuronides (Kassahun et al., 1997; Prakash et al., 2007a). Dansyl chloride has been used for the derivatization of polar metabolites such as phenol derivatives and the separation of metabolites from the interfering endogenous compounds to enhance the MS response (Prakash et al., 2007b). Miao et al. (2005) have reported the characterization of an unstable metabolite, a 2-mercaptophenyl-amidine analog of the antipsychotic ziprasidone, by chemical derivatization with *N*-dansylaziridine. Derivatization with hexafluoroacetylacetone was proved effective in identification of an unusual metabolite that resulted by scission of the pyrimidine ring (Prakash and Cui, 1997). Several quaternary nitrogen and triphenyl phosphonium compounds have been used for the derivatization of low-molecular-weight primary and secondary amines and carboxylic acids to enhance their detection by LC–MS. In addition, $TiCl_3$ has been used to differentiate *N*-oxides from hydroxylated metabolites (Prakash and Cui, 1997).

9.3 IN SILICO TOOLS

Modern data-processing software packages from manufactures such as Metabolynx (Waters, Manchester, UK), Lightsight (Applied Biosystem, Concord, ON, Canada), Metworks (Thermo Fisher Scientific, San Jose, CA), and Mass Hunter (Agilent Technologies, Santa Clara, CA) are capable of performing target-list-dependent detection of potential drug-related components from complex data. The subsequent list contains common biotransformation pathways and is used to search for the predicted metabolites in full-scan MS based on the structure of the parent drugs (Anari and Baillie, 2005). Commercial software, which allows generation of this knowledge-based metabolite prediction lists, can be divided into two categories. The early generation of software is knowledge-based expert systems such as META, MetabolExpert, and METEOR, which require the analyst to have biotransformation knowledge and provide structure-dependent rules for each metabolite prediction. Another category includes database-based systems to predict metabolite-based published literature and similarity to parent structures and include MDL metabolite (www.mdl.com), Accelrys metabolism (www.accelrys.com) database, and recent MetaDrug (www.genego.com). The software and knowledge-based metabolic predictions have been used in many metabolite studies (Anari et al., 2004; Boyer et al., 2009). While this approach is likely to be reliable for common biotransformation pathways, unusual or multiple-step metabolic reactions may be overlooked.

9.4 CONCLUSIONS AND FUTURE TRENDS

Metabolism studies are an integral part of drug discovery and development, and LC–MS/MS has now become the most powerful tool for the rapid detection and structure elucidation of drug metabolites from biological fluids.

This chapter discussed the common strategies used for characterization of metabolites of small-molecule compounds. In low-resolution MS such as triple quadrupole (QqQ) the detection of metabolite often starts with CNL or PI as survey scans. The general approach for metabolite identification using triple quad MS involves the following steps:

1. Obtain a product ion spectrum of the parent compound: interpret the spectrum.
2. Identify major fragment ion and possible neutral loss.
3. Run precursor ions (PI), neutral loss scans (CNL), or MRM (MIM) of biological samples.
4. Run product ion scans for all possible metabolites identified from step 3 plus expected metabolites.
5. Interpret the spectra and assign structures of metabolites.

Because the ion trap MS is not capable to conduct CNL or PI acquisition, detection of metabolites will rely on post-data-processing of CNL or product ion scans. However, MS^n function of ion trap MS provides its advantage on structural assignment. The application of quadrupole–linear ion trap (QqLIT) on metabolite identification is described in more details in Chapter 15.

In high-resolution MS such as Orbitrap and QTOF, metabolite identification involves the following steps:

1. Obtain a full scan of the compound in question facilitated with background subtraction and MDF post-data-processing.
2. Obtain product ion spectra of the parent and all possible metabolites identified.
3. Interpret the spectra facilitated with elemental analysis and assign structures of metabolites.

Chapter 13 describes the details on applying high-resolution MS and post-acquisition data mining for metabolite identification.

The newer MS experiments in a data-dependent acquisition mode provide the MS and MS^n data from a single injection. Accurate mass measurements, software-assisted data acquisition, and processing methods have been very useful for metabolite detection and identification. In addition, when MS is combined with other analytical techniques such as derivatization, H/D exchange, and stable isotope labeling have been proven very useful for structural characterization of unusual, uncommon, and difficult metabolites. Further, the flexibility and broad applications of mass spectrometry have allowed for the creation of hybrid instruments and coupling to other powerful analytical techniques, most notably nuclear magnetic resonance (NMR), to further enhance the utility in the field of drug metabolism.

Recently, significant progress has been made in the development of software for the prediction of drug metabolites and the automated identification of metabolite peaks from the complex mixtures. The next step, automated assignment of biotransformations and fragment ions, will be a major focus for scientists involved in drug metabolism studies. In addition, the attempts will be made in the development of in silico tools to predict the rate of metabolism and drug metabolism enzymes involved in the metabolism.

ACKNOWLEDGMENT

We would like to thank Dr. Natalia Penner for her excellent assistance in the review of the manuscript.

REFERENCES

Anari MR, Baillie TA. Bridging cheminformatic metabolite prediction and tandem mass spectrometry. *Drug Discov Today* 2005;10:711–717.

Anari MR, Sanchez RI, Bakhtiar R, Franklin RB, Baillie TA. Integration of knowledge-based metabolic predictions with liquid chromatography data-dependent tandem mass spectrometry for drug metabolism studies: Application to studies on the biotransformation of indinavir. *Anal Chem* 2004;76:823–832.

Baillie TA, Cayen MN, Fouda H, Gerson RJ, Green JD, Grossman SJ, Klunk LJ, LeBlanc B, Perkins DG, Shipley LA. Drug metabolites in safety testing. *Toxicol Appl Pharmacol* 2002;182:188–196.

Baillie TA, Davis MR. Mass spectrometry in the analysis of glutathione conjugates. *Biol Mass Spectrom* 1993;22:319–325.

Baranczewski P, Stanczak A, Kautiainen A, Sandin P, Edlund PO. Introduction to early in vitro identification of metabolites of new chemical entities in drug discovery and development. *Pharmacol Rep* 2006;58:341–352.

Bateman KP, Castro-Perez J, Wrona M, Shockcor JP, Yu K, Oballa R, Nicoll-Griffith DA. MSE with mass defect filtering for in vitro and in vivo metabolite identification. *Rapid Commun Mass Spectrom* 2007;21:1485–1496.

Bjorge SM. Identification and characterization of drug metabolites using stable isotope techniques. *Pharmacochem Libr* 1997;26:233–242.

Boyer D, Bauman JN, Walker DP, Kapinos B, Karki K, Kalgutkar AS. Utility of MetaSite in improving metabolic stability of the neutral indomethacin amide derivative and selective cyclooxygenase-2 inhibitor 2-(1-(4-chlorobenzoyl)-5-methoxy-2-methyl-1*H*-indol-3-yl)-*N*-phenethyl-acetamide. *Drug Metab Dispos* 2009;37:999–1008.

Castro-Perez J, Plumb R, Granger JH, Beattie I, Joncour K, Wright A. Increasing throughput and information content for in vitro drug metabolism experiments using ultra-performance liquid chromatography coupled to a quadrupole time-of-flight mass spectrometer. *Rapid Commun Mass Spectrom* 2005a;19:843–848.

Castro-Perez J, Plumb R, Liang L, Yang E. A high-throughput liquid chromatography/tandem mass spectrometry method for screening glutathione conjugates using exact mass neutral loss acquisition. *Rapid Commun Mass Spectrom* 2005b;19:798–804.

Chowdbury SK, Gopaul VS, Blumenkrantz N, Zhong R, Kulmatycki KM, Alton KB. Detection and characterization of highly polar metabolites by LC-MS: Proper selection of LC column and use of stable isotope-labeled drug to study metabolism of ribavirin in rats. *Prog Pharm Biomed Anal* 2005;6:277–293.

Clayton E, Bateman RH, Preece S, Sinclair I. Precursor ion and neutral loss scanning on quadrupole and sector time of flight mass spectrometers in real time LC-MS. *Adv Mass Spectrom* 2001;15:403–404.

Dai H, Wang M, Li X, Wang L, Li Y, Xue M. Structural elucidation of in vitro and in vivo metabolites of cryptotanshinone by HPLC-DAD-ESI-MSn. *J Pharm Biomed Anal* 2008;48:885–896.

Dybala-Defratyka A, Rostkowski M, Paneth P. Enzyme mechanisms from molecular modeling and isotope effects. *Arch Biochem Biophys* 2008;474:274–282.

Food and Drug Administration (FDA). Guidance for Industry: Safety Testing of Drug Metabolites, 2008, available: http//www.fda.gov/cder/guidance/index.htm.

Fura A, Shu YZ, Zhu M, Hanson RL, Roongta V, Humphreys WG. Discovering drugs through biological transformation: Role of pharmacologically active metabolites in drug discovery. *J Med Chem* 2004;47:4339–4351.

Gao H, Materne OL, Howe DL, Brummel CL. Method for rapid metabolite profiling of drug candidates in fresh hepatocytes using liquid chromatography coupled with a hybrid quadrupole linear ion trap. *Rapid Commun Mass Spectrom* 2007;21:3683–3693.

Grillo MP, Hua F, Knutson CG, Ware JA, Li C. Mechanistic studies on the bioactivation of diclofenac: Identification of diclofenac-*S*-acyl-glutathione in vitro in incubations with rat and human hepatocytes. *Chem Res Toxicol* 2003;16:1410–1417.

Hager JW, Le Blanc JC. High-performance liquid chromatography-tandem mass spectrometry with a new quadrupole/linear ion trap instrument. *J Chromatogr A* 2003;1020:3–9.

Herrin GL, McCurdy HH, Wall WH. Investigation of an LC-MS-MS (QTrap) method for the rapid screening and identification of drugs in postmortem toxicology whole blood samples. *J Anal Toxicol* 2005;29:599–606.

Hobby K, Gallagher RT, Caldwell P, Wilson ID. A new approach to aid the characterisation and identification of metabolites of a model drug; partial isotope enrichment combined with novel formula elucidation software. *Rapid Commun Mass Spectrom* 2009;23:219–227.

Holcapek M, Kolarova L, Nobilis M. High-performance liquid chromatography-tandem mass spectrometry in the identification and determination of phase I and phase II drug metabolites. *Anal Bioanal Chem* 2008;391:59–78.

Huang Y, Liu S, Miao S, Jeanville PM. Using multiple mass defect filters and higher energy collisional dissociation on an LTQ Orbitrap XL for fast, sensitive, and accurate metabolite ID, 2008, available: www.thermo.com/appnotes.

Jemal M, Ouyang Z, Zhao W, Zhu M, Wu WW. A strategy for metabolite identification using triple-quadrupole mass spectrometry with enhanced resolution and accurate mass capability. *Rapid Commun Mass Spectrom* 2003;17:2732–2740.

Jensen Berit P, Smith C, Wilson ID, Weidolf L. Sensitive sulphur-specific detection of omeprazole metabolites in rat urine by high-performance liquid chromatography/inductively coupled plasma mass spectrometry. *Rapid Commun Mass Spectrom* 2004;18:181–183.

Johnson BM, Kamath AV, Leet JE, Liu X, Bhide RS, Tejwani RW, Zhang Y, Qian L, Wei DD, Lombardo LJ, Shu YZ. Metabolism of 5-isopropyl-6-(5-methyl-1,3,4-oxadiazol-2-yl)-*N*-(2-methyl-1*H*-pyrrolo[2,3-*b*]pyridin-5-yl)pyrrolo[2,1-*f*][1,2,4]triazin-4-amine (BMS-645737): Identification of an unusual *N*-acetylglucosamine conjugate in the cynomolgus monkey. *Drug Metab Dispos* 2008;36:2475–2483.

Kalgutkar AS, Gardner I, Obach RS, Shaffer CL, Callegari E, Henne KR, Mutlib AE, Dalvie DK, Lee JS, Nakai Y, O'Donnell JP, Boer J, Harriman SP. A comprehensive listing of bioactivation pathways of organic functional groups. *Curr Drug Metab* 2005;6:161–225.

Kamel A, Prakash C. High performance liquid chromatography/atmospheric pressure ionization/tandem mass spectrometry (HPLC/API/MS/MS) in drug metabolism and toxicology. *Curr Drug Metab* 2006;7:837–852.

Kassahun K, Mattiuz E, Nyhart E Jr, Obermeyer B, Gillespie T, Murphy A, Goodwin RM, Tupper D, Callaghan JT, Lemberger L. Disposition and biotransformation of the antipsychotic agent olanzapine in humans. *Drug Metab Dispos* 1997;25:81–93.

Kostiainen R, Kotiaho T, Kuuranne T, Auriola S. Liquid chromatography/atmospheric pressure ionization-mass spectrometry in drug metabolism studies. *J Mass Spectrom* 2003;38:357–372.

Lampinen-Salomonsson M, Bondesson U, Petersson C, Hedeland M. Differentiation of estriol glucuronide isomers by chemical derivatization and electrospray tandem mass spectrometry. *Rapid Commun. Mass Spectrom* 2006;20:1429–1440.

Leclercq L, Mortishire-Smith RJ, Huisman M, Cuyckens F, Hartshorn MJ, Hill A. IsoScore: Automated localization of biotransformations by mass spectrometry using product ion scoring of virtual regioisomers. *Rapid Commun Mass Spectrom* 2009;23:39–50.

Li W, Zhang D, Wang L, Zhang H, Cheng PT, Zhang D, Everett DW, Humphreys WG. Biotransformation of carbon-14-labeled muraglitazar in male mice: Interspecies difference in metabolic pathways leading to unique metabolites. *Drug Metab Dispos* 2006;34:807–820.

Lim HK, Chen J, Sensenhauser C, Cook K, Subrahmanyam V. Metabolite identification by data-dependent accurate mass spectrometric analysis at resolving power of 60,000 in external calibration mode using an LTQ/Orbitrap. *Rapid Commun Mass Spectrom* 2007;21:1821–1832.

Ling KH, Hanzlik RP. Deuterium isotope effects on toluene metabolism. Product release as a rate-limiting step in cytochrome P-450 catalysis. *Biochem Biophys Res Commun* 1989;160:844–849.

Liu DQ, Hop CE, Beconi MG, Mao A, Chiu SH. Use of on-line hydrogen/deuterium exchange to facilitate metabolite identification. *Rapid Commun Mass Spectrom* 2001;15:1832–1839.

Liu Y, Wang M, Xue M, Li Y, Li X, Ruan J, Liu K. Structural elucidation of in vivo metabolites of penehyclidine in rats by the method of liquid chromatography-mass spectrometry, gas chromatography-mass spectrometry and isotope ion cluster. *J Chromatogr B Anal Technol Biomed Life Sci* 2008;873:41–50.

Ma S, Chowdhury SK, Alton KB. Application of mass spectrometry for metabolite identification. *Curr Drug Metab* 2006;7:503–523.

Mahajan MK, Evans CA. Dual negative precursor ion scan approach for rapid detection of glutathione conjugates using liquid chromatography/tandem mass spectrometry. 2008;*Rapid Commun Mass Spectrom* 22:1032–1040.

Mauriala T, Chauret N, Oballa R, Nicoll-Griffith DA, Bateman KP. A strategy for identification of drug metabolites from dried blood spots using triple-quadrupole/linear ion trap hybrid mass spectrometry. *Rapid Commun Mass Spectrom* 2005;19:1984–1992.

Miao Z, Kamel A, Prakash C. Characterization of a novel metabolite intermediate of ziprasidone in hepatic cytosolic fractions of rat, dog, and human by ESI-MS/MS, hydrogen/deuterium exchange, and chemical derivatization. *Drug Metab Dispos* 2005;33:879–883.

Mirica LM, McCusker KP, Munos JW, Liu H-w, Klinman JP. 18O Kinetic isotope effects in non-heme iron enzymes: Probing the nature of Fe/O_2 intermediates. *J Am Chem Soc* 2008;130:8122–8123.

Mortishire-Smith RJ, Castro-Perez JM, Yu K, Shockcor JP, Goshawk J, Hartshorn MJ, Hill A. Generic dealkylation: A tool for increasing the hit-rate of metabolite rationalization, and automatic customization of mass defect filters. *Rapid Commun Mass Spectrom* 2009;23:939–948.

Mutlib A, Lam W, Atherton J, Chen H, Galatsis P, Stolle W. Application of stable isotope labeled glutathione and rapid scanning mass spectrometers in detecting and characterizing reactive metabolites. *Rapid Commun Mass Spectrom* 2005;19: 3482–3492.

Mutlib A, Shockor J. Application of LC/MS, LC/NMR, NMR and stable isotopes in identifying and characterizing metabolites. In *Drug Metabolizing Enzymes: Cytochrome P450 and Other Enzymes in Drug Discovery and Development*, Lee J, RS O, MB F, Eds. Marcel Dekker, New York, 2003, pp. 33–86.

Nakazawa T, Miyata K, Omura K, Iwanaga T, Nagata O. Metabolic profile of FYX-051 (4-(5-pyridin-4-yl-1*h*-[1,2,4]triazol-3-yl)pyridine-2-carbonitrile) in the rat, dog, monkey, and human: Identification of *N*-glucuronides and *N*-glucosides. *Drug Metab Dispos* 2006;34:1880–1886.

Paglia G, D'Apolito O, Corso G. Precursor ion scan profiles of acylcarnitines by atmospheric pressure thermal desorption chemical ionization tandem mass spectrometry. *Rapid Commun Mass Spectrom* 2008;22:3809–3815.

Prakash C, Chen W, Rossulek M, Johnson K, Zhang C, O'Connell T, Potchoiba M, Dalvie D. Metabolism, pharmacokinetics, and excretion of a cholesteryl ester transfer protein inhibitor, torcetrapib, in rats, monkeys, and mice: Characterization of unusual and novel metabolites by high-resolution liquid chromatography-tandem mass spectrometry and 1H nuclear magnetic resonance. *Drug Metab Dispos* 2008a;36:2064–2079.

Prakash C, Cui D. Metabolism and excretion of a new antianxiety drug candidate, CP-93,393, in cynomolgus monkeys: Identification of the novel pyrimidine ring cleaved metabolites. *Drug Metab Dispos* 1997;25:1395–1406.

Prakash C, Cui D, Potchoiba MJ, Butler T. Metabolism, distribution and excretion of a selective *N*-methyl-D-aspartate receptor antagonist, traxoprodil, in rats and dogs. *Drug Metab Dispos* 2007a;35:1350–1364.

Prakash C, O'Donnell J, Khojesteh C. Excretion, pharmacokinetics and metabolism of the substance P receptor antagonist, CJ-11,974, in humans: Identification of polar metabolites by LC/MS/MS and chemical derivatization *Drug Metab Dispos* 2007b;35: 1071–1080.

Prakash C, Shaffer CL, Nedderman A. Analytical strategies for identifying drug metabolites. *Mass Spectrom Rev* 2007c;26:340–369.

Prakash C, Sharma R, Gleave M, Nedderman A. In vitro screening techniques for reactive metabolites for minimizing bioactivation potential in drug discovery. *Curr Drug Metab* 2008b;9:952–964.

Rock D, Wahlstrom J, Wienkers L. Cytochrome P450s: Drug–drug interactions. Methods Principles Med Chem 2008;38:197–246.

Rousu T, Pelkonen O, Tolonen A. Rapid detection and characterization of reactive drug metabolites in vitro using several isotope-labeled trapping agents and ultra-performance liquid chromatography/time-of-flight mass spectrometry. *Rapid Commun Mass Spectrom* 2009;23:843–855.

Ruan Q, Peterman S, Szewc MA, Ma L, Cui D, Humphreys WG, Zhu M. An integrated method for metabolite detection and identification using a linear ion trap/Orbitrap mass spectrometer and multiple data processing techniques: Application to indinavir metabolite detection. *J Mass Spectrom* 2008;43:251–261.

Sang S, Lee M-J, Yang I, Buckley B, Yang CS. Human urinary metabolite profile of tea polyphenols analyzed by liquid chromatography/electrospray ionization tandem mass spectrometry with data-dependent acquisition. *Rapid Commun Mass Spectrom* 2008;22:1567–1578.

Scholz K, Dekant W, Volkel W, Pahler A. Rapid detection and identification of *N*-acetyl-L-cysteine thioethers using constant neutral loss and theoretical multiple reaction monitoring combined with enhanced product-ion scans on a linear ion trap mass spectrometer. *J Am Soc Mass Spectrom* 2005;16:1976–1984.

Sinz M, Podoll T. The mass spectrometer in drug metabolism. In *Mass Spectrometry in Drug Discovery*, Rossi D, Sinz M, Eds. Marcel Dekker, New York, 2002, pp. 271–336.

Song R, Lin H, Zhang Z, Li Z, Xu L, Dong H, Tian Y. Profiling the metabolic differences of anthraquinone derivatives using liquid chromatography/tandem mass spectrometry with data-dependent acquisition. *Rapid Commun Mass Spectrom* 2009;23:537–547.

Sun J, Yang M, Han J, Wang B, Ma X, Xu M, Liu P, Guo D. Profiling the metabolic difference of seven tanshinones using high-performance liquid chromatography/multi-stage mass spectrometry with data-dependent acquisition. *Rapid Commun Mass Spectrom* 2007;21:2211–2226.

Tarning J, Bergqvist Y, Day NP, Bergquist J, Arvidsson B, White NJ, Ashton M, Lindegardh N. Characterization of human urinary metabolites of the antimalarial piperaquine. *Drug Metab Dispos* 2006;34:2011–2019.

Tiller PR, Yu S, Castro-Perez J, Fillgrove KL, Baillie TA. High-throughput, accurate mass liquid chromatography/tandem mass spectrometry on a quadrupole time-of-flight system as a "first-line" approach for metabolite identification studies. *Rapid Commun Mass Spectrom* 2008;22:1053–1061.

Triolo A, Altamura M, Dimoulas T, Guidi A, Lecci A, Tramontana M. In vivo metabolite detection and identification in drug discovery via LC-MS/MS with data-dependent scanning and postacquisition data mining. *J Mass Spectrom* 2005;40:1572–1582.

Watt AP, Mortishire-Smith RJ, Gerhard U, Thomas SR. Metabolite identification in drug discovery. *Curr Opin Drug Discov Devel* 2003;6:57–65.

Wienkers LC, Steenwyk RC, Sanders PE, Pearson PG. Biotransformation of tirilazad in human: 1. Cytochrome P450 3A-mediated hydroxylation of tirilazad mesylate in human liver microsomes. *J Pharmacol Exp Ther* 1996;277:982–990.

Wrona M, Mauriala T, Bateman KP, Mortishire-Smith RJ, and O'Connor D. "All-in-one" analysis for metabolite identification using liquid chromatography/hybrid quadrupole time-of-flight mass spectrometry with collision energy switching. *Rapid Commun Mass Spectrom* 2005;19:2597–2602.

Xu L, Adams B, Jeliazkova-Mecheva VV, Trimble L, Kwei G, Harsch A. Identification of novel metabolites of colchicine in rat bile facilitated by enhanced online radiometric detection. *Drug Metab Dispos* 2008;36:731–739.

Yan Z, Caldwell GW. Stable-isotope trapping and high-throughput screenings of reactive metabolites using the isotope MS signature. *Anal Chem* 2004;76:6835–6847.

Yan Z, Caldwell GW, Maher N. Unbiased high-throughput screening of reactive metabolites on the linear ion trap mass spectrometer using polarity switch and mass tag triggered data-dependent acquisition. *Anal Chem* 2008;80:6410–6422.

Yao M, Ma L, Humphreys WG, Zhu M. Rapid screening and characterization of drug metabolites using a multiple ion monitoring-dependent MS/MS acquisition method on a hybrid triple quadrupole-linear ion trap mass spectrometer. *J Mass Spectrom* 2008;43:1364–1375.

Zhang H, Ma L, He K, Zhu M. An algorithm for thorough background subtraction from high-resolution LC/MS data: Application to the detection of troglitazone metabolites in rat plasma, bile, and urine. *J Mass Spectrom* 2008a;43:1191–1200.

Zhang H, Yang Y. An algorithm for thorough background subtraction from high-resolution LC/MS data: Application for detection of glutathione-trapped reactive metabolites. *J Mass Spectrom* 2008;43:1181–1190.

Zhang H, Zhang D, Ray K. A software filter to remove interference ions from drug metabolites in accurate mass liquid chromatography/mass spectrometric analyses. *J Mass Spectrom* 2003;38:1110–1112.

Zhang H, Zhu M, Ray KL, Ma L, Zhang D. Mass defect profiles of biological matrices and the general applicability of mass defect filtering for metabolite detection. *Rapid Commun Mass Spectrom* 2008b;22:2082–2088.

Zhang Z, Chen Q, Li Y, Doss GA, Dean BJ, Ngui JS, Silva Elipe M, Kim S, Wu JY, Dininno F, Hammond ML, Stearns RA, Evans DC, Baillie TA, Tang W. In vitro bioactivation of dihydrobenzoxathiin selective estrogen receptor modulators by cytochrome P450 3A4 in human liver microsomes: Formation of reactive iminium and quinone type metabolites. *Chem Res Toxicol* 2005;18:675–685.

Zheng J, Ma L, Xin B, Olah T, Humphreys WG, Zhu M. Screening and identification of GSH-trapped reactive metabolites using hybrid triple quadruple linear ion trap mass spectrometry. *Chem Res Toxicol* 2007;20:757–766.

10 Mass Spectral Interpretation

LI-KANG ZHANG, and BIRENDRA N. PRAMANIK
Global Analytical Chemistry, Merck Research Laboratories, Kenilworth, New Jersey

Mass spectrum interpretation is essential to solve one or more of the following problems: establishment of molecular weight and of empirical formula; detection of functional groups and other substituents; determination of molecular skeleton (atom connectivity); elucidation of precise structure and, even in favorable cases, certain stereochemical features. As discussed in the previous chapters, electrospray (ESI) and atmospheric pressure chemical ionization (APCI) are two of the most effective and successful interfaces for the liquid chromatography–mass spectrometry (LC–MS) that have been developed. Thus, we will focus on how to interpret the mass spectral data generated by either ESI or APCI in this section.

Mass Spectrometry in Drug Metabolism and Disposition: Basic Principles and Applications, First Edition. Edited by Mike S. Lee and Mingshe Zhu.

10.1 MOLECULAR WEIGHT AND EMPIRICAL FORMULA DETERMINATION

All qualitative applications of mass spectrometry are based on the determination of the mass-to-charge (m/z) ratio of an analyte. ESI and APCI MS are well known to produce abundant molecular ions and are consequently known as "soft" ionization techniques. This means that a relatively small amount of energy is transferred to the molecule as a result of vaporization and ionization. The efficiency of ion formation depends, in part, on a molecule's ability to carry a charge (e.g., its proton affinity). The positive-ion mode ionization process can be defined by the following simple protonation reaction:

$$M + H^+ \rightarrow MH^+$$

Mobile-phase additives are used in high-pressure liquid chromatography (HPLC) to control the pH and ensure efficient and reliable separations. They need to be compatible with ESI or APCI conditions. The most suitable additives in LC–MS are acetic acid (0.05–0.1%), formic acid (0.05–0.1%, and ammonia acetate (2–10 m*M*). Trifluoroacetic acid (TFA) has been used in the LC analysis of basic compounds extensively as an ion-pair agent to improve peak shapes of basic compounds on silica-based columns. On the other hand, TFA is known to suppress the ESI signals of analytes and reduce assay sensitivity when used in the LC–MS mobile phase. This is probably due to the ability of TFA to form gas-phase ion pairs with positively charged analyte ions. It has been reported by Shou and co-workers (Shou and Weng, 2005) that by addition of 0.5% acetic acid to the mobile phase containing either 0.025 or 0.05% TFA could minimize the negative effect of TFA in LC–MS analysis.

10.1.1 Formation of Adduct Ions

In contrast to traditional MS, the highest mass peaks in ESI or APCI spectra are not always the molecular ion of interest. Instead, noncovalent complex ions are commonly observed. The adduct ions are generally formed by an analyte–adduct interaction in solution that is preserved as a result of the soft ionization of the ESI/APCI process (Bose et al., 1978, 1982; Cerny et al., 1994; Pramanik et al., 1984, 1989; Pettit et al., 1983; Bartner et al., 1997). To our knowledge, the first report of the formation of the adduct ions in mass spectra came from the work performed in the laboratory of Prof. A. K. Bose at the Stevens Institute of Technology (Bose et al., 1978). The authors reported (Bose et al., 1978) that a series of multifunctional polar and thermolabile compounds displayed abundant ammoniated adducts of the molecules ($[M + NH_4^+]$) in chemical ionization (CI) MS, when the experiment was conducted by adding a trace level of NH_4Cl to the analyte before analysis. The use of salts (i.e., NH_4Cl, NaCl, KCl) in MS was expanded quite rapidly to cover electron ionization (EI), liquid secondary ion mass spectrometry (LSIMS), and fast atom bombardment (FAB) MS for the enhancement of the relative abundances of the molecular ions as ammonium, sodium, or potassium ion adducts in mass spectra of a wide range of organic molecules. This has now become a widely used technique in a MS laboratory (Cerny et al., 1994; Pramanik et al., 1984, 1989; Pettit et al,, 1983, Bose et al., 1982; Bartner et al., 1997). The adduct ions are generally formed by analyte–adduct gas-phase collisions in the spray chamber (Ikonomou et al., 1991). On the other hand, the exact mechanism of the formation of analyte–adducts in ESI or APCI still has not been established. More often than not, the adduct–ion formation is a major cause for low detection limit and poor mass resolving power for ESI or APCI. These associative processes, however, have also created interest in the study of drug–protein/drug–oligonucleotide gas-phase complexes that benefit from the ability of ESI to APCI MS analysis.

The formation of adduct ions depends explicitly on the structure and functionality of the species involved in the resulting complex as well as the instrument conditions (Daniel et al., 2002, Brodbelt, 1997, 2000). The $[M + X]^+$ formation in ESI–MS was first reported by Yamashita and Fenn (1984a, b). They showed the effect of flow rate, source temperature, and probe voltage on methanol, water, and acetonitrile solvent systems containing various additives such as LiCl, NaCl, $(CH_3)_4NI$, and HCl. Kebarle and co-workers (Ikonomou et al., 1990, 1991; Kebarle, 1992; Tang and Kebarle, 1991), in early examples, recorded the adduct ions of 30 organic compounds where association with NH_4^+, Na^+, K^+, Cs^+ and Ca^{2+} occurred. They reported that the analyte ion sensitivities decrease with analyte ion concentration and the presence of salts in the solution. They found the detection limits of those organic bases to be in the subfemtomole to attomole range when the salt concentration was below 10^{-4} *M*. Leize et al. (1996) also demonstrated that Cs^+ showed the greatest propensity to form adducts relative to Li^+, Na^+, K^+, and Rb^+ in equimolar mixtures of LiCl, NaCl, KCl, RbCl, and CsCl solution. They found that salvation energy played an important role in the determination of the ionization efficiency (Leize et al., 1996).

The binding of Na^+, NH_4^+ and other background species to an analyte is often seen in the ESI–MS analysis (Table 10.1). To demonstrate adduct ion

TABLE 10.1 Common Adduct Ions

Pseudo-Molecular Ions	Mass (*m/z*)
$[M + Na]^+$	M + 23
$[M + K]^+$	M + 39
$[M + Li]^+$	M + 7
$[M + Na + K - H]^+$	M + 61
$[M + H + NH_3]^+$	M + 18
$[M + H + ACN]^+$	M + 42
$[M + H + MeOH]^+$	M + 33
$[M + Na + ACN]^+$	M + 64
$[M + K + ACN]^+$	M + 80
$[M + H + CH_3CH_2NH_2]^+$	M + 46
$[M + Cl]^-$	M + 35
$[M + CH_3COO]^-$	M + 59
$[M + CF_3COO]^-$	M + 113

STRUCTURE 1 Verapamil.

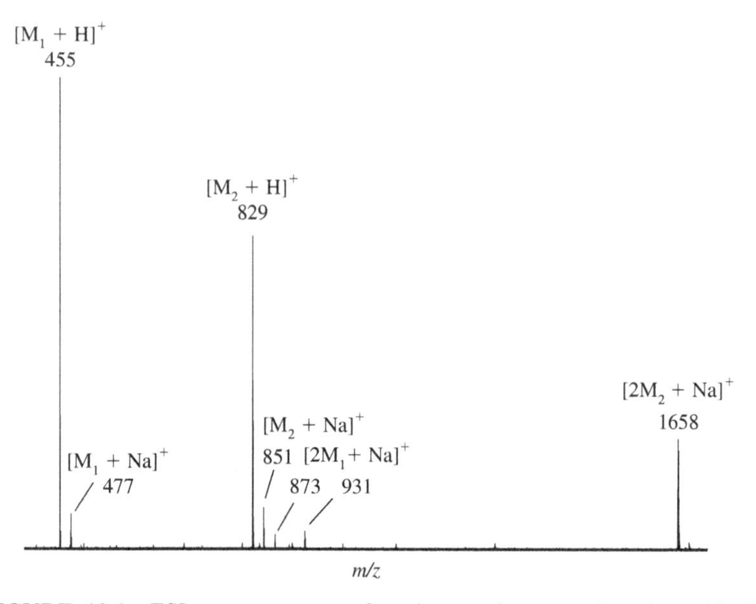

FIGURE 10.1 ESI mass spectrum of a mixture of verapamil and peptide A.

formation, we analyzed a mixture of verapamil (Structure 1) and a small peptide A (ALILTLVS) by ESI–MS operated in the positive-ion mode. ESI of the mixture produced the molecular ions of verapamil ($[M_1 + H]^+$) and peptide A ($[M_2 + H]^+$) at m/z 455 and 829, respectively (Fig. 10.1). The spectrum also shows three peaks at m/z 477, m/z 851, and m/z 873 for the sodium adduct of verapamil ($[M_1 + Na]^+$) and peptide A ($[M_2 + Na]^+$, $[M_2 - H + 2Na]^+$), respectively. The adduct ion at m/z 931 was confirmed to be of verapamil ($[2M_1 + Na]^+$), likely a sodim-bridged dimer ion, by the high mass resolving power data. A moderately abundant noncovalent dimer ion of peptide A ($[2M_2 + H]^+$) was also observed in the ESI spectrum at m/z 1658.

Sometimes, the $[M + Na]^+$, $[M + NH_4]^+$ and other adduct ions (Table 10.1) were found to be prominent in the mass spectra. An example is reported by Zhou and Hamburger (1996) for the characterization of secondary metabolites of plant and microbial origin by ESI–MS. A number of different compounds, including macrolides, peptides, aminoglycosides, polyethers, polyenes, alkaloids, terpenoids, purines, amindes, phenolics, and glycosides, were characterized based on the dominant adduct ions observed. They found that with an APCI interface, most of the compounds produced $[M + H]^+$ and/or $[M - H]^-$ ions instead of adducts with various metal ions (Zhou and Hamburger, 1996), except some poorly functionalized and very thermolabile compounds. More recently, Schug and McNair (2002, 2003) investigated the adduct ion formation by ESI–MS operated in the negative-ion mode. They performed extensive ESI–MS experiments designed to determine the underlying principles in the formation of hydrogen-bound dimer ions and sodium-bridged dimer ions from the halide-substituted benzoic acid derivatives (Schug and McNair, 2003) and six acidic pharmaceuticals (Schug and McNair, 2002).

The formation of solvent–analyte noncovalently bound complexes (e.g., $[M + H + ACN]^+$, $[M + H + MeOH]^+$, etc.) complicates molecular ion determination in ESI–APCI MS (Table 10.1). The relative abundances of these solvent cluster ions depend on the components in the solution phase, ionization mode, spray voltage, capillary temperature, sheath gas pressure, as well as auxiliary gas flow. Zhao et al. (2004) recently reported that acetonitrile could be reduced to ethyl amine under ESI conditions (Scheme 1). They demonstrated that the "M + 46" ion in the mass spectrum represented the ethyl amine adduction ($[M + H + CH_3CH_2NH_2]^+$) when the ESI–MS was performed by infusion of the compound in acetonitrile and water (1% HCOOH + 1% NH_4OH) (1 : 1; v : v). Moreover, they showed the same analyte produced a moderate $[M + H + CD_3CH_2NH_2]^+$ (M + 49) signal when acetonitrile-d_3 was used as the organic solvent (Scheme 1).

Adduct ion formation is a common feature of most ESI or APCI–MS analysis. Although these ions can complicate the mass spectra, once understood, they are useful for the confirmation of the molecular weight of an analyte of interest.

10.1.2 Isotopic Clusters

Most of the common elements encountered in organic molecules have more than one isotope, except fluorine, iodine, and phosphorus (Table 10.2). This results in "isotopic clusters" produced in the mass spectrum. One example is

$$CH_3CN \xrightarrow[ESI]{+4H} CH_3CH_2NH_2$$

$$CD_3CN \xrightarrow[ESI]{+4H} CD_3CH_2NH_2$$

SCHEME 1

TABLE 10.2 Natural Isotope Abundance and Exact Masses of Common Elements

Element	Symbol	Nominal Mass (Da.)	Exact Mass	Abundance
Hydrogen	H	1	1.00783	99.99
	D or ^{2}H	2	2.01410	0.01
Carbon	^{12}C	12	12.0000	98.91
	^{13}C	13	13.0034	1.09
Nitrogen	^{14}N	14	14.0031	99.6
	^{15}N	15	15.0001	0.37
Oxygen	^{16}O	16	15.9949	99.76
	^{17}O	17	16.9991	0.037
	^{18}O	18	17.9992	0.20
Fluorine	F	19	18.9984	100
Silicon	^{28}Si	28	27.9769	92.28
	^{29}Si	29	28.9765	4.70
	^{30}Si	30	29.9738	3.02
Phosphorus	P	31	30.9738	100
Sulphur	^{32}S	32	31.9721	95.02
	^{33}S	33	32.9715	0.74
	^{34}S	34	33.9679	4.22
Chlorine	^{35}Cl	35	34.9689	75.77
	^{37}Cl	37	36.9659	24.23
Bromine	^{79}Br	79	78.9183	50.5
	^{81}Br	81	80.9163	49.5
Iodine	I	127	126.9045	100

chlorine, which has two isotopes of 35 and 37 Da with a characteristic isotopic ratio of 3 : 1. On ionization of an organic compound containing one atom of chlorine (e.g., 4-chloro-phenlyamine), 75.77% of the molecular ions will have a mass of 127 Da and 24.23% will have a mass of 129 Da. For the mass spectral analysis of a small molecule, the molecular ion is generally assigned to the peak representing the more or most abundant isotope.

Although the isotopic pattern may complicate the molecular weight assignment, the pattern usually will also provide valuable reference for recognition of the type and number of elements in a molecule. The characteristic patterns resulting from multiple isotopic contributions of the chlorine, bromine, and sulfur isotopes are shown graphically in Figure 10.2. Mometasone furoate (Structure 2) is a topical glucocorticosteroid with local anti-inflammatory and anti-allergic properties. The mass spectrum of mometasone furoate (Fig. 10.3) shows prominent molecular ions of m/z 521 ($[M+H]^{+}$), 523, and 525, of relative abundances 100 : 70 : 14. The isotopic intensity pattern is similar to that of the second entry of the reference bar graphs, namely that for Cl_2 in Figure 10.2, and the pattern arises from the ions $C_{27}H_{31}O_6{}^{35}Cl_2$, $C_{27}H_{31}O_6{}^{35}Cl^{37}Cl$ and $C_{27}H_{31}O_6{}^{37}Cl_2$. Note the good agreement among the chlorine-isotope pattern regardless of the other elements (i.e., C, H, O) present. Often, the

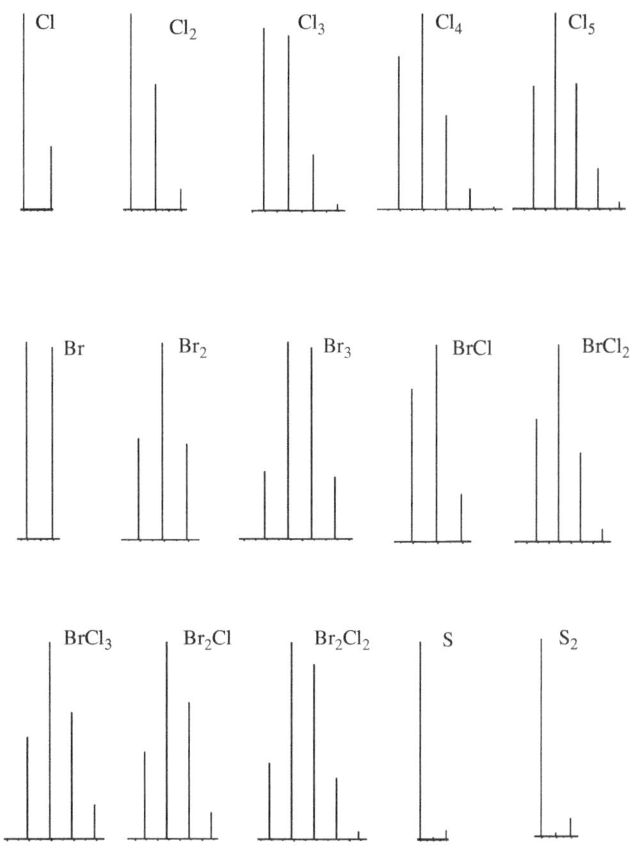

FIGURE 10.2 Isotopic abundances for ions containing different numbers of sulfur, chlorine, and/or bromine atoms.

STRUCTURE 2 Mometasone furoate.

characteristic patterns (Fig. 10.2) resulting from combinations of the isotope peaks can be used to ascertain or confirm the elemental composition of the corresponding ion.

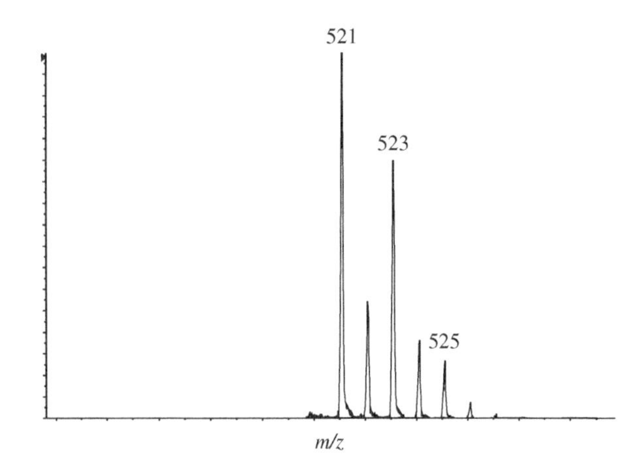

FIGURE 10.3 ESI mass spectrum of mometasone furoate.

10.1.3 Nitrogen Rule

If the nominal molecular weight (MW) of an analyte appears to be an even mass number, the compound contains an even number of nitrogen atoms (or no nitrogen atoms). On the other hand, if a compound contains an odd number of nitrogen atoms, then its molecular weight will be an odd mass number. This rule is very useful for determining the nitrogen content of an unknown compound. In the case of verapamil (Structure 1), the molecular ion appears at $m/z = 421$ (MW = 420 Da), indicating an even number of nitrogen atoms in the structure. The nitrogen rule is also useful in interpreting spectra. The neutral loss of 29 Da in ESI is more likely to mean CH_2NH than C_2H_5 given that even-electron ions are preferred in ESI and APCI.

10.1.4 Accurate Mass Measurement

The development of ionization techniques and mass analyzers has enabled mass spectrometrists to determine the accurate mass of small molecules as well as biomolecules that are present at low levels. The modem quadrupole time of flight (Q-TOF), Fourier transform ion cyclotron resonance MS (FT-ICR MS), and Orbitrap instruments allow determination of the mass of an ion with accuracies and precisions beyond the decimal point. Mass accuracy measurement, typically measured and reported as parts per million (ppm), is essential for elemental composition assignment.

An internal mass calibration is generally needed to achieve mass measurement accuracy of 5–10 ppm with a Q-TOF mass spectrometer (Clauwaert et al., 2003; Wolff et al., 2001; Pilard et al., 2000). Internal calibration is based on mixing one or several internal standards or calibrants of known molecular weight with the analyte and then using the known masses to calibrate the mass measurements of unknown ones that coexist in the sample mixture. With modern instruments, the number of calibrant ions can be small because the mass calibration law is well known (consider FT-ICR or TOF instruments) compared to the older sector mass spectrometers where scores of reference ions were needed to calibrate the mass range of the instrument.

Currently, FT-ICR MS provides the highest mass resolving power and mass accuracy among all the mass spectrometric methods. Using external calibration, FT-ICR MS is capable of achieving mass measurement accuracies (MMA) of 1 ppm or better. Internal calibration can provide an order of magnitude greater mass accuracy than external calibration. The mass accuracy of the newer Orbitrap instruments is comparable although perhaps slightly lower. Time-of-flight instruments are capable of 2–5 ppm of accuracy and precision.

One example is the high resolution ESI-FT-ICR MS analysis of a mixture of verapamil and peptide A. Abbreviated segment of the mass spectrum of verapamil and peptide A (Fig. 10.1) shows the molecular ion peak of peptide A, as well as its accompanying isotope peaks (Fig. 10.3). The following values were obtained for the monoisotopic molecular ion and the isotopic molecular ions using an external calibration (calculated mass; mass error in ppm): (829.5393; 0.1 ppm), (830.5423; 1.3 ppm), (831.5450; 1.0 ppm), and (832.5476; 1.2 ppm). Similar results were obtained for verapamil (data not shown).

Accurate mass measurement is essential in establishing compound identity. For example, an unknown with an m/z of 122.0606 ($[M+H]^+$) must be C_7H_8NO; it cannot be $C_5H_{13}NCl$ (m/z 122.0731), $C_3H_8NO_4$ (m/z 122.0447), $C_4H_{12}NO_3$ (m/z 122.0811), C_4H_9NOCl (m/z 122.0367), or $C_8H_{12}N$ (m/z 122.0964). A mass measurement accuracy of only 102 ppm is required to distinguish these elemental compositions. Thus, the MMA needed to make an unequivocal elemental composition assignment depends on the problem and the mass of the analyte. For most small-molecule analyses (MW $<$ 500 Da), a compound can be identified if the mass measurement accuracy is 5 ppm or better. The accurate mass measurement of the relative intensities of the isotope peaks, as described in the previous section, has also been used as a powerful additional metric for obtaining the correct elemental compositions for Br, Cl, or S containing compounds with their distinctive isotope patterns.

The newer LTQ/Orbitrap MS instrument offers an alternative to the generally more expensive FT MS system with a lower resolving power (up to 100,000). The LTQ/Orbitrap is often more compatible with HPLC or ultra-pressure liquid chromatography (UPLC) flow rates and also provides data-dependant MS^n. The MS^n and accurate mass capabilities (better than 2 ppm MMA) make it capable for metabolite characterization and structural elucidation.

10.1.5 Double-Bond Equivalency (DBE)

To evaluate whether a formula, $C_xH_yN_zO_n$, is a reasonable elemental composition for a certain mass, one can calculate the DBE (numbers of rings and double bonds) by using the formula: $x - y/2 + z/2 + 1$ as shown in Eq. 10.1. A more general case $I_yII_nIII_zIV_x$ was suggested by McLafferty (1980), where I = H, F, Cl, Br, I; II = O, S; III = N, P; and IV = C, Si.

$$\text{DBE(R + DB)} = x - y/2 + z/2 + 1 \tag{10.1}$$

The value 12 found for mometasone furoate ($C_{27}H_{30}O_6Cl_2$; Structure 2) represents five rings and seven double bonds of this molecule (Eq. (10.2)]. Notice that even though oxygen is present in the formula of mometasone furoate, it does not contribute to the DBE. The abundance of the molecular ion usually parallels the chemical stability of the molecule, and compounds with large numbers of rings and double bonds (DBE) show higher molecular ion abundance than those with low DBE. This is consistent with the abundant molecular ion peak we observed in the ESI–MS spectrum of mometasone furoate (Fig. 10.3).

$$\text{DBE(R + DB)} = 27 - 32/2 + 1 = 12 \tag{10.2}$$

10.2 COMMON FRAGMENTATION REACTIONS

Fragmentation patterns of analytes, after ionization in the mass spectrometer, can provide structural information. As discussed in the previous sections, tandem mass spectrometers are the manifestation of true MS/MS operation whereby a precursor ion can be selected and all the product ions resulting from its collisionally induced dissociation (CID) are analyzed. The fragmentation chemistry and mechanisms are reasonably well understood. Favorable fragmentation processes naturally occur more often, and the ions thus formed always dominate the product ion spectrum. The prominent fragment ions are usually the most stable fragments, although one should not forget that fragmentation is a kinetic phenomenon and rates are determined by the heights of transition states and abundances by both the rates of formation and of decomposition of a product ion.

Many of the fragment ions observed in the product ion spectra are formed by collision-induced heterolytic cleavage. For example, the formation of the product ion $[M + H - HX]^+$ can be explained by a 1,4 hydrogen rearrangement mechanism (Scheme 2). The product ion is formed by the neutral loss of HX, where X can be a heteroatom or a more electronegative group. For instance, the formation of stable product ions, including acylium ion, benzylic ion, and allylic carbonium ion are able to promote heterolytic cleavage. A less common fragmentation mechanism by homolytic cleavage is also observed in tandem MS experiments. In this case, the driving force for the fragmentation of an ion is dependent on the stabilities of the resulting ion and the radical species relative to the energy of the initial ionic species; often the product ions are distonic species.

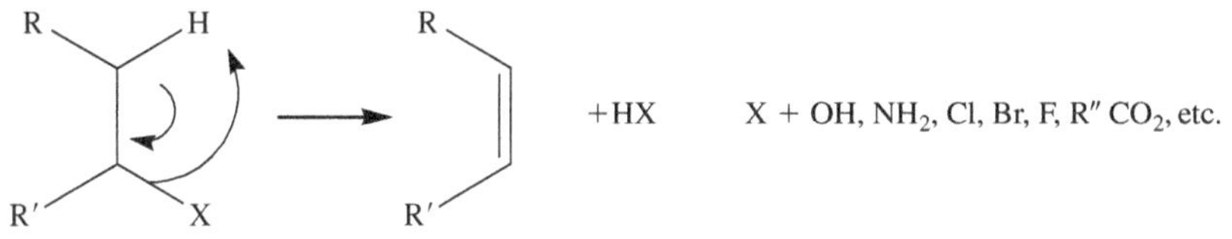

$+HX$ X + OH, NH_2, Cl, Br, F, R″ CO_2, etc.

SCHEME 2

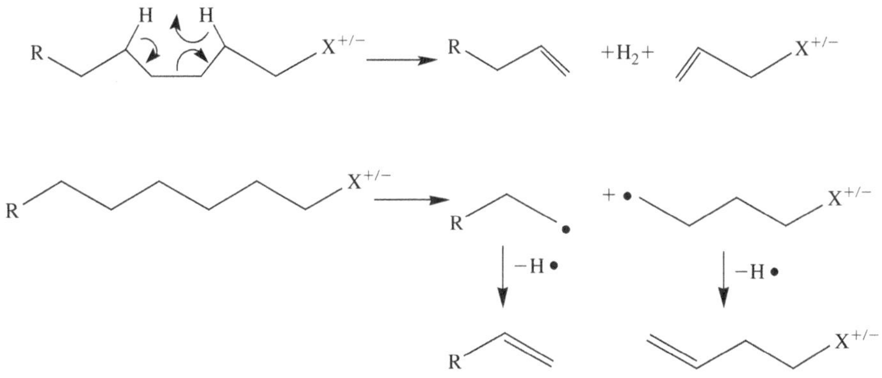

SCHEME 3

Many fragmentation processes are driven by charge. Charge remote fragmentation is defined as a class of gas-phase decompositions that occur physically remote from the charge site (Gross, 1992, 2000; Cheng and Gross, 2000, Adams and Songer, 1993; Adams, 1990). Although the mechanism of charge remote fragmentation is still debated (Scheme 3)(Cheng and Gross, 2000), it has proven useful in the structural determination of long-chain or poly-ring molecules, including fatty acids, phospholipids, glycolipids, triacylglycerols, steroids, peptides, ceramides, and the like. These studies have been thoroughly reviewed (Adams and Songer, 1993; Adams, 1990; Gross, 1992; 2000; Cheng and Gross, 2000).

It is possible to derive structural information from the fragmentation pattern in a number of ways. First, the appearance of prominent peaks at certain mass numbers is empirically correlatable with certain structural elements. For example, the mass spectrum of an aromatic compound is usually dominated by a peak at m/z 91, corresponding to the tropylium ion. In the case of mass spectrum, information can also be obtained from the differences between the masses of two peaks. For example, a fragment ion occurring 20 Da below the molecular ion suggests strongly a loss of a HF, and therefore that a fluorine group is present in the substance examined.

Second, the knowledge of the principles governing the mode of fragmentation of ions makes it possible to confirm the structure assigned to a compound

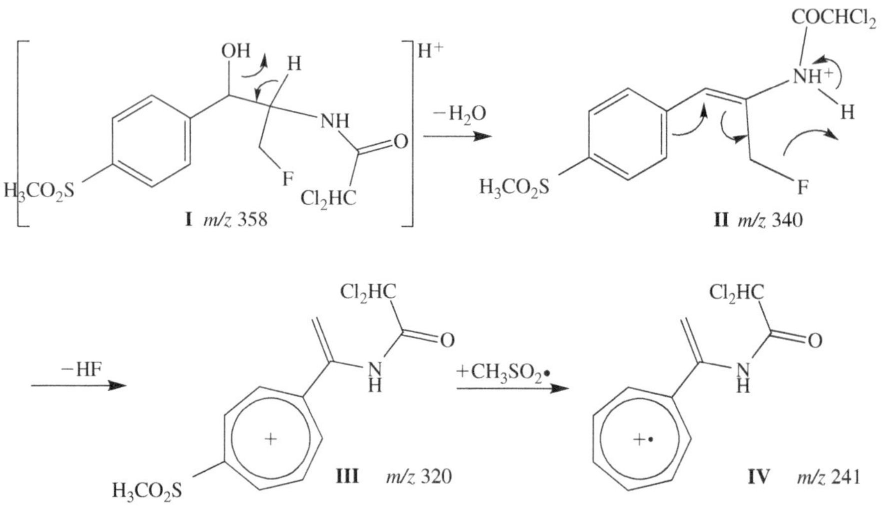

$-H_2O$

I m/z 358

II m/z 340

$-HF$

$+CH_3SO_2\bullet$

III m/z 320

IV m/z 241

SCHEME 4

and, quite often, to determine the juxtaposition of structural fragments and, thus, distinguish between isomeric substances. That is, the analyst can make reasonable guesses as to which fragment peaks to expect in a mass spectrum if the isomeric possibilities are known.

The ESI product ion spectrum of florfenicol (Scheme 4, I) does not show a molecular ion peak at m/z 358 (Fig. 10.4). The significant peak corresponding to the loss of water is formed by heterolytic fragmentation as shown in Scheme 4 (II, m/z 340). The II decomposes to III (m/z 320) by a neutral loss of HF, and this fragmentation is promoted by the formation of a substituted tropylium ion (Scheme 4). The product ion spectrum of florfenicol is characterized by the unusual feature of a most abundant peak occurring at odd mass, namely at m/z 241. The IV, a radical ion, is formed by homolytic cleavage of the sulfur–carbon bond in III, and loss of the methanesulfinic radical (Scheme 4).

10.3 PRACTICAL APPLICATIONS

Metabolism studies play an important role in the drug discovery and development process (Borchardt et al., 1998; Liu and Hop, 2005; Naganeo and Iwasaki, 2004; Baillie, 2004; Hop, 2004; Korfmacher, 2003, 2005). The metabolite characterization of a new chemical entity (NCE) in various drug discovery stages is crucial in assessment of the safety of a drug for human use. The identification of metabolites may reveal the metabolically labile portions of a molecule in a particular drug series. This information can be used by the synthetic chemists to synthesize compounds that are less susceptible to metabolism and, consequently, have a lower elimination rate and a longer

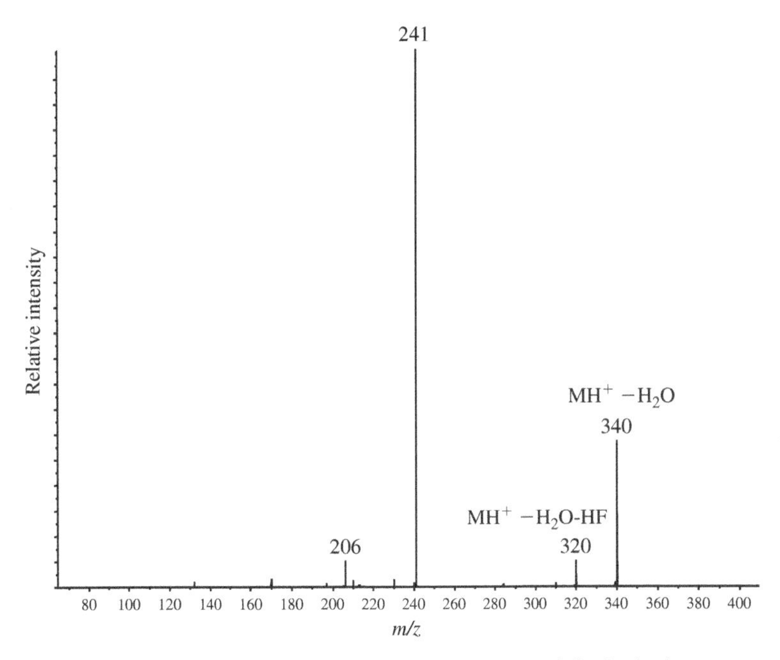

FIGURE 10.4 ESI product ion spectrum of florfenicol.

half-life. In addition, early metabolite identification studies are useful to minimize potential safety concerns that may arise from bioactivation of lead compounds to electrophilic reactive intermediates.

In general, drugs are metabolized to more polar, hydrophilic entities, and so facilitate their elimination from the body. There are two major pathways for metabolism. Biotransformation reactions catalyzed by enzymes (i.e., cytochrome P450), including oxidation, reduction, and hydrolysis, are usually referred to as phase I metabolism. On the other hand, the reactions that involve the addition of bulky and polar groups through conjugation to a nucleophilic site on the drug molecule, are referred to as phase II metabolism (Liu and Hop, 2005; Korfmacher, 2005; Kostiainen et al., 2003). Both phase I and phase II metabolism may occur in parallel for particular compounds (Liu & Hop, 2005; Korfmacher, 2005; Kostiainen et al., 2003). The application of LC–MS/(MS) to elucidate metabolite structures is illustrated using a number of examples here.

10.3.1 Metabolite Profiling by LC–MS

Mass spectrometry in general and LC–MS/MS in particular have played a key role in supporting the lead optimization phase of drug discovery. The high sensitivity, selectivity, and mass accuracy of LC–MS established it as a routine analytical tool for drug metabolism studies (Liu and Hop, 2005; Naganeo and Iwasaki, 2004; Hop, 2004; Korfmacher, 2003, 2005; Baillie, 2004). A number of review articles and books were devoted to metabolite characterization (Borchardt et al., 1998; Liu and Hop, 2005; Korfmacher, 2005; Evans et al., 2004; Oliveira and Watson; 2000; Kobayashi, 1997; Ekins et al., 2001), and the reader is directed to these reviews for more information. In this section, we will focus on the most recent application of LC–MS to provide metabolite profiling.

The metabolites generated from the parent compound through phase I and II metabolism could be viewed as simply incurring an addition and/or subtraction of functional groups of the parent structure. The comparison of the tandem MS spectra between the parent compound and the generated metabolites will easily highlight the structural changes while confirming the parts of the structure that are unaffected. On the other hand, to completely characterize the unknown structure of metabolites usually requires combination of HPLC, MS, and nuclear magnetic resonance (NMR) spectroscopy analysis. For example, Hop et al. (2002) used a combination of LC–MS, LC–MS/MS, and NMR techniques to identify metabolites of a substance P (Neurokinin 1 receptor) antagonist, compound A (Scheme 5), in rat hepatocytes and rat plasma. In both in vitro and in vivo studies, the samples were prepared for analysis by protein precipitation using acetonitrile followed by centrifugation. The supernatant was profiled with LC–MS and LC–MS/MS. The analysis involved the use of the product ion spectra of compound A and ^{14}C-labeled compound A as the structural templates for the identification of metabolite structures. A comparison of the product ion spectrum of compound A with that of the ^{14}C-labeled compound A suggested that the fragments at *m/z* 231, 215, 203, 191, and 175 were associated with the trifluoromethoxy phenyl moiety, whereas the fragments at *m/z* 184, 172, 159, 131, 91, and 56 were associated with the phenyl piperidine moiety (Table 10.6). The assignment of the fragment ions were also confirmed by high-resolution LC–MS/MS analysis. These product ions served as diagnostic markers for structural modification. Based on their product ion spectra, nine major metabolites, the *O*-dealkylated metabolite (Scheme 5, B), the hydroxylamine metabolite (Scheme 5, C), the nitrones (D and E), the lactam metabolite (Scheme 5, F), the hydroxylated *O*-dealkylated metabolite (G), the glucuronide of metabolite G (Scheme 5, H), the oxime metabolite (Scheme 5, I), and the keto acid metabolite (Scheme 5, J), were characterized by high-resolution LC–MS and LC–MS/MS (Table 10.7). The authors found that the major circulating metabolite observed in vivo was generated by oxidative deamination of the piperidine ring yielding a keto acid metabolite, J (Scheme 6) (Hop et al., 2002). Other metabolites, which might be the intermediates for formation of the keto acid, were also observed in the radiochromatogram of rat plasma (spectrum not shown) (Hop et al., 2002).

One of the goals of metabolite characterization is to identify the metabolic pathways and to determine whether or not any potentially reactive or toxic metabolites are formed (Liu and Hop, 2005; Korfmacher, 2005; Evans et al., 2004; Evans and Baillie, 2005). It is generally accepted that toxicities can stem from drug bioactivation in vivo, thus, identifying the potential toxic

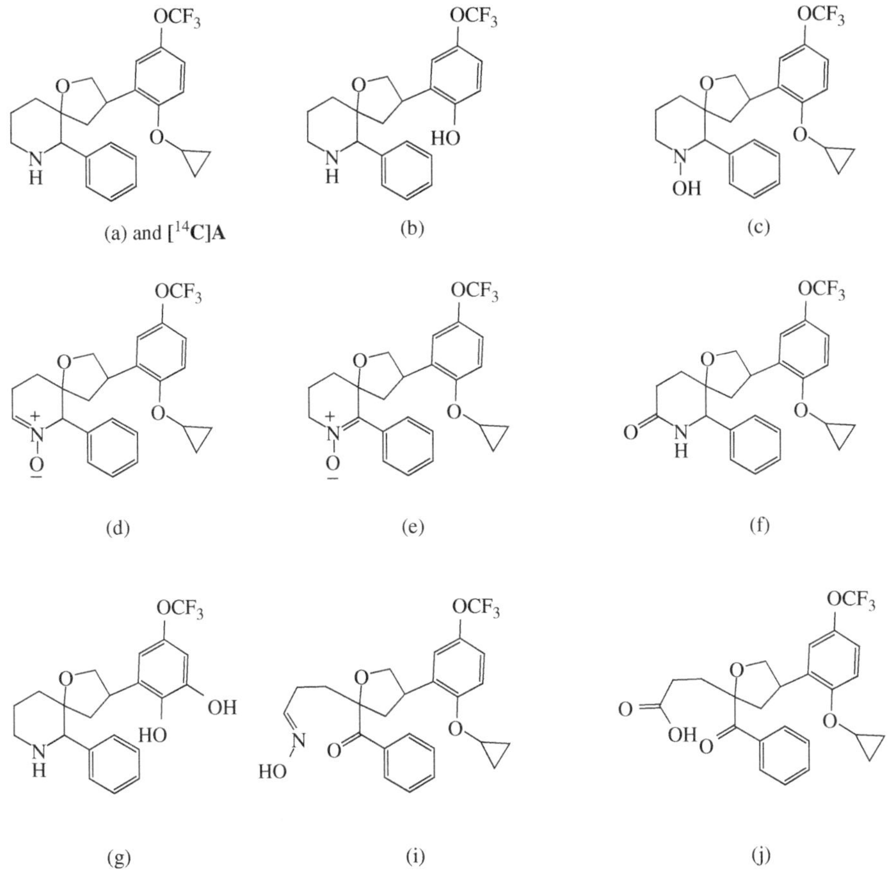

OCF_3
O
O
N
H

(a) and [^{14}C]A

OCF_3
O
HO
N
H

(b)

OCF_3
O
O
N
OH

(c)

OCF_3
O
O
+
N
O

(d)

OCF_3
O
O
+
N
O

(e)

OCF_3
O
O
O
N
H

(f)

OCF_3
O
OH
HO
N
H

(g)

OCF_3
O
N
HO
O
O

(i)

OCF_3
O
O
OH
O
O

(j)

SCHEME 5 Structure of compound A and its metabolites.

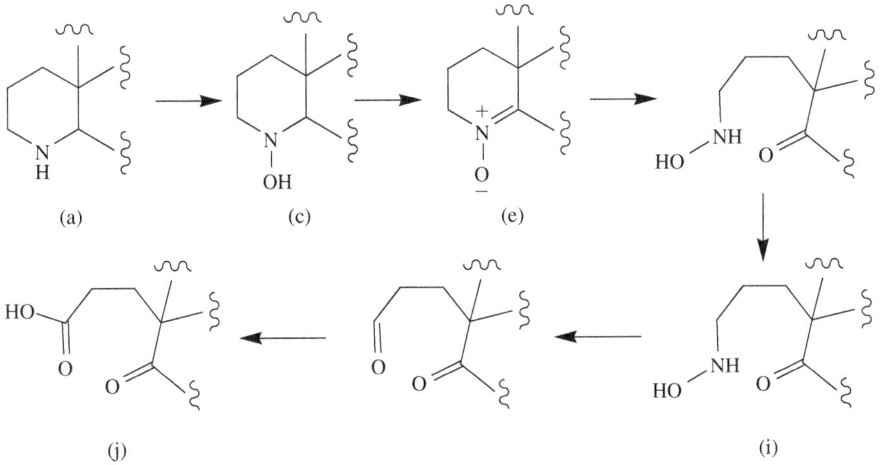

N
H

(a)

N
OH

(c)

+
N
O

(e)

HO
NH
O

HO
NH
O

(i)

O
O

HO
O
O

(j)

SCHEME 6 Mechanism for the formation of metabolite J via oxidative deamination.

metabolites is crucial in the lead optimization process (Liu and Hop, 2005; Korfmacher, 2005; Evans et al., 2004; Evans and Baillie, 2005). The generation of acyl glucuronide and glutathione (GSH) metabolites are important biotransformation pathways for many drugs and xenobiotics (Evans et al., 2004; Evans and Baillie, 2005; Ghosal et al., 2004). Recently, the formation of acyl glucuronide conjugate had forced the withdraw of four drugs, bromofenac, benoxaprofen, ibufenac, and Zomepirac, from market due to hepatotoxicity (Lasser et al., 2002).

The hydroxylation or oxidation of heterocyclic atoms (i.e., nitrogen) is a common phase I oxidative reaction. The metabolites generated by the oxidation at the N atom are known as N-oxides. Several N-oxides are reported to be carcinogenic and/or to exhibit toxicological effects (Sugimura et al., 1966; Bosin and Maickel, 1973; Kiese et al. 1966). Chowdhury and co-workers demonstrated the application of LC–APCI–MS(/MS) to distinguishing N-oxide metabolites from hydroxylated metabolites (Ramanathan et al., 2000; Tong et al., 2001).

In LC–APCI–MS, the molecular ions of N-oxides are found to undergo thermal deoxygenation at elevated temperatures (Ramanathan et al., 2000; Tong et al., 2001). The resulted product ion ($[MH^+ - 16]$) had been attributed to the loss of elemental oxygen from the protonated N-oxide. On the other hand, the $[MH^+ - 16]$ ions were not producted in the LC–APCI–MS spectra of hydroxylated metabolites (Ramanathan et al., 2000; Tong et al., 2001). Since this thermal deoxygenation is unique to N-oxide metabolites, it can also be used to differentiate *N*-oxides from other hydroxylated metabolites.

More sensitive capillary LC–MS/MS technology had been applied to investigate whether metabolite-mediated genotoxicity of estrogens may play an important role in the initiation of mammary tumors (Chakravarti et al., 2001). Studies have shown that exposure to long-term and high-level estrogen or estrogen replacement therapy increases the risk of women developing breast cancer (Colditz, 1998; Service et al., 1998; Feigelson and Henderson, 1996; Colditz et al., 1995; Henderson et al., 1991). One hypothesis for lesion formation in breast cancer is that metabolic activation of estrogen leads to deoxyribonucleic acid (DNA) damage via abasic site formation (Nutter et al., 1991; Cavalieri and Rogan, 1996; Cavalieri et al., 1997, 2000). One carcinogenic pathway of interest to us is the metabolism of estradiol (E_2) and estrone (E_1) to 4-hydroxyestradiol (4-OHE_2) and 4-hydroxyestrone (4-OHE_1)(Scheme 7) (Cavalieri et al., 1997). These catechols may be further oxidized enzymatically to *o*-quinones. As good electrophiles, the *o*-quinones can react with purine bases of DNA via a Michael addition reaction to form depurinating adducts.

A highly sensitive and specific capillary LC–ESI–MS/MS method was developed by Gross's group (Chakravarti et al., 2001; Zhang, 2001) for the direct analysis of biological extracts generated by in vitro or in vivo experiments. The depurinating adducts, 4-OHE2-N7Gua and 4-OHE2-N3-Ade, formed in the in vitro experiments were determined by this method. The presence of the adducts was confirmed by comparing both the HPLC retention

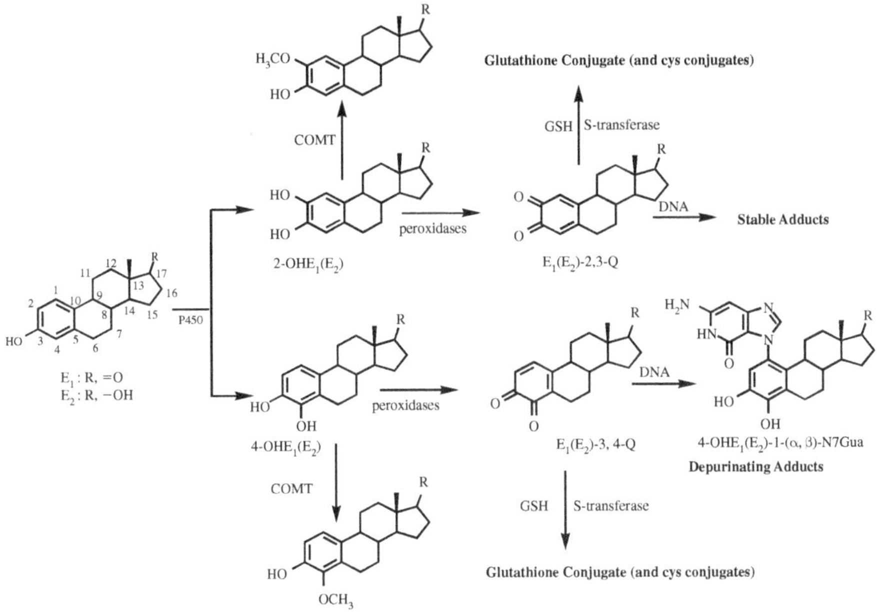

SCHEME 7 Proposed mechanisms for DNA damage by estrogen quinones.

time and the full-scan, product ion mass spectra of the adducts to those of synthetic standards (Fig. 10.5). The low detection limit (~100 fmol) of the LC–MS/MS method also allows the identification and quantification of catechol–estrogen adducts formed in vivo from mouse skin tissue and other biosources (Fig. 10.5). The results are consistent with the hypothesis that estrogen metabolites react with both model oligodeoxynucleotides (ODNs) and DNA to form depurinating adducts, which may result in mutations.

Exact mass measurements and elemental composition assignment are essential for the characterization of metabolites (Zhang et al., 2005). It is well accepted that the accurate mass measurement of the product ions, formed in an MS/MS experiment facilitates the structure elucidation of new or unknown materials (Zhang et al., 2005). Hybrid quadrupole/time-of-flight MS had been employed for accurate mass measurement for well over a decade (Mihaleva et al., 2008; Bristow et al., 2008; Tolmachev et al., 2006). The mass measurement precision/accuracy can be ±5 ppm with internal calibration methods. One limitation of the instruments of this type was the narrow ion abundance range over which accurate mass measurements could be made with a high degree of certainty (Mihaleva et al., 2008; Bristow et al., 2006; Koefeler and Gross, 2005). On the other hand, by using external calibration, FT-ICR MS is capable of achieving mass measurement accuracies of 1 ppm or better (Zhang et al., 2005). Smith and co-workers recently reported that by using a new trapped-ion cell with improved dc potential harmonicity, they achieved under 0.05 ppm

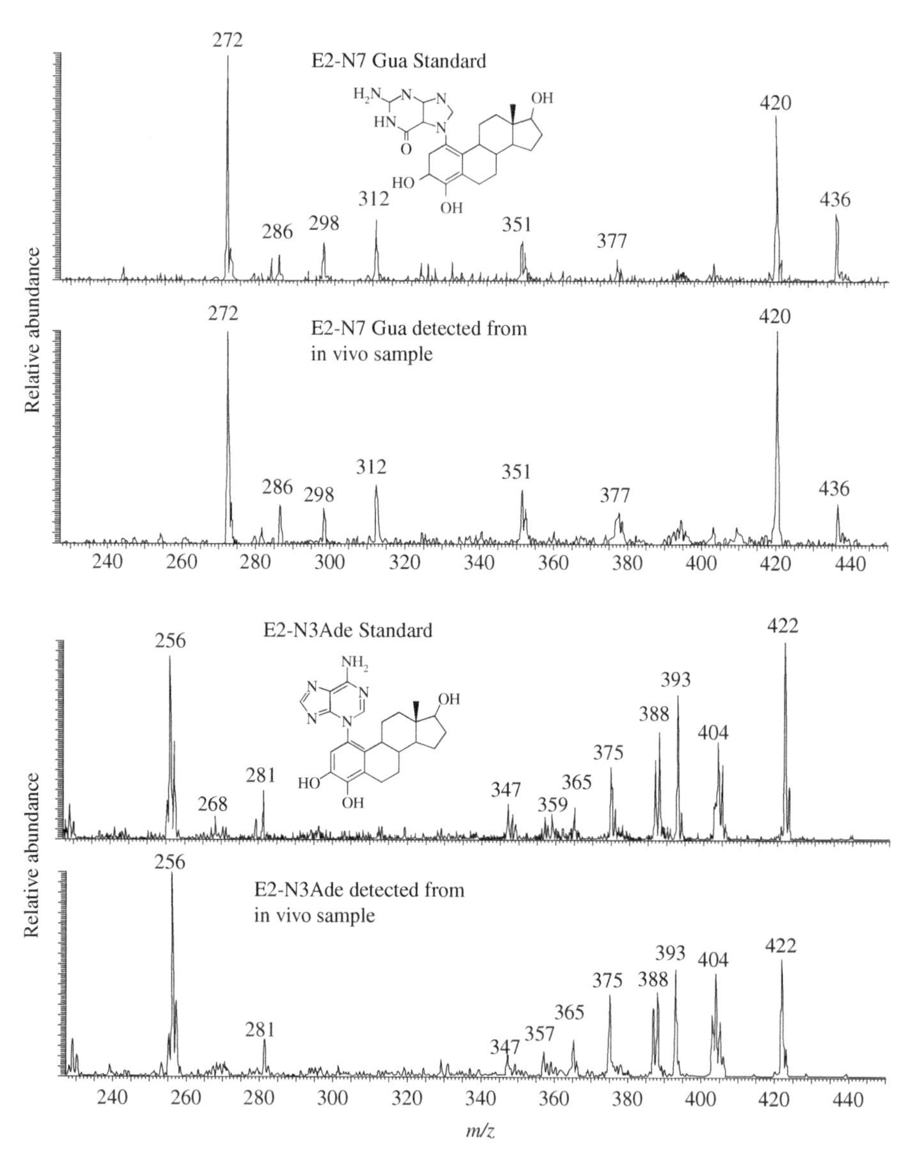

FIGURE 10.5 Detection of estrogen quinone-induced DNA adduct in mice skin tissue samples by capillary HPLC–MS/MS.

root-mean-square precision (Tolmachev et al., 2008). This is an impressive application that is not possible with any other known mass spectrometric equipment. Although FT MS has rapidly developed as one of the most effective techniques for the determination of unknowns, its application to drug metabolite screening is much less routine than for the drug impurities identification (Williams and Muddiman, 2007; Williams et al., 2008; Tolmachev et al., 2006;

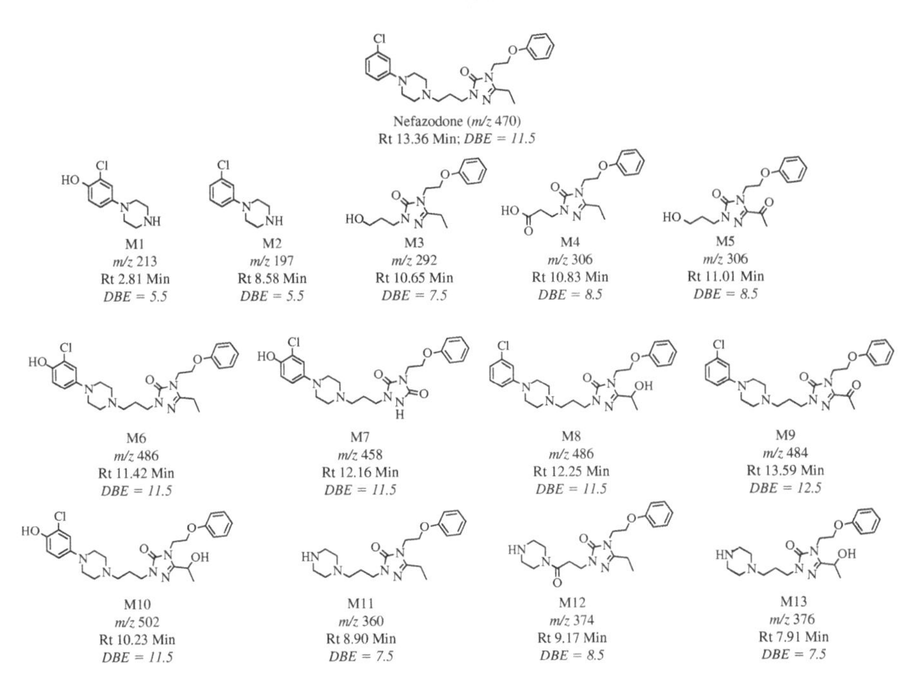

SCHEME 8 Metabolic pathways of the antidepressant nefazodone in humans.

Maziarz et al., 2005; Li et al., 2008; Herniman et al., 2005; Feng and Siegel, 2007; Chen et al., 2007). Bai et al. (2004) evaluated the application of MALDI–FT-ICR MS for the identification of the degradation products of surfactant proteins. They reported that owing to the high resolving power, FT-ICR MS was found to provide substantial advantages for the structural identification of surfactant proteins and their degradants from complex biological matrixes with high mass accuracy (Bai et al., 2004).

The relatively recent introduction of high-resolution mass spectrometers and associated high-speed data processing systems has enhanced the value of MS-based technologies in metabolite detection and identification through the provision of accurate mass data and corresponding elemental composition information. The application of the new high-resolution FT MS and linear ion trap/Orbitrap mass spectrometer allows unambiguous structural characterization of metabolites (Peterman et al., 2006; Ruan et al., 2008; Sanders et al., 2006; Yoo et al., 2007). Nefazodone is a psychoactive drug and antidepressant. Peterman et al. (2006) demonstrated the characterization of human liver microsomal metabolites of nefazodone, using a LTQ-Orbitrap instrument. A total of 13 metabolites were identified in the studies (Scheme 8) (Fig. 10.6). They found the full-scan accurate mass analysis was useful for generating the chemical formula and rings plus double bonds of the metabolites. The mass measurement errors for all full-scan MS and MS/MS spectra was less than 3 ppm, regardless of the relative ion abundance. The molecular weight of the

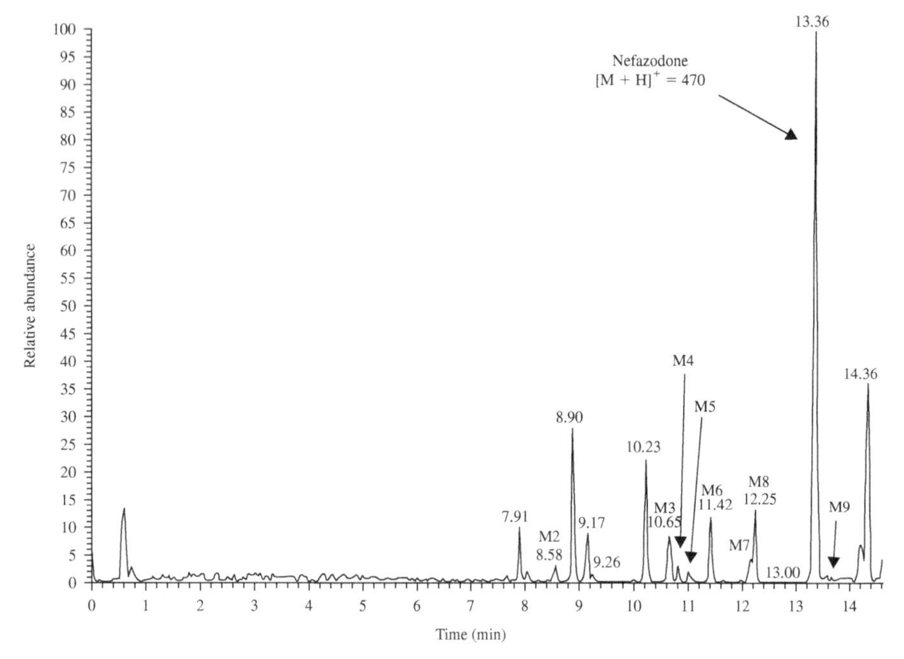

FIGURE 10.6 Total ion chromatogram of an NADPH-supplemented human liver microsomal incubation of nefazodone (10 μ*M*).

nefazodone, which eluted at 13.36 min, was confirmed to be 469 Da. An initial step in elucidating the metabolite structures of nefazodone is to understand the fragmentation pattern of the drug substance. Extensive tandem mass spectrometry analysis of the fragmentation pattern of the nefazodone provides a basis for assessing structural assignment for the metabolites. The product ion spectrum is dominated by the fragment ion of *m*/*z* 274 (measured ion intensity of $2.1 \times e^{07}$ counts and mass error of −1.4 ppm). This most abundant product ion observed in the spectrum corresponded to the phenoxyethyl-5-ethyltriazolone propyl motif (Scheme 9). In addition, the collision-induced decomposition of the ESI-produced MH^+ precursor also generates ions at *m*/*z* 317.19855, 246.12343, 180.1128, and 140.08145 (Fig. 10.7) (Scheme 9). The ability of LTQ-Orbitrap MS to provide an exact mass measurement for each of the ions produced in the MS/MS experiment assists greatly in the investigation of the fragmentation mechanism of the nefazodone (Fig. 10.7). According to the nominal mass of the *m*/*z* 180 product ion, two possible structures can be proposed as shown in the Scheme 9. However, the measured mass of 180.1128 (calculated: 180.1131 Da, error −2 ppm) unambiguously proves that the charge-driven cleavage on the right-hand side of the nefazodone structure is the most likely route to the *m*/*z* 180 product ion given that the resulting mass error is only 2 ppm compared to 308 ppm for the other possible product ion. Peterman et al. (2006) also found that despite large differences in measured ion

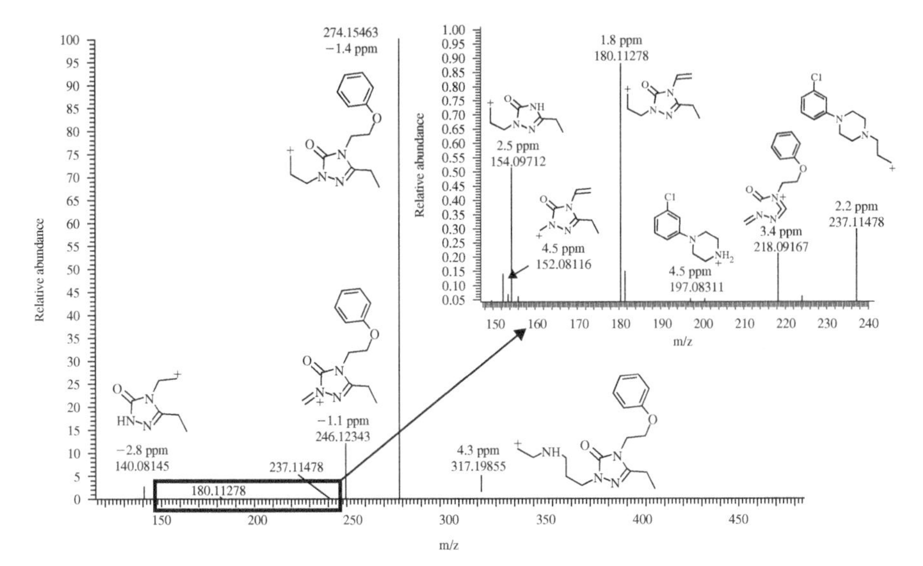

FIGURE 10.7 Full-scan accurate mass spectrum of nefazodone (*m*/*z* 470) acquired in the Orbitrap mass spectrometer using external mass calibration. The expanded spectrum showed the low-intensity product ions. Product ion assignment resulted from predictive fragmentation using Mass Frontier 4.0.

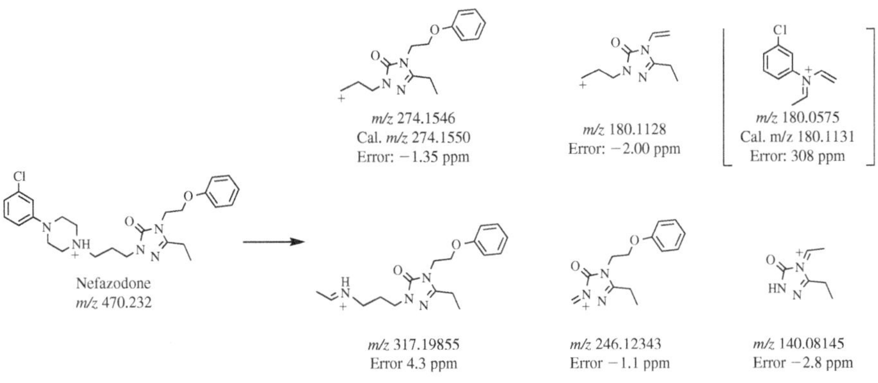

Nefazodone
m/z 470.232

m/z 274.1546
Cal. *m/z* 274.1550
Error: −1.35 ppm

m/z 180.1128
Error: −2.00 ppm

m/z 180.0575
Cal. m/z 180.1131
Error: 308 ppm

m/z 317.19855
Error 4.3 ppm

m/z 246.12343
Error −1.1 ppm

m/z 140.08145
Error −2.8 ppm

SCHEME 9 Proposed fragmentation mechanism for the nefazodone.

intensity between product ions, the measured mass accuracy is quite comparable. For example, there is a 100-fold difference between the measured ion intensities of the product ions at *m*/*z* 140 ($5.7 \times e^{05}$) and at *m*/*z* 274 ($2.1 \times e^{07}$); however, the errors of the accurate mass accuracy are found to be −2.8 ppm and −1.4 ppm for the two product ions, respectively (Fig. 10.7).

The extensive tandem MS studies allow the structural characterization of isomeric metabolites of nefazodone. For example, two monohydroxy metabolites of nefazodone (*m*/*z* 486) elute at 11.42 (M6) and 12.25 min (M8), respectively. The mass fragmentation pattern for M6 (Fig. 10.8*a*) is quite

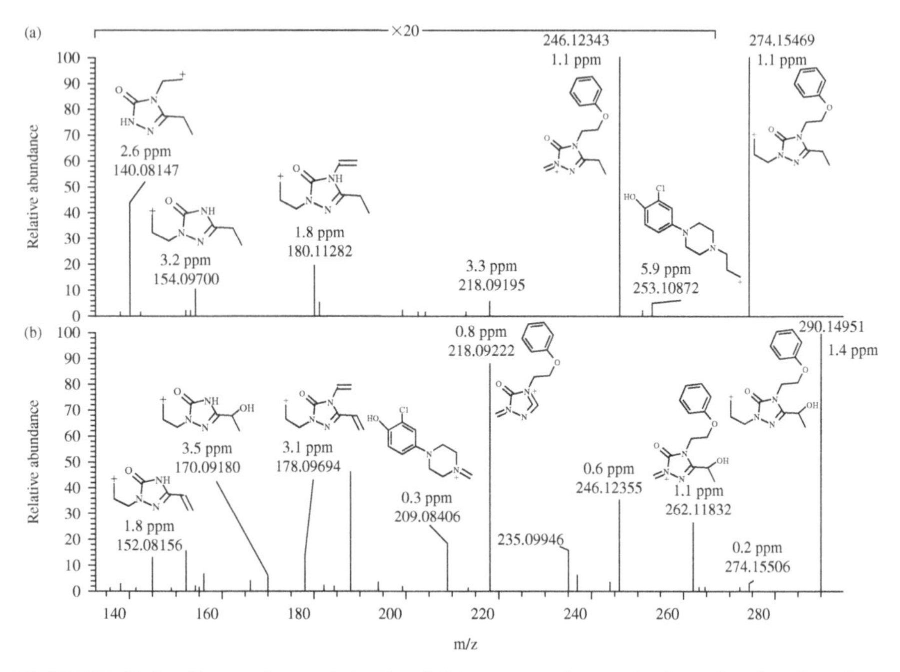

FIGURE 10.8 Comparison of the MS/MS spectra of monohydroxylated nefazodone metabolites M_6 ($R_t = 11.42$ min) and M_8 ($R_t = 12.25$ min).

different from that for M8 (Fig. 10.8*b*). Collisional activation of the precursor ion of M6 results in a major fragment ion of *m/z* 274.15469, which was also observed in the product ion spectrum of the nefazodone. This indicates this part of nefazodone is intact for M6. The product ion spectrum of the M6 (Fig. 10.8*a*) also reveals a fragment ion at *m/z* 253 representing an addition of a mass of 16 to the fragment ion at *m/z* 237 in the nefazodone parent ion spectra (Fig. 10.7), implicating the 3-chlorophenylpiperazine ring as the site of hydroxylation. On the other hand, the product ion spectrum (Fig. 10.8*b*) of M8 depicts diagnostic fragment ions at *m/z* 170.09180, 262.11832, and 290.14951, which represent the additions of a mass of 16 to the fragment ions at *m/z* 154, 246, and 274, respectively, in the product ion spectrum of nefazodone. These product ions indicate that the ethyl group in the 5-ethyl-triazolone motif is the site of monohydroxylation. Thus, the two monohydroxy metabolites, M6 and M8, can be easily distinguished by the LC–MS/MS analysis.

10.3.2 Special Strategies for Characterization of Drug Metabolites by LC–MS

Special strategies, such as online hydrogen/deuterium (H/D) exchange LC–MS/MS and chemical derivatization, were well-established techniques for characterization of drug metabolites (Liu and Hop, 2005; Liu et al., 2001;

Lam and Ramanathan, 2002; Ohashi et al., 1998). Recently, Liu and Hop (2005) thoroughly reviewed the chemical derivatization strategies for the characterization of drug metabolites by LC–MS, and the reader is directed to the review for more information. In this part, we will focus on the online H/D exchange techniques, particularly those that utilize LC–MS as the mode of metabolite analysis.

The H/D exchange for structural elucidation had been used for many years by mass spectrometrists (Biemann, 1962). Recently, this technology had also been applied for determination of the affinity constants for protein–ligand interactions and for quantifying the conformational changes associated with ligand binding to proteins (Zhu et al., 2003). In principle, the labile protons in the side chains and amide hydrogen, which are not protected from solution, generally exchange rapidly with deuterium. The determination of the number of exchangeable hydrogen atoms in a structure can provide additional information for structural characterization.

The combination of H/D exchange and tandem MS experiments had been used to discriminate between N- or S-oxide formation and hydroxylation in drug metabolism studies (Liu and Hop, 2005; Liu et al., 2001; Lam and Ramanathan, 2002; Ohashi et al., 1998). Ohashi et al. (1998) demonstrated the online H/D exchange for characterization of metabolites of promethazine (MW = 284 Da) (Scheme 14). M1 and M2, metabolites of promethazine, gave rise to pseudomolecular ions at *m/z* 301 (Fig. 10.9*a* and Fig. 10.10*a*),

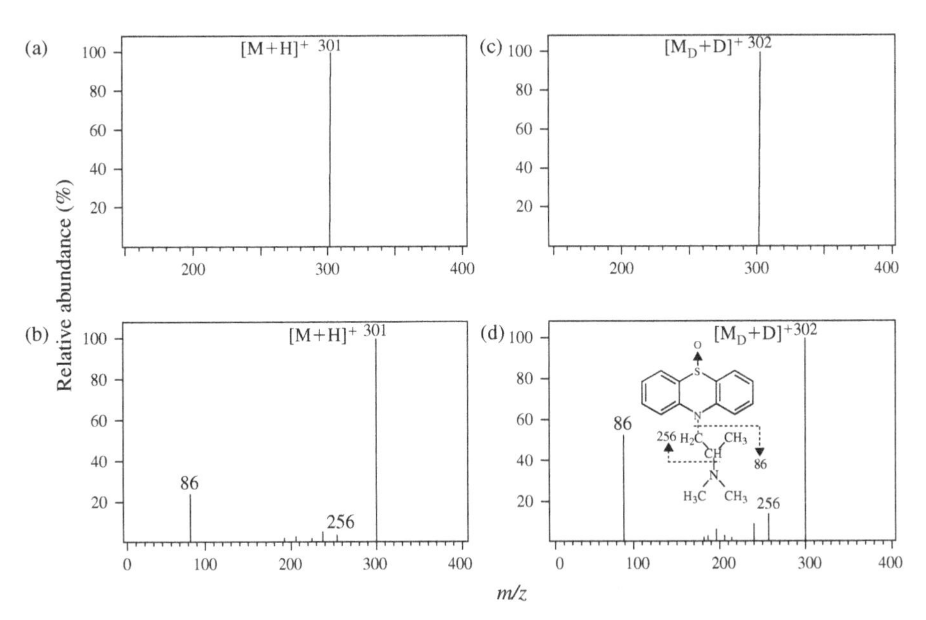

FIGURE 10.9 (*a*) ESI mass spectrum of M_1, a metabolite of promethazine; (*b*) product ion spectrum of M_1; (*c*) ESI mass spectrum of M_1 in D_2O; (*d*) product ion spectrum of M_1 in D_2O.

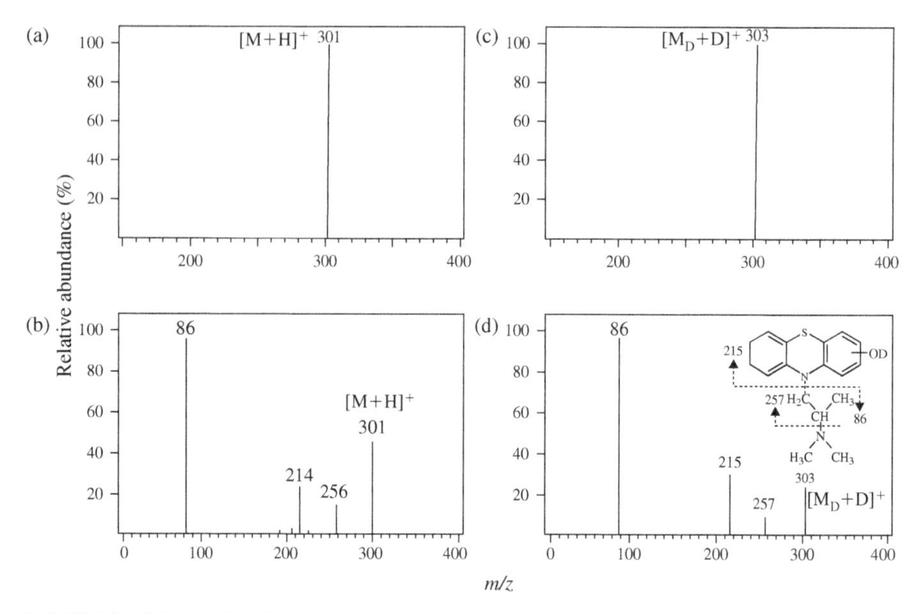

FIGURE 10.10 (*a*) ESI mass spectrum of M_2, a metabolite of promethazine; (*b*) product ion spectrum of M_2; (*c*) ESI mass spectrum of M_2 in D_2O; (*d*) Product ion spectrum of M_2 in D_2O.

SCHEME 10 Promethazine.

which are 16 Da higher than the parent compounds (Scheme 10). This suggested M1 and M2 might be oxidation metabolites of promethazine, that is, addition of oxygen or a hydroxyl group to the phenothiazine. However, the discrimination between the oxidation types could not be achieved by analysis of the product ion spectra (Fig. 10.9*b* and Fig. 10.10*b*).

Therefore, the H/D exchange method was evaluated by the authors to differentiate between these possibilities. The online H/D exchange LC–MS experiment using D_2O in the mobile phase afforded $[M_D + D^+]$ ions at m/z 302 and 303 for M1 (Fig. 10.9*c*) and M2 (Fig. 10.10*c*), respectively. This revealed that M1 had no exchangeable hydrogen atom in its structure, which ruled out the possibility of the hydroxyl structure. The authors assumed that M1 to be an S-oxidated metabolite of promethazine. On the other hand, M2 had one exchangeable hydrogen atom; thus it was assumed to be a hydroxylated metabolite. The fragmentation patterns of the product ion spectra (Fig. 10.9*d* and Fig. 10.10*d*) were found to be consistent with the proposed structures.

Similar approaches had been developed by Liu et al. (2001) to facilitate metabolite identification of an investigational compound (R_1–CH(NH_2)–CO–N(R_2)–CH_2–S–R_3). A disadvantage of the online H/D exchange technique is that it requires relatively expensive deuterated solvents. To solve this problem, Lam and Ramanathan (2002) developed a modified method employing H/D exchange in the electrospray ionization source to characterize metabolites. However, the authors found that the exchange efficiencies of ion-source H/D exchange were not as high as those of solution phase H/D exchange (Lam and Ramanathan, 2002).

10.4 CONCLUSION

Liquid chromatography–mass spectroscopy has become one of the most powerful analytical techniques in the drug discovery and development process. LC tandem MS is obviously the technique of choice for the identification of metabolites as illustrated in this chapter. Exact mass measurements and elemental composition assignment are essential for the characterization of the metabolites. Accurate mass measurements of the product ions, formed in an MS/MS experiment, greatly facilitate the structure elucidation of the metabolites. With the ever evolving technological advances in mass spectrometry and separation science, LC–MS will continue to play an important role in metabolite identification in the future.

ACKNOWLEDGMENT

The authors thank Professor Michael L. Gross (Washington University in St. Louis) for proofreading the manuscript. The authors also acknowledge Dr Malcolm MacCoss for his support on the project.

REFERENCES

Adams J. Charge-remote fragmentations: Analytical applications and fundamental studies. Mass Spectrom Rev 1990;9:141–186.

Adams J, Songer MJ. Charge-remote fragmentations for structural determination of lipids. TrAC, Trends Anal Chem 1993;12:28–36.

Bai Y, Galetskiy D, Damoc E, Paschen C, Liu Z, Griese M, Liu S, Przybylski M. High resolution mass spectrometric alveolar proteomics: Identification of surfactant protein SP-A and SP-D modifications in proteinosis and cystic fibrosis patients. Proteomics 2004;4:2300–2309.

Baillie TA. Drug discovery and development in the post-genome era. Can we rationally design safer drugs? Adv in Mass Spectrom 2004;16:1–18.

Bartner P, Pramanik BN, Saksena AK, Liu Y-H, Das PR, Sarre O, Ganguly AK. Structure elucidation of everninomicin-6, a new oligosaccharide antibiotic, by chemical degradation and FAB-MS methods. J Am Soc Mass Spectrom 1997;8:1134–1140.

Biemann K. Mass Spectrometry: Organic Chemical applications. McGraw-Hill, New York, 1962.

Borchardt RT, Freidinger RM, Sawyer TK, Smith PL, Eds. Integration of pharmaceutical discovery and development: Case histories. Pharm Biotechnol 1998;11.

Bose AK, Fujiwara H, Pramanik BN, Lazaro E, Spillert CR. Mass spectral studies. VIII. Some aspects of chemical ionization mass spectroscopy using ammonia as reagent gas: A valuable technique for biomedical and natural products studies. Anal Biochem 1978;89:284–291.

Bose AK, Pramanik BN, Bartner PL. Mass spectral studies. Part 120. Facile ionization induced by ammonium salts: Mass spectra of nonvolatile compounds using unmodified electron-impact mass spectrometers. J Org Chem 1982;47:4008–4010.

Bosin TR, Maickel RP. Mass spectra of carcinogenic 4-hydroxylaminoquinoline-*N*-oxides. Res Commun Chem Pathol and Pharmacol 1973;6:813–820.

Bristow T, Constantine J, Harrison M, Cavoit F. Performance optimisation of a new-generation orthogonal-acceleration quadrupole-time-of-flight mass spectrometer. Rapid Commun Mass Spectrom 2008;22:1213–1222.

Brodbelt JS. Analytical applications of ion-molecule reactions. Mass Spectrom Rev 1997;16:91–110.

Brodbelt JS. Probing molecular recognition by mass spectrometry. Int J Mass Spectrom 2000;200:57–69.

Cavalieri E, Rogan E. The primary role of apurinic sites in tumor initiation. Polycyclic Aromat Compd 1996;10:251–258.

Cavalieri E, Frenkel K, Liehr JG, Rogan E, Roy D. Chapter 4: Estrogens as endogenous genotoxic agents—DNA adducts and mutations. J Natl Cancer Inst Monogr 2000;27:75–93.

Cavalieri EL, Stack DE, Devanesan PD, Todorovic R, Dwivedy I, Higginbotham S, Johansson SL, Patil KD, Gross ML, Gooden JK, Ramanathan R, Cerney RL, Rogan EG. Molecular origin of cancer: Catechol estrogen-3,4-quinones as endogenous tumor initiators. Proc Natl Acad Sci 1997;94:10937–10942.

Cerny RL, MacMillan DK, Gross ML, Mallams AK, Pramanik BN. Fast-atom bombardment and tandem mass spectrometry of macrolide antibiotics. J Am Soc Mass Spectrom 1994;5:151–158.

Chakravarti D, Mailander PC, Li K-M, Higginbotham S, Zhang HL, Gross ML, Meza JL, Cavalieri EL, Rogan EG. Evidence that a burst of DNA depurination in SENCAR mouse skin induces error-prone repair and forms mutations in the H-ras gene. Oncogene 2001;20:7945–7953.

Chen G, Zhang L-K, Pramanik BN. LC/MS: Theory, instrumentation, and applications to small molecules. In HPLC for Pharmaceutical Scientists. Yuri Kazakevich, Rosario Lobrutto, Editors, Wiley Hoboken, NJ, 2007, pp. 281–346.

Cheng C, Gross ML. Applications and mechanisms of charge-remote fragmentation. Mass Spectrom Rev 2000;19:398–420.

Clauwaert K, Vande Casteele S, Sinnaeve B, Deforce D, Lambert W, Van Peteghem C, Van Bocxlaer J. Exact mass measurement of product ions for the structural confirmation and identification of unknown compounds using a quadrupole time-of-flight spectrometer: A simplified approach using combined tandem mass spectrometric functions. Rapid Commun Mass Spectrom 2003;17:1443–1448.

Colditz GA. Relationship between estrogen levels, use of hormone replacement therapy, and breast cancer. J Nat Cancer Inst 1998;390:814–823.

Colditz GA, Hankinson SE, Hunter DJ, Willett WC, Manson JE, Stampfer MJ, Hennekens C, Rosner B, Speizer FE. The use of estrogens and progestins and the risk of breast cancer in postmenopausal women. N Engl J Med 1995;15332:1589–1593.

Daniel JM, Friess SD, Rajagopalan S, Wendt S, Zenobi R. Quantitative determination of noncovalent binding interactions using soft ionization mass spectrometry. Int J Mass Spectrom 2002;216:1–27

Ekins S, Ring BJ, Grace J, McRobie-Belle DJ, Wrighton SA. Present and future in vitro approaches for drug metabolism. J Pharmacol Toxicol Methods 2001;44:313–324.

Evans DC, Baillie TA. Minimizing the potential for metabolic activation as an integral part of drug design. Curr Opin Drug Discov Devel 2005;8:44–50.

Evans DC, Watt AP, Nicoll-Griffith DA, Baillie TA. Drug-protein adducts: An industry perspective on minimizing the potential for drug bioactivation in drug discovery and development. Chem Res Toxicol 2004;17:3–16.

Feigelson HS, Henderson BE. Estrogens and breast cancer. Carcinogenesis 1996; 17:2279–2284.

Feng X, Siegel MM. FTICR-MS applications for the structure determination of natural products. Anal Bioanal Chem 2007;389:1341–1363.

Ghosal A, Hapangama N, Yuan Y, Achanfuo-Yeboah J, Iannucci R, Chowdhury S, Alton K, Patrick JE, Zbaida S. Identification of human UDP-glucuronosyltransferase enzyme(S) responsible for the glucuronidation of ezetimibe (ZETIA). Drug Metab Dispos 2004;32:314–320.

Gross ML. Charge-remote fragmentations: Method, mechanism and applications. Int J Mass Spectrom Ion Process 1992;118–119:137–165.

Gross ML. Charge-remote fragmentation: An account of research on mechanisms and applications. Int J Mass Spectrom 2000;200:611–624.

Henderson BE, Ross RK, Pike MC. Toward the primary prevention of cancer. Science 1991;22254:1131–1138.

Herniman JM, Langley GJ, Bristow TWT, O'Connor G. The validation of exact mass measurements for small molecules using FT-ICRMS for improved confidence in the selection of elemental formulas. J Am Soc Mass Spectrom 2005;16: 1100–1108.

Hop CECA. Applications of quadrupole-time-of-flight mass spectrometry to facilitate metabolite identification. Am Pharm Rev 2004;7:76,78–79,48.

Hop CECA, Wang Y, Kumar S, Elipe MVS, Raab CE, Dean DC, Poon GK, Keohane C-A, Strauss J, Chiu S-HL, Curtis N, Elliott J, Gerhard U, Locker K, Morrison D, Mortishire-Smith R, Thomas S, Watt AP, Evans DC. Identification of metabolites of a substance P (neurokinin 1 receptor) antagonist in rat hepatocytes and rat plasma. Drug Metab Dispos 2002;30:937–943.

Ikonomou MG, Blades AT, Kebarle P. Investigations of the electrospray interface for liquid chromatography/mass spectrometry. Anal Chem 1990;62:957–967.

Ikonomou MG, Blades AT, Kebarle P. Electrospray-ion spray: A comparison of mechanisms and performance. Anal Chem 1991;63:1989–1998.

Kebarle P. Ion-molecule equilibria, how and why. J Am Soc Mass Spectrom 1992;3:1–9.

Kiese M. The biochemical production of ferrihemoglobin-forming derivatives from aromatic amines, and mechanisms of ferrihemoglobin formation. Pharmacol Rev 1966;18:1091–1161.

Kobayashi N. Quantitative analysis of pharmaceuticals in clin. Samples by LC/MS/MS method. Gendai Kagaku Zokan 1997;31:284–289.

Koefeler HC, Gross ML. Correction of accurate mass measurement for target compound verification by quadrupole time-of-flight mass spectrometry. J Am Soc Mass Spectrom 2005;16:406–408.

Korfmacher WA. Lead optimization strategies as part of a drug metabolism environment. Curr Opin Drug Discov Devel 2003;6:481–485.

Korfmacher WA, Editor. Using Mass Spectrometry for Drug Metabolism Studies. Taylor & Francis CRC Press, Boca Raton, FL, 2005.

Kostiainen R, Kotiaho T, Kuuranne T, Auriola S. Liquid chromatography/atmospheric pressure ionization-mass spectrometry in drug metabolism studies. J Mass Spectrom 2003;38:357–372.

Lam W, Ramanathan R. In electrospray ionization source hydrogen/deuterium exchange LC-MS and LC-MS/MS for characterization of metabolites. J Am Soc Mass Spectrom 2002;13:345–353.

Lasser KE, Allen PD, Woolhandler SJ, Himmelstein DU, Wolfe SM, Bor David H. Timing of new black box warnings and withdrawals for prescription medications. J Am Med Assoc 2002;1287:2215–2220.

Leize E, Jaffrezic A, Van Dorsselaer A. Correlation between solvation energies and electrospray mass spectrometric response factors. Study by electrospray mass spectrometry of supramolecular complexes in thermodynamic equilibrium in solution. J Mass Spectrom 1996;31:537–544.

Li M, Chen B, Lin M, Rustum A. Application of LC-MSn in conjunction with mechanism-based stress studies in the elucidation of drug impurity/degradation product structures and degradation pathways. Am Pharm Rev 2008;11:98–103.

Liu DQ, Hop CECA. Strategies for characterization of drug metabolites using liquid chromatography-tandem mass spectrometry in conjunction with chemical derivatization and on-line H/D exchange approaches. J Pharm Biomed Anal 2005;37:1–18.

Liu DQ, Hop CECA, Beconi MG, Mao A, Chiu S-HL. Use of on-line hydrogen/deuterium exchange to facilitate metabolite identification. Rapid Commun Mass Spectrom 2001;15:1832–1839.

Maziarz EP III, Baker GA, Wood TD. Electrospray ionization Fourier transform mass spectrometry of polycyclic aromatic hydrocarbons using silver(I)-mediated ionization. Can J Chem 2005;83:1871–1877.

McLafferty FW. Interpretation of Mass Spectra, 3rd ed. University Science Books, Mill Valley, CA, 1980.

Mihaleva VV, Vorst O, Maliepaard C, Verhoeven HA, de Vos RCH, Hall RD, van Ham RCHJ. Accurate mass error correction in liquid chromatography time-of-flight mass spectrometry based metabolomics. Metabolomics 2008;4:171–182.

Naganeo F, Iwasaki K. Application of high throughput LC/MS on drug metabolism and pharmacokinetics in drug discovery. J Mass Spectrom Soc Jpn 2004;52:137–141.

Nutter LM, Ngo EO, Abul-Hajj YJ. Characterization of DNA damage induced by 3,4-estrogen-*o*-quinone in human cells. J Biol Chem 1991;266:16380–16386.

Ohashi N, Furuuchi S, Yoshikawa M. Usefulness of the hydrogen-deuterium exchange method in the study of drug metabolism using liquid chromatography-tandem mass spectrometry. J Pharm Biomed Anal 1998;18:325–334.

Oliveira EJ, Watson DG. Liquid chromatography-mass spectrometry in the study of the metabolism of drugs and other xenobiotics. Biomed Chromatogr 2000; 14:351–372.

Peterman SM, Duczak N, Kalgutkar AS, Lame ME, Soglia JR. Application of a linear ion trap/orbitrap mass spectrometer in metabolite characterization studies: Examination of the human liver microsomal metabolism of the non-tricyclic antidepressant nefazodone using data-dependent accurate mass measurements. J Am Soc Mass Spectrom 2006;17:363–375.

Pettit GR, Holzapfel CW, Cragg GM, Herald CL, Williams P. Antineoplastic agents. 92. Broad scope secondary ion mass spectrometry. J Nat Prod 1983;46:917–922.

Pilard S, Caradec F, Jackson P, Luijten W. Identification of an *N*-(hydroxysulfonyl)oxy metabolite using in vitro microorganism screening, high-resolution and tandem electrospray ionization mass spectrometry. Rapid Commun Mass Spectrom 2000;14:2362–2366.

Pramanik BN, Das PR, Bose AK. Molecular ion enhancement using salts in FAB matrixes for studies on complex natural products. J Nat Prod 1989;52:534–546.

Pramanik BN, Mallams AK, Bartner PL, Rossman RR, Morton JB, McGlotten JH. Special techniques of fast atom bombardment mass spectrometry for the study of oligosaccharide containing macrotetronolide antibiotic, kijanimicin. J Antibiot 1984;37:818–821.

Ramanathan R, Su AD, Alvarez N, Blumenkrantz N, Chowdhury SK, Alton K, Patrick J. Liquid chromatography/mass spectrometry methods for distinguishing *N*-oxides from hydroxylated compounds. Anal Chem 2000;72:1352–1359.

Ruan Q, Peterman S, Szewc MA, Ma L, Cui D, Humphreys WG, Zhu M. An integrated method for metabolite detection and identification using a linear ion trap/Orbitrap mass spectrometer and multiple data processing techniques: Application to indinavir metabolite detection. J Mass Spectrom 2008;43:251–261.

Sanders M, Shipkova PA, Zhang H, Warrack BM. Utility of the hybrid LTQ-FTMS for drug metabolism applications. Curr Drug Metab 2006;7:547–555.

Schug K, McNair HM. Adduct formation in electrospray ionization. Part 1: Common acidic pharmaceuticals. J Separ Sci 2002;25:760–766.

Schug K, McNair HM. Adduct formation in electrospray ionization mass spectrometry II. Benzoic acid derivatives. J Chromatogr, A 2003;985:531–539.

Service RF. New role for estrogen in cancer? Science 1998;13279:1631–1633.

Shou WZ, Weng N. Simple means to alleviate sensitivity loss by trifluoroacetic acid (TFA) mobile phases in the hydrophilic interaction chromatography-electrospray tandem mass spectrometric (HILIC-ESI/MS/MS) bioanalysis of basic compounds. J Chromatogr B 2005;825:186–192.

Sugimura T, Okabe K, Nagao M. The metabolism of 4-nitroquinoline-1-oxide, a carcinogen. 3. An enzyme catalyzing the conversion of 4-nitroquinoline-1-oxide to 4-hydroxyaminoquinoline-1-oxide in rat liver and hepatomas. Cancer Res 1966;26:1717–1721.

Tang L, Kebarle P. Effect of the conductivity of the electrosprayed solution on the electrospray current. Factors determining analyte sensitivity in electrospray mass spectrometry. Anal Chem 1991;63:2709–2715.

Tolmachev AV, Monroe ME, Jaitly N, Petyuk VA, Adkins JN, Smith RD. Mass measurement accuracy in analyses of highly complex mixtures based upon multidimensional recalibration. Anal Chem 2006;78:8374–8385.

Tolmachev AV, Robinson EW, Wu S, Kang H, Lourette NM, Pasa-Tolic L, Smith RD. Trapped-ion cell with improved DC potential harmonicity for FT-ICR MS. J Am Soc Mass Spectrom 2008;19:586–597.

Tong W, Chowdhury SK, Chen J-C, Zhong R, Alton KB, Patrick JE. Fragmentation of *N*-oxides (deoxygenation) in atmospheric pressure ionization: Investigation of the activation process. Rapid Commun Mass Spectrom 2001;15:2085–2090.

Williams DK Jr, Chadwick MA, Williams TI, Muddiman DC. Calibration laws based on multiple linear regression applied to matrix-assisted laser desorption/ionization Fourier transform ion cyclotron resonance mass spectrometry. J Mass Spectrom 2008;43:1659–1663.

Williams DK Jr, Muddiman DC. Parts-per-billion mass measurement accuracy achieved through the combination of multiple linear regression and automatic gain control in a Fourier transform ion cyclotron resonance mass spectrometer. Anal Chem 2007;79:5058–5063.

Wolff J-C, Eckers C, Sage AB, Giles K, Bateman R. Accurate mass liquid chromatography/mass spectrometry on quadrupole orthogonal acceleration time-of-flight mass analyzers using switching between separate sample and reference sprays. 2. Applications using the dual-electrospray ion source. Anal Chem 2001;73:2605–2612.

Yamashita M, Fenn JB. Negative ion production with the electrospray ion source. J Phys Chem 1984a;88:4671–4675.

Yamashita M, Fenn JB. Electrospray ion source. Another variation on the free-jet theme. J Phys Chem 1984b;88:4451–4459.

Yoo HJ, Liu H, Hakansson K. Infrared multiphoton dissociation and electron-induced dissociation as alternative MS/MS strategies for metabolite identification. Anal Chem 2007;79:7858–7866.

Zhang L-K. Mass spectrometric method development and application for elucidation of structure, sequence, and reactivity of oligodeoxynucleotides and their carcinogen adducts. Washington University, St. Louis, MO, 2001.

Zhang L-K, Rempel D, Pramanik Birendra N, Gross Michael L. Accurate mass measurements by Fourier transform mass spectrometry. Mass Spectrom Rev 2005;24:286–309.

Zhao X-G, Ma J, Feng H, Wu J, Gu Z-M. Loss of hydroxyl radical in CAD MS/MS for identification of oxidized drug metabolites. In proceedings of the 52nd ASMS Conference on Mass Spectrometry and Allied Topics, May 23–27, In 2004, Nashville, TN, 2004.

Zhou S, Hamburger M. Application of liquid chromatography-atmospheric pressure ionization mass spectrometry in natural product analysis. Evaluation and optimization of electrospray and heated nebulizer interfaces. J Chromatogr A 1996;755: 189–204.

Zhu M, Rempel DL, Du Z, Gross ML. Quantification of protein-ligand interactions by mass spectrometry, titration, and H/D exchange: PLIMSTEX. J Am Soc Mass Spectrom 2003;125:5252–5253.

11 Techniques to Facilitate the Performance of Mass Spectrometry: Sample Preparation, Liquid Chromatography, and Non-Mass-Spectrometric Detection

MARK HAYWARD, MARIA D. BACOLOD, QING PING HAN, MANUEL CAJINA, AND ZACK ZOU

Department of Chemistry, Lundbeck Research USA, Paramus, New Jersey

Mass Spectrometry in Drug Metabolism and Disposition: Basic Principles and Applications, First Edition. Edited by Mike S. Lee and Mingshe Zhu.

11.1 INTRODUCTION

Although high specificity and selectivity is often achieved with modern liquid chromatography–mass spectrometry (LC/MS), there are other factors, including sample preparation, chromatographic separation, and chemical detection techniques, that can greatly affect the performance of LC/MS in metabolite profiling and bioanalysis. First, sample preparation to remove or minimize interference from biological matrices is an important factor that affects the limit of detection (LOD) and accuracy of LC/MS measurements. The substances that interfere with mass spectrometry performance could be, but are not limited to, biomatrices such as proteins, peptides, lipids, and inorganic salts as well as added surfactants from the formulation of compounds. In addition, some samples contain proteins or other substances that will precipitate during an LC/MS or liquid chromatography–tandem mass spectrometry (LC/MS/MS) analysis, which must be removed prior to LC/MS/MS analysis. Sample preparation is a frequent key contributor to the overall analytical performance (smooth baseline, lower noise or interference peaks, less matrix effect) allowing more robust LC/MS/MS methods that achieve lower LOD.

Quality of liquid chromatographic performance is another key factor contributing to sensitivity, accuracy, and throughput of LC/MS analysis. Optimization of LC operations is often accomplished by selecting the correct LC columns, elution solvents, and conditions and instrumentation. Finally, use of additional online or offline liquid chromatographic detection in conjunction with LC/MS can significantly enhance the results of LC/MS analysis by providing additional quantitative or qualitative information that is not available from LC/MS data. These liquid chromatographic detection technologies include ultraviolet (UV) detection, radioactivity detection, and nuclear magnetic resonance (NMR) spectroscopy. In this chapter, the various sample preparation technologies that are used for either drug metabolite profiling or quantitative analysis of a drug and its metabolites in plasma (bioanalysis) are reviewed. Additionally, various liquid chromatographic technologies applied for both metabolite separation and bioanalysis and detection technologies such as UV, radiodetection, and NMR are discussed.

11.2 SAMPLE PREPARATION FOR BIOANALYSIS

Although the main purposes of sample preparation for bionalysis and metabolite profiling are the same, namely removal of biological matrices and concentration of analytes, the nature of analytes and requirements for sample preparations associated with the two types of studies are fundamentally different (McDowall, 1989; McDowall et al., 1989). In the bioanalysis profiling, one or two known analytes that are usually the parent drug and one or a few metabolites are involved. An internal standard is often added to samples prior to sample preparation so that sample preparation recovery that can be easily monitored is not a concern. Protein precipitation (PPT) based methods, solid-phase extraction (SPE), and liquid–liquid extraction (LLE) are the most commonly used approaches in bioanalysis sample preparation (Herman, 2002). Occasionally, ion-exchange chromatography is also used.

Sensitivity, selectivity, and ruggedness are a few factors that help to determine the appropriate sample preparation method. Sample preparation requirements also may depend on the specific stage of drug development, drug discovery, or clinical development. Often times drug discovery methods demand quick turn-around time, and, consequently, there is minimal time devoted to method development activities. As a result, some of these methods may achieve the highest level of performance in terms of sensitivity. In contrast, ruggedness and selectivity are typical requirements for a method in a clinical development setting and thus require meeting these requirements regardless of the effort involved.

11.2.1 Protein Precipitation

PPT-based sample preparation methods are commonly used in drug discovery (Biddlecombe and Pleasance, 1999; Watt et al., 2000; Ma et al., 2008). These methods require minimal to no method development. Generally, PPT techniques are easy, fast, relatively inexpensive, and can be easily automated. On the other hand, PPT approaches dilute the sample that will affect the LOD. Furthermore, some salts and endogenous molecules still remain in the supernatant of the sample, which could lead to a matrix effect. PPT-based methods, which create a low-polarity environment, such as a 4/1 dilution with acetonitrile, are often very thorough at crashing out proteins and other polar endogenous interferences while being highly effective at keeping less polar drugs dissolved. However, some drugs are quite polar and metabolites often can be significantly more polar than the drug itself. Thus, experimentally determined lower ratios and/or strengths of organic solvents may be necessary to avoid also precipitating the drug and metabolites as well as achieve sufficient retention when injected on the LC column.

11.2.2 Solid-Phase Extraction

Solid-phase extraction (SPE) is a very useful technique employed for sample preparation. Its advantages are that it can be applied to many drug compounds while increasing sensitivity by concentrating the sample and reducing matrix effects (Simpson et al., 1998; Venn et al., 2005; Zang et al., 2005). However, SPE is not generally fast, requires a large volume of solvent, and tends to be labor intensive unless automation tools are added to the process. The basic approach is that a sample, in liquid form, is loaded to a cartridge (packed sorbent particles), binding the analyte to the solid sorbent. Then, a wash step (normally 100% water or water containing 4% methanol) is used to remove unwanted proteins while ensuring your analyte is still retained, followed by an elution step (normally high organic of methanol or acetonitrile) used to release your analyte from the sorbent particles. Sometimes evaporation and reconstitution steps are added after the elution. These extra measures facilitate concentrating the sample and achieving increased precision resulting from better control of the final sample volume.

Solid-phase extraction techniques tend to require method development. These development activities can be fairly straightforward and often times generic approaches work when reverse-phase sorbents are chosen. One disadvantage of reverse-phase methods is that they usually are very similar to the high-resolution LC separation used for analysis. Thus, these methods may not remove all interfering components that a truly orthogonal approach may be able to deliver. Ion-exchange-based sorbents offer a truly orthogonal cleanup that often is very effective at removing interferences. The drawback with ion-exchange approaches is the time spent screening sorbents, solvents, and buffers for the initial method development as well as perhaps desalting the sample after elution. Nevertheless, this time commitment can still be worthwhile when other approaches have not been fruitful.

11.2.3 Turbulent Flow Chromatography

Turbulent flow chromatography (TFC) is another popular method for sample preparation (Xin et al., 2010; Herman, 2002; Du et al., 2005; Turnpenny et al., 2007). The authors suggest that TFC should be viewed simply as an online form of SPE. Like SPE, TFC features the use of larger particles (ca. 50 μm) and most of the common sorbents types are readily available. This online approach has significant advantages over traditional SPE in that a separate gradient pump is used at 3–5 mL/min for rapid sample cleanup without a PPT step, which facilitates a high degree of automation for both sample processing and method development. Like SPE, TFC allows direct injection of plasma or serum sample, which means that only a small amount of sample is needed to run the method. Direct injection of plasma or serum samples is ideal when working with small quantities of sample volume (e.g., mouse plasma). In practice, samples are often diluted to avoid fouling the injection port. Samples are then mixed and injected directly for analysis. This procedure generally does not reduce sensitivity because large injection volumes can be used since the analyte can be refocused when the cleanup column is backflushed with the high-resolution analytical gradient. Furthermore, analyte response can be increased with multiple injections on the cleanup column, allowing as much as 500 μL of plasma to be injected for a single analysis.

In the most common approach to using TFC, dual columns are used where the sample is injected and extracted in the large-particle (ca. 50 μm) extraction column, then eluted into a small-particle (1.5–5 μm) analytical column for

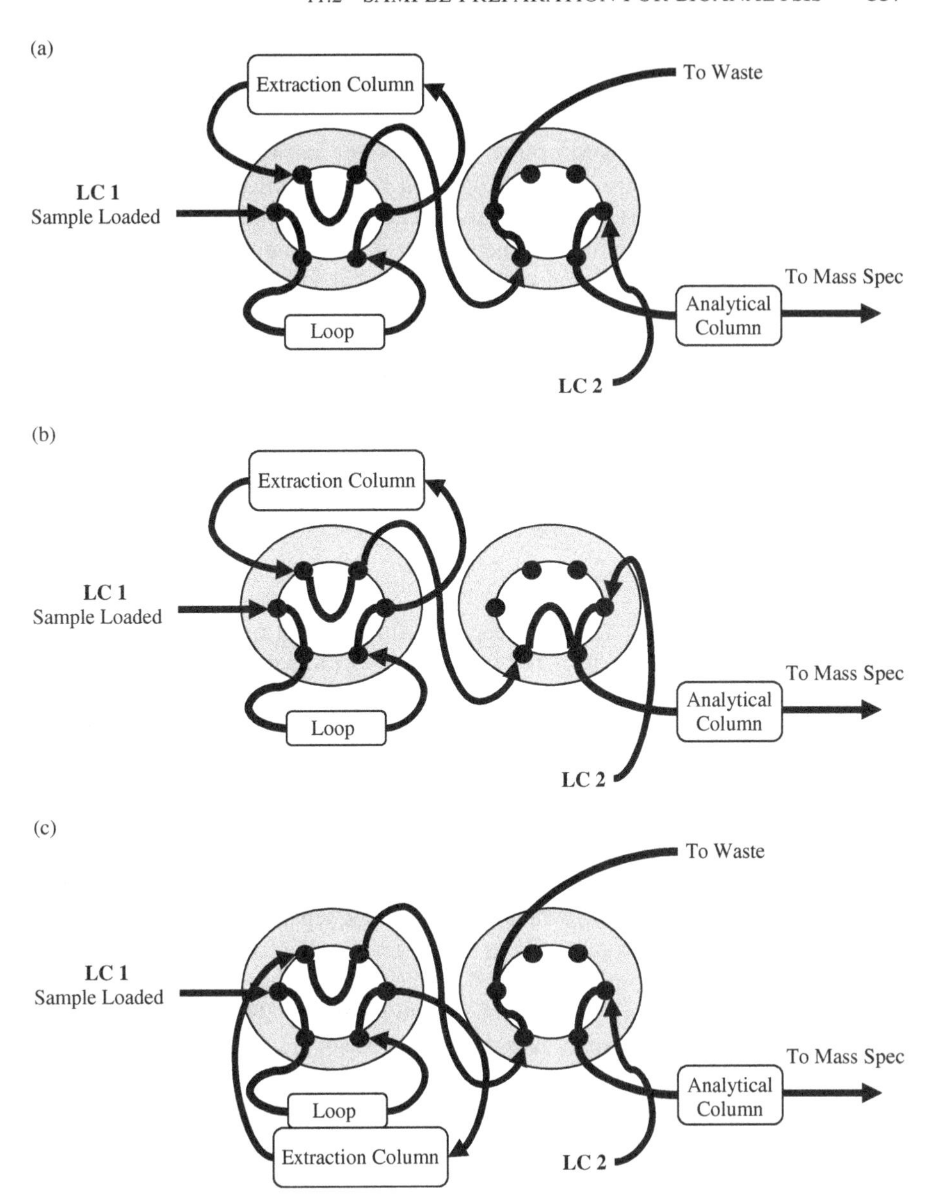

FIGURE 11.1 Three steps [(a)–(c)] of performing TFC sample introduction into an MS for bioanalysis as it is most commonly practiced.

high-resolution separation prior to elution into the MS. The beginning of this process is illustrated in Figure 11.1*a* where LC 1 loads the sample at a high flow rate (2–5 mL of 90+% aqueous) into the extraction column with the valve diverted to waste. The sample is retained in the extraction column while larger unwanted proteins are washed away. Simultaneously, LC 2 is cleaning and conditioning the analytical column in anticipation of the sample. In the next step of the process, LC 1 flows are ramped or stepped toward high organic content, then the divert valve is switched from flow to waste to flow into the analytical column, allowing sample to flow from the extraction to the analytical column as shown in Figure 11.1*b*. In the third step, LC 1 flow switches again to high aqueous, with divert valve back to waste, allowing the extraction column to be washed and prepared for the next sample. Simultaneously, LC 2 starts a gradient to elute the analyte from the analytical column into the MS for measurement as shown in Figure 11.1*c*. Method development for TFC is completely analogous to SPE except that the experimental work can leverage the automation aspect to minimize the labor involved. The automation features of TFC can be expanded even further after method development by utilizing the capability to run up to four systems in parallel into a single mass spectrometer. Also, when reverse-phase sorbents are used for both the cleanup and analytical separations, generic and relatively all-purpose methods can be utilized.

11.2.4 Liquid–Liquid Extraction

Liquid–liquid extraction (LLE) is a powerful purification technique that is commonly practiced in most of the basic sciences (Xue et al., 2006, 2007). As applied to bioanalysis, LLE is frequently regarded as among the more effective as well as laborious approaches for sample cleanup (McDowall, 1989; Wille and Lambert, 2007). The classic approach with LLE is to use pH to drive the analyte of interest away from the sample matrix as well as other analytes (Hendriks et al., 2007). For example, a sample containing a basic drug may be double washed in high pH followed by low pH in a water–heptane mixture. This approach will completely remove phospholipids and often provides samples clean enough for single-stage MS detection (MS/MS not needed). The separation work flow can be simplified by freezing the aqueous layer in dry ice, but the technique still requires significant labor. This approach has led to the development of semi-automated robotically driven approaches (Steinborner and Henion, 1999; Zhang et al., 2000) and fully automated approaches using a filter (Peng et al., 2001) to help separate layers or hollow fiber membranes to accomplish the same in liquid-phase microextraction (LPME) (Pedersen-Bjergaard and Rasmussen, 1999; Psillakis and Kalogerakis, 2003). Despite this progress, LLE (or LPME) has not gained widespread use in bioanalysis because it is still deemed to be insufficiently automated for true high-throughput work flows.

Nevertheless, there is another way to perform LLE that can achieve ultra-high throughput and be completely automated without sacrificing the highly efficient separation. This is achieved by the use of low volatility halogenated solvent analogs as the organic layer. Halogenated solvents are used to increase the density of the solvent so it is easily centrifuged to the bottom of the well for sampling by standard autosamplers. Another benefit of added halogenated solvents is that the vapor pressure can be greatly reduced to avoid evaporation while causing only small incremental changes in the solvents' dissolution properties.

This approach requires a bit more method development, but it is clearly well worth the effort when the same method needs to be used for thousands or hundreds of thousands of samples. Method development starts with purchasing 6–12 halogenated solvent analogs and empirically measuring (screening) extraction recoveries. Once a high-recovery solvent is found, execution of the cleanup is simple: add solvent and mix by pipette, then centrifuge.

The beauty of this method is that it is easy. The resulting samples are very clean, and when the volatility of the solvent is low enough, the samples can be highly stabilized to the point of being able to sit open on the bench for weeks without changing the analysis results when remeasured. Achieving this level of sample stability is very important because it allows complete decoupling of sample generating and sample measurement processes. The aqueous layer on top causes no problems because it only comes in contact with the outside of the autosampler needle. Eventually, the aqueous layer dries to a white ring of solid around the rim of the well and then is physically removed from coming in contact with the autosampler needle. Bromobenzene (BB) has been found to be a very efficient LLE solvent and works well for greasy, steroidlike analytes. BB is surprisingly efficient at extraction and often requires little or no mixing, however it does require the use of glass or poly(tetrafluoroethylene) (PTFE)-coated plates for containers. 222-Trichloroethanol (TCE) is a suitable LLE solvent for analytes containing both significant lipid properties and a highly charged ionic site. However, TCE has high surface tension with water and requires thorough mixing (10 pipette aspirate/dispenses). While using either BB or TCE, sample cleanup often is so effective that LC column lifetimes can reach as high as 10,000 injections and single-stage MS detection often suffices (MS/MS is not needed).

11.2.5 Plasma and Blood Sample Preparation

In the bioanalytical experiments that support pharmacokinetic (PK) evaluations of drug candidates in drug discovery and development, more than 90% of uncounted samples are plasma or blood samples (Hopfgartner et al., 2002). PPT is the most often used sample preparation technique in drug discovery as it offers minimal method development and fast turnaround times. The sample is normally diluted multiple times with an organic solvent, mixed, centrifuged, and the supernatent is removed from the precipitated proteins (Du et al., 2005; Wu et al., 2007). Not all endogenous molecules and salts are removed, but this still provides a fast and efficient way to analyze a large number of samples, which is a huge requirement in drug discovery support. TFC is another method that meets the demand of fast turnaround time in drug discovery. It allows direct injection of plasma or serum sample, which means that only a small amount of sample is needed to run the method. This direct injection format is ideal when working with small quantities of sample volume (e.g., mouse PK) as samples are normally diluted, mixed, and injected directly for analysis. SPE is typically the preferred choice for the preparation of blood due to the complexity of the sample and the single use nature of the cleanup sorbent. SPE is not ideal when supporting drug discovery as fast turnaround time is not the strength of this method. When time is the driving force, PPT often is used as an alternative.

11.3 SAMPLE PREPARATION FOR METABOLITE PROFILING AND IDENTIFICATION

In the metabolite or glutathione profiling experiment, a large number of unknown analytes are often encountered, some of which could be very polar components, such as glucuronide conjugates. Additionally, analyte recovery from a sample cleanup procedure cannot be monitored. Low recovery of some metabolites in the sample preparation, which could lead to totally different outcomes of a metabolite profiling experiment, is a significant problem. Therefore, a good sample cleanup procedure for metabolite profiling should be able to preserve unknown metabolites with high and consistent recoveries, especially for polar constituents.

11.3.1 In Vitro Sample Preparation

In vitro metabolism samples from liver microsome or hepatocyte incubations have limited amounts of biological matrices. The main objective of in vitro incubation sample preparation is to remove proteins and buffer salts, which is routinely carried out using PPT methods with two or three volumes of acetonitrile or methanol. After centrifugation, supernatants can be either directly injected into LC/MS or concentrated using nitrogen or vacuum evaporation prior to injection. For processing incubation samples in 96-well format filtration plates are often used for removal of proteins precipitated by organic solvent (Yao et al., 2007). In some cases, incubations in test tubes followed by SPE are conducted in determining in vitro metabolite profiles (Zhu et al., 2005a).

11.3.2 Plasma, Urine, and Bile Sample Preparation

Radiolabeled drug metabolism and disposition studies in preclinical species and humans are routinely conducted to support drug development and registration. In these studies, plasma samples are analyzed for determining exposures of the drug and its metabolites (Zhu et al., 2009). In addition, metabolite profiles in urine, feces, and/or bile are analyzed for determining drug clearance pathways (Zhang et al., 2009b). In some cases, profiles and quantities of metabolites in animal tissue samples are determined to address toxicology or pharmacology issues after administration of radiolabeled or nonlabeled drugs. Plasma samples containing huge amounts of proteins and metabolite concentrations are usually lower than in vitro incubation samples (D'Souza et al., 2007; Hosoda et al., 2008). Therefore, sample preparation processes are carried out to accomplish two objectives: removal of proteins and enrichment of analytes. The PPT processes used in the treatment of in vitro samples, which are described above, are applicable to plasma or blood samples. Additionally, SPE methods are often employed for the preparation of plasma samples in determining metabolite profiles. After loading a plasma sample that is often diluted with water at a 1 : 1 ratio, water or an aqueous solution with less than 5% of organic solvent is added to SPE cartridges to remove salts and proteins. Then methanol is added to elute all drug-related components. The methanol fraction is concentrated and reconstituted with high-pressure liquid chromatography (HPLC) solvents prior to LC/MS analysis. Urine and bile samples contain very limited amounts of proteins or other substances that may precipitate out during LC/MS analysis. Therefore sample preparation of urine may be simply performed by centrifugation or filtration to remove solid particles. In some cases, urine samples are processed by SPE, which not only cleans up the samples but also concentrates analytes. Metabolite concentrations in bile are usually much higher than those in urine. For processing bile samples, organic solvents can be added to precipitate proteins and inorganic salts prior to injection, and in many cases, these samples can be injected directly onto the LC column for analysis.

11.3.3 Fecal and Tissue Sample Preparation

Preparation of fecal and tissue samples usually starts with homogenizing of tissues, which are suspended in about three to four fold (v/w) liquid using a homogenizer (Monleon et al., 2009; Zhang et al., 2009a). This liquid can be water or a solution that contains a low percentage of organic solvents and volatile pH buffers for increasing the analytes' solubility and stability or stopping the potential enzyme reactions. The percentage of organic solvents should not be too high to avoid protein precipitation, which would clog the homogenizer. PPT is performed using organic solvents in multiple times. After centrifugation, combined supernatants are concentrated to enrich analytes.

11.4 LIQUID CHROMATOGRAPHIC SEPARATION IN BIOANALYSIS

11.4.1 Basic Approach and Method Development

The most common way to perform method development is to infuse the analyte and allow software to optimize all tuning parameters that will give the best intensity and resolution. However, even though most vendors offer the ability to automatically tune many different compounds simultaneously, this process could take a long time, which is not ideal for drug discovery applications. An alternative is to use a low concentration, ~250 μg/mL of analyte with a generic LC method and application of a full data dependent mass spectrometer scan (Toll et al., 2005; Saunders et al., 2009). First, a full Q_1 or Q_3 scan verifies that the mass of the parent compound is detected, followed by dependent scans of increased collision energies that will generate fragmentation. Note that collision energy is not the only parameter that is needed to develop a good method, but in supporting drug discovery generally generic methods are used for speed. However, one of the key parameters that is extremely compound dependent is the collision energy, and this combined approach is not only faster than infusion but also provides an ability to determine if the sample will be retained in a specific column as well as the retention time. Fragments or product ions of the parent compound are obtained at different collision energy levels. This information can be used to approximate the optimal collision energy required to generate the most abundant product ion.

Liquid chromatography is a vital component of bioanalysis (Knox and Jurand, 1973; Pedersen-Bjergaard and Rasmussen, 2005). If quality LC conditions cannot be achieved, then sample preparation may be irrelevant (Jemal and Xia, 2006). Numerous factors must be considered when choosing the correct LC conditions: type of solvents used, buffers, pH adjustments, isocratic versus gradient elution, reverse versus normal phase, and columns, just to name a few (Jemal and Ouyang, 2000; Gritti and Guiochon, 2006; Wirth, 2007). As previously stated, fast turnaround time is a major requirement in the drug discovery field. Therefore, the overall analysis time must be weighed when choosing the correct LC method. The following set up of mobile phases is a logical starting point: 0.1% formic acid in water, for a weak solvent, and 0.01% formic acid in acetonitrile for a strong solvent. A gradient elution is used that results in a faster elution rate than with isocratic elution. Finally, a small column with small particle size is chosen, C18 Gemini column, 3×2.00 mm 3μ (Phenomenex); both will help minimize the run time and generate narrower peaks. Note that this method may not be optimal for the analysis of every compound. As stated previously, this is a representative or generic method that is currently used to support drug discovery, and its goal is delivering reliable results with a quick turnaround time.

An obstacle that one could encounter is that of matrix effects and vehicle effects from the dosing formulation. Unfortunately, this may be unavoidable when choosing PPT as the sample preparation method. Not all salts and endogenous molecules are removed with PPT; therefore, an extraction that would yield a cleaner extract, such as LLE or SPE, should be used. However, the use of either LLE or SPE will increase turnaround time. Furthermore, LLE and SPE are more laborious methods than PPT. One way to deal with matrix effects and at the same time maintain the speed of PPT sample preparation is to vary the gradient. Normally, the salts and endogenous polar molecules tend to elute along with the aqueous mobile phase. Thus, delaying when a high percentage of organic phase is introduced to the gradient can help with the separation. The result is that unwanted polar components of the matrix elute earlier and matrix effects are minimized when the gradient rate is reduced. The primary matrix effect from the phospholipids also can be minimized with slower gradients, causing them to elute later than the drug, although run time is increased.

Analysis of low-molecular-weight (MW) compounds (<200) is also challenging because of the difficulty of retaining these analytes in the HPLC column. In this situation, two parameters can be changed, the length of the column and/or the run time of the gradient. When compounds have a very low MW, analysis with a short column is very difficult; therefore, the length of the column must be increased. Again, it should be noted that this method may not be optimal for the analysis of every compound in a drug discovery environment. Of course, better chromatography could be achieved if buffers were used, or if the pH was adjusted, or a different column or modified gradient were used. But optimizing all of the above factors will lead to longer method development time, which again is not optimal for drug discovery support.

11.4.2 Splitting LC Flow for Introduction into MS

It is a common situation for those coupling LC with MS to achieve an optimal separation at a different flow than that which yields optimum sensitivity and lowest maintenance. All too often in bioanalysis, the separation is sacrificed to accommodate the MS. Sometimes the MS performance is sacrificed for analysis speed. A common, readily achievable combination for speed is 0.5–1.0+ mL/min through a 2.1 × 50 mm column, while the optimum for the ion source is usually 100–400 μL/min. If electrospray ionization (ESI) is used, then ion intensity is concentration (not mass) driven, (Ardrey, 2003) and there is no reason not to split in order to operate under optimal conditions for both the separation and the detection.

It is well known there are suitable highly tuned, off-the-shelf splitters available based on long fused-silica capillaries, but these are best used for high split ratios to split/sample purification streams. While lesser known, a rugged, low ratio splitter also is available (Analytical Sales, Pompton Lakes, NJ). This splitter uses constant-pressure regulation and is much better suited for the task of high-throughput bioanalysis (Fig. 11.2). In this approach, waste flow is made variable using a pressure regulator. Thus, the setup shown can be used to deliver the same optimum flow to the MS over any LC flow range that is

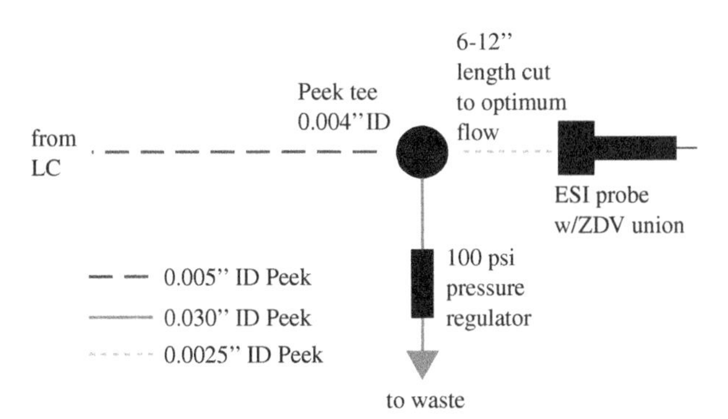

FIGURE 11.2 Rugged constant-pressure flow splitter for LC introduction to MS.

greater than the target flow into the MS. Flow is regulated with the length of the tube connecting tee to MS ion source. Thus, one simply performs the tubing length adjustment once to match the MS (no more adjustments!). When good plumbing technique is used, transit times are very low (<1 s). Narrow peaks can be achieved ($2\sigma \leq 1.5$ s) without negatively impacting sensitivity while providing reproducibility and ruggedness that are completely solid (lifetime easily $>10^5$ injections).

11.4.3 Stepping up Productivity with Fast LC Separations

A wealth of advanced information is available on the fundamentals of LC separations and perhaps even more information is available on practical LC/MS. Recently, there has been a real shift in focus toward very fast separations (Hsieh and Korfmacher, 2006). A significant portion of this focus has been directed toward the measurement of productivity in terms of the plate height of the column, which in many ways may not provide a true representation of productivity. In reality, resolution per unit time is a more accurate representation of separation and productivity. Therefore, if the analyte is resolved, then the work is done.

The first, most essential part of achieving a high-resolution separation is sweeping the column volume with a sufficient amount of eluent. If the analyte elutes in one column volume ($k' = 1$), then a separation was not likely achieved because there is no retention. Often, optimal resolution is achieved at $5 < k' < 10$ (sometimes higher). Thus, increased productivity without the loss of resolution involves the use of $k' = 5–10$, or more generally, a higher k' per unit time. Thus, the use of a higher k' requires higher eluent velocities.

More velocity, quite simply, requires more flow (and thus pressure). Increased velocity can also be achieved with smaller column inner diameter (ID) at the same flow. Similarly, higher productivity (k'/min) can be achieved with the use of shorter columns at the same or preferably increased flow. All of these changes are straightforward; however, these changes also result in the use of velocities that are further away from the optimum velocity and, furthermore, add other departures from the ideal separation. For example, to avoid significant loss in separation efficiency, LC columns must be operated with infinite diameter (Snyder LR and Kirkland JJ, 1974). Thus, the column ID must be large enough to ensure that the analyte never reaches the column wall where the packing is less dense and the flow higher than the center of the column. The bottom line is that the column ID must be matched to the delivered injection volume. This is best measured with a UV detector connected directly to the injection valve and can be as much as 50 times the selected sample volume (Gritti et al., 2006 & Scott, 2002). Table 11.1 (page 355) provides some optimal column ID ranges for different instrument sample injection capabilities.

The use of a shorter column is the classic way to increase speed, but there is an inherent loss in separation efficiency. The classic solution to this problem has been to reduce the particle diameter to restore efficiency due to shorter

TABLE 11.1 Common Examples of Matching Column ID to Autosampler Performance

Delivered Injection Volume[a]	Column ID Range (mm)	Example Instrument
50	4–6	Any ordinary HPLC
10	1.5–2.1	Waters Acquity UPLC
0.2	0.2–0.3	Eksigent Express

([a]2σ in μL with UV connected to valve).

column length. However, there are limits on what can be accomplished with particle diameter reduction, which is also likely to originate in the autosampler. With a reasonably optimized LC system, approximately 80% of the observed variance (peak dispersion) is due to the injection process. So, even if a dramatic 50% reduction in variance from the column can be achieved (e.g., from 3.5–1.7 μm particle diameter), the result is only a 10% improvement in the overall observed peak variance. This practical limitation has led to a loosely defined empirically derived consensus that column lengths of at least 50 mm are required to obtain acceptable separations instead of the 10 mm lengths once envisioned (Knox and Saleem, 1969).

In many ways, column dimensions are predefined by the required resolution. Thus, the goal is narrowed to increase velocity and the ability to reoptimize the separation back to its original resolution. The two known approaches to increase optimum velocity are to use smaller particle diameters (optimum velocity proportional to 1/particle diameter) (Knox and Saleem, 1969 and Nguyen et al., 2006) and increase the column temperature (optimum velocity proportional to $e^{-k/RT}$) (Antia and Horvath, 1988; Yan et al., 2000). The use of a smaller particle diameter is straightforward; decreasing the particle diameter in half allows for the velocity to be doubled (e.g., 3.5–1.7 μm allows shifting from 1.5 to 3 mm/s at 30°C). However, this increased velocity is accompanied by huge increases in pressure (four fold for the 3.5- to 1.7-μm example), which can ultimately limit velocity and productivity (k'/min). Temperature is more complex because the rate of change can vary with choice of solvent or column particle type as well as the mode of operation (isocratic or gradient). Conveniently, in all cases, as temperature is increased to raise optimum velocity, pressure is reduced to allow more velocity to be achieved. Interestingly, choosing the solvent, column, and mode of operation has become easier as more data is compiled and it is clear that silica based (or silica hybrid) columns used with gradient acetonitrile (ACN) elution allows much higher velocities than other options when operated at elevated temperatures.

11.4.4 Using SFC and MS for Chiral Bioanalysis

Historically, pharmaceutical researchers mostly ignored stereochemistry (prior to the 1980s). Eventually in the 1980s, it became clear that individual stereoisomers can have unique pharmacological profiles and hence medical benefits. Success at meeting previously unmet medical needs with individual stereoisomers has resulted in pharma integrating the study of drug candidate stereochemistry into research and development (R&D) (Agranat et al., 2002 and Caner et al., 2004). This pharma need has led to expanded use of packed-column-based chiral separations. Chiral separations are used in a variety of ways to support pharma R&D. First, in drug discovery there is the usual enantiomeric excess (%ee) and basic compound purification needs required to test enantiomers in a variety of biological assays.

Then, as a compound progresses, we may want to perform chiral bioanalysis of the drug (Li and Shieh, 2008 and Brock, 2006) to answer questions such as reactivity reasons that the in vivo and in vitro assays do not agree on which isomer is most potent. One hypothesis might lead one to test if isomer interconversion occurred. There are significant examples of enantiomeric interconversion such as thalidomide, where one enantiomer is therapeutic and another has negative side effects, but neither can be prevented from racemizing (Eriksson et al., 1995). Another example is the analgesic profens, where the R enantiomers are a prodrug to the active S enantiomers, leaving little benefit for the S-only drug (Hutt and Caldwell, 1983). Experimentally, these situations are readily tested by dosing individual stereoisomers (ee = 99+%) and measuring %ee in plasma and target organs. If the %ee changes (beyond the small statistical variation, i.e., other isomers appear), then interconversion has occurred.

Certainly, there are other possible ways to explain anomalous biological data. Different mechanisms and, hence, different rates of metabolism may provide rational explanations. Still another possibility may be that the stereospecific potency is target related (Baumann et al., 2002) instead of absorption, distribution, metabolism, and elimination (ADME) related. These possibilities can be tested by dosing a known mixture of the individual enantiomers and measure ee in plasma and target organs. If the preferentially available stereoisomer is the less potent isomer, then the drug–target protein combination may be the possible source of stereospecific behavior. This would have to be validated with further experiments. If the preferentially available stereoisomer also is the more potent isomer, then drug availability may be the source of stereospecific behavior (commonly observed).

Once one has decided that chiral bioanalysis is needed (Bamba, 2008), then the question becomes: What technique should be used? Reverse-phase (RP) LC perhaps is an automatic comfortable choice because it is used so frequently in ordinary bioanalysis. While using acetonitrile gradients is great for rapidly spanning a range of polarities in one chromatogram, it is not so simple for chiral separations. Addressing different polarities does not address stereoisomers. In practice, RP chiral separations often are very pH dependent, and this makes method development difficult. Normal-phase (NP) LC works and is the well-proven approach for chiral resolution. Difficult stereoisomers can usually be well resolved, but gradients do not tend to work well (functional range of gradient is too small), separations are not MS compatible, and retention times usually are long (30+ min). Nevertheless, NP method development still is easier/faster than RP where pH sometimes must be stepped 0.1 pH units. This is because NP usually only requires screening three or fewer different buffers to achieve satisfactory conditions.

The technology alternative to chiral LC is supercritical fluid chromatography (SFC) (Liu et al., 2002; Mangelings and Vander Heyden, 2008a, 2008b). SFC is essentially an NP format where CO_2 is substituted for hexane. Gradients work well with SFC and usually span 5–50% B (alcohol) and facilitates fast method development. The combination of gradient and low viscosity often facilitates fast separations (usually at least three times faster than NP-LC). SFC works when combined with MS detection, but historically has not been easy to perform and has not worked as well as RP-LC/MS (White and Burnett, 2005).

Interfacing SFC to MS has most frequently been accomplished by splitting flow and using make-up solvent (Combs et al., 1997). Despite an eluent being mostly gas at atmospheric pressure (AP), full flow (1–5 mL/min) into the source, ESI has not worked particularly well (high background, low response) but has been functional. Interestingly, atmospheric pressure chemical ionization (always heated) has tended to give better response. After trying many different configurations, the authors concluded that temperature was the primary missing ingredient in interfacing SFC to cold ESI-MS. This was based on simply observing the eluent stream as the pressure goes from supercritical (nearly) to atmospheric. When this happens, three distinct phases are formed: gas, large liquid droplets, and small solid snowflakes. Spraying this three-phase mixture into a cold ESI source still gives ions, but the signal is usually low and noisy.

Heating the eluent stream while under pressure (100 bar) is among the most effective ways to maintain a homogeneous fluid during the expansion to AP and stabilizes the ESI–MS signal. Use of a simple mobile-phase preheater for temperature control (Selerity, Salt Lake City, UT) allows full flow (3–4 mL/min, 30% isopropyl alchol) into a cold ESI source while achieving good sensitivity (ng/mL) and high signal to noise (>1000). Perhaps the simplest way to direct full flow in the interface is to eliminate the back pressure regulator (Pinkston, 2005) and replace it with an empirically calibrated piece of Peek (3–5 ft of 0.004–0.005 inches ID). This yields performance that is at least as good as with well heated interfaces and shows an optimum temperature setting of ≥90°C (under 100 bar pressure), which is consistent with earlier custom-split interfaces (Sadoun, 1993). Now with sufficient heating, using ordinary ESI sources, functionally equivalent bioanalysis performance levels (precision, sensitivity, dynamic range) as well as robustness can be achieved with LC–MS/MS or SFC–MS/MS (Fig. 11.3).

Over time, performing many chiral bioanalyses in the drug discovery setting leads to a few observations that are unique and merit discussion. First, chiral separations are no where near as effective at separating the analyte of interest from the biological matrix as is the usual gradient reverse-phase approach used

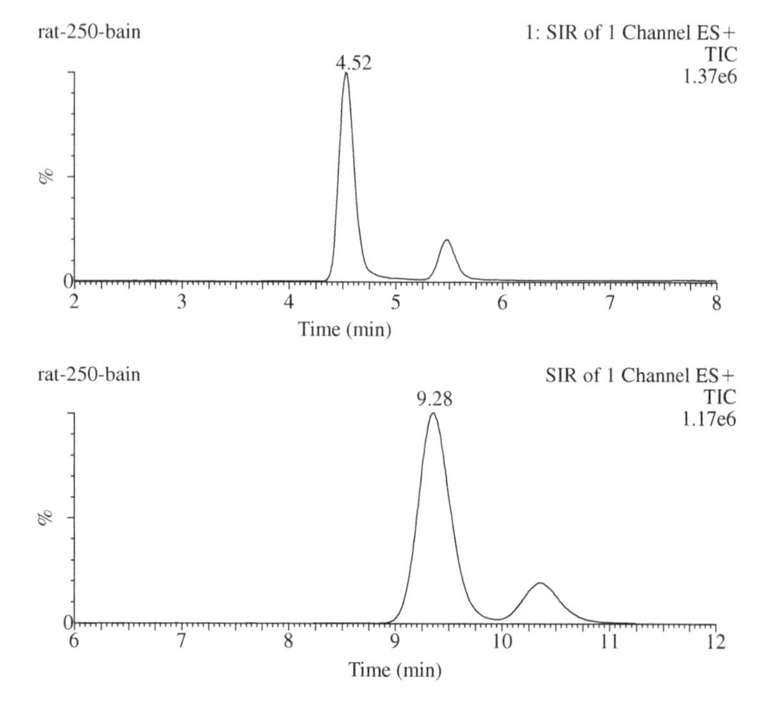

FIGURE 11.3 Chiral bioanalysis in rat brains using SFC/MS and LC/MS.

in achiral separations. Thus, samples for chiral separation tend to need more effective sample cleanup SPE or (LLE) and the commonly used ACN precipitation often is insufficient if a robust, sensitive, and high dynamic range method is needed. Once the sample is cleaned up, single-stage MS is generally sufficient as shown in Figure 11.3 and MS/MS is not needed.

Two additional interesting trends are observed when dosing mixed isomers. The isomers always (for all studies performed in our labs) are observed in a different ratio than dosed, and these ratios are always very constant between animals given the same dose measured at same time point. In a range of studies from small (2 animal) to large (>40), animal-to-animal variation has been 2% or better, even though total drug concentrations tended to span a range of a factor of 2. For the same animal, plasma and brain ratios tend to agree within 1% or better. These observations lead us to hypothesize that *all* ADME aspects of enantiomers are identical (i.e., not stereospecific) except for the mechanism(s) and hence rates of metabolism. This hypothesis (and our observed basis data) is entirely consistent with what is known in general (Lu, 2007 and Brock, 2006) and for the well-studied drug citalopram (Sidhu et al., 1997 and Kugelberg et al., 2003).

Finally, one more interesting trend is how different the perceptions of drug chirality are relative to the actual observations. By far, the most common hypothesis that we focus our experiments on is the in vivo interconversion of enantiomers. Despite our awareness of this process in the literature (Eriksson et al., 1995; Hutt and Caldwell, 1983) we have yet to observe in vivo interconversion first hand over several years of studying many drug candidates. Instead, perhaps not surprisingly, in the overwhelming majority of the cases, we see the more in vivo active enantiomer is the more available enantiomer. More specifically, the more in vivo active enantiomer is the enantiomer with lower clearance. While it is certainly well known that stereospecific target activity is not unusual, it seems to generally align with greater drug availability. There is one rare drug we have studied, citalopram, where the more active enantiomer is less available (proven to be cleared faster) (Bezchlibnyk et al., 2000) but that surely seems to be the exception rather than the rule.

11.5 LIQUID CHROMATOGRAPHIC SEPARATION TECHNOLOGIES IN METABOLITE PROFILING

Although LC/MS analytical tools are overwhelmingly used in both bioanalysis and metabolite identification, the nature and requirements of the two types of experiments are significantly different. First, bioanalysis is quantitative analysis of one or a few known compound(s) mainly in plasma, while metabolite identification involves detection and structural characterization of unknown metabolites in various biological matrices, including in incubations, plasma, urine, bile, and feces. Second, bioanalysis deals with a very large number of samples, for example, a full PK study in a preclinical species often has more than 100 samples, including biological, quality control, and standard samples. In a clinical trial, a bioanalysis method can be used repeatedly for up to more than 10,000 samples. Data acquisition of LC/MS analysis constitutes the majority of time spent in the bioanalysis experiment. In contrast, an in vitro metabolite profiling across four species, or a radiolabeled ADME study in human, or a preclinical species usually has less than a total of 20 samples for analysis. Data analysis and interpretation instead of data acquisition accounts for a majority of time of a metabolite profiling and identification study. Finally, method development and validation that are accomplished by using synthetic standards are crucial steps in bioanalysis, especially in bioanalytical procedures involved in safety toxicology and clinical studies. In drug metabolite identification studies, metabolite standards are not available and regulated procedures are not required so that very limited method development and validation are performed. Because of these differences, analytical strategies and liquid chromatographic methodologies employed in metabolite profiling are significantly different from those applied in regulated bioanalysis.

11.5.1 LC Methods for Metabolite Profiling of Nonradiolabeled Compounds

In the drug discovery stage, most drug metabolite profiling studies are conducted with nonradiolabeled lead compounds by using LC/UV/MS

(Wong et al., 2008). These studies include metabolic soft spot determination, reactive metabolite screening, and comparative metabolism across species in incubations of liver microsomes and/or hepatocytes (Zhang et al., 2009b). Liquid chromatographic tools most frequently employed in this type of experiment are reverse LC column technology (2 mm ID column) at flow rates of approximately 0.25 mL/min with run times of 10–60 min. To avoid early elution of very polar metabolites with endogenous components, such as proteins, LC gradient systems often start with a low percentage of organic solvent and are often isocratic in the first 3 min. In many drug metabolism laboratories, generic LC methods are used initially for compounds from discovery programs with no or very limited efforts in LC method development or validation. If major metabolites are co-eluted or have similar retention times, which may lead to the difficulty in determination of relative quantification of metabolites, then different generic methods can be tested (Zhu et al., 2009) or existing ones adapted. Figure 11.4 (page 361) illustrates a good example of in vitro metabolite profiling using LC/UV/MS. The analysis was accomplished using a generic LC method at a 10-min run time. Major metabolites were separated in the UV profile (Fig. 11.4*a*), which provided quantitative information on both the test drug and metabolites. The glutathione (GSH) adduct elutes at 3.5 min and does not overlap with proteins or other polar endogenous components so that potential matrix effects are minimized. On the other hand, LC/MS data acquired from the same run provided useful information on metabolite structures (Fig. 11.4*c* and 11.4*d*). Recently, ultra performance liquid chromatography (UPLC) coupled with UV and MS has been used to improve metabolite separation and shorten LC run time to 5–20 min.

11.5.2 LC Methods for Metabolite Profiling of Radiolabeled Compounds

Most metabolite profiling and identification studies in drug development involve radiolabeled drugs and in vivo ADME samples. To attain a confident quantitative estimation of the parent drug and its radiolabeled metabolites, baseline separation of radioactivity components are often required (Athersuch et al., 2008; Cuyckens et al., 2008). In addition, to be able to load a large amount of sample into the LC, which can improve sensitivity of radioactivity detection, conventional reversed LC (with 3.9- or 4.6-mm columns and at 1-mL flow rate) are routinely employed. The run times are usually from 30 to 60 min. Usually, limited method development work (one or few days) that focuses on metabolite separation is performed via selecting suitable LC columns and optimizing elution conditions. Figure 11.5 highlights both radiochromatographic and LC/MS profiles of metabolites of a test radiolabeled drug in monkey plasma. Multiple radioactive metabolites were separated well with excellent peak shapes.

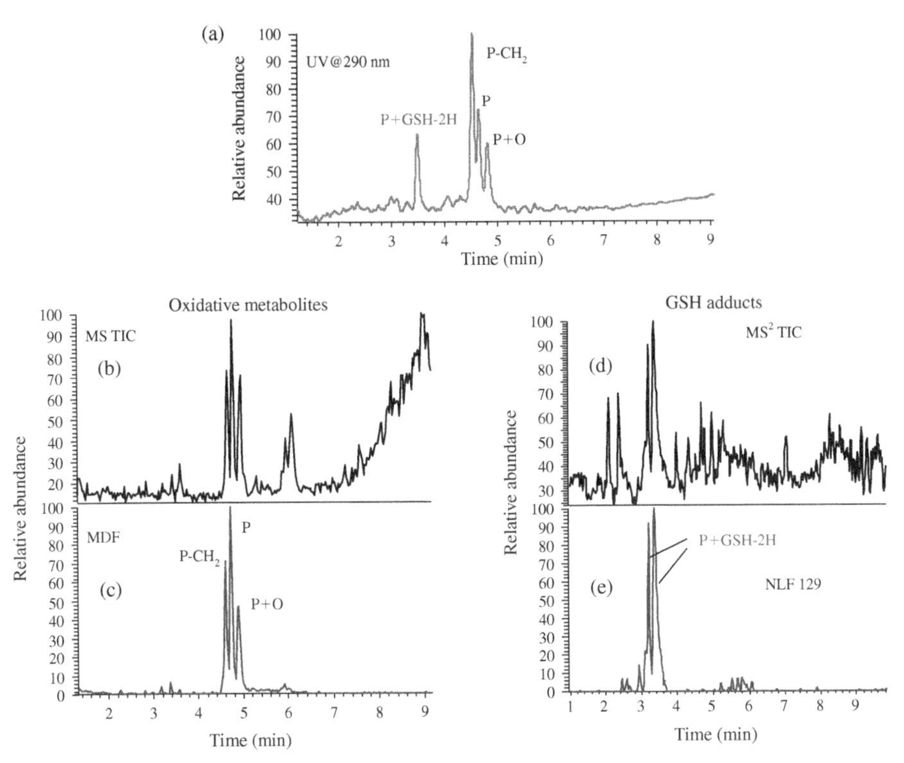

FIGURE 11.4 Metabolite profiles of clozapine in rat liver microsome incubation in the presence of glutathione. (*a*) UV profile of clozapine that displays four drug-related peaks, P (parent drug), P-CH_2 (demethylated metabolite), P+O (N-oxide metabolite) and P+GSH-2H (a GSH adduct). (*b*) TIC of full-scan MS analysis by LTQ-Obitrap mass spectrometry. (*c*) MDF processed profiling using the parent drug as a filter template. (*d*) TIC of data-dependent MS/MS scan. (*e*) Processed MS/MS data with neutral loss of 129 Da.

11.6 LIQUID CHROMATOGRAPHIC DETECTION

In LC/MS metabolite profiling and identification analysis, other liquid chromatographic detection formats, including UV, radiodetection, and NMR, are often employed to facilitate metabolite detection, quantification, or structural characterization. These liquid chromatographic detection formats provide high analytical values that are complementary to mass spectrometry and critical to accomplishing objectives of drug metabolism studies (Zhu et al., 2009).

11.6.1 UV Absorbance Detection

UV detectors can be easily coupled to LC/MS systems and operated with very low cost. UV spectra of parent drugs are not the same as the UV spectra of their metabolites; however, metabolites generally retain most of the core structure of the parent drug, and the differences of UV responses between a drug and its metabolites are smaller than the corresponding mass spectrometric responses (Yau et al., 2007). Therefore, UV detection has been widely employed in quantitative estimation of metabolites in biological samples without analytical standards (Yau et al., 2007). The major application of UV detection in drug metabolism studies is analysis of in vitro metabolite profiles using LC/UV/MS in drug discovery (Zhang et al., 2009b; Zhu et al., 2009). These studies, including metabolic soft spot determination, reactive metabolite screening, and in vitro metabolism across species, are usually conducted at 10–30 μM of a test compound in liver microsomes and hepatocytes. Mass spectral data acquired from the analysis is crucial in structural elucidation of metabolites, while UV responses provide quantitative information on the disappearance of the parent drug and metabolite formation. The utility of LC/UV/MS in determination of metabolic soft spots and reactive metabolite formation is illustrated in Figure 11.4 (page 361). The UV chromatogram (Fig. 11.4*a*) displays the formation of two major oxidative metabolites and one glutathione adduct in the incubation of clozapine (10 μM) with human liver microsomes in the presence of GSH, a reactive metabolite trapping agent. The LC/MS data acquired with an online high-resolution LTQ-Orbitrap mass spectrometer includes total ion chromatograms of accurate full-scan MS data (Fig. 11.4*b*) and MS/MS data (Fig. 11.4*d*). Post-acquisition processing of these LC/MS data files with mass detect filters and neutral loss filtering are able to selectively detect these oxidative metabolites and GSH adducts that showed the same LC retention times and peak shapes as those displayed in UV chromatogram (Fig. 11.4*a*). The same approach can be applied to in vivo metabolite profiling when metabolite concentrations are relatively higher. However, UV detection does not work well with in vitro incubation samples with lower substrate concentrations (<1 μM), or for metabolites in plasma, urine, or feces where many endogenous components and metabolite concentrations are low. In these cases, estimation of metabolite quantities can be carried out based on mass spectrometric responses related to those of metabolite standards that are either chemically synthesized or generated in incubations in the presence of high concentrations of metabolizing enzymes and substrates.

11.6.2 Radioactivity Detection

Liquid chromatography is routinely used for the quantitative analysis of a drug and its metabolites in radiolabeled in vitro metabolism experiments and in humans and animal ADME studies. Online radio-flow detection (RFD) and offline microplate scintillation counting (MSC) are the most commonly used radiochromatographic techniques in radiolabeled metabolite profiling and quantification (Boernsen et al., 2000; Nassar et al., 2003; Bruin et al., 2006). LC/RFD is compatible with ESI mass spectrometry and provides high analytical speed and excellent separation resolution (Athersuch et al., 2008).

The flow rate of LC/RFD/MS is usually at 1 mL/min. After splitting, a small portion (≈20%) of the HPLC effluent is introduced into a mass spectrometer, and the rest of the effluent goes to the liquid detection cell of the RFD. This setup provides flow rates suitable for both mass spectrometry and RFD and a large sample loading capacity suitable for in vivo sample analysis. The major limitation of LC/RFD is its poor radiodetection sensitivity (Zhu et al., 2005b). The radiochromatographic technology is not suitable for analysis of low-level radioactivity such as plasma metabolites from radiolabeled ADME studies. To improve the sensitivity of LC/RFD, a few new RFD technologies have developed and been evaluated, such as stop-flow RFD and dynamic flow RFD. With relatively short counting times, the stop-flow LC/RFD is approximately 10-fold more sensitive than traditional LC/RCD (Nassar et al., 2003; Zhao et al., 2008). Dynamic-flow RFD is a more advanced online radiochromatographic technique and operates at a consistent HPLC flow rate with variable ratios of liquid scintillation cocktail to HPLC flow. Dynamic-flow LC/RFD dose not change or stop the HPLC flow, but it is comparable to the stopped-flow technology with respect to sensitivity and is easy to couple with a mass spectrometer. If even greater sensitivity is needed, increased counting times often can provide another 3-fold improvement in sensitivity in the stop flow experiment.

In contrast to LC/RFD, radioactivity profiling by offline LC/MSC provides radiodetection sensitivity 50- to 100-fold better than LC/RFD (Zhu et al., 2005b). In the LC/MSC analysis, HPLC effluent is collected into 96-well microplates and then evaporated to dryness followed by radioactivity counting of microplates with multiple detectors. Figure 11.5 (page 364) demonstrates the utility of LC/MS coupled with offline MSC (Zhu et al., 2009). The radiochromatogram of monkey plasma metabolites displayed very minor metabolite 15 (20 cpm; Fig. 11.5*a*). At the same time LC/MS profile (Fig. 11.5*b*), which was obtained after mass defect filtering of accurate mass spectral data, showed the metabolites corresponding to this radioactivity. Thus, the analysis was able to provide both quantitative and structural information of metabolites in plasma. The major limitation of LC/MSC is that it is a slow speed process. It requires multiple manual operation processes such as fraction collection and a long time to count radioactivity. Recently, a novel fraction collector was introduced to increase the productivity of metabolite profiling by LC/MSC. It can handle up to 20 microplates and allow automated performance of multiple injections.

11.6.3 Nuclear Magnetic Resoance

Although LC/MS/MS is the most widely used analytical technique for drug metabolism studies, this analytical platform may not provide complete molecular structure information in certain circumstances. As a complementary analytical tool, offline or online nulear magnetic resonance (NMR) is usually able to elucidate the precise structures of drug metabolites (Martinez-Granados et al., 2006; Lenz et al., 2007; Trivedi et al., 2009). NMR techniques are much less

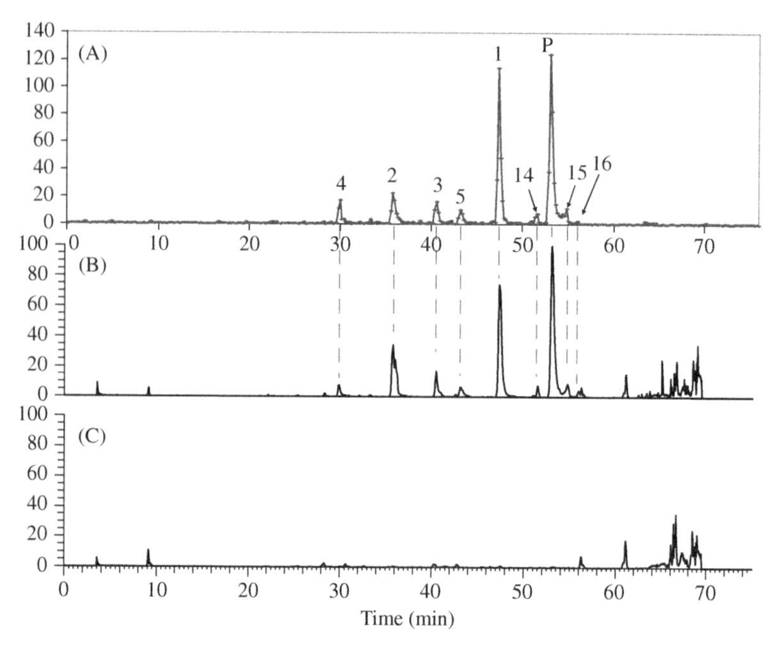

FIGURE 11.5 Comparison of plasma metabolite profiles of a drug between a radiochromatogram and the LC–MS profiles to illustrate the comprehensiveness of high-resolution mass spectrometry-based profiling techniques. The sample was obtained from monkey dosed with a radiolabeled proprietary compound. High-resolution LC–MS data were acquired using LTQ-Orbitrap and were processed with multiple mass defect filters. (*a*) Radiochromatogram determined by offline LC–MSC, (*b*) > MDF-processed LC–MS profile, and (*c*) MDF-processed LC–MS profile of a predose sample.

sensitive than mass spectrometry and require relatively pure analytes that are isolated from biological samples or separated online by liquid chromatography. Recently, advances in NMR technology, including high-field magnets (1 GHz), cryogenic probes, miniaturized probes, and software, allow NMR to be much more sensitive. Pure metabolite in the submicrogram range is adequate to obtain a high-quality ^{1}H spectrum in minutes. Furthermore, a 1-μg sample summed for a little longer can yield a ^{13}C spectrum using modest 400- to 500-MHz instruments. The ^{13}C spectrum allows for the determination of the number of carbons and can be used in combination with the accurate mass MS to conclusively determine the elemental composition. The combination of the elemental composition with the ^{1}H- and ^{13}C-NMR spectra can then be used to obtain the three-dimensional (3D) structure in a very high percentage of cases.

As a result, NMR is increasingly used in metabolic studies (Coen et al., 2008, Sandvoss et al., 2005 and Ceccarelli et al., 2008). In addition, NMR can be used for quantitative analysis to obtain quantitative excretive balance data and avoid costly radiolabelling studies (Zhu et al., 2009 and Wilson et al., 1999). This approach is particularly useful for the analysis of drugs/metabolites that contain ^{19}F. Coupled with other techniques (Lenz et al., 2007) such as SPE, HPLC, and MS, methods based on high-throughput NMR have been established for metabolic study, fingerprinting and profiling. The HPLC–SPE–NMR technique is also commercially available (Sandvoss et al., 2005). The frequently used nuclei for metabolic study using NMR are ^{1}H (most useful), ^{13}C, ^{19}F, ^{31}P, and ^{15}N. Some metabolites present in cleaner biofluids such as cerebrospinal fluid (CSF), and urine can be analyzed directly with NMR without sample cleanup with ^{19}F-NMR (Spraul et al., 1993; Ellis et al., 1995; Orhan et al., 2004). The use of magic angle spinning (MAS) NMR allows metabolic information to be acquired from intact tissues (Mao et al., 2007). NMR-based metabolic studies have also been reported in the related fields of live imaging (Henry et al., 2006) and disease biomarkers (Bollard et al., 2008).

Often, the main issue with NMR for metabolite identification (primary use of NMR in metabolism) is the ability to successfully obtain or collect 1 μg or more of a pure sample. The most common approaches for sample collection involves scale-up of a microsomal incubation or to dose a rat and collect urine. However, at times, it may be necessary to extract tissues. In any of these cases, given the dilute nature of the samples, a 1-μg metabolite single injection for online LC–NMR tends to be unlikely to succeed (1-μg metabolite likely to require >1-mL sample injection). High-resolution analytical-scale LC with fraction collection is one way to collect 1 μg, but this approach requires 50–100 injections. Interestingly, the same LC/MS purification platforms used for medicinal chemistry compounds are particularly well suited to this task. This is because the metabolite is most frequently originally known based on RP LC/MS detection and these analytical methods are readily scaled up. The result is that sufficient pure metabolite can be readily obtained in 5 or fewer injections and often in a single injection. With some systems/applications, tens or even hundreds of milligrams can be purified in a single injection. This also can be a convenient first source of standard. As a consequence, laboratories that often rely on NMR as a tool for metabolite identification are starting to gravitate toward the productivity offered by the use of such preparative LC/MS systems (Cai et al., 2007).

REFERENCES

Agranat I, Caner H, Caldwell J. Putting chirality to work: The strategy of chiral switches. Nat Rev Drug Discov 2002;1:753–768

Antia FD, Horvath C. High performance liquid chromatography at elevated temperatures: examination of conditions for the rapid separations of large molecules. J Chromatogr 1998;435:1–15

Ardrey RE. Liquid chromatography mass spectrometry: An introduction. 2003;Wiley: New York

Athersuch TJ, Sison RL, Kenyon AS, Clarkson-Jones JA, Wilson ID. Evaluation of the use of UPLC-TOFMS with simultaneous [14C]-radioflow detection for drug metabolite profiling: Application to propranolol metabolites in rat urine. J Pharm Biomed Anal 2008;48:151–157.

Bamba T. Application of supercritical fluid chromatography to the analysis of hydrophobic metabolites. J Sep Sci 2008;31:1274–1278

Baumann P, Zullino DF, Eap CB. Enantiomers' potential in psychopharmacology—a critical analysis with special emphasis on the antidepressant escitalopram. Europ Neuro-psychopharmcol 2002;12:433–444

Bezchlibnyk-Butler K, Aleksic L, Kennedy SH. Citalopram—a review of pharmacological and clinical effects. J Psychiatry Neurosci 2000;25:241–253

Biddlecombe RA, Pleasance S. Automated protein precipitation by filtration in the 96-well format. J Chromatogr B Biomed Sci Appl 1999;734:257–265.

Boernsen KO, Floeckher JM, Bruin GJ. Use of a microplate scintillation counter as a radioactivity detector for miniaturized separation techniques in drug metabolism. Anal Chem 2000;72:3956–3959.

Bollard ME, Contel NR, Ebbels TMD, Smith L, Beckonert O, Cantor GH, Mckeeman LL, Holmes EC, Lindon JC, Nicholson JK, Keun HC. NMR-Based Metabolic Profiling Identifies Biomarkers of Liver Regeneration Following Partial Hepatectomy in the Rat. J proteome Res 2009;8:59–69

Brock DR. Drug disposition in three dimensions: an update on Stereoselectivity in pharmacokinetics. Biopharm Drug Dispos 2006:27:387–406

Bruin GJ, Waldmeier F, Boernsen KO, Pfaar U, Gross G, Zollinger M. A microplate solid scintillation counter as a radioactivity detector for high performance liquid chromatography in drug metabolism: Validation and applications. J Chromatogr A 2006;1133:184–194.

Cai P, Tsao R, and Ruppen ME. In Vitro Metabolic Study of Temsirolimus: Preparation, Isolation, and Identification of the Metabolites. Drug Metab and Dispos 2007;35:1554–1563

Caner H, Groner E, Levy L, Agranat I. Trends in the development of chiral drugs. Drug Discov Today 2004;9(3):105–110

Ceccarelli SM, Schlotterbeck G, Boissin P, Binder M, Buettelmann B, Hanlon S, Jaeschke G, Kolczewski S, Kupfer E, Peters J, Porter RHP, Prinssen EP, Rueher M, Ruf I, Spooren W, Stämpfli A, Vieira E. Metabolite Identification via LC-SPE-NMR-MS of the In vitro Biooxidation Products of a Lead mGlu5 Allosteric Antagonist and Impact on the Improvement of Metabolic Stability in the Series. ChemMedChem 2008;3:136–144

Coen M, Holmes E, Lindon JC. Nicholson JK. NMR-based metabolic profiling and metabonomic approaches to problems in molecular toxicology. Chem Res Toxicol 2008;21:9–27

Combs MT, Ashraf-Khorassani M, Taylor LT. Packed Column Supercritical Fluid Chromatography-Mass Spectroscopy: A Review. J Chromatragr A 1997;785:85

Cuyckens F, Koppen V, Kembuegler R, Leclercq L. Improved liquid chromatography—Online radioactivity detection for metabolite profiling. J Chromatogr A 2008;1209:128–135.

D'Souza RA, Partridge EA, Roberts DW, Ashton S, Ryan A, Patterson AB, Wilson Z, Thurrell CC. Distribution of radioactivity and metabolite profiling in tumour and plasma following intravenous administration of a colchicine derivative (14C-ZD6126) to tumour-bearing mice. Xenobiotica 2007;37:328–340.

Du L, Musson DG, Wang AQ. High turbulence liquid chromatography online extraction and tandem mass spectrometry for the simultaneous determination of suberoylanilide hydroxamic acid and its two metabolites in human serum. Rapid Commun Mass Spectrom 2005;19:1779–1787.

Ellis MK, Naylor JL, Green T, Collins MA. Identification and quantification of fluorine-containing metabolites of 1–chloro-2,2,2-trifluoroethane (HCFC133A) in the rat by 19F-NMR spectroscopy. Drug Metab Dispos 1995;23:102–106.

Eriksson T, Bjorkman S, Roth B, Fyge A, Hoglund P. Stereospecific determination, chiral inversion in vitro and pharmacokinetics in humans of the enantiomers of thalidomide. Chirality 1995;7:44–52.

Gritti F, Felinger A, Guiochon G. Influence of the errors made in the measurement of the extra-column volume on the accuracies of estimates of the column efficiency and the mass transfer kinetics. J Chromatogr A. 2006;1136:57–72.

Gritti F, Guiochon G. General HETP equation for the study of mass-transfer mechanisms in RPLC. Anal Chem 2006;78:5329–5347.

Hendriks G, Uges DR, Franke JP. Reconsideration of sample pH adjustment in bioanalytical liquid-liquid extraction of ionisable compounds. J Chromatogr B Anal Technol Biomed Life Sci 2007;853:234–241.

Henry PG, Adriany G, Deelchand D, Gruetter R, Marjanska M, Öz G, Seaquist ER, Shestov A, Uğurbil K. In vivo 13C NMR spectroscopy and metabolic modeling in the brain: a practical perspective. MRI 2006;24(4):527–539

Herman JL. Generic method for on-line extraction of drug substances in the presence of biological matrices using turbulent flow chromatography. Rapid Commun Mass Spectrom 2002;16:421–426.

Hopfgartner G, Husser C, Zell M. High-throughput quantification of drugs and their metabolites in biosamples by LC-MS/MS and CE-MS/MS: Possibilities and limitations. Ther Drug Monit 2002;24:134–143.

Hosoda K, Furuta T, Yokokawa A, Ogura K, Hiratsuka A, Ishii K. Plasma profiling of intact isoflavone metabolites by high-performance liquid chromatography and mass spectrometric identification of flavone glycosides daidzin and genistin in human plasma after administration of kinako. Drug Metab Dispos 2008;36:1485–1495.

Hsieh Y, Korfmacher WA. Increasing speed and throughput when using HPLC-MS/MS systems for drug metabolism and pharmacokinetic screening. Curr Drug Metab 2006;7:479–489.

Hutt AJ, Caldwell J. The metabolic chiral inversion of 2-arylpropionic acids—a novel route with pharmacological consequences. J Pharm Pharmacol 1983:35: 693–704

Jemal M, Ouyang Z. The need for chromatographic and mass resolution in liquid chromatography/tandem mass spectrometric methods used for quantitation of lactones and corresponding hydroxy acids in biological samples. Rapid Commun Mass Spectrom 2000;14:1757–1765.

Jemal M, Xia YQ. LC-MS development strategies for quantitative bioanalysis. Curr Drug Metab 2006;7:491–502.

Knox JH, Jurand J. Application of high-speed liquid chromatography to the analysis of morphine, heroin, 6-(*O*-acetyl)morphine and methadone. J Chromatogr 1973;87: 95–108.

Knox JH, Saleem M. High speed liquid chromatography. J Chromatogr Sci 1969;7:745–752

Kugelberg FC, Carlsson B, Ahlner J, Bengtsson F, Stereoselective single-dose kinetics of citalopram and its metabolities in rats. Chirality 2003;15:622–629

Lenz EM, D'Souza RA, Jordan AC, King CD, Smith SM, Phillips PJ, McCormick AD, Roberts DW. HPLC–NMR with severe column overloading: Fast-track metabolite identification in urine and bile samples from rat and dog treated with [14C]-ZD6126. J Pharm Biomed Anal 2007;43:1065–1077.

Li F, Shieh Y. Supercritical fluid chromatography-mass spectrometry for chemical analysis. J Sep Sci 2008;31:1231–1237

Liu Y, Berthod A, Mitchell CR, Xiao TL, Zhang B, Armstrong DW. Super/subcritical fluid chromatography chiral separations with macrocyclic glycopeptide stationary phases. J Chromatogr A 2002;978:185–204.

Lu H. Stereoselectivity in drug metabolism. Expert Opin. Drug Metab Toxicol 2007; 3(2):149–158.

Ma J, Shi J, Le H, Cho R, Huang JC, Miao S, Wong BK. A fully automated plasma protein precipitation sample preparation method for LC-MS/MS bioanalysis. J Chromatogr B Anal Technol Biomed Life Sci 2008;862:219–226.

Mangelings D, Vander Heyden Y. Chiral separations in sub- and supercritical fluid chromatography. J Sep Sci 2008a;31:1252–1273.

Mangelings D, Vander Heyden Y. Screening approaches for chiral separations. Adv Chromatogr 2008b;46:175–211.

Mao H, Toufexia D, Wang X, Lacreuse A and Wu S. Changes of metabolite profile in kainic acid induced hippocampal injury in rats measured by HRMAS NMR Exp. Brain Res 2007;183(4):447–485

Martinez-Granados B, Monleon D, Martinez-Bisbal MC, Rodrigo JM, del Olmo J, Lluch P, Ferrandez A, Marti-Bonmati L, Celda B. Metabolite identification in human liver needle biopsies by high-resolution magic angle spinning 1H NMR spectroscopy. NMR Biomed 2006;19:90–100.

McDowall RD. Sample preparation for biomedical analysis. J Chromatogr 1989; 492:3–58.

McDowall RD, Doyle E, Murkitt GS, Picot VS. Sample preparation for the HPLC analysis of drugs in biological fluids. J Pharm Biomed Anal 1989;7: 1087–1096.

Monleon D, Morales JM, Barrasa A, Lopez JA, Vazquez C, Celda B. Metabolite profiling of fecal water extracts from human colorectal cancer. NMR Biomed 2009; 22:342–348.

Nassar AE, Bjorge SM, Lee DY. On-line liquid chromatography-accurate radioisotope counting coupled with a radioactivity detector and mass spectrometer for metabolite identification in drug discovery and development. Anal Chem 2003;75:785–790.

Nguyen DTT, Guillarme D, Rudaz S, Veuthey JL. Fast analysis in liquid chromatography using small particle size and high pressure. J Sep Sci 2006; 29:1836–1848.

Orhan H, Commandeur JN, Sahin G, Aypar U, Sahin A, Vermeulen NP. Use of 19F-nuclear magnetic resonance and gas chromatography-electron capture detection in the quantitative analysis of fluorine-containing metabolites in urine of sevoflurane-anaesthetized patients. Xenobiotica 2004;34:301–316.

Pedersen-Bjergaard S, Rasmussen KE. Liquid-liquid-liquid microextraction for sample preparation of biological fluids prior to capillary electrophoresis. Anal Chem 1999;71:2650–2656.

Pedersen-Bjergaard S, Rasmussen KE. Bioanalysis of drugs by liquid-phase microextraction coupled to separation techniques. J Chromatogr B Anal Technol Biomed Life Sci 2005;817:3–12.

Peng SX, Branch TM, King SL. Fully automated 96 well liquid–liquid extraction for analysis of biological samples by liquid chromatography with tandem mass spectrometry 2001;73:708–714.

Pinkston, JD. Advantages and drawbacks of popular supercritical fluid chromatography-mass interfacing approaches—a user's perspective. Eur J Mass Spectrom 2005;11:189.

Psillakis E, Kalogerakis N. Hollow-fibre liquid-phase microextraction of phthalate esters from water. J Chromatogr A 2003;999:145–153.

Sadoun F, Virlizer H, Arpino PJ. Packed-column supercritical fluid chromatography coupled with electrospray ionization mass spectrometry J Chromatogr A 3;647: 351–359

Sandvoss M, Bardsley B, Beck TL, Smith EL, North SE, Moore PJ, Edwards AJ, Smith RJ. HPLC–SPE–NMR in pharmaceutical development: Capabilities and applications. Magn Reson Chem 2005;43:762–770.

Saunders KC, Ghanem A, Boon Hon W, Hilder EF, Haddad PR. Separation and sample pre-treatment in bioanalysis using monolithic phases: A review. Anal Chim Acta 2009;652:22–31.

Scott RPW. Extra-column dispersion in liquid chromatography systems. J Liq Chromatogr Related Technol 2002;25(17):2567–2587

Sidhu J, Priskorn M, Poulsen M, Segonzac A, Grollier G, Larsen F. Steady-state pharmacokinetics of the enantiomers of citalopram and its metabolites in humans, Chirality 1997;9:686–692

Simpson H, Berthemy A, Buhrman D, Burton R, Newton J, Kealy M, Wells D, Wu D. High throughput liquid chromatography/mass spectrometry bioanalysis using 96-well disk solid phase extraction plate for the sample preparation. Rapid Commun Mass Spectrom 1998;12:75–82.

Synder LR, Kirkland JJ. Modern liquid chromatography. Wiley: New York; 1974; p411.

Spraul M, Hofmann M, Wilson ID, Lenz E, Nicholson JK, Lindon JC. Coupling of HPLC with 19F-and 1H-NMR spectroscopy to investigate the human urinary excretion of flurbiprofen metabolites. J Pharm Biomed Anal 1993;11: 1009–1015.

Steinborner S, Henion J. Liquid-liquid extraction in the 96-well plate format with SRM LC/MS quantitative determination of methotrexate and its major metabolite in human plasma. Anal Chem 1999;71:2340–2345.

Toll H, Oberacher H, Swart R, Huber CG. Separation, detection, and identification of peptides by ion-pair reversed-phase high-performance liquid chromatography—

Electrospray ionization mass spectrometry at high and low pH. J Chromatogr A 2005;1079:274–286.

Trivedi A, Kaushik P, Pandey A. Identification and metabolite profiling of *Sitophilus oryzae* L. by 1D and 2D NMR spectroscopy. Bull Entomol Res 2009;100:287–296.

Turnpenny P, Fraier D, Chassaing C, Duckworth J. Development of a micro-turbulent flow chromatography focus mode method for drug quantitation in discovery bioanalysis. J Chromatogr B Anal Technol Biomed Life Sci 2007;856:131–140.

Venn RF, Merson J, Cole S, Macrae P. 96-Well solid-phase extraction: A brief history of its development. J Chromatogr B Anal Technol Biomed Life Sci 2005;817:77–80.

Watt AP, Morrison D, Locker KL, Evans DC. Higher throughput bioanalysis by automation of a protein precipitation assay using a 96-well format with detection by LC-MS/MS. Anal Chem 2000;72:979–984.

White C, Burnett J. Integration of supercritical fluid chromatography into drug discovery as a routine support tool. II. investigation and evaluation of supercritical fluid chromatography for achiral batch purification. J Chromatogr A 2005;1074: 175–185.

Wille SM, Lambert WE. Recent developments in extraction procedures relevant to analytical toxicology. Anal Bioanal Chem 2007;388:1381–1391.

Wilson ID, Nicholson JK, Lindon JC. The role of nuclear magnetic resonance spectroscopy in drug metabolism. Handbook of drug metabolism, edited by Thomas L. Woolf. Marcel Dekker Inc., New York, Basel, 1999;523–550

Wirth MJ. Mass transport in sub-2-microm high-performance liquid chromatography. J Chromatogr A 2007;1148:128–130.

Wong MC, Lee WT, Wong JS, Frost G, Lodge J. An approach towards method development for untargeted urinary metabolite profiling in metabonomic research using UPLC/QToF MS. J Chromatogr B Anal Technol Biomed Life Sci 2008;871:341–348.

Wu ST, Schoener D, Jemal M. Plasma phospholipids implicated in the matrix effect observed in liquid chromatography tandem mass spectrometry bioanalysis: evaluation of the use of colloidal silica in combination with divalent or trivalent for selective removal of phospholipids from plasma. Rapid Comm. Mass Spectrom 2008;22(18): 2873–2881.

Wu TL, Tsai IC, Chang PY, Tsao KC, Sun CF, Wu LL, Wu JT. Establishment of an in-house ELISA and the reference range for serum amyloid A (SAA): Complementarity between SAA and C-reactive protein as markers of inflammation. Clin Chim Acta 2007;376:72–76.

Xin GZ, Zhou JL, Qi LW, Li CY, Liu P, Li HJ, Wen XD, Li P. Turbulent-flow chromatography coupled on-line to fast high-performance liquid chromatography and mass spectrometry for simultaneous determination of verticine, verticinone and isoverticine in rat plasma. J Chromatogr B Anal Technol Biomed Life Sci 2010;878:435–441.

Xue YJ, Liu J, Unger S. A 96-well single-pot liquid-liquid extraction, hydrophilic interaction liquid chromatography-mass spectrometry method for the determination of muraglitazar in human plasma. J Pharm Biomed Anal 2006;41:979–988.

Xue YJ, Pursley J, Arnold M. Liquid-liquid extraction of strongly protein bound BMS-299897 from human plasma and cerebrospinal fluid, followed by high-performance liquid chromatography/tandem mass spectrometry. J Pharm Biomed Anal 2007;43: 1728–1736.

Yao M, Zhu M, Sinz MW, Zhang H, Humphreys WG, Rodrigues AD, Dai R. Development and full validation of six inhibition assays for five major cytochrome P450 enzymes in human liver microsomes using an automated 96-well microplate incubation format and LC-MS/MS analysis. J Pharm Biomed Anal 2007;44:211–223.

Yan B, Znqo J, Brown JS, Blackwell J, Carr PW. High temperature ultra fast liquid chromatography. Anal Chem 2000;72:1253–1262

Yau WP, Vathsala A, Lou HX, Zhou SF, Chan E. Simple reversed-phase liquid chromatographic assay for simultaneous quantification of free mycophenolic acid and its glucuronide metabolite in human plasma. J Chromatogr B Anal Technol Biomed Life Sci 2007;846:313–318.

Zang X, Luo R, Song N, Chen TK, Bozigian H. A novel on-line solid-phase extraction approach integrated with a monolithic column and tandem mass spectrometry for direct plasma analysis of multiple drugs and metabolites. Rapid Commun Mass Spectrom 2005;19:3259–3268.

Zhang D, He K, Raghavan N, Wang L, Mitroka J, Maxwell BD, Knabb RM, Frost C, Schuster A, Hao F, Gu Z, Humphreys WG, Grossman SJ. Comparative metabolism of 14C-labeled apixaban in mice, rats, rabbits, dogs, and humans. Drug Metab Dispos 2009a;37:1738–1748.

Zhang N, Hoffman KL, Li W, Rossi DT. Semi-automated 96-well liquid-liquid extraction for quantitation of drugs in biological fluids. J Pharm Biomed Anal 2000;22:131–138.

Zhang Z, Zhu M, Tang W. Metabolite identification and profiling in drug design: Current practice and future directions. Curr Pharm Des 2009b;15:2220–2235.

Zhao W, Wang L, Zhang D, Zhu M. Rapid and sensitive characterization of the metabolite formation enzyme kinetics of radiolabeled drugs using stop-flow liquid radiochromatography. Drug Metab Lett 2008;2:41–46.

Zhu M, Zhang D, Zhang H, Shyu WC. Integrated strategies for assessment of metabolite exposure in humans during drug development: analytical challenges and clinical development considerations. Biopharm Drug Dispos 2009;30:163–184.

Zhu M, Zhao W, Jimenez H, Zhang D, Yeola S, Dai R, Vachharajani N, Mitroka J. Cytochrome P450 3A-mediated metabolism of buspirone in human liver microsomes. Drug Metab Dispos 2005a;33:500–507.

Zhu M, Zhao W, Vazquez N, Mitroka JG. Analysis of low level radioactive metabolites in biological fluids using high-performance liquid chromatography with microplate scintillation counting: Method validation and application. J Pharm Biomed Anal 2005b;39:233–245.

PART III
Applications of New LC–MS Techniques in Drug Metabolism and Disposition

12 Quantitative In Vitro ADME Assays Using LC–MS as a Part of Early Drug Metabolism Screening

WALTER KORFMACHER
Exploratory Drug Metabolism, Merck Research Laboratories, Kenilworth, New Jersey 07033

12.1 INTRODUCTION

Over the last two decades, there has been a major shift in how mass spectrometry (MS) is used for drug metabolism applications. Previously, MS was confined to the narrow role of metabolite identification for compounds in the development cycle. At this time, most quantitative assays were performed by using high-performance liquid chromatography (HPLC) combined with an ultraviolet (UV) detection system. Currently, HPLC–MS is the primary tool for both quantitative and qualitative drug metabolism applications. This shift occurred primarily for two reasons: (1) the commercialization of the electrospray ionization (ESI) source and the ability to perform HPLC–ESI–MS provided a whole new set of analytical capabilities for both quantitative and qualitative applications; and (2) at the same time, drug metabolism

Mass Spectrometry in Drug Metabolism and Disposition: Basic Principles and Applications, First Edition. Edited by Mike S. Lee and Mingshe Zhu.

departments were required to provide data on various absorption, distribution, metabolism, and excretion (ADME) properties for new chemical entities (NCEs) prior to the development phase.

Once the MS instrument vendors understood that this interest in HPLC–ESI–MS would lead to new customers, they responded with improvements in both hardware and software capabilities. These enhancements led to a series of new MS systems. Many of these systems were found to be very useful by scientists who worked in the early drug metabolism (EDM) phase of new drug discovery. Fig. 12.1 illustrates various types of mass spectrometers that have been found to be useful for various ADME assays. Table 12.1 shows a listing of the most common ADME assays that are used as part of new drug discovery screening. The iterative nature of the new drug discovery paradigm is shown in Fig. 12.2. Many of these steps include assays that are performed in the EDM phase (as part of lead optimization). There have been multiple review articles and book chapters in the last few years that cover the general topics of the utility of HPLC–MS for EDM assays (Korfmacher et al., 1997; Ackermann et al., 2002, 2008; Hopfgartner et al., 2002, 2004; Hopfgartner and Bourgogne, 2003; Korfmacher, 2003, 2005a, 2005b, 2005c, 2008, 2009; Hopfgartner and Zell, 2005; Kassel, 2005; Chu and Nomeir, 2006; Hsieh et al., 2006b; Hsieh and Korfmacher, 2006, 2009; Jemal and Xia, 2006; Hsieh, 2008; Kerns and Di, 2008). This chapter will focus primarily on how HPLC–MS can be used to perform metabolic stability, drug permeability and absorption, and cytochrome P450 (CYP) inhibition studies. The chapter will conclude with a

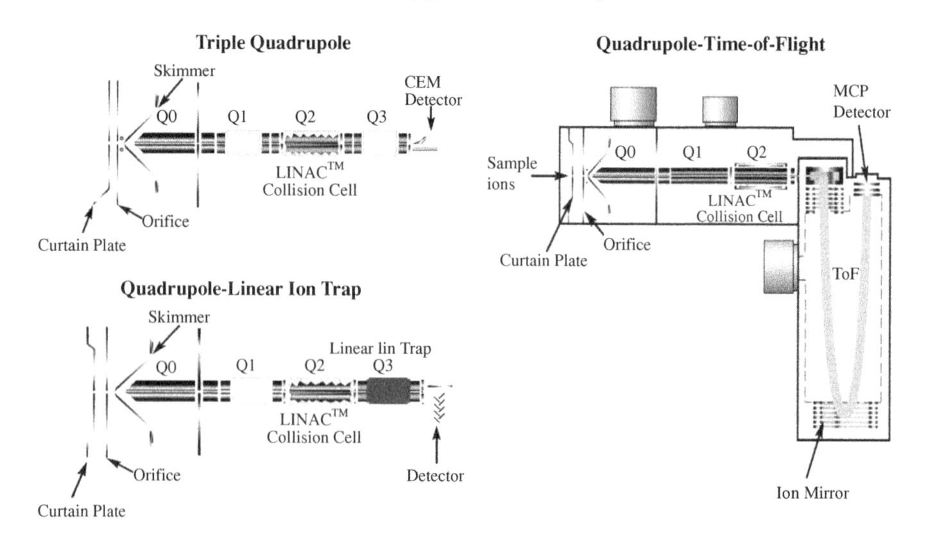

FIGURE 12.1 Various types of mass spectrometers used for EDM studies. (Courtesy of ABI-Sciex.).

TABLE 12.1 ADME Screening Assays Used for New Drug Discovery

Assay Name[a]	Assay Type	MS Method Development Needed for Test Compounds?	Assay Purpose	References
Caco-2 cell	In vitro	Yes	Measures drug permeability	Fung et al., 2003; Briem et al., 2005; Smalley et al., 2006
PAMPA	In Vitro	Yes	Measures drug permeability	Avdeef, 2005; Avdeef and Tsinman; 2006; Avdeef et al., 2007; Chen et al., 2008
CYP P450	In vitro	No	Measures CYP inhibition potential	Chu et al., 2000; Peng et al., 2003; Zientek et al., 2008
Metabolic stability	In vitro	Yes	Used to predict in vivo Cl	Korfmacher et al., 1999; Drexler et al., 2007; Di et al., 2008
Rodent oral PK sreen	In vivo	Yes	Used to assess oral bioavailability	Korfmacher et al., 2001; Liu et al., 2008

[a] Each of these assays are discussed in this chapter.

short discussion on some novel MS technology that might allow for higher throughput in the future.

12.2 METABOLIC STABILITY ASSAYS

One of the early in vitro assays that is used as part of the new drug discovery paradigm is the metabolic stability assay. This assay is also referred to as the microsomal stability assay or the hepatocyte stability assay or sometimes simply the in vitro stability assay (Thompson, 2000, 2001; Xu et al., 2002; Jenkins et al., 2004; Baranczewski et al., 2006). As these names suggest, there are various ways to perform the "in-life" part of the assay; some departments prefer to screen with liver microsomes, while others find that hepatocytes provide more meaningful data (Lau et al., 2002). Regardless of the in-life part of the in vitro stability assay, the analytical part represents a significant challenge because the analyst must have compound-specific methods in order to properly analyze the samples.

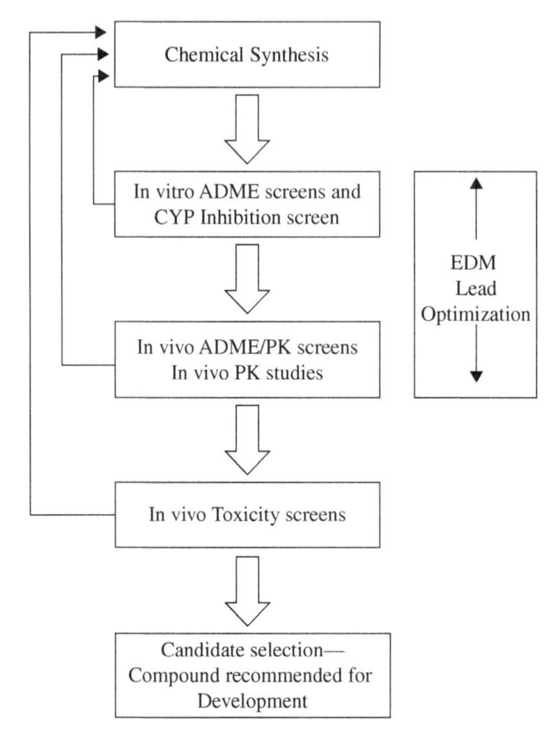

FIGURE 12.2 Iterative nature of the new drug discovery process. [Adapted from Korfmacher (2003) with permission from The Thomson Corporation.].

Di et al. (2008) discussed the use of the microsomal stability assay for drug discovery applications. The study involved the in vitro–in vivo correlation of 306 discovery compounds. A high-throughput microsomal stability assay was performed as well as a rat pharmacokinetic (PK) screen in which three rats were dosed intravenously with one of the test compounds. In this way, the in vitro half-life ($t_{1/2}$) was compared with the in vivo clearance (CL) for each of the compounds. Since microsomal stability measures only one part of the total CL that is observed in vivo, one should not expect a complete correlation. These studies indicated that microsomal stability measurements were very useful for the identification of compounds that had a high (in vivo) CL. Thus, compounds that were not very stable in microsomes were typically not good candidates for dosing in vivo. For compounds that were relatively stable in microsomes, about half showed high CL in vivo and half were low CL (so no correlation). These findings suggest that the proper use of microsomal stability data is to filter out compounds that are likely to have high CL in vivo, but the assay is not useful for predicting which compounds will have low CL in vivo.

One early example of a semiautomated assay for microsomal stability was described by Korfmacher et al. (1999). This procedure was performed with an LC–MS system based on a single quadrupole MS. There was no MS method development needed because the MS system was operated in the "full-scan" mode. The software tools were designed to find the expected protonated molecule ($[M+H]^+$) and use this information to get the retention time (t_R) of the compound as well as the mass spectrum for the compound in an automated fashion from the time zero sample. The software was used to assay the time zero sample and the 20-min sample for the expected compound. A peak area was obtained from the appropriate extracted ion chromatogram from each sample. The peak area ratio of these two samples provided a measure of the metabolic stability for the compound.

One recent example of a high-throughput assay for metabolic stability was reported by Fonsi et al. (2008). In this report, the authors described the use of a robotic platform to prepare the compounds for a microsomal stability assay. The authors selected five time points (20, 30, 45, 60, and 90 min) so the intrinsic clearance (CL_{int}) could be calculated with an Excel spreadsheet. The assay was performed with a triple quadrupole tandem mass spectrometer (MS/MS) system. The system included software tools for automated MS/MS method development—an important requirement for any MS/MS system that will be used for high-throughput assays for microsomal stability assays. The authors also made use of a cassette assay system where samples were pooled after the incubation step was completed. Up to four analytes were pooled and were assayed in one HPLC–MS/MS procedure with a generic chromatographic gradient system.

Another recent example of an automated liver microsomal stability assay was reported by Drexler et al. (2007). In this report, the authors described a two-step assay, which they named the "MetFast assay." The goal of this assay was to serve as the first-tier metabolic stability screen to flag compounds as highly metabolically unstable. The first step of their process is a structural integrity module based on LC–UV–MS where the goal is to test the compound for purity (by UV) as well as structural integrity by MS. In order to be incubated in the microsomal stability step, a compound had to have a purity of at least 50% and be confirmed by MS (in terms of expected molecular weight). This first step also provided the opportunity to perform MS/MS method development for the compound in preparation for the second step. The MS/MS system used for these assays was an ion trap system. The authors stated that the advantage of the ion trap MS system was that there was no need for collision energy optimization as there would be on a triple stage quadrupole MS/MS system that is commonly used for this assay. For the metabolic stability assay, the authors noted that a typical test set of 45 test compounds plus 3 control compounds resulted in 144 assays and 624 samples—based on 3 microsomal species (e.g., mouse, rat, and human) and 2 time points (zero and 10 min) in duplicate. This full set of samples could be assayed with a single HPLC–MS/MS system in less than 24 h. The time zero time point was used to compare to the 10 min sample and calculate the percent remaining at 10 min; the authors noted that an acceptable "% remaining" value was between 0 and 120%. The overall speed of the assay was impressive—45 compounds every

48 h (based on one HPLC–UV–MS/MS instrument), from compound receipt to data upload to the database.

McNaney et al. (2008) described a second-tier metabolic stability assay that was used to further characterize compounds and provide a metabolic stability half-life measurement. The main change for this assay versus the MetFast assay was the measurement of the substrate depletion of a compound for up to six time points in addition to the zero time point. These additional time points allowed for the calculation of the metabolic stability half-life for each compound.

Di et al. (2006) have described a high-throughput stability assay specifically for insoluble compounds. The authors noted that the typical high-throughput stability assay works well for commercial drugs and NCEs that have good druglike properties. However, many NCEs exhibit poor solubility and were found to provide incorrect metabolic stability results. For this reason, a method referred to as the "cosolvent method" was developed; in this method, compounds were diluted into working solutions that contained higher organic solvent content (than their typical method). This solution was added directly to the microsomes to provide an increased solubility. The method was found to also reduce nonspecific binding to various plastic surfaces. Overall, the procedure was stated to be very useful for early microsomal stability screening in a drug discovery arena.

Hsieh et al. (2006a) demonstrated the use of packed-column supercritical fluid chromatography (SFC) combined with atmospheric pressure chemical ionization (APCI)–tandem mass spectrometry (MS/MS) for the analysis of metabolic stability samples. For the SFC step, the mobile phase is liquid carbon dioxide with some organic modifiers added to adjust the retention time of the analytes. In this report, the primary organic modifier was methanol, and the authors described how the percentage of methanol in the mobile phase varied the retention of the analytes and affected the relative response of the test compounds in the APCI source. The authors also demonstrated that the results obtained by the SFC–APCI–MS/MS assay were equivalent to those obtained by a conventional HPLC–APCI–MS/MS assay, as shown in Figure 12.3.

In another study Hsieh et al. (2005) reported that SFC–APCI–MS/MS could be used for a chiral assay of metabolic stability samples. The test compounds for this study were pindolol and propranolol; these drugs are available as racemic mixtures. A Chiralcel OD-H (250 × 4.6 mm, 5 μm) column was used, and the percentage of methanol used as the mobile-phase modifier was adjusted to achieve baseline resolution of the two isomers with assay times of less than 3 min. The chromatographic separation obtained for these two sets of isomers with either 30 or 50% of the modifier in the mobile phase is shown in Fig. 12.4. It can also be seen in this figure that increasing the percentage of the modifier reduced the response for the compounds in the APCI source.

Jacobson et al. (2007) described an automated assay for the determination of metabolic stability based on rat or human hepatocytes. The automated system was based on a 96-well-plate system and was designed for 40 compounds in a

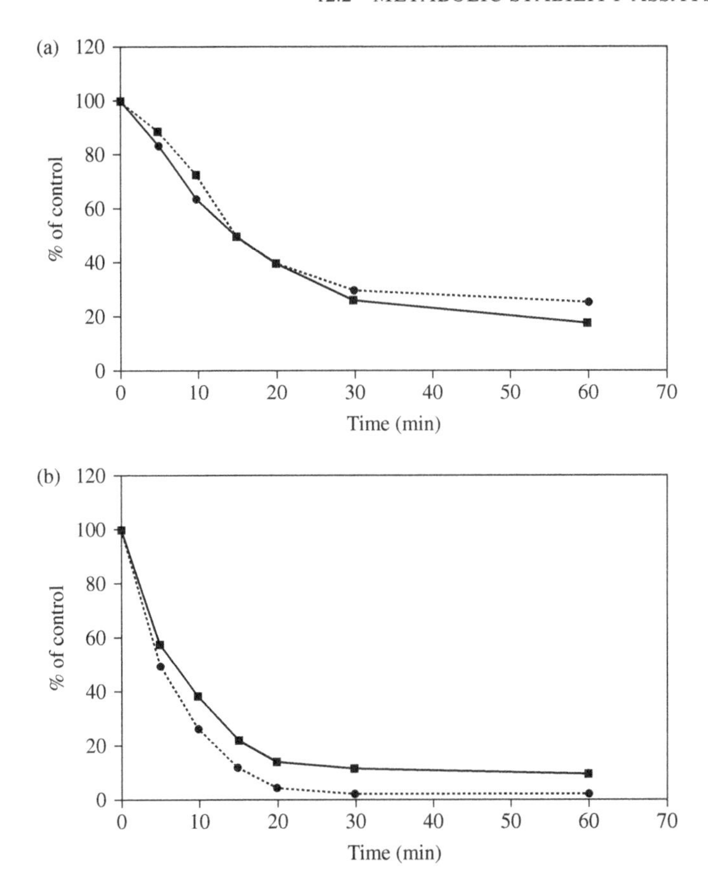

FIGURE 12.3 Equivalent results were obtained when rat microsomal samples were assayed by either SFC–APCI–MS/MS (dotted line) or HPLC–APCI–MS/MS (solid line) for two test compounds: (*a*) ondansetron and (*b*) clozapine. [Reprinted from Hsieh et al. (2006a) with permission.]

batch and a zero time sample plus 5 additional samples at time points out to 60 min. The assay was based on a robotic sample preparation and incubation system combined with HPLC–MS/MS assays for the samples. The assay provided a measure of the intrinsic clearance (CLint) for a molecule. This measurement could be used to further calculate a predicted in vivo CL in rats or humans. The authors tested the in vitro–in vivo correlation of the assay by testing a series of commercial drugs and comparing the predicted in vivo CL to the measured in vivo CL in either rats or humans. The authors found that the predicted in vivo CL was within twofold of the actual in vivo CL of 86 and 77% of the time for human and rat hepatocytes, respectively.

Another example of a metabolic stability assay using hepatocytes was provided by Reddy et al. (2005). This assay was based on a 96-well-plate format and made use of robotic sample preparation. The assay was designed to

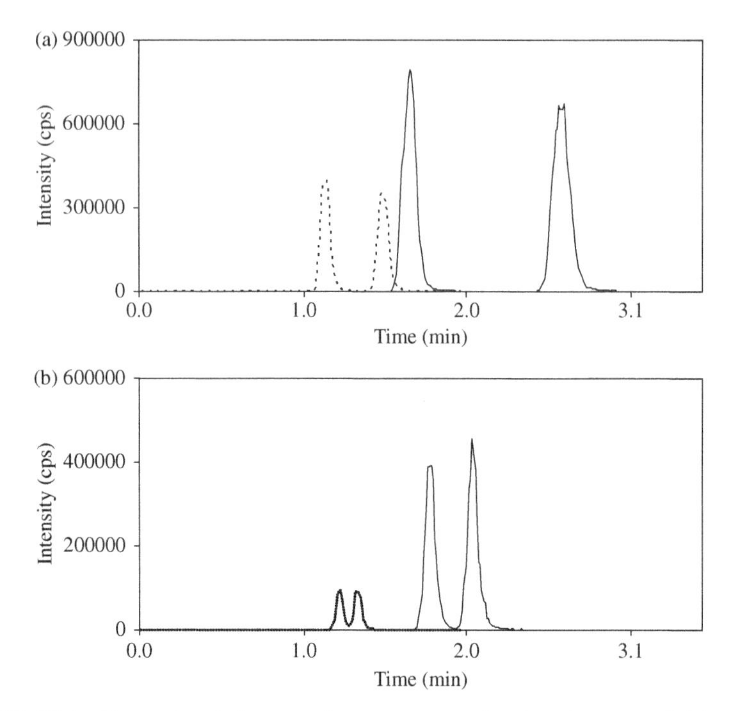

FIGURE 12.4 Reconstructed mass chromatograms for two isomers of (*a*) pindolol and (*b*) propranolol when assayed by SFC–APCI–MS/MS using either a 30% mobile-phase modifier (solid line) or a 50% mobile-phase modifier (dotted line). [Reprinted from Hsieh et al. (2005) with permission.]

handle batches of 48 discovery compounds and included multiple time points for each compound. The authors chose to use cryopreserved human hepatocytes (instead of human liver microsomes) to demonstrate both phase I and phase II metabolic activity (while microsomes are generally limited to phase I activity). The robotic sample preparation handled the incubation and sampling steps. The sample time points included the zero time points as well as 6 additional sample times out to 240 min. Each set of 48 test compounds resulted in approximately 600 samples. The HPLC–MS/MS system was based on a standard triple quadrupole MS system. The sample analysis times varied based on the conditions needed for various compounds, but ranged from 2.5 to 3.5 min for the test compounds.

12.3 DRUG ABSORPTION AND PERMEABILITY ASSAYS

The importance of oral bioavailability screens early in the lead optimization process has been discussed in multiple articles and book chapters (Bohets et al., 2001; Chaturvedi et al., 2001; Mandagere et al., 2002; Korfmacher, 2003, 2009;

Pintore et al., 2003; van de Waterbeemd and Jones, 2003; Crowley and Martini, 2004; Thompson, 2005; Avdeef et al., 2007; Cheng et al., 2007; Liu et al., 2008; Nomeir et al., 2009). Oral bioavailability can be viewed as a combination of two factors: absorption and first-pass metabolism. Indeed, Cheng et al. (Lau et al., 2004; Cheng et al., 2006; Li et al., 2007) described a novel high-throughput in vitro assay based on a hybrid system that involved a combination of an absorption screen as well as metabolism screen in one system; this system was used to evaluate multiple test compounds and has the potential to be used as a discovery oral bioavailability screening tool in the future.

A standard model for higher-throughput oral screening is shown in Fig. 12.5. As shown in Fig. 12.5, a common scenario for an oral bioavailability screen would be to combine the results of an in vitro absorption screen with the results on an in vitro metabolic stability screen to select compounds to go into an in vivo oral PK screen. The NCEs that still appeared to be promising lead compounds after these various screens could then be selected for "full PK" studies (oral and intravenous dosing in three to four animals for each dose route). In this way, only those compounds that were likely to exhibit good oral bioavailability would be tested in the full PK studies.

Two good examples of in vivo oral bioavailability screens are the cassette-accelerated rapid rat screen (CARRS) described by Korfmacher et al. (2001) and the "snapshot PK" screen described by Liu et al. (2008). Both of these rodent oral PK screens were found to be very helpful for filtering out compounds that had poor oral bioavailability. In a retrospective analysis,

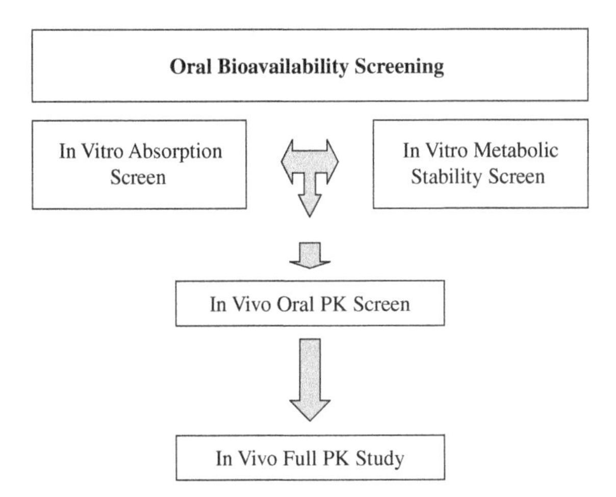

FIGURE 12.5 Schematic representation of a typical process for oral bioavailability screening in a drug discovery setting. The results of both an in vitro absorption screen as well as an in vitro metabolic stability screen are combined to select compounds for an in vivo oral PK screen. The compounds that survive these various screens can then be assayed in the standard in vivo full PK study.

Mei et al. (2006) demonstrated that the CARRS screen could be used to predict rat oral bioavailability and was therefore a useful in vivo discovery screen.

The most widely used in vitro absorption screen is the Caco-2 cell (obtained from human adenocarcinoma cells) assay. The Caco-2 cell assay provides a measure of permeability and has been demonstrated to have a reasonable correlation with in vivo human absorption (Artursson and Karlsson, 1991; Camenisch et al., 1998; Artursson et al., 2001; Korfmacher, 2003; Stoner et al., 2004; Thompson, 2005; Uchida et al., 2009). The assay for samples from Caco-2 cell studies is fairly simple—the analysis of two samples from the two sides (basolateral and apical) of the Caco-2 cell system. There are two challenges for this assay. The first challenge is the need to perform MS/MS method development on the test compound before it can be assayed. The second challenge is the large number of test compounds that may need to be screened for Caco-2 cell permeability. Thus, a typical HPLC–MS/MS system might be overwhelmed by the sample demands. Various techniques have been used to try to meet these challenges. In one example, Fung et al. (2003) described a higher-throughput Caco-2 screening method based on using the four-way multiplexed (MUX) electrospray interface, which was able to assay four different HPLC systems with one MS/MS system. Fung et al. (2003) also demonstrated that the Caco-2 screen assay could be performed without a standard curve; the concentration ratio of the basolateral and apical samples was calculated by simply using the ratio of the MS response ratios for the two samples—this approach greatly reduced the number of samples that were required to assay for each NCE. Together, these changes allowed the researchers to screen over 100 compounds per week for Caco-2 cell permeability. Briem et al. (2005) described another approach for using a four-channel HPLC system to assay Caco-2 samples. Their method made use of the staggered injection system, which allows one to select a small part of the chromatographic time for MS/MS analysis while the major portion of the HPLC effluent is diverted to waste. The utility of the staggered injection system is illustrated in Fig. 12.6. By making more efficient use of the MS/MS system, the staggered injection system can increase the throughput for the HPLC–MS/MS assay by threefold.

Smalley et al. (2006) published a method that increased the throughput of a Caco-2 cell screening assay by using an online turbulent flow extraction system as part of the HPLC–MS/MS configuration. In this method, the authors used calibration curves for the assay but reduced the sample load by mixing 10 compounds together for the Caco-2 screen as well as the calibration curves. Both Laitinen et al. (2003) and Koljonen et al. (2006) also described the utility of the "N-in-one," or cocktail, approach for screening NCEs in a Caco-2 cell permeability assay. Their reports demonstrated good correlation between the single and the cocktail assay results for the described test compounds.

Another approach commonly used to measure permeability is the parallel artificial membrane permeability assay (PAMPA) (Avdeef, 2005; Avdeef and Tsinman, 2006; Avdeef et al., 2007; Chen et al., 2008). The main advantage of PAMPA is that it is an artificial membrane, so it is easier to do the test. Once

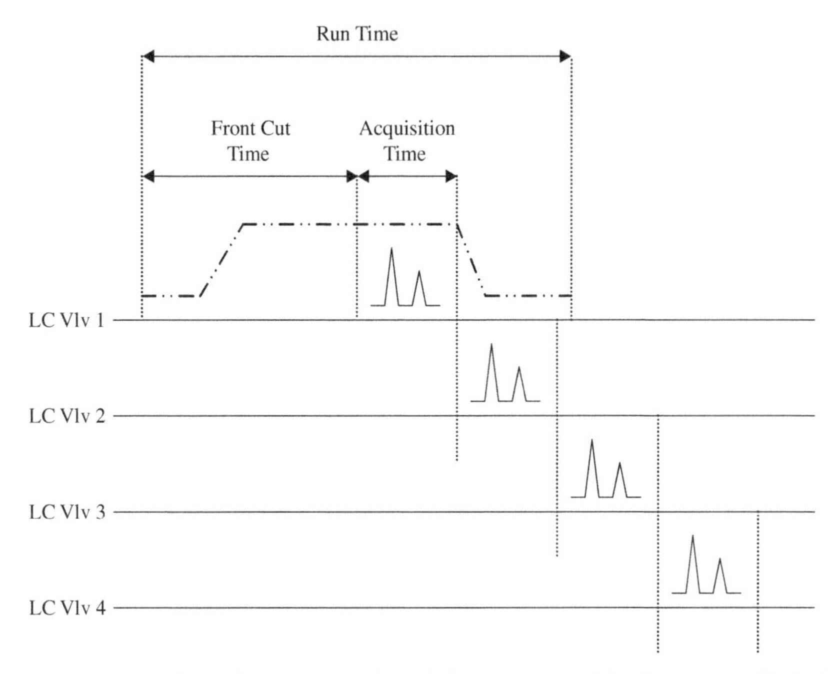

FIGURE 12.6 Schematic representation of the process used in the staggered injection technique. This process can provide a threefold increase in sample throughput by making more efficient use of the MS/MS system. [Reprinted with permission from Briem et al. (2005).].

the test samples are obtained, the analytical step is basically the same as for the Caco-2 assay. Kerns et al. (2004) compared the results of the Caco-2 cell assay with PAMPA for a series of compounds and made multiple observations and suggestions based on their results:

1. In many cases the results from the two assays were equivalent; for these compounds, both systems measured passive diffusion.
2. For compounds where PAMPA showed much lower permeability than the Caco-2 cell assay, it can be concluded that absorption transport is the primary permeability mechanism for these compounds.
3. For compounds where PAMPA showed much higher permeability than the Caco-2 cell assay, it can be concluded that secretory transport is the primary mechanism for these compounds.

In order to best use the capabilities of these two permeability assays, Kerns et al. (2004) suggested that one should use the PAMPA for early discovery screening and Caco-2 cell screening for compounds that are likely to be recommended for development. More recently, Masungi et al. (2008) have also suggested that it is useful to use both the PAMPA and the Caco-2 cell assays for screening NCEs in a new drug discovery setting.

12.4 CYTOCHROME P450 (CYP) ASSAYS

The cytochrome P450 (CYP) assays are very important for early drug discovery screening of new compounds. The CYP assays provide data that can be used to determine whether a compound has the potential to cause drug–drug interactions (DDIs) in a clinical setting (Li, 2001, 2004; Di and Kerns, 2003; Houston and Galetin, 2003; Kerns and Di, 2003, 2008; Hutzler et al., 2005; Zlokarnik et al., 2005; Venkatakrishnan and Obach, 2007; McGinnity et al., 2008; Youdim et al., 2008; Grime et al., 2009; Grimm et al., 2009). Typically, these screening assays are used to measure a test compound's IC_{50} (concentration that inhibits activity of the isozyme by 50%) values for various CYP isozymes and are based on incubations with either recombinant human CYP isozymes (rhCYP) or human liver microsomes (HLM). In many cases, CYP assays rely on HPLC–MS/MS for the analytical step and most of the variations in the published assays are based on the number of CYP isozymes that are tested (Rodrigues and Lin, 2001).

Over the last 10 years, multiple reports of various CYP assays have been published. During this time, the assays have become more complex. One of the earlier publications to describe an HPLC–MS/MS-based assay was provided by Chu et al. (2000). This assay featured the use of HLM and measured the inhibition of two CYPs (CYP2D6 and CYP3A4). The analysis time was less than 1 min per sample. Peng et al. (2003) described a higher-throughput assay that tested NCEs for their inhibition potential for five major CYP isozymes (CYP1A2, CYP2C9, CYP2C19, CYP2D6, and CYP3A4). Their HPLC–MS/MS method used a monolithic HPLC column in order to reduce the assay time, which was less than 30 s per sample. Kim et al. (2005) described a high-throughput assay based on HPLC–MS/MS in which the inhibition potential of NCEs was measured for a total of nine CYP isozymes (CYP1A2, CYP 2A6, CYP2B6, CYP2C8, CYP2C9, CYP2C19, CYP2D6, CYP2E1, and CYP3A4).

The latest trend with CYP assays is to modify them for higher throughput by adapting them to work with 384-well-plate formats. One example of this trend is highlighted in the recent publication by Zientek et al. (2008). The authors noted that five major CYP isozymes account for 70% of all drug clearance in humans (Zientek et al., 2008); therefore, a high-throughput screen was designed to measure the inhibition potential of NCEs for these five CYPS (CYP1A2, CYP2C9, CYP2C19, CYP2D6, and CYP3A4). The HPLC–MS/MS assay for the analysis had been described previously by Smith et al. (2007). The authors studied various test parameters in order to optimize the assay. For example, dimethyl sulfoxide (DMSO) which is commonly used for stock solutions of standards, had a significant effect on some of the isozyme assays at levels of 0.2–0.5% in the incubation system. Based on these findings, the final assay was designed so that the amount of DMSO in the final "cocktail" (substrate mixture) was not greater than 0.1%. All of the experiments to test various parameters of the assay were completed with a standard 96-well-plate-based robotic system. The authors also converted the assay to one based on 384 well plates and demonstrated that a library of 9494 compounds could be tested using this higher-throughput assay.

Youdim et al. (2008) reported on a method to measure CYP inhibition potential of NCEs that was designed to be an automated high-throughput assay based on using 384 well plates. The assay made use of HLM for the incubation step, which included a cocktail of probe substrates designed to measure inhibition of the five major CYP isozymes (CYP1A2, CYP2C9, CYP2C19, CYP2D6, and CYP3A4). The authors performed a series of tests to confirm that the same results could be obtained by either the cocktail probe assay method or the individual probe assay method. The advantage of the cocktail assay was that a single incubation mixture (NCE + probe substrates) could be used to assess the inhibition potential of the NCE. The challenge was to ensure that the HPLC–MS/MS assay could measure the multiple probes in less than 1 min. The HPLC–MS/MS step used a fast gradient with a cycle time of less than 1 min. The HPLC column (Phenomenex Synergi Fusion column, 20 mm × 2.0 mm, 2.5 μm particle size) was operated at a column temperature of 45°C and a flow rate of 1.0 mL/min. The authors also added stable-label internal standards (when available) to help ensure that the HPLC–MS/MS assay was accurate.

There continues to be some discussion on how to use the CYP inhibition potential data (Yan and Caldwell, 2001; Bjornsson et al., 2003a, 2003b; Korfmacher, 2003). A typical use for these data is described by Yan and Caldwell (2001) who classify a compound's potential for inhibition based on the measured IC_{50} value; a strong inhibitor has an $IC_{50} < 1\ \mu M$, while a moderate inhibitor has an IC_{50} from 1 to 10 μM, and a weak inhibitor has an $IC_{50} > 10\ \mu M$. Based on this classification scheme, Yan and Caldwell (2001) described a multistep approach [see Figure 5 in Yan and Caldwell (2001)] for utilizing CYP inhibition data.

12.5 NEW TECHNOLOGY FOR HIGH-THROUGHPUT ASSAYS

Various new assay technologies have become available in the last several years. While some of these technologies can be viewed as improvements on the HPLC–MS/MS platform, other new technologies are MS–based but do not have a chromatographic step. It is important to be aware of these new techniques because in some cases, they may be a significant improvement over our current technology.

One new technology that has been found to be very useful for various HPLC–MS/MS applications is ultra-high-performance liquid chromatography (UPLC) (Yu et al., 2006, 2007). UPLC takes advantage of the improvement in chromatographic speed and resolution that can be obtained by using sub-2 μm particles for the HPLC column (Guillarme et al., 2007; Nguyen et al., 2007). Recently, Rainville et al. (2008) demonstrated the utility of UPLC–MS/MS as the analytical method for the rapid analysis of six CYP probe substrates. The rapid assay for the six compounds was completed in 30 sec. In another report by Plumb et al. (2008), the use of UPLC–MS/MS was demonstrated for a series of ADME screening assays. In one example, the authors described how UPLC–MS/MS could be used for metabolic stability assays. In this example, the UPLC system provided very sharp chromatographic peaks (peak widths of 2–3 s) and the assay time was less than 1.5 min per sample.

Another new technology that has the potential to be useful for various EDM assays is direct analysis in real time (DART) MS (Cody et al., 2005; Cody, 2009). DART is an "open-air ionization" technique that can be adapted to standard atmospheric pressure ionization (API) sources as part of a triple quadrupole MS system. The main advantages of DART are no sample preparation and fast assays. Yu et al. (2009) have recently published a report showing that DART–MS/MS could be applied to metabolic stability assay samples. The main limitation of DART is that because there is no chromatographic step, there is the potential for interferences from metabolites that are labile in this source (e.g., glucuronide metabolites of the test compound); this issue could lead to a false-positive result in some cases. In spite of this issue, DART may become a useful technique for some special EDM assays.

Another technique that shows potential for use in the EDM assay arena is the nanoelectrospray technology. Nanoelectrospray has been available for many years but has recently become available in a convenient chip-based system that can be used for both quantitative and qualitative applications (Dethy et al., 2003; Ackermann and Dethy, 2005; Wickremsinhe et al., 2005, 2006). The main advantage of this technique is that it is fairly easy to implement (it can be interfaced directly to many triple quadrupole MS systems) and it provides a fast assay time (about 20–30 s/sample). Van Pelt et al. (2003) described the use of the nanoelectrospray system for the analysis of Caco-2 samples. More recently, Balimane et al. (2005) demonstrated that a chip-based nanoelectrospray system could be used to assay PAMPA samples in a high-throughput manner. In this report, the samples for 10 test compounds were assayed by HPLC–UV or nanoelectrospray and the permeability results from the two assays were found to be similar for this test set.

Another new technology that shows promise as a high-throughput screening tool is FlashQuant. FlashQuant is a triple quadrupole MS/MS system that has a high-speed matrix-assisted laser desorption/ionization (MALDI) source. The MALDI source can provide for the very rapid assay of samples on a plate format. Because there is no chromatography, the sample assay speed is about 6 s/sample (10 samples/min)—this is 10–20 times a typical HPLC–MS/MS assay speed. The technique has potential for use in early in vitro assays that have simple matrices. At this time, it is too early to state whether or not this technique will be useful for EDM screens.

12.6 CONCLUSIONS

Advances in both chromatography and mass spectrometry have allowed HPLC–MS/MS to remain the premier analytical tool for the analysis of samples from various in vitro and in vivo EDM ADME screens as part of the new drug discovery process. In spite of these advances, there is continuing pressure to develop new ways to assay samples faster. Novel technologies now being tested by various groups may well lead the way forward to a time when sample assay speed is not the rate-limiting step for scientists who work in this demanding environment.

REFERENCES

Ackermann B, Berna M, Eckstein JA, Ott LW, Chaudhary AG. Current applications of liquid chromatography/mass spectrometry in pharmaceutical discovery after a decade of innovation. Annu Rev Anal Chem 2008;1:357–396.

Ackermann B, Dethy JM. Understanding the role and potential of infusion nanoelectrospray ionization for pharmaceutical analysis. In *Using Mass Spectrometry for Drug Metabolism Studies*, Korfmacher W, Ed. CRC Press, Boca Raton, FL, 2005, pp. 329–356.

Ackermann BL, Berna MJ, Murphy AT. Recent advances in use of LC/MS/MS for quantitative high-throughput bioanalytical support of drug discovery. Curr Top Med Chem 2002;2(1):53–66.

Artursson P, Karlsson J. Correlation between oral drug absorption in humans and apparent drug permeability coefficients in human intestinal epithelial (Caco-2) cells. Biochem Biophys Res Commun 1991;175(3):880–885.

Artursson P, Palm K, Luthman K. Caco-2 monolayers in experimental and theoretical predictions of drug transport. Adv Drug Deliv Rev 2001;46(1–3):27–43.

Avdeef A. The rise of PAMPA. Expert Opin Drug Metab Toxicol 2005;1(2):325–342.

Avdeef A, Bendels S, Di L, Faller B, Kansy M, Sugano K, Yamauchi Y. PAMPA—Critical factors for better predictions of absorption. J Pharm Sci 2007;96 (11):2893–2909.

Avdeef A, Tsinman O. PAMPA—A drug absorption in vitro model 13. Chemical selectivity due to membrane hydrogen bonding: In combo comparisons of HDM-, DOPC-, and DS-PAMPA models. Eur J Pharm Sci 2006;28(1–2):43–50.

Balimane PV, Pace E, Chong S, Zhu M, Jemal M, Pelt CK. A novel high-throughput automated chip-based nanoelectrospray tandem mass spectrometric method for PAMPA sample analysis. J Pharm Biomed Anal 2005;39(1–2):8–16.

Baranczewski P, Stanczak A, Sundberg K, Svensson R, Wallin A, Jansson J, Garberg P, Postlind H. Introduction to in vitro estimation of metabolic stability and drug interactions of new chemical entities in drug discovery and development. Pharmacol Rep 2006;58(4):453–472.

Bjornsson TD, Callaghan JT, Einolf HJ, Fischer V, Gan L, Grimm S, Kao J, King SP, Miwa G, Ni L, Kumar G, McLeod J, Obach RS, Roberts S, Roe A, Shah A, Snikeris F, Sullivan JT, Tweedie D, Vega JM, Walsh J, Wrighton SA. The conduct of in vitro and in vivo drug-drug interaction studies: A Pharmaceutical Research and Manufacturers of America (PhRMA) perspective. Drug Metab Dispos 2003a;31(7):815–832.

Bjornsson TD, Callaghan JT, Einolf HJ, Fischer V, Gan L, Grimm S, Kao J, King SP, Miwa G, Ni L, Kumar G, McLeod J, Obach SR, Roberts S, Roe A, Shah A, Snikeris F, Sullivan JT, Tweedie D, Vega JM, Walsh J, Wrighton SA. The conduct of in vitro and in vivo drug-drug interaction studies: A PhRMA perspective. J Clin Pharmacol 2003b;43(5):443–469.

Bohets H, Annaert P, Mannens G, Van Beijsterveldt L, Anciaux K, Verboven P, Meuldermans W, Lavrijsen K. Strategies for absorption screening in drug discovery and development. Curr Top Med Chem 2001;1(5):367–383.

Briem S, Pettersson B, Skoglund E. Description and validation of a four-channel staggered LC-MS/MS system for high-throughput in vitro screens. Anal Chem 2005;77(6):1905–1910.

Camenisch G, Alsenz J, van de Waterbeemd H, Folkers G. Estimation of permeability by passive diffusion through Caco-2 cell monolayers using the drugs' lipophilicity and molecular weight. Eur J Pharm Sci 1998;6(4):317–324.

Chaturvedi PR, Decker CJ, Odinecs A. Prediction of pharmacokinetic properties using experimental approaches during early drug discovery. Curr Opin Chem Biol 2001; 5(4):452–463.

Chen X, Murawski A, Patel K, Crespi CL, Balimane PV. A novel design of artificial membrane for improving the PAMPA model. Pharm Res 2008;25(7):1511–1520.

Cheng KC, Korfmacher W, White RE, Njoroge FG. Lead optimization in discovery drug metabolism and pharmacokinetics/case study: The hepatitis C virus (HCV) protease inhibitor SCH 503034. Perspect Med Chem 2007;1(1):1–9.

Cheng KC, Li C, Hsieh Y, Montgomery D, Liu T, White RE. Development of a high-throughput in vitro assay using a novel Caco-2/rat hepatocyte system for the prediction of oral plasma area under the concentration versus time curve (AUC) in rats. J Pharmacol Toxicol Methods 2006;53(3):215–218.

Chu I, Favreau L, Soares T, Lin C, Nomeir AA. Validation of higher-throughput high-performance liquid chromatography/atmospheric pressure chemical ionization tandem mass spectrometry assays to conduct cytochrome P450s CYP2D6 and CYP3A4 enzyme inhibition studies in human liver microsomes. Rapid Commun Mass Spectrom 2000;14(4):207–214.

Chu I, Nomeir AA. Utility of mass spectrometry for in-vitro ADME assays. Curr Drug Metab 2006;7(5):467–477.

Cody RB. Observation of molecular ions and analysis of nonpolar compounds with the direct analysis in real time ion source. Anal Chem 2009;81(3):1101–1107.

Cody RB, Laramee JA, Durst HD. Versatile new ion source for the analysis of materials in open air under ambient conditions. Anal Chem 2005;77(8):2297–2302.

Crowley PJ, Martini LG. Enhancing oral absorption in animals. Curr Opin Drug Discov Devel 2004;4(1):73–80.

Dethy JM, Ackermann BL, Delatour C, Henion JD, Schultz GA. Demonstration of direct bioanalysis of drugs in plasma using nanoelectrospray infusion from a silicon chip coupled with tandem mass spectrometry. Anal Chem 2003;75(4):805–811.

Di L, Kerns EH. Profiling drug-like properties in discovery research. Curr Opin Chem Biol 2003;7(3):402–408.

Di L, Kerns EH, Li SQ, Petusky SL. High throughput microsomal stability assay for insoluble compounds. Int J Pharm 2006;317(1):54–60.

Di L, Kerns EH, Ma XJ, Huang Y, Carter GT. Applications of high throughput microsomal stability assay in drug discovery. Comb Chem High Throughput Screen 2008;11(6):469–476.

Drexler DM, Belcastro JV, Dickinson KE, Edinger KJ, Hnatyshyn SY, Josephs JL, Langish RA, McNaney CA, Santone KS, Shipkova PA, Tymiak AA, Zvyaga TA, Sanders M. An automated high throughput liquid chromatography–mass spectrometry process to assess the metabolic stability of drug candidates. Assay Drug Dev Technol 2007;5(2):247–264.

Fonsi M, Orsale MV, Monteagudo E. High-throughput microsomal stability assay for screening new chemical entities in drug discovery. J Biomol Screen 2008;13 (9):862–869.

Fung EN, Chu I, Li C, Liu T, Soares A, Morrison R, Nomeir AA. Higher-throughput screening for Caco-2 permeability utilizing a multiple sprayer liquid chromatography/tandem mass spectrometry system. Rapid Commun Mass Spectrom 2003;17 (18):2147–2152.

Grime KH, Bird J, Ferguson D, Riley RJ. Mechanism-based inhibition of cytochrome P450 enzymes: An evaluation of early decision making in vitro approaches and drug-drug interaction prediction methods. Eur J Pharm Sci 2009;36(2–3):175–191.

Grimm SW, Einolf HJ, Hall SD, He K, Lim HK, Ling KH, Lu C, Nomeir AA, Seibert E, Skordos KW, Tonn GR, Van Horn R, Wang RW, Wong YN, Yang TJ, Obach RS. The conduct of in vitro studies to address time-dependent inhibition of drug metabolizing enzymes: A perspective of the Pharmaceutical Research and Manufacturers of America (PhRMA). Drug Metab Dispos 2009;37(7):1355–1370.

Guillarme D, Nguyen DT, Rudaz S, Veuthey JL. Recent developments in liquid chromatography—Impact on qualitative and quantitative performance. J Chromatogr A 2007;1149(1):20–29.

Hopfgartner G, Bourgogne E. Quantitative high-throughput analysis of drugs in biological matrices by mass spectrometry. Mass Spectrom Rev 2003;22 (3):195–214.

Hopfgartner G, Husser C, Zell M. High-throughput quantification of drugs and their metabolites in biosamples by LC-MS/MS and CE-MS/MS: Possibilities and limitations. Ther Drug Monit 2002;24(1):134–143.

Hopfgartner G, Varesio E, Tschappat V, Grivet C, Bourgogne E, Leuthold LA. Triple quadrupole linear ion trap mass spectrometer for the analysis of small molecules and macromolecules. J Mass Spectrom 2004;39(8):845–855.

Hopfgartner G, Zell M. Q Trap MS: A new tool for metabolite identification. In *Using Mass Spectrometry for Drug Metabolism Studies*, Korfmacher W, Ed. CRC Press, Boca Raton, Fl, 2005, pp. 277–304.

Houston JB, Galetin A. Progress towards prediction of human pharmacokinetic parameters from in vitro technologies. Drug Metab Rev 2003;35(4):393–415.

Hsieh Y. HPLC-MS/MS in drug metabolism and pharmacokinetic screening. Expert Opin Drug Metab Toxicol 2008;4(1):93–101.

Hsieh Y, Favreau L, Cheng KC, Chen J. Chiral supercritical fluid chromatography/ tandem mass spectrometry for the simultaneous determination of pindolol and propranolol in metabolic stability samples. Rapid Commun Mass Spectrom 2005;19 (21):3037–3041.

Hsieh Y, Favreau L, Schwerdt J, Cheng KC. Supercritical fluid chromatography/ tandem mass spectrometric method for analysis of pharmaceutical compounds in metabolic stability samples. J Pharm Biomed Anal 2006a;40(3):799–804.

Hsieh Y, Fukuda E, Wingate J, Korfmacher WA. Fast mass spectrometry-based methodologies for pharmaceutical analyses. Comb Chem High Throughput Screen 2006b;9(1):3–8.

Hsieh Y, Korfmacher W. The role of hyphenated chromatography–mass spectrometry techniques in exploratory drug metabolism and pharmacokinetics. Curr Pharm Des 2009;15(19):2251–2261.

Hsieh Y, Korfmacher WA. Increasing speed and throughput when using HPLC-MS/MS systems for drug metabolism and pharmacokinetic screening. Curr Drug Metab 2006;7(5):479–489.

Hutzler M, Messing DM, Wienkers LC. Predicting drug-drug interactions in drug discovery: Where are we now and where are we going? Curr Opin Drug Discov Devel 2005;8(1):51–58.

Jacobson L, Middleton B, Holmgren J, Eirefelt S, Frojd M, Blomgren A, Gustavsson L. An optimized automated assay for determination of metabolic stability using hepatocytes: Assay validation, variance component analysis, and in vivo relevance. Assay Drug Dev Technol 2007;5(3):403–415.

Jemal M, Xia YQ. LC-MS development strategies for quantitative bioanalysis. Curr Drug Metab 2006;7(5):491–1502.

Jenkins KM, Angeles R, Quintos MT, Xu R, Kassel DB, Rourick RA. Automated high throughput ADME assays for metabolic stability and cytochrome P450 inhibition profiling of combinatorial libraries. J Pharm Biomed Anal 2004;34(5):989–1004.

Kassel DB. High Throughput strategies for *in vitro* ADME assays: How fast can we go? In *Using Mass Spectrometry for Drug Metabolism Studies*, Korfmacher W, Ed. CRC Press, Boca Raton, FL, 2005, pp. 83–102.

Kerns E, Di L Drug-like Properties: Concepts, Structure Design and Methods. Academic, Burlington, MA, 2008.

Kerns EH, Di L. Pharmaceutical profiling in drug discovery. Drug Discov Today 2003;8 (7):316–23.

Kerns EH, Di L, Petusky S, Farris M, Ley R, Jupp P. Combined application of parallel artificial membrane permeability assay and Caco-2 permeability assays in drug discovery. J Pharm Sci 2004;93(6):1440–1453.

Kim MJ, Kim H, Cha IJ, Park JS, Shon JH, Liu KH, Shin JG. High-throughput screening of inhibitory potential of nine cytochrome P450 enzymes in vitro using liquid chromatography/tandem mass spectrometry. Rapid Commun Mass Spectrom 2005;19(18):2651–2658.

Koljonen M, Hakala KS, Ahtola-Satila T, Laitinen L, Kostiainen R, Kotiaho T, Kaukonen AM, Hirvonen J. Evaluation of cocktail approach to standardise Caco-2 permeability experiments. Eur J Pharm Biopharm 2006;64(3):379–387.

Korfmacher W. Bioanalytical assays in a drug discovery environment. In *Using Mass Spectrometry for Drug Metabolism Studies*, Korfmacher W, Ed. CRC Press, Boca Raton, FL, 2005a, pp. 1–34.

Korfmacher W Strategies for increasing throughput for PK samples in a drug discovery environment. In *Identification and Quantification of Drugs, Metabolites and Metabolizing Enzymes by LC-MS*, Chowdhury S, Ed. Elsevier, Amsterdam, The Netherlands, 2005b, pp. 7–34.

Korfmacher W Strategies and techniques for higher throughput ADME/PK assays. In *High-Throughput Analysis in the Pharmaceutical Industry*, Wang PG, Ed. CRC Press, Boca Raton, FL, 2008, pp.205–232.

Korfmacher WA. Lead optimization strategies as part of a drug metabolism environment. Curr Opin Drug Discov Devel 2003;6(4):481–485.

Korfmacher WA. Principles and applications of LC-MS in new drug discovery. Drug Discov Today 2005c;10(20):1357–1367.

Korfmacher WA. Advances in the integration of drug metabolism into the lead optimization paradigm. Mini Rev Med Chem 2009;9(6):703–716.

Korfmacher WA, Cox KA, Bryant MS, Veals J, Ng K, Lin CC. HPLC-API/MS/MS: A powerful tool for integrating drug metabolism into the drug discovery process. Drug Discov Today 1997;2:532–537.

Korfmacher WA, Cox KA, Ng KJ, Veals J, Hsieh Y, Wainhaus S, Broske L, Prelusky D, Nomeir A, White RE. Cassette-accelerated rapid rat screen: A systematic procedure for the dosing and liquid chromatography/atmospheric pressure ionization tandem mass spectrometric analysis of new chemical entities as part of new drug discovery. Rapid Commun Mass Spectrom 2001;15(5):335–340.

Korfmacher WA, Palmer CA, Nardo C, Dunn-Meynell K, Grotz D, Cox K, Lin CC, Elicone C, Liu C, Duchoslav E. Development of an automated mass spectrometry system for the quantitative analysis of liver microsomal incubation samples: a tool for rapid screening of new compounds for metabolic stability. Rapid Commun Mass Spectrom 1999;13(10):901–907.

Laitinen L, Kangas H, Kaukonen AM, Hakala K, Kotiaho T, Kostiainen R, Hirvonen J. *N*-in-one permeability studies of heterogeneous sets of compounds across Caco-2 cell monolayers. Pharm Res 2003;20(2):187–197.

Lau YY, Chen YH, Liu TT, Li C, Cui X, White RE, Cheng KC. Evaluation of a novel in vitro Caco-2 hepatocyte hybrid system for predicting in vivo oral bioavailability. Drug Metab Dispos 2004;32(9):937–942.

Lau YY, Sapidou E, Cui X, White RE, Cheng KC. Development of a novel in vitro model to predict hepatic clearance using fresh, cryopreserved, and sandwich-cultured hepatocytes. Drug Metab Dispos 2002;30(12):1446–1454.

Li C, Liu T, Cui X, Uss AS, Cheng KC. Development of in vitro pharmacokinetic screens using Caco-2, human hepatocyte, and Caco-2/human hepatocyte hybrid systems for the prediction of oral bioavailability in humans. J Biomol Screen 2007;12 (8):1084–1091.

Li AP. Screening for human ADME/Tox drug properties in drug discovery. Drug Discov Today 2001;6(7):357–366.

Li AP. In vitro approaches to evaluate ADMET drug properties. Curr Top Med Chem 2004;4(7):701–706.

Liu B, Chang J, Gordon WP, Isbell J, Zhou Y, Tuntland T. Snapshot PK: A rapid rodent in vivo preclinical screening approach. Drug Discov Today 2008;13 (7–8):360–367.

Mandagere AK, Thompson TN, Hwang KK. Graphical model for estimating oral bioavailability of drugs in humans and other species from their Caco-2 permeability and in vitro liver enzyme metabolic stability rates. J Med Chem 2002;45(2):304–311.

Masungi C, Mensch J, Van Dijck A, Borremans C, Willems B, Mackie C, Noppe M, Brewster ME. Parallel artificial membrane permeability assay (PAMPA) combined with a 10-day multiscreen Caco-2 cell culture as a tool for assessing new drug candidates. Pharmazie 2008;63(3):194–199.

McGinnity DF, Waters NJ, Tucker J, Riley RJ. Integrated in vitro analysis for the in vivo prediction of cytochrome P450-mediated drug-drug interactions. Drug Metab Dispos 2008;36(6):1126–1134.

McNaney CA, Drexler DM, Hnatyshyn SY, Zvyaga TA, Knipe JO, Belcastro JV, Sanders M. An automated liquid chromatography–mass spectrometry process to determine metabolic stability half-life and intrinsic clearance of drug candidates by substrate depletion. Assay Drug Dev Technol 2008;6(1):121–129.

Mei H, Korfmacher W, Morrison R. Rapid in vivo oral screening in rats: Reliability, acceptance criteria, and filtering efficiency. Aaps J 2006;8(3):E493–500.

Nguyen DT, Guillarme D, Heinisch S, Barrioulet MP, Rocca JL, Rudaz S, Veuthey JL. High throughput liquid chromatography with sub-2 microm particles at high pressure and high temperature. J Chromatogr A 2007;1167(1):76–84.

Nomeir AA, Morrison R, Prelusky D, Korfmacher W, Broske L, Hesk D, McNamara P, Mei H. Estimation of the extent of oral absorption in animals from oral and intravenous pharmacokinetic data in drug discovery. J Pharm Sci 2009;98 (11):4027–4038.

Peng SX, Barbone AG, Ritchie DM. High-throughput cytochrome p450 inhibition assays by ultrafast gradient liquid chromatography with tandem mass spectrometry using monolithic columns. Rapid Commun Mass Spectrom 2003;17(6):509–518.

Pintore M, van de Waterbeemd H, Piclin N, Chretien JR. Prediction of oral bioavailability by adaptive fuzzy partitioning. Eur J Med Chem 2003;38(4):427–431.

Plumb RS, Potts WB 3rd, Rainville PD, Alden PG, Shave DH, Baynham G, Mazzeo JR. Addressing the analytical throughput challenges in ADME screening using rapid ultra-performance liquid chromatography/tandem mass spectrometry methodologies. Rapid Commun Mass Spectrom 2008;22(14):2139–2152.

Rainville PD, Wheaton JP, Alden PG, Plumb RS. Sub one minute inhibition assays for the major cytochrome P450 enzymes utilizing ultra-performance liquid chromatography/tandem mass spectrometry. Rapid Commun Mass Spectrom 2008;22(9):1345–1350.

Reddy A, Heimbach T, Freiwald S, Smith D, Winters R, Michael S, Surendran N, Cai H. Validation of a semi-automated human hepatocyte assay for the determination and prediction of intrinsic clearance in discovery. J Pharm Biomed Anal 2005;37 (2):319–326.

Rodrigues AD, Lin JH. Screening of drug candidates for their drug–drug interaction potential. Curr Opin Chem Biol 2001;5(4):396–401.

Smalley J, Kadiyala P, Xin B, Balimane P, Olah T. Development of an on-line extraction turbulent flow chromatography tandem mass spectrometry method for cassette analysis of Caco-2 cell based bi-directional assay samples. J Chromatogr B Anal Technol Biomed Life Sci 2006;830(2):270–277.

Smith D, Sadagopan N, Zientek M, Reddy A, Cohen L. Analytical approaches to determine cytochrome P450 inhibitory potential of new chemical entities in drug discovery. J Chromatogr B Anal Technol Biomed Life Sci 2007;850 (1–2):455–463.

Stoner CL, Cleton A, Johnson K, Oh DM, Hallak H, Brodfuehrer J, Surendran N, Han HK. Integrated oral bioavailability projection using in vitro screening data as a selection tool in drug discovery. Int J Pharm 2004;269(1):241–249.

Thompson TN. Early ADME in support of drug discovery: The role of metabolic stability studies. Curr Drug Metab 2000;1(3):215–241.

Thompson TN. Optimization of metabolic stability as a goal of modern drug design. Med Res Rev 2001;21(5):412–49.

Thompson TN. Drug metabolism i*n vitro and in vivo* results: How do these data support drug discovery. In *Using Mass Spectrometry for Drug Metabolism Studies*, Korfmacher W, CRC Press, Boca Raton, FL, 2005, pp. 35–82.

Uchida M, Fukazawa T, Yamazaki Y, Hashimoto H, Miyamoto Y. A modified fast (4 day) 96-well plate Caco-2 permeability assay. J Pharmacol Toxicol Methods 2009;59 (1):39–43.

van de Waterbeemd H, Jones BC. Predicting oral absorption and bioavailability. Prog Med Chem 2003;41:1–59.

Van Pelt CK, Zhang S, Fung E, Chu I, Liu T, Li C, Korfmacher WA, Henion J. A fully automated nanoelectrospray tandem mass spectrometric method for analysis of Caco-2 samples. Rapid Commun Mass Spectrom 2003;17(14):1573–1578.

Venkatakrishnan K, Obach RS. Drug-drug interactions via mechanism-based cytochrome P450 inactivation: Points to consider for risk assessment from in vitro data and clinical pharmacologic evaluation. Curr Drug Metab 2007;8(5):449–462.

Wickremsinhe ER, Ackermann BL, Chaudhary AK. Validating regulatory-compliant wide dynamic range bioanalytical assays using chip-based nanoelectrospray tandem mass spectrometry. Rapid Commun Mass Spectrom 2005;19(1):47–56.

Wickremsinhe ER, Singh G, Ackermann BL, Gillespie TA, Chaudhary AK. A review of nanoelectrospray ionization applications for drug metabolism and pharmacokinetics. Curr Drug Metab 2006;7(8):913–928.

Xu R, Nemes C, Jenkins KM, Rourick RA, Kassel DB, Liu CZ. Application of parallel liquid chromatography/mass spectrometry for high throughput microsomal stability screening of compound libraries. J Am Soc Mass Spectrom 2002;13(2):155–165.

Yan Z, Caldwell GW. Metabolism profiling, and cytochrome P450 inhibition & induction in drug discovery. Curr Top Med Chem 2001;1(5):403–425.

Youdim KA, Zayed A, Dickins M, Phipps A, Griffiths M, Darekar A, Hyland R, Fahmi O, Hurst S, Plowchalk DR, Cook J, Guo F, Obach RS. Application of CYP3A4 in vitro data to predict clinical drug-drug interactions; predictions of compounds as objects of interaction. Br J Clin Pharmacol 2008;65(5):680–692.

Yu K, Di L, Kerns E, Li SQ, Alden P, Plumb RS. Ultra-performance liquid chromatography/tandem mass spectrometric quantification of structurally diverse drug mixtures using an ESI-APCI multimode ionization source. Rapid Commun Mass Spectrom 2007;21(6):893–902.

Yu K, Little D, Plumb R, Smith B. High-throughput quantification for a drug mixture in rat plasma—A comparison of ultra performance liquid chromatography/tandem mass spectrometry with high-performance liquid chromatography/tandem mass spectrometry. Rapid Commun Mass Spectrom 2006;20(4):544–552.

Yu S, Crawford E, Tice J, Musselman B, Wu JT. Bioanalysis without sample cleanup or chromatography: The evaluation and initial implementation of direct analysis in real time ionization mass spectrometry for the quantification of drugs in biological matrixes. Anal Chem 2009;81(1):193–202.

Zientek M, Miller H, Smith D, Dunklee MB, Heinle L, Thurston A, Lee C, Hyland R, Fahmi O, Burdette D. Development of an in vitro drug-drug interaction assay to simultaneously monitor five cytochrome P450 isoforms and performance assessment using drug library compounds. J Pharmacol Toxicol Methods 2008;58(3):206–214.

Zlokarnik G, Grootenhuis PD, Watson JB. High throughput P450 inhibition screens in early drug discovery. Drug Discov Today 2005;10(21):1443–1450.

13 High-Resolution Mass Spectrometry and Drug Metabolite Identification

RUSSELL J. MORTISHIRE-SMITH

Janssen Pharmaceutical Companies of Johnson & Johnson,
B-2340 Beerse, Belgium

HAIYING ZHANG

Biotransformation, Bristol-Myers Squibb Research & Development,
Pennington, New Jersey

KEVIN P. BATEMAN

Drug Metabolism and Pharmacokinetics, Merck Frosst Canada,
Kirkland, Quebec H9H 3L1, Canada

Mass Spectrometry in Drug Metabolism and Disposition: Basic Principles and Applications, First Edition. Edited by Mike S. Lee and Mingshe Zhu.

13.1 INTRODUCTION

Drug metabolite identification is a critical component of drug discovery and development and provides a variety of inputs, including in vitro metabolite profiling, in vitro/in vivo correlation, cross-species comparison, characterization of major circulating metabolites, rationalizing drug–drug interactions, and identifying pharmacologically active or toxic metabolites and the mechanisms by which they are formed.

Drug metabolite identification is technically challenging, primarily because of the vast diversity of possible drug metabolites, the demand for ever-increasing throughput in drug discovery, and the challenge of characteristically low concentrations of drug metabolites in the presence of a high abundance of endogenous matrix background. A variety of approaches have been proposed for circumventing, in particular, this latter challenge. Here, we focus on the significant advantages of high-resolution liquid chromatography–mass spectrometry (LC–MS) technology as a platform for drug metabolite characterization and describe a variety of powerful data acquisition and data-mining techniques that take advantage of high mass resolution data.

Given that the biotransformation reactions for forming different phase I and phase II metabolites are covered in Chapters 2 and 3 and experimental models for studying drug metabolism are covered in Chapter 5, we will provide only an overview of the challenges presented by different sample types in metabolite identification studies, highlighting the throughput and sample matrix concerns. This chapter will provide the background for understanding the advantages and practicalities of high-resolution mass spectrometry over the traditional nominal mass resolution technologies for drug metabolite identification.

13.2 CHALLENGES PRESENTED BY DIFFERENT SAMPLES

The most common samples for in vitro metabolite identification are obtained from liver microsomal incubations. Microsomes are a relatively simple and straightforward system with which to assess phase I and some phase II enzyme activities (with appropriate cofactors or fortification). In recent years, the ready commercial availability of hepatocytes has led to an increasing use of this system. Hepatocytes are a more complete system than liver microsomes and contain phase I and phase II enzyme activities, as well as cellular transporters. In addition, samples from tissue slice incubations are used for some in vitro metabolism investigations. Tissue slices are the most complete in vitro system that contain not only enzymes but also intercellular membranes and transporter systems.

Typically, crude in vitro samples are pretreated prior to injection on an LC–MS system to reduce the concentrations of matrix components such as salts, proteins, lipids, cofactors, cellular components, and other incubation components. This pretreatment is necessary not only to avoid column fouling, but also to improve the selectivity for drug metabolite analysis.

Samples used for in vivo metabolite identification are typically derived from body fluids such as plasma, bile, and urine. Sometimes, homogenized samples obtained from solids (e.g., feces or tissues) are also used. The major challenge for in vivo samples is the matrix content, which is substantially higher than for in vitro samples. For in vivo samples, as with in vitro samples, pretreatment is a necessity to extract drug metabolites and minimize the amount of endogenous components. However, even with sample pretreatment, the presence of endogenous components can still be overwhelming and may suppress the mass spectrometric response of certain drug metabolites. For this reason, a good LC separation is desirable to ensure the sensitivity and reliability of metabolite identification.

Typically, sample pretreatment techniques (James, 2008; Kallury, 2009) include protein precipitation (Polson et al., 2003), liquid–liquid extraction (Pedersen-Bjergaard and Rasmussen, 2005), and solid-phase extraction (Fedeniuk and Shand, 1998). Protein precipitation with a denaturing organic solvent is a general approach that is applicable to most samples and lends itself to automated high-throughput sample handling (Watt et al., 2000). Liquid–liquid extraction and solid-phase extraction typically generate cleaner samples, but these approaches are less generic and, depending on the nature of the metabolites, may need specific steps to optimize the recovery. These procedures are often challenging for metabolite identification tasks when drug metabolites are unknown and their recoveries are difficult to monitor. In any event, no matter what sample pretreatment technique is used, the samples inevitably contain substantial matrix components (e.g., lipids, fatty acids, endogenous peptides, and residual proteins) that will complicate the LC–MS data, making drug metabolite identification challenging. Aside from the sample matrix concerns, the limited availability of sample volumes in some studies (e.g., plasma samples) also imposes a demand on the methodology to generate largest amount of useful information from a single LC–MS experiment.

In brief, the throughput demand of drug metabolite analysis, the diversity of potentially unknown metabolite structures, and the inevitable presence of matrix components interfering with drug metabolite identification are factors that drive a need for methodology that is fast, efficient, and robust and requires minimal up-front investment to be successful. Fortunately, high-resolution mass spectrometry offers an opportunity to improve the throughput, efficiency, and richness of drug metabolite identification.

13.3 FUNDAMENTAL ADVANTAGE OF HIGH-RESOLUTION MASS SPECTROMETRY: SPECIFICITY/SELECTIVITY IN A SINGLE GENERIC METHOD

The foremost advantage of high-resolution mass spectrometry over nominal mass resolution mass spectrometry is the fact that a complete, unbiased metabolism data set can be acquired in a single injection without the requirement to invest upfront resources in method development.

Because of the complexity of drug metabolite samples, the task of metabolite identification with LC–MS has been conventionally conceived as two parts: the detection of molecular ions of drug metabolite species as the first part and the localization of the site of oxidation in the second part. This approach is achievable due to soft ionization techniques (Yamashita and Fenn, 1984a, 1984b, Karas et al., 1987; Koichi et al., 1988) that allow for nonspecific generation of molecular ions for components in a sample and tandem mass spectrometric technologies that allow for fragmentation of selected ions to generate structure-related information (Kerns et al., 1994, 1995; Yu et al., 1999; Gangl et al., 2002; Lafaye et al., 2003; Anari et al., 2004; Peterman et al., 2006; Lim et al., 2007).

The first approach, which we term "theory down," uses a knowledge of the fragmentation pattern of the parent drug, and the biotransformations likely to occur, to build a library of MS transitions that are expected to capture the majority of drug metabolites present in the sample (Tiller et al., 2002; Yao et al., 2008). The generation of transitions is tractable to automated method creation, and to ensure adequate coverage, a reasonably large number of transitions must be acquired. Selectivity over matrix background is achieved on the basis that endogenous compounds are unlikely to fragment in the same way as the parent compound, and metabolites are identified on the basis that they will (a) have a similar fragmentation pathway to the parent and (b) have a defined mass change. An advantage of this approach is that some localization information (product ions) is obtained in the same data set. Coverage of most common oxidations can be easily achieved, but the approach has a single, major drawback—it is predicated on limiting assumptions that may miss the most interesting metabolites (novel pathways and/or rearrangements that lead to metabolites that do not look particularly similar to the parent compound, at least, in mass spectrometric terms). A more generic approach on quadrupole-based instruments is to use precursor ion scanning, which is able to identify a greater proportion of novel biotransformations (i.e., is not list dependent), but which is inherently less sensitive and still assumes that the major product ions of metabolites will be common with those of the parent compound.

In the absence of compound-specific fragmentation to drive selectivity, metabolite detection on nominal mass instruments becomes significantly more difficult. Typically, significant ambiguity exists in nominal MS data due to the presence of endogenous interferences, isobaric with metabolites, which cannot be distinguished. Fortunately, high-resolution mass spectrometric data provide a perfect solution to address this issue. Indeed, the advance of high-resolution mass spectrometry (HRMS) technology has enabled a number of techniques to improve the selectivity in metabolite detection, and the sensitivity limitations of early high-resolution instruments have largely been overcome.

The key advantage of a high-resolution-based approach is that it is a data-up, rather than theory-down, approach. The basic data set acquired is a full mass range, high-resolution scan. Some preacquisition optimization of source conditions can be valuable but is by no means obligatory for most druglike compounds. No considerations need be given to the data that should be acquired. The critical decisions about what metabolites to look for, and how best to find them, are left to the post-data-acquisition stage. On the other hand, with modern mass spectrometers, data acquisition can include either data-independent (MS^E) (Wrona et al., 2005; Bateman et al., 2007) or data-dependent (MS^n) high-energy channels (Kantharaj et al., 2003; Lim et al., 2007; Ruan et al., 2008) without significantly compromising the quality of data obtained in the basic low-energy channel. Indeed, almost any data-dependent high-energy scan function can be acquired, triggered by isotope pattern, mass defect, specific mass, or (pseudo)-constant neutral loss, and, if interleaved, will still leave the low-energy channel with sufficient data points for a full chromatogram analysis.

As a consequence of these features, the high-resolution MS approach has become a popular option for metabolite detection in recent years (Zhang et al., 2000b; Chernushevich et al., 2001; Mortishire-Smith et al., 2005; Wrona et al., 2005; Tiller et al., 2008; Bateman et al., 2009; Peters et al., 2009; Holman et al., 2009).

13.4 HIGH-RESOLUTION MASS SPECTROMETRY: IMPORTANT CONCEPTS

Fundamentally, a mass spectrometer is an apparatus with which to separate ions according to their mass-to-charge (m/z) ratios. The term *mass resolution* is used to describe the mass resolving power according to the degree to which close m/z values can be separated. A practical and convenient way of evaluating the mass resolution of an instrument is the use of the full width at half maximum (FWHM) definition in which the $m/z\ /\Delta m/z$ ratio is calculated, where m/z is the mass-to-charge value of an ion peak and $\Delta m/z$ is the full width at half the maximal height of the peak (Fig. 13.1). Nominal mass resolution instruments (e.g., quadrupoles and ion traps) typically have resolution in the low-thousands range and are not capable of resolving isobaric ions with the same nominal m/z value. High-resolution instruments such as time-of-flight (TOF) (Pilard et al., 2000; Sin et al., 2002; Nikolic et al., 2004; Castro-Perez et al., 2005; Ferrer and Thurman, 2005) and Fourier transform (Godejohann et al., 2004; Peterman et al., 2006; Marshall and Hendrickson, 2008) have mass

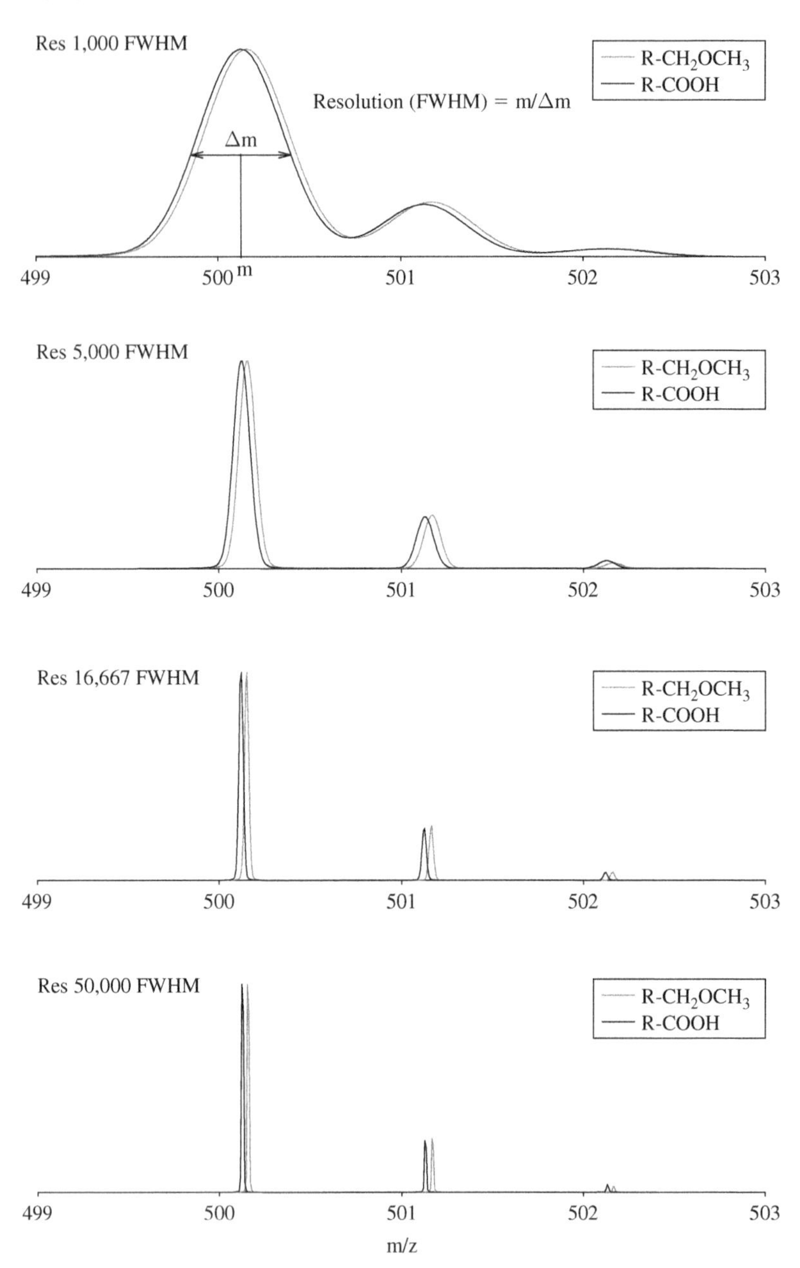

FIGURE 13.1 Definition of FWHM and comparison of mass spectral data simulated for two nominally isobaric metabolites (500 Th) but with a decimal mass difference of 36.4 mDa using FWHM resolutions of 1,000, 5,000, 16,667, and 50,000 (top to bottom).

resolutions that approach 10,000–20,000 or higher. As a consequence of this higher mass resolution, the degree to which drug-related entities can be distinguished from nominally isobaric endogenous species is significantly higher (Fig. 13.1). For example, with a TOF instrument, spectral data measured with resolution of 5,000–8,000 over a typical scan range for metabolite analysis of 80–1,000 Th can be extracted using a 0.1 mass unit window (Zhang et al., 2000b). A visualization of the practical consequence of using narrower mass windows on component discrimination is shown in Figure 13.2. As the mass window is reduced, the number of peaks present in the chromatogram falls until only the metabolite signals are left, with no consequent loss in peak intensity.

In principle, as long as data can be measured precisely, high-resolution data are sufficient to demonstrate the presence or absence of defined species. Fortunately, modern high-resolution LC–MS technology also significantly improves the accuracy of the mass measurements. The term *mass accuracy* is used to define how close the mass measured by the mass spectrometer is to the theoretical exact mass of an ion. Mass accuracy is typically expressed as a relative mass error using the ratio of the difference between the experimental and theoretical m/z values over the theoretical m/z value of an ion. Since the ratio is always a small fractional value, it is commonly expressed as parts per million (ppm, 10^6). Under certain circumstances, good accurate mass measurements of metabolites can be achieved with even a low-resolution triple quadrupole instrument (Kostiainen et al., 1997); but, in general, high mass accuracy measurements can only be routinely achieved with values below 5 ppm with modern high-resolution mass spectrometry technologies with the aid of LC separation. For example, with TOF technologies, accurate mass measurements of fragment ions of two glucuronides dissociated in the ion source can be made with an accuracy of better than 2.1 ppm and with an interday relative standard deviation within ±0.00049 % (Sundstrom et al., 2002). Such sub-part-per-million mass accuracy allows for the differentiation for components with milli-Th mass differences in the typical mass range of 80–1,000 Th used for metabolite identification studies. Such reliable accurate measurement significantly improves the specificity of a number of data-dependent acquisitions and post-acquisition data processing techniques for the detection of metabolites (as will be further elaborated in this chapter).

13.5 HIGH-RESOLUTION INSTRUMENTATION

Magnetic sector instruments have long been used for metabolite identification studies and reveal the power of accurate mass for unambiguous determination of elemental composition of metabolites (e.g., see Yergey et al., 2001). In recent years, though, this type of mass analyzer has fallen out of favor for metabolism studies with the advent of easier to operate and maintain mass analyzers. Currently, there is a range of instrumentation for HRMS data collection, based

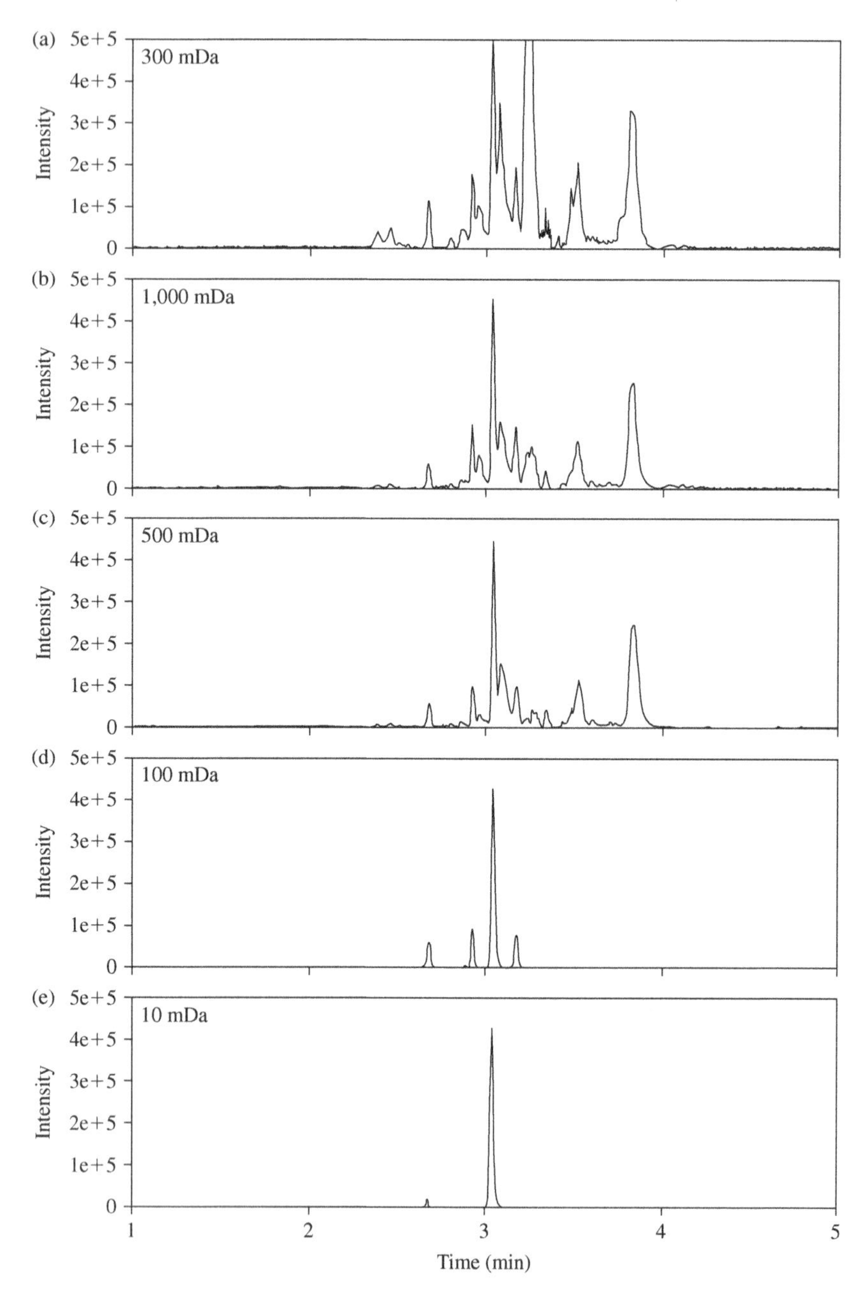

FIGURE 13.2 Extracted ion chromatograms with the same data file using decreasing extraction windows at (*a*) 3,000 mDa, (*b*) 1,000 mDa, (*c*) 500 mDa, (*d*) 100 mDa, and (*e*) 10 mDa. Improved selectivity is obtained without loss of sensitivity using high-resolution full-scan data acquisition. [Bateman et al. (2009); © Wiley.]

either on TOF or Fourier transform (FT) iontrap (Orbitrap or ion cyclotron resonance (ICR)) mass analyzers. A variety of different configurations of hybrid systems using either TOF (quadrupole TOF, iontrap TOF, TOF–TOF, quadrupole-ion mobility–TOF) or FT-based analyzers have evolved (iontrap Orbitrap, iontrap FTICR, and quadrupole FTICR). Each instrument has distinct advantages and limitations with respect to data collection and data quality. Schematics of three commercial implementations of the main three HRMS analyzers are shown in Figure 13.3.

TOF-based systems (Chernushevich et al., 2001) have been used for many years, and performance has improved substantially over time. Modern commercial TOF mass analyzers can achieve mass resolution of >40,000 FWHM and mass accuracies of <1 ppm with internal calibration. Improvements in detector technologies and high-speed digitizers have enabled within-spectrum dynamic ranges of >10,000. Spectral acquisition rates of up to 20 Hz enable compatibility with fast chromatography and complex data acquisition procedures such as data-dependent or nonspecific fragmentation-based approaches. TOF-based analyzers operate at full sensitivity all the time; that is, they are not limited by high ion counts that could prevent detection of low abundance ions.

TOF-based instruments do have some limitations that can impact data quality and subsequent data processing and interpretation. Mass accuracy is limited by ion statistics, so low abundance ions will have poorer mass accuracy than high abundance ions. Resolution, while much improved over older TOF instruments, is still lower than that achievable with FT-based mass analyzers. In some circumstances, this can be critical, since higher resolutions can be used to good effect in very complex matrices in which it is necessary to differentiate between closely related compounds and/or fragment ions. Detection systems on TOF-based analyzers have traditionally suffered from poor dynamic range and poor mass accuracy due to saturation of the detector. Current generation systems, based on analog-to-digital conversion with fast digitizers, do not saturate as readily and have much wider dynamic range (see above). To achieve good mass accuracy (<5 ppm) on TOF analyzers, internal calibration is required. Commercial instruments have a variety of approaches for automatically providing an internal calibrant to overcome this limitation.

Direct TOF-based systems are not typically used for metabolite identification studies as they lack the ability to generate mass-directed tandem mass spectra (MS/MS), thus the majority of published metabolite identification studies using TOF mass analyzers have been accomplished with hybrid quadrupole TOF (QTOF) systems. This configuration allows for a wider range of data collection scan functions to be performed. The instrument consists of an ion path based on quadrupole (or related) technology transmitting ions into a TOF mass analyzer orthogonal to the ion beam. When operated in radio frequency (RF)–only mode, the mass filter passes a wide m/z range for mass analysis in the TOF analyzer. When operated as a mass-resolving quadrupole in tandem with the collision cell, fragment ions of selected precursor ions can be generated prior to mass analysis.

(a)

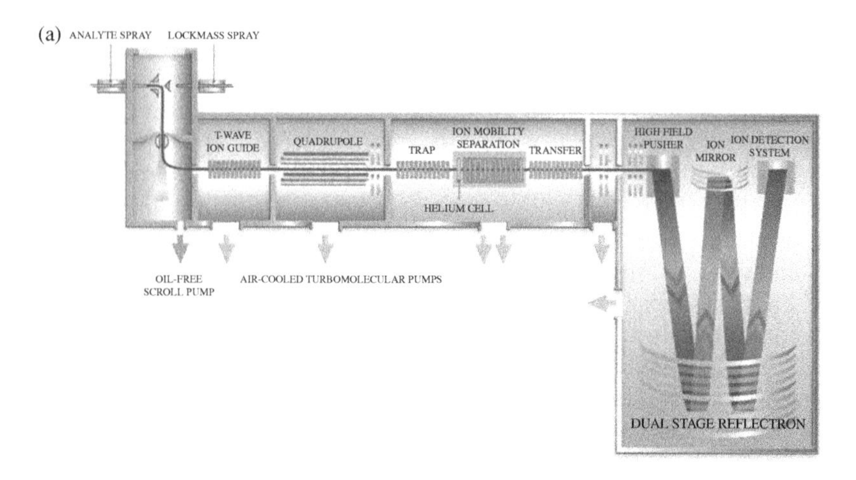

(b)

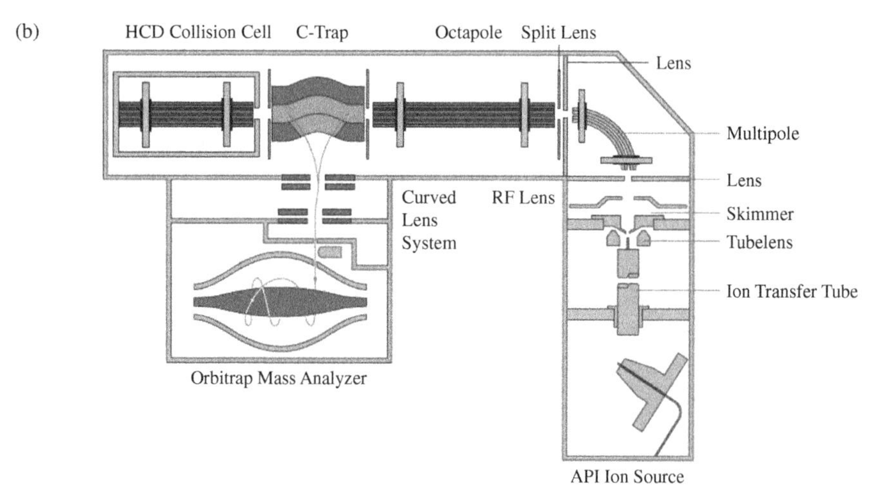

(c)

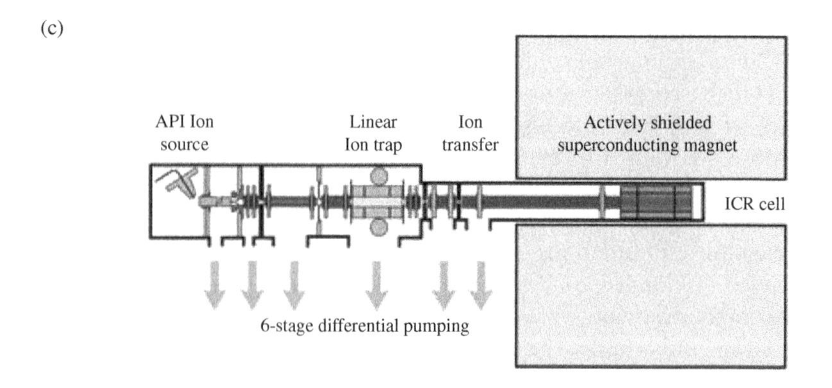

FIGURE 13.3 Instrument schematics for (*a*), Waters Synapt QTOF (© Waters), (*b*) Thermo Finnigan LTQ-Orbitrap, and (*c*) Thermo Finnegan LTQ FTICR mass spectrometers (© Thermo Finnegan).

Several modes of data acquisition are available from this type of mass analyzer beyond those described above. These modes either take advantage of the high resolution and accurate mass directly in the acquisition experiment or rely on the HRMS data to facilitate postacquisition processing. Several data-dependent modes of operation have been described using QTOF technologies that leverage accurate mass data. Both precursor ion scanning and pseudo neutral loss scanning are possible on a QTOF instruments. These scan functions use a full-scan low-collision energy scan, followed by a full-scan high-collision energy scan, to generate a data set that is processed in real time to look for specific precursors or neutral losses to trigger a mass-directed MS/MS scan function (Bateman et al., 2002; Castro-Perez et al., 2005). The *m/z* values of precursor ions and neutral losses are defined to a narrow tolerance to leverage the accurate mass capability of the instrument. A novel approach leveraging mass-defect-based data-dependent acquisition has been described using QTOF technology (Bloomfield, 2007). A more comprehensive discussion of the mass defect filter and its utility in HRMS approaches for metabolite identification is presented later in this chapter. Both these approaches rely on the accurate mass data to improve data acquisition of HRMS data.

Alternatively, an approach described by Wrona et al. (2005) building on earlier work (Carr et al., 1993; Allen et al., 1997; Bateman et al., 2002) leverages nonspecific collisionally induced dissociation and postacquisition data processing of the HRMS data to enable a simple and generic approach to data acquisition. Based on the neutral loss scan function described above, this approach collects two scan functions, one low-collision energy, one high-collision energy, without triggering a data-dependent MS/MS scan. This approach provides fragment ions in a nonselective manner and has the advantage that the likelihood of missing a metabolite is reduced relative to a data-dependent approach where rules and thresholds are required. However, to be truly effective, both good chromatographic resolution and data processing tools are required.

An alternative hybrid TOF instrument is the ion trap TOF, which uses a three-dimensional (3D) ion trap linearly coupled to a TOF mass analyzer. This configuration couples the ion trap's ability to do multiple stages of fragmentation with the strength of a TOF analyzer for fast data acquisition, high resolution, and accurate mass. The commercial implementation of this design uses highly stable, fast switching power supplies to enable positive–negative ion switching at rates up to 2.5 Hz. The performance of the TOF on the commercial implementation is not at a level comparable to the most recent TOF systems, and as such lags in performance characteristics such as resolution, dynamic range, and mass accuracy.

The most recent implementation of hybrid TOF technology is the MS–ion mobility spectrometry (IMS)–TOF mass spectrometer. This instrument uses a series of three stacked ring ion guides that serve several functions, including ion trapping, collision cell, IMS device, and ion transfer. When not using ion mobility, the device functions as a normal QTOF instrument with the added advantage of having the ability to do two stages of fragmentation in the ion guides. When using IMS, multidimensional data sets are obtained with chromatographic retention time, drift time, m/z, and intensity dimensions. Fragmentation can be done before and/or after the IMS device giving information-rich spectra with accurate mass on the fragment ions to aid in interpretation of metabolite structures.

Orbitrap-based mass spectrometers (Hu et al., 2005; Scigelova and Makarov 2009) have seen rapid uptake into metabolite identification laboratories, mainly in the hybrid iontrap–Orbitrap configuration discussed below. The Orbitrap mass analyzer provides a resolution of up to 100,000 FWHM at m/z 200 at a 1 Hz acquisition rate. Mass accuracy is <2 ppm with internal calibration and is <5 ppm with external calibration. It is stable over extended periods (days) when using external calibration. The mass accuracy is also very good for low abundance ions and is reproducible from scan to scan. Dynamic range is >4,000 within a spectrum and acquisition rates of 10 Hz are possible at reduced resolution.

While providing excellent resolution and mass accuracy, the Orbitrap is still a trap-based instrument and has the limitations associated with this type of mass analyzer. Resolution is scan speed dependent requiring longer total cycle times to achieve higher resolutions. This dependency can be a limiting factor when using fast high-resolution chromatography with narrow chromatographic peak widths. Traps are limited by space charging at high ion counts, and this can be problematic when working with complex matrices such as bile, urine, and plasma. High abundance background ions can fill the trap to the exclusion of low abundance ions of interest.

The uptake of accurate mass instrumentation into drug metabolism studies has been accelerated by the development of the linear trap quadrupole (LTQ)-Orbitrap (Makarov et al., 2006a, 2006b). This instrument combines what has been the traditional workhorse for metabolite identification studies, the quadrupole ion trap, with the high-resolution Orbitrap. The ion trap benefits from versatility for data-dependent experiments and the ability to run MS^n experiments. The configuration allows for flexible data acquisition using the high-resolution Orbitrap to collect an accurate mass full-scan spectrum and the fast scanning linear ion trap to rapidly collect MS/MS or MS^n spectra in low-resolution mode (Li et al., 2007). Alternatively, all the data can be collected using the Orbitrap by transmitting the fragment ions from the LTQ into the Orbitrap for mass measurement (Peterman et al., 2006). The addition of a collision cell to the Orbitrap to enable higher energy collisional dissociation (HCD) has enabled further data acquisition modes (Olsen et al., 2007). One approach uses the HCD cell to collect wide mass range MS/MS in the Orbitrap giving nonselective data similar to the MS^E described above. In addition to this data stream, a series of data-dependent low-resolution scans on the LTQ are also collected that are triggered from a high-resolution full-scan MS in the Orbitrap.

Most recently, a single-stage Orbitrap mass spectrometer has been introduced, known commercially as the Exactive (Wieghaus et al., 2008). This instrument, when coupled with a HCD cell, enables MS^E data collection but does not have a mass selective MS/MS capability. The instrument provides high performance with respect to stable and accurate mass measurement and resolution desirable for metabolite identification studies (Bateman et al., 2009; Koulman et al., 2009). With such feature-rich data sets, it is critical that fully featured software tools be developed to leverage the complex data sets generated by this instrument.

Not to be forgotten in the list of instrumentation for HRMS and metabolite identification studies is the FTICR mass analyzer (Sanders et al., 2006; Shipkova et al., 2009). This type of mass analyzer offers the ultimate in performance with respect to resolving power, with specifications >1,000,000 being cited on commercial instruments. Predating the LTQ-Orbitrap, the LTQ-FTICR provides similar capabilities with the added benefit of higher resolution. The limitation of this mass analyzer is the initial expense and that of maintaining the cryogenically cooled magnet needed to provide the high performance. However, this type of instrument has not gained wide acceptance for metabolite identification studies given the performance, lower cost, and ease, of use of the LTQ-Orbitrap instrument.

13.6 ADVANTAGES OF HIGH-RESOLUTION MS: THE CONCEPT OF MASS DEFECT FILTRATION

Aside from improving the selectivity and accuracy of the existing metabolite identification techniques, the introduction of high-resolution mass spectrometry into the drug metabolism work flow has also catalyzed new approaches to the process by which metabolites can be identified in complex data sets. One of the most important of these is mass defect filtration [also referred to as "fractional mass filtration" by some groups; for good recent reviews see Huang (2009), and Zhang et al. (2009b)].

The atomic *mass defect* is defined as the amount by which the mass of an atomic nucleus is less than the sum of the masses of its constituent particles, typically fractions of a dalton. This amount of mass difference is equivalent to the energy released when the nucleons are bound to form an atomic nucleus. For example, the hydrogen atom, with an atomic mass of 1.0084 Da, has a mass defect of 5.8 mDa. The mass defect of a molecule is a simple summation of the mass defects of all its constituent atoms and is therefore a characteristic property that reflects the empirical formula of the molecule. This is the basis of accurate mass measurement for structural characterization, as described above. A *mass defect filter* is a software-based data filter that leverages high-resolution data acquired by accurate mass LC–MS instruments with the predictable mass defects of drug metabolites (Zhang et al., 2003). The novelty and attractiveness of mass defect filtering is that since drug metabolites rarely have elemental compositions comparable with those of the endogenous components of biological matrices, this affords a method of eliminating ions

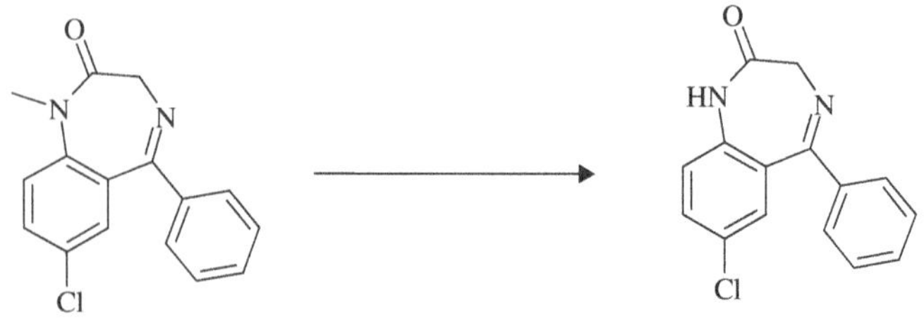

Exact Mass = 284.0716
Molecular Formula = C16H13ClN2O

Exact Mass = 270.0560
Molecular Formula = C15H11ClN2O

Mass change is −14.0156

f.m.c. = −15.6 mDa

FIGURE 13.4 Mass defect [fractional mass change (fmc)] consequent upon a typical biotransformation of diazepam.

that derive from the latter. Practically speaking, the approach involves the removal of all ions from the data set that have observed mass defects that fall outside of an operator-determined range. With the majority of interference ions thus removed, the filtered data set should be substantially enriched with ions that correspond to drug metabolites.

An important reason for the success of mass defect filtration as a generic approach derives from the observation that, irrespective of the gross integral mass change associated with a biotransformation, mass defects of phase I and phase II metabolite ions typically fall within 50 mDa relative to that of the parent drug (Fig.13.4 and Tables 13.1 and 13.2), (Mortishire-Smith et al., 2005); for example, hydroxylation changes the mass defect by −5 mDa, dehydrogenation by −16 mDa, demethylation by −23 mDa, glucuronidation by +32 mDa, sulfation by −43 mDa, with the exception of glutathione, which results in a +68 mDa change to the mass defect. Therefore, it is possible to define a mass defect window (e.g., ±50 mDa around the mass defect of the parent drug), which should retain most drug-related ions yet exclude most matrix-related ions.

Conceptually, the value of mass defect filters can be visualized using a mass defect plot (Zhang et al., 2008b) (Fig.13.5) in which the intensity of the mass spectrometry data is ignored, and the *m*/*z* values of all ions in the data are separated into two dimensions: the mass defect is plotted along the *y* axis and the nominal mass is plotted along the *x* axis. For mass defect filter design, one or more regions (the rectangular boxes in Fig.13.5) may be defined in both the *y* (mass defect) and *x* (nominal mass) dimensions. For example, a filter can be designed by defining its range in the *y* and *x* dimensions for a parent drug with a mass of 500 Da and a mass defect around 0.3 Da (see the middle rectangular box for an example). If glutathione conjugates are expected, then a filter around the appropriate mass range can be set (rightmost box in Fig.13.5).

Even if a drug molecule is expected to undergo a metabolic cleavage, thus changing the mass and mass defect to another range, yet another filter can be

TABLE 13.1 Mass and Fractional Mass (Mass Defect) Changes for Phase 1 Biotransformations

Description	Mass Change (μ)	Fractional Mass Change (mμ)	Formula Change
tert-Butyl to acid	−12.0726	−72.6	−C3−H8+O2
Propyl ketone to acid	−40.0675	−67.5	−C4−H8+O
tert-Butyl to alcohol	−40.0675	−67.5	−C4−H8−O
Propyl ether to acid	−28.0675	−67.5	−C3−H8+O
tert-Butyl dealkylation	−56.0624	−62.4	−C4−H8
Isopropyl to acid	1.943	−57	−C2−H6+O2
Ethyl ketone to acid	−26.0519	−51.9	−C3−H6−O
Isopropyl to alcohol	−26.0519	−51.9	−C3−H6+O
Ethyl ether to acid	−14.0519	−51.9	−C2−H6+O
Debenzylation	−90.0468	−46.8	−C7−H6
Isopropyl dealkylation	−42.0468	−46.8	−C3−H6
Ethyl to carboxylic acid	15.9586	−41.4	−C−H4+O2
Demethylation and methylene to ketone	−0.0365	−36.5	−C−H4−O
Methyl ketone to acid	−12.0363	−36.3	−C2−H4+O
Ethyl to alcohol	−12.0363	−36.3	−C2−H4+O
2-Ethoxyl to acid	−0.0363	−36.3	−C−H4+O
Oxidative deamination to ketone	−1.0316	−31.6	−N−H3+O
Two sequential desaturations	−4.0314	−31.4	−H4
Deethylation	−28.0312	−31.2	−C2−H5
Demethylation and two hydroxylations	17.9741	−25.9	−C−H2+O2
Hydroxylation and ketone formation	29.9741	−25.9	−H2+O2
Quinone formation	29.9741	−25.9	−H2+O2
Demethylation to carboxylic acid	29.9742	−25.8	−H2+O2
Demethylation and hydroxylation	1.9792	−20.8	−C−H2+O
Methylene to ketone	13.9792	−20.8	−H2+O
Hydroxylation and desaturation	13.9792	−20.8	−H2+O
Oxidative deamination to alcohol	0.984	−16	−N−H+O
Demethylation	−14.0157	−15.7	−C−H2
Hydroxylation and dehydration	−2.0157	−15.7	−H2
First/second alcohols to aldehyde/ketone	−2.0157	−15.7	−H2
Desaturation	−2.0157	−15.7	−H2
1, 4-Dihydropyridines to pyridines	−2.0157	−15.7	−H2

(Continued)

TABLE 13.1 (Continued)

Description	Mass Change (μ)	Fractional Mass Change (mμ)	Formula Change
3x Hydroxylation	47.9847	−15.3	+O3
Aromatic thiols to sulfonic acids	47.9847	−15.3	+O3
Hydroxymethylene loss	−30.0106	−10.6	−C−H2−O
Alcohols dehydration	−18.0105	−10.5	−2H−O
Dehydration of oximes	−18.0105	−10.5	−2H−O
2x-Hydroxylation	31.9898	−10.2	+O2
Thioether to sulfone	31.9898	−10.2	+O2
Hydroxylation	15.9949	−5.1	+O
Second/third amine to hydroxylamine/N-oxide	15.9949	−5.1	+O
Thioether to sulfoxide, sulfoxide to sulfone	15.9949	−5.1	+O
Aromatic ring to arene oxide	15.9949	−5.1	+O
Alkene to epoxide	15.9949	−5.1	+O
Trifluoromethyl loss	−67.9984	1.6	−C−F3+H
Oxidative defluorination	−1.9957	4.3	−F+H+O
Decarbonylation	−27.9949	5.1	−C−O
Sulfoxide to thioether	−15.9949	5.1	−O
Alkenes to dihydrodiol	34.0054	5.4	+H2+O2
Reductive defluorination	−17.9906	9.4	+H−F
Decarboxylation	−43.9898	10.2	−C−O2
Hydration, hydrolysis (internal)	18.0106	10.6	+H2+O
Hydrolysis of aromatic nitriles	18.0106	10.6	+H2+O
Hydrolysis of nitrate esters	−44.9851	14.9	−N−O+H
Ketone to alcohol	2.0157	15.7	+H2
2x Reductive defluorination	−35.9811	18.9	+H2−F2
Thioureas to ureas	−15.9772	22.8	+O−S
Nitro reduction	−29.9742	25.8	+H2−O2
Oxidative dechlorination	−17.9662	33.8	+O+H−Cl
Reductive dechlorination	−33.9611	38.9	+H−Cl
2x Reductive defluorination	−67.9222	77.8	+H2−Cl2
Oxidative debromination	−61.9156	84.4	+H+O−Br
Reductive debromination	−78.9105	89.5	+H−Br

[a] *Source:* Mortishire-Smith et al. (2005); © John Wiley & Sons.

set for that range based on the expected fragments (leftmost box in Fig.13.5). After applying the designed mass defect filters, all the ions outside of the filter ranges, which are expected to be ions of no interest, will be removed. It is also possible to apply multiple mass defect filters across selected mass ranges based on the mass defects of the parent drug ion and its observed fragments, as well as

TABLE 13.2 Mass Change and Fractional Mass (Mass Defect) Changes for Phase II Biotransformations

Description	Mass Change (μ)	Fractional Mass Change (mμ)	Formula Change
2x Sulfate conjugation	159.9136	−86.4	+6O+2S
Hydroxylation and sulfation	95.9517	−48.3	+4O+S
Sulfate conjugation	79.9568	−43.2	+3O+S
Taurine conjugation	107.0041	4.1	+2C+5H+N+2O+S
Cysteine conjugation	119.0041	4.1	+3C−5H+N+2O+S
Hydroxylation and methylation	30.0105	10.5	+C+2H+O
Acetylation	42.0106	10.6	+2C+2H+O
N-acetylcysteine conjugation	161.0147	14.7	+6C+8H+N+3O+S
(O_r N_r S) methylation	14.0157	15.7	+C+2H
Glycine conjugation	57.0215	21.5	−2C+3H+N+O
Hydroxylation + glucuronide	192.027	27	+6C+8H+7O
Glucuronide conjugation	176.0321	32.1	+6C+8H+6O
Decarboxylation and glucuronidation	148.0372	37.2	+5C+8H+5O
Desaturation + GSH conjugation	303.0525	52.5	−10C+13H+3N+6O+S
Epoxidation + GSH conjugation	321.0631	63.1	−10C+15H+3N+7O+S
2x Glucuronide conjugation	352.0642	64.2	+12C+16H+12O
GSH conjugation	305.0682	68.2	+10C+15H+3N+6O+S

[a] *Source:* Mortishire-Smith et al. (2005); © John Wiley & Sons.

possible metabolic reactions (e.g., glutathione conjugation), to retain as many metabolite ions as possible (Zhang et al., 2003; Zhu et al., 2006; Bateman et al., 2007). The necessity of creating mass defect filters in a chemically intelligent manner is highlighted in Figure 13.6, in which the likely metabolic cleavages of four typical drugs have been combined with classical phase 1 biotransformations to create the likely metabolic mass defect spaces for each compound. This is critically important for compounds that are asymmetric with respect to hydrogen content (e.g., where a molecule contains multiple aromatic rings on one half and alkyl groups on the other). Where a number of compounds need to be screene,d for their metabolism, or when time is of the essence, automated mass defect filtration is a valuable addition. To this end, an approach to automated, generic mass defect filter composition has been described (Mortishire-Smith et al., 2009) in which all bonds likely to undergo metabolic cleavage in a molecule are systematically identified, disconnected, and convoluted with sets of phase 1 and/or phase 2 biotransformations to generate the mass defect space of all likely metabolites from which any number of subfilters can be created and applied to the data set.

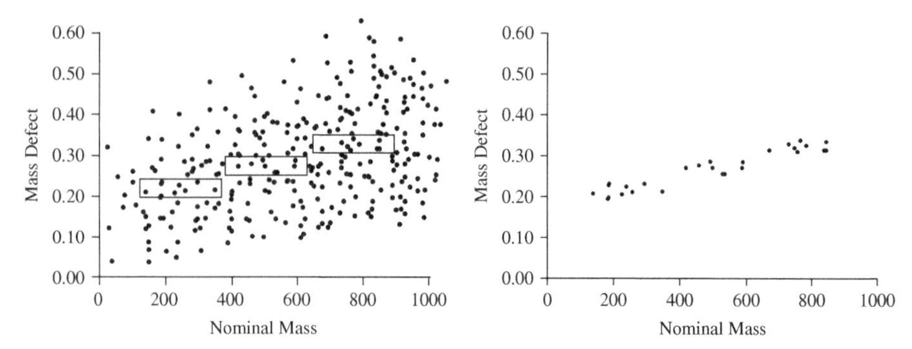

FIGURE 13.5 Illustration of the mass defect filter concept for the removal of interference ions from drug metabolites. Mass defects of ions in mass spectra from high-resolution LC–MS data are represented in mass defect plots. The left panel illustrates the mass defect plot of an unprocessed spectrum. The rectangular boxes in the plot illustrate multiple mass defect filters over distinct mass ranges. The filters can be designed with a narrow width (e.g., ± 50 mDa) around the mass defect(s) of a parent drug, of its glutathione conjugate, or of its anticipated breakdown metabolite. The right panel illustrates the mass defect plot of a mass spectrum after such filtering, which contains fewer ions and thus facilitates metabolite identification [Zhang et al. (2008b); © Wiley.]

It is noteworthy that the selectivity of the MDF approach depends not only on the mass defect similarities between drug metabolites but also the mass defect differences between drug metabolites and endogenous components in a biological matrix. This relationship is illustrated with the mass defect profiles of typical biological matrices for drug metabolite identification shown in Figure 13.7, (Zhang et al., 2008c). The efficiency of a mass defect filter in a given biological matrix is affected primarily by how close it is in a mass defect profile to the dense clusters of endogenous components. The latter, in turn, is dependent on the kind of matrix under investigation. In general, the efficiency of a MDF is better in plasma, bile, and fecal matrices than in urine. This effiency is evident in Figure 13.7 where the majority of 115 commonly prescribed drugs, as represented by the oval defined in Figure 13.7*e*, show a pattern that is largely distinct from the dense clusters of endogenous components in plasma, bile, and feces but overlapped with many of the components in urine. Significantly, mass defect filters for detecting glutathione (GSH) conjugates can be more efficient than mass defect filters for oxidative metabolites. This is because the positions of GSH conjugates in a mass defect profile move to the right relative to the parent drug (in the integral mass dimension) without much increase in the decimal mass dimension, while the mass defects of matrix clusters generally increase monotonically with mass. This behavior gives rise to a better separation of these conjugated metabolites from the interference clusters than that predicted for oxidative metabolites. This advantageous selectivity of MDF toward GSH conjugates has been demonstrated in application by the analysis of diclofenac acid–GSH adducts in rat bile (Zhu et al., 2007).

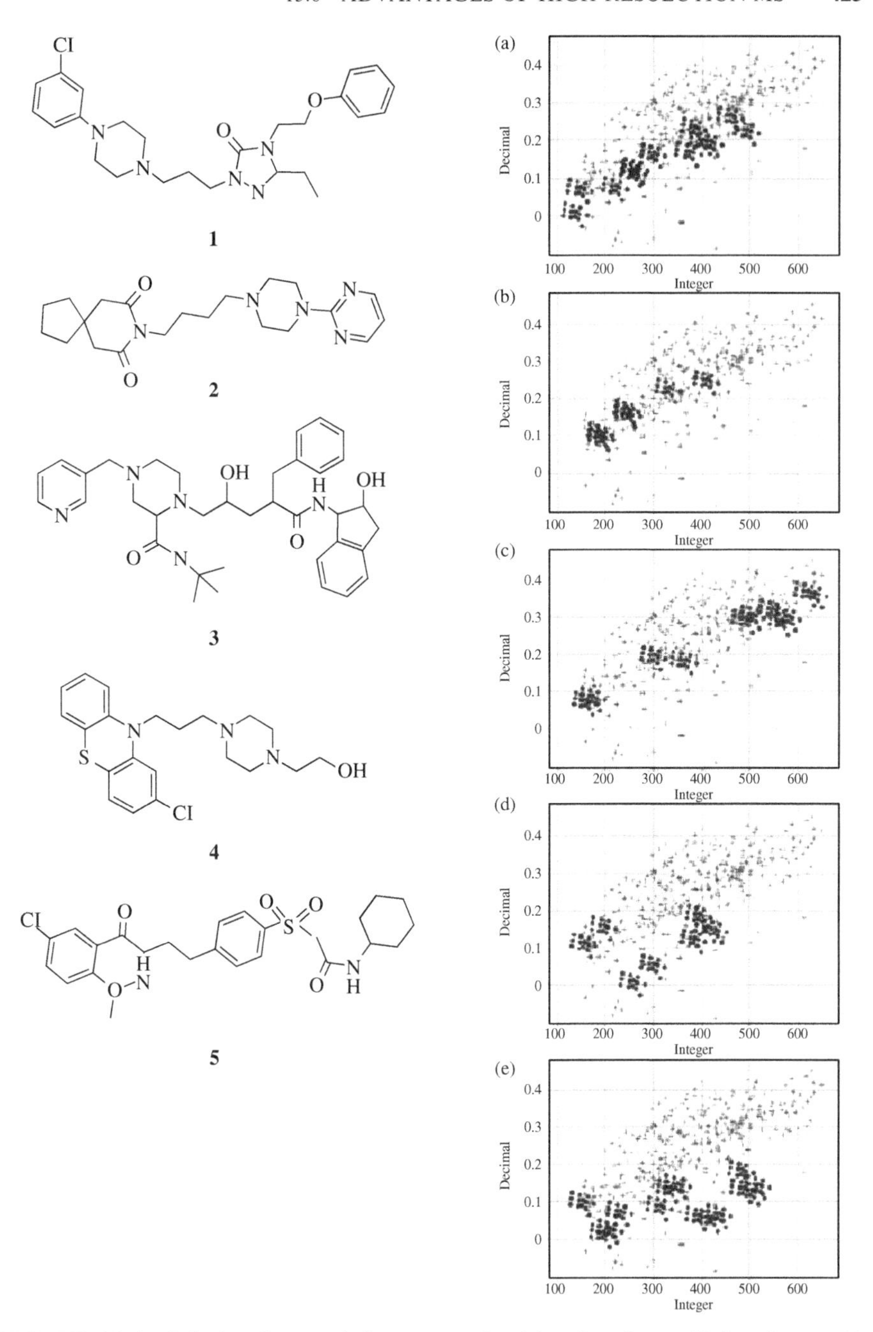

FIGURE 13.6 Calculated mass defect spaces for (*a*) nefazodone, (*b*) buspirone, (*c*) indinavir, (*d*) perphenazine, and (*e*) glyburide. Each solid circle denotes the integer/decimal pairing for a disconnection combined with a phase I biotransformation. Crosses represent integer/decimal pairings for background ions observed in analysis of a control microsomal incubation. [Mortishire-Smith et al. (2009); © Wiley.]

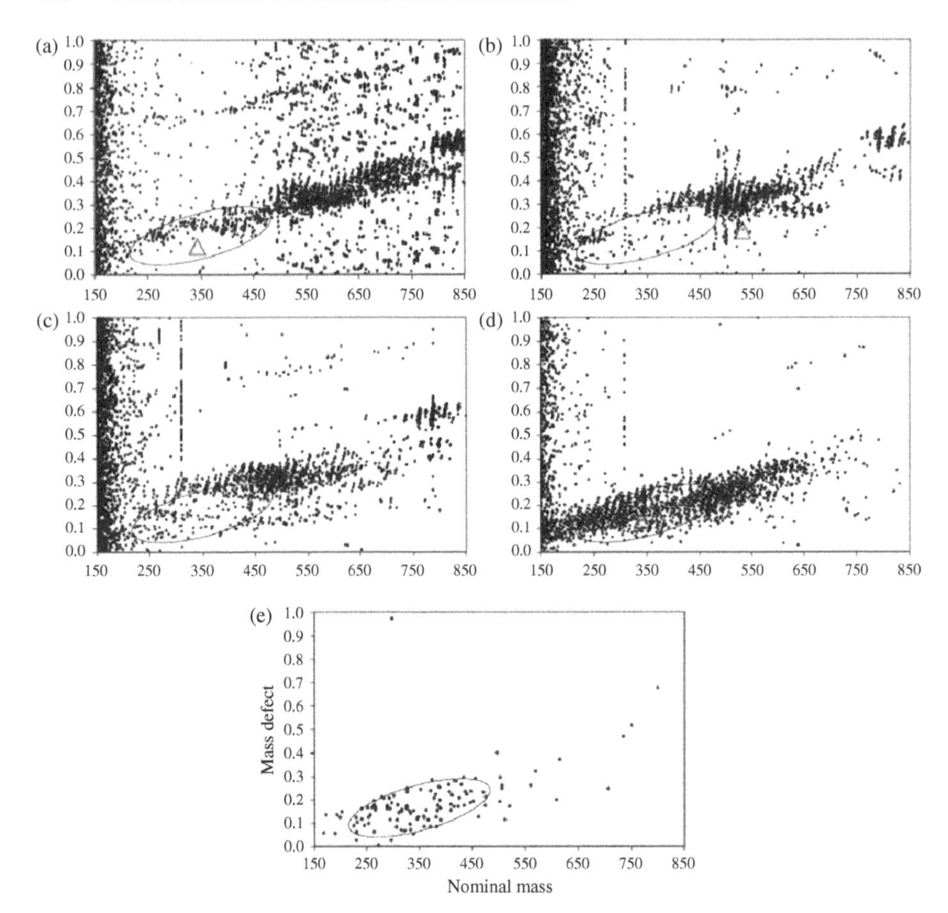

FIGURE 13.7 Mass defect plots of human biological matrices of (*a*) plasma, (*b*) bile, (*c*) feces, and (*d*) urine from high-resolution LC–MS analysis compared with those of (*e*) 115 marked drugs. Full-scan LC–MS data from each biological matrix were combined into one spectrum to generate the mass defect plots. The oval in each plot represents the mass defect–nominal mass coordinates of the majority of the 115 commonly prescribed drugs. [Zhang et al. (2008c); © Wiley.]

13.7 POSTPROCESSING STRATEGIES FOR IDENTIFYING METABOLITES IN COMPLEX HIGH-RESOLUTION DATA SETS

13.7.1 Classical Metabolites

The purist's approach to data-up metabolite identification might be to evaluate each analyte-specific peak in turn, determining the biotransformation from the accurate *m*/*z* obtained for the metabolite, in practice, a significant proportion of metabolites—those formed on the basis of classical phase 1 or phase 2 biotransformations—can be retrieved on the basis of narrow-window extracted ion chromatograms (nwXICs) (Mortishire-Smith et al., 2005; O'Connor et al., 2006; Bateman et al., 2007; Koulman et al., 2009; Zhang et al., 2009c). Here, a chromatogram is reconstructed from the full-scan low-energy channel based on the presence of ions within a defined tolerance of the predicted *m/z* of the metabolite (typically, a few parts per million). The reduction of background signal with increasing mass resolution and decreasing tolerance is dramatic (Fig.13.2). A comparison with a control sample still allows false-positive peaks to be identified, though at high mass resolution, false-positive peaks in these data sets most commonly arise from minor oxidative impurities in the compound batch with which the incubations are spiked, rather than mass-coincident species in the matrix background.

Given the ease with which nwXICs can be generated and peak picked, it makes little sense not to pick these proverbial "low-hanging fruit," leaving the analyst to focus on more challenging metabolites.

13.7.2 Identifying All Other Analyte-Specific Peaks

While classical metabolites can be identified with relative ease, the purpose of drug metabolism studies is to characterize the metabolic fate of a compound to a level appropriate to the development stage of the compound and to ensure that there are no surprises later on. Practically speaking, this means identifying all (i.e., not only simple phase 1) metabolites present above a defined threshold, understanding the nature of the metabolic pathways involved, and then reporting metabolites of interest to the project team.

To this end, the limiting issue in analyzing high-resolution LC–MS data sets is identifying all analyte-specific, drug-related entities in the presence of large amounts of matrix background signal. A variety of algorithms are available to facilitate this process, some of which use the properties of the parent compound.

13.8 CONTROL COMPARISON

In the ideal situation the control sample will be perfectly matched to the analyte sample such that a simple subtraction of control from analyte leaves only the analyte-specific peaks behind. In reality, of course, such a perfect match can rarely be achieved. Nonetheless, a careful choice of the control sample can reduce the number of analyte-specific false positives. For in vitro samples, a control incubation performed under identical conditions to the analyte, but with the test compound spiked at termination, was demonstrated to the best matched to the analyte, compared with an incubation lacking cofactor or a generic blank incubation (Mortishire-Smith et al., 2005). For in vivo samples, the best control sample is likely to be one taken from a vehicle-dosed animal and sampled at equivalent time points to the dosed animals. One-to-one comparison of control and analyte data sets is relatively easy to implement and understand, and is the most widely used. More complex multivariate statistical algorithms are now also appearing in the literature, for example, Sun et al. (2009) identified metabolites of tolcapone excreted in rat urine using QTOF technology and principal component analysis.

13.9 BACKGROUND SUBTRACTION

A second, compound-independent, strategy for the removal of false-positive peaks from consideration is that of background subtraction. Algorithms to accomplish this are applicable to both nominal and high mass resolution data sets, have been in use for many years, and are commonly embedded in the operating software of most instruments. A useful comparison of four commonly used filters is described by Fredriksson et al. (2007). More recently, publications have appeared in which novel systematic background subtraction strategies are applied to high-resolution data sets (Zhang and Yang 2008; Zhang et al., 2008a, 2009a; Zhu et al., 2009a), and include those from in vitro, fortified in vitro and in vivo sampling. In most of these, ions in a peak of interest in the analyte sample are evaluated for their presence in a broader time window in the control sample.

Control comparison and background subtraction approaches have the advantage of being truly objective with respect to enriching analyte data sets—they do not rely on knowledge of the structure of the test compound. Where relatively few analyte-specific peaks are left for review following these cleanup steps, there is no necessity to resort to more subjective strategies. However, where a control sample is poorly matched with the analyte sample (with respect either to content or retention times), the number of peaks presented for review may be unmanageably large. Under these circumstances, compound-dependent approaches are required.

13.10 ISOTOPE FILTRATION

Where a drug contains a halogen atom or an isotopic label at less than 100% isotopic enrichment, the result is a set of isotope pattern ratios that is not commonly present in the biological background. This phenomenon has been used for many years to highlight drug metabolites (Leal et al., 1992; Lanting et al., 1993). Essentially, all ions in the analyte sample (either before or after control comparison/background subtraction) are evaluated for the presence of a partner ion separated by a defined mass and with a defined intensity ratio. This procedure can still be accomplished at nominal mass resolution (Yan et al., 2008), but, using an accurate mass difference (e.g., 1.9970 for chlorine, 1.0062 for deuterium, 2.0032 for ^{14}C) to filter the data results in a significantly improved enrichment of the analyte data set with respect to drug metabolites (Cuyckens et al., 2008, 2009; Hobby et al., 2009; Zhu et al., 2009b). Where the drug molecule contains the label to start with, the subjective assumption is made that all metabolites will also contain the label, which may not be the case when oxidative dehalogenation or an internal cleavage of the molecule has occurred. In the specific situation where the metabolite search is directly toward conjugates of reactive intermediates, the use of an isotopically labeled trapping agent such as glutathione or cyanide (Zhu et al., 2007) can be used to good advantage.

13.11 "ALL-IN-ONE" DATA ANALYSIS

All-in-one data acquisition, first described in its application to drug metabolism using QTOF technology (Wrona et al., 2005; Bateman et al., 2007), and subsequently also implemented on Orbitrap instrumentation (Zhang et al., 2009a; Bateman et al., 2009), is a powerful, generic approach for facilitating metabolite finding and characterization. Two scan functions are acquired, one at low collision energy and the second at high collision energy, both without preselection of a target mass. Together, these two channels provide complementary data that can be mined ad libitum for metabolites. The low-energy channel is typically mined as described above—by generating nwXICs for individual biotransformations, and further identifying analyte-specific peaks and determining their drug relatedness. The high-energy channel can be mined using a wider variety of techniques, bearing in mind that since the fragmentation is pseudo-MS/MS, with no preselection of a target ion, care must be taken with interpretation. A comparison between high-energy all-in-one data and directed MS/MS data for selected microsomal metabolites of verapamil is shown in Figure 13.8.

Generation of Product Ion Data for the Parent Compound and Metabolites Identified in Low-energy Trace High-energy ions with a retention time and chromatographic peak shape similar to that of the selected low-energy peaks are likely to be product ions of the low-energy species.

Parent Precursor Ion Scanning Generate nwXICs from the high energy channel for all parent compound product ions. This result will highlight all metabolite peaks which have product ions in common with the parent.

Modified Precursor Ion Scanning Generate nwXICs from the high-energy channel for m/z values calculated by shifting the m/z values of the parent compound product ions by a defined amount (typically, the biotransformation mass shift). This result identifies all metabolites that have shifted product ions relative to the parent drug.

Defined Precursor Ion Scanning A compound-independent method of identifying glutathione conjugates has previously been described by Dieckhaus et al.

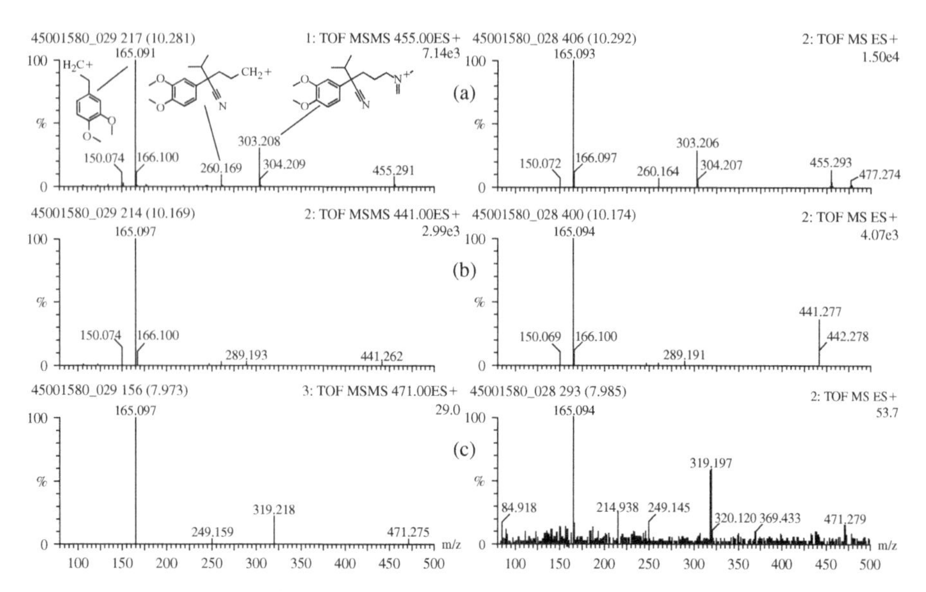

FIGURE 13.8 Comparison of targeted MS–MS (left-hand panel) and nonselective MS/MS (right-hand panel) for (*a*) verapamil, (*b*) *N*-demethylated verapamil, and (*c*) a minor hydroxylated metabolite of verapamil. (Wrona et al., 2005 © Wiley.)

(2005) on nominal mass instrumentation, which takes advantage of the observation that in negative ion experiments, glutathione conjugates frequently fragment to yield a 272.0888-Th ion. Where all-in-one experiments have been acquired in negative ion mode, searching the high-energy function for this ion retrieves glutathione conjugates that fragment by this pathway.

Neutral Loss Scanning In a conventional neutral loss scan function on a triple quadrupole instrument, Q_1 and Q_3 scan in synchrony with an offset corresponding to the neutral loss of interest. In all-in-one data analysis, a similar effect can be achieved by identifying all high-energy scans in which two ions separated by a designated mass difference appear (Fig 13.9). For example, glucuronides can be identified by searching for pairs of ions separated by 176.0321 Th and glutathione conjugates by 129.0426 Th. This feature has been implemented as a data-dependent acquisition method on a variety of instrumentation. More generally, though, a key advantage of all-in-one data is that any neutral loss of interest can be trivially recreated as a chromatogram, for example, where a novel biotransformation is identified with a characteristic fragmentation pathway, one can search for related metabolites.

Cross-Correlation Analysis Cross-correlation analysis is an established methodology for determining the likelihood that two spectral data sets are related to each other (Owens, 1991). Where novel analyte-specific peaks are identified in

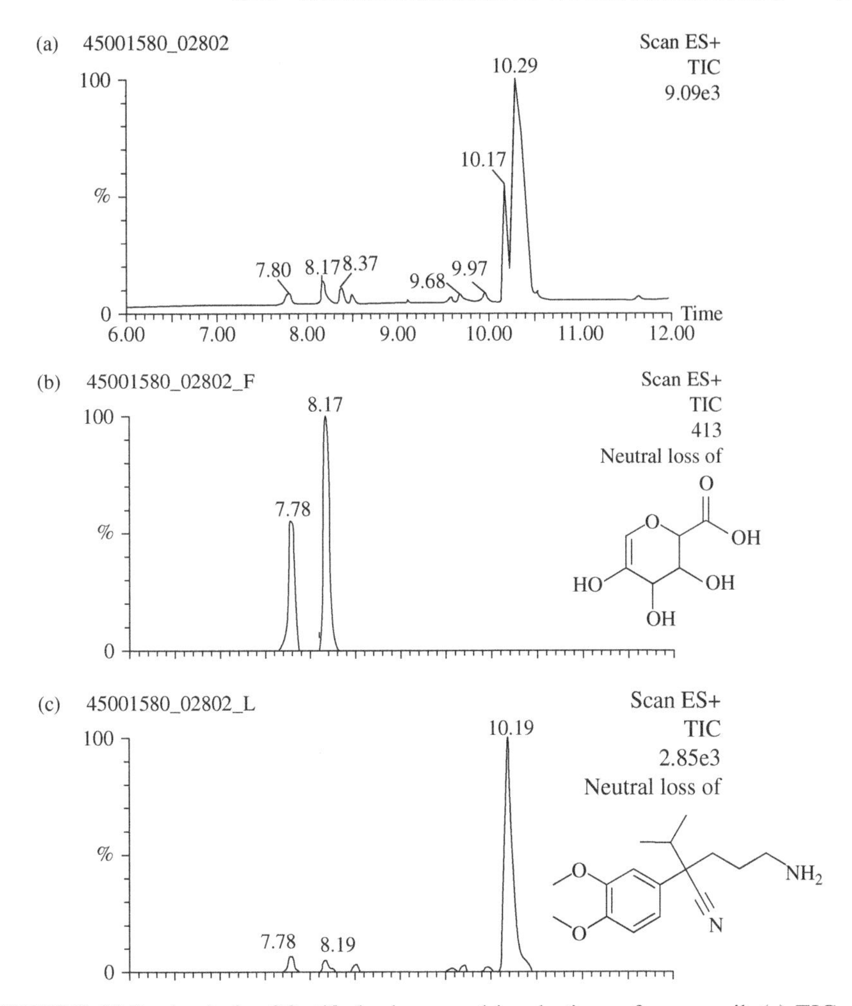

FIGURE 13.9 Analysis of fortified microsomal incubations of verapamil; (*a*) TIC of higher energy scan, and reconstructed neutral loss chromatograms for (*b*) glucuronidation (176.034 ± 0.01 Da), and (c) 276.184 ± 0.01 Da. Bateman et al., 2009, (© Wiley.)

the low-energy channel, it is possible to use coeluting high-energy product ions to rank the peaks on the basis of parent relatedness.

13.12 RATIONALIZATION OF NOVEL METABOLITES

Metabolites that have "conventional" mass changes relative to parents are usually readily identified and require little additional investment of time to be understood, beyond localization of the site of metabolism. Metabolites in which the mass change is unexpected are often of greater interest since they may be formed by a nonstandard (e.g., non-cytochrome P450 (CYP)) enzyme, or else reflect an intrinsic structural liability of the molecule toward oxidation. Many of these novel mass changes are likely to be secondary metabolites (i.e., downstream biotransformations of a primary oxidation, either oxidative or conjugative). To this end, where software is able to take advantage of chemically intelligent inputs, an increased hit rate for metabolite rationalization is the result. For example, using a simple set of biotransformations, and no compound-dependent cleavages, only 10 out of 26 metabolites of indinavir present in the LC–MS data set were rationalized by Metabolynx (Mortishire-Smith et al., 2009). Including all likely metabolic cleavages in the input list resulted in a correct assignment for all 26 metabolites. Similarly, for nefazodone, of 46 metabolites known to be present in a microsomal incubation, only 17 were assigned using the simple standard set, but all were assigned when compound-specific cleavages were used.

A critical component in confirming the nature of these unusual/novel metabolites is the determation of the likely elemental composition of the metabolite, and this is where high-resolution mass spectrometry has a dramatic advantage over nominal mass instrumentation. Historically, elemental composition determination was done in a more tedious way with double-focusing magnetic sector instruments, but nowadays it can be routinely achieved with all of the instrumentation described in this chapter (Godejohann et al., 2004; Nikolic et al., 2004; Castro-Perez et al., 2005). The ability to determine elemental compositions of molecular adducts and product ions is important in metabolite identification, as it facilitates structural elucidation for metabolites detected as well as elimination of some false positives. This advantage was recognized very early on, and a number of applications of the QTOF instrument around the turn of the century described the use of high mass accuracy to assign empirical formulas to molecular and product ions for structural elucidation (Chernushevich et al., 2001). For example, Eckers et al. (2000) reported accurate mass measurements for determination of both protonated and deprotonated molecules and product ions. Hop et al. (2001) used accurate mass measurement of product ions of metabolites to investigate a novel fragmentation mechanism. In addition, the accurate mass data also allowed for the identification of six pharmaceuticals (Zhang et al., 2000b) and a hydroxylamide sulfoconjugated metabolite (Pilard et al., 2000).

The key to successful determination of elemental compositions is the mass accuracy of the data. A high confidence in the accuracy of the measured mass enables one to determine the right elemental composition among the wealth of different possibilities at a given nominal m/z or at least to limit the number of possible compositions to a few of the most probable ones (Ferrer and Thurman, 2005). For example, good mass accuracy allowed for definitive determination of the parent and fragment ions of some mercapturic acid derivatives of hydroxydiclofenac (Poon et al., 2001). In addition to mass accuracy, the molecular mass of an ion also affects the elemental composition determination. Assuming a potential elemental composition range encompassing C0-100, H3-74, O0-4, and N0-4, there will be only one possible formula within 34 ppm for a molecular mass of 118 Da. However, there will be 626 probable formulas that are within 5 ppm for a molecular mass of 750.4 Da. Therefore, for small molecules up to 300 Da a mass accuracy of 5 ppm is typically sufficient for nonambiguous elemental composition determination, whereas for higher molecular masses better mass accuracy (<1 ppm) is desirable (Gross, 1994). In practicality, a third consideration that may be useful for further reducing the number of probable formulas is to utilize the knowledge of a potential metabolite candidate, for example, the number of its N, O, S, or halogen atoms, the limit of its double-bond equivalents, and/or its possible isotope patterns.

Characterization of Novel Metabolites One of the strengths of high-resolution mass spectrometry is its ability to facilitate the detection and identification of unusual drug metabolites in complex samples. The value of this ability is demonstrated in the detection of two unpredicted oxazole ring-opened metabolites (M9 and M16) of muraglitazar in the feces of humans (Zhang et al., 2007). The unusual nature of the two metabolites, first of all, made the extracted ion chromatogram approach inapplicable. Second, the traditional way of locating their metabolite ions based on radiochromatogram peak reference had also failed primarily because of the low intensities of the metabolite ions and the significant interferences from the fecal matrix. Third, unlike other muraglitazar metabolites, M9 and M16 formed few or no protonated species but instead were observed as NH_4^+ or Na^+ adducts. Without prior knowledge, such adduct ions made it even more challenging to identify drug-related ions amid fecal interferences. With mass defect filtering on high-resolution LC–MS data, however, matrix interference ions were significantly removed; the resulting total ion chromatogram was comparable to the radiochromatogram profile and clearly displayed drug metabolite peaks including M9 and M16. The corresponding mass spectra revealed the mutually confirming proton/sodium or ammonium/sodium adducts. The accurate mass information associated with each of these ions facilitated the determination of their elemental compositions, which led to the rationalization of the identity of the metabolites. The facile detection and identification of the metabolite ions allowed for follow-up studies to definitely confirm the metabolite structures.

13.13 ASSIGNING PRODUCT ION SPECTRA USING THE POWER OF ACCURATE MASS

Despite the many and significant advances made in recent years toward fully efficient strategies for the identification of drug metabolites in complex biological matrices, and subsequent assignment of a specific biotransformation to the mass change, ultimately the information that this affords is often of relatively little use to drug discovery scientists. The primary reasons for characterizing drug metabolites in the discovery phase are to rationalize high plasma clearance, to better understand potential mechanisms of toxicity, or to fully characterize human-specific metabolites in order to be able to synthesize authentic reference standards for quantitation. Where the formula change of the biotransformation by itself is sufficient to completely define the structure of the metabolite (e.g., demethylation in a compound with one methyl group), the job of the drug metabolism scientist is then complete. But these represent a relatively small proportion of metabolites. In other cases, the nature of the biotransformation gives a strong direction as to the site of metabolism—for example, oxidative defluorination in a compound that contains one aromatic fluorine atom is likely have only one or two regioisomeric explanations. For all remaining metabolites, biotransformations must be localized as far as possible in order to direct the next stages of research on development.

And herein lies a key challenge (or opportunity, for the optimistic, or those who have been on management training courses)! The key resource in the toolbox of the mass spectrometrist is the product ion spectrum, enhanced if possible with data on the pathways by which the product ions have been formed (i.e., $MS^{2,\ 3\ \cdots\ n}$ experiments). The interpretation and analysis of product ion spectra still lies primarily in the realm of the experienced (and in many cases, wizened) expert. Where the structure of the parent ion is defined (e.g., the drug under investigation), assignment of the product ion spectrum is facilitated—but is nonetheless tedious when undertaken manually. However, with the majority of drug metabolites, the structure of the metabolite is not known or at best is described by a set of between 2 and *n* plausible regioisomers. In some cases, the product ion spectrum of the metabolite under investigation is closely related to that of the parent compound, with only one or two mass changes that point directly to localisation on a narrowly defined substructure. In most other situations, the challenge is to determine the minimal substructure to which the biotransformation can be localized from a set of product ions that may be substantially different to those obtained for the parent drug.

What, then, are the options currently available to the scientist, and is accurate mass measurement a help? The latter question is largely rhetorical, especially for anyone who has tried to make sense of product ion spectra obtained on instrumentation capable only of generating nominal mass data. Trivial neutral losses can usually be assigned on the latter instruments, but not always, for example, distinguishing between a neutral loss of methane from that of an oxygen atom is not possible. The presence or absence of a halogen isotope pattern can still be used to limit possible explanations for a product ion. On the other hand, the benefit of accurate mass measurement is dramatic, as noted earlier, in that it affords a dramatic reduction in elemental compositions that are possible explanations of the measured m/z value—and thus limits the number of possibilities that must be considered by the analyst.

Indeed, the literature is replete with carefully determined assignments of product ion spectra across a broad range of compound classes, many of which have been undertaken using high-resolution instrumentation and the introduction of QTOF mass spectrometers has had a particularly significant impact in this regard. Relatively sparse, however, are publications describing formal strategies by which product ions can be assigned with confidence.

This situation, however, appears to be changing, as increased attention is being paid to the challenges associated with localization. For example, an approach entitled EPIC (Elucidation of Product Ion Connectivity) was described in 2005 (Hill and Mortishire-Smith, 2005) and is currently commercially available both in stand-alone form and integrated with instrumentation software. The approach uses a brute force, rather than rule-based, approach to generate substructures whose elemental compositions are compared with all compositions possible for the observed product ion (within defined tolerances). Input to the software comprises a set of experimentally measured m/z values and a structure. The first step in the process is a systematic n-ply ($n \leq 5$) disconnection of all heavy atom–heavy atom bonds in the molecule to generate all fragments that are accessible by disconnecting up to n bonds. Then, for each observed m/z value, compatible elemental compositions are calculated and matched against the virtual fragment library. Where multiple substructures are feasible within defined Th tolerances, each is assigned a score that captures the likelihood that the substructure can be formed from the parent compound (i.e., number of bonds broken, types of bonds broken, hydrogen deficit, etc.). The user is then presented, for each observed ion, with a ranked set of possible fragments. A number of alternative software packages are available that use proprietary bond disconnection approaches, for example, MassFragmenter (Advanced Chemical Development) was used by Pelander et al. (2009) in association with instrumentation software to identify and localize a number of in vivo urinary metabolites of quetiapine and Fragment Identificator (Heinonen et al., 2008).

13.14 LOCALIZATION: THE FINAL FRONTIER

Clearly, high-resolution mass spectrometric data dramatically simplifies the problem of product ion assignment. The routine, automatic translation of product ion data into a maximally constrained oxidation (i.e., the substructure identified as containing the site of oxidation is the smallest possible set of atoms consistent with the observed product ions) remains an open problem at the moment, at least in the sense that no robust, generic, open-source, or commercially available solution exists. A small number of publications have presented proof-of-concept or company-internal approaches to the localization problem, either with or without high-resolution data as input. In the simplest situations, either (a) the product ion spectrum of the metabolite is identical to that of the parent drug, in which case the biotransformation is likely to be on the smallest possible neutral which is lost, or (b) the product ion spectrum of the metabolite is identical to that of the parent compound, with the exception of one, attributable, low m/z product ion shifted by the mass of the biotransformation. Both these situations are relatively straightforwardly interpreted by inspection, for example, using the systematic approach first described by Kerns et al. (1994, 1995) for analyzing product ion spectra of natural products. More complex are the situations where intermediate product ions are shifted in mass (in which case, the open question is whether using other ion assignments can localize the biotransformation further), or the case in which the product ion spectrum of the metabolite has very few ions (and still fewer biotransformation-shifted ions) in common with that of the parent compound. MS^n approaches (either using accurate or nominal mass spectra) in which major product ions are themselves subjected to further fragmentation to generate product ion "trees" do ultimately afford a significant advantage for data analysis since defining the elemental composition of a precursor ion automatically reduces the number of elemental compositions possible for smaller product ions (see, e.g., the elaborated product ion tree for carvelilol in Fig. 13.10).

Cross-correlation approaches have long been used for estimating the likelihood that a novel species is in fact drug related (Fernandez-Metzler et al., 1999; Crockford et al., 2008) but are by themselves limited when authentic standards with known structures are not available. An interesting concept, metabolite identification by comparison of correlated analogs (MICCA) presented (but not subsequently published in manuscript format) by Johnson et al. (Johnson et al., 2006) takes advantage of the fact that corporate compound databases commonly contain large numbers of structural analogs synthesized to explore structure–activity relationships, thus to a degree circumventing the lack of availability of metabolite standards. Given the product ion spectrum of a metabolite of unknown structure, the MICCA algorithm involves cross correlating this spectrum with that of the parent compound, and of appropriate structural analogs (Owens, 1991), identifying the maximum common structure of each analog with the parent, and finally identifying substructures that match the correlated ions. As described by Johnson et al. (2006), the approach is not based on high-resolution product ion spectra, although conceptually there is no reason why this could not be so.

More recently, Leclercq et al. (2009) described a systematic approach to the localization problem called IsoScore, which does use high-resolution data, is somewhat related to the MICCA approach in that it invokes a systematic evaluation of analog data, but differs significantly in that the analogs are created virtually. The IsoScore algorithm takes as input the high-resolution product ion spectra of the parent compound and the metabolite under investigation. First, all possible regioisomers of the biotransformation are created (i.e., for an M+O metabolite, all plausible sites of hydroxylation). Then, iterating through each virtual regioisomer, each ion in the experimentally observed product ion spectrum is scored using the EPIC engine (Hill et al., 2005), for the likelihood that it can be formed from that structure. At the simplest level, the most likely solution for the localization is that with the

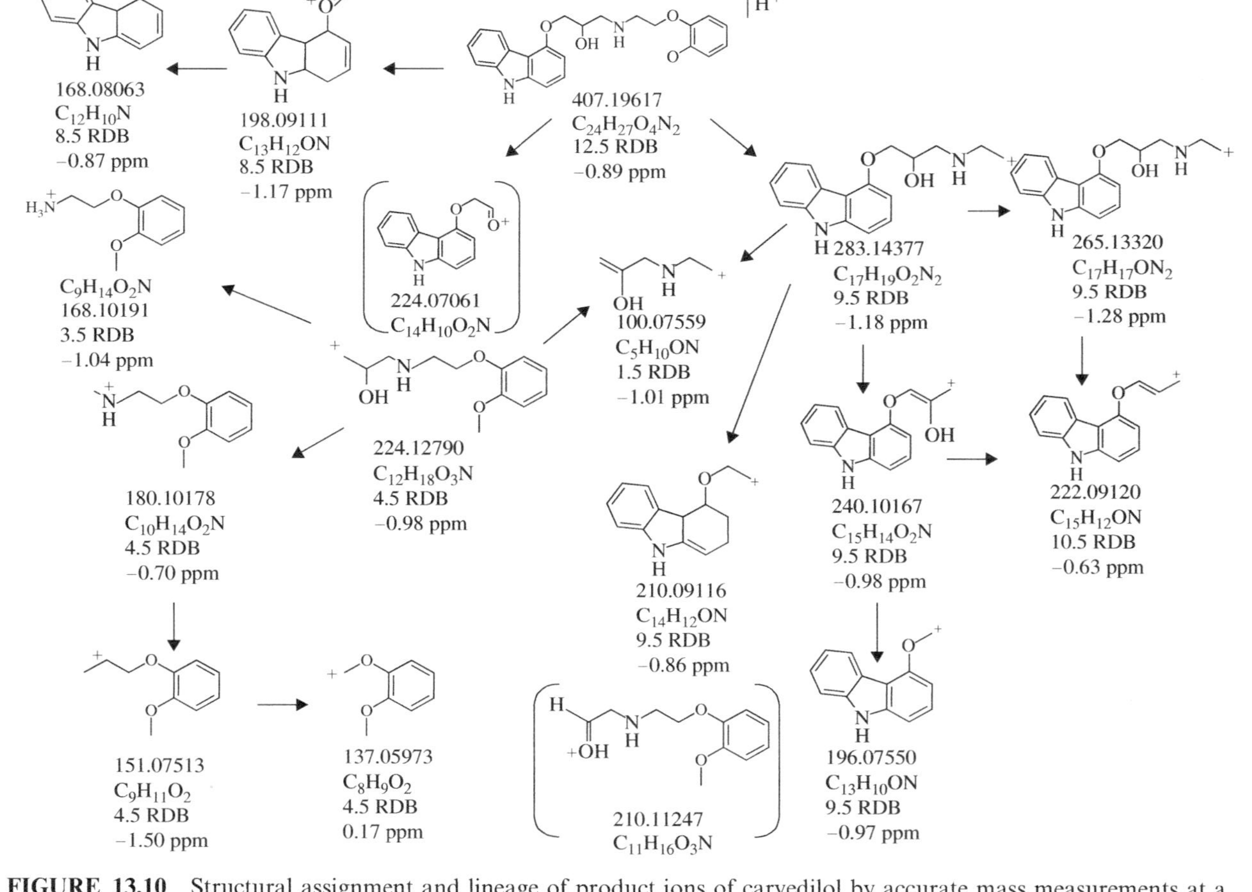

FIGURE 13.10 Structural assignment and lineage of product ions of carvedilol by accurate mass measurements at a resolving power of 60,000 including structural elucidation of three sets of isobaric product ions. [Lim et al. (2007); © Wiley.]

lowest cumulated score, however, a number of refinements are included to improve the prediction fidelity. A key point is that high-resolution data are critical since this results in a limitation of the number of possible elemental compositions.

All the previously described analytical strategies for high-resolution data analysis are combined in the flowchart shown in Figure 13.11.

13.15 QUANTITATIVE AND QUALITATIVE IN VIVO PHARMACOKINETIC DATA FROM A SINGLE INJECTION PER SAMPLE

As previously noted, the advantages of high-resolution mass spectrometry for qualitative studies of drug metabolism are significant. An obvious question is whether there is any reason not to combine the qualitative component of the experiment with the quantitative deliverable. To do so would be to add value to the most commonly performed in vivo adsorption, distribution, metabolism, and excretion (ADME) experiment—determination of plasma pharmacokinetics—at no cost to the scientist beyond the time taken to analyze the data. The limiting issue here is that such a large investment has been made in triple-quadrupole-based bioanalysis that accurate mass-based approaches are considered relatively specialized. Yet, there is no particular reason why this should be the case. A number of studies have been published that demonstrate quite clearly that a combined quantitative/qualitative endpoint is achievable on real-world samples, using a variety of instrumentation (Zhang et al., 2000a, 2000b; Williamson et al., 2008). A number of these more recent studies are highlighted below.

- O'Connor et al., (2006a) acquired full-range scan data using a triple quadrupole–time of flight instrument on plasma obtained from rats following dosing of an experimental drug that had previously been demonstrated to have highly variable exposure in rats. The methodology used was a combined quantitative/qualitative experiment in which the plasma concentrations of the parent compound were obtained via narrow-window XICs for the parent drug. Both the individual plasma concentrations and the derived pharmacokinetic parameters were comparable with those obtained using a conventional triple quadrupole instrument. The QTOF data was also analyzed for circulating metabolites, and the interanimal variability determined to be a consequence of an oxidation catalyzed by aldehyde oxidase, an enzyme known to be highly polymorphic in the rat O'Connor et al. (2006b).
- Zhang et al. (2009c) compared the quantitative performance of a current-generation Orbitrap instrument with that of one of the most commonly used instruments in bioanalysis, an Applied Biosystems API-4000 triple quadrupole system. The study was conducted using a test set of 15 compounds under evaluation in drug discovery, with the quantitation

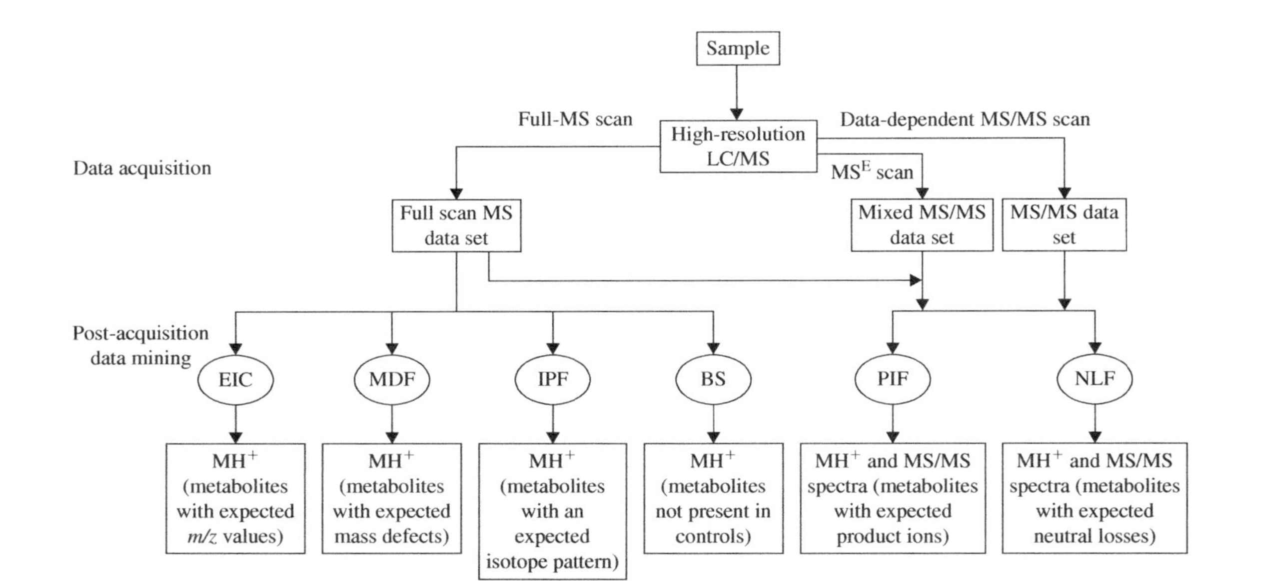

FIGURE 13.11 Analytical strategies for drug metabolite detection and identification using high-resolution mass spectrometry and multiple postacquisition data-mining techniques. The accurate mass full-scan MS and MS/MS data sets are acquired with data-dependent methods, including intensity-dependent MS/MS scan and simultaneous acquisition of exact mass at high and low collision energy (MS^E). Extracted ion chromatography (EIC) process, mass defect filter (MDF), background subtraction (BS), product ion filter (PIF), neutral loss filter (NLF), and isotope pattern filter (IPF) can be employed for data mining. Additionally, isotope pattern-dependent MS/MS scanning can be used in data acquisition. [Zhang et al. (2009b); © Wiley.]

on the Orbitrap being performed using a mass window of +5–10 mTh to extract analyte peaks of interest in the full-scan HRMS data and the quantitation on API-4000 being selective reaction monitoring (SRM) based. They concluded that the performance of the two systems was not distinguishable and demonstrated that information on circulating metabolites was directly obtainable from the same data sets used for quantitation. As noted previously, one of the biggest advantages of the full-scan HRMS method is that the data on metabolites can be retrieved at any time postacquisition should the need arise to investigate additional species of interest. For example, they demonstrated that although both the full-scan HRMS and the SRM methods produced mutually agreeable plasma concentration–time profiles for a drug candidate, only the full-scan HRMS data allowed for further evaluation on whether the metabolites observed in vitro would present in the plasma samples. Based on the evaluation, they were able to determine without any additional experiments that there were no oxidative metabolites present in the plasma samples and that a glucuronide metabolite was the major circulating metabolite of the drug candidate in rats.

- Bateman et al. (2009) describe the use of a nonhybrid Orbitrap mass spectrometer for both qualitative and quantitative analysis of in vitro and in vivo samples. Data collection used nonselective MS/MS data capture with the HCD-cell-enabled system. Metabolic stability was determined for a series of compounds with simultaneous determination of metabolites and their structures from the MS and MS/MS data sets. Plasma drug levels were measured using full-scan HRMS data collected using a FWHM resolution of 25,000. It was found that this was the minimum resolution required for acceptable quantification at low levels in plasma. Circulating metabolites were also detected and time profiles plotted to show relative abundance of several metabolites.

A key issue for the successful future use of this strategy for data collection is the need for continuing improvements to software tools to allow metabolites to be identified, localized, and reported efficiently.

13.16 FUTURE OPPORTUNITIES

The uptake of HRMS instrumentation into metabolite identification laboratories points to future developments and uses for these types of mass analyzers. From a hardware perspective, developments in TOF and Orbitrap technology will be needed to address lingering issues specific to these analyzers. In the TOF area, current commercial systems just coming to the market have addressed historical limitations of resolution, dynamic range, and mass accuracy. It remains to be seen if these specifications are sufficient or if further gains are needed. These instruments are not yet as sensitive as high-end triple quadrupole mass spectrometers, and improvements are needed in this area. For Orbitrap-based systems, the main limitations are scan speed and sensitivity. As chromatography is pushed to provide gas-chromatographic-like peak widths, will the Orbitrap be able to keep pace? Recent patents issued in this area, such as parallel detection systems (Makarov et al., 2008) based on the Orbitrap mass analyzer, suggest ways in which the duty cycle issues may be addressed.

Having found and assigned metabolites, the question always remains "What is the concentration of the metabolite?" While some approaches described in the literature show promise (e.g., the use of nanospray LC–MS), and calculation of relative response factors from LC–ultraviolet), there is no universal solution available at the moment, and this remains a ripe area for future development in the field of metabolite identification.

More generally, the most pressing need is for continued software development to streamline data analysis and to identify new, robust strategies for extracting and reporting all the information present in these complex data sets. There is a pressing need to build more chemical and metabolic intelligence into such algorithms, so that, for example, only reasonable rationalizations are made for novel metabolites. And if the rate-limiting step of metabolite identification is not to become the documentation of final results in a format suitable for discussion by project teams, some attention needs to be given to the need for automatic creation of camera-ready summaries, including annotated chromatograms, metabolite schemes, and product ion assignments.

REFERENCES

Allen M, Anacleto J, Bonner R, Bonnici P, Shushan B, Nuwaysir L. Characterization of protein digests using novel mixed-mode scanning with a single quadrupole instrument. Rapid Commun Mass Spectrom 1997;11:325–329.

Anari MR, Sanchez RI, Bakhtiar R, Franklin RB, Baillie TA. Integration of knowledge-based metabolic predictions with liquid chromatography data-dependent tandem mass spectrometry for drug metabolism studies: Application to studies on the biotransformation of indinavir. Anal Chem 2004;76:823–832.

Bateman RH, Carruthers R, Hoyes JB, Jones C, Langridge JI, Millar A, et al. A novel precursor ion discovery method on a hybrid quadrupole orthogonal acceleration time-of-flight (Q-TOF) mass spectrometer for studying protein phosphorylation. J Am Soc Mass Spectrom 2002;13:792–803.

Bateman KP, Castro-Perez J, Wrona M, Shockcor JP, Yu K, Oballa R et al. MSE with mass defect filtering for in vitro and in vivo metabolite identification. Rapid Commun Mass Spectrom 2007;21:1485–1496.

Bateman KP, Kellmann M, Muenster H, Papp R, Taylor L. Quantitative-qualitative data acquisition using a benchtop Orbitrap mass spectrometer. J Am Soc Mass Spectrom 2009;8:1441–1450.

Bloomfield, N. Mass defect triggered information dependent acquisition. MDS Sciex, Can. and Applera Corporation, is available at http://www.wipo.int/pctdb/en/wo.jsp?WO=2007076606, 2007.

Carr SA, Huddleston MJ, Bean MF. Selective identification and differentiation of N- and O-linked oligosaccharides in glycoproteins by liquid chromatography–mass spectrometry. Protein Sci 1993;2:183–196.

Castro-Perez J, Plumb R, Liang L, Yang E. A high-throughput liquid chromatography/tandem mass spectrometry method for screening glutathione conjugates using exact mass neutral loss acquisition. Rapid Commun Mass Spectrom 2005;19:798–804.

Chernushevich IV, Loboda AV, Thomson BA. An introduction to quadrupole-time-of-flight mass spectrometry. J Mass Spectrom 2001;36:849–865.

Crockford DJ, Maher AD, Ahmadi KR, Barrett A, Plumb RS, Wilson ID, et al. 1H NMR and UPLC-MS(E) statistical heterospectroscopy: Characterization of drug metabolites (xenometabolome) in epidemiological studies. Anal Chem 2008; 80:6835–6844.

Cuyckens F, Balcaen LIL, Wolf K, Samber B, Looveren C, Hurkmans R, et al. Use of the bromine isotope ratio in HPLC-ICP-MS and HPLC-ESI-MS analysis of a new drug in development. Analy Bioanal Chem 2008;390:1717–1729.

Cuyckens F, Hurkmans R, Castro-Perez JM, Leclercq L, Mortishire-Smith RJ. Extracting metabolite ions out of a matrix background by combined mass defect, neutral loss and isotope filtration. Rapid Commun Mass Spectrom 2009;23:327–332.

Dieckhaus CM, Fernandez-Metzler CL, King R, Krolikowski PH, Baillie TA. Negative ion tandem mass spectrometry for the detection of glutathione conjugates. Chem Res Toxicol 2005;18:630–638.

Eckers C, Wolff JC, Haskins NJ, Sage AB, Giles K, Bateman R. Accurate mass liquid chromatography/mass spectrometry on orthogonal acceleration time-of-flight mass analyzers using switching between separate sample and reference sprays. 1. Proof of concept. Anal Chem 2000;72:3683–3688.

Fedeniuk RW, Shand PJ. Theory and methodology of antibiotic extraction from biomatrices. J Chromatogr A 1998;812:3–15.

Fernandez-Metzler CL, Owens KG, Baillie TA, King RC. Rapid liquid chromatography with tandem mass spectrometry-based screening procedures for studies on the biotransformation of drug candidates. Drug Metab Dispos 1999;27:32–40.

Ferrer I, Thurman EM. Measuring the mass of an electron by LC/TOF-MS: A study of "twin ions". Analy Chem 2005;77:3394–3400.

Fredriksson M, Petersson P, Jornten-Karlsson M, Axelsson BO, Bylund D. An objective comparison of pre-processing methods for enhancement of liquid chromatography–mass spectrometry data. J Chromatogr A 2007;1172:135–150.

Gangl E, Utkin I, Gerber N, Vouros P. Structural elucidation of metabolites of ritonavir and indinavir by liquid chromatography–mass spectrometry. J Chromatogr A 2002;974:91–101.

Godejohann M, Tseng LH, Braumann U, Fuchser J, Spraul M. Characterization of a paracetamol metabolite using on-line LC-SPE-NMR-MS and a cryogenic NMR probe. J Chromatogr A 2004;1058:191–196.

Gross ML. Accurate masses for structure confirmation. J Am Soc Mass Spectrom 1994;5:57.

Heinonen M, Rantanen A, Mielikainen T, Kokkonen J, Kiuru J, Ketola RA, et al. FiD: A software for ab initio structural identification of product ions from tandem mass spectrometric data. Rapid Commun Mass Spectrom 2008;22:3043–3052.

Hill AW, Mortishire-Smith RJ. Automated assignment of high-resolution collisionally activated dissociation mass spectra using a systematic bond disconnection approach. Rapid Commun Mass Spectrom 2005;19:3111–3118.

Hobby K, Gallagher RT, Caldwell P, Wilson ID. A new approach to aid the characterisation and identification of metabolites of a model drug; partial isotope enrichment combined with novel formula elucidation software. Rapid Commun Mass Spectrom 2009;23:219–227.

Holman SW, Wright P, Langley GJ. High-throughput approaches towards the definitive identification of pharmaceutical drug metabolites. 2. An example of how unexpected dissociation behaviour could preclude correct assignment of sites of metabolism. Rapid Commun Mass Spectrom 2009;23:2017–2025.

Hop CE, Yu X, Xu X, Singh R, Wong BK. Elucidation of fragmentation mechanisms involving transfer of three hydrogen atoms using a quadrupole time-of-flight mass spectrometer. J Mass Spectrom 2001;36:575–579.

Hu Q, Noll RJ, Li H, Makarov A, Hardman M, Cooks RG. The Orbitrap: A new mass spectrometer. J Mass Spectrom 2005;40:430–443.

Huang Y. Massive accomplishment. Drug Discov Devel 2009;12:29–31.

James CA. Sample preparation. In *Principles and Practice of Bioanalysis*, Second ed.; Venn RF, Ed.; CRC Press: Boca Raton, USA. 2008, pp 19–39.

Johnson SR, Josephs JL, Claus B, Langish RA. Automated regional assignment of metabolic modification using cross correlation algorithms, maximum common substructure analysis, and MS/MS spectral libraries. Abstracts of Papers. Paper presented at the 232nd ACS National Meeting, San Francisco, CA, September, 10–14, 2006.

Jones P, Atack JR, Braun MP, Cato BP, Chambers MS, O'Connor D, et al. Pharmacokinetics and metabolism studies on (3-*tert*-butyl-7-(5-methylisoxazol-3-yl)-2-(1-methyl-1*H*-1,2,4-triazol-5-ylmethoxy)pyrazolo[1,5-d][1,2,4]triazine), a functionally selective GABAA alpha 5 inverse agonist for cognitive dysfunction. Bioorg Med Chem Lett 2006;16:872–875.

Kallury K. High-throughput sample preparation techniques and their application to bioanalytical protocols and purification of combinatorial libraries. In *Critical Reviews in Combinatorial Chemistry*, Perry Wang, Ed.; CRC Press: Boca Raton, 2009, pp. 1–72.

Kantharaj E, Tuytelaars A, Proost PE, Ongel Z, Van Assouw HP, Gilissen RA. Simultaneous measurement of drug metabolic stability and identification of metabolites using ion-trap mass spectrometry. Rapid Commun Mass Spectrom 2003;17:2661–2668.

Karas M, Bachmann D, Bahr U, Hillenkamp F. Matrix-assisted ultraviolet laser desorption of non-volatile compounds. Int J Mass Spectrom Ion Process 1987; 78:53–68.

Kerns EH, Volk KJ, Hill SE, Lee MS. Profiling taxanes in Taxus extracts using lc/ms and lc/ms/ms techniques. J Nat Prod 1994;57:1391–1403.

Kerns EH, Volk KJ, Hill SE, Lee MS. Profiling new taxanes using LC/MS and LC/MS/MS substructural analysis techniques. Rapid Commun Mass Spectrom 1995;9:1539–1545.

Koichi T, Hiroaki W, Yutaka I, Satoshi A, Yoshikazu Y, Tamio Y, et al. Protein and polymer analyses up to *m/z* 100,000 by laser ionization time-of-flight mass spectrometry. Rapid Commun Mass Spectrom 1988;2:151–153.

Kostiainen R, Tuominen J, Luukkanen L, Taskinen J, Green BN. Accurate mass measurements of some glucuronide derivatives by electrospray low resolution quadrupole mass spectrometry. Rapid Commun Mass Spectrom 1997;11:283–285.

Koulman A, Woffendin G, Narayana VK, Welchman H, Crone C, Volmer DA. High-resolution extracted ion chromatography, a new tool for metabolomics and lipidomics using a second-generation orbitrap mass spectrometer. Rapid Commun Mass Spectrom 2009;23:1411–1418.

Lafaye A, Junot C, Ramounet-Le Gall B, Fritsch P, Tabet JC, Ezan E. Metabolite profiling in rat urine by liquid chromatography/electrospray ion trap mass spectrometry. Application to the study of heavy metal toxicity. Rapid Commun Mass Spectrom 2003;17:2541–2549.

Lanting AB, Bruins AP, Drenth BF, de Jonge K, Ensing K, de Zeeuw RA, et al. Identification with liquid chromatography–ionspray mass spectrometry of the metabolites of the enantiomers *N*-methyl dextrorphan and *N*-methyl levorphanol after rat liver perfusion. Biol Mass Spectrom 1993;22:226–234.

Leal M, Hayes MJ, Powell ML. The metabolism of CGS 15873 in man using stable isotope pattern recognition techniques. Biopharm Drug Dispos 1992;13:617–628.

Leclercq L, Mortishire-Smith RJ, Huisman M, Cuyckens F, Hartshorn MJ, Hill A. IsoScore: Automated localization of biotransformations by mass spectrometry using product ion scoring of virtual regioisomers. Rapid Commun Mass Spectrom 2009;23:39–50.

Li AC, Shou WZ, Mai TT, Jiang XY. Complete profiling and characterization of in vitro nefazodone metabolites using two different tandem mass spectrometric platforms. Rapid Commun Mass Spectrom 2007;21:4001–4008.

Lim HK, Chen J, Sensenhauser C, Cook K, Subrahmanyam V. Metabolite identification by data-dependent accurate mass spectrometric analysis at resolving power of 60,000 in external calibration mode using an LTQ/Orbitrap. Rapid Commun Mass Spectrom 2007;21:1821–1832.

Makarov A, Denisov E, Kholomeev A, Balschun W, Lange O, Strupat K et al. Performance evaluation of a hybrid linear ion trap/Orbitrap mass spectrometer. Anal Chem 2006a; 78:2113–2120.

Makarov A, Denisov E, Lange O, Horning S. Dynamic range of mass accuracy in LTQ Orbitrap hybrid mass spectrometer. J Am Soc Mass Spectrom 2006b; 17:977–982.

Makarov AA, Horning S. A system and method of mass spectrometry. Thermo Fisher Scientific, Bremen, Germany, Patent is available at http://www.ipo.gov.uk/p-find-publication-getPDF.pdf? PatentNo=GB2445169&DocType=A&JournalNumber=6215, 2008.

Marshall AG, Hendrickson CL. High-resolution mass spectrometers. Annu Rev Anal Chem 2008;1:579–599.

Mortishire-Smith RJ, O'Connor D, Castro-Perez JM, Kirby J. Accelerated throughput metabolic route screening in early drug discovery using high-resolution liquid chromatography/quadrupole time-of-flight mass spectrometry and automated data analysis. Rapid Commun Mass Spectrom 2005;19:2659–2670.

Mortishire-Smith RJ, Castro-Perez JM, Yu K, Shockcor JP, Goshawk J, Hartshorn MJ, et al. Generic dealkylation: A tool for increasing the hit-rate of metabolite rationalization, and automatic customization of mass defect filters. Rapid Commun Mass Spectrom 2009;23:939–948.

Nikolic D, Li Y, Chadwick LR, Grubjesic S, Schwab P, Metz P, et al. Metabolism of 8-prenylnaringenin, a potent phytoestrogen from hops (*Humulus lupulus*), by human liver microsomes. Drug Metab Dispos 2004;32:272–279.

O'Connor D, Jones P, Chambers MS, Maxey R, Szekeres HJ, Szeto N, et al. Aldehyde oxidase and its contribution to the metabolism of a structurally novel, functionally selective GABA-A-alpha5-subtype inverse agonist. Xenobiotica: The Fate and Safety Evaluation of Foreign Compounds in Biological Systems 2006a;36:315–330.

O'Connor D, Mortishire-Smith R. High-throughput bioanalysis with simultaneous acquisition of metabolic route data using ultra performance liquid chromatography coupled with time-of-flight mass spectrometry. Anal Bioanal Chem 2006b;385: 114–121.

Olsen JV, Macek B, Lange O, Makarov A, Horning S, Mann M. Higher-energy C-trap dissociation for peptide modification analysis. Nat Methods 2007;4:709–712.

Owens KG. Application of correlation analysis techniques to mass spectral data. Appl Spectrosc Rev 1991;27:1–49.

Pedersen-Bjergaard S, Rasmussen KE. Bioanalysis of drugs by liquid-phase microextraction coupled to separation techniques. J Chromatogr B Analyt Technol Biomed Life Sci 2005;817:3–12.

Pelander A, Tyrkko E, Ojanpera I. In silico methods for predicting metabolism and mass fragmentation applied to quetiapine in liquid chromatography/time-of-flight mass spectrometry urine drug screening. Rapid Commun Mass Spectrom 2009;23:506–514.

Peterman SM, Duczak N, Kalgutkar AS, Lame ME, Soglia JR. Application of a linear ion trap/Orbitrap mass spectrometer in metabolite characterization studies: Examination of the human liver microsomal metabolism of the non-tricyclic antidepressant nefazodone using data-dependent accurate mass measurements. J Am Soc Mass Spectrom 2006;17:363–375.

Peters RJB, van Engelen MC, Touber ME, Georgakopoulus C, Nielen MWF. Searching for in silico predicted metabolites and designer modifications of (cortico)steroids in urine by high-resolution liquid chromatography/time-of-flight mass spectrometry. Rapid Commun Mass Spectrom 2009;23:2329–2337.

Pilard S, Caradec F, Jackson P, Luijten W. Identification of an *N*-(hydroxysulfonyl)oxy metabolite using in vitro microorganism screening, high-resolution and tandem electrospray ionization mass spectrometry. Rapid Commun Mass Spectrom 2000;14:2362–2366.

Polson C, Sarkar P, Incledon B, Raguvaran V, Grant R. Optimization of protein precipitation based upon effectiveness of protein removal and ionization effect in liquid chromatography–tandem mass spectrometry. J Chromatogr B Analyt Technol Biomed Life Sci 2003;785:263–275.

Poon GK, Chen Q, Teffera Y, Ngui JS, Griffin PR, Braun MP, et al. Bioactivation of diclofenac via benzoquinone imine intermediates-identification of urinary mercapturic acid derivatives in rats and humans. Drug Metab Dispos 2001;29:1608–1613.

Ruan Q, Peterman S, Szewc MA, Ma L, Cui D, Humphreys WG, et al. An integrated method for metabolite detection and identification using a linear ion trap/Orbitrap mass spectrometer and multiple data processing techniques: application to indinavir metabolite detection. J Mass Spectrom 2008;43:251–261.

Sanders M, Shipkova PA, Zhang H, Warrack BM. Utility of the hybrid LTQ-FTMS for drug metabolism applications. Curr Drug Metab 2006;7:547–555.

Scigelova M, Makarov A. Advances in bioanalytical LC-MS using the Orbitrap mass analyzer. Bioanalysis 2009;1:741–754.

Shipkova PA, Josephs JL, Sanders M. Changing role of FTMS in drug metabolism. In Mass Spectrometry in Drug Metabolism and Disposition Wiley, NJ, Hoboken, 2008b, pp. 191–121.

Sin CH, Lee ED, Lee ML. Atmospheric pressure ionization time-of-flight mass spectrometry with a supersonic ion beam. Anal Chem 2002;63:2897–2900.

Sun J, Von Tungeln LS, Hines W, Beger RD. Identification of metabolite profiles of the catechol-*O*-methyl transferase inhibitor tolcapone in rat urine using LC/MS-based metabonomics analysis. J Chromatogr B Anal Technol Biomed Life Sci 2009;877:2557–2565.

Sundstrom I, Hedeland M, Bondesson U, Andren PE. Identification of glucuronide conjugates of ketobemidone and its phase I metabolites in human urine utilizing accurate mass and tandem time-of-flight mass spectrometry. J Mass Spectrom 2002;37:414–420.

Tiller PR, Romanyshyn LA. Liquid chromatography/tandem mass spectrometric quantification with metabolite screening as a strategy to enhance the early drug discovery process. Rapid Commun Mass Spectrom 2002;16:1225–1231.

Tiller PR, Yu S, Castro-Perez J, Fillgrove KL, Baillie TA. High-throughput, accurate mass liquid chromatography/tandem mass spectrometry on a quadrupole time-of-flight system as a "first-line" approach for metabolite identification studies. Rapid Commun Mass Spectrom 2008;22:1053–1061.

Watt AP, Morrison D, Locker KL, Evans DC. Higher throughput bioanalysis by automation of a protein precipitation assay using a 96-well format with detection by LC-MS/MS. Anal Chem 2000;72:979–984.

Wieghaus A, Makarov A, Froehlich U, Kellmann M, Denisov E, Lange O. Development and applications of a new benchtop Orbitrap mass spectrometer. In Proceedings of the 56th ASMS Conference on Mass Spectrometry and Allied Topics. 2008.

Williamson LN, Zhang G, Terry AV Jr, Bartlett MG. Comparison of time-of-flight mass spectrometry to triple quadrupole tandem mass spectrometry for quantitative bioanalysis: Application to antipsychotics. J Liq Chromatogr Relat Technol 2008;31:2737–2751.

Wrona M, Mauriala T, Bateman KP, Mortishire-Smith RJ, O'Connor D. "All-in-one" analysis for metabolite identification using liquid chromatography/hybrid quadrupole time-of-flight mass spectrometry with collision energy switching. Rapid Commun in Mass Spectrom 2005;19:2597–2602.

Yamashita M, Fenn JB. Electrospray ion source. Another variation on the free-jet theme. J Phys Chem 1984a;88:4451–4459.

Yamashita M, Fenn JB. Negative ion production with the electrospray ion source. J Phys Chem 1984b;88:4671–4675.

Yan Z, Caldwell GW, Maher N. Unbiased high-throughput screening of reactive metabolites on the linear ion trap mass spectrometer using polarity switch and mass tag triggered data-dependent acquisition. Anal Chem 2008;80:6410–6422.

Yao M, Ma L, Humphreys WG, Zhu M. Rapid screening and characterization of drug metabolites using a multiple ion monitoring-dependent MS/MS acquisition method on a hybrid triple quadrupole–linear ion trap mass spectrometer. J Mass Spectrom 2008;43:1364–1375.

Yergey JA, Trimble LA, Silva J, Chauret N, Li C, Therien M, et al. In vitro metabolism of the COX-2 inhibitor DFU, including a novel glutathione adduct rearomatization. Drug Metab Dispos 2001;29:638–644.

Yu X, Cui D, Davis MR. Identification of in vitro metabolites of Indinavir by "intelligent automated LC-MS/MS" (INTAMS) utilizing triple quadrupole tandem mass spectrometry. J Am Soc Mass Spectrom 1999;10:175–183.

Zhang H, Yang Y. An algorithm for thorough background subtraction from high-resolution LC/MS data: Application for detection of glutathione-trapped reactive metabolites. J Mass Spectrom 2008;43:1181–1190.

Zhang H, Henion J, Yang Y, Spooner N. Application of atmospheric pressure ionization time-of-flight mass spectrometry coupled with liquid chromatography for the characterization of in vitro drug metabolites. Anal Chem 2000a;72: 3342–3348.

Zhang N, Fountain ST, Bi H, Rossi DT. Quantification and rapid metabolite identification in drug discovery using API time-of-flight LC/MS. Anal Chem 2000b;72:800–806.

Zhang H, Zhang D, Ray K. A software filter to remove interference ions from drug metabolites in accurate mass liquid chromatography/mass spectrometric analyses. J Mass Spectrom 2003;38:1110–1112.

Zhang D, Cheng PT, Zhang H. Mass defect filtering on high resolution LC/MS data as a methodology for detecting metabolites with unpredictable structures: Identification of oxazole-ring opened metabolites of muraglitazar. Drug Metab Lett 2007;1:287–292.

Zhang H, Ma L, He K, Zhu M. An algorithm for thorough background subtraction from high-resolution LC/MS data: Application to the detection of troglitazone metabolites in rat plasma, bile, and urine. J Mass Spectrom 2008a;43:1191–1200.

Zhang H, Zhang D, Zhu M, Ray K. High-resolution LC-MS-based mass defect filter approach: Basic concept and application in metabolite detection. In Mass Spectrometry in Drug Metabolism and Pharmacokinetics, ed Ramanathan, R. Wiley, NJ, Hoboken, 2008b, pp. 223–251.

Zhang H, Zhu M, Ray KL, Ma L, Zhang D. Mass defect profiles of biological matrices and the general applicability of mass defect filtering for metabolite detection. Rapid Commun Mass Spectrom 2008c;22:2082–2088.

Zhang H, Grubb M, Wu W, Josephs J, Humphreys WG. Algorithm for thorough background subtraction of high-resolution LC/MS data: Application to obtain clean product ion spectra from nonselective collision-induced dissociation experiments. Anal Chem 2009a;81:2695–2700.

Zhang H, Zhang D, Ray K, Zhu M. Mass defect filter technique and its applications to drug metabolite identification by high-resolution mass spectrometry. J Mass Spectrom 2009b;44:999–1016.

Zhang NR, Yu S, Tiller P, Yeh S, Mahan E, Emary WB. Quantitation of small molecules using high-resolution accurate mass spectrometers—A different approach for analysis of biological samples. Rapid Commun Mass Spectrom 2009c;23: 1085–1094.

Zhu M, Ma L, Zhang D, Ray K, Zhao W, Humphreys WG, et al. Detection and characterization of metabolites in biological matrices using mass defect filtering of liquid chromatography/high resolution mass spectrometry data. Drug Metab Dispos 2006;34:1722–1733.

Zhu M, Ma L, Zhang H, Humphreys WG. Detection and structural characterization of glutathione-trapped reactive metabolites using liquid chromatography-high-resolution mass spectrometry and mass defect filtering. Anal Chem 2007;79: 8333–8341.

Zhu P, Ding W, Tong W, Ghosal A, Alton K, Chowdhury S. A retention-time-shift-tolerant background subtraction and noise reduction algorithm (BgS-NoRA) for extraction of drug metabolites in liquid chromatography/mass spectrometry data from biological matrices. Rapid Commun Mass Spectrom 2009a;23: 1563–1572.

Zhu P, Tong W, Alton K, Chowdhury S. An accurate-mass-based spectral-averaging isotope-pattern-filtering algorithm for extraction of drug metabolites possessing a distinct isotope pattern from LC-MS data. Anal Chem 2009b;81:5910–5917.

14 Distribution Studies of Drugs and Metabolites in Tissue by Mass Spectrometric Imaging

RICHARD F. REICH, DANIEL P. MAGPARANGALAN, TIMOTHY J. GARRETT, and RICHARD A. YOST
Department of Chemistry and Medicine University of Florida
Gainesville, Florida

Mass Spectrometry in Drug Metabolism and Disposition: Basic Principles and Applications, First Edition. Edited by Mike S. Lee and Mingshe Zhu.

14.1 INTRODUCTION

Pharmaceutical development depends on understanding the pharmacological activity as well as the ADMET (adsorption, distribution, metabolism, elimination, and toxicity) properties of drug candidates in vivo. Compounds that show high potency in vitro may display no in vivo efficacy as a result of poor pharmacokinetic (PK) properties such as low absorption rates or short half-life (Davis and Riley, 2004). The compound may even prove to be highly toxic for in vivo models due to the formation of reactive metabolites or from the target molecule itself. Drug-induced toxicity has been reported to be the major cause of attrition in drug discovery and development (Boguslavsky, 2001). A major determinant of drug-induced organ toxicity is the local concentration of the chemical at a particular organ. Understanding the localization of the drug and its metabolites in tissue in context with the dose, formulation, and ADMET–PK properties will serve a critical role in drug development.

Conventional drug analysis in tissue involves homogenization of the tissue prior to subsequent chromatographic analysis (Drummer and Gerstamoulos, 2002). Such sample pretreatments are known to introduce variation in detection due to inhomogeneity of the analyte within the sample matrix (Stimpfl and Reichel, 2007). Also, homogenization of tissue eliminates the opportunity to acquire detailed anatomical and histological information for in situ drug distribution. Imaging techniques that include mass spectrometric imaging (MSI) can help provide this information.

14.2 TISSUE IMAGING TECHNIQUES

A number of analytical techniques are capable of imaging drugs in vivo and ex vivo, including positron emission tomography (PET) (Schiffer et al., 2007), single photon emission computed tomography (Gatley and Volkow, 1998), magnetic resonance imaging (Sosnovik and Weissleder, 2007), X-ray computed tomography (Wang et al., 2008), optical fluorescence imaging (Gumbleton and Stephens, 2005), optical bioluminescence imaging (Contag and Ross, 2002), ultrasound (Denoyer et al., 2008), whole-body autoradiography (WBA) (Som et al., 1994), infrared imaging (Garidel and Boese, 2007), and magnetically labeled nanoparticles (Mahmoudi et al., 2009). However, disadvantages of these imaging methods include low sensitivity, low specificity, limited functional and molecular information, poor spatial resolution, and the need for the drug to be labeled with either a radioactive isotope or a fluorescent tag, which can be time consuming and costly (Niu and Chen, 2008). A specific disadvantage for those techniques that require a chemical tag is the need to monitor the tag rather than the intact drug, and, therefore, the ability to differentiate the drug from a metabolite that may have retained the tag is difficult. In addition, a chemical tag may alter the pharmacological properties of the compound, which could affect both bioavailability and localization within the tissue.

Mass spectrometric imaging (MSI) has higher molecular specificity compared to other tissue imaging techniques, particularly when used in combination with tandem mass spectrometry (MS/MS). The high selectivity of the instrument eliminates the need for labeling because the ion (or product ion as in tandem mass spectrometry) is monitored directly and leaves the drug molecule of interest functionally unmodified. An unmodified drug compound also removes the potential interference of fluorescent/radioactive labels with the biological function (e.g., when the drug must pass through the blood–brain barrier). This analyte specificity of the instrument also provides the ability to simultaneously image drugs and their metabolites due to the parallel detection of multiple analytes. With MSI, an image can be produced for each of the hundreds of detected analytes within the mass spectral data set. Another advantage is its high sensitivity. Unfortunately, MSI is a destructive imaging technique, although only a few molecular monolayers of sample are affected by the analysis. This characteristic precludes MSI from being used for in vivo studies and, therefore, limits the flexibility of ADMET–PK studies.

MSI collects chemical data normally associated with mass spectrometry, but in a spatially defined manner, and processes that information into chemical image maps. Secondary ion mass spectrometry (SIMS) (Pacholski and Winograd, 1999) and matrix-assisted laser desorption/ionization (MALDI) mass spectrometry (MS) (Caprioli et al., 1997) are the two main techniques used with MSI. MALDI–MS has been shown to be very effective for the direct analysis of drugs and their metabolites in tissues (Troendle et al., 1999; Reyzer et al., 2003; Bunch et al., 2004; Wang et al., 2005; Cristoni et al., 2006; Crossman et al., 2006; Hsieh et al., 2006, 2007; Khatib-Shahidi et al., 2006; Drexler et al., 2007; Stoeckli et al., 2007; Chen et al., 2008; Cornett et al., 2008; Hopfgartner et al., 2009; Li et al., 2009). MALDI–MS is currently the most common MSI technique used for mapping pharmaceuticals in tissue, although new MSI techniques may emerge as new surface ionization methods are developed. Ambient ionization methods such as desorption electrospray ionization (DESI) (Takats et al., 2004) and laser ablation electrospray ionization (LAESI) (Nemes and Vertes, 2007) show potential for the in vivo analysis on the surface of skin of organisms with high specificity. DESI has been used for the in vivo detection of the antihistamine loratadine from the finger of a person who had taken 10 mg of the drug, 40 min prior to analysis (Takats et al., 2004). DESI has also been used to localize clozapine directly from histological sections of brain, lung, kidney, and testis without prior chemical treatment (Cooks et al., 2006).

14.3 MASS SPECTROMETRIC IMAGING BACKGROUND

Historically, MSI of pharmaceuticals in biological tissues began with the analysis of elemental ions. Ions are created with SIMS as energetic particles of several kiloelectron volts of kinetic energy bombard solid surfaces. Due to the high degree of fragmentation caused by the ionization process, cellular and subcellular monitoring of pharmaceuticals relies upon the detection of a drug that contains an atom that is not natively found in cells or tissues. Thus, the distribution of the drug, and in some cases its metabolites, can be determined by the analysis of a specific atom. SIMS was used to characterize the distribution of *p*-boronphenylalanine-fructose (BPA-F), a boron-containing drug used in boron neutron capture therapy in brain tumors. (Clerc et al., 1993). Boron captures a neutron and emits a heavy, high-energy-charged particle, which causes cell death. BPA-F was determined by SIMS imaging to be strongly localized in tumor tissue (Clerc et al., 1993). In another study, SIMS was used to identify the distribution of the fluorine-containing drug mefloquine in tissue, which was infected with the malaria parasite. The fluorine signal appears homogeneously in uninfected red blood cells but is localized in the food vacuole, cytoplasm, and nucleus of the parasite in infected cells (Adovelande et al., 1994). With lower primary ion fluences (so-called static SIMS), molecular ion signals can be observed, but with relatively low sensitivity due in part to the shallow (monolayer) sampling depth of SIMS.

Photons are used to desorb and ionize molecules in laser desorption ionization (LDI). The laser microprobe mass analyzer (LAMMA) (Hillenkamp et al., 1975) was an early LDI-based commercial instrument, which provided the ability to spatially focus a laser at a particular location and acquire a mass spectrum from the ions in the resulting plume. The LAMMA instrument combined a high-resolution optical microscope with a time-of-flight (TOF) mass spectrometer. The LAMMA was originally developed for high sensitivity molecular analysis of thin histological sections but was primarily used for in situ determination of physiological cations in organ tissues. Similar to SIMS, LAMMA was restricted to elemental analysis because of the high degree of fragmentation caused by the LDI process. An example of a LAMMA analysis involved the study of the localization and concentration of copper-containing compounds in liver biopsy samples from rats with an inherited disorder that caused copper toxicosis (Iancu et al., 1996). LAMMA was also a useful technique for the study of the localization of trace elements in various tissues that contained excessive amounts of metals (Iancu et al., 1996). However, this approach did not serve as an effective tool for pharmaceutical analysis.

Another LDI instrument that was similar in principle to LAMMA was developed by Perchalski (1985) that featured the additional selectivity of two stages of mass analysis provided by a triple quadrupole mass spectrometer (QqQ). The LDI–QqQ was shown to have potential for use as a probe-type analyzer for molecular analysis of mixtures, as demonstrated by the detection of a mixture of nine antiepileptic drugs by monitoring the precursor ion/product ion pair for each drug (Perchalski et al., 1983). The LDI–QqQ, however, was determined to be too slow to adequately characterize molecules ionized by cationization or anionization after desorption by a single-shot laser. Also, the vaporization/ionization process on the LDI–QqQ was unable to ionize polar, nonvolatile, and/or thermally unstable molecules (Perchalski, 1985).

One limitation of SIMS and LDI is that higher molecular weight compounds (>1000 Da) do not survive the ionization process. The introduction of MALDI (Karas and Hillenkamp, 1988) enabled the analysis of intact higher molecular weight molecules in tissue (Caprioli et al., 1997). Incorporation of a matrix compound during MALDI sample preparation reduces the fragmentation observed in SIMS and LDI. MALDI will be discussed further in Section 14.4.2.

14.4 MSI METHODOLOGY

In order to use a mass spectrometer as an imaging instrument, it is necessary for the mass spectrometer to be equipped with a rastering function, which allows the sample (or the laser beam) to be rastered in a precise manner to move the laser beam across the tissue surface. A sophisticated data acquisition system is necessary to handle the large amount of data acquired during analysis, and visualization software is essential to create images from the ion intensities of selected ions. Several commercial mass spectrometric imaging instruments are now available with these features. Almost all manufacturers have developed in-house software and have included methods for instrument and image reconstruction, but there are some noncommercial products as well.

14.4.1 Mass Analyzers

A variety of mass analyzers have been used in MSI experiments such as TOF, quadrupole ion trap (QIT), linear ion trap (LIT), QqQ, Fourier transform ion cyclotron resonance (FTICR), and Orbitrap. Also, various tandem configurations of these mass analyzers such as QqTOF, QqLIT, and TOF/TOF have been used. Due to the many possible interferences from endogenous compounds or from the matrix, the use of tandem mass spectrometry (MS/MS or MS^n) or the ability to perform high-resolution and accurate mass measurements for the analysis of drugs by MALDI is essential. An overview of some of the established instrumentation and their respective capabilities is presented below.

14.4.1.1 Time of Flight (TOF) The majority of SIMS and MALDI imaging experiments are performed using TOF mass analyzers. TOF-based instruments are ideally suited to the pulsed nature of these ionization techniques, which have a precisely defined ionization time. Also, the TOF has the highest scan rate (microseconds versus milliseconds) compared to the other types of mass analyzers. This fast scan rate makes it well suited for use with high-frequency lasers, thus reducing the time needed to acquire images. TOF instruments have high detection efficiency, excellent mass resolution (>10,000), and accurate mass capabilities (below 20 ppm).

Measurements on TOF instruments can be performed using linear or orthogonal extraction. The linear geometry, in which the ions are accelerated directly into the mass analyzer, is more common for MSI because this configuration provides the highest sensitivities due to higher transmission efficiencies. Linear extraction MALDI–TOF imaging experiments are often performed using delayed-extraction, and a reflectron is used for small molecules (< 10,000 Da). It has been shown that a short time delay between firing the laser and extracting the ions can significantly improve the sensitivity and mass resolution of a MALDI experiment and minimize the kinetic energy distribution of the analyte (Brown and Lennon, 1995). Delayed extraction is seldom used with measurement that involves SIMS because the kinetic energy distribution of molecular ions is narrower due to the much lower density of the particles sputtered from the sample (Garrison et al., 2003). An integral part of TOF–SIMS instrumentation is the electrostatic sector analyzer (ESA), which compensates for the difference in ion flight time due to variations in the initial kinetic energies and emission angles of secondary ions.

The selectivity and sensitivity of MSI measurements using TOF mass analyzers can be increased with tandem mass spectrometry (MS/MS or MS^2). Analyte fragmentation is usually obtained using a collision-induced dissociation (CID) with tandem quadrupole/time-of-flight (QqTOF) or TOF/TOF instruments. In-source decay is another approach used to obtain analyte fragmentation with a single TOF analyzer. In this case, the laser energy is increased to observe in-source degradation of an analyte ion or any other ion that may be present. MALDI–TOF finds much use in the analysis of macromolecules; the technique is less widely used to analyze small molecules (< 1000 Da) because of its inability to resolve low mass analyte ions from the intense interfering MALDI matrix background. MS/MS is used to overcome this challenge (Troendle et al., 1999; Garrett, 2006; Garrett and Yost, 2006; Garrett et al., 2007).

14.4.1.2 Triple Quadrupole (QqQ) The combination of MALDI with a triple quadrupole mass spectrometer (MALDI–QqQ) allows for the use of the MS/MS technique of selected reaction monitoring (SRM). SRM monitors the ion current associated with the transition of a given precursor ion to a particular product ion during CID of the precursor ion. In most cases, the high degree of specificity inherent to SRM allows the differentiation of analyte signal from the MALDI matrix background. Unfortunately, quadrupole scan rates are too slow (time scale of hundreds of milliseconds) to obtain complete MS/MS scans with a pulsed laser.

14.4.1.3 Ion Trap The first demonstration of full-scan MS/MS data for MALDI imaging was obtained with a three-dimesional (3D) quadrupole ion trap (QIT) (Troendle et al., 1999; Garrett et al., 2005). These experiments were performed on an instrument constructed at the University of Florida and clearly demonstrated the critical role of MS/MS for imaging small molecules in tissue. The use of ion traps for MSI has been reviewed (Garrett and Yost, 2009). The QIT can trap and store ions from a pulsed laser desorption event; in contrast, quadrupole mass spectrometers and other scanning instruments are not able to obtain spectra over a wide mass range for a pulsed ionization event. Ions produced by the laser pulse can be detected with the QIT regardless of the duration of the laser desorption event (Brodbelt et al., 1995).

The two-dimensional (2D) linear ion trap (LIT) offers higher sensitivity, better mass resolution, and higher scan speed than the QIT. The LIT can be used to perform multiple stages of mass analysis (MS^n) with a faster full-scan speed than the QqQ. Furthermore, the multiple stages of tandem mass spectrometry can be used to generate more detailed structure information and enable the identification of unknowns. In addition, because of the ability to collect full-scan product ion spectra for a given isolated *m/z*, isobaric compounds can be identified based on unique fragment pathways. Sometimes more stages of mass analysis are required above MS/MS to provide the necessary specificity to identify compounds. MSI has been performed on a triple quadrupole linear ion trap (QqLIT) mass spectrometer (Hopfgartner et al., 2009). This instrument has the high-throughput capability of SRM for the analysis of large tissue sections. Furthermore, the enhanced trap ion mode and the precursor ion mode can also be used to confirm the presence of the parent drug or to screen for the presence of structural analogs (Hopfgartner et al., 2009).

14.4.1.4 Fourier Transform Ion Cyclotron Resonance (FTICR) As discussed above, MALDI–MSI analyses of pharmaceutical compounds are often performed with MS/MS. Mass analyzers that can provide high resolution (>100,000) and accurate mass experiments (sub-ppm) such as FTICR instruments offer an alternate strategy for imaging small molecules and rely on differences in the exact mass of analyte ions and matrix ions. The measurement of isobaric ions from tissue is possible with a high-resolution mass spectrometer. Thus, there is no need to specify individual MS/MS precursor and product ion masses. A disadvantage of FTICR imaging, however, is that it does not provide the same molecular specificity as MS/MS or MS^n. Isomeric ions (with the same elemental composition) will have the same exact mass, and some isobaric ions may have masses too close to distinguish. In these situations, subsequent MS/MS measurements can be made at discrete pixel locations to confirm the presence of a molecular structure and to determine if multiple isomers contribute to the selected ion (Kutz et al., 2004; Taban et al., 2007; Cornett et al., 2008). This procedure is not always straightforward, however, since the mass resolution of precursor ion isolation for MS/MS is limited to unit resolution.

A key factor that has hindered the use of FTICR for imaging applications is the inherently low throughput. For example, the typical imaging speed for FTICR–MSI is 4 pixels/min, which is dramatically slower than imaging systems based on TOF mass analyzers or ion traps (30–50 pixels/min). Due to the extended measurement times involved with FTICR, it is recommended to use these other faster imaging methods to preselect areas of interest before the ultra-high-mass accuracy of FTICR is used.

14.4.1.5 Orbitrap A new approach to Fourier transform mass spectrometry is the Orbitrap analyzer (Makarov, 2000), which uses electric fields rather than magnetic fields to trap the ions. Ions orbit around an axial electrode with a specific frequency of oscillation that is inversely proportional to m/z. Mass analysis is based on detecting the frequency of oscillation using image current and transforming these oscillations into mass spectra using fast Fourier transform, similar to FTICR. The orbitrap mass analyzer is now commercially available as a hybrid linear ion trap/Orbitrap mass spectrometer and as a standalone liquid chromatography–mass spectrometry (LC–MS) system (Makarov et al., 2006a). This novel instrument combines the MS^n capability of the linear ion trap mass spectrometer with the high-resolution and mass accuracy capability of the Orbitrap. All measurements on the LIT/Orbitrap (including ion accumulation and image current detection) were performed in less than 1 second at a resolving power of 30,000. Also, 5-ppm mass accuracy of the Orbitrap mass analyzer is attained with >95% probability at a dynamic range of more than 5000, which is at least an order of magnitude higher than typical values for TOF instruments (Makarov et al., 2006b). The Orbitrap also takes advantage of much greater ion numbers obtained by each laser shot in each single scan than provided for typical TOF instrumentation. This increase is accomplished by decoupling the MALDI process from the analysis by employing the LIT as a trapping buffer to accumulate ions (Strupat et al., 2009). The MALDI–LIT/Orbitrap mass spectrometer was recently demonstrated for the imaging of lipids in spinal cord (Landgraf, 2009).

14.4.2 Ionization Sources

14.4.2.1 Secondary Ion Mass Spectrometry For SIMS analysis, a sample surface is bombarded with a beam of primary ions. Primary ion sources for SIMS require high vacuum conditions ($< 10^{-6}$ torr) and typically use monoatomic (Ar^+, Xe^+, Au^+, Ga^+, Cs^+), polyatomic (He_3^+, SF_6^+, CO_2^+, $C_2F_6^+$), or cluster (gold, C_{60}^+) ion beams (Appelhans and Delmore, 1989; Hand et al., 1990; Tempez et al., 2004; Xu et al., 2004) with kinetic energies in the kilo- to megaelectron volt regime. These primary ions are accelerated and focused by an electric field to form a pulsed ion beam. On impact with the sample surface, the primary ions induce a chain of binary collisions in which the analyte atoms and molecules are ejected from the surface of the sample into the gas phase in the form of charged and neutral particles.

Damage cross section is an important SIMS parameter and provides a measure of the surface area affected by the impact of a single primary ion. The magnitude of the incident ion dose establishes whether a dynamic or static SIMS experiment is performed. For dynamic SIMS, the number of incident ions exceeds the number of surface atoms on the sample and results in erosion due to sputtering and chemical damage to the surface. Dynamic SIMS is primarily used in quantitative elemental imaging applications. In contrast, static SIMS measurements are performed so that the number of incident ions is about an order of magnitude less than the number of surface atoms. This situation results in less than 1% of the top surface atoms and molecules interacting with the primary ion beam, so that a primary ion is likely to strike a fresh surface region. Consequently in static SIMS, significantly less fragmentation of sample molecules occurs and allows the technique to be used to obtain images of small organic compounds [molecular weight (MW) <1000]. Surface modifications can also be implemented to enhance the ionization yield for larger intact molecular ions by static SIMS, such as matrix-enhanced (ME) (Wu and Odom, 1996) and metal-assisted (MetA) SIMS (Delcorte et al., 2003). Unfortunately, static SIMS has poor molecular sensitivity and limited use for the detection of trace molecular analytes in tissue.

Spatial resolution for MSI is usually limited by the spot size of the ionization source. Small spot sizes are routinely obtained with SIMS using bright point sources, ion beam collimation, and ion-optical focusing lenses. Spot sizes as low as 50 nm are commercially available and smaller spots have been obtained (30 nm) (Slodzian et al., 1992). However, ultra-high-resolution images have only been recorded for atomic ions and low mass molecular fragments. This limitation is due in part to the collective motion of the particles removed from the surface of the sample as a result of the ionization process. The larger the collective motion, the larger the analytes that can be desorbed (Garrison et al., 2003). Unfortunately, SIMS has a smaller collective motion and higher molecular internal energy than MALDI (Luxembourg and Heeren, 2006). This combination of properties results in more extensive fragmentation than MALDI.

14.4.2.2 Desorption Electrospray Ionization DESI is carried out by directing electrosprayed charged droplets and ions of solvent onto the surface to be analyzed. The impact of the charged particles on the surface produces gaseous ions of material originally present on the surface. The resulting mass spectra are similar to normal electrospray ionization (ESI) mass spectra in that they show singly or multiply charged molecular ions of the analytes (Takats et al., 2004). An advantage of DESI is that no matrix is required, which eliminates the interference from matrix and analyte-matrix cluster ions (<750 Da) when analyzing low-molecular-weight drug and metabolite compounds. DESI is also performed under ambient (atmospheric pressure) conditions. In addition, the production of multiply charged biological ions is advantageous in extending the mass range, as is the case with ESI.

The primary disadvantage of DESI is the low spatial resolution when compared to MALDI and SIMS. DESI achieved a lateral resolution of approximately 400 μm when imaging specific lipids in coronal sections of rat brain tissue using a 50-μm spray tip (Ifa et al., 2007). The spot size in DESI–MS imaging depends on the spray tip diameter and the tip-to-surface distance, among other parameters (Takats et al., 2005). However, the achievable spatial resolution appears to be significantly poorer than the impacting spot diameter (Wiseman et al., 2006). Microphotography of the impact site shows streams of solvent rolling along the surface for a distance equal to about twice the spray diameter (Venter et al., 2006), which may explain the disparity between spatial resolution and spot diameter.

14.4.2.3 ***Matrix-Assisted Laser Desorption/Ionization*** The most commonly used ionization source for mass spectrometric imaging is MALDI. Unlike SIMS and DESI, MALDI requires the addition of a matrix to the sample. An advantage of the matrix is that the solvent used to apply the matrix is used to extract the analyte out of the tissue, not just analyte at the surface. However, the matrix solvent allows the potential for analyte migration. MALDI is a soft ionization technique in which laser energy is applied for an instant to a co-crystallized mixture of a compound (called a matrix) and the analyte molecules. A typical matrix is a small organic compound that absorbs at the wavelength of the laser and consequently promotes desorption of the analyte. The ionization mechanisms of MALDI are not fully understood but have been critically reviewed (Zenobi and Knochenmuss, 1998). In brief, the chromophore of the matrix couples with the laser energy and causes a rapid vibrational excitation that desorbs matrix and analyte molecules from the solid solution. The photoexcited matrix molecules are then stabilized through proton transfer to the analyte. The first step in sample preparation for MALDI–MSI involves application of a homogeneous layer of matrix to the sample. The sample is then analyzed by moving it stepwise beneath a pulsed laser beam and MALDI mass spectra are acquired from each point. Two-dimensional images may then be obtained by plotting the relative or absolute ion abundance (considered to be proportional to analyte concentration) versus spatial dimensions of X and Y. Figure 14.1 illustrates the overall protocol of a MALDI–MSI experiment.

Laser wavelength is an important parameter in MALDI. The most commonly used wavelength is 337 nm from the nitrogen laser, but harmonics of the Nd:YAG laser 1065 nm fundamental (3×, 355 nm and 4×, 266 nm), various excimer laser lines that include XeCl (308 nm), KrF (248 nm), and ArF (193 nm) and infrared lasers such as carbon dioxide (10.6 μm) and Er:YAG (2.94 μm) lasers have been employed (Zenobi and Knochenmuss, 1998). It has been shown that MALDI mass spectra obtained from ultraviolet (UV) and infrared (IR) laser wavelengths are similar (Niu et al., 1998). However, IR-MALDI requires higher laser pulse energy due to lower MALDI matrix absorption, and the sample consumption is also higher (Berkenkamp et al., 1997). Characteristics of IR-MALDI that have been reported include a greater tendency to form multiply charged high mass ions, less metastable fragmentation, and adduct ion formation (Zhang et al., 1998).

Spatial resolution for MALDI–MSI experiments is limited by laser spot size, laser step size, matrix crystal size, and analyte migration. Spatial resolution increases with decreasing laser spot size, but MALDI mass spectrometers are usually equipped with N_2 (337 nm) or tripled Nd:YAG (355 nm) lasers having relatively large spot sizes (about 100 μm diameter). The rate of energy redistribution rapidly increases with smaller laser spot sizes and higher laser

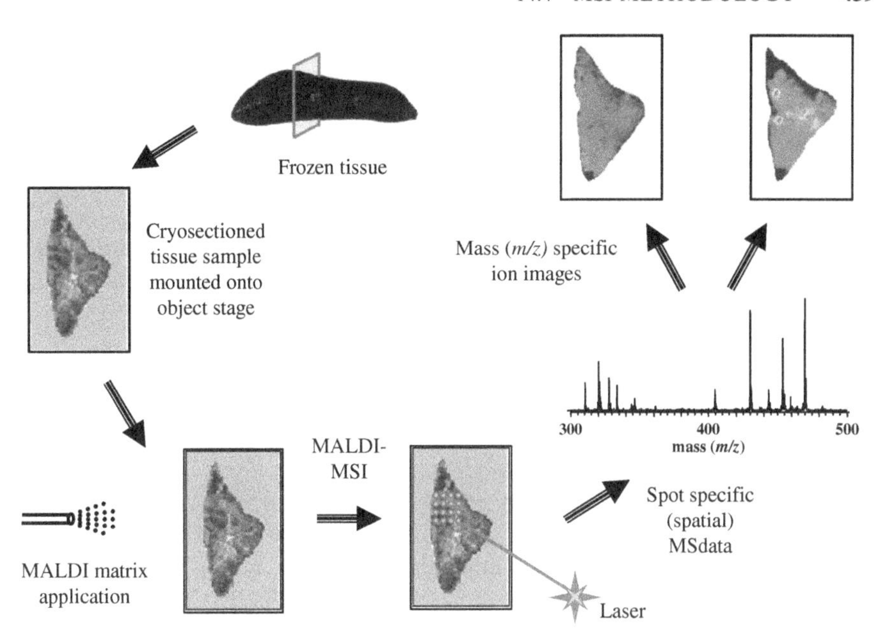

FIGURE 14.1 Schematic of the process flow of MSI by MALDI–MS. Frozen tissue samples are sectioned with a cryostat, after which the sections are mounted onto a sample stage (e.g., glass microscope slide or MALDI target plate). A suitable MALDI matrix is applied to the tissue surface using a dispensing technique (e.g., airbrush or microspotter). The analytes then migrate to the tissue surface and co-crystallize with the MALDI matrix. The sample is analyzed by rastering the tissue surface beneath the laser generating a spot specific (spatial) array of mass spectra that afford mass (m/z) specific density maps. [Adapted from Drexler et al. (2007).]

fluences are required for MALDI to occur (Dreisewerd, 2003). These higher laser fluences can cause extensive fragmentation. Laser spot sizes focused to 7–8 μm in diameter lead to a decrease in ion yields of two orders of magnitude compared to a normal (100-μm) laser spot (Schriver et al., 2003), since the cross-sectional area analyzed is approximately 100 times smaller. In addition, high spatial resolution experiments are more sensitive to analyte migration during the matrix application step and dramatically increase analysis time for whole tissue section analysis.

An alternate approach to increase spatial resolution is by oversampling (using a step size that is smaller than the laser spot width). This method involves, first, the complete ablation of the MALDI matrix coating the sample at each sample position and, second, moving the sample target a distance less than the diameter of the laser beam before repeating the process. The reported method enabled commercial MALDI instruments with large laser spots sizes (100 μm) to image with approximately 25 μm imaging spatial resolution (Jurchen et al., 2005).

Another factor that determines spatial resolution for MALDI–MSI is the size of the matrix crystals formed during the matrix deposition process. The size of the sample matrix co-crystals grown is strongly dependent on the sample matrix solution composition and the rate at which the crystals are grown (Cohen and Chait, 1996). For the majority of MALDI–MSI experiments, the spot size of the laser is such that multiple crystals are sampled in each laser shot, thus the spatial resolution is limited by the laser spot size and not the crystals formed. However, it is still important to avoid nonuniformities in the matrix layer (crystals), which can cause ionization yields to vary across the sample and hinder the interpretation of spatial information. Some approaches for MALDI matrix application, such as inkjet printing (Nakanishi et al., 2005; Baluya et al., 2007), can produce a uniform coating of small crystals. Different approaches for the application of MALDI matrix will be discussed further in Section 14.4.4.2.

In conclusion, MALDI is the most common approach for the MSI of drugs and their metabolites in tissue. SIMS–MSI is capable of imaging surfaces with a lateral resolution of approximately 50 nm; however, it is limited to the analysis of atomic ions and small molecular species (< 500 Da) due to its low sensitivity for high molecular species and extensive fragmentation. DESI is an ionization technique like MALDI that produces abundant molecular ions with minimal fragmentation and has the additional benefits of no sample preparation and atmospheric sampling; however, MALDI and SIMS offer much higher spatial resolution than DESI. Since the vast majority of MSI studies of drugs and metabolites employ MALDI, only MALDI–MSI will be discussed for the remainder of this chapter.

14.4.3 Tissue Preparation

MALDI–MSI of intact tissue involves preparation procedures with minimal sample handling, which decreases analyte losses compared to analyses that involve the preparation of tissue homogenates followed by extraction. Nonetheless, tissue preparation for MALDI–MSI is critical to maintain the integrity of the spatial arrangement of drug and metabolite compounds within tissue. Mishandling or improperly storing tissue samples in the early sample preparation steps may cause delocalization or degradation of the analytes. Experimental procedures that should be considered include excision of tissue, tissue sectioning, sample transfer to MALDI target plate or glass microscope slide, matrix application, and tissue storage after sectioning.

14.4.3.1 Excision of Tissue Tissue samples should be surgically removed so that the original shape of the tissue is retained. Immediately after removal, the tissue may be loosely wrapped in aluminum foil and frozen in liquid nitrogen by gently lowering the tissue into the liquid nitrogen over a period of 30–60 seconds (Schwartz et al., 2003). Immersing the tissue into the liquid nitrogen too quickly can lead to cracking and brittle edges. The foil acts to provide support for more malleable tissue and prevents adhesion of the tissue to the sides of the liquid nitrogen dewar. Freshly excised tissue that is placed into small plastic tubes may mold to the shape of the tube when frozen. Whole tissues may remain frozen in a freezer at −80°C for at least a year with little to no degradation of the sample (Schwartz et al., 2003).

14.4.3.2 Tissue Sectioning and Mounting Frozen tissue samples are cut into thin sections in a cryostat, which allows for accurate sectioning to be accomplished at subfreezing temperatures with minimal sample contamination. It is recommended that tissue samples be attached to the sample stage of the cryostat by freeze mounting with a few drops of deionized water at the interface between the tissue and the stage (Landgraf, 2009). It is not advised to use an embedding medium such as agar or OCT (optimal cutting temperature polymer) to mount the tissue to the sample stage because these compounds could suppress ion formation in MALDI–MS analysis (Schwartz et al., 2003). Tissue samples, once mounted to the cryostat sample stage, are sliced with a stainless steel microtome blade. The sample stage temperature is typically maintained between −5 and −25°C, depending on the tissue type. Tissues that have a higher fat content require lower temperatures to avoid tearing during sectioning. Although tissue thickness is not critical for most studies, 10- to 20-μm-thick tissue sections are optimal for handling. Analyte signal intensity has previously been shown to increase with increasing tissue section thickness; it was hypothesized that, during matrix application, matrix solvent may obtain access to the interior of the tissue to extract more analyte (Crossman et al., 2006).

14.4.3.3 Sample Transfer The tissue section can be transferred with a thin artist's brush and carefully positioned onto a cold MALDI target plate or glass microscope slide. (Conductive microscope slides must be used when performing experiments on a TOF instrument. MSI on LIT and other instruments may use either nonconductive or conductive slides.) Care should be taken during the transfer to avoid folding or tearing the thinly sliced tissue. Tears or rips distort the tissue section and create holes or gaps, which were not present in the native tissue. All equipment that will come into contact with the frozen tissue, including the plate or glass slides, should be kept in the cold box of the cryostat during sectioning. Once the tissue slice is positioned on the cold target plate or glass slide, they are removed from the cold box and quickly warmed, thus thaw mounting the tissue onto the sample plate or slide. Thaw-mounted tissue samples should be stored in a freezer at −80°C until analyzed.

When tissue samples are ready to be analyzed, they are dehydrated in a vacuum desiccator at room temperature to remove moisture and avoid lateral migration of analytes before application of MALDI matrix. Traditional low-pressure (~10^{-6} torr) MALDI requires samples to be dried completely (~2 h) before exposure to vacuum conditions. This prohibits the analysis of freshly cut tissue and reduces sample throughput. In addition, low-pressure MALDI has been shown to produce in-source fragmentation of lipids in tissue, which makes low-level detection difficult (unpublished results). MALDI operated at an intermediate pressure (IP) of 0.17 torr (100,000 times higher than traditional vacuum MALDI) has been shown to reduce the degree of source fragmentation by collisional cooling (Garrett and Yost, 2006). Tissue drying times with IP-MALDI can be reduced to 30 min, which will increase sample throughput and allow for the analysis of tissue samples shortly after dissection.

14.4.4 MALDI Matrix

14.4.4.1 Matrix Selection The success of MALDI–MSI for the analysis of drugs in tissue is dependent on the choice of matrix. The common UV-absorbing molecules used as matrices for MALDI analysis are benzoic-acid-based components with low molecular weights ($<$ 500 Da) such as sinapinic acid (SA, 3,5-dimethoxy-4-hydroxycinnamic acid), α-cyano-4-hydroxycinnamic acid (CHCA), and 2,5-dihydroxybenzoic acid (DHB). Various MALDI matrices, including organic, solid ionic, liquid, and liquid/solid two-phase matrices, have been reviewed (Zenobi and Knochenmuss, 1998). Unfortunately, ions formed from most matrix compounds dominate the low mass range background for a typical MALDI–MS spectrum, making MS/MS or high–resolution MS critical for the analysis of small molecules.

One approach to circumvent matrix interference is to use a higher molecular weight matrix, which does not interfere in the low mass region. To this end, some porphyrins have been employed as MALDI matrices (Ayorinde et al., 1999). Although the porphyrin matrices have been shown to be valuable for the detection of low mass analytes with minimum mass interference from matrix signals (Ayorinde et al., 1999; Cohen and Gusev, 2002), poor ion production yield for drug molecules in tissues was observed when these porphyrin matrices were employed (Karas and Kruger, 2003).

Because of the nature of biological tissues, the growth of matrix crystals is more complicated on tissue than on an inert plate where a small volume of matrix is mixed with a neat drug solution. For example, on tissue, the matrix solvent not only plays a role in the co-crystallization of the matrix and analyte molecules, but the solvent composition also facilitates the extraction of analyte molecules to the surface of the tissue. Therefore, selecting a solvent composition that can readily dissolve the analyte is critical for crystal formation as well as analyte extraction. Solvent composition can also play an active role in protonation of the analyte and result in higher ionization efficiency. Strong acids such as 0.1% trifluoroacetic acid (TFA) are normally added to the matrix solution to assist protonation of proteins but have been found to have a marginal effect on the ionization efficiency for small molecules (Hsieh et al., 2007). For small molecules, a higher matrix concentration (matrix-to-analyte ratio) can also produce better quality mass spectra (Hsieh et al., 2007).

14.4.4.2 Matrix Deposition The analyte signal intensity, suppression of the matrix signal, and laser shot-to-shot reproducibility can be affected by the distribution of matrix and analyte during crystallization (Chapman, 1998). Crystal irregularities can occur when the matrix–analyte mixture partitions during the slow crystallization process (Hensel et al., 1997); thus, it is very important that the solubilities of all components are suitably matched. Many sample preparation procedures for improved co-crystallization of matrix and analyte have been reported and include electrospraying (Hensel et al., 1997), fast evaporation (Nicola et al., 1995), pneumatic spraying (Garrett et al., 2007), spray-droplet method (Sugiura et al., 2006), sublimation (Hankin et al., 2007), inkjet printing (Baluya et al., 2007), acoustic drop ejection (Aerni et al., 2006), and solvent-free matrix dry coating (Puolitaival et al., 2008). By far the most common matrix deposition approach for MALDI–MSI of drugs in tissue is pneumatic spraying (Reyzer et al., 2003; Bunch et al., 2004; Crossman et al., 2006; Hsieh et al., 2006; Khatib-Shahidi et al., 2006; Atkinson et al., 2007; Drexler et al., 2007; Chen et al., 2008; Trim et al., 2008; Hopfgartner et al., 2009; Li et al., 2009) with either CHCA, SA, or DHB matrix. Pneumatic spraying is an inexpensive and easy technique of applying MALDI matrix that is effective at depositing a homogeneous layer of small matrix crystals across the entire tissue sample.

14.4.4.3 Tissue Washing To optimize matrix crystallization, a washing step is sometimes performed before matrix deposition, which allows the majority of salts to be removed from the surface of the tissue (Schwartz et al., 2003). Recent studies have shown that matrix crystallization and analyte incorporation are hampered by the presence of high concentrations of salt, which can result in an inhomogeneous sample surface and lead to high signal variability (Luxembourg et al., 2003). The removal of salt from tissue sections is typically performed by rinsing in 70–80% ethanol (Schwartz et al., 2003). Improved peptide and protein signals were demonstrated with tissue washing in organic solvents traditionally used for lipid extraction (i.e., methylene chloride, hexane, toluene, acetone, and xylene), especially from older or even archived tissue sections (Lemaire et al., 2006). However, great care must be taken to prevent migration of analyte molecules or even the loss of analyte; thus, tissue washing is not recommended for small-molecule applications.

14.4.5 Quantitative MALDI–MS

Although MALDI–MS is an established method for qualitative analysis, quantitative analysis is more difficult because MALDI exhibits irreproducible analyte signals as a result of inhomogeneous crystal formation, inconsistent sample preparation, and laser shot-to-shot variability (Onnerfjord et al., 1999). Indeed, relative standard deviations can be higher than 50% (Cohen and Gusev, 2002; Sleno and Volmer, 2006). The addition of an internal standard can compensate for several of these experimental factors that seriously complicate quantitative MALDI–MS (Nicola et al., 1996; Ling et al., 1998; Hatsis et al., 2003; Cui et al., 2004; Sleno and Volmer, 2006).

An appropriate internal standard for MALDI must compensate not only for any crystallization irregularities but also for subsequent desorption and gas-phase effects. In choosing an internal standard, the relative polarities of the analytes and internal standard as well as their solvent solubilities should be considered (Sleno and Volmer, 2006). Structural similarities should reflect the gas-phase behavior of the involved molecules and extend to solubility. Naturally, an isotope-labeled standard is the ideal choice since its chemical behavior is nearly identical to its unlabeled counterpart (Gusev et al., 1996). Such a standard guarantees identical crystallization and gas-phase behavior of the analyte and internal standard (Kang et al., 2001).

Traditional MALDI experiments demonstrate that using the ratio of the analyte peak intensities to those of a deuterated internal standard can improve signal reproducibility (Gusev et al., 1996). Upon direct analysis of tissue sections, a variety of compounds are ionized, which can interfere with a targeted analysis of exogenous compounds such as drugs; therefore, it is critical to utilize MS/MS and MS^n in order to differentiate exogenous ion signals from endogenous ones (e.g., lipids) as well as matrix ions. An approach to combine the use of an internal standard with MS/MS would involve two alternating MS/MS experiments: first perform MS/MS of the analyte ion, then MS/MS of the internal standard to calculate the ratios of the intensities of the targeted product ion to the internal standard fragment ion. An alternative approach would feature the use of a wide isolation window that includes both the targeted compound and internal standard ions, performing a single MS/MS experiment. In the wide isolation window approach, a wide mass range is chosen in order to include both the precursor ion m/z value and the internal standard m/z value in the same isolation event. The center of the isolation window is an m/z value between those of the precursor ion and the internal standard. Preliminary results have shown that using a single isolation method (the wide isolation window) compared to two alternating MS/MS experiments improves precision dramatically (10–20 times reduction in the percent relative standard deviation) (Reich et al., 2008; Reich et al., 2010).

An example of the application of the wide isolation window in tandem mass spectrometry is the detection and quantitative imaging of cocaine in post-mortem human brain tissue (Reich et al., 2008; Reich et al., 2010). In this instance, it was determined necessary to develop an MS^3 wide isolation window method because of interfering background ions. Cocaine detected in brain tissue was confirmed by matching six MS^3 product ions with those from a cocaine standard. For the quantitation of cocaine analyzed from the tissue, the trideuterated ($^2H_3$) internal standard was spiked beneath the tissue at known concentrations before matrix application to develop a calibration curve. The MS^3 wide isolation method was then employed for the analysis of cocaine and cocaine-d_3. Cocaine was analyzed successfully and quantified using this approach.

14.5 APPLICATIONS OF MSI FOR DETECTION OF DRUG METABOLITES IN TISSUE

The analysis of drugs and their metabolites in tissue is of great importance and can grant insight to their circulatory pathway. From this information, drug distribution, areas of drug metabolism, and the rate of metabolism, and eventual excretion from the body can be determined. All of these aspects are important to drug development as they can determine the efficacy and effectiveness of a potential drug. This section will describe several examples that demonstrate the utility of MSI for analysis of drugs and their metabolites in tissue.

14.5.1 Localizing Drugs and Their Metabolites to Verify Targeted Drug Distribution

After a drug has been administered, it is important to determine its circulatory pathway and site of metabolism. Drugs to treat neurological diseases have little efficacy if they cannot pass through the blood–brain barrier, whereas anti-cancer agents can wreak havoc in the rest of the body if they are widely distributed. An article by Li et al. (2009) demonstrated the use of MSI on a MALDI–QqTOF in the MS/MS mode to examine the distribution of an antihistamine (astemizole) and one of its metabolites in rat brain sections. The drug was equally distributed throughout the brain section, although the major metabolite was found only in the ventricle regions. From this result, they inferred that the metabolite did not pass through the blood–brain barrier. This application illustrates the utility of MSI for the determination of the distribution of both the drug and corresponding metabolite.

An example of determining the site of metabolism was described by Atkinson et al. (2007), where they demonstrated imaging of a prodrug (AQ4N) and its active metabolite (AQ4) in solid cancer tissue utilizing a MALDI–QqTOF in the MS mode. Effective cancer treatments attempt to maximize the delivery of the active drug species to the cancerous cells, while minimizing distribution to healthy cells. An advantage of using a bioreductive drug such as AQ4N is that they are administered as inactive prodrugs with little to no cytotoxicity. Once they reach the targeted site, the reductase enzymes inside the solid tumors convert the prodrug into the active metabolite. In regions of hypoxia (such as those found in solid tumors), AQ4N has been found to be reduced to AQ4, a compound found to sensitize tumors to existing cancer treatments (McKeown et al., 1995, 1996; Friery et al., 2000; Gallagher et al., 2001). MSI was particularly well suited to verifying the metabolism of AQ4N to AQ4 in solid tissue; in contrast, autoradiography techniques would be unable to distinguish between the bioreductive prodrug and its active metabolite due to the monitoring of the tag and not the molecule. This set of experiments examined sections of H460 human tumor xenografts that were treated with AQ4N. MS images were

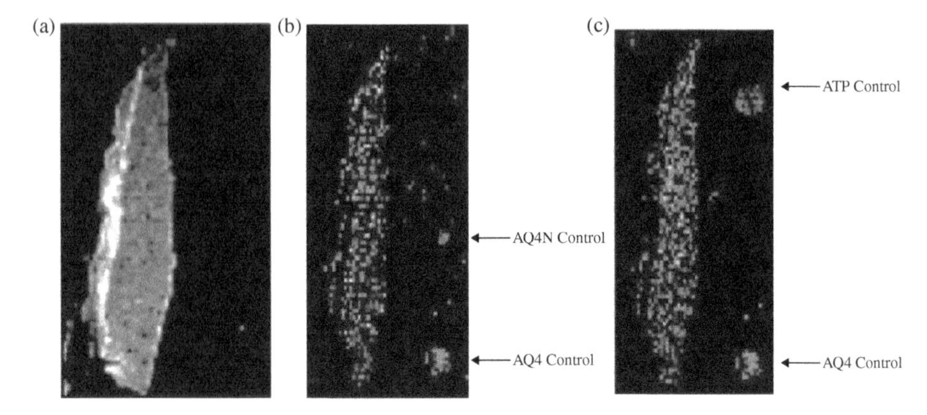

FIGURE 14.2 MALDI–MS images of the distribution of AQ4N, AQ4, and ATP in tumor tissue. Signal has been normalized to the respective matrix ions. To the right of (*b*) and (*c*) are the control spots. (*a*) MS image of *m*/*z* 184 (phosphatidyl choline head group). (*b*) MS image of $[M+Na]^+$ ion 467 (AQ4N) combined with the image of $[M+H]^+$ ion 413 (AQ4). (c) MS image of $[M+Na]^+$ ion 467 (AQ4N) combined with the image of ATP $[M\text{-}H_3PO_4]^-$ (*m*/*z* 409). [Reprinted from Atkinson et al. (2007).]

able to demonstrate a highly accurate localization of the AQ4N prodrug and its active metabolite AQ4, Figures 14.2*a*–14.2*c*. Figure 14.2*a* is an image of the phosphatidylcholine head group (*m*/*z* 184) given to illustrate the position of the tumor. Figure 14.2*b* is an image of the AQ4N and AQ4. The authors suggest that AQ4N is confined to the oxygenated regions of the tissue, whereas AQ4 is localized in areas of hypoxic tissue. To confirm this hypothesis, adenosine 5′-triphosphate (ATP) was imaged. It has been shown that levels are reduced in areas of hypoxia (Kribben et al., 2003). Figure 14.2*c* shows the images of ATP and AQ4. The areas of overlap are again minimal with areas of low ATP levels (and thus indicating a hypoxic environment) dominated by relatively high levels of AQ4. In this fashion, areas of drug metabolism can be identified with confidence that the active metabolite is targeting the cancerous tissue. This result is in contrast with autoradiography, which would be unable to distinguish between metabolite or drug compound. Moreover, the detection of ATP to verify areas of hypoxia would be impossible with autoradiography.

14.5.2 Analysis of Whole-Body Tissue Sections Utilizing Mass Spectral Imaging

While imaging an individual organ or tumor is useful for determining local distribution and concentration, it is important to determine the circulatory pathway over the entire body. Whole-body autoradiography (WBA) is one technique that can display drug distribution postdrug administration.

However, one disadvantage is its inability to distinguish between a labeled drug and a drug metabolite that may have retained the radioactive label; furthermore, since detection is based on the radioactive label, metabolites that lose the label are effectively invisible with this technique.

Mass spectral imaging of whole-body tissue sections would appear to be well suited to tackle the detection of both a drug and its metabolite in this type of tissue section. An early study that examined whole-body sections by MSI using a TOF/TOF instrument in MS/MS mode was a work by Khatib-Shahidi et al. (2006). Physiologically equivalent amounts of olanzapine (OLZ) were orally dosed in 10-week-old male rats; the animals were sacrificed at 2 and 4 h postdose and whole-body sections were then prepared and imaged by MSI. Since the whole-body sections were larger than the MALDI target plates, the sections were divided into four pieces and then placed onto separate target plates. After analysis in the mass spectrometer, image processing was used to piece the four images together to generate a single whole-body tissue section image. Using this method, they were able to detect OLZ and its *N*-desmethyl and 2-hydroxymethyl metabolites from the whole-body tissue sections. (Fig. 14.3) The antipsychotic drug was found to be distributed throughout the entire tissue section in varying degrees of concentration. The metabolites were found in lower concentration; however, neither was present in the brain and spinal cord and verified earlier work that demonstrated that the metabolites do not act within the central nervous system.

Along with the distribution of a drug, ADMET studies examine the metabolism and excretion of a drug and metabolite. Stoeckli et al. (2007) demonstrated the applicability of MSI for determining the excretion pathway of a labeled drug and metabolite in mice using a TOF/TOF instrument. Chen et al. (2008) described the use of MSI on a QqTOF in MS/MS mode to determine the cause of poor oral bioavailability for terfenadine in rats. In this case, MSI on a QqTOF in MS/MS mode was used on whole-body sections of Sprague–Dawley rats. The 1 h postdosed sections showed that terfenadine localized in the stomach and the intestine, whereas the metabolite, fexofenadine, was found in the liver, intestine, and stomach. More notably, the metabolite was absent from the systemic circulation and suggested that the liver metabolized the drug prior to entering the bloodstream. In this fashion, the low bioavailability of the drug was determined.

These three examples demonstrate the applicability of MSI for monitoring drugs and metabolites at all stages of ADMET studies. While WBA is a commonly used technique, MSI provides a complementary technique that avoids the complications of WBA. The drug distribution of a ^{14}C-labeled compound is compared in Figure 14.4 using WBA and MSI in which the 5-min and 1-h drug distributions for both techniques are very similar (Stoeckli et al., 2007). The elimination of a costly synthesis of a radio-labeled drug compound, the long exposure time for the WBA experiment, and the lack of analyte–metabolite specificity are obstacles that can be avoided if MSI is used.

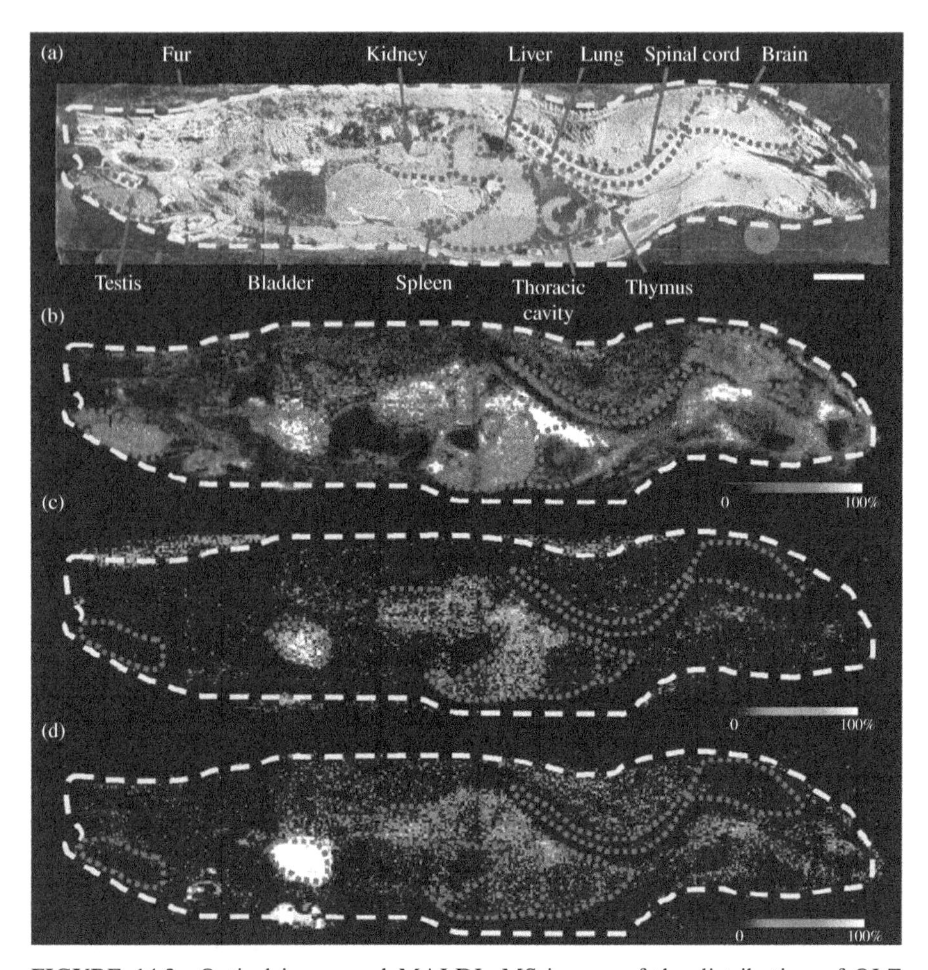

FIGURE 14.3 Optical image and MALDI–MS images of the distribution of OLZ and its metabolites 2 h postdose in whole-body sagittal rat sections. Each section was divided across 4 MALDI target plates and reprocessed postacquisition. (*a*) Optical image of a rat whole-body section 2 h post-OLZ dose. (*b*) MS/MS image of OLZ m/z 313 → 256. (*c*) MS/MS image of *N*-desmethyl OLZ m/z 299 → 256. (*d*) MS/MS image of 2-hydroxymethyl OLZ m/z 329 → 272. [Adapted from Khatib-Shahidi et al. (2006).]

14.5.3 Increasing Analyte Specificity for Mass Spectral Images

While MSI can provide images that allow the user to discriminate between drug and metabolite, it is far more difficult to differentiate between isobaric compounds. Due to the complex nature of tissue, many different lipids, peptides, and other biological compounds (not to mention MALDI matrix clusters) can complicate the MSI analysis for drug compounds. Several

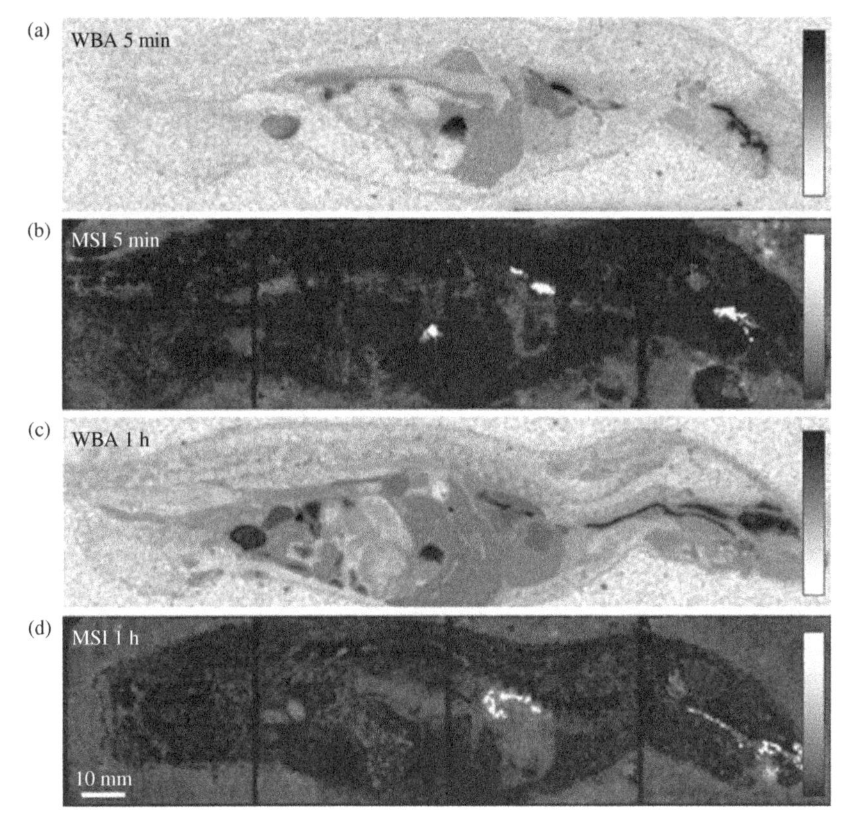

FIGURE 14.4 Comparison of WBA and MSI images using whole-body sections of rats sacrificed either 5 min or 1 h postdose. The 5-min and 1-h sections are from the same animal, respectively. [Adapted from Stoeckli et al. (2007).]

techniques to improve analyte specificity for MSI analysis have been developed, as described below.

Many of the previously described applications use MS/MS to achieve analyte specificity for the drug–metabolite compound in tissue. Additional stages of mass spectrometry (MS^n) afford increased analyte specificity by monitoring an analyte's product ion(s). Use of an ion trap (either 2D or 3D) is ideal for MSI–MS^n experiments (Garrett et al., 2007). Drexler et al. (2007) demonstrated the use of an LIT for the determination of unknown crystalline features found in rat spleens undergoing high dosages of a proprietary drug candidate.

Microcrystals were found in spleen sections after a high-dose prodrug was administered for 2 weeks. As Figure 14.5 shows, MSI was able to confirm the identity of the birefringent crystals as the active drug by comparing the MS/MS product ion scans from microcrystalline areas to microcrystalline-free areas and verifying that both regions demonstrated similar mass spectra for the

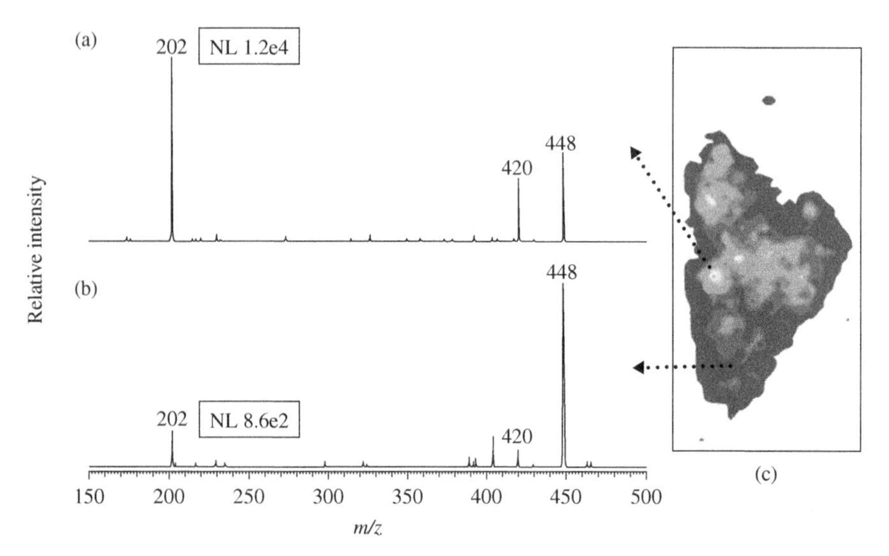

FIGURE 14.5 Comparison of two areas on a spleen section containing microcrystalline material. (*a*) MS/MS image of the active drug (BMS-X, MW 447 Da) m/z 448 → 202. (*b*) MS/MS spectrum (m/z 448 →) at an area rich in microcrystalline material. (*c*) MS/MS spectrum (m/z 448 →) at an area free of the microcrystalline material. The fragmentation between the two spectra is similar enough to confirm the presence of the active drug in both areas. [Adapted from Drexler et al. (2007).]

fragmentation of the active drug (m/z 448) (although varied intensities were noted). The results were further verified by using laser capture microdissection (Emmert-Buck et al., 1996) on both microcrystalline and microcrystalline-free sections and then comparing each sample's MS^3 spectrum to that of a standard.

Albeit MS/MS analysis of drug compounds and metabolites within tissue provides excellent compound specificity, MS/MS is limited by the number of analytes that are imaged per experiment. Although MS/MS can be employed to differentiate target compounds from other interfering isobaric species (e.g., matrix clusters or endogenous biological species), another strategy to increase specificity is to employ high-resolution mass spectrometry. For example, Cornett et al. (2008) demonstrated the use of FTICR mass spectrometry to distinguish between a drug metabolite of OLZ and three isobaric species in kidney tissue. At the nominal mass m/z 329, four different species are detected, but only the species at m/z 329.069 is related to OLZ (2-hyrdoxymethyl OLZ) (Fig. 14.6). Thus, the mass spectral image for the nominal mass at m/z 329 would not reflect just OLZ but the summation of the intensities for the four ions present.

One drawback to using high-resolution mass spectrometry is long analysis times. MSI of individual rat brain sections is typically performed within a couple of hours on a unit resolution system, whereas the same analysis on

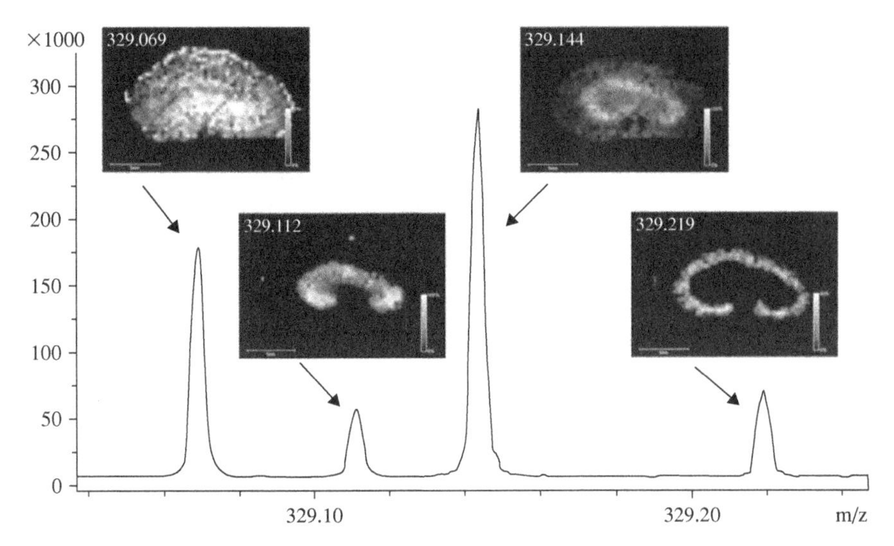

FIGURE 14.6 FTICR mass spectrum of *m/z* 329 and the MS images of *m/z* 329.069, 329.112, 329.144, and 329.219. The ion at *m/z* 329.069 is 2-hydroxymethyl OLZ. The remaining three ions (*m/z* 329.112, 329.144, and 329.219) are isobaric interferants. Without the use of high-resolution MS, the MS image of nominal mass *m/z* 329 would be incorrect due to the interfering peaks. [Adapted from Cornett et al. (2008).]

a high-resolution instrument would take several times that. Nevertheless, high-resolution instruments such as FTICR and LIT-Orbitrap provide excellent specificity for analysis of isobaric species in full-scan mass spectrometric data.

Another strategy to improve analyte specificity is to couple a separation device prior to mass analysis. Much as adding an LC or gas chromatography (GC) prior to the mass spectrometer, the addition of another stage of analysis can enhance the signal-to-noise (S/N) ratio of the experiment. However, the incorporation of LC or GC into an imaging experiment is impractical. An alternate separation device to provide additional analyte specificity in the gas phase is ion mobility spectrometry (IMS). IMS separates ions on the basis of their size (cross section) in the gas phase. An early example of using IMS with mass spectral imaging was work performed by McLean et al. (2007). IMS was used to differentiate between a lipid and a peptide deposited on mouse brain tissue that formed isobaric ions in MALDI.

In the same fashion, IMS can be used to discriminate against potential isobaric species to improve analyte specificity in MSI experiments. Trim et al. (2008) demonstrated the utility of IMS to enhance detection of vinblastine in whole-body tissue sections. Sprague–Dawley rats were dosed intravenously with vinblastine, after which they were sacrificed 1 h postdose. After preparing the tissue section, the samples were analyzed on a QqTOF in MS/MS mode. CID occurred after the ions were separated by IMS, thus product ions can be assigned to their precursor ions since they have the same drift time. Upon examining spiked tissue sections with IMS–MS/MS, the IMS plot showed the presence of multiple ions at *m/z* 811. For example, one of the product ions detected with a longer drift time than vinblastine was *m/z* 184, likely the head group of a phosphatidylcholine lipid. If IMS is used to extract ions with a drift time that match vinblastine, then other interfering signals can be eliminated. The image in Figure 14.7 illustrates the power of adding an additional stage of analysis (IMS separation) for MS and MS/MS imaging experiments. At the regions indicated by the white arrow, signal is reduced in the IMS images compared to the sections without the IMS separation. Thus, with the addition of IMS to the mass spectrometry experiment, the IMS–MS and IMS–MS/MS images show a reduction of interferants at *m/z* 811 and *m/z* 811–751, therefore producing a more accurate image in the renal pelvis region.

Endogenous interferences and MALDI matrix cluster can complicate the analysis of xenobiotic drug compounds and their metabolites in MSI. The use of hyphenated techniques (either MS^n or IMS) or high-resolution mass spectrometry can serve to increase analyte specificity. Although instrument

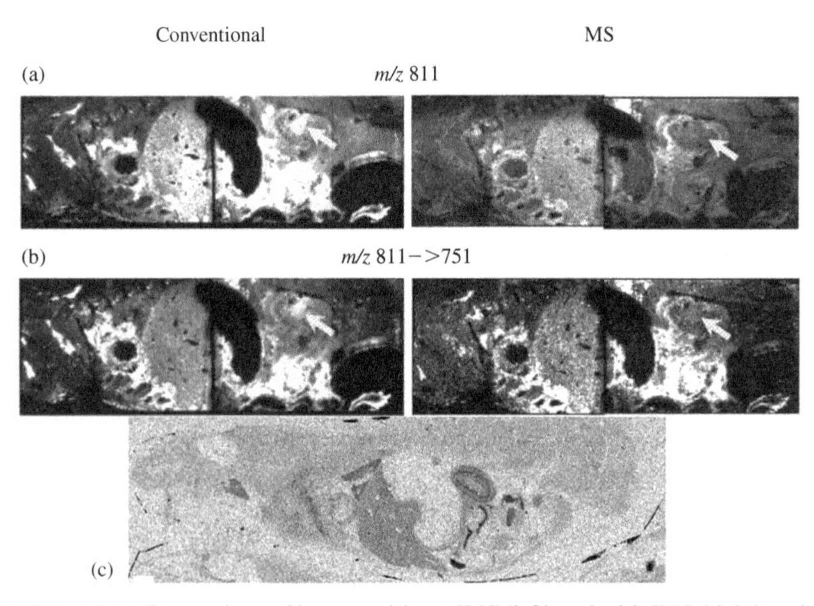

FIGURE 14.7 Comparison of images without IMS (left) and with IMS (right) and the WBA image (bottom). Rats were dosed with ^{3}H-vinblastine and whole-body sections taken. (*a*) Comparison of conventional MSI–MS of ^{3}H-vinblastine (*m/z* 811) (left) and MSI with ion mobility (right). (*b*) MSI–MS/MS image of ^{3}H-vinblastine *m/z* 811 → 751, comparing conventional MSI–MS/MS (left) and ion mobility MSI–MS/MS (right). (*c*) WBA of the rat section. A white arrow in (*a*) and (*b*) highlights a region where IMS eliminated interferants from the image. [Adapted from Trim et al. (2008).]

complexity and/or analysis time may increase, the overall image quality is enhanced by the utilization of these techniques.

14.5.4 Alternative Source Options for Mass Spectral Imaging

MALDI is the dominant ionization source choice for MSI; however, it is not without drawbacks. The application of MALDI matrix on top of the tissue surface complicates the analysis by adding potentially isobaric matrix cluster ions that may obscure the drug compound. Careful selection of MALDI matrix can reduce the matrix effect. An alternative would be to either use the water native to the tissue as a matrix such as with IR-MALDI or altogether eliminate the need for matrix. DESI is an atmospheric pressure technique that permits the direct analysis of surface samples, including tissue sections, with minimal sample preparation (Takats et al., 2004; Cooks et al., 2006). In contrast to MALDI–MSI, no matrix is required; however, the spatial resolution for DESI–MSI is worse when compared to MALDI or SIMS imaging experiments.

Despite the limited spatial resolution, DESI–MSI technique could prove to be complementary to current MALDI–MSI techniques. Studies have been performed describing the use of DESI–MSI for the detection of drug compounds and metabolites in tissue sections (Kertesz et al., 2008; Wiseman et al., 2008). Wiseman et al. (2008) described the detection of clozapine and its major metabolites from individual rat organs using DESI–MSI with an LIT mass analyzer. Clozapine was detected in the brain, lung, kidney, and testis, whereas the *N*-desmethyl clozapine metabolite was detected only in lung tissue. A separate LC–MS/MS analysis determined that clozapine had the highest concentrations in the lung tissue, which could explain the detection of the metabolite only in the lung sections.

Another example of using DESI–MSI was demonstrated by Kertesz et al. (2008) with whole-body sections using a quadrupole ion trap mass analyzer. They examined rats that were dosed with propranolol (an antihypertensive and antiarrhythmic agent). Although the mass spectral images of the dosed drug correlated well with the WBA results (Fig. 14.8), the signal of the known metabolites of propranolol did not rise above the background. In a separate experiment using a liquid microjunction surface sampling probe/electrospray emitter coupled to a quadrupole trap instrument (Van Berkel et al., 2008), the presence of the metabolite was confirmed. As in the analysis of individual organs, whole-body tissue analysis showed similar results with respect to the ability to detect the parent drug, but difficulty in detecting the metabolite.

DESI–MSI is a potentially complementary imaging technique to MALDI–MSI due to the lack of pretreatment set required and similar analysis times. However, difficulties observed to date in detection of metabolites within tissue could prove to be hindrance to wide acceptance as a tool for pharmacological studies.

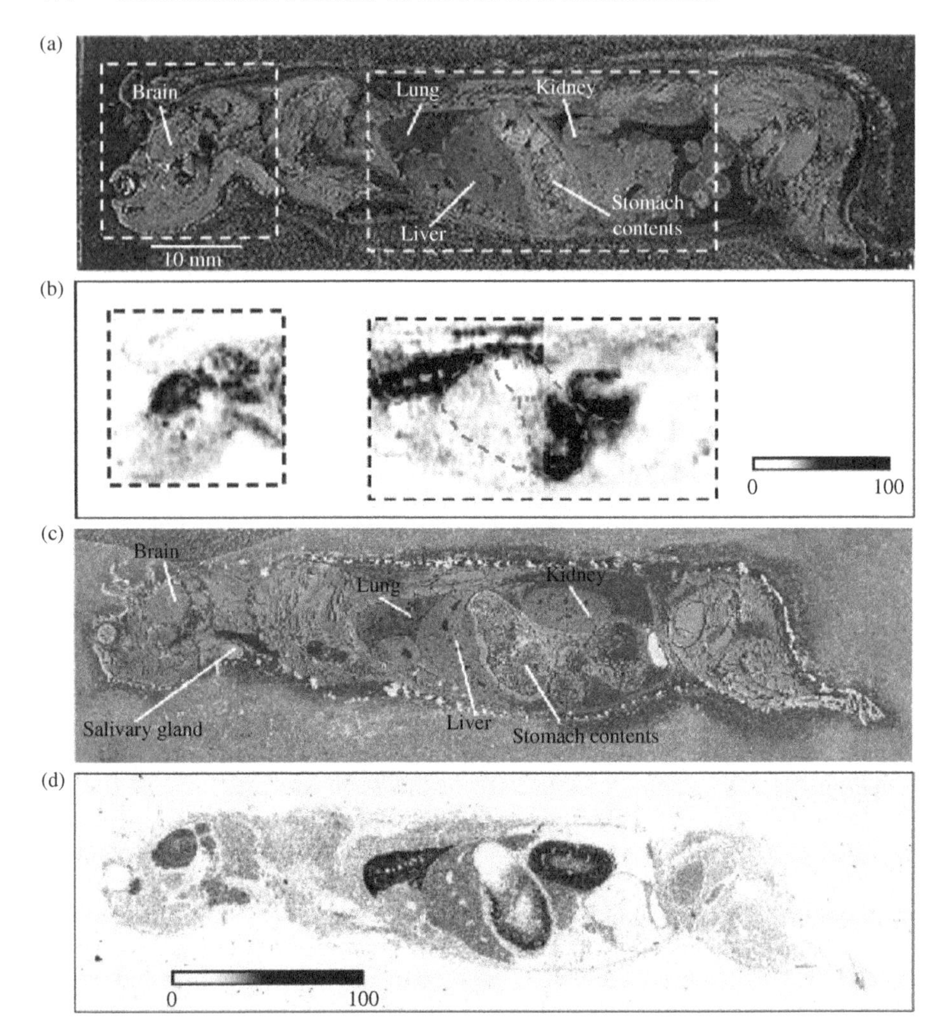

FIGURE 14.8 Comparison of optical, MSI, and WBA images for the detection of propranolol in whole-body sagittal rat sections (all animals sacrificed 1 h postdose). (*a*) Optical image of the whole-body rat section with organs labeled from a rat dosed with propranolol. (*b*) DESI–MS/MS image of the drug propranolol (m/z 260 → 116) from the section displayed in (*a*). (*c*) Optical image of ^{3}H-propranolol. (*d*) WBA image of ^{3}H-propranolol in the whole-body section in (*c*). [Adapted with permission from Kertesz et al. (2008).]

14.6 CONCLUSIONS

Monitoring a drug and its metabolite(s) throughout the body aids a pharmaceutical drug's development. Although several tools exist to observe the circulatory pathway of drugs and metabolites, they are not without their drawbacks. For example, LC–MS analysis of tissue homogenates provides sensitive, selective, and quantitative results for both the drug and metabolite, but spatial information is lost. Autoradiography retains the spatial information of a radio-labeled drug, but this technique cannot distinguish between a radio-labeled drug and a drug metabolite that retained the radioactive tag. Moreover, the cost of synthesizing the radio-labeled drug may be cost prohibitive.

MSI offers a complementary technique as mass spectrometry can differentiate between various ions on the basis of their molecular weight. Additional selectivity is achieved by utilizing MS^n, high-resolution mass spectrometry, or orthogonal separation techniques, either separately or combined. Each additional stage of analysis serves to reduce the interfering chemical noise that originates from the tissue without tedious sample preparation.

There is still much room for growth in this field. Image resolution, for instance, is limited by the laser spot and matrix crystal size. In addition, MSI is primarily a qualitative technique. At best, semiquantitative results are achieved since, among other things, it is difficult to apply an internal standard to tissue. Despite these challenges, mass spectral imaging offers a complementary approach for the detection of drugs and their metabolites in tissue when used with other established methods.

14.7 ACKNOWLEDGMENTS

The authors thank Dr. Walter Korfmacher of Schering-Plough for valuable assistance with the preparation of this work. The authors gratefully acknowledge the support of Bristol-Myers Squibb Company and Thermo Scientific. In addition, DPM acknowledges support from the Alumni Fellowship from the University of Florida.

REFERENCES

Adovelande J, Boulard Y, Berry J-P, Galle P, Slodzian G, Schrevel J. Detection and cartography of the fluorinated antimalarial drug mefloquine in normal and *Plasmodium falciparum* infected red blood cells by scanning ion microscopy and mass spectrometry. Biol Cell 1994;81:185–192.

Aerni H-R, Cornett DS, Caprioli RM. Automated acoustic matrix deposition for MALDI sample preparation. Anal Chem 2006;78:827–834.

Appelhans AD, Delmore JE. Comparison of Polyatomic and atomic primary beams for secondary ion mass spectrometry of organics. Anal Chem 1989;61:1087–1093.

Atkinson SJ, Loadman PM, Sutton C, Patterson, LH, Clench MR. Examination of the distribution of the bioreductive drug AQ4N and its active metabolite AQ4 in solid tumours by imaging matrix-assisted laser desorption/ionisation mass spectrometry. Rapid Commun Mass Spectrom 2007;21:1271–1276.

Ayorinde FO, Hambright P, Porter TN, Keith QL, , Jr. Use of meso-tetrakis (pentafluorophenyl) porphyrin as a matrix for low molecular weight alkylphenol ethoxylates in laser desorption/ionization time-of-flight mass spectrometry. Rapid Commun Mass Spectrom 1999;13:2474–2479.

Baluya DL, Garrett TJ, Yost RA. Automated MALDI matrix deposition method with inkjet printing for imaging mass spectrometry. Anal Chem 2007;79:6862–6867.

Berkenkamp S, Menzel C, Karas M, Hillenkamp F. Performance of infrared matrix-assisted laser desorption/ionization mass spectrometry with lasers emitting in the 3 micro m wavelength range. Rapid Commun Mass Spectrom 1997;11:1399–1406.

Boguslavsky J. Drug Discovery Dev. 2001; 4: 26-30.

Brodbelt JS, Vargas RR, Yost RA. Laser desorption in a quadrupole ion trap. In Practical Aspects of Ion Trap Mass Spectrometry, Vol. 2, Ion Trap Instrumentation, March RE, Todd JFJ, Eds. CRC Press, Boca Raton, FL, 1995, pp. 205–234.

Brown RS, Lennon JJ. Mass resolution improvement by incorporation of pulsed Ion extraction in a matrix-assisted laser desorption/ionization linear time-of-flight mass spectrometer. Anal Chem 1995;67:1998–2003.

Bunch J, Clench MR, Richards DS. Determination of pharmaceutical compounds in skin by imaging matrix-assisted laser desorption/ionisation mass spectrometry. Rapid Commun Mass Spectrom 2004;18:3051–3060.

Caprioli RM, Farmer TB, Gile J. Molecular imaging of biological samples: Localization of peptides and proteins using MALDI-TOF MS. Anal Chem 1997; 69:4751–4760.

Chapman JR. Practical Organic Mass Spectrometry: A Guide for Chemical and Biochemical Analysis, Wiley, New York, 1998.

Chen J, Hsieh Y, Knemeyer I, Crossman L, Korfmacher WA. Visualization of first-pass drug metabolism of terfenadine by MALDI-imaging mass spectrometry. Drug Metab Lett 2008;2:1–4.

Clerc J, Halpern S, Fourre C, Omri F, Briancon C, Jeusset J, Fragu P. SIMS microscopy imaging of the intratumor biodistribution of metaiodobenzylguanidine in the human SK-N-SH neuroblastoma cell line xenografted into nude mice. J Nucl Med 1993;34: 1565–1570.

Cohen LH and Gusev AI. Small molecule analysis by MALDI mass spectrometry. Anal Bioanal Chem 2002;373:571–586.

Cohen SL, Chait BT. Influence of matrix solution conditions on the MALDI-MS analysis of peptides and proteins. Anal Chem 1996;68:31–37.

Contag CH, Ross BD. It's not just about anatomy: In vivo bioluminescence imaging as an eyepiece into biology. J Magn Reson Imaging 2002;16:378–387.

Cooks RG, Ouyang Z, Takats Z, Wiseman JM. Ambient mass spectrometry. Science 2006;311:1566–1570.

Cornett DS, Frappier SL, Caprioli RM. MALDI-FTICR imaging mass spectrometry of drugs and metabolites in tissue. Anal Chem 2008;80:5648–653.

Cristoni S, Brioschi M, Rizzi A, Sironi L, Gelosa P, Tremoli E, Bernardi LR, Banfi C. Analysis of rosuvastatin by imaging mass spectrometry. Rapid Commun Mass Spectrom 2006;20:3483–3487.

Crossman L, McHugh NA, Hsieh Y, Korfmacher WA, Chen J. Investigation of the profiling depth in matrix-assisted laser desorption/ionization imaging mass spectrometry. Rapid Commun Mass Spectrom 2006;20:284–290.

Cui M, McCooeye MA, Fraser C, Mester Z. Quantitation of lysergic acid diethylamide in urine using atmospheric pressure matrix-assisted laser desorption/ionization ion trap mass spectrometry. Anal Chem 2004;76:7143–7148.

Davis AM, Riley RJ. Predictive ADMET studies, the challenges and the opportunities. Curr Opin Chem Biol 2004;8:378–386.

Delcorte A, Bour J, Aubriet F, Muller J-F, Bertrand P. Sample metallization for performance improvement in desorption/ionization of kilodalton molecules: Quantitative evaluation, imaging secondary ion MS, and laser ablation. Anal Chem 2003; 75:6875–6885.

Denoyer A, Ossant F, Arbeille B, Fetissof F, Patat F, Pourcelot L, Pisella P-J. Very-high-frequency ultrasound corneal imaging as a new tool for early diagnosis of ocular surface toxicity in rabbits treated with a preserved glaucoma drug. Ophthalm Res 2008;40:298–308.

Dreisewerd K. The desorption process in MALDI. Chem Rev 2003;103:395–425.

Drexler DM, Garrett TJ, Cantone JL, Diters RW, Mitroka JG, Prieto Conaway MC, Adams SP, Yost RA, Sanders M. Utility of imaging mass spectrometry (IMS) by matrix-assisted laser desorption ionization (MALDI) on an ion trap mass spectrometer in the analysis of drugs and metabolites in biological tissues. J Pharmacol Toxicol Methods 2007;55:279–288.

Drummer OH, Gerstamoulos J. Postmortem drug analysis: Analytical and toxicological aspects. Ther Drug Monitor 2002;24:199–209.

Emmert-Buck MR, Bonner RF, Smith PD, Chuaqui RF, Zhang Z, Goldstein SR, Weiss RA, Liotta LA. Laser capture microdissection. Science 1996;274:998–1001.

Friery OP, Gallagher R, Murray MM, Hughes CM, Galligan ES, McIntyre IA, Patterson LH, Hirst DG, McKeown SR. Enhancement of the antitumor effect of cyclophosphamide by the bioreductive drugs AQ4N and tirapazamine. Br J Cancer 2000;82:1469–1473.

Gallagher R, Hughes CM, Murray MM, Friery OP, Patterson LH, Hirst DG, McKeown SR. The chemopotentiation of cisplatin by the novel bioreductive drug AQ4N. Br J Cancer 2001;85:625–629.

Garidel P, Boese M. Mid infrared microspectroscopic mapping and imaging: A bio-analytical tool for spatially and chemically resolved tissue characterization and evaluation of drug permeation within tissues. Microsc Res Tech 2007;70:336–349.

Garrett TJ. Imaging small molecules in tissue by matrix-assisted laser desorption/ionization tandem mass spectrometry. PhD dissertation. University of Florida, Gainesville, FL, 2006, p. 166.

Garrett TJ, Kovtoun V, Prieto-Conaway MC, Miller K, Stafford G. Imaging mass spectrometry with LD/MS^n at different pressure regimes. Paper presented at the 53rd Annual Conference on Mass Spectrometry and Allied Topics, 2005.

Garrett TJ, Prieto-Conaway MC, Kovtoun V, Bui H, Izgarian N, Stafford G, Yost RA. Imaging of small molecules in tissue sections with a new intermediate-pressure MALDI linear ion trap mass spectrometer. Int J Mass Spectrom 2007;260:166–176.

Garrett TJ, Yost RA. Analysis of intact tissue by intermediate-pressure MALDI on a linear ion trap mass spectrometer. Anal Chem 2006;78:2465–2469.

Garrett TJ, Yost RA. The role of trapped ion mass spectrometry for imaging. In Practical Aspects of Trapped Ion Mass Spectrometry, Vol. 5, Applications of Ion Trapping Devices, March RE, Todd JFJ, Eds. CRC Press, Boca Raton, FL, 2009.

Garrison BJ, Delcorte A, Zhigilei LV, Itina TE, Krantzman KD, Yingling YG, McQuaw CM, Smiley EJ, Winograd N. Big molecule ejection-SIMS vs. MALDI. Appl Surf Sci 2003;203–204:69–71.

Gatley SJ, Volkow ND. Addiction and imaging of the living human brain. Drug Alcohol Dependence 1998;51:97–108.

Gumbleton M, Stephens DJ. Coming out of the dark: The evolving role of fluorescence imaging in drug delivery research. Adv Drug Deliv Rev 2005;57:5–15.

Gusev AI, Wilkinson WR, Proctor A, Hercules DM. Direct quantitative analysis of peptides using matrix-assisted laser desorption/ionization. Fresenius J Anal Chem 1996;354:455–463.

Hand OW, Majumdar TK, Cooks RG. Effects of primary ion polyatomicity and kinetic energy on secondary ion yield and internal energy in SIMS. Int J Mass Spectrom 1990;97:35–45.

Hankin JA, Barkley RM, Murphy RC. Sublimation as a method of matrix application for mass spectrometric imaging. J Am Soc Mass Spectrom 2007;18:1646–1652.

Hatsis P, Brombacher S, Corr J, Kovarik P, Volmer DA. Quantitative analysis of small pharmaceutical drugs using a high repetition rate laser matrix-assisted laser/desorption ionization source. Rapid Commun Mass Spectrom 2003;17:2303–2309.

Hensel RR, King RC, Owens KG. Electrospray sample preparation for improved quantitation in matrix-assisted laser desorption/ionization time-of-flight mass spectrometry. Rapid Commun Mass Spectrom 1997;11:1785–1793.

Hillenkamp F, Unsoeld E, Kaufman R, Nitsche R. High-sensitivity laser microprobe mass analyzer. Appl Phys 1975;8:341–348.

Hopfgartner G, Varesio E, Stoeckli M. Matrix-assisted laser desorption/ionization mass spectrometric imaging of complete rat sections using a triple quadrupole linear ion trap. Rapid Commun Mass Spectrom 2009;23:733–736.

Hsieh Y, Casale R, Fukuda E, Chen J, Knemeyer I, Wingate J, Morrison R, Korfmacher W. Matrix-assisted laser desorption/ionization imaging mass spectrometry for direct measurement of clozapine in rat brain tissue. Rapid Commun Mass Spectrom 2006;20:965–972.

Hsieh Y, Chen J, Korfmacher WA. Mapping pharmaceuticals in tissues using MALDI imaging mass spectrometry. J Pharmacol Toxicol Methods 2007;55:193–200.

Iancu TC, Perl DP, Sternlieb I, Lerner A, Leshinsky E, Kolodny EH, Hsu A, Good PF. The application of laser microprobe mass analysis to the study of biological material. BioMetals 1996;9:57–65.

Ifa DR, Wiseman JM, Song Q, Cooks RG. Development of capabilities for imaging mass spectrometry under ambient conditions with desorption electrospray ionization (DESI). Int J Mass Spectrom 2007;259:8–15.

Jurchen JC, Rubakhin SS, Sweedler JV. MALDI-MS imaging of features smaller than the size of the laser beam. J Am Soc Mass Spectrom 2005;16:1654–1659.

Kang M-J, Tholey A, Heinzle E. Application of automated matrix-assisted laser desorption/ionization time-of-flight mass spectrometry for the measurement of enzyme activities. Rapid Commun Mass Spectrom 2001;15:1327–1333.

Karas M, Hillenkamp F. Laser desorption ionization of proteins with molecular masses exceeding 10,000 daltons. Anal Chem 1988;60:2299–301.

Karas M, Kruger R. Ion formation in MALDI: The cluster ionization mechanism. Chem Rev 2003;103:427–439.

Kertesz V, Van Berkel GJ, Vavrek M, Koeplinger KA, Schneider BB, Covey TR. Comparison of drug distribution images from whole-body thin tissue sections obtained using desorption electrospray ionization tandem mass spectrometry and autoradiography. Anal Chem 2008;80:5168–5177.

Khatib-Shahidi S, Andersson M, Herman JL, Gillespie TA, Caprioli RM. Direct molecular analysis of whole-body animal tissue sections by imaging MALDI mass spectrometry. Anal Chem 2006;78:6448–6456.

Kribben A, Feldkamp T, Horbelt M, Lange B, Pietruck F, Herget-Rosenthal S, Heemann U, Philipp T. ATP protects, by way of receptor-mediated mechanisms, against hypoxia-induced injury in renal proximal tubules. J Lab Clin Med 2003; 141:67–73.

Kutz KK, Schmidt JJ, Li L. In situ tissue analysis of neuropeptides by MALDI FTMS in-cell accumulation. Anal Chem 2004;76:5630–5640.

Landgraf RR. Analysis of lipids in nerve tissue by MALDI tandem mass spectrometric imaging. PhD dissertation. University of Florida, Gainesville, FL, 2009, p. 135.

Lemaire R, Wisztorski M, Desmons A, Tabet JC, Day R, Salzet M, Fournier I. MALDI-MS direct tissue analysis of proteins: Improving signal sensitivity using organic treatments. Anal Chem 2006;78:7145–7153.

Li F, Hsieh Y, Kang L, Sondey C, Lachowicz J, Korfmacher WA. MALDI-tandem mass spectrometry imaging of astemizole and its primary metabolite in rat brain sections. Bioanalysis 2009;1:299–307.

Ling Y-C, Lin L, Chen Y-T. Quantitative analysis of antibiotics by matrix-assisted laser desorption/ionization time-of-flight mass spectrometry. Rapid Commun Mass Spectrom 1998;12:317–327.

Luxembourg SL, Heeren RMA. Fragmentation at and above surfaces in SIMS: Effects of biomolecular yield enhancing surface modifications. Int J Mass Spectrom 2006;253:181–192.

Luxembourg SL, McDonnell LA, Duursma MC, Guo X, Heeren RMA. Effect of local matrix crystal variations in matrix-assisted ionization techniques for mass spectrometry. Anal Chem 2003;75:2333–2341.

Mahmoudi M, Simchi A, Imani M, Hafeli UO. Superparamagnetic iron oxide nanoparticles with rigid cross-linked polyethylene glycol fumarate coating for application in imaging and drug delivery. J Phys Chem C 2009;113:8124–8131.

Makarov A. Electrostatic axially harmonic orbital trapping: A high-performance technique of mass analysis. Anal Chem 2000;72:1156–1162.

Makarov A, Denisov E, Kholomeev A, Balschun W, Lange O, Strupat K, Horning S. Performance evaluation of a hybrid linear ion trap/Orbitrap mass spectrometer. Anal Chem 2006a;78:2113–2120.

Makarov A, Denisov E, Lange O, Horning S. Dynamic range of mass accuracy in LTQ Orbitrap hybrid mass spectrometer. J Am Soc Mass Spectrom 2006b;17:977–982.

McKeown SR, Friery OP, McIntyre IA, Hejmadi MV, Patterson LH, Hirst DG. Evidence for therapeutic gain when AQ4N or tirapazamine is combined with radiation. Br J Cancer Suppl 1996;74:S39–S42.

McKeown SR, Hejmadi MV, McIntyre IA, McAleer JJ, Patterson LH. AQ4N: An alkylaminoanthraquinone *N*-oxide showing bioreductive potential and positive interaction with radiation in vivo. Br J Cancer 1995;72:76–81.

McLean JA, Ridenour WB, Caprioli RM. Profiling and imaging of tissues by imaging ion mobility–mass spectrometry. J Mass Spectrom 2007;42:1099–1105.

Nakanishi T, Ohtsu I, Furuta M, Ando E, Nishimura O. Direct MS/MS analysis of proteins blotted on membranes by a matrix-assisted laser desorption/ionization-quadrupole ion trap-time-of-flight tandem mass spectrometer. J Proteome Res 2005;4:743–747.

Nemes P, Vertes A. Laser ablation electrospray ionization for atmospheric pressure, in vivo, and imaging mass spectrometry. Anal Chem 2007;79:8098–8106.

Nicola AJ, Gusev AI, Hercules DM. Direct quantitative analysis from thin-layer chromatography plates using matrix-assisted laser desorption/ionization mass spectrometry. Appl Spectrosc 1996;50:1479–1482.

Nicola AJ, Gusev AI, Proctor A, Jackson EK, Hercules DM. Application of the fast-evaporation sample preparation method for improving quantification of angiotensin II by matrix-assisted laser desorption/ionization. Rapid Commun Mass Spectrom 1995;9:164–1171.

Niu G, Chen X. Has molecular and cellular imaging enhanced drug discovery and drug development? Drugs in R&D 2008;9:351–368.

Niu S, Zhang W, Chait BT. Direct comparison of infrared and ultraviolet wavelength matrix-assisted laser desorption/ionization mass spectrometry of proteins. J Am Soc Mass Spectrom 1998;9:1–7.

Onnerfjord P, Ekstrom S, Bergquist J, Nilsson J, Laurell T, Marko-Varga G. Homogeneous sample preparation for automated high throughput analysis with matrix-assisted laser desorption/ionization time-of-flight mass spectrometry. Rapid Commun Mass Spectrom 1999;13:315–322.

Pacholski ML, Winograd N. Imaging with mass spectrometry. Chem Rev 1999;99: 2977–3005.

Perchalski R. Characteristics and application of a laser ionization/evaporation source for tandem mass spectrometry. PhD dissertation. University of Florida, Gainesville, FL, 1985, p. 195.

Perchalski RJ, Yost RA, Wilder BJ. Laser desoprtion chemical ionization mass spectrometry/mass spectrometry. Anal Chem 1983;55:2002–2005.

Puolitaival SM, Burnum KE, Cornett DS, Caprioli RM. Solvent-free matrix dry-coating for MALDI imaging of phospholipids. J Am Soc Mass Spectrom 2008;19: 882–886.

Reich RF, Cudzilo K, Yost RA. Quantitative imaging of cocaine and its metabolites in postmortem brain tissue by intermediate-pressure MALDI/linear ion trap tandem mass spectrometry. Paper presented at the 56th ASMS Conference on Mass Spectrometry and Allied Topics, Denver, CO, 2008.

Reich RF, Cudzilo K, Levisky JA, Yost RA. Quantitative MALDI-MS^n analysis of cocaine in the autopsied brain of a human cocaine user employing a wide isolation window and internal standards. J Am Soc Mass Spectrom 2010;21:564–571.

Reyzer ML, Hsieh Y, Ng K, Korfmacher WA, Caprioli RM. Direct analysis of drug candidates in tissue by matrix-assisted laser desorption/ionization mass spectrometry. J Mass Spectrom 2003;38:1081–1092.

Schiffer WK, Liebling CNB, Patel V, Dewey SL. Targeting the treatment of drug abuse with molecular imaging.Nucl Med Biol 2007;34 833–847.

Schriver KE, Chaurand P, Caprioli RM. High resolution imaging mass spectrometry: characterization of ion yields and laser spot sizes. Paper presented at the 51st ASMS

Conference on Mass Spectrometry and Allied Topics, Montreal, Canada, 2003;231–232:485–489.

Schwartz SA, Reyzer ML, Caprioli RM. Direct tissue analysis using matrix-assisted laser desorption/ionization mass spectrometry: Practical aspects of sample preparation. J Mass Spectrom 2003;38:699–708.

Sleno L, Volmer DA. Assessing the properties of internal standards for quantitative matrix-assisted laser desorption/ionization mass spectrometry of small molecules. Rapid Commun Mass Spectrom 2006;20:1517–1524.

Slodzian G, Daigne B, Girard F, Boust F, Hillion F. Scanning secondary ion analytical microscopy with parallel detection. Biol Cell 1992;74:43–50.

Som P, Oster ZH, Wang G-J, Volkow ND, Sacker DF. Spatial and temporal distribution of cocaine and effects of pharmacological interventions: Wholebody autoradiographic microimaging studies. Life Sci 1994;55:1375–1382.

Sosnovik DE, Weissleder R. Emerging concepts in molecular MRI. Curr Opin Biotechnol 2007;18:4–10.

Stimpfl T, Reichel S. Distribution of drugs of abuse within specific regions of the human brain. Forensic Sci Int 2007;170:179–182.

Stoeckli M, Staab D, Schweitzer A. Compound and metabolite distribution measured by MALDI mass spectrometric imaging in whole-body tissue sections. Int J Mass Spectrom 2007;260:195–202.

Strupat K, Kovtoun V, Bui H, Viner R, Stafford G, Horning S. MALDI produced ions inspected with a linear ion trap-Orbitrap hybrid mass analyzer. J Am Soc Mass Spectrom 2009;20:1451–1463.

Sugiura Y, Shimma S, Setou M. Two-step matrix application technique to improve ionization efficiency for matrix-assisted laser desorption/ionization in imaging mass spectrometry. Anal Chem 2006;78:8227–8235.

Taban IM, Altelaar AFM, van der Burgt YEM, McDonnell LA, Heeren RMA, Fuchser J, Baykut G. Imaging of peptides in the rat brain using MALDI-FTICR mass spectrometry. J Am Soc Mass Spectrom 2007;18:145–151.

Takats Z, Wiseman JM, Cooks RG. Ambient mass spectrometry using desorption electrospray ionization (DESI): instrumentation, mechanisms and applications in forensics, chemistry, and biology. J Mass Spectrom 2005;40:1261–1275.

Takats Z, Wiseman JM, Gologan B, Cooks RG. Mass spectrometry sampling under ambient conditions with desorption electrospray ionization. Science 2004;306: 471–473.

Tempez A, Schultz JA, Della-Negra S, Depauw J, Jacquet D, Novikov A, Lebeyec Y, Pautrat M, Caroff M, Ugarov M, Bensaoula H, Gonin M, Fuhrer K, Woods A. Orthogonal time-of-flight secondary ion mass spectrometric analysis of peptides using large gold clusters as primary ions. Rapid Commun Mass Spectrom 2004;18:371–376.

Trim PJ, Henson CM, Avery JL, McEwen A, Snel MF, Claude E, Marshall PS, West A, Princivalle AP, Clench MR. Matrix-assisted laser desorption/ionization-ion mobility separation-mass spectrometry imaging of vinblastine in whole body tissue sections. Anal Chem 2008;80:8628–8634.

Troendle FJ, Reddick CD, Yost RA. Detection of pharmaceutical compounds in tissue by matrix-assisted laser desorption/ionization and laser desorption/chemical ionization tandem mass spectrometry with a quadrupole ion trap. J Am Soc Mass Spectrom 1999;10:1315–1321.

Van Berkel GJ, Kertesz V, Koeplinger KA, Vavrek M, Kong A-NT. Liquid microjunction surface sampling probe electrospray mass spectrometry for detection of drugs and metabolites in thin tissue sections. J Mass Spectrom 2008;43:500–508.

Venter A, Sojka PE, Cooks RG. Droplet dynamics and ionization mechanisms in desorption electropray ionization mass spectrometry. Anal Chem 2006;78:8549–8555.

Wang G, Yu H, De Man B. An outlook on x-ray CT research and development. Med Phys 2008;35:1051–1064.

Wang H-YJ, Jackson SN, McEuen J, Woods AS. Localization and analyses of small drug molecules in rat brain tissue sections. Anal Chem.2005;77:6682–6686.

Wiseman JM, Ifa DR, Song Q, Cooks RG. Tissue imaging at atmospheric pressure using desorption electrospray ionization (DESI) mass spectrometry. Angew Chem Int Ed 2006;45:7188–7192.

Wiseman JM, Ifa DR, Zhu Y, Kissinger CB, Manicke NE, Kissinger PT, Cooks RG. Desorption electrospray ionization mass spectrometry: Imaging drugs and metabolites in tissues. Proc Natl Acad Sci USA 2008;105:18120–18125.

Wu KJ, Odom RW. Matrix-enhanced secondary ion mass spectrometry: A method for molecular analysis of solid surfaces. Anal Chem 1996;68:873–882.

Xu J, Szakal CW, Martin SE, Peterson BR, Wucher A, Winograd N. Molecule-specific imaging with mass spectrometry and a buckminsterfullerene probe: Application to characterizing solid-phase synthesized combinatorial libraries. J Am Chem Soc 2004;126:3902–3909.

Zenobi R, Knochenmuss R. Ion formation in MALDI mass spectrometry. Mass Spectrom Rev 1998;17:337–366.

Zhang W, Niu S, Chait BT. Exploring infrared wavelength matrix-assisted laser desorption/ionization of proteins with delayed-extraction time-of-flight mass spectrometry. J Am Soc Mass Spectrom 1998;9:879–884.

15 Use of Triple Quadrupole–Linear Ion Trap Mass Spectrometry as a Single LC–MS Platform in Drug Metabolism and Pharmacokinetics Studies

WENYING JIAN

BA/DMPK, Pharmaceutical Research and Development,
Johnson & Johnson, Raritan, New Jersy

MING YAO

Department of Biotransformation, Bristol-Myers Squibb Research and Development, Princeton, New Jersy

BO WEN

Department of Drug Metabolism and Pharmacokinetics,
Hoffmann-La Roche, Nutley, New Jersey

ELLIOTT B. JONES

Applied Biosystems Inc., Foster City, California

MINGSHE ZHU

Department of Biotransformation, Bristol-Myers Squibb Research and Development, Princeton, New Jersy

Mass Spectrometry in Drug Metabolism and Disposition: Basic Principles and Applications, First Edition. Edited by Mike S. Lee and Mingshe Zhu.

15.1 INTRODUCTION

Drug metabolism and pharmacokinetics (DMPK) research is an integral part of drug discovery and development (Zhang et al., 2009b; Zhu et al., 2009). The DMPK organization in the modern pharmaceutical industry can be categorized into three distinct groups: drug metabolism, pharmacokinetics (PK), and bioanalysis. Drug metabolism scientists study drug disposition and enzymes and transporters involved in drug metabolism and disposition. Bioanalysis scientists are mainly responsible for quantitative analysis of drugs and their metabolites in PK and toxicokinetics (TK) studies as well as some in vitro absorption, distribution, metabolism, and elimination (ADME) studies. The main responsibilities of PK scientists include PK study design, PK parameter calculation, and data interpretation. Scientists in each of these groups usually have relatively unique skill sets, although some DMPK scientists work in two or more DMPK functions.

As summarized in Fig. 15.1, various DMPK studies are performed through the entire drug discovery and development process. In the lead selection and optimization stages of drug discovery, metabolic stability analysis, cytochrome P450 (CYP) inhibition, parallel artificial membrane permeability assay (PAMPA), metabolic soft spot determination, reactive metabolite screening, and animal PK studies are conducted in optimizing DMPK properties of lead compounds. Radiolabeled ADME studies in animals and humans and PK studies in a large number of healthy subjects, patients, and special populations are carried out in the stage of drug development. From an analytical methodology perspective, all DMPK studies (Fig. 15.1 and Table 15.1) are divided into four main types: (1) quantitative analysis of drugs and their metabolites in

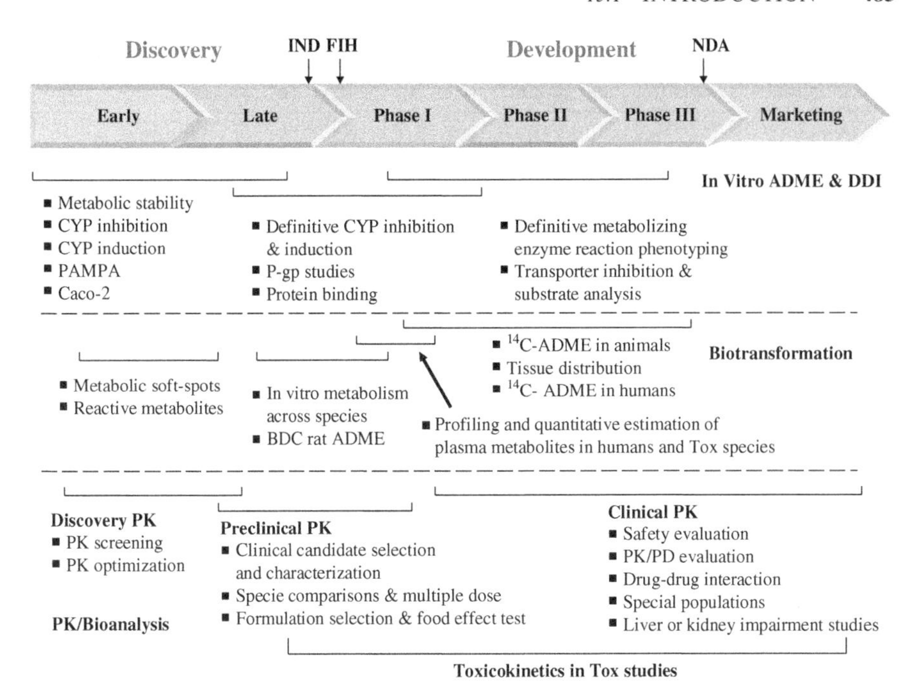

FIGURE 15.1 Summary of drug metabolism and pharmacokinetics studies in drug discovery and development. (Abbreviations: **ADME**, absorption, distribution, metabolism and elimination; **BDC**, bile duct cannulated; **CYP**, cytochrome P450; **DDI**, drug-drug interaction; **DMPK**, drug metabolism and pharmacokinetics; **FIH**, first in human; **IND**, investigative new drug; **NDA**, new drug application; **PAMPA**, parallel artificial membrane permeability assay; **P-gp**, P-glycoprotein; **PK/PD**, pharmacokinetics/pharmacodynamatics; **TK**, toxicokinetics.

the circulation of animals and humans, namely bioanalysis, in support of PK and TK studies, (2) metabolite profiling and identification, (3) quantitative analysis of drugs (such as metabolic stability, protein binding, and transporter studies), metabolites or substrates in in vitro systems (such as CYP inhibition and induction using probe substrates), and (4) radioactivity analysis in tissue distribution and radiolabeled ADME studies. Liquid chromatography coupled with mass spectrometry (LC–MS) plays a predominant role in all ADME studies in both metabolite identification and quantitative analysis of small molecules, including drugs, metabolites, and substrates of metabolizing enzymes and transporters.

Traditionally, quantitative analysis of drugs and their metabolites both in vivo and in vitro is mainly dependent on multiple reaction monitoring (MRM) with triple quadruple instruments. In contrast, drug metabolite detection and identification often involve various types of mass spectrometers (Ma and Chowdhury, 2007; Prakash et al., 2007). Full-scan MS experiments followed by data-dependent tandem MS/MS acquisition with ion trap or linear ion trap mass spectrometers are carried out to identify common metabolites whose

TABLE 15.1 Preferred and Optional Methods for In Vitro ADME, Metabolite Profiling and Identification, and Bioanalysis Using QTRAP Instruments

Experiments	Nature and Objective of Experiment	Preferred Method	Optional Method
Metabolic stability, transporter, PAMPA, and protein binding	Rapidly measure the disappearance of a drug or drug concentrations in incubations	• MRM • Require predetermination of MS/MS spectra	
Metabolism rates, CYP inhibition, induction, and reaction phenotyping	Determine the formation rates of metabolites of substrates in incubations	• MRM • Very sensitive and selective	
Metabolic soft spots	Rapidly determine structures of major metabolites in incubations	• MIM–EPI followed by data mining • More sensitive than EMS–EPI	• EMS–EPI followed by data mining • Generic data acquisition method
Metabolite profiles across species in vitro	Determine human metabolite profiles and confirm the presence of human metabolites in animal species	• MIM–EPI followed by data mining • Confirmation of human metabolites using MRM–EPI	• pMRM–EPI • Very sensitive • Suited for common metabolites
GSH adduct screening	Detect and structurally characterize reactive metabolites trapped by GSH	• PI–EPI with polarity switching • Sensitive and selective	• pMRM–EPI • Very sensitive • Suited for common metabolites
NAC adduct screening	Detect and structurally characterize reactive metabolites trapped by NAC	• NL–EPI with polarity switching • Fast analysis	• pMRM–EPI with polarity switching • Suited for common NAC adducts
GSH ethyl ester adduct screening	Detect and structurally characterize reactive metabolites trapped by GSH ethyl ester	• PI-EPI with polarity switching • Very sensitive and selective	

(*Continued*)

TABLE 15.1 (Continued)

Experiments	Nature and Objective of Experiment	Preferred Method	Optional Method
D-GSH adduct screening and semiquantitiation	D-GSH is quantitatively determined by fluoresce detection and structurally characterized by MS	• pMRM–EPI with polarity switching	
Identification of oxidative metabolites in vivo	Detect and identify metabolites in complex matrices	• pMRM–EPI for common metabolites • PI-EPI and NL-EPI for uncommon metabolites	• EMS–EPI for high level metabolites
Bioanalysis in PK and TK studies and tissue imaging	Determine drug and metabolite concentrations in biological matrices such as plasma, urine, or tissue	• MRM with synthetic standards	

molecular weights are predictable based on mass shifts from the parent drugs. Precursor ion (PI) scan and neutral loss (NL) scan with triple quadrupole mass spectrometers are commonly employed to detect common and uncommon metabolites based on their predicable fragmentation patterns regardless of their molecular weights. Recently, high-resolution mass spectrometry (HRMS) has been extensively used for drug metabolite detection and identification, in which mass defect filtering and other data-mining technologies play a key role in searching for metabolites and recovering MS/MS spectra of detected metabolites (Zhang et al., 2009a).

Recently introduced hybrid triple quadruple–linear ion trap (QTRAP) mass spectrometer has scan functions identical to those of classical triple quadruple (PI scan, NL scan, and MRM). It can be used for small-molecule quantification with MRM in both bioanalysis and in vitro ADME studies (Hopfgartner et al., 2004). The QTRAP also has scan functions (full MS scan and MS/MS scan) similar to an ion trap mass spectrometer. Most importantly, MS, MRM, PI, and NL scans on QTRAP instruments can serve as survey scans to trigger the information-dependent acquisition (IDA) of enhanced product ion spectra (EPI) with polarity switching, which provide product ion spectra with rich fragments with no low-mass cutoff (Xia et al., 2003; Zheng et al., 2007; Wen et al., 2008b; Yao et al., 2008; Jian et al., 2009). The information-dependent acquisitions have been widely applied to reactive metabolite screening and metabolic soft spot determination. The QTRAP mass spectrometer is a unique LC–MS platform and well suited for both quantitative analysis and structural elucidation in DMPK studies. In this chapter, instrumentation and scan functions of newly introduced 5500 QTRAP instruments are described. Utilities and examples of the QTRAP in various types of DMPK studies are discussed.

15.2 INSTRUMENTATION AND SCAN FUNCTIONS

15.2.1 Instrumentation

The QTRAP is a unique LC–MS system and can function either as a triple quadrupole or ion trap within a given scan cycle (Hopfgartner et al., 2004; King and Fernandez-Metzler, 2006). This versatility allows for an extensive array of quantitative and qualitative experiments to support various DMPK studies (Fig. 15.1). The ability of a QTRAP to perform triple quadrupole MS/MS-based survey scans as a trigger for MS^2 acquisition by linear ion trap allows for sensitive and selective detection of metabolite ions and recording of their product ion spectra. The recent development of the 5500 QTRAP further enhances these unique scan functions with increased quadrupole and ion trap sensitivity and improved scan rates (Fig. 15.2). Compared to 4000 QTRAP, 5500 QTRAP allows for four times greater sampling over a chromatographic

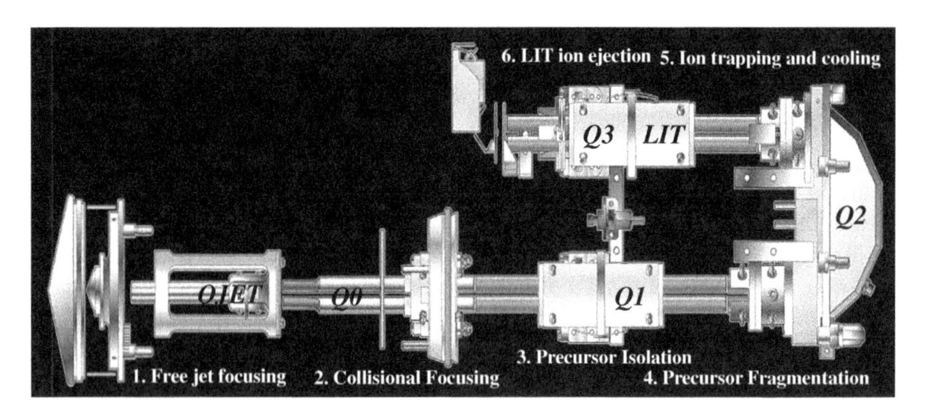

FIGURE 15.2 Diagram of complete ion path of the 5500 QTRAP. The QJet focuses the ions as they enter the main vacuum chamber. Q_0 further focuses the ions to improve sensitivity. Q_1 can act as a mass filter for MS or MS/MS experiments. Q_2 produces ion fragmentation for both trap and quadrupole analysis. Q_3 can act as a quadrupole or as an ion trap depending upon the experimental need. The Q_3 linear ion trap is capable of switching between these modes in less then 10 ms.

peak, which is critical for quantitative determination of narrow chromatographic peaks and for use of the broad array of scan modes and IDA.

The ion path of 5500 QTRAP is similar to other triple quadrupole instruments (Fig. 15.2). An improved ion focusing element, QJet, increases ion transmission from the interface to the Q_0 quadrupole and leads to increased sensitivity. The ion path contains two mass resolving radio frequency (RF)/direct current (DC) quadrupoles. A 180° curved collision cell allows for submillisecond ion transfer, which improves the rate of PI and NL scanning as well as sensitivity of MRM mode at sub-10-ms dwell times. The Q_3 quadrupole has been modified to act as a linear ion trap (LIT). In the quadrupole MS/MS mode, the ions are filtered by the mass-to-charge ratio m/z in Q_1, fragmented in Q_2, and filtered again in Q_3. In MS/MS LIT mode, filtering is also done in Q_1 with fragmentation in Q_2. The dynamic fill time (DFT) algorithm sets the fill time during which the trap will collect fragment ions for MS^1, MS^2, and MS^3 to allow for maximum sensitivity, without overfilling and producing space charge effects. For any trap mode, a short burst of N_2 gas is injected into the trap once it is filled to aid in the radial cooling of the ions. Radial cooling increases sensitivity and resolution. After a ~10 ms cooling, the ions are resonantly excited by a dipolar excitation of two of the quadrupole rods. A mass-calibrated amplitude increase over the trap scan time causes the ions to pass through the exit barrier field at the rear of the trap in order of increasing m/z. A linear ion accelerator rod set pushes the ions toward the exit barrier and improves ejection efficiency. Since the trap loses very few ions compared to a sequential quadrupole scan, sensitivity is very good for MS^2 in this mode compared to a quadrupole product ion scan.

Quadrupole single MS can be run using a Q_1 or Q_3 scan. Q_3 is typically preferred because in this mode N_2 is used in Q_2 to focus the ions into the quadrupole. The higher pressure of N_2 in Q_2 makes this mode compatible with other MS/MS-based IDA cycles. In this experiment, either Q_1 or Q_3 is operated in the RF/DC filtering mode, which allows the transmission of a single ion (at unit resolution, 0.7 m/z at half-height) at a single quadrupole step. The quadrupole is swept or, peak hopped, by the step size, from the low to high mass range specified in the acquisition method (e.g., 100–500 m/z, 0.2 amu step). Because each quadrupole step is mass calibrated, the system can record the number of ions/second that are detected for that step and produce a mass spectrum.

Quadrupoles transmit a narrow width of ions in the RF/DC mode at a given step, making them less efficient than time-of-flight (TOF) mass spectrometer or traditional trap instruments for single MS. Trap sensitivity can be significantly reduced for low abundance ions due to fill time constraints necessary to avoid space charge effects. These competing effects limit the single MS sensitivity difference between traps and quadrupoles, but traps generally have a scan rate and sensitivity advantage over quadrupoles, although both can be useful for qualitative applications. Single MS experiments can be useful for general high-level metabolite identification and in vitro profiling. However, all quadrupole and trap varieties of single MS are less effective for low-level and/or in vivo metabolite identification because of the low selectivity (for IDA) of single MS and high general background of this mode.

15.2.2 Scan Functions

A QTRAP instrument is capable of a wide variety of scan modes: full MS scan (MS), product ion scan (MS^2 and MS^3), PI scan, NL scan, MRM, predictive MRM (pMRM), multiple ion monitoring (MIM), enhanced multiply charged (EMC) scan and time-delayed fragmentation (TDF) (Hopfgartner et al., 2004; King and Fernandez-Metzler, 2006). Most of these scan modes can be incorporated into an IDA experiment to achieve metabolite detection and MS/MS spectral recording in a single run.

15.2.2.1 Enhanced MS and Enhanced Product Ion Scans The enhanced MS (EMS) scan and enhanced product ion scan (EPI) on QTRAP instruments are performed in the ion trap. Ions produced in the source undergo free-jet expansion when entering the vacuum chamber and are immediately focused into a tight stream by the QJet entrance quadrupole to reduce ion loss, which improves sensitivity. The ion beam is collisionally focused further by Q_0 before entry into Q_1 for precursor selection. The selected ion is then accelerated into Q_2 using the collision energy (CE) defined in the method. The CE is often modulated between low, medium, and high values using the collision energy spread (CES) to allow for a wider range of fragmentation. The CES is especially useful for phase II conjugate confirmation, where low-energy conjugate fragments and high-energy parent compound fragments are necessary in one spectrum. The fragmented ions are ejected in <1 ms into the LIT for trap analysis, as described previously. EMS mode differs from EPI as Q_1 and Q_2 are operated in RF-only (full ion transmission) mode. All of the ions generated in the source are passed through Q_1 and Q_2 in the RF mode to the trap for collection and analysis. In both cases, DFT can be used to determine the optimum fill time for the trap to avoid space charge effects but maximize sensitivity. DFT utilizes a short quadrupole scan to determine the total ion abundance before a given trap scan. The fill time is then dynamically adjusted to ensure optimal scan quality for each MS^1, MS^2, MS^3 scan individually.

15.2.2.2 MS^3 Scan QTRAP performs MS^3 differently than a traditional ion trap. In a QTRAP instrument, the steps for MS^3 are the same as those for an EPI scan, up to and including the point of the fragments being cooled in the LIT. After cooling, the fragments are isolated in the trap using an RF/DC field, similar to what is used during a normal quadrupole m/z transmission. The main difference is that the entrance and exit trapping potentials are activated to prevent the isolated ions from being lost. The RF/DC isolation allows a narrow band of $\sim 0.5–5$ m/z to be stabilized while other ions are radially ejected, in a similar manner to a normal quadrupole scan, leaving only the fragment of interest in the trap. The same dipolar energy used to excite ions during an LIT scan is applied at a lower energy to create collision-induced dissociation (CID) with N_2 injected into the trap. A distribution of mainly low-energy fragments is produced (similar to a standard ion trap) above the ~1/3 low mass cutoff limitation of in trap fragmentation. Detailed fragment analysis and quantitation are the two main reasons for using MS^3 on the QTRAP. Use of MS^3 is not generally necessary for structural elucidation because EPI can produce a complete fragmentation and MS^3 is often only used when very detailed fragment investigation is necessary. Quantitation by MS^3 has become more popular as sensitivity and scan speed have increased and is particularly useful when background signal in MRM mode becomes prohibitively high. Background signal is dramatically reduced in MS^3 because interferences must pass two stages of fragmentation and mass selection. Space charge tends to limit the high end of a quantitation range, which means MS^3 has a relatively low dynamic range of 2.5–3 orders of magnitude. This limitation can be compensated for by running MRM to extend the dynamic range at higher levels.

15.2.2.3 Precursor Ion and Neutral Loss Scans The PI scan and NL scan are two classic quadrupole scan modes used for metabolite detection. During a PI scan, Q_1 scans from a low to high m/z. The ions are fragmented in Q_2 and a single, specified fragment is allowed to pass through Q_3 to the detector. PI scan generates a very low background spectrum, which is only made up of parent ions that generate the desired fragment. The constant NL scan is complimentary to a PI scan and detects parent ions that lose a common neutral fragment in Q_2 as opposed to generating a common charged fragment. In this mode, Q_1 scans from a low to high m/z, Q_2 fragments the ions, and Q_3 scans from the same low to high mass range of Q_1 *offset* by the neutral fragment of interest. The only masses detected are those that lose the specified neutral fragment in Q_2 to produce the charged product ion. PI and NL scans are particularly effective for the detection of phase II metabolites [glucuronides, sulfates, cyanide, and glutathione (GSH) conjugates]. For comprehensive detection of various types of metabolites, both PI and NL scans are needed, each of which may be effective to detect certain oxidative metabolites. The 5500 QTRAP has a faster scan rate that allows for multiple PI and NL scans to be performed in a single experiment.

15.2.2.4 Multiple Reaction Monitoring Quantitative multiple reaction monitoring (MRM) is the most popular quantitative LC–MS analytical technique. Figure 15.3a illustrates the principle of the MRM mode. During an MRM scan, Q_1 operates in RF/DC mode and allows the desired parent mass to enter and fragment in Q_2. Q_3 is also set to RF/DC mode and allows only the specified product ion to pass from Q_2 through to the detector. This process produces a trace of the intensity for the transition over the course of the chromatographic run that will show a peak at a specific retention time. This trace can then be integrated and compared to a known standard. Modern triple quadrupole

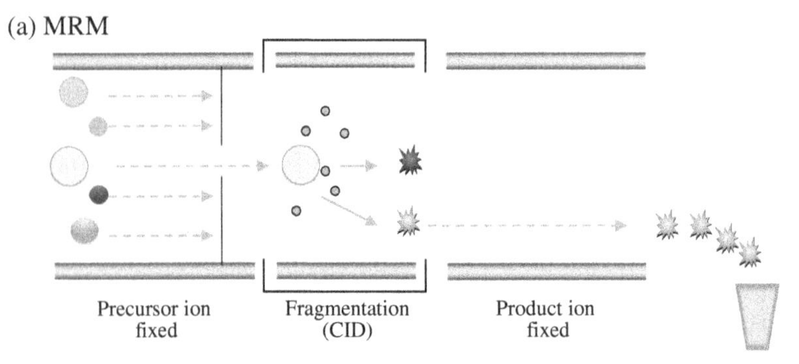
(a) MRM
Precursor ion
fixed
Fragmentation
(CID)
Product ion
fixed
MRM is characterized by parent ion selection in Q1, fragmentation in Q2, and key fragment isolation in Q3. The resultant trace can be integrated in a quantitative manner.

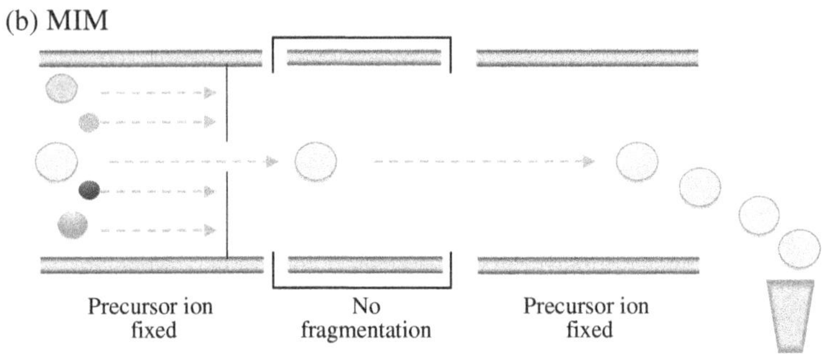
(b) MIM
Precursor ion
fixed
No
fragmentation
Precursor ion
fixed
MIM is characterized by parent ion selection in Q1, no fragmentation in Q2, and parent ion isolation in Q3.

FIGURE 15.3 (*a*) MRM and (*b*) MIM scan modes on QTRAP. MRM can be operated either MRM-only for quantification or pMRM–EPI for metabolite detection and MS/MS spectra recording. MIM is mainly used as MIM-EPI in either a targeted or screening method for metabolite profiling and identification.

instruments can rapidly switch between many MRM transitions within a single acquisition cycle.

15.2.2.5 Predictive Multiple Reaction Monitoring Unlike traditional triple quadrupole instruments, the 4000 and 5500 QTRAP instruments allow the performance of over 600 pMRM on an LC time frame. This novel scan function, namely predictive MRM (pMRM), has been employed for targeted analysis of drug metabolites and screening for drugs and toxic chemicals based on their predicable molecular weights and fragmentation patterns. The example shown in Figure 15.4*a* illustrates the principle of pMRM for detection of oxidative metabolites. The protonated molecular ion $[M+H]^+$ of the drug is observed at m/z 455 and a major product ion is observed at m/z 165. To target a potential monooxidation product at m/z 471, in which the location of the oxidation could be on either side of the fragmentation cleavage, two MRM

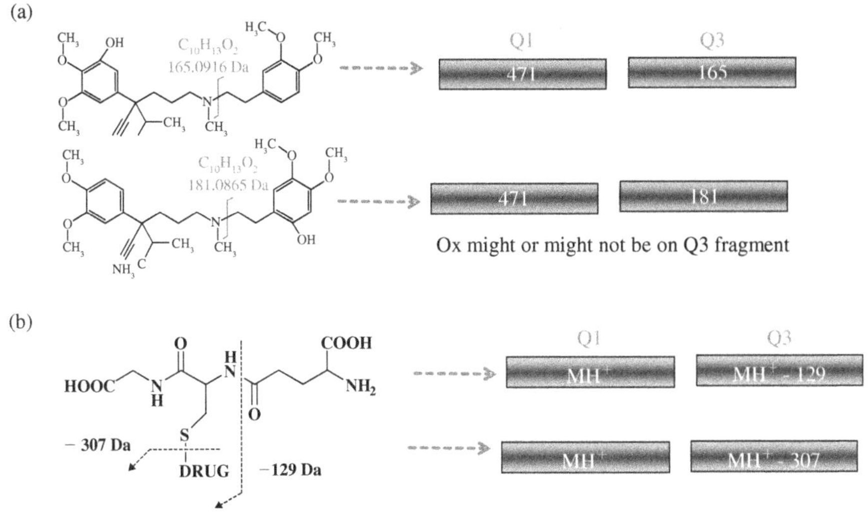

FIGURE 15.4 MRM transitions in pMRM screening for (*a*) a monooxidation product and (*b*) a GSH conjugate. The $[M+H]^+$ and a major product ion of the oxidative metabolite is *m*/*z* 471 and 165/181. The major product ions of the GSH adduct are $[M+H]^+$ −129 or/and $[M+H]^+$ −307.

transitions from *m*/*z* 471 to *m*/*z* 165 and *m*/*z* 471 to *m*/*z* 181 are required. To target a conjugated metabolite that produces a common neutral loss, a single MRM from predicted $[M+H]^+$ to a product ion derived from the neutral loss is effective. For example, glucuronide conjugates usually undergo the neutral loss of 176 Da. An MRM transition from $[M+H]^+$ to $[M+H]^+$ −176 can be employed for searching for a glucuronide of a parent drug. All classes of GSH conjugates, including aliphatic, benzylic, and aromatic thioethers, give the NL of either 129 and/or 307. Therefore, two MRM transitions from $[M+H]^+$ to the ions derived from neutral losses of 129 and 307 are used to search for a potential GSH adduct (Fig. 15.4*b*). The target analysis of drug metabolites with the QTRAP instrument can be very effective. For example, a 400 pMRM scan on 5500 QTRAP can target up to 200 potential GSH adducts with the cycle time for only 1200 ms (with a 2-ms dwell and 1-ms pause). This pMRM technique is highly selective and sensitive, which is especially useful in detecting trace amounts of metabolites in complex biological matrices; however, this technique is not capable of detecting drug metabolites with unpredictable molecular weights or fragmentation pathways.

15.2.2.6 Multiple Ion Monitoring Recently, MIM has emerged as a new and promising scanning method for rapid screening of metabolites formed in the in vitro experiments (Yao et al., 2008, 2009). MIM is derived from MRM. In the MIM experiment, the same molecular ions are monitored in Q_1 and Q_3, with the collision energy in Q_2 set at a minimal value so that fragmentation is not produced (Fig. 15.3). The molecular ions isolated in Q_1 pass through Q_2 with no fragmentation and are monitored in Q_3. MIM is mainly employed as a survey scan to trigger EPI acquisition. MIM–EPI retains the sensitivity and duty cycle similar to pMRM–EPI for detection of in vitro metabolites but requires no knowledge of the fragmentation pattern of the metabolites. Therefore, MIM–EPI is applicable to fast metabolite profiling without construction of compound-dependent acquisition protocols and suitable in detecting metabolites that have no predicable fragmentation patterns.

15.2.2.7 Information-Dependent Acquisition and Polarity Switching One of the QTRAP's unique abilities is that it can perform IDA cycles that use both quadrupole and LIT based scans. The first stage of the IDA cycle is the survey scan, which can be any combination of MS, pMRM, PI, NL, or MIM in either polarity. The second stage is creation of a mass peak list from a combination of the results of the survey scans. The IDA then evaluates the peak list based on mass range, intensity, dynamic background subtraction, isotope pattern, dynamic exclusion, and any user-generated inclusion/exclusion lists. The selected candidates are then analyzed by data-dependent MS^2 scans with or without polarity switch. The major fragments from MS^2 can be fed to another level of logic for further MS^3 acquisition.

15.3 IN VITRO AND IN VIVO METABOLITE PROFILING AND IDENTIFICATION

15.3.1 Metabolic Stability Analysis

Fast metabolism of a drug candidate often leads to poor bioavailability, undesirable PK properties, such as fast drug clearance and short half-life values, and drug–drug interactions. Additionally, drug metabolism could result in the formation of toxic metabolites or chemically reactive intermediates. Therefore, high metabolic rates, especially those associated with CYP-mediated reactions, are considered as an undesirable property of pharmaceutical compounds. A metabolic stability test, which is conducted by monitoring the disappearance of parent compounds over time in liver microsome incubations, is widely employed as one of the first-tier in vitro ADME assays in drug discovery (Fig. 15.1) The MRM mode on a classical triple quadrupole or QTRAP instrument provides excellent sensitivity and selectivity to obtain the quantitative information regarding the turnover rate of the parent drug (Table 15.1). If a high metabolic rate is observed for a tested compound, a follow-up experiment that determines metabolic soft spots is conducted in some cases. Results from a metabolic stability assay and/or metabolic soft spot analysis are useful to rank and select compounds for further evaluation, design more metabolically stable compounds, and predict PK in animals and humans (Zhang et al., 2009b).

The QTRAP instrument has recently been employed to simultaneously determine metabolic rates and monitor the formation of specific metabolites (Shou et al., 2005). In this approach, the MRM of the parent drug and those of the predicted common metabolites are used as survey scans to trigger the data-dependent EPI scan. It has been demonstrated that quantitation of the parent drug using MRM is not negatively impacted by the addition of pMRM channels that trigger EPI acquisition. Due to the highly specific nature of MRM and the large variety of metabolites, the pMRM could miss important, uncommom metabolites. Additionally, metabolic stability assays are usually conducted at substrate concentrations of approximately 0.5–2 μ*M*, in which metabolites detected by LC–MS are often not visible in corresponding ultraviolet (UV) chromatograms due to their lower concentrations. Therefore, these incubation samples may not be suited for metabolic soft spot determination which usually involves UV analysis.

15.3.2 Metabolic Soft Spot Determination

A metabolic soft spot is a substructure or function group of a drug that is mostly susceptible to biotransformation that leads to high turnover rate of this drug in vitro or in vivo. A drug could have one or a few metabolic soft spots. Metabolic soft spot information can help medicinal chemists to design more metabolically stable drug candidates (Zhang et al., 2009b). The most common experiment to determine metabolic soft spots involves the incubation of a compound at 5–30 μ*M* in liver microsomes followed by LC–UV–MS analysis (Fig. 15.1 and Table 15.1). Metabolite UV chromatograms provide semiquantitative information of the relative abundance of the metabolites generated in the incubation, while metabolite mass spectral data are crucial in structural elucidation. The key to a successful metabolic soft spot assay is to rapidly identify the major metabolites in a high-throughput fashion. Therefore, an ideal approach for metabolic soft spot determination should use a generic data acquisition method with no need for predetermining the compound-specific information or constructing the compound-dependent acquisition protocol prior to sample analysis. A QTRAP that provides various information-dependent MS/MS acquisition functions is well suited for this application. EMS–EPI followed by data mining has been developed for fast metabolite profiling (Fig. 15.5) (Li et al., 2007). In the experiment, a generic full MS scanning method serves as the survey scan. Once an ion above a preset intensity threshold is detected, the corresponding MS/MS acquisition is triggered. This approach allows for collection of both MS and MS/MS spectral data in a single injection and metabolite detection via postacquisition data mining. Date mining and spectral interpretation can be conducted in parallel without interruption of data acquisition. Neutral loss filter (NLF) and product ion filter (PIF) are also used in post-acquisition data mining for metabolic soft spot determination (Fig 15.5). They utilize the same mechanisms as NL and PI performed by a classical triple quadrupole instrument to search for metabolites but do not require predetermination of MS/MS

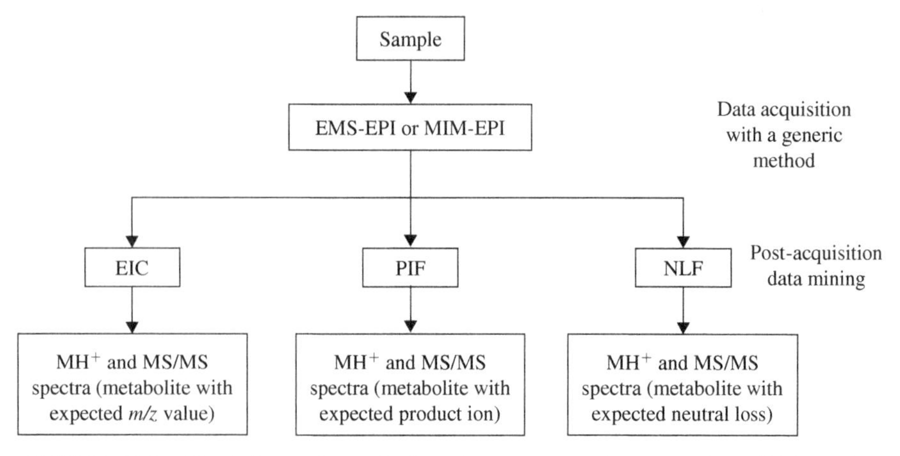

FIGURE 15.5 Work flow of metabolic soft spot determination using EMS–EPI or MIM–EPI; EIC, extracted ion chromatography; PIF, product ion filter; NLF, neutral loss filter.

spectra of the parent drugs and compound-dependent experimental protocols. In addition, NLF and PIF approaches do not need repeated injections, which are usually required when different NL scans or PI scans are performed on a classical triple quadrupole instrument.

As an alternative, MIM–EPI has been recently employed for fast metabolic soft spot determination (Fig. 15.5) (Yao et al., 2009). In this approach, MIM–EPI scans with selected *m*/*z* ranges that cover the expected metabolites are utilized to acquire MS/MS spectral data. Metabolite detection and recovery of their MS/MS spectra are achieved via post-acquisition data mining using EIC, PIF, and NLF. This approach is similar to the EMS–EPI-based data-mining approach and allows for high-throughput drug metabolite without prior knowledge of the molecular weights and fragmentation patterns of the metabolites. MIM–EPI has better sensitivity and selectivity than EMS–EPI. Figure 15.6 illustrates the comparison of the ritonavir metabolite profiles in rat hepatocytes using different scanning methods (Yao et al., 2008). The total ion current (TIC) of EPI for ritonavir metabolites determined by EMS–EPI displayed only four metabolites (M1, M4, M5, and M6 in Fig. 15.6a) while most other metabolite ions were not visible due to higher levels of background noise and endogenous components from rat hepatocytes. In comparison, the ritonavir metabolite profile determined by MIM–EPI revealed 11 metabolites (Fig. 15.6*b*). It is also evident that MIM–EPI had comparable sensitivity and selectivity as the pMRM–EPI (Fig. 15.6*c*) at least for detecting in vitro metabolites. Overall, the MIM–EPI-based post-acquisition data-mining approach provides efficiency and flexibility for rapid screening of the major metabolites formed in the in vitro incubation experiments and, therefore, are the preferred approaches for metabolic soft spot determination (Table 15.1).

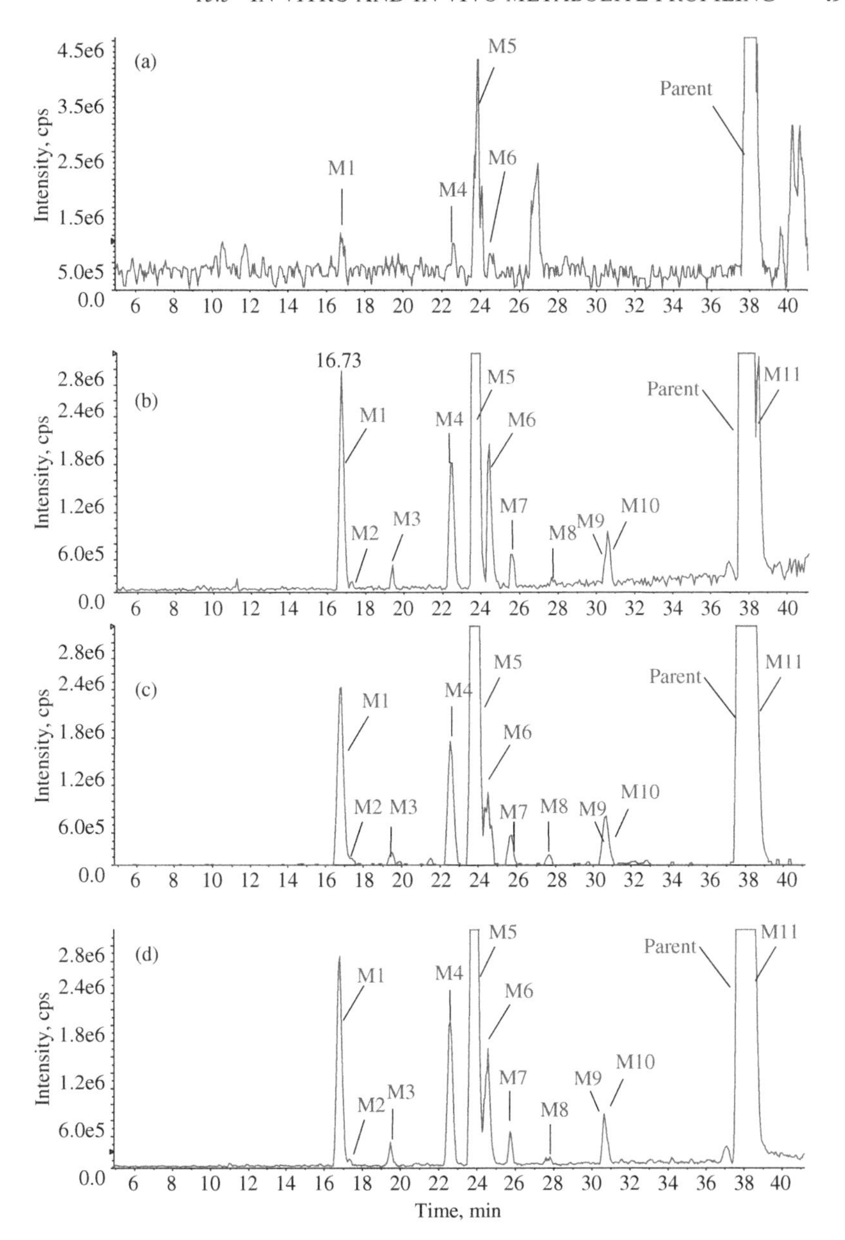

FIGURE 15.6 Detection of ritonavir metabolites in rat hepatocytes by the 4000 QTRAP. (*a*) TIC of EPI from EMS–EPI Scan. (*b*) TIC of EPI from MIM–EPI scan with 24 MIM transitions. (*c*) TIC of EPI from MRM–EPI scan with 22 transitions. (*d*) TIC of EPI from the MIM–MRM–EPI scan. The MRM–MIM survey monitored either 22 MRM transitions or 24 MIM transitions. [Reprinted from Yao et al. (2008) with permission of John Wiley & Sons.]

EMS–EPI is an also effective method for metabolic soft spot analysis, when analytical sensitivity is not an issue.

15.3.3 In vitro Species Comparison

Understanding in vitro metabolism of animal species relevant to that in humans is important in the selection of animal toxicological species and prediction of human PK from animal PK. This information also provides initial examination for the presence of unique human metabolites that would be an issue in drug development as indicated in recently published Food and Drug Administration (FDA) guidance on safety testing of drug metabolites (FDA, 2008). In the late drug discovery and early development stages, in vitro comparative metabolism studies are often conducted in liver microsomes and hepatocytes across species. Unlike metabolic soft spot determination experiments that focus on the analysis of major metabolites, comparative metabolite profiling requires exhaustive identification of all the potential human metabolites even at low concentrations because minor in vitro metabolites can potentially be major circulation metabolites in vivo. Due to the superior sensitivity, selectivity, and automatic acquisition of informative MS/MS spectra, pMRM–EPI scan on QTRAP has been applied in profiling of hepatocyte metabolites such as oxidative metabolites, glucuronide conjugates and sulfate conjugate cross species. In one study, 48 pMRM transitions designed to cover the common phase I and phase II metabolites of the test drug were used as survey scan to trigger product ions scan (Gao et al., 2007). This method provided the necessary sensitivity to detect minor metabolites in a relevant therapeutic concentration of the drug (1–10 μ*M*). In addition, the peak area ratios of individual identified metabolites to that of internal standards provided semiquantitative data about the levels of metabolite in different species.

The key to effectively identify metabolites using pMRM–EPI is to choose the theoretical metabolite ions and to predict their product ions based on the similarity of the fragmentation to that of the parent drugs. However, certain types of metabolites such as those formed via amide hydrolysis or dealkylation are significantly smaller than the parent drugs and often give unpredictable fragmentation patterns totally different from that of the parent. As a result, pMRM survey scans could miss uncommon metabolites that have unpredictable molecular weights. In contrast, the MIM–EPI method does not need the prediction of the fragmentation patterns of the metabolites and has comparable sensitivity and selectivity as pMRM–EPI for detecting in vitro metabolites (Yao et al., 2009). Therefore, the MIM–EPI method is considered to be a preferred method for the comprehensive detection of both common and uncommon metabolites in comparative metabolism studies (Fig. 15.5 and Table 15.1). In addition, MIM–EPI can be combined with pMRM–EPI (Fig. 15.6*d*). This approach takes advantages of both MRM and MIM scanning methods and can be especially useful when dealing with low levels of metabolites in complex biological matrix such as plasma (Yao et al., 2008).

15.3.4 Identification of In Vivo Oxidative Metabolites

Screening of in vivo metabolites in nonclinical and clinical studies provides critical information for understanding of metabolic clearance pathways and exposure of metabolites (Zhang et al., 2009b). Detection and identification of in vivo metabolites in complex biological matrices such as plasma, urine, bile, and tissues present analytical challenge. Endogenous component could give high background noise that obscures the metabolite-related signals, especially when less selective acquisition methods such as EMS–EPI and MIM–EPI scans are performed. Therefore, pMRM–EPI is often chosen as the primary method for screening of in vivo metabolites due to its inherent sensitivity and selectivity (Li et al., 2005; Mauriala et al., 2005; Jian et al., 2009). As complementary techniques, NL-EPI and PI-EPI are suitable for screening for uncommon oxidative metabolites based on their predicted fragmentation patterns (Table 15.1). In one example, nonclinical plasma samples collected for PK analysis were screened for circulating metabolites (Li et al., 2005). pMRM–EPI experiments were incorporated into the conventional MRM method that was used for quantitation of the parent drug. As a result, the concentration data for the parent drug and the structural information of the metabolites were obtained in the same injection. In addition, the relative quantitative estimation of plasma metabolites cross species can be readily accomplished using pMRM–EPI.

15.4 REACTIVE METABOLITE SCREENING AND CHARACTERIZATION

15.4.1 In Vitro Reactive Metabolite Screening

Most reactive metabolites are electrophiles in nature. Owing to the inherent reactivity and instability, they can be readily trapped with a suitable nucleophile to form stable adducts, such as protein adducts and GSH conjugates. The naturally occurring tripeptide glutathione (GSH) contains a free sulfydryl group, a soft nucleophile capable of reacting with soft electrophiles, including quinones, epoxides, nitrenium ions, alkyl halides, Michael acceptors, and forming stable GSH adducts. Formation of GSH adducts in vivo is typically considered as a detoxification mechanism. However, in cases of drug overdose where cellular stores of GSH are depleted, biological macromolecules are subjected to covalent binding of survived reactive electrophiles leading to tissue injury (Gibson et al., 1996; Hinson et al., 2004). A recent study showed that the rates of GSH adduct formation of 10 structurally diverse compounds correlated well with the protein covalent binding of radioactivity in human and rat liver microsomes (Takakusa et al., 2008). In addition, a study to quantitatively estimate formation of reactive metabolites of commonly prescribed 50 drugs found a group of drugs that did not induce liver toxicity had a significantly lower load of reactive metabolites in humans than that in a group of drugs that expressed hepatotoxicity (Gan et al., 2009). These examples support the notions that screening for GSH adducts would identify a significant portion of reactive metabolites formed from a drug candidate, and lower exposure of reactive metabolites would have a lower risk of hepatotoxicity in humans. Hence, screening and minimization of reactive metabolite formation in lead optimization are common practice in many drug discovery organizations (Fig. 15.1) (Ma et al., 2009; Wen and Fitch, 2009a).

15.4.1.1 Glutathione as a Trapping Agent GSH serves as a natural trapping agent for chemically reactive metabolites and has been routinely used in vitro to screen and evaluate reactive metabolite formation. As illustrated in Figure 15.7, characteristic fragmentation of GSH conjugates under both positive and negative ion modes following CID has been extensively investigated (Haroldsen et al., 1988; Murphy et al., 1992; Baillie and Davis, 1993; Dieckhaus et al., 2005). Similar to CID of peptides, the fragmentation of GSH conjugates is mainly resulted from the cleavage of the peptide backbone of the GSH moiety in the positive ion mode (Fig. 15.7*a*). Although the abundance of these fragment ions can vary remarkably and is governed by the structural nature, GSH conjugates generally undergo an NL of 129 Da (pyroglutamic acid) to produce *e*-type fragment ions (Fig. 15.7*a*). Therefore, GSH conjugates can be readily detected by a constant NL scan of 129 Da (Baillie and Davis, 1993; Chen et al., 2001; Yan and Caldwell, 2004). On the other hand, CID of the $[M - H]^-$ ions of GSH conjugates affords a series of preeminent anions at *m/z* 128, 143, 160, 179, 210, 254, and 272 in the negative ion mode (Dieckhaus et al., 2005) (Fig. 15.7*b*). First-generation product ions from the $[M - H]^-$ species include *m/z* 160 and 272 and are presumably resulted from elimination of the elements of glutamate and H_2S, respectively. Further loss of water from the fragment anion at *m/z* 272 affords the ion at *m/z* 254. Based on these characteristic fragmentations, different analytical methodologies have been developed recently, including NL, PI, and MRM approaches, to achieve improved sensitivity, selectivity, and throughput for the screening and characterization of GSH conjugates.

Structurally different classes of GSH conjugates behave differently upon CID in the positive ion mode (Baillie and Davis, 1993). It has been recognized that not all classes of GSH conjugates afford an NL of 129 Da as the primary fragmentation pathway. Aliphatic and benzylic thioether conjugates may fragment by a neutral loss of GSH (307 Da) and/or afford a protonated GSH product ion (*m/z* 308). Thioester conjugates typically fragment by loss of pyroglutamic acid and water (147 Da). Additionally, many GSH adducts yield doubly charged $[M + 2H]^{2+}$ ions under the positive ion mode, which typically do not fragment by a neutral loss of 129 Da under CID (Dieckhaus et al., 2005). This limited coverage of NL scanning suggests that a broader MS/MS survey scan is needed for the detection of different structural classes of GSH conjugates. To overcome this drawback, Dieckhaus et al. demonstrated that precursor ion (PI) scanning of *m/z* 272 (deprotonated γ-glutamyl-dehydroalanyl-glycine; Fig. 15.7*b*) in the negative ion mode provided a general survey scan

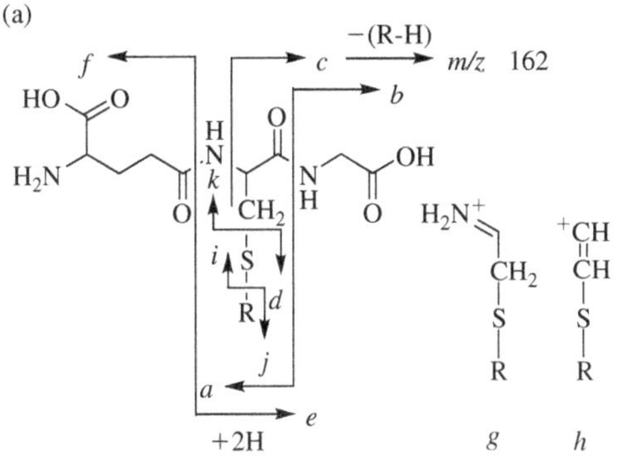

$[M + H]^+$	
Fragmention	*m/z*
a	$MH^+ - 75$
b	76
c	$MH^+ - 146$
c–(R–H)	162
c–H_2O	$MH^+ - 164$
d	$MH^+ - 273$
e	$MH^+ - 129$
e–H_2O	$MH^+ - 147$
f	130
g	$MH^+ - 232$
g–H_2O	$MH^+ - 250$
h	$MH^+ - 249$
i	308
j	$MH^+ - 305/307$
k	274

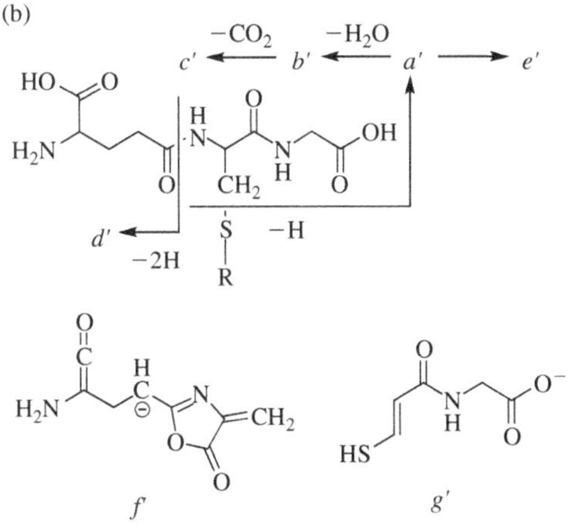

$[M - H]^-$	
Fragmention	*m/z*
a'	272
b'	254
c'	210
d'	128
e'	143
f'	179
g'	160

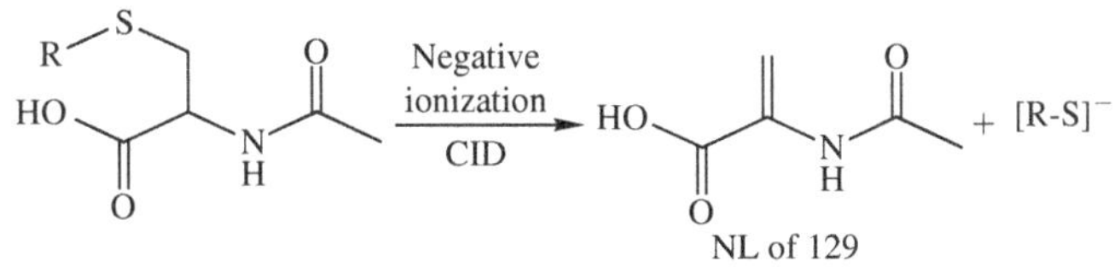

FIGURE 15.7 Fragmentation patterns of GSH conjugates in (*a*) the positive mode and (*b*) negative mode, and fragmentation pattern of NAC conjugates in (*c*) the negative mode.

for the successful detection of GSH conjugates of different classes, including benzylic, aromatic, aliphatic, and thioester GSH conjugates (Dieckhaus et al., 2005). In comparison, NL scanning of either *m*/*z* 129 or *m*/*z* 307 in the positive ion mode did not enable the detection of all four types of GSH adducts in a single LC–MS/MS run. Analysis of in vitro and in vivo GSH adducts have also demonstrated that the PI scanning method detected various classes of GSH adducts with improved selectivity and sensitivity compared to NL scanning.

To further improve the selectivity, an extension of this methodology was recently reported by incorporating a simultaneous dual negative precursor ion scan for *m/z* 272 and 254 (the dehydrated form of *m/z* 272) (Mahajan and Evans, 2008). Both anions are abundant and characteristic fragment ions of GSH conjugates and originate from the glutathionyl moiety regardless of structural classes. With both fragment anions scanned in parallel, this method achieved a further increase in selectivity.

To overcome a major drawback of classical triple-quadrupole-based methods that need additional runs to record MS/MS spectra of detected GSH conjugates, a high-throughput method was recently developed for screening and characterization of reactive metabolites using polarity switching on a quadrupole–linear ion trap mass spectrometer (Wen et al., 2008b). As illustrated in Figure 15.8*a*, this approach uses a PI scan of *m/z* 272 in the negative ion mode as the survey scan for unambiguous detection of GSH adducts, with polarity switching to the positive ion mode to acquire EPI spectra for structural elucidation in a single LC–MS/MS run. Results from analysis of reactive metabolites of several model compounds suggest that this PI-EPI approach is a feasible high-throughput method that can be used to screen for a large number of structurally diversified compounds with reliable sensitivity and selectivity in a drug discovery setting. In addition to previously reported major reactive metabolites, multiple minor GSH adducts were also detected and characterized by this approach. This polarity-switching approach was applied to the detection of several novel reactive metabolites of trazodone (Wen et al., 2008c), flutamide (Wen et al., 2008a), amitriptyline, and its metabolite nortriptyline (Wen et al., 2008d) in human liver microsome (HLM) incubations. An example of using this approach for detecting and structural characterizing reactive metabolites is presented in Figure 15.9. The negative PI chromatogram of a flutamine (FLU) incubation sample displayed multiple GSH adducts with no false positives (Fig. 15.9*b*) (Wen et al., 2008a). Additionally, the positive MS/MS spectrum of an adduct, FLU-G1, showed informative product ions that facilitated structural elucidation (Fig. 15.9*a*). As a result, complicated bioactivation pathways of flutamine in human liver microsomes, leading to multiple GSH-trapped intermediates, were identified (Fig. 15.9*c*).

As an alternative approach, pMRM–EPI was developed for detection and characterization of GSH-trapped reactive metabolites using a 4000 QTRAP (Zheng et al., 2007). In the analysis, pMRM transitions are constructed from the protonated molecules of potential GSH adducts to their product ions derived from neutral losses of 129 and 307 Da in the positive ion mode. Once a predicted GSH adduct is detected, acquisition of its MS/MS spectrum in the positive ion mode is triggered (Fig. 15.8*b*). The MRM survey scan can be set to monitor up to 100 transitions with the 4000 QTRAP without a significant loss of sensitivity. The effectiveness and reliability of this approach were evaluated using several model compounds known to undergo bioactivation in HLM. Results showed that the MRM-based approach

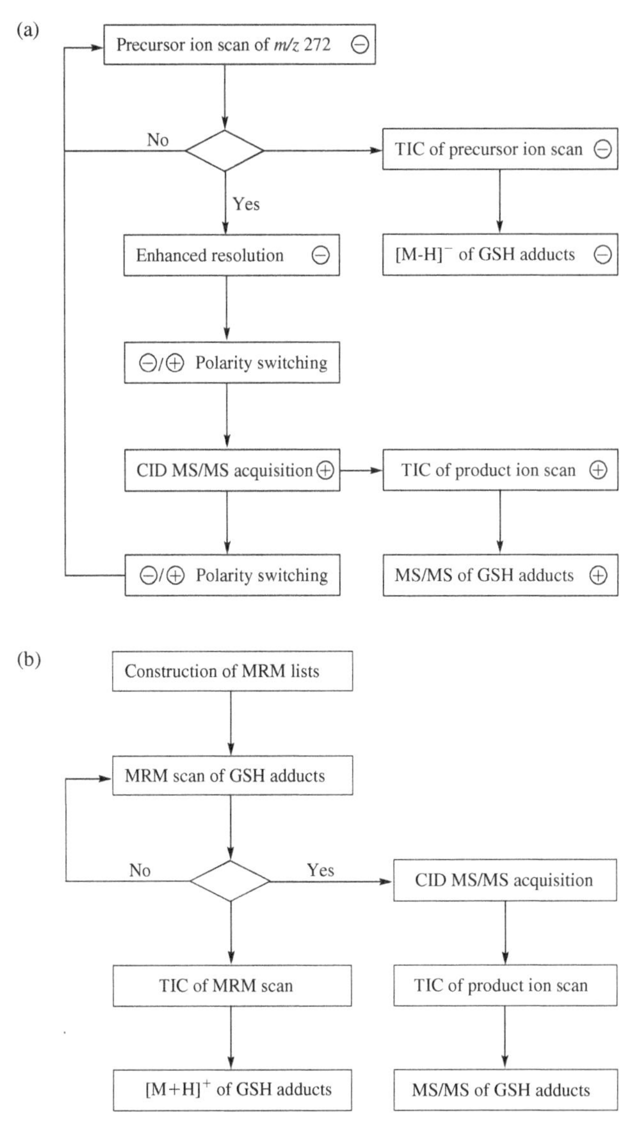

FIGURE 15.8 Work flow of (*a*) PI-EPI with polarity switching and (*b*) pMRM–EPI for the detection and structural characterization of GSH conjugates. (*c*) Work flow of NL-EPI or pMRM–EPI with polarity switching for the detection and structural characterization of NAC adducts. (*d*) Work flow of pMRM–EPI with polarity switch for the detection and structural characterization of dGSH adducts.

provided superior sensitivity and selectivity than NL and PI methods. This pMRM–EPI methodology is well suited for detection of low levels of GSH adducts in complex biological samples as demonstrated in the sensitive detection of trace amounts of GSH and cyanide adducts of hepatotoxin

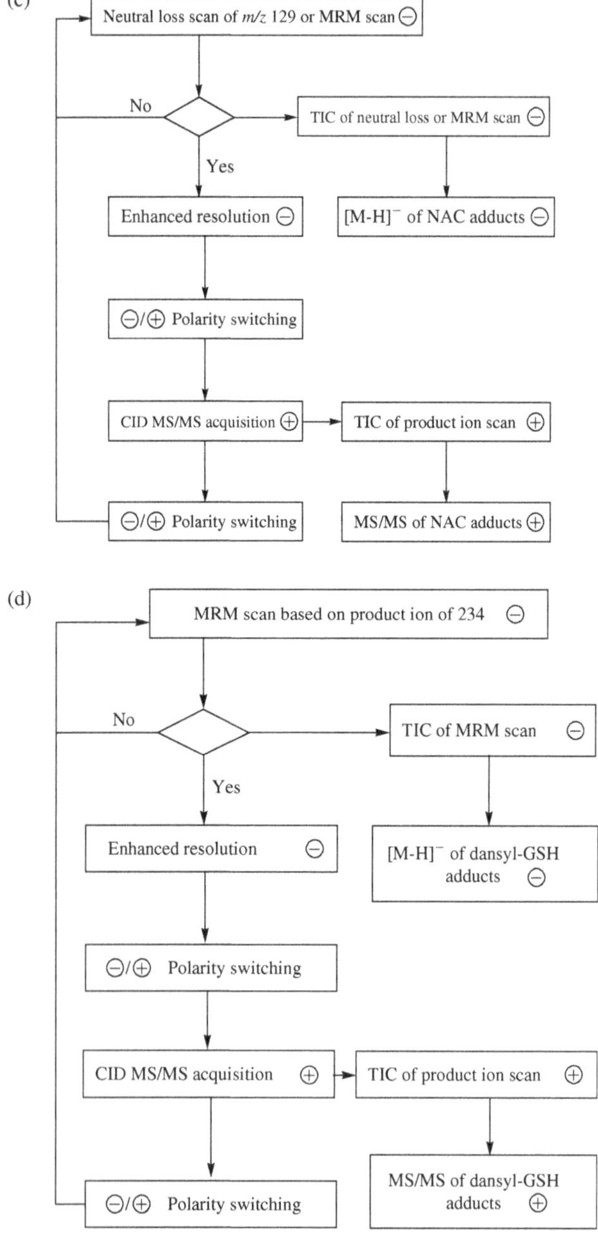

FIGURE 15.8 (*Continued*)

nefazodone (Bauman et al., 2008). Limitations of this pMRM-based approach include the lack of high-throughput screening and incapability of detecting GSH adducts that have unpredicted molecular weights.

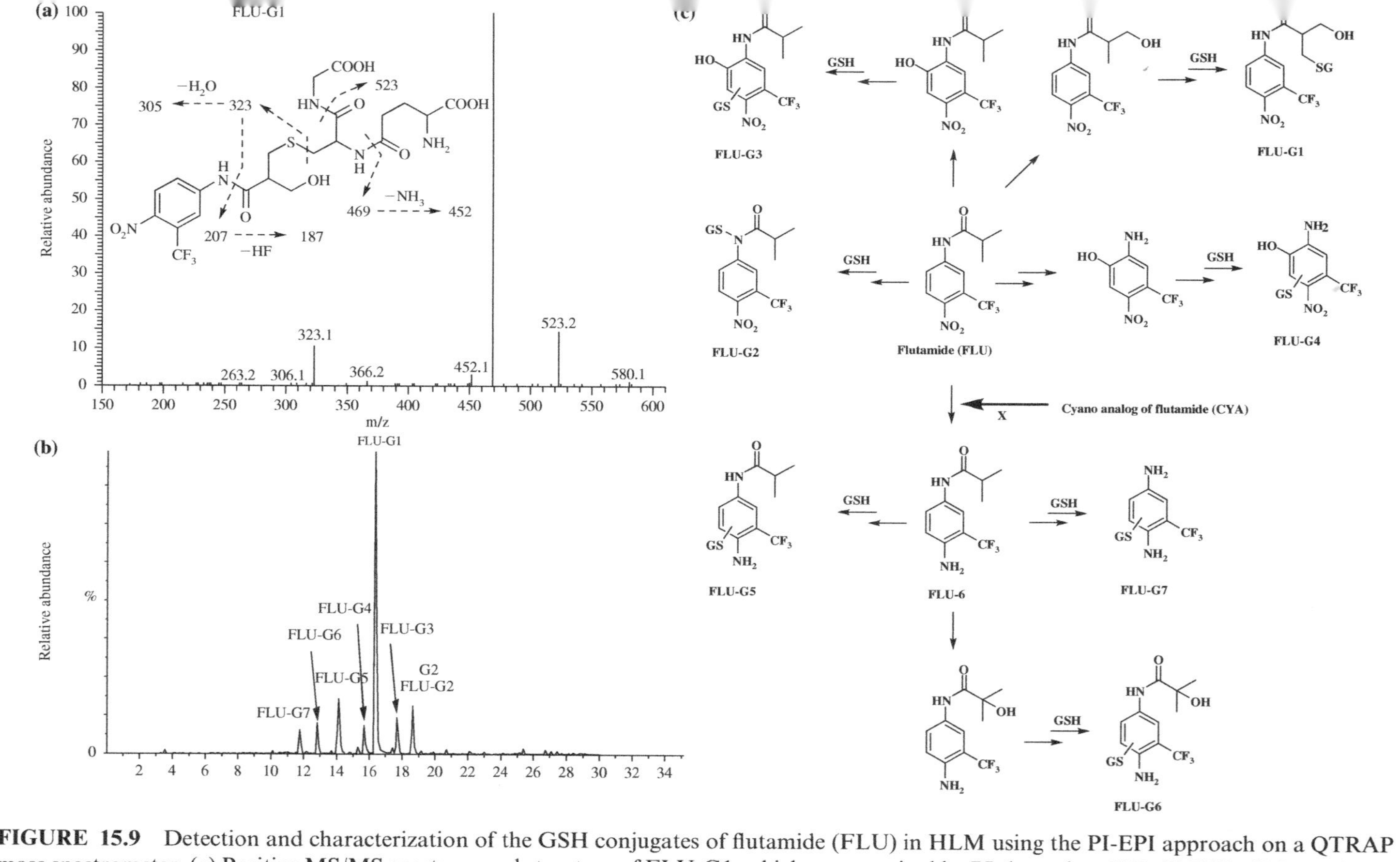

FIGURE 15.9 Detection and characterization of the GSH conjugates of flutamide (FLU) in HLM using the PI-EPI approach on a QTRAP mass spectrometer. (*a*) Positive MS/MS spectrum and structure of FLU-G1, which was acquired by PI-dependent EPI. (*b*) TIC of PI scan in the negative mode. (*c*) Proposed bioactivation pathways of FLU in HLM. [Reprinted from Wen et al. (2008a) with permission of American Chemical Society.]

15.4.1.2 Glutathione Ethyl Ester as a Trapping Agent GSH analogs have been described as trapping agents to evaluate reactive metabolite formation (Soglia et al., 2004, 2006; Gan et al., 2005). The use of the close analog of GSH, GSH ethyl ester (GSH-EE), was previously shown to increase the MS sensitivity by an approximate 10-fold in the detection of reactive metabolites using a NL-based approach (Soglia et al., 2004). However, like other NL-based assays, the main drawback of this approach is that not all GSH conjugates afford a neutral loss of 129 Da in the CID MS/MS analysis. Similarly, the PI-EPI methodology was applied to the detection and characterization of reactive metabolites using glutathione ethyl ester (GSH-EE) as the trapping agent (Wen and Fitch, 2009b). Unambiguous detection of GSH-EE conjugates were achieved by the negative PI scanning of *m/z* 300 (deprotonated γ, glutamyl-dehydroalanyl-glycine ethyl ester), while simultaneous CID MS/MS acquisition in the positive ion mode provides structural elucidation, which greatly enhances the throughput of reactive metabolite screening (Wen and Fitch, 2009b).

15.4.1.3 N-Acetyl-L-Cysteine as a Trapping Agent GSH adducts undergo a series of in vivo biotransformation to give rise to *N*-acetyl-L-cysteine (NAC) adducts (mercapturic acid adducts) (Dickinson and Forman, 2002*a*, 2002*b*). It is well documented that NAC adducts undergo a single primary NL of 129 Da in the negative ion mode regardless of their structure. Based on this knowledge, Scholz et al. (2005) developed an NL-EPI (129 Da in the negative mode) for detection and MS/MS acquisition of synthetic NAC adducts in vitro and those spiked in rat urine with a hybrid triple quadrupole instrument (2000 QTRAP). The methodology was also applied to metabonomics and biomarker discovery studies via analyzing patterns of endogenous NAC adducts in urinary samples from human subjects (Wagner et al., 2006, 2007). Results from these studies proved that the NL or pMRM analysis is a particularly useful tool for the confirmation of targeted NAC adducts. However, MS/MS spectra of NAC adducts acquired in the negative ion mode displayed the same NL fragmentation with little fragment information on drug moiety.

In addition, NAC can be used as an in vitro trapping agent to screening for reactive metabolites. Resultant NAC adducts can be used as standards for in vivo NAC adducts. Recently, a new method with the 4000 QTRAP was developed for fast analysis of NAC adducts (Jian et al., 2009). The 4000 QTRAP is capable of selectively detecting unknown NAC adducts and acquiring fragment-rich MS/MS spectra for structural elucidation by using the polarity switch capabilities of the QTRAP. The work flow of the novel method is illustrated in Figure 15.8*c*. This method uses a negative NL scan of 129 Da or pMRM from predicted *m/z* values to product ions derived from the NL of 129 Da as a survey scan to trigger acquisition of EPI spectra in the positive ion mode. Thus, selective detection of NAC adducts and acquisition of fragment-rich MS/MS spectra were accomplished in a single LC–MS run. The pMRM–EPI approach provides excellent sensitivity for detection of NAC adducts generated from common reactive metabolites. The major limitation of the pMRM–EPI approach is the lack of fast analysis capability. In the pMRM–EPI experiment, data acquisition protocols are compound dependent, in which MRM transitions are calculated based on the molecular mass of each test compound. Therefore, the set up of acquisition protocols for a large number of samples is a time-consuming process. In comparison to the pMRM–EPI approach, the NL–EPI approach is better suited for rapid screening and structural characterization of NAC adducts formed in NAC trapping experiments. In the NL–EPI analysis, a simple, generic acquisition method is employed so that the set up of an acquisition protocol for sample analysis takes minimal time and mass spectral data acquisition can be performed continuously. In addition, NL-EPI provides a similar level of sensitivity as the pMRM–EPI method for detecting of NAC adducts formed in the in vitro trapping experiments. Overall, the NL-EPI method uses a generic data acquisition protocol and enables the detection of various types of NAC adducts without multiple injections. Therefore, the NL-EPI method is well suited for rapid screening of major NAC adducts in vitro and in urine.

The major advantage of the NAC trapping experiment over the GSH trapping experiment is that NAC adducts may have the potential to generate more fragments from cleavages of the drug moiety than those from GSH adducts and provide critical information about the structures of the reactive metabolites (Jian et al., 2009). This advantage is due to the fact that NAC is a smaller molecule with fewer bonds susceptible to CID fragmentation than GSH, and the collision energy has more chance to distribute to the drug moiety itself instead of the trapping group. For example, as shown in Figure 15.10*a* (Jian et al., 2009), an NAC adduct of *N*-oxide clozapine formed in an NAC trapping experiment generated more informative fragmentations in EPI scan than the GSH adduct of the same *N*-oxide clozapine formed in the GSH trapping experiment (Fig. 15.10*b*). The product ion at *m*/*z* 331 derived from NAC adduct confirmed unambiguously that the monooxidation occurred on the nitrogen of the methylpiperazine ring, and the product ion at *m*/*z* 285 suggested that NAC was attached on the chlorine-substituted phenyl ring (Fig. 15.10*a*). However, these key fragmentation pathways were not observed in the MS/MS spectrum of the GSH adduct analog (Fig. 15.10*b*).

15.4.1.4 Dansyl-Glutathione as Trapping Agent Dansyl-glutathione (dGSH) was recently used as a trapping agent of reactive metabolites. dGSH adducts can be semiquantitatively determined using high-performance liquid chromatography (HPLC)–fluorescence analysis (Gan et al., 2005). Results from this analysis allow to rank drug candidates with respect to reactive metabolite formation when radiolabels are not available, which is very useful in lead optimization. To be able to provide the structural information of dGSH adducts, an LC–fluoresce detection–QTRAP method was developed recently (Fig. 15.8*d*) (Gan et al., 2009). In the analysis, the negative MRM is used to monitor a characteristic product ion of dGSH (5-dimethylamino-1-naphthalenesulfinic acid, *m*/*z* 234) that is common to all dGSH adducts (Fig. 15.11*c*).

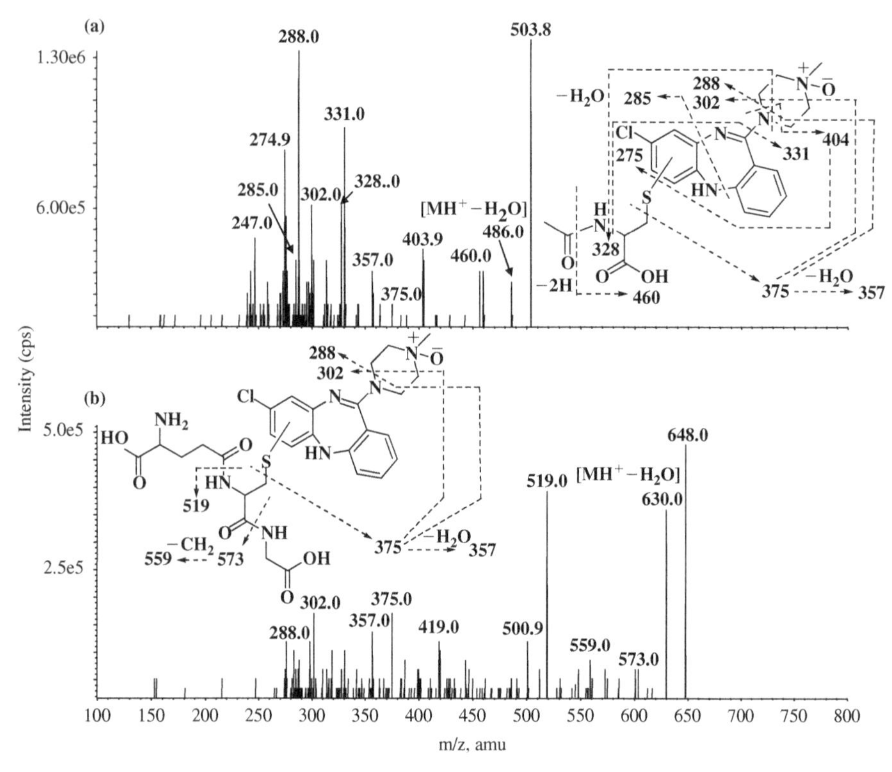

FIGURE 15.10 Product ion spectra and proposed structures of clozapine adducts formed in rat liver microsome incubation with (*a*) NAC and (*b*) GSH. (*a*) Positive product ion spectra acquired by negative NL-directed EPI with polarity switch. (*b*) Positive product ion spectra acquired by negative PI-directed EPI with polarity switch. [Reprinted from Jian et al. (2009) with permission of American Chemical Society.]

It has been demonstrated that negative MRM provides better sensitivity than positive MRM in detection of dGSH adduct, as shown in an example of an analysis of dGSH conjugates of omeprazole formed human liver microsome incubation (Fig. 15.11*a* vs. Fig. 15.11*b*) (Gan et al., 2009). Once a dGSH adduct is detected, then a positive EPI spectrum is acquired via information-dependent polarity switching (Fig. 15.8*d*), which provide more informative product ions (Fig. 15.11*d*) than those recorded in the negative mode (Fig. 15.11*c*). Overall, dGSH as a trapping agent combined with LC–fluorescence detection–QTRAP mass spectrometry provides a sensitive and selective method for rapid detection, semiquantiation, and characterization of reactive metabolites formed in the in vitro trapping experiments (Table 15.1).

15.4.2 Analysis of Adducts of Reactive Metabolites In Vivo

Analysis of adducts of reactive metabolites or other stable products of reactive metabolites in vivo is often conducted in pharmaceutical research. The purpose

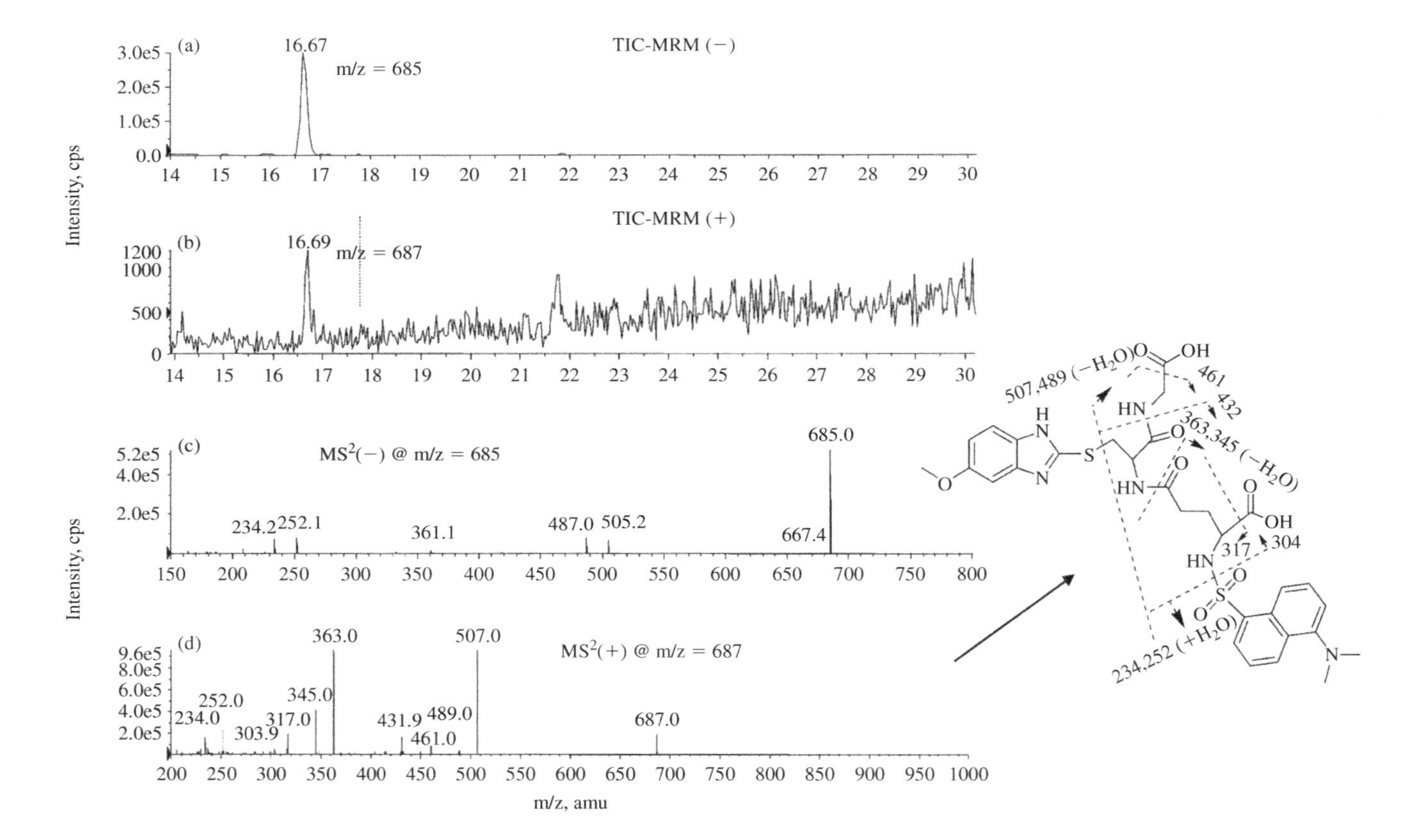

FIGURE 15.11 LC–MS/MS analysis of omeprazole incubation with dGSH. (*a*) TIC of the negative MRM. (*b*) TIC of the positive MRM. (*c*) Negative product ion spectrum of the dGSH adduct peak (*m/z* 685) eluted at 16.7 min. (*d*) Positive product ion spectrum of the dGSH adduct peak (*m/z* 687) eluted at 16.7 min. [Reprinted from Gan et al. (2009) with permission of American Chemical Society.]

of these studies include (1) screening for reactive metabolites that may not be formed in liver microsomes or other in vitro systems, (2) correlation of in vitro bioactivation to reactive metabolite formation in vivo, and (3) quantitative estimation of bioactivation for the assessment of exposure of reactive metabolites in animals and humans (the total load of reactive metabolites in vivo and ratios of bioactivation to the total metabolites) (Ma and Zhu, 2009). Although there are large amounts of GSH (5–10 m*M*) in the liver and GSH can readily react with most types of reactive metabolites, GSH adducts are usually not present in the circulation. GSH adducts formed in liver are either directly excreted in bile and/or transformed to NAC adducts (mercapturic acid adducts), which are mainly excreted into urine. Thus, the evaluation of reactive metabolite formation in animals is often accomplished by direct analysis of GSH adducts in bile collected from bile-duct cannulated (BDC) rats or other animal species. Both PI scan of *m/z* 272 in negative mode or NL scan of 129 Da in positive mode have been used for detection of GSH adducts formed in vivo in bile samples (Table 15.1). A study for analysis of GSH adducts in the bile samples from a rat dosed with troglitazone indicated that PI scan of *m/z* 272 offered better selectivity than the NL scan of 129 Da and resulted in less false-positive peaks (Dieckhaus et al., 2005). PI-EPI combined with polarity switching methods on a QTRAP instrument can provide rich structural information in addition to good selectivity for analysis of the GSH adducts in bile. For detection of low-level GSH adducts, a negative pMRM-directed positive EPI approach in which *m/z* 272 are used as the product ions in the pMRM transitions could be employed for better sensitivity.

In clinical studies, collection of human bile samples in short periods of time (2–6 h postdose) followed by LC–MS analysis is conducted in some cases in order to have better assessment of GSH adducts or metabolites that may not be stable such as glucuronides or subjected to further metabolism in feces (Wang et al., 2006). On the other hand, it is much more practical to analyze drug–NAC adducts excreted in urine for the assessment of the structures and exposure levels of reactive metabolites in humans, especially after multiple doses. Urinary NAC conjugates are considered as a biomarker of exposure to toxic chemicals, sediment contaminators, and drugs that form reactive metabolites. The majority of LC–MS methods employed so far in the analysis of urinary NAC adducts are associated with target analysis and quantification of specific NAC adducts using MRM methods on triple quadrupole instruments (Table 15.1). The recently developed NL-EPI/pMRM–EPI approach with polarity switch on QTRAP instruments (Fig. 15.8*c*) enabled the simultaneous detection of the presence of NAC adducts and acquisition of the product ion spectra (Jian et al., 2009). This approach was applied to the analysis of the urine samples collected from a health volunteer after an oral administration of acetaminophen. The NL-EPI method revealed a single predominant component, AM2, as shown in Figure 15.12*a*. AM2 was identified as the NAC derivative of direct conjugation between acetaminophen and GSH (Fig. 15.12*g*). Compared to the NL-EPI approach, the pMRM–EPI method provided better sensitivity and selectivity for analyzing NAC adducts

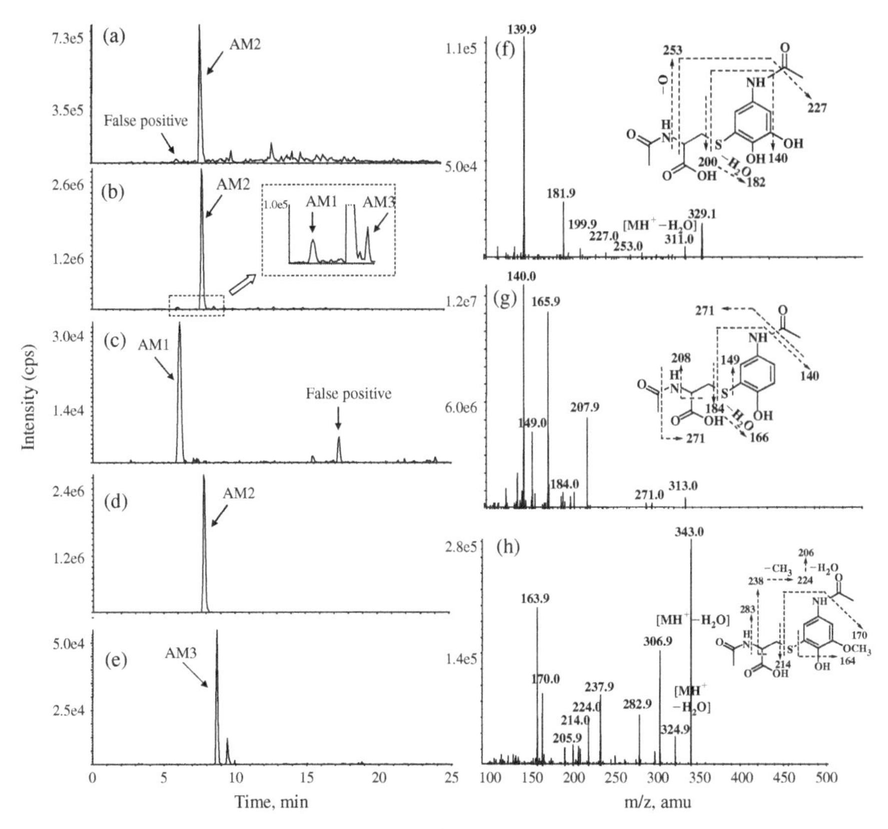

FIGURE 15.12 LC–MS/MS analysis of acetaminophen NAC adducts in human urine. (*a*) TIC of negative NL scan of 129 Da. (*b*) TIC of negative MRM with 40 transitions. (*c*) Extracted MRM transition for AM1 (*m*/*z* 327>198). (*d*) Extracted MRM transition for AM2 (*m*/*z* 311>182). (*e*) Extracted MRM transition for AM3 (*m*/*z* 341>212). (*f*, *g*, *h*) Positive product ion spectra and proposed structures of acetaminophen NAC conjugate AM1, AM2, and AM3, respectively, acquired by negative MRM-directed EPI with polarity switch. [Reprinted from Jian et a. (2009) with permission of American Chemical Society.]

in urine. As illustrated in Figures 15.12*b*–*e*, the pMRM–EPI analysis clearly determined the identities of AM1 and AM3, two minor NAC adducts of acetaminophen. These NAC adducts were identified as NAC adducts of 3-hydroxyl acetaminophen and 3-methoxyl acetaminophen, respectively (Figs. 15.12*f* and 15.12*h*). The pMRM–EPI approach provides superior sensitivity in analysis of NAC adducts in urine samples but requires a compound-dependent acquisition protocol. Therefore, this method is especially useful in the target analysis or confirmation of low levels of urinary NAC adducts of drugs or environmental contaminants when high analytical sensitivity and selectivity are required. In addition, an MRM scan can be applied to quantitative or semiquantitative estimation of NAC adducts with or without synthetic standards, which is essential for evaluation of exposure level of reactive metabolites in human relative to toxicological species.

15.5 IN VITRO DRUG INTERACTION STUDIES

Metabolically mediated drug–drug interactions (DDI) can cause severe side effects and reduced efficacy. DDI may potentially lead to early termination of development, refusal of approval, severe prescribing restrictions, and withdrawal of a drug from the market. DDI in humans are often associated with alternations of major metabolic clearance pathways of a drug via inhibition or induction of the corresponding CYP enzyme(s) by a co-administrated drug. As listed in Table 15.1, in vitro DDI studies include CYP inhibition, CYP induction, metabolizing reaction phenotyping, and in vitro studies associated with transporters. Unlike metabolic stability, protein binding, PAMPA, and in vitro transporter studies (Table 15.1) that involve the determination of concentrations of test compounds (either drug candidates or enzyme/transporter substrates), CYP inhibition, induction, and reaction phenotyping experiments often require the measurement of concentrations of the metabolites of test compounds and CYP substrates. The requirement for high analytical sensitivity and the lack of metabolite standards are the two major challenges involved in these studies.

15.5.1 Enzyme Kinetics Analysis

Prior to P450 reaction phenotyping or inhibition experiments, it is important to determine enzyme kinetic parameters such as K_m and V_{max} for the formation of selected metabolites that are subjected to quantitative analysis by LC–MS. For example, *S*-warfarin is catalyzed by CYP2C9 to a specific metabolite, 7-hydroxy-*S*-warfarin (Fig. 15.13). Thus, a CYP2C9 inhibition assay is developed based on the reaction. In the assay, *S*-warfarin is incubated with HLM in the presence of a test compound, followed by quantification of 7-hydroxy-*S*-warfarin by LC–MS (Zhang et al., 2001). To set up this assay in our lab, enzyme kinetics for the formation of 7-hydroxy-*S*-warfarin in HLM was determined. In this experiment, warfarin was incubated at concentrations from 0 to 250 μM with HLM at optimized conditions. Rates of 7-hydroxy-*S*-warfarin formation at various substrate concentrations were determined as shown in Figure 15.13*a*, from which K_m and V_{max} values were calculated. The warfarin assay represented an analytical challenge since the turnover of warfarin in the HLM system was extremely low. To be able to quantitatively determine low concentrations of 7-hydroxy-*S*-warfarin in the incubations, a very sensitive LC–MS method that used MRM with a 4000 QTRAP has been developed (Fig. 15.13*a*).

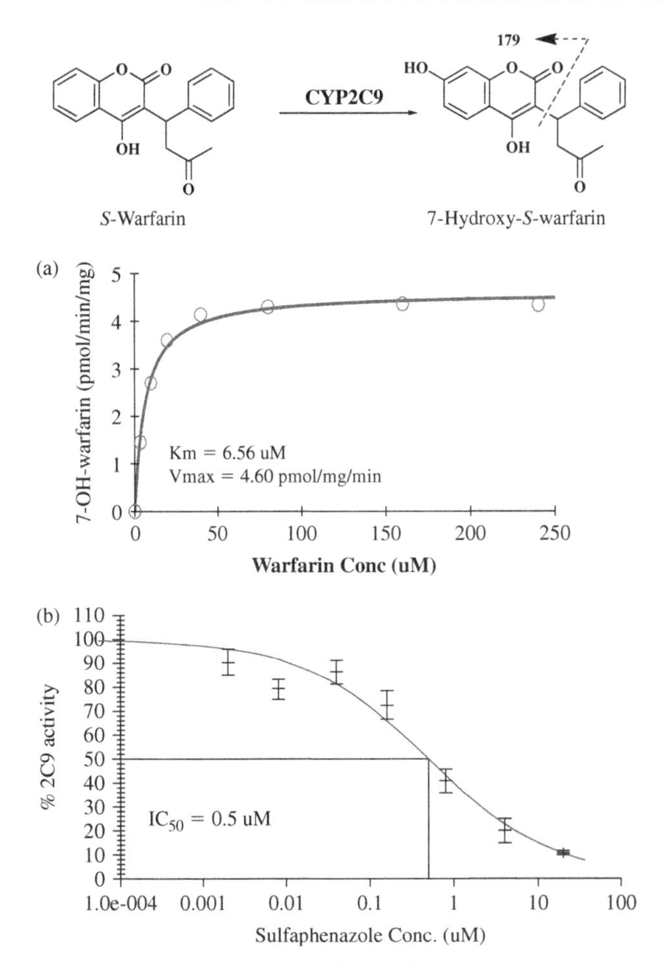

FIGURE 15.13 Analysis of enzyme kinetics of *S*-warfarin 7-hydroxylase (CYP2C9) by QTRAP. (*a*) ($K_m/V_{\max}$) determination for the formation of 7-hydroxy-*S*-warfarin from *S*-warfarin in HLM (7-hydroxy-*S*-warfarin was monitored with MRM transition from m/z 325>179). (*b*) IC_{50} determination for inhibition of CYP2C9 by sulfaphenazole.

15.5.2 Metabolizing Enzyme Reaction Phenotyping

The goal of metabolizing enzyme reaction phenotyping is to determine enzyme (s) that is/are responsible for a specific metabolic reaction (Zhang et al., 2007; Harper and Brassil, 2008). These studies are often accomplished by incubating test compounds with individual CYP enzymes or HLM in the presence of specific chemical inhibitors of various CYP enzymes (Zhu et al., 2005). The specific metabolite formed from the reaction is quantitatively determined using the metabolite standard and LC–MS (Raghavan et al., 2005). Since metabolite standards are not readily available in drug discovery and early development, relative ratios of metabolite peaks are measured for reaction phenotyping of GSH adducts (Wen et al., 2008c) and stable oxidative metabolites. In the late stages of drug development, metabolism enzyme reaction phenotyping is routinely conducted after a significant metabolic clearance in humans is identified in a radiolabeled human ADME study. Like any triple quadruple instruments, the MRM mode of the QTRAP is well suited for the measurement of both absolute concentrations (with metabolite standard) of metabolites and peak ratios of metabolites (without metabolite standard).

15.5.3 CYP Inhibition Assays

HLM assays are often carried out in HLM incubations with a CYP probe substrate and a test compound followed by MRM analysis of a metabolite of the probe substrates. These assays are frequently employed from lead optimization to preclinical development (Fig. 15.1) (Di et al., 2007; Yao et al., 2007; Mori et al., 2009; Perloff et al., 2009). To improve throughput of the CYP inhibition evaluation, HLM assays with multiple probe substrates are performed in many drug metabolism labs (Mori et al., 2009). In this case, MRM analysis of multiple metabolites of multiple probe substrates can be readily carried with a QTRAP. The FDA recently renewed DDI guidelines that recommend the use of probe substrates, inhibitors, and inducers of each individual CYP enzymes. For example, *S*-warfarin is a preferred probe substrate of CYP2C9, and sulfaphenazole is a recommended selective chemical inhibitor of CYP2C9 (Fig. 15.13). In a routine CYP2C9 inhibition experiment, sulfaphenazole is used as a positive control of CYP2C9 inhibition. In the experiment, sulfaphenazole is incubated at range of 0–20 μM. After incubation, proteins are precipitated by one volume of acetonitrile and then supernatants are collected by 96-well-plate filtration and analyzed with a 4000 QTRAP. As displayed in Figure 15.13*b*, the CYP2C9 IC_{50} value for sulfaphenazole (0.5 μM) is calculated based on the formation rates of 7-hydroxy-*S*-warfarin at the presence of various concentrations of sulfaphenazole.

15.6 QUANTIFICATION AND SCREENING OF DRUGS AND SMALL MOLECULES

15.6.1 PK and TK Sample Analysis

The QTRAP instrument has the standard MRM function that is identical to that of classical triple quadrupole instruments such as the API 4000 triple quadrupole mass spectrometer and can be used to routinely quantify parent drug and their metabolites in PK and TK bioanalytical quantitative work. In addition, pMRM can be used as a survey scan to trigger the information-dependent EPI to confirm the identity of the observed peaks. Studies have shown that quantitation data obtained by MRM-only and pMRM–EPI were equivalent and demonstrated comparable accuracy and precision from standards and quality control (QC) samples. For PK sample analysis, concentration–time profiles of parent drugs determined by both methods correlated well and produced overlapping curves (Li et al., 2005).

In addition to the MRM scan, full-scan data can be obtained using the EMS function on the QTRAP to provide valuable information for evaluation of the bioanalysis method. For example, dosing vehicle polyethylene glycol (PEG) has been found to cause ion suppression in electrospray ionization, and chromatographic resolution of PEG from the analytes of interest is crucial for reliable performance of a quantitative bioanalytical assay. The EMS scan of the incurred samples could provide valuable information on the presence and the retention time of PEG, which cannot be obtained by MRM (King et al., 2003). Phospholipids have been shown to cause ion suppression, especially when a generic extraction method such as protein precipitation is used. The EMS function on QTRAP can be used to elucidate the phospholipid profile and to monitor the separation of phospholipids from the analyte and the internal standard (King et al., 2003).

15.6.2 Tissue Imaging of Drugs

Tissue distribution information of a drug candidate can provide insight into mechanistic questions during drug development, such as tissue PK, route of elimination, CYP or P-gp mediated drug–drug interactions, site-specific drug localization and retention, and penetration into specific targets (Solon et al., 2002). Currently, whole-body autoradiography (WBA) using radiolabeled drugs is a standard method for quantitative imaging of the distribution of a drug and its metabolites in thin-layer tissue sections. This method suffers from several disadvantages, including the requirement for radiolabeled drug, the inability to differentiate the parent drug and the metabolites, and the labor and time intensity (4–6 weeks turnaround time). In comparison, MS-based tissue imaging does not require labeled drug and can provide structure-related information with much shorter turnaround time, usually within 24 h or less.

Recently, a QTRAP instrument was used in an imaging study to demonstrate the feasibility of using desorption electrospray ionization (DESI)–MS/MS to obtain tissue distribution information following drug administration at a pharmacologically relevant level (Kertesz et al., 2008). In this study, propranolol was monitored with a QTRAP using DESI–MRM during the scanning of the whole-body tissue section of a mouse dosed intravenously with propranolol. Figures 15.14*a* and 15.14*b* contains the optical image and distribution profile of propranolol measured by DESI–MS/MS, respectively. The images revealed that the method confidently detected propranolol in the brain, lung, stomach, and kidney regions. This study demonstrated that the QTRAP is well suited for MS-based tissue imaging studies. It is worthwhile to mention that QTRAP provides additional scan functions such as PI, NL,

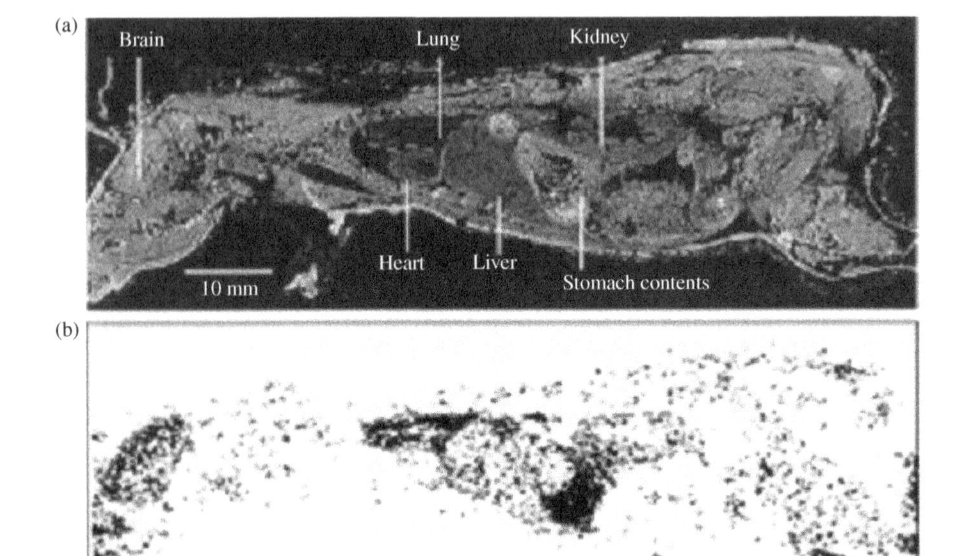

FIGURE 15.14 Tissue imaging using DESI–MS/MS and autoradiography in mice dosed with propranolol. (*a*) Scanned optical image of a 40-μm-thick sagittal whole-body tissue section of a mouse dosed intravenously with 7.5 mg/kg propranolol and euthanized 60 min after dose. (*b*) Distribution of propranolol in the 94 mm × 30 mm tissue section presented in (*a*) measured by DESI–MS/MS. [Reprinted from Kertesz et al. (2008) with permission of American Chemical Society.]

and EPI that can be further explored for the detection, confirmation, and characterization of the parent drugs and their metabolites in the whole-body tissue sections.

15.6.3 Screening of Drugs and Toxic Chemicals in Biological Samples

General unknown screening (GUS) of drugs and toxic compounds in biological specimens is a crucial part of clinical toxicology analysis, doping drug surveillance, and forensic detection. In recent years, LC–MS has become a popular complement to enzyme immunoassay (EIA) techniques, HPLC–diode array detection (DAD), and gas chromatography (GC)–MS for GUS (Marquet 2002a, 2002b; Saint-Marcoux et al., 2003; Maurer, 2005). The very first utilization of QTRAP in GUS adopted EMS as the survey scan and EPI as the dependent scan to analyze test compounds spiked in serum (Marquet et al., 2003). A comparison to HPLC–DAD and GC–MS for the analysis of clinical samples including serum, whole blood, urine, or gastric content showed that QTRAP has the capability to identify most of the compounds tested using the EMS–EPI-based method (94% of the total) (Sauvage et al., 2006).

In comparison to EMS-based methods, MRM provides superior selectivity and increased sensitivity, especially in the presence of biological matrices. Therefore, MRM-based approach is a more popular method of choice for GUS (Herrin et al., 2005; Mueller et al., 2005; Politi et al., 2007; Stanley et al., 2007). In one study, a multitarget screening (MTS) procedure for drugs in blood and urine for toxicological analysis was developed for the fast detection and identification of 301 forensically important drugs (Mueller et al., 2005). A multiple MRM survey for 301 drugs and three internal standards was conducted with IDA criteria that select MRM signal intensity above a certain threshold. The EPI spectrum was searched against the mass spectral library for the facile confirmation of the identified peaks. This method is useful for rapid detection or exclusion of the multiple target compounds in forensic or clinical cases. The number of target compounds that can be screened using MRM–EPI approach is limited by the maximum cycle time of the experiment and the peak width of the chromatographic peak. In order to increase the number of target compounds, the experiment could be split into several retention windows (periods) containing MRMs for the compounds that are expected to be eluted during the specific window. The recently released Analysis Software 1.5 has the Scheduled MRM algorithm that automatically maximizes the dwell time and cycle time for 1000 or more MRM in a single LC run. In addition, the 5500 QTRAP has the capability to conduct MRM with much faster speed (2 ms dwell time). These advances in instrumentation would significantly increase the number of the target compounds that can be included in the method, and therefore, meet the needs for high-throughput screening of the ever-increasing numbers of drugs.

In another MRM-based work, separate modes were used for the screening of 24 diuretics in human urine using both positive and negative modes (Politi et al., 2007). Two individual injections were required for each sample and led to the increased cycle times. A single polarity switch on the 4000 QTRAP would require a settling time of 700 ms, extending the cycle time significantly, which would be too high to allow a minimum of 12 points per peak. The splitting of the acquisition into different periods could not solve this problem due to closely eluted compounds with opposite modes along the whole run. This problem could be potentially solved by 5500 QTRAP, which has a polarity switch speed only of 50 ms and enables the simultaneous acquisition in both positive and negative modes without sacrificing the acquisition time.

15.6.4 Analysis of Pharmaceuticals in Wastewater

Nowadays, contamination of urban wastewater has caused growing social concern and is demanding prompt actions. Even though subjected to treatment, wastewater has been demonstrated to contain multiple organic compounds, such as pharmaceuticals, personal care products, hormones, and other disrupting compounds, some of which are becoming ubiquitous in the environment (Carballa et al., 2004; Jones-Lepp et al., 2007). In order to improve the wastewater treatment technologies, suitable analytical methodology that would allow an exhaustive characterization of wastewater effluents would be necessary. LC–MS has arisen in the recent years as the techniques of choice for wastewater analysis. Target analysis (MRM) based on LC–MS/MS provides excellent performance due to inherent selectivity and sensitivity.

In a recent study, the QTRAP instrument was used in the MRM mode for quantitative analysis of a selected group of 56 organic pollutants in wastewater, including 38 pharmaceuticals and 10 of their most frequent metabolites, 6 pesticides, and 2 disinfectants (Bueno et al., 2007). Three MRM transitions were used for each compound to provide unequivocal identification, except for ibuprofen, which has only one major suitable transition. An additional EPI was developed for ibuprofen to obtain structural information for confirmation purposes. Linearity, precision, matrix effect, detection limit, and recovery were evaluated for all the tested compounds, and the result indicated that the performance on the QTRAP was, in general, comparable with those of other triple quadrupole instrument. This study demonstrated that the MRM scan on the QTRAP is well suited for high-throughput quantitative analysis of wastewater and the EPI function could provide further structure confirmation. The new advances in the QTRAP instrument will show further promise to increase the capability of the assay and to improve the efficiency and performance of wastewater analysis.

15.7 SUMMARY

Although a mature form of hybrid triple quadrupole–linear ion trap mass spectrometry (4000 QTRAP) was introduced only in the mid-2000s, the QTRAP has emerged a very unique and attractive LC–MS platform for drug metabolism and pharmacokinetics research in the pharmaceutical industry (Fig. 15.1). The QTRAP offers versatile scanning functions that are identical to those of classical triple quadrupole instruments (NL, PI, and MRM) and similar to those of linear ion trap instruments (EPI and MS^3). The QTRAP can be employed as a triple quadrupole instrument in bioanalytical experiments to support in vivo pharmacokinetics studies as well as in vitro ADME screening to support metabolic stability, PAMPA, protein binding, and transporter studies. In addition, the QTRAP enables the performance of various combinations of triple quadrupole and ion trap scanning functions via information-dependent acquisition and polarity switching (EMS–EPI, NL-EPI, PI-EPI, pMRM–EPI, and MIM–EPI). The recently introduced 5500 QTRAP has made these information-dependent acquisitions much quicker to implement and the pMRM–EPI, or MIM–EPI to cover a much wider mass range. These unique scan modes not only provide highly valuable results for the high-throughput detection and structural characterization of drug metabolites (Fig. 15.8) but also meet specific needs for the analysis of various types of metabolites in different biological matrices

(Table 15.1). As a multiple-task instrument, the QTRAP may serve as the LC–MS platform of choice in certain DMPK and bioanalysis laboratories. Specific examples of these laboratories would include: (1) bioanalytical laboratories where detection of plasma metabolites or in vitro ADME screening are also performed; (2) drug metabolism laboratories where metabolite identification and/or in vitro ADME screening are conducted, and (3) small bioanalysis and drug metabolism laboratories where quantitative and qualitative analyses of drugs and/or metabolites are routinely conducted with limited numbers of LC–MS instruments and scientists.

REFERENCES

Baillie TA, Davis MR. Mass spectrometry in the analysis of glutathione conjugates. Biol Mass Spectrom 1993;22:319–325.

Bauman JN, Frederick KS, Sawant A, Walsky RL, Cox LM, Obach RS, Kalgutkar AS. Comparison of the bioactivation potential of the antidepressant and hepatotoxin nefazodone with aripiprazole, a structural analog and marketed drug. Drug Metab Dispos 2008;36:1016–1029.

Bueno MJ, Aguera A, Gomez MJ, Hernando MD, Garcia-Reyes JF, Fernandez-Alba AR. Application of liquid chromatography/quadrupole-linear ion trap mass spectrometry and time-of-flight mass spectrometry to the determination of pharmaceuticals and related contaminants in wastewater. Anal Chem 2007;79:9372–9384.

Carballa M, Omil F, Lema JM, Llompart M, Garcia-Jares C, Rodriguez I, Gomez M, Ternes T. Behavior of pharmaceuticals, cosmetics and hormones in a sewage treatment plant. Water Res 2004;38:2918–2926.

Chen WG, Zhang C, Avery MJ, Fouda HG. Reactive metabolite screen for reducing candidate attrition in drug discovery. In Biological Reactive Intermediates, Vol. 6, Dansette PM, Delaforge RSM, Gilbon GG, Greim H, Jollow DJ, Monks TJ, Sipes IG. Eds. Kluwer Academic/Plenum, New York, 2001, pp.521–534.

Di L, Kerns EH, Li SQ, Carter GT. Comparison of cytochrome P450 inhibition assays for drug discovery using human liver microsomes with LC-MS, rhCYP450 isozymes with fluorescence, and double cocktail with LC-MS. Int J Pharm 2007;335:1–11.

Dickinson DA, Forman HJ. Cellular glutathione and thiols metabolism. Biochem Pharmacol 2002a;64:1019–1026.

Dickinson DA, Forman HJ. Glutathione in defense and signaling: Lessons from a small thiol. Ann NY Acad Sci 2002b;973:488–504.

Dieckhaus CM, Fernandez-Metzler CL, King R, Krolikowski PH, Baillie TA. Negative ion tandem mass spectrometry for the detection of glutathione conjugates. Chem Res Toxicol 2005;18:630–638.

Food and Drug Administration (FDA). Guidance for Industry–Satety Testing of Drug Metabolites, FDA, Washington, DC, 2008.

Gan J, Harper TW, Hsueh MM, Qu Q, Humphreys WG. Dansyl glutathione as a trapping agent for the quantitative estimation and identification of reactive metabolites. Chem Res Toxicol 2005;18:896–903.

Gan J, Ruan Q, He B, Zhu M, Shyu WC, Humphreys WG. In vitro screening of 50 highly prescribed drugs for thiol adduct formation—Comparison of potential for drug-induced toxicity and extent of adduct formation. Chem Res Toxicol 2009;22:690–698.

Gao H, Materne OL, Howe DL, Brummel CL. Method for rapid metabolite profiling of drug candidates in fresh hepatocytes using liquid chromatography coupled with a hybrid quadrupole linear ion trap. Rapid Commun Mass Spectrom 2007;21:3683–3693.

Gibson JD, Pumford NR, Samokyszyn VM, Hinson JA. Mechanism of acetaminophen-induced hepatotoxicity: Covalent binding versus oxidative stress. Chem Res Toxicol 1996;9:580–585.

Haroldsen PE, Reilly MH, Hughes H, Gaskell SJ, Porter CJ. Characterization of glutathione conjugates by fast atom bombardment/tandem mass spectrometry. Biomed Environ Mass Spectrom 1988;15:615–621.

Harper TW, Brassil PJ. Reaction phenotyping: Current industry efforts to identify enzymes responsible for metabolizing drug candidates. Aaps J 2008;10:200–207.

Herrin GL, McCurdy HH, Wall WH. Investigation of an LC-MS-MS (QTrap) method for the rapid screening and identification of drugs in postmortem toxicology whole blood samples. J Anal Toxicol 2005;29:599–606.

Hinson JA, Reid AB, McCullough SS, James LP. Acetaminophen-induced hepatotoxicity: Role of metabolic activation, reactive oxygen/nitrogen species, and mitochondrial permeability transition. Drug Metab Rev 2004;36:805–822.

Hopfgartner G, Varesio E, Tschappat V, Grivet C, Bourgogne E, Leuthold LA. Triple quadrupole linear ion trap mass spectrometer for the analysis of small molecules and macromolecules. J Mass Spectrom 2004;39:845–855.

Jian W, Yao M, Zhang D, Zhu M. Rapid detection and characterization of in vitro and urinary *N*-acetyl-L-cysteine conjugates using quadrupole-linear ion trap mass spectrometry and polarity switching. Chem Res Toxicol 2009;22:1246–1255.

Jones-Lepp TL, Stevens R. Pharmaceuticals and personal care products in biosolids/sewage sludge: The interface between analytical chemistry and regulation. Anal Bioanal Chem 2007;387:1173–1183.

Kertesz V, Van Berkel GJ, Vavrek M, Koeplinger KA, Schneider BB, Covey TR. Comparison of drug distribution images from whole-body thin tissue sections obtained using desorption electrospray ionization tandem mass spectrometry and autoradiography. Anal Chem 2008;80:5168–5177.

King R, Fernandez-Metzler C. The use of Qtrap technology in drug metabolism. Curr Drug Metab 2006;7:541–545.

King RC, Gundersdorf R, Fernandez-Metzler CL. Collection of selected reaction monitoring and full scan data on a time scale suitable for target compound quantitative analysis by liquid chromatography/tandem mass spectrometry. Rapid Commun Mass Spectrom 2003;17:2413–2422.

Li AC, Alton D, Bryant MS, Shou WZ. Simultaneously quantifying parent drugs and screening for metabolites in plasma pharmacokinetic samples using selected reaction monitoring information-dependent acquisition on a QTrap instrument. Rapid Commun Mass Spectrom 2005;19:1943–1950.

Li AC, Gohdes MA, Shou WZ. "*N*-in-one" strategy for metabolite identification using a liquid chromatography/hybrid triple quadrupole linear ion trap instrument using multiple dependent product ion scans triggered with full mass scan. Rapid Commun Mass Spectrom 2007;21:1421–1430.

Ma S, Chowdhury S. Application of liquid chromatography/mass spectrometry for metabolite identification. In Drug Metabolism in Drug Design and Development: Basic Concepts and Practice, Zhang D, Zhu M, Humphreys WG, Eds. Wiley, Hoboken, NJ, 2007; pp.319–367.

Ma S, Zhu M. Recent advances in applications of liquid chromatography–tandem mass spectrometry to the analysis of reactive drug metabolites. Chem Biol Interact 2009;179:25–37.

Mahajan MK, Evans CA. Dual negative precursor ion scan approach for rapid detection of glutathione conjugates using liquid chromatography/tandem mass spectrometry. Rapid Commun Mass Spectrom 2008;22:1032–1040.

Marquet P. Is LC-MS suitable for a comprehensive screening of drugs and poisons in clinical toxicology? Ther Drug Monit 2002a;24:125–133.

Marquet P. Progress of liquid chromatography–mass spectrometry in clinical and forensic toxicology. Ther Drug Monit 2002b;24:255–276.

Marquet P, Saint-Marcoux F, Gamble TN, Leblanc JC. Comparison of a preliminary procedure for the general unknown screening of drugs and toxic compounds using a quadrupole-linear ion-trap mass spectrometer with a liquid chromatography–mass spectrometry reference technique. J Chromatogr B Anal Technol Biomed Life Sci 2003;789:9–18.

Maurer HH. Multi-analyte procedures for screening for and quantification of drugs in blood, plasma, or serum by liquid chromatography–single stage or tandem mass spectrometry (LC-MS or LC–MS/MS) relevant to clinical and forensic toxicology. Clin Biochem 2005;38:310–318.

Mauriala T, Chauret N, Oballa R, Nicoll-Griffith DA, Bateman KP. A strategy for identification of drug metabolites from dried blood spots using triple-quadrupole/linear ion trap hybrid mass spectrometry. Rapid Commun Mass Spectrom 2005;19:1984–1992.

Mori K, Hashimoto H, Takatsu H, Tsuda-Tsukimoto M, Kume T. Cocktail-substrate assay system for mechanism-based inhibition of CYP2C9, CYP2D6, and CYP3A using human liver microsomes at an early stage of drug development. Xenobiotica 2009;39:415–422.

Mueller CA, Weinmann W, Dresen S, Schreiber A, Gergov M. Development of a multi-target screening analysis for 301 drugs using a QTrap liquid chromatography/tandem mass spectrometry system and automated library searching. Rapid Commun Mass Spectrom 2005;19:1332–1338.

Murphy CM, Fenselau C, Gutierrez PL. Fragmentation characteristic of glutathione conjugates activated by high-energy collision. J Am Soc Mass Spectrom 1992;3:815–822.

Perloff ES, Mason AK, Dehal SS, Blanchard AP, Morgan L, Ho T, Dandeneau A, Crocker RM, Chandler CM, Boily N, Crespi CL, Stresser DM. Validation of cytochrome P450 time-dependent inhibition assays: A two-time point IC50 shift approach facilitates kinact assay design. Xenobiotica 2009;39:99–112.

Politi L, Morini L, Polettini A. A direct screening procedure for diuretics in human urine by liquid chromatography–tandem mass spectrometry with information dependent acquisition. Clin Chim Acta 2007;386:46–52.

Prakash C, Shaffer CL, Nedderman A. Analytical strategies for identifying drug metabolites. Mass Spectrom Rev 2007;26:340–369.

Raghavan N, Zhang D, Zhu M, Zeng J, Christopher L. Cyp2D6 catalyzes 5-hydroxylation of 1-(2-pyrimidinyl)-piperazine, an active metabolite of several psychoactive drugs, in human liver microsomes. Drug Metab Dispos 2005;33:203–208.

Saint-Marcoux F, Lachatre G, Marquet P. Evaluation of an improved general unknown screening procedure using liquid chromatography–electrospray–mass spectrometry by comparison with gas chromatography and high-performance liquid-chromatography—Diode array detection. J Am Soc Mass Spectrom 2003;14:14–22.

Sauvage FL, Saint-Marcoux F, Duretz B, Deporte D, Lachatre G, Marquet P. Screening of drugs and toxic compounds with liquid chromatography–linear ion trap tandem mass spectrometry. Clin Chem 2006;52:1735–1742.

Scholz K, Dekant W, Volkel W, Pahler A. Rapid detection and identification of *N*-acetyl-L-cysteine thioethers using constant neutral loss and theoretical multiple reaction monitoring combined with enhanced product-ion scans on a linear ion trap mass spectrometer. J Am Soc Mass Spectrom 2005;16:1976–1984.

Shou WZ, Magis L, Li AC, Naidong W, Bryant MS. A novel approach to perform metabolite screening during the quantitative LC-MS/MS analyses of in vitro metabolic stability samples using a hybrid triple-quadrupole linear ion trap mass spectrometer. J Mass Spectrom 2005;40:1347–1356.

Soglia JR, Contillo LG, Kalgutkar AS, Zhao S, Hop CE, Boyd JG, Cole MJ. A semiquantitative method for the determination of reactive metabolite conjugate levels in vitro utilizing liquid chromatography–tandem mass spectrometry and novel quaternary ammonium glutathione analogues. Chem Res Toxicol 2006;19:480–490.

Soglia JR, Harriman SP, Zhao S, Barberia J, Cole MJ, Boyd JG, Contillo LG. The development of a higher throughput reactive intermediate screening assay incorporating micro-bore liquid chromatography–micro-electrospray ionization–tandem mass spectrometry and glutathione ethyl ester as an in vitro conjugating agent. J Pharm Biomed Anal 2004;36:105–116.

Solon EG, Balani SK, Lee FW. Whole-body autoradiography in drug discovery. Curr Drug Metab 2002;3:451–462.

Stanley SM, Wee WK, Lim BH, Foo HC. Direct-injection screening for acidic drugs in plasma and neutral drugs in equine urine by differential-gradient LC-LC coupled MS/MS. J Chromatogr B Anal Technol Biomed Life Sci 2007;848:292–302.

Takakusa H, Masumoto H, Yukinaga H, Makino C, Nakayama S, Okazaki O, Sudo K. Covalent binding and tissue distribution/retention assessment of drugs associated with idiosyncratic drug toxicity. Drug Metab Dispos 2008;36:1770–1779.

Wagner S, Scholz K, Donegan M, Burton L, Wingate J, Volkel W. Metabonomics and biomarker discovery: LC-MS metabolic profiling and constant neutral loss scanning combined with multivariate data analysis for mercapturic acid analysis. Anal Chem 2006;78:1296–1305.

Wagner S, Scholz K, Sieber M, Kellert M, Voelkel W. Tools in metabonomics: An integrated validation approach for LC-MS metabolic profiling of mercapturic acids in human urine. Anal Chem 2007;79:2918–2926.

Wang L, Zhang D, Swaminathan A, Xue Y, Cheng PT, Wu S, Mosqueda-Garcia R, Aurang C, Everett DW, Humphreys WG. Glucuronidation as a major metabolic clearance pathway of 14c-labeled muraglitazar in humans: Metabolic profiles in subjects with or without bile collection. Drug Metab Dispos 2006;34:427–439.

Wen B, Coe KJ, Rademacher P, Fitch WL, Monshouwer M, Nelson SD. Comparison of in vitro bioactivation of flutamide and its cyano analogue: Evidence for reductive activation by human NADPH:cytochrome P450 reductase. Chem Res Toxicol 2008a;21:2393–2406.

Wen B, Fitch WL. Analytical strategies for the screening and evaluation of chemically reactive drug metabolites. Expert Opin Drug Metab Toxicol 2009a;5:39–55.

Wen B, Fitch WL. Screening and characterization of reactive metabolites using glutathione ethyl ester in combination with Q-trap mass spectrometry. J Mass Spectrom 2009b;44:90–100.

Wen B, Ma L, Nelson SD, Zhu M. High-throughput screening and characterization of reactive metabolites using polarity switching of hybrid triple quadrupole linear ion trap mass spectrometry. Anal Chem 2008b;80:1788–1799.

Wen B, Ma L, Rodrigues AD, Zhu M. Detection of novel reactive metabolites of trazodone: Evidence for CYP2D6-mediated bioactivation of *m*-chlorophenylpiperazine. Drug Metab Dispos 2008c;36:841–850.

Wen B, Ma L, Zhu M. Bioactivation of the tricyclic antidepressant amitriptyline and its metabolite nortriptyline to arene oxide intermediates in human liver microsomes and recombinant P450s. Chem Biol Interact 2008d;173:59–67.

Xia YQ, Miller JD, Bakhtiar R, Franklin RB, Liu DQ. Use of a quadrupole linear ion trap mass spectrometer in metabolite identification and bioanalysis. Rapid Commun Mass Spectrom 2003;17:1137–1145.

Yan Z, Caldwell GW. Stable-isotope trapping and high-throughput screenings of reactive metabolites using the isotope MS signature. Anal Chem 2004;76:6835–6847.

Yao M, Ma L, Duchoslav E, Zhu M. Rapid screening and characterization of drug metabolites using multiple ion monitoring–depedent MS/MS scan and postacquisition data mining on a hybrid quadrupole–linear ion trap mass spectrometer. Rapid Commun Mass Spectrom 2009;23:1683–1693.

Yao M, Ma L, Humphreys WG, Zhu M. Rapid screening and characterization of drug metabolites using a multiple ion monitoring–dependent MS/MS acquisition method on a hybrid triple quadrupole–linear ion trap mass spectrometer. J Mass Spectrom 2008;43:1364–1375.

Yao M, Zhu M, Sinz MW, Zhang H, Humphreys WG, Rodrigues AD, Dai R. Development and full validation of six inhibition assays for five major cytochrome P450 enzymes in human liver microsomes using an automated 96-well microplate incubation format and LC-MS/MS analysis. J Pharm Biomed Anal 2007;44: 211–223.

Zhang H, Davis CD, Sinz MW, Rodrigues AD. Cytochrome P450 reaction-phenotyping: An industrial perspective. Expert Opin Drug Metab Toxicol 2007;3:667–687.

Zhang H, Zhang D, Ray K, Zhu M. Mass defect filter technique and its applications to drug metabolite identification by high-resolution mass spectrometry. J Mass Spectrom 2009a;44:999–1016.

Zhang Z, Zhu M, Tang W. Metabolite identification and profiling in drug design: Current practice and future directions. Curr Pharm Des 2009b;15:2220–2235.

Zhang ZY, King BM, Wong YN. Quantitative liquid chromatography/mass spectrometry/mass spectrometry warfarin assay for in vitro cytochrome P450 studies. Anal Biochem 2001;298:40–49.

Zheng J, Ma L, Xin B, Olah T, Humphreys WG, Zhu M. Screening and identification of GSH-trapped reactive metabolites using hybrid triple quadruple linear ion trap mass spectrometry. Chem Res Toxicol 2007;20:757–766.

Zhu M, Zhang D, Zhang H, Shyu WC. Integrated strategies for assessment of metabolite exposure in humans during drug development: Analytical challenges and clinical development considerations. Biopharm Drug Dispos 2009;30:163–184.

Zhu M, Zhao W, Jimenez H, Zhang D, Yeola S, Dai R, Vachharajani N, Mitroka J. Cytochrome P450 3A-mediated metabolism of buspirone in human liver microsomes. Drug Metab Dispos 2005;33:500–507.

16 Quantitative Drug Metabolism with Accelerator Mass Spectrometry

JOHN S. VOGEL, PETER LOHSTROH, BRAD KECK, and STEPHEN R. DUEKER
Vitalea Science, Davis, California

16.1 RELEVANCE OF AMS TO DRUG METABOLISM

Accelerator mass spectrometry (AMS) has long been used for quantifying concentrations of metabolites in humans, primarily for activated metabolites of environmental compounds that form genotoxic macromolecular adducts (Turteltaub et al., 1993, 1997, 1999; Dingley et al., 1999; Mauthe et al., 1999; Lightfoot et al., 2000; Cupid et al., 2004) agricultaural chemicals that chronically

Mass Spectrometry in Drug Metabolism and Disposition: Basic Principles and Applications, First Edition. Edited by Mike S. Lee and Mingshe Zhu.

expose workers at low levels (Gilman et al., 1998; Buchholz et al., 1999), and vitamins at physiologically relevant doses (Dueker et al., 2000; Hickenbottom et al., 2002; Ho et al., 2009). Prior to AMS, these metabolities and molecular interactions were quantified at well above relevant biological concentrations due to the lack of sufficient analytical sensitivity. The first application of AMS to biochemical tracing showed an immediate detection increase of three orders of sensitivity and a large improvement in precision over previous methods and added precise quantitation to metabolic studies at compound exposures relevant to environmental toxicology and trace nutrition (Turteltaub et al., 1990). Nutrient metabolism has particularly advanced under the use of AMS for sensitive isotopic tracing because vitamins are potent therapeutic compounds that are ingested at microgram per day doses into large endogenous pools of otherwise indistinguishable chemicals (Lin et al., 2004; Lemke et al., 2003; Hickenbottom et al., 2002). The isotopic label assures that all circulating forms of the nutrient dose are detected and quantified, barring poorly designed labels in labile metabolic positions. The distinguishable dosed compound is then available for quantifying effects of unlabeled interacting doses (Lemke et al., 2003; Dingley et al., 2003; Ebeler et al., 2005) Thus, quantitation of drug and drug candidate compound metabolism is a direct extension of developed methods.

The 2008 FDA (US FDA, 2008) Guidance on Metabolite Safety In Toxicology (MIST) has heightened the need for precise and accurate quantitation, if not always identification, of all human transformations of compounds in drug development. Quantitation must not only be independent of metabolizing pathways and biological matrices but also independent of animal species, since MIST requires comparison across humans and toxicology species to determine that animal toxicity or safety is relevant. Major advantages of sensitive AMS isotope quantitation include the universal applicability of isotopic quantitation without extensive specific method development and the assurance that all derivatives of properly labeled compounds will be quantifiable due to the wide linear range of AMS quantitation.

16.2 INTRODUCTION TO AMS

Since the time of Scholenheimer (1938), isotopic molecular labels have revealed the pathways and transformation of nutritive, toxic, and therapeutic compounds within living animals and humans. With its discovery in 1940 by Kamen (1949), radiocarbon, ^{14}C, quickly became the most useful isotopic tracer for organic compounds because carbon has maximum potential for stable molecular inclusion but also because its radioactive decay provides an inherent means of quantitation through counting the energetic electrons that emerge from a sample. Radioisotopes have a further benefit over the stable isotopic labels in their greater absolute sensitivity, a result of their natural rarity.

Quantitation by decay counting is inefficient for isotopes with lives as long as radiocarbon, for which only 1 in 4.4 billion atoms decay per minute.

Accelerator mass spectrometry was initially developed for radiocarbon dating (Nelson et al., 1977; Bennett et al., 1977) to directly count rare isotopes without their decay. AMS is an isotope ratio mass spectrometry (IRMS) method for counting individual ions of the rare isotope against the abundance of common stable isotopes of the element after equivalent ion mass separations. No solely stable isotopes are quantified by AMS, per se, although AMS spectrometers measure stable isotope ratios concomitant with counting the rare isotope. AMS was designed to quantify particularly long-lived radioisotopes that occur naturally at parts per trillion (ppt) to quintillion and found rapid acceptance in archaeological and earth science studies (Taylor et al., 1992). The initial concentration of natural ^{14}C to be measured is 1.18 parts per trillion, a defined unit internationally termed "Modern," which then decays away within any previously living plant or animal once life no longer recycles fresh biosphere material. Decay-counted carbon dating required large samples and long count times but seldom achieved precise dates much older than 10,000 years. This was the initial impetus for direct counting of long-lived isotopes. Direct ion counting depends on engineered quantities such as ionization efficiency and ion optics independent of the physical constraint of half-life. Kutschera (2005) provides a review of the many isotopes and applications that have benefited from the high sensitivity of AMS measurements.

This inherent inefficiency in decay counting requires large doses of isotope for effective quantitative tracing of labeled compounds, including therapeutic drug candidates. The fundamental description of radioactive decay is that the decay rate (the measured quantity) is proportional to the number of the decaying isotopes present, with the proportionality constant being the mean isotopic lifetime [$\tau = t_{1/2} \ln(2)$]:

$$A\,(\text{dpm}) = -\frac{dN}{dt} = \frac{N}{\tau} \tag{16.1}$$

where the mean life of ^{14}C is 4.35×10^9 minutes (8270 years), meaning that 1 dpm (disintegrations per minute) represents 4.35×10^9 ^{14}C atoms or 7.2 fmol tracer in the sample. One milligram of carbon (mg C) contains 83 μmol of carbon, and the isotope ratio of 1 dpm ^{14}C in 1 mg C is thus 87ppt, 75 times the natural concentration of ^{14}C in the living biosphere. AMS thus quantifies tracers well below the femtomole levels of dpm samples down into subattomole (amol) amounts and has a range 4–5 orders of magnitude above its sensitivity limit. AMS counting is a Poisson process, as is decay counting, whose precision of measurement is proportional to the number of counts. AMS provides a large gain in measurement precision by directly counting ^{14}C ions instead of decays (Vogel et al., 2004). In biochemistry and drug development, this sensitivity is applied to increasing precision, to lowering radiactive doses presented to human volunteers or animal subjects, and to isolating ever more specific metabolic products to obtain more quantitative detail in absorption, distribution, metabolism, and elimination (ADME) studies.

Although the possibility of using AMS for biochemical research was suggested shortly after its development, the first AMS quantification of macromolecular binding by reactive metabolites of a ^{14}C-labeled compound was published in 1990 (Turteltaub et al., 1990). Since that time, multiple research groups used AMS in biological studies of ^{3}H (Dingley et al., 1998b), ^{10}Be (Chiarappa-Zucca et al., 2004), ^{26}Al (Flarend et al., 1997), ^{36}Cl (Baxter et al., 2009), ^{41}Ca (Denk, et al. 2006, 2007), as well as ^{14}C. This chapter confines itself to ^{14}C isotopic labels and reviews the basic structure of ^{14}C AMS and the fundamentals of interpreting an isotope ratio measurement versus the more familiar decay counting technique. The method and concepts are then applied to quantitative metabolite measurements with ultra-high-performance liquid chromatography (UPLC)–AMS compared to competing ^{14}C counting technologies. Particular emphasis is placed on the importance of accurate universal quantitative assay of drug metabolites possible with AMS in light of the 2008 FDA Guidance on Metabolites in Safety Toxicology (MIST).

16.3 FUNDAMENTALS OF AMS INSTRUMENTS

Accelerator mass spectrometry, like all IRMS, must distinguish the desired isotopic signal (e.g., ^{14}C) from both nuclear (^{14}N) and molecular ($^{12}CH_2$) isobars. AMS uses two fundamental capabilities, both based on accelerating the sample ions with a high kinetic potential (100–10,000 kV). At these energies, a collision cell of confined gas or a thin solid film disrupts even highly bound molecules such as hydrides to the constituent ions, removing molecular isobars. AMS is a type of tandem mass spectrometry with a low-energy mass selection that directs ions of interest to the dissociation cell at high potential followed by a high-energy mass selection that sorts and quantifies the atomic debris as shown in the schematic of the Vitalea spectrometer in Figure 16.1 and pictured in Figure 16.2. The high energy of the resultant atomic ions identifies individual ions through the dependence of ions energy loss on nuclear charge (Z) in an ion detector, distinguishing nuclear isobars. ^{14}C-AMS was particularly favored by the inability of its nuclear isobar, ^{14}N, to retain an extra electron to become a negative atomic ion (Middleton, 1974). Thus, tandem van de Graaff style electrostatic accelerators, which are injected with low-energy negative ions, became the standard for ^{14}C-AMS, with detailed spectrometer research finally leading to quite low energy "accelerators" such as the 200-kV spectrometer produced by the Paul Scherrer Institute (Zurich, Switzerland) for Vitalea Science shown in Figure. 16.2 (Synal et al., 2004, 2007; Schultz-Konig et al., 2010). The name MICADAS stands for MIniature CArbon DAting Spectrometer and was, as are most AMS, initially designed for ^{14}C dating. This spectrometer is modified for the higher counting intensities of tracer samples. This is a sequential ion spectrometer in which ion masses 12,13, and 14 are alternately selected electrostatically through the low-energy mass spectrometer at repetition rates of 5–10 times per second.

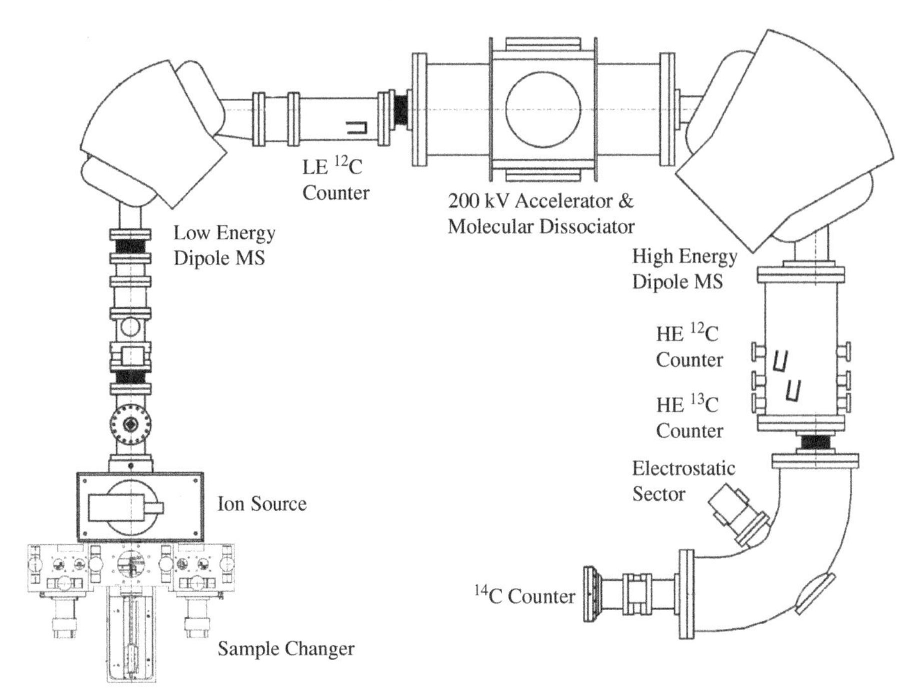

FIGURE 16.1 Schematic of Bio-MICADAS. Negative carbon ions travel counter-clockwise from the ion source, are selected to 1-amu resolution at the low-energy dipole, are accelerated to 200 kV, and pass through argon gas to lose electrons and are accelerated again across 200 kV, resolved again to 1 amu by the high-energy dipole. ^{14}C ions then are filtered by electrostatic sector into their counter. Ion beams of ^{12}C and ^{13}C are measured in indicated Faraday cups.

Low- and high-energy ^{12}C current pulses and high-energy ^{13}C current pulses are quantified by integrating the amount of ion beam entering respective Faraday cups, while individual ions of ^{14}C are summed in an ion counter. The validated specifications for this particular spectrometer for ^{14}C are listed in Table 16.1. Other AMS spectrometers arrange the separation sectors differently and use a variety of accelerating voltages, but all are essentially the same in accelerating ^{14}C for direct counting with one or both stable carbon ions accelerated to normalize the collision cell transmission (Barker and Garner 1999; Skog 2007; Zoppi et al., 2007; Young et al., 2008).

Instrument specificity refers to counts in the ^{14}C detector, that are ^{12}C or ^{13}C ions scattered off residual gas in the instrument at low probabilities (10^{-12}–10^{-15}). This instrument background is insignificant compared to natural and processing backgrounds discussed below. The spectrometers themselves operate robustly in that multiple operators obtain the same raw isotope ratio for National Institute of Standards and Technology (NIST)–traceable standards to within a few percent over periods of years in routine operation. These external reference standards normalize even this imprecision from

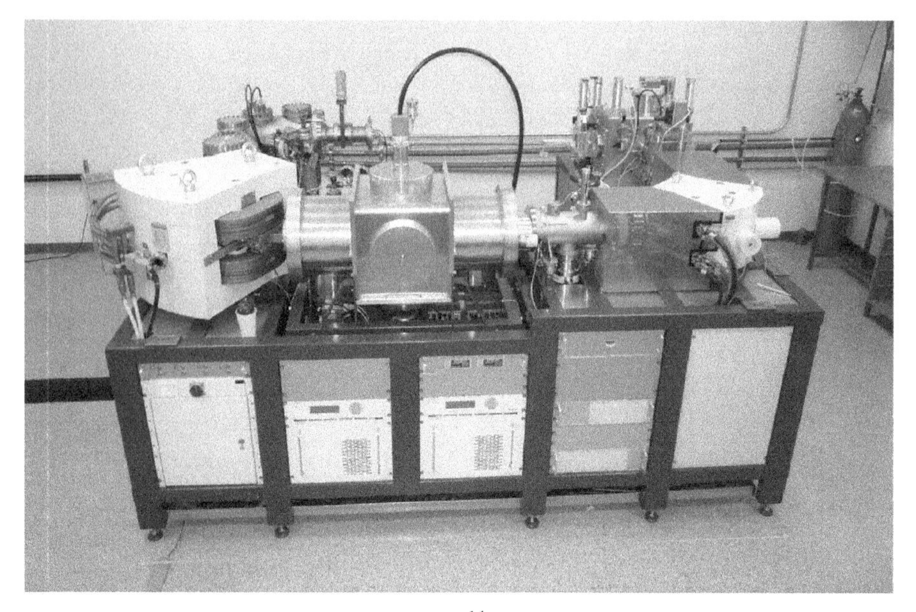

FIGURE 16.2 Image of Bio-MICADAS ^{14}C spectrometer at Vitalea Science. The instrument is 3 m on a side.

TABLE 16.1 Validated Performance of Bio-MICADAS

	^{14}C Spectrometer
Specificity	1:10^{14}
Robustness	2.5%
Reproducibility	<1% single sample
Precision	1.5% multiple samples
Stability	<3% per year
LLOQ	0.01 Modern
ULOQ	100 Modern
Accuracy/linearity	0.99 ± 0.01
Throughput	150 samples/day

multiple measures of a single sample to less than 1%. Precision for one material prepared into multiple samples over a period of time is 1.5% in routine operation. The isotope decay is an insignificant fraction of the isotope concentration for ^{14}C over many years, and no biological, chemical, or physical (short of nuclear irradiation) effects change the isotope ratio of a defined sample. AMS quantitation of well-mixed stored samples shows stability of <3% over long periods and multiple freeze–thaw cycles. The Vitalea spectrometer is validated over a 10^4 range from 0.01 Modern (approximately 1 amol ^{14}C per sample) to 100 Modern (10 fmol ^{14}C per sample). Over this validated range, AMS is linear and accurate to 1%. Throughput depends on multiple choices in performing a study (compound's label concentration, dose, etc.), the resolution of metabolite isolation, operating characteristics of the ion source, desired precision, and the like. With automated operation requiring little or no personal attendance, 150 or more samples are quantified per day. Finally, AMS is sufficiently robust and reliable that a series of "suitability" tests to assure quality in precision, accuracy, and limits of quantitation are only needed weekly or less often.

16.4 SAMPLE DEFINITION AND INTERFACES

AMS quantifies the isotope ratio of a homogenized discrete sample. Homogenizatation for ^{14}C analysis occurs at the combustion of sample material to carbon dioxide. This combustion is also the "definition" of the sample: All the combusted carbon becomes equivalent and indistinguishable. Samples and standards are presented in a uniform chemical form that efficiently produces negative carbon ions, so that all carbon in all samples is equivalently ionized and measured. The most common form for this material is a fullerene, which is rapidly and inexpensively reduced from carbon dioxide over an iron group catalyst (Vogel et al., 1984, 1987; Vogel, 1992; Ognibene et al., 2003). The process for reducing CO_2 to a fullerence is designed to be performed in "batch" mode with the throughput limited only by the capacities of the batch combustion furnace(s) and the reduction heater(s). The reduction process is robust against exact temperatures and catalyst ratios to precisions of 1% desired in biological measurements and can be pushed to even higher precision (Xu et al., 2007; Kim et al., 2009). Ognibene et al.'s (2003) refinement of the method into septa-seal reduction vessels enhanced throughput and eliminated chances of sample cross contamination by using inexpensive individual gas transfer manifolds produced from rapidly assembled Luer-style components. An operator can convert over >100 samples in a shift, working in continuous mode, so that samples combusted one day are reduced the following day while another set is combusted. This fullerene product has long been recognized as giving the highest ion output efficiency in AMS (Bronk Ramsey and Hedges, 1997; Fallon et al., 2007), producing the most rapid measurements for high throughput. The nonvolatile material protects the laboratory and spectrometer from sample cross contamination. The fullerene is known stable within a percent over at least a 20-year period, as shown by remeasuring long archived fullerene samples. This extreme stability in the quantified material uncouples the analytical production and definition of samples from the actual quantitation by the spectrometer, freeing both the AMS and the chromatography for maximum independent scheduling.

Direct feed of CO_2 to a negative ion source was demonstrated in 1984 (Middleton, 1984), but subsequent attempts to use this gas in carbon dating (Bronk Ramsey and Hedges, 1997) and in biochemistry (Liberman et al., 2004) suffered from reduced spectrometer throughput or decreased precision, to the point that CO_2 is seldom used in carbon spectrometers except for particular systems designed for molecule-specific gas chromatography–AMS (GC–AMS) (Flarakos et al., 2008; McIntyre et al., 2009). The variable emission of ions from a gas-fed source modifies transmission of ions through the accelerator collision cell and requires acceleration of one or more stable isotopes for normalizing the ^{14}C counts. One chromatography spectrometer is better described as an accelerator-aided ion detector because an isotope abundance ratio is not derived from quantitatively equivalent ion trajectories. This spectrometer achieves precisions of 20–30% and accuracies of 15–20% through interpolated response calibrations over changing baselines (Liberman et al., 2007; Flarakos et al., 2008) but may require analyte-specific internal standards to approach the performance of isotope ratio AMS. The calculations and characteristics of AMS in this chapter do not pertain to this instrument.

Research to link chromatography directly to AMS continues at several institutions (Liberman et al., 2007; Ognibene et al., 2007; McIntyre et al., 2009) and has great analytical appeal because chromatographic separations are continuous processes whose resolution is potentially diminished by collection of discrete time fractions. Better quantitative precision arises now from defined fractions whose isotope abundance is fixed by combustion and has no bleed effects from volumetric movement of fluids or gases. No "baseline" effects are seen for solid-sample AMS and the discrete fullerene samples do not cross contaminate. While direct LC connection is a goal, we show here that batch-produced chromatography samples from isolated fractions presents comparatively little loss in metabolite resolution.

16.5 AMS QUANTITATION

AMS measures the ratio of all ^{14}C to total carbon in the sample, which is then compared to the same ratio from an external standard prepared and measured in identical circumstances. The most commonly used external standard is SRM-4990-C, oxalic acid ^{14}C reference standard from the U.S. National Institute of Standards and Technology (NIST) at a raw isotope ratio of 1.1361 Modern with a ^{13}C content of 1.106%. Other international standards are available from the International Atomic Energy Agency (Vienna). The Center for AMS at Lawrence Livermore National Laboratory (Livermore, California) has generated secondary standards of oxalic acid at 12 and 100 Modern.

Vogel and Love (2005) provide a tutorial on the contributions and quantifications of AMS results. Briefly, the ratio of all the ^{14}C to all the carbon in the presented material is expressed as

$$R_s = \frac{^{14}C_n + {}^{14}C_t + {}^{14}C_c + {}^{14}C_x}{C_n + C_t + C_c + C_x} \tag{16.2}$$

Both the ^{14}C and the carbon that provide a sample's isotope ratio (R_s) arise from multiple sources (n, natural; t, tracer; c, carrier; x, unknown or "contamination") that must be understood or estimated for accurate AMS quantitation. The amount of tracer ^{14}C used for radioactive tracing was so much greater than the natural content that the latter did not need consideration in decay counting. Low-dose pharmacokinetic tracing with AMS uses whole isotopic doses that are equal or only a few factors higher than the natural human content so that the previously ignorable sources of ^{14}C become important in calculating net tracer (Garner et al., 2002). All non-specimen-derived carbon in the sample must be recorded and reported as essential "process knowledge" throughout a study from clinic to analysis. Again, this differs from radioactive measurements because the stable carbon content was of no consequence unless its form and concentration impeded the collection and counting of the scintillation flashes from decay products. Too often, contamination by materials or surfaces elevated in ^{14}C are considered and emphasized, to the oversight of ^{14}C-free "dead" forms of carbon, which also contribute contamination, C_x, that renders the measured ratio inconclusive. These "unknown" contributions are often intentional additions that enter the sample as anticoagulants or solubility aids.

$$[^{14}C_t] = (R_s - R_n)[C_n] \tag{16.3}$$

Assuming no added carrier carbon or unknown contributions and that the tracer compound itself adds insignificant carbon, the tracer concentration $[^{14}C_t]$ is obtained in moles per milliliter from the predose sample ratio, R_n, the measured sample ratio, R_s, and the carbon concentration in milligram/milliliter of the measured fluid, $[C_n]$, as shown in Eq. (16.3). The same equation holds true for tracer concentrations in tissues, precipitated proteins, cell isolates, and so on, where the natural carbon concentration, $[C_n]$, is expressed as the carbon content of the wet or dry isolate, whichever is relevant, providing moles of tracer per gram of specimen. The particular volume or mass of the sample is not important since the isotopic concentration provides the tracer, hence tracee, concentration.

The result is expressed in any units of the ^{14}C: C ratio, including parts per trillion. The international unit of ^{14}C concentration is Modern (the units of R), which corresponds to 1.180 parts ^{14}C per trillion parts carbon, or one of the equivalent numerical definitions shown in Table 16.2. The mole equivalent

TABLE 16.2 Equivalent Definitions of Modern

Modern	Units
0.01356	dpm/mg C
6.11	fCi/mg C
0.814	Bq/mg C
97.9	amol/mg C

definition of Modern is favored because AMS measures concentrations of atoms rather than decays as implied by the units of radioactivity. The approximation of Modern at 100 amol ^{14}C per milligram carbon eases rough mental calculations of isotope ratios. ^{14}C tracer concentration is converted to moles of the traced compound using the fraction of the compound's molecules that are ^{14}C labeled, the "isomolar" fraction, which is the direct counting equivalent of the decay counting "specific activity" of the compound. The isomolar fraction is readily expressed, with many compounds labeled to an isomolar fraction of 0.8 mol ^{14}C/mol compound (50 Ci/mol, 1 mol of $^{14}C = 62.4Ci$). Familiar units such as gram-equivalents per milliliter are then available by multiplication by the molecular weight. Reporting the AMS analysis as moles ^{14}C per milliliter or milligram allows the analytical calculations to be standardized until a final calculation introduces the particular isomolar fraction and molecular weight of the compound under study. This allows the analysts to expect similar AMS quantitation limits for every study, since AMS limit of detection (LOD) and lower limit of quantitation (LLOQ) of ^{14}C do not depend on the compound under study, its label level, or its molecular weight.

The LLOQ for a tracer concentration from Eq. (16.3) is simply six times the standard deviation (SD) in natural background, following the FDA Guidance on Bioanalytical Analysis (FDA, 2001). This natural background is best defined as the predose samples in a clinical cohort. A cohort of over 100 volunteers in one study had a coefficient of variance (CV) of 1.2% around the near Modern predose level, giving an LLOQ of 7.2% above the natural ^{14}C concentration (Keck, et al. 2010). Ten percent Modern above natural ^{14}C is a good approximation for rough LLOQ estimates for biological fluids and tissue samples. This 10 amol/mg C (10% of 100 amol/mg C) is about 0.4 pM ^{14}C in plasma or 0.1 pM in urine, depending on the urine's carbon concentration.

Biochemical isolates may have too little natural carbon for efficient manipulation through combustion, reduction, and measurement. These samples are bulked by addition of a known amount of carrier carbon. Natural ^{14}C and the carrier ^{14}C contents are inserted into Eq. (16.3) as multiples of their measurable isotope ratios and their masses of carbon in the mixed sample, resulting in Eq. (16.4) for the amount of ^{14}C tracer in the measured isolate:

$$^{14}C_t = (R_s - R_c)C_c + (R_s - R_n)C_n \tag{16.4}$$

Liquid chromatography fractions are quantified using carrier carbon and report amount of ^{14}C per fraction. The fractions are collected as eluates in solvents derived from ^{14}C-free petrochemicals, with small amounts of natural solutes and the labeled compound(s) of interest. High-performance liquid chromatography (HPLC) fractions often have milliliter volumes, from which tens of microliters are pipetted as the fraction sample (Creek et al., 1994) UPLC sepraration is well suited to high-throughput AMS because the solvent volumes are much smaller than from HPLC flow rates. The solvents are evaporated just to dryness in a clean vacuum centrifuge, leaving at most tens of micrograms of carbon in the residue. (This drying process is found to be the most common source of contamination in AMS since stringent precautions are needed to prevent backstreaming of contamination from the vacuum pump and solvent condensers. Aspiration pumps with refrigerated water have the least problems.) The small amount of sample carbon is bulked with about 0.5–1 mg carbon carrier compound prior to combustion and reduction. High sensitivity and low quantitation limits suggest that a ^{14}C-free carrier compound would be best, but one that is around 10–20% Modern in ^{14}C permits rapid determination of LLOQ from a set of quantified blanks, as well as alerting the analyst to the presence of a solvent component (such as acetate buffer) that did not fully evaporate, confounding quantitation (Miyashita et al., 2001) One or two microliters of a low-vapor-pressure, nitrogen-free fluid, such as tributyrin, are easily added to sample vials as filled capillary pipettes, providing highly reproducible ($\pm 2\%$) carrier additions.

Equation (16.4) converts the isotope ratios to an amount (e.g. moles) of tracer eluted per fraction or per unit elution time. The ^{14}C molar concentrations for fraction contents are obtained by dividing by the injected sample-equivalent volume corrected for quantitative recovery discussed below. The fractions' natural carbon contents, C_n, are approximated by apportioning the injected sample fluid's carbon content over the total collected fractions. For example, plasma proteins constitute about 70 mg/mL of human plasma, or about 37 mg C/mL, given 0.53 mg C/mg protein (Rouwenhorst et al., 1991), while human plasma contains 43 mg C/mL, leaving 6 mg C/mL in the protein-depleted fluid used for metabolic studies of plasma. A 50-μL aliquot of this isolate has 300 μg of carbon that would be bulked with carrier for a direct measure of tracer applied to the column. The quantity of tracer is found from Eq. (16.4) with R_s being the measured sample ratio in Modern, R_c and C_c being the added carrier ratio and mass (e.g., 0.1 Modern and 620 μg), while R_n is about 1.05 Modern (the predose level) with C_n being the 300 μg from protein-depleted human plasma. Carbon content measurements are an important factor in making high-precision AMS quantitations. These fraction concentrations are converted to gram equivalents of traced compound using the particular isomolar fraction and the molecular weight of the traced compound.

A pair of cross-validated Waters Acuity UPLC instruments (Waters Corp., Milford MA) are used at Vitalea Science for metabolic analyses and have very high reproducibility. Figure 16.3 shows AMS quantitation for a single metabolite peak from urine separated on different days. AMS provides an uncertainty estimate for each measurement based on the number of recorded ^{14}C atoms ($CV = 100/\sqrt{N}$), and multiple measures (>3) are made for each fraction providing a normal standard deviation that often equals the counting statistics. There should be two of the six data replicates that do not overlap at the 1σ uncertainties; but there is only one, and that one occurs at the sharply rising start of the peak. Thus, these UPLCs appear reproducible with AMS quantitation to "better than statistical" accuracy. The LLOQ of chromatographic fractions is discussed in detail below.

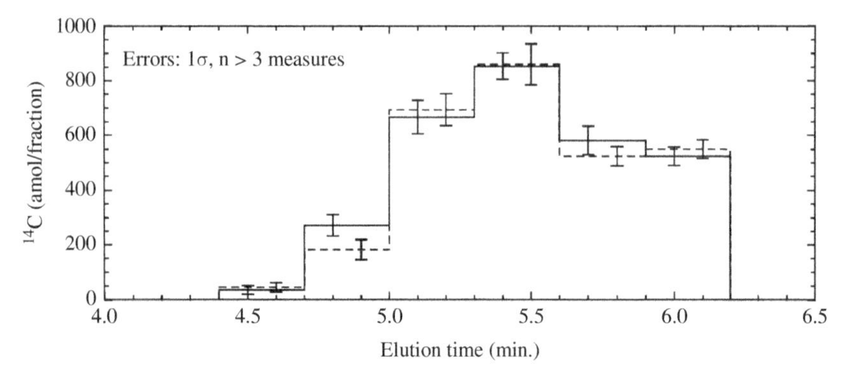

FIGURE 16.3 A Test of UPLC–AMS quantitative reproducibility shows a metabolite peak in urine samples separated by UPLC and measured by AMS on different days. Error bars represent the AMS counting uncertainty at 1σ and should overlap only at 4 of the 6 points. The only nonoverlap occurs at the rising edge of the peak.

AMS quantification of concentrations or contents for ^{14}C-labeled compounds using Eqs. (16.3) and (16.4) are absolute. There are no internal reference standards in AMS quantifications. The AMS ratios for sample, background, and carrier are determined to almost any desired precision by extended counting. Accuracy of the ratios is assured by normalization to external reference standards. All other components of data transformation (carrier mass, isomolar fraction, molecular weight, etc.) are externally measured and assured independently of the sample being quantified by AMS. This universal precision and accuracy is particularly useful for assaying drug metabolism because there are no quantitative surprises even if unexpected metabolites appear in one or more species during drug development and clinical trials. AMS is a particularly robust platform for the quantitation of metabolism in drug development.

16.6 LC–AMS ANALYSIS OF DRUG METABOLITES

This section reviews and defines for this chapter how discrete time fractions of ^{14}C-labeled metabolic traces are presented and analyzed. Chromatographic parameters originated for continuous detectors in which peaks areas represent the mass or concentration of the eluate. This is maintained in fraction-collected quantitation using a "step" plot such that equal areas on linear axes represent the quantified signal.(e.g., quantity of ^{14}C) within the fraction collected, as shown in Figure 16.4. The integral [area under the curve (AUC)] of the plot or portion of the plot must then be equal to the total signal over the time period integrated. This specification requires the y axis to be "quantity per unit time," often represented as "^{14}C per fraction," if the fraction collection times are equal. Either fixed or variable fraction widths are represented by plotting the response per unit elution time versus that elution time. The data in Figure 16.4

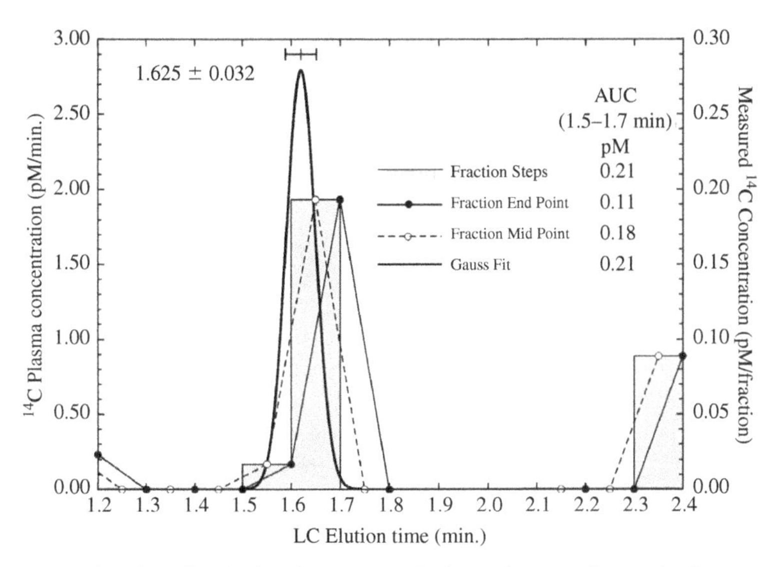

FIGURE 16.4 Plot of an isolated UPLC peak shown in step-plot, endpoint connected plot, midpoint connected plot. Point plotting shows signal area where none occurred and produces incorrect integrals over defined times. The Gaussian function was fit to the data requiring an area equal to the step-plot rectangles.

was collected in 0.1-min fractions, and the ^{14}C content in fractions was converted to concentration in plasma using the known injection volume of protein-depleted plasma. The data was normalized to eluted concentration per minute (left axis) so that integration of area results in picomolar ^{14}C in plasma for the integrated peak. Note that units of radioactivity (Bq, Ci, dpm, etc.) are units of isotope content, as defined in Eq. (16.1) and are completely equivalent to moles or milligrams of ^{14}C. A fraction will have 1 dpm whether it is counted for 10, 100, or 1000 min and whether the fraction is collected over 0.1, 1, or 10 min of LC elution. Thus, 1 min fractions reported in units of dpm should plot "dpm per fraction" or "dpm per minute" (dpmpm?) to fulfill the tradition of chromatography for quantitative area.

Point-to-point plots do not maintain the integrity of area as a measure of analyte quantity or mass. These plots transfer area into adjacent fractions at the expense of the resolved fraction peak, as seen by the point plots in Figure 16.4, which show the commonly applied endpoint trace (light solid line) and the more correct midpoint line trace (dashed line). Integrals (AUCs) over short periods of point-represented LC fractions can underestimate the fraction areas by as much as 50% and displace the visual peak location. The step plots in this chapter were created in pro-Fit for Macintosh (www.quantsoft.com), but step plots are an option within plotting programs such as Dplot, (www.dplot.com), Kaleidagraph (www.synergy.com), and Grapher (www.ssg-surfer.com), as well as being part of statistical packages such as JMP (www.jmp.com) and SAS-Graph (www.sas.com).

$$G(X) = Ae^{-(X-X_0)^2/2\sigma^2} \tag{16.5}$$

$$\int_{-\infty}^{\infty} G(X) = \mathrm{AUC(peak)} = \sum \frac{^{14}\mathrm{C}}{\mathrm{fraction}} = A\sigma\sqrt{2\pi} \tag{16.6}$$

$$G(X) = \mathrm{AUC}e^{-(X-X_0)^2/2\sigma^2} \Big/ \sigma\sqrt{2\pi} \tag{16.7}$$

Liquid chromatography peaks are nearly Gaussian, to a few parts per 100 at least, due to the high multiples of eluate adhesion and re-solution between the solid and liquid phases through the column, with greater or lesser tailing toward longer retention. The location and quantitation of even a single isolated fraction is sufficient to completely define a Gaussian elution shape, although such manipulations are anathema to dedicated chromatography scientists. A Gaussian peak [Eq. (16.5)] was fit to the 2 fraction data in Figure 16.4 using the precisely quantified total ^{14}C content (area) of the two fractions to constrain the height and width of the shape using Eq. (16.6). The Gaussian fit then solves only for the width, σ, and the centroid, x_0, as in Eq. (16.7). The amplitude A, the centroid, and the width are tightly constrained by the requirement that the Gaussian area exactly fit and maximally overlap the fraction rectangle areas. The fit is shown as the the heavy line in Figure 16.4, and the integral of the function between 1.5 and 1.7 min is 97.5% of the quantified ^{14}C, with a slight error in the tail beyond 1.7 min. Note that this approach more precisely defines the peak elution time (1.63 ± 0.03 min), with the small fraction from 1.5 to 1.6 min helping to refine the peak location so that better quantitative alignments with LC–MS/MS identification can be obtained even from fraction-collected data.

Figure 16.5 quantifies metabolites of ^{14}C-testosterone in excreta of Japanese quail in a chromatogram containing 92 fractions of 1 min each (B. Faulkner, personal communication). A log scale expansion of the fractions' ^{14}C content is also shown for details at low levels.

The LLOQ ($10 \times \sigma_{\mathrm{bgnd}}$) for this data set was 8 amol (10^{-18} mol) per fraction (dotted line in the log trace) with a dynamic range of at least 30:1. Precision of better than 3% is obtained across the applicable dynamic range, and all metabolites are quantifiable to equivalent precisions. Gaussian functions were fit to 8 peaks in which at least 3 fractions defined the peak location and width. The Gaussian area constraint as described above was used even for overlapping peaks seen at early elution times. The integral (AUC) of the fitted multipeak function accounted for 96% of the ^{14}C in the covered fractions. These 8 peaks had an average standard deviation of under 0.7 min for an average peak full width at half maximum (WHM) of 1.5 min, only 50% wider than the fraction width. Resolution for Gaussian peaks is defined as the time difference between peak centers divided by the average tangential width of the peaks [Eq. (16.8)]. For true Gaussians, this tangential width is 4 times the standard deviation, σ. We define a fraction resolution (R_f) equal to the

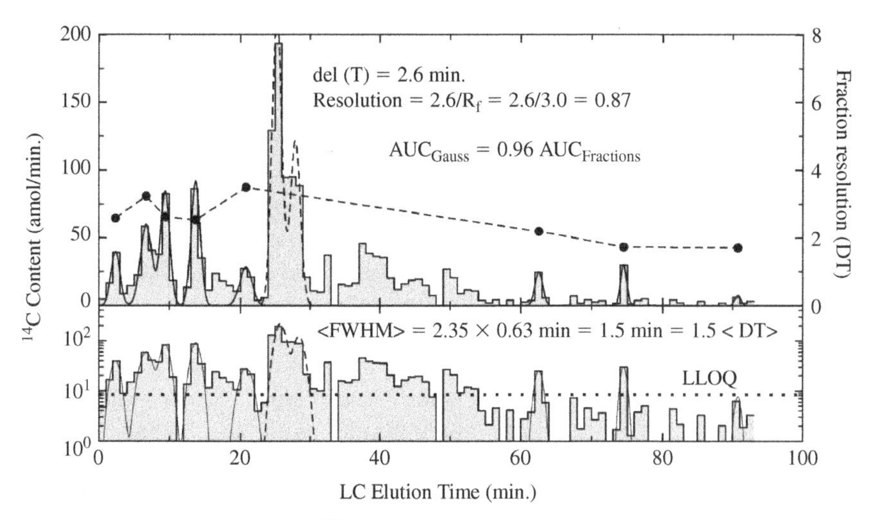

FIGURE 16.5 Metabolites of ^{14}C testosterone in Japanese quail show the resolution of LC–AMS used in defining the optimal metabolite for developing a field enzyme-linked immunosorbent assay (ELISA) test. Fractional resolution is shown for 8 Gaussian-fitted peaks. Two unresolved peaks have dashed Gaussian fits that predict a 2.6-min separation for a resolution of 0.93.

tangential width of a Gaussian peak, 4σ, divided by the fraction's time width or dwell time (DT) of the quantified fractions. This fraction resolution is plotted in Figure 16.5 referring and to the right y axis. Since the width of the Gaussian is strongly constrained by the area quantitation and the dwell time, this fraction resolution need not be much larger than unity for "peaks" in a fraction step plot to represent resolved Gaussians.

$$\text{Res} = \frac{0.5 \times (W_2^t + W_1^t)}{(t_2 - t_1)} \tag{16.8}$$

$$R_f = \frac{4\sigma}{\text{DT}} \tag{16.9}$$

A double Gaussian was fit to the unresolved peaks at 28 min which are defined by 7 fractions using the average peak width, σ, of 0.7 min (fraction resolution of 2.8 fractions), shown as a thin dashed line in Figure 16.5. The fit shows a 2.6-min separation between the component peaks for a resolution of 0.93 ($\Delta t/R_f$), which means that the two peaks are almost resolved, although there are no low fractions between the centroids. Despite apparently coarse fraction collections, precise quantitation of the area of fractions produces width and location estimates based on the expected Gaussian shape of elution peaks. Such finer estimates are not needed in comparative metabolism studies but may assist in linking quantification by LC–AMS with identification by LC–MS/MS.

The resolution of fraction collection can be dynamically controlled to lower sample number and costs of measurement. Variable width fractions are possible with programmable fraction collectors. Alternatively, high time resolution fractions are collected in well plates, and equivolumes of fractions are combined over the wells expected to have low-resolution features. Elution periods needing higher resolution are available after initial measurements using the unused portions of the fine fraction wells. Alternatively, a repeat separation provides higher detail by normalizing finer fraction quantitations through equivalent fractions adjoining the expanded resolution. This capability depends on the high reproducibility shown in Figure 16.3 and provides the analyst with the freedom to adjust fraction resolution (and AMS load) as needed.

UPLC offers advantages over HPLC for AMS quantitation. The delivery of compound (parent or metabolite) occurs in a small volume. For example, a 12 s peak at a flow rate of 600 μL per minute is contained with an eluent volume of only 120 μL. This provides compatibility with automated well plate collection, a minimal amount of extraneous solvent carbon, and an optimum compound-to-eluent ratio. Second, UPLC column recovery is generally better than for HPLC, and it is often possible to get 95–100% of the radiocarbon through the column. This is difficult to compare to HPLC separation, as this parameter is often not assessed or reported, but our experience indicates an 80–90% recovery to be realistic for HPLC. Quantified high column recovery assures that no unknown metabolites have been lost in separation. The high throughput available from short elution times of UPLC is not accompanied by a loss peak resolution, as seen in the next section.

16.7 COMPARATIVE RESOLUTION OF FRACTION LC MEASUREMENTS

^{14}C-labeled compounds and metabolites are quantified in fractions after chromatography, with the traditional method being liquid scintillation counting (LSC) of the fractions. Flow scintillation detectors for LC appear to produce continuous output, but the finite volume of the scintillating chamber imposes a limit on temporal resolution due to flow rates and peak mixing in the detection volume. Dwell time is more explicit in an LC–ARC instrument in which stop-flow LC is used to count the radioisotopic decays in essentially "isolated" flow fractions to higher precision than in continuous-flow counters (Nassar et al., 2003). Another counting technology, Top-Count, places scintillating material at the bottom of fraction-collecting well plates. These count methods are limited to high concentrations of the isotope label and/or to large fraction volumes, which are available from HPLC. We compare the fraction resolutions of these technologies to LC–AMS results and to our use of UPLC–AMS in this section.

Figure 16.6 compares six technologies for high-sensitivity metabolic chromatography using ^{14}C molecular labels. Data plots for HPLC–ARC of human fecal metabolites (Colizza et al., 2007), HPLC–LSC and HPLC–Top-Count of human liver microsome metabolites (Zhu et al., 2005), UPLC–RAM

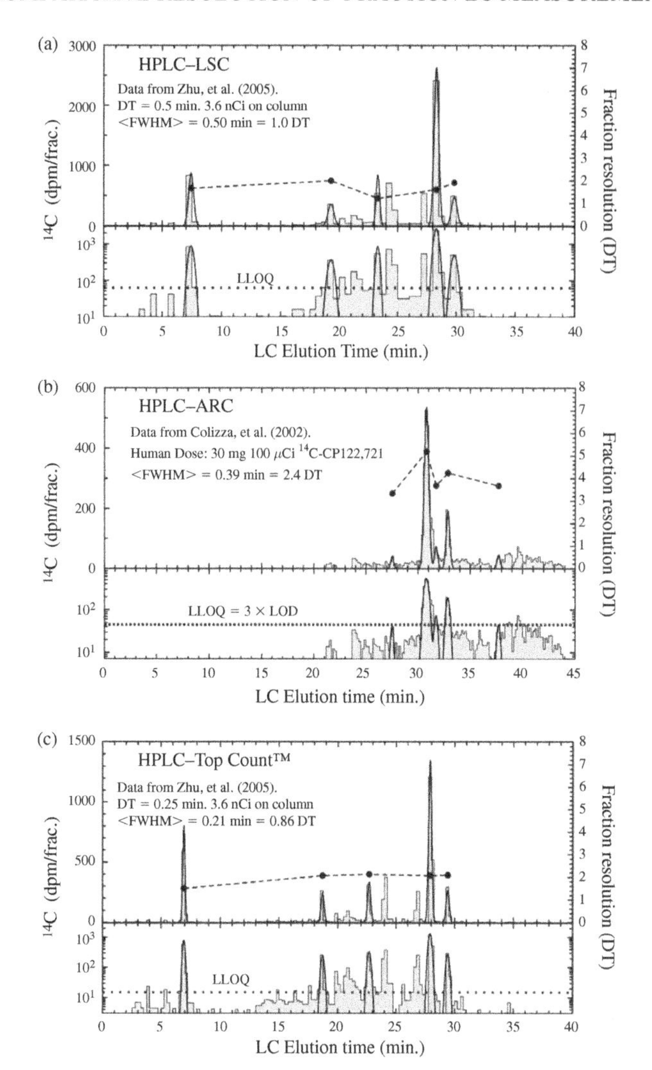

FIGURE 16.6 Equivalently plotted and analyzed results form ^{14}C-metabolite LC analyses are shown for: (*a*) HPLC–LSC counted fractions of liver microsome metabolites with a dwell time (DT) of 30 s over a 40-min elution; (*b*) HPLC–ARC stop-flow decay counting of fecal extract metabolites with a DT of 10 s counted for 60 s over a 45-min elution; (*c*) Top-Count well plate counting of HPLC fractions of liver microsome metabolites with a DT of 15 s over a 40-min elution; (*d*) UHPLC–RAM continuous decay counting of dog urine metabolites from paired columns through a 200-μL detector at flows of 1 mL/min over 28-min; (*e*) HPLC–AMS fractions of plasma with a DT of 60 s over a 106-min elution; and (*f*) UPLC–AMS quantitation of fecal extract metabolites with variable DT down to 6 s over a 10-minute elution. LLOQ's are noted as dotted lines in log-scaled frames where available. The average FWHM of the fitted peaks are given as a multiplier of DT to indicate peak shape resolution. Peak resolution (tangential width) is divided by dwell time and plotted versus elution time as solid points connected by dashed lines referring to the right *y* axis. Dose sizes and dosed ^{14}C are given where known. Amounts of ^{14}C placed on column are given where known.

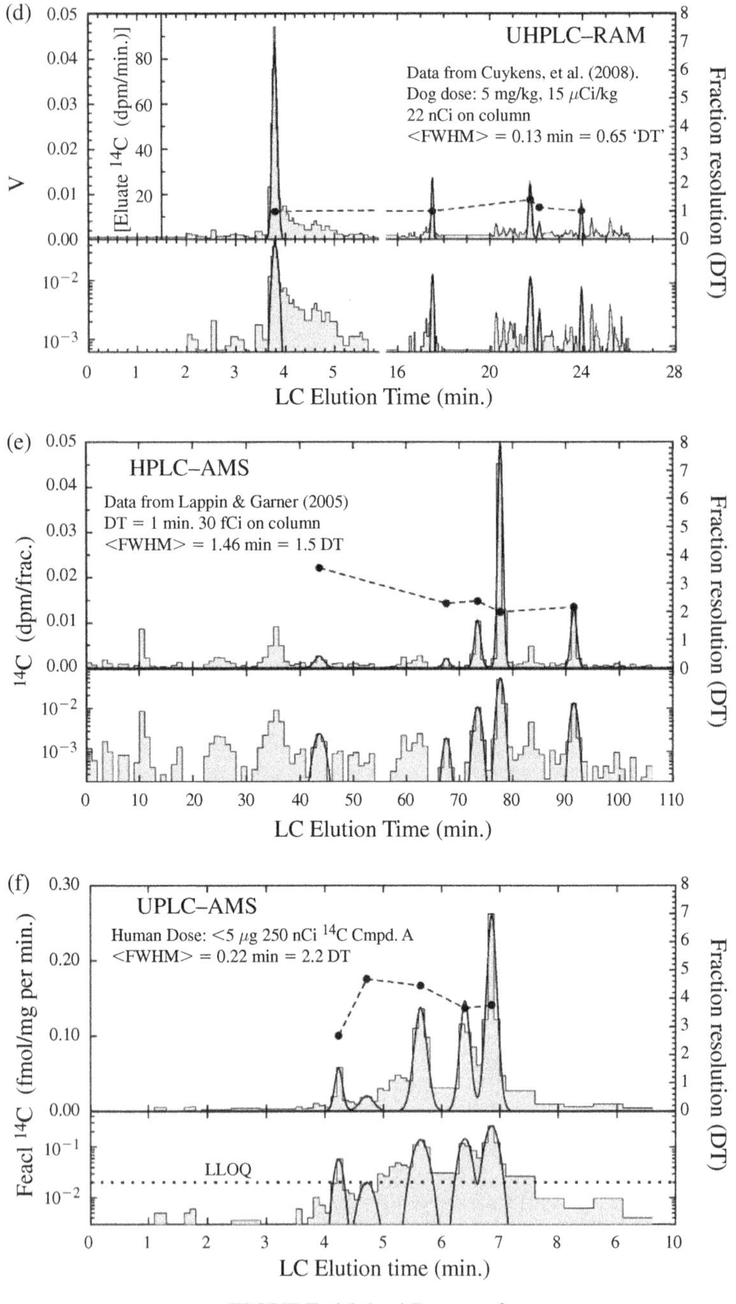

FIGURE 16.6 (*Continued*)

(radioactive monitor) of dog urine (Cuyckens et al., 2008), and HPLC–AMS of human plasma metabolites (Lappin and Stevens, 2008) were digitized to extract data for completely equivalent presentation and analysis along with UPLC–AMS data. Data are plotted as step plots of ^{14}C concentration per unit time (dpm per fraction for Figs. 16.6*a*, *b*, *c*, and *e*) using the dwell times per fraction stated in the publications. UPLC–RAM does not have stationary dwell time and was originally published with the voltage, V, from the radioactivity monitor's count rate circuit. A nominal dwell time of 0.2 min was defined as the active counting volume (200 μL) divided by the flow rate (1 mL/min). Total injection load and response integral were used to approximate a decay rate per elution minute shown on the inset left axis. UPLC–RAM requires very high levels of ^{14}C but is included for peak separation comparisons. Only the UPLC–AMS data used variable-width fractions, which are plotted as fecal ^{14}C concentration per unit elution time. Each plot in Figure 16.6 has a frame with the data on a linear scale, as commonly expected, and a lower frame expanded on a log scale to show the finer detail available from these high-precision counting methods and to allow visible plotting of the LLOQ as a heavy dotted line when available. AMS shares with TopCount the need to dry the sample, with potential loss of volatile components. These issues have been addressed in studies of animal metabolism and distribution of volatile toxins such as benzene (Creek et al. 1997), (Mani et al., 1999) and trichloroethylene (Kautiainen et al., 1997) but have not yet been fully addressed for metabolic separations.

Gaussian peaks were fitted to five definable peaks for each technique using a Levenberg–Marquadt algorithm after initial visual adjustment to center the fit and to approximate the proper width, assuming a 3% uncertainty in quantitation and a 10% DT uncertainty in fraction width. Without initial visual adjustment, the algorithm is unable to estimate parameter uncertainties with such minimal data (sometimes only the peak fraction and two side fractions). Note that software easily fits the sum of peaks and does not rely on interpolated peak baselines, adding the appropriate amount of each peak to a shared fraction when the tails overlap. Concepts such as "tangential width" arose in a time when the best analysis of a chromatogram consisted of drawing straight lines to the axis from the sides of peaks and then cutting out the peak for weighing. Such methods were not amenable to proportionate summing of incompletely resolved peaks. The sum response of five peaks were fit in each trace using Eq. (16.7) to force the summed peaks to have the same quantitative area as the quantified fractions. This summation corrects for overlapping peaks, as seen in Figure 16.6*b* at 32 min and Figure 16.6*f* at 6.5 min. Fitted Gaussians gave complete parameter specification, despite the sparse data, and provided peak width error estimates of ± 5–30% (CV) on individual peaks. Fitted peaks contained 96–102% of the quantified fraction contents. Quality of fits for even the very small peaks are visible in the log scale frames.

The average peak width was found for each trace and the average FWHM (2.35σ) is shown in each plot as a multiplier of the dwell time. The higher this

TABLE 16.3 Summary of Fitted Peak Widths and Fraction Resolution Derived from ^{14}C-LC Techniques Shown in Figure 16.6

	Peak Width (min)	DT (min)	R_f (DT)	Total Elution (min)
HPLC–LSC	0.209 ± 0.038	1	1.67 ± 0.31	40
HPLC–ARC	0.168 ± 0.030	0.16	4.03 ± 0.73	45
LC–Top Count	0.123 ± 0.016	0.25	1.97 ± 0.26	40
UHPLC–RAM	0.056 ± 0.009	0.2	1.11 ± 0.18	28
HPLC–AMS	0.62 ± 0.15	1	2.47 ± 0.62	106
UPLC–AMS	0.096 ± 0.020	0.1	3.84 ± 0.79	10

multiplier, the more the fitted peak width reflects the chromatographic width rather than the dwell time. The fine time resolution of the LC–ARC system gave the highest FWHM of 2.4 times DT, reflecting the 270:1 ratio of the total elution (TE) to dwell time. AMS uses 50:1 up to 100:1 elution times versus DT to avoid having more than 100 fractions per trace, providing FWHM of 1.5–2.2 DT. LC–TopCount had a TE-to-DT ratio of 160:1 but had a FWHM 0.86 DT, reflecting very fine chromatographic separation of peaks for that particular analyte and method as well as emphasizing the quantitative value of precise peak area measurement in determining peak width. UHPLC–RAM also had peak FWHM, less than the defined DT, supporting the very narrow peak demonstrated in that work using LC–MS/MS detection (Cuyckens et al., 2008). The LSC data unsurprisingly had the coarsest relative resolution with an TE-to-DT ratio of 80:1. Prakash et al. (2007) show well-resolved peaks using an unconventional accelerator system, but the quantitative comparison of the diluted urine measured by accelerator agrees with the undiluted measurements by HPLC–RAM only to 50% root-mean-square (rms) accuracy. Without the precise quantitation of conventional AMS, this comparison of forced Gaussian areas fails and was not further analyzed.

The fraction resolution, R_f, is derived from each peak width using Eq. (16.9) and plotted against the right-hand linear axis. Average peak widths and fraction resolution for the six technologies are in Table 16.2, along with the total elution time. This analysis shows that peaks can be resolved at tangential width in only two or three fractions, as observed between the resolved peaks in Figure 16.6*a* at 28.5 and 30 min and in Figure 16.6*c* at 26 and 28 min. UPLC–AMS has equivalent peak resolution to the HPLC–ARC system, both of which apply to fecal extractions that may have wider chromatographic peaks than the urine and microsome analytes of the other method examples. The tangential width resolution equals four fractions in these two techniques, suggesting that coarser fractions could be taken with no loss in peak quantitation. This high resolution is augmented in UPLC–AMS by chromatographic throughput that is increased by 4 or more, sample volumes reduced by 20, and column recovery that is often 100%, while the AMS sensitivity reduces chemical dose by 6000 and radiative dose by 400 from other examples of Figure 16.6. Thus, any fear that metabolite resolution might be lost in using UPLC–AMS is unfounded.

Accurate quantitative metabolism is now an important factor in drug development, given the FDA MIST requirements on metabolites reaching a certain threshold of concentration compared to parent or total compound. Relative metabolic profiles were sufficiently informative for medicinal chemistry investigations of pathways, but quantitative biological concentrations of metabolites are now available using complete quantitative analysis for high precision investigation of compound physiochemistry. Quantitative recovery analysis is measurable throughout the process of defining, isolating, and separating the matrix sample, providing biological concentrations as easily as the measured ^{14}C response after chromatography.

16.8 QUANTITATIVE EXTRACTION AND RECOVERY

Quantitative power (and present cost) of AMS is wasted unless processes are quantitative from the drawn specimen to measurement of the isolated metabolite. Quantification of losses throughout analytical preparation determines biological metabolite concentrations from final measured tracer concentrations, and isotopic labeling is the most direct method of obtaining reliable recovery factors. The first loss of analyte can occur the moment that a specimen is placed in a storage vessel. Storage recovery is determined with multiple samples of a known amount of labeled analyte spiked into matrices within the containers to be used. These spiked materials are measured after multiple freeze-thaw cycles to obtain a storage recovery factor that is usually in the range of 90–100% but may be much less for distinct hydrophobes. Alcohol or other organic solvents may be added to sample vials prior to specimen collection to improve recovery and are removed through evaporation once transferred to the final sample vial.

Whole blood is preferably collected into heparinzed tubes because low levels of natural heparin affect isotope ratios less than larger amounts of petrochemically derived ethylenediaminetetraacetic acid (EDTA). If whole blood is measured, an aliquot is frozen immediately after drawn for storage and shipping, requiring a complete thaw and homogenization prior to analysis. Alternatively, a spot of whole blood is dried into a glass filter "paper" for transport to the laboratory. A 15 to 20-μL aliquot is pipetted into a combustion sample tube. The specific volume of blood is not critical since the isotope ratio normalizes the sample size using the carbon content of blood. The sample is dried within a jacketing tube in a vacuum centrifuge. The dried sample is then combusted to CO_2 in a sealed quartz tube and reduced to a fullerene on an iron group catalyst in a septa-seal vial as described by Ognibene et al. (2003).

Plasma is separated from cellular contents by immediate centrifugation, and the cellular pellet may be retained for cell content determination. Aliquots of plasma separated by blood centrifugation at the clinical site are frozen at −70°C and shipped to the analysis laboratory. A 25-μL aliquot of the thawed vortexed plasma is pipetted to a sample combustion vial and reduced to carbon for AMS analysis as above. This provides the total ^{14}C measure of all compounds, metabolites, and bound fractions in the plasma to quantify the recovery in following analytical procedures. Again, the exact volume of quantified plasma need not be known since the tracer concentration is measured by AMS in reference to the known or measurable volume concentration plasma carbon. Predose plasma constitutes the background signal to be subtracted for label quantitation and the precision of multiple predose samples determines the LLOQ for plasma quantification.

Plasma proteins are precipitated using volumes of precipitation solvent separated by vortexing. The proteins are concentrated by centrifugation and an aliquot of the supernatant dried in the vacuum centrifuge. This sample is resuspended in appropriate solvent and filtered at 0.45 μm before injection on the LC. A second aliquot is placed into a sample combustion vial for separate direct measurement of ^{14}C content in the injectate to quantify recovery off the column. Tens of micrograms of the unlabeled parent compound and/or specific metabolites can be added during the protein precipitation or resuspension to act as ultraviolet (UV) markers of specific LC fractions. This is useful for a parent versus burden measure in which the parent fraction is collected separate form a pool of the remaining eluate and requires only 2 AMS analyses for each plasma sample. Alternatively, the captured parent concentration is subtracted directly from the plasma's total isotope concentration if quantitative column recovery is assumed. A portion of the protein pellet is also dried into a sample combustion vial and prepared as above for AMS measurement to assess quantitative recoveries and/or to measure protein binding. The protein carbon content of plasma is used with the measured tracer concentration in an arbitrary amount of the protein pellet to obtain the concentration of protein-bound ^{14}C concentration in plasma.

The number of AMS measurements per metabolite profile is reduced by using variable fraction widths without loss of quantitative yield since these fractions are still accurately quantified to reveal unexpected contributions of ^{14}C metabolites. A rat plasma metabolite profile with variable fraction widths is shown in Figure 16.7. A single fraction between 4 and 5 min quantified an unexpectedly high concentration whose proper relation to other quantified peaks is maintained by plotting ^{14}C tracer eluted per minute in the gray-shaded isometric step plot of Figure 16.7. This wide fraction contained 2.6% of the injected ^{14}C signal, as does the isolated peak in the next minute of elution, both delineated by double arrow lines in the lower trace of Figure 16.7. Localization of this metabolite is possible using finer fractions in a repeat measure, given the instrument reproducibility revealed in Figure 16.3.

Table 16.4a shows the quantitative recovery of the rat plasma preparation for the UPLC trace in Figure 16.7. Injected plasma supernatant was quantified with 87.6% of the ^{14}C concentration, 35.9 fmol ^{14}C per milliliter plasma (p*M*), of the initial plasma concentration 41 p*M*, while the precipitated protein pellet contained 13.4% of the original label concentration with 5.5 p*M* plasma equivalent concentration, showing 100% process retention prior to LC separation. The column recovery of this hydrophobic compound was only 85 ± 6%,

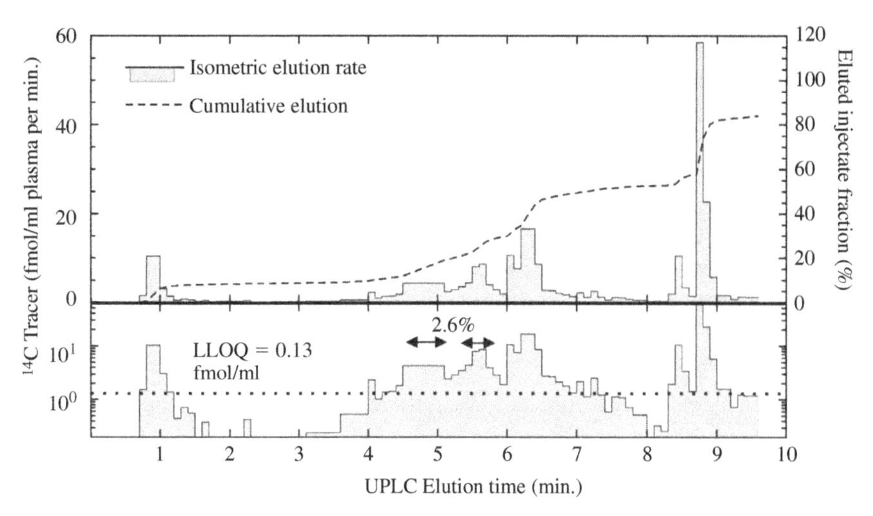

FIGURE 16.7 Quantitative UPLC trace of rat plasma metabolites is shown in femtomoles of ^{14}C label per milliliters (p*M*) plasma for variable width fractions. The concentrations are shown as an isometric step plot of concentration per elution minute that normalizes response over fraction widths. Cumulative column recovery (dashed line, right axis) is shown as a fraction of the injected ^{14}C quantity. LLOQ (dotted line) as 10 times the standard deviation of the fractions' carrier background is shown in the lower panel using a log axis for visibility.

TABLE 16.4 Quantitation of Analytical Recovery of a Hydrophobic Compound in (a) Rat plasma and a Hydrophilic Compound in (b) Human Plasma

	[C] (mg/ml)	R_s (Modern)	R_{bgnd} (Modern)	LLOQ (pM ^{14}C)	Tracer (pM^{14}C)	Fraction (%)
Plasma	32	14.2	1.079 ± 0.019	0.6	41.0	100
Pellet	26	3.2	1.079 ± 0.019	0.5	5.5	13
Injectate	6	15.2	0.096 ± 0.008	0.04	35.4	86
Column			0.135 ± 0.015	0.08	30.1 ± 2.0	85 ± 6
Human						
Plasma	43	2.91	1.061 ± 0.019	0.8	7.8	100
Pellet	37	1.06	1.061 ± 0.019	0.7	0	0
Injectate	6	3.73	0.096 ± 0.008	0.12	7.1	91
Column			0.096 ± 0.008	0.12	7.0 ± 0.1	98.5 ± 1.5

with uncertainty calculated from the quadrature summed uncertainties in the background for each fraction. Methanol (25 μL) was added to the collection wells prior or LC separation to decrease loss of nonpolar signal to well walls, but the 16% loss in process recovery was most likely surface losses from the very dilute eluate in the well plate. This can be avoided by lining the wells with tin cups from CHN analyzer supplies. The tin cups are then transferred completely to the combustion vial for AMS preparation. An overall correction to all circulating metabolic fractions of 1.18(1/85%) is not warranted since wall loss is probably not constant with polarity. The LLOQ for individual fractions (0.13 fmol ^{14}C per milliliter plasma equivalent per minute) is shown in the lower trace of Figure 16.7 as 10 times the uncertainty in the ^{14}C concentration found in multiple fractions of pre-injection mobile-phase applied to the volume equivalent of injected plasma.

Figure 16.8*a* shows the UPLC profile in linear and log scales from a pooled plasma set from humans 4 h after oral dosing with a hydrophilic compound. The analytical recoveries are summarized in Table 16.8, and the cumulative

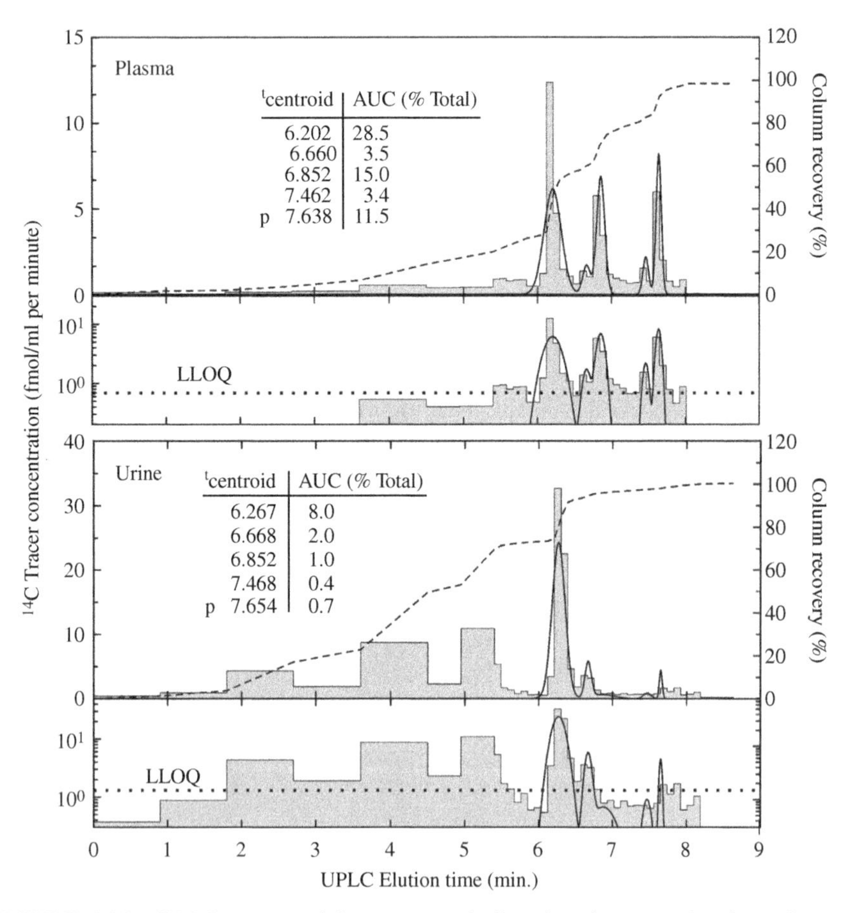

FIGURE 16.8 UPLC traces of human metabolites in plasma and urine after an exposure to a hydrophilic compound. The shaded step plot shows the isometric eluate per minute where equal areas represent equal concentrations of the isotopic label in the separate plasma and urine traces. The column recovery is shown in dashed line and the dotted line in the lower frame gives the LLOQ of individual fractions at 10 times background uncertainty. Centroids of metabolites and parent are tabulated.

recovery is plotted as a dashed line in the linear plot of response (upper frame, right axis). In this instance, there was no retention of the compound or metabolites in the plasma protein pellet. The UPLC injectate has a ^{14}C concentration that was 91% that of the unprocessed plasma. Again, variable-width fractions were quantified by AMS because polar eluates were less important. The column recovery adds up 100%. In this instance, one would correct the quantified metabolite profiles by a factor of 1.10 to account for the 9.4% loss in processing the plasma prior to LC separation.

Urine is prepared for LC following procedures much like plasma without a protein precipitation. A known volume of urine is dried and resuspended in LC mobile phase and filtered at 0.45 μm before aliquoting to a sample vial for direct measure and to the LC injector. Figure 16.8*b* shows a UPLC trace of urine from pooled 12-h collections of 6 volunteers who were dosed at 2–3 kBq of ^{14}C. The urine contains more polar components and essentially none of the parent compound, as expected. Column recovery of the injected signal was $100 \pm 2\%$ for this compound in urine.

Feces is the dominant excretion route for many hydrophobic compounds that undergo biliary excretion and enterohepatic recirculation. These compounds and metabolites present multiple analytical challenges, not the least of which are (a) maintaining overall mass balance through multiple procedures of dilution, extraction, and solubilization from the matrix; (b) obtaining quantitative removal from storage and shipping vessels; and (c) retaining solubility throughout the chromatography solvent phases and loss of very dilute nonpolar metabolite eluates to the collection well walls. The analytical procedures must address these issues for each compound separately and will depend heavily on experience from preclinical analyses. For example, Dueker et al. (2000) described the procedures for extracting beta carotene and metabolites from human excreta. Essentially, chemical extraction, perhaps with acidified acetonitrile or some suitable solvent must be coupled with physical dispersion of the matrix to obtain a fluid that can be sampled. Further chemical extractions may be required to release recalcitrant compounds from fibrous or fatty residues prior to centrifugation, supernatant collection, lyophilization, and recovery in the chromatographic mobile phase. Some laboratories promote lyophilization prior to extraction, but this method is seen as dangerous from a cross-contamination point of view. As with the vacuum centrifuge, caution is required to assure proper trapping of vapors. Additionally, the drying process could confound subsequent removal of the drug from the complex matrix.

Our data are presented as moles of isotopic tracter ^{14}C so that the LLOQs and ranges of quantitation are directly pertinent to future studies. The molar concentrations of tracer label are easily converted to hypothetical sensitivities for planning a trial by dividing the plotted or tabled quantities by the isomolar fraction of a compound (moles ^{14}C per mole compound) and then multiplying by the molecular weight to obtain the familiar units of gram-equivalent per milliliter. The LLOQ for metabolite profiles depends on the particular compound chosen as a carrier carbon, and different facilities may quantify this carrier background to higher or lower precisions. There are other quantifiable effects on LLOQ as discussed below. However, the AMS sensitivity (limit of detection, LOD) for a single fraction will remain 0.1–0.5 amol of ^{14}C (or 100–500 zeptomoles to use the next named unit), which is a plasma sensitivity of 5 fM ^{14}C for a 50 μL column injection. The quantitative LLOQ lies about a factor of 10 above this and is related to precision of measurement rather than sensitivity.

The above examples include human plasma and urine and rat plasma for both a hydrophilic compound and a hydrophobic compound, yet the AMS quantitation methods remained the same for all matrices, species, and compounds. There are no compound- or species-specific method developments to be done for AMS quantitation. The analytical method for AMS is summarized: define the sample, burn it, measure it. Thus, these specific examples provide universal estimating tools for planning UPLC–AMS analyses.

16.9 LC–AMS BACKGROUND AND SENSITIVITY

This section discusses UPLC–AMS measurement background and sensitivity in more detail and may not be of interest to all. It is included to show that these quantities submit to rigorous analysis.

An AMS "instrument" background exists in the form of scattered ions that reach the ^{14}C detector. This background is 0.1% Modern or less in well-designed and properly operated biochemical spectrometers. A "preparation" background arises from natural ^{14}C adhering to combustion and reduction glassware or reactants and is inseparably incorporated into the sample, but this level is less than 0.1% Modern in routine sample production. The primary background in an AMS measurement arises from the well-defined natural or carrier ^{14}C in samples, which is much greater than these smaller uncontrolled sources. This differs from other bioanalytical assays in which detection response (UV absorbance, the area of a specific mass peak, luminescent intensity, etc.) has a finite range at all times in relation to quantified analytes but no defined isolatable background component. Calibration curves are needed to define the LLOQ for those continuous responses. The FDA Guidance on Bioanalytical Validation (2001), calling for an LLOQ that is at least 5 times the response compared to a blank response, was written with these methods in mind but does not translate well to AMS. A multiplier of 6 represents a likelihood of only 9 ppm that normal distributions of the background and the sample with equivalent uncertainties would overlap at the 3σ levels. The second approach provided in the FDA Guidance for finding the concentration at which uncertainty rises above 20% in repeated measures of calibrations near the background is also not amenable for AMS since the uncertainty in any one AMS measure is derived from the Poisson statistics of the ^{14}C counts ($\sqrt{N}$). Distributions of repeated measures have widths consistent with this uncertainty (Vogel et al., 2004). In a Poisson measurement, the uncertainty in both the background and the sample are reduced merely by continuation of measurement, given sufficient spectrometer time and sample survival. AMS produces homoscedastic analysis, in which the variance in the measurement is independent of the sample size or the analyte (^{14}C) content, although the content will control the length of time required to attain a given variance. Background measurements and low concentration samples have equivalent variances because the background and analyte are fully additive within the sample. The ICH Guidance, Q2B Validation of Analytical Procedures (FDA, 1996) took a broader view on analytical, rather than strictly bioanalytical, methods and defined an LLOQ as a multiplier of the standard deviation in the background quantitation (perhaps corrected for slope in heteroscedastic data). This definition better fits AMS data in which background samples are measured independently of chromatographic analyte. An LLOQ of 10 times the standard deviation of repetitive measures of true background samples was suggested.

Solid AMS samples use carrier carbon in LC fractions having low natural carbon. One microliter of tributyric acid (tributyrin) contains 615 μg of carbon and is a suitable carrier additive. The compound can be purchased with between 9 and 35% natural biospheric carbon, apparently introduced by ingredient stocks obtained from natural lipids. A set of independently quantified background carrier samples taken with the UPLC trace shown in Figure 16.9 averaged 9.65 ± 0.29% Modern, or an uncertainty of 0.18 amol on the 6 amol

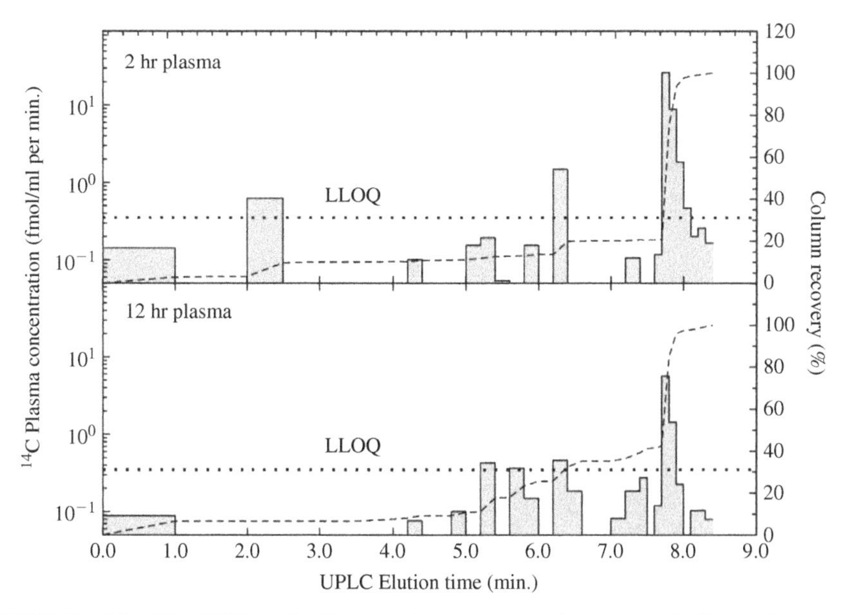

FIGURE 16.9 The UPLC–AMS trace form human plasma metabolites of a compound labeled at less than 1 molecule per 100 at 2h (upper) and 12h (lower) postdose. Due to the light label content, many fractions are well below the LLOQ and had no net tracer ^{14}C.

^{14}C per fraction containing 0.615 mg carrier carbon. Figure 16.9 shows human plasma UPLC separations at 2 and 12 h postdose with a compound of very low isomolar fraction (<1%, <0.6 Ci/mol) for which the 0.04 p*M* LLOQ of AMS converts to 4 p*M* molecular concentration. This corresponds to a mass-equivalent LLOQ of 1.4 pg-eq/mL if the compound has a molecular weight of 350 g/mol. A more typically labeled compound with an 80% isomolar fraction (50 Ci/mol) would have an LLOQ of 0.018 pg-eq/mL for metabolite LC fractions.

The compound in Figure 16.9 was so lightly labeled that many of the collected fractions may contain no compound-related ^{14}C, permitting determination of the full UPLC process background directly. The ^{14}C contents of fractions corrected for both carrier and uniform natural background ^{14}C are plotted on a normal probability plot as amol of ^{14}C per fraction in Figure 16.10. A Gaussian distribution of measurements forms a straight line on a probability plot, as appears for 50% of the data set. There is a disjoint above this set, perhaps the initial concentrations of true net compound ^{14}C in collected fractions, reenforcing the suspicion that a large number of fractions did not contain net ^{14}C. The average of the encircled data is 0.23 ± 0.33 amol ^{14}C, as also determined by

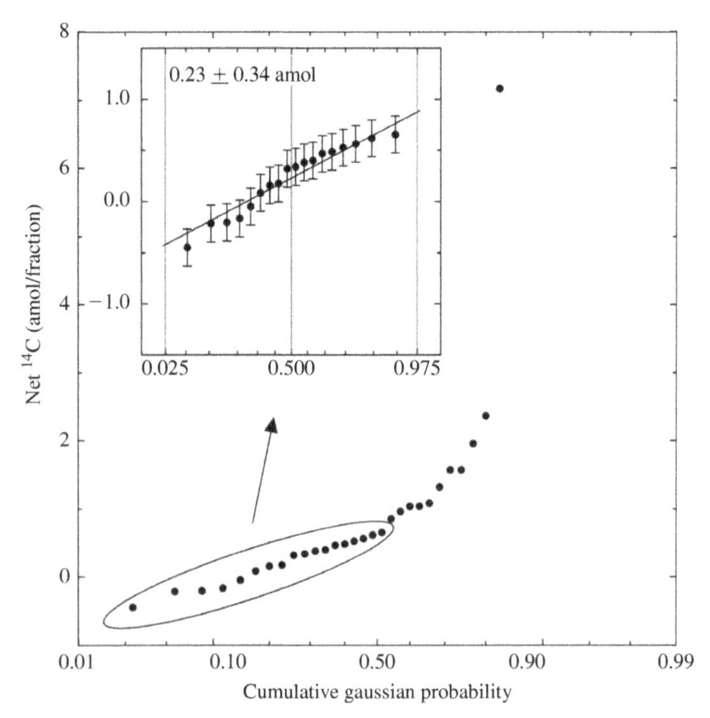

FIGURE 16.10 Probability plot of background-subtracted data from UPLC analyses of a lightly labeled compound reveal a normal distribution of data that quantifies the uncertainty in the complete analytical background for UPLC–AMS.

the linear fit to the inset plot. The 0.18 amol ^{14}C, uncertainty in the carrier carbon backgrounds is applied to the measurements of the individual fractions, as befits homoscedastic data, and are plotted on the inset data, showing that the data well approximate a normal distribution within the uncertainties. This normal distribution of the lowest UPLC concentrations implies a standard deviation in complete UPLC background of 0.33 amol ^{14}C per fraction, nearly twice the 0.18 amol found from repeated measures of the carrier compound alone. The wider total background distribution may arise from low levels of column retention and bleed between separations, but is more likely from the uncertain amount of biogenic plasma carbon in the fraction. A 50-μL injection of protein-free plasma contains almost 300 μg of biogenic carbon with 30 amol ^{14}C. In the simplest assumption, this ^{14}C is distributed evenly across the 10-min LC elution, although the natural constituents are primarily polar molecules that elute in the earlier fractions. The additional 0.15 amol in background distribution width is not unreasonable given this biogenic background that contributes between 3 amol ^{14}C to the 1-min fractions and 0.3 amol to the 6-s fractions. The effect of this biogenic fraction is proportionally reduced by using a larger (but still defined) amount of carrier, such as a 2-μL capillary pipette of tributyrin (1.23 mg C).

An LLOQ for UPLC–AMS fractions at 6 times this comprehensively measured standard deviation is 2.0 amol ^{14}C per fraction or 0.04 p*M* ^{14}C for a 50-μL injection. This is nearly twice the 1.1 amol ^{14}C per fraction uncertainty (6×97.9 amol/mgC $\times 0.615$ mgC $\times 0.0029$ Modern) determined solely from carrier background distribution width, which suggests a 0.02-p*M* ^{14}C LLOQ for a 50-μL plasma equivalent injection. In this instance, the low label content of the compound afforded sufficient fractions with no net ^{14}C with which to measure this complete measurement uncertainty. More generally, there are few such fractions, as seen in the log frames of Figures 16.7 and 16.8 and Figures 16.6*e* and 16.6*f*. One solution to determining LLOQ when there is a possibility of unaccountable natural ^{14}C contributions is to use the larger multiplier of 10 times measurable background uncertainties, as has been done here for all UPLC–AMS plots. The lower factor of 6 is still appropriate for samples not supplemented with carrier compound because the natural ^{14}C concentration in living subjects is well defined and universal.

16.10 CLINICAL ASPECTS OF AMS METABOLITE STUDIES

Clinical procedures must assure that no unaccounted sources of ^{14}C or carbon enter the defined sample, introducing a C_x in (Eq. 16.4) that is not easily understood in postclinical analysis. AMS sensitivity is specific for the particular isotope tracer, helping to focus procedures for maintaining sample integrity. AMS quantitation is not affected by the presence of other radioisotopes, such as tritium, when ^{14}C is used as the isotopic tracer. Much of the early work developing AMS for biochemical studies was performed in laboratories that had long histories of ^{14}C tracing, but procedures isolated sample materials from contact with surfaces and instruments that had remnant traces of radioisotope that would overwhelm the tracer levels in samples (Buchholz et al., 2000). The guiding principle is one of assuring that the sample contacts only new clean surfaces. The most contaminating surface in a laboratory or clinic is the hand (door knobs in isotope labs are often the "hottest" surface in the room), and a primary lesson is that gloves are changed to protect the samples and not the person wearing them. A discussion with the AMS provider is often sufficient to learn of potential procedures that need modifying for AMS quantitation in a typical clinical site.

A predose sample of blood is required from every clinical volunteer as soon as clinically possible, not only to establish the background ^{14}C concentration to be subtracted form the later measurements but also as an integrated test of the isotope cleanliness of the volunteers, the clinical setting with its food and entertainment amenities, and the handling and shipping procedures. The "contamination" to be guarded against is often not the ^{14}C but the unnoted addition of "dead" carbon, often from a process (e.g., ethanol addition to enhance solubility) or labware (e.g., new plastic collection vessels) that had never presented difficulties in previous isotope or bioanalytical tracing because the analysis by decay counting or MS was insensitive to the presence of such added carbon. For example, a cohort of volunteers had plasma ^{14}C at the expected 1.09 ± 0.01 Modern (in 2001), while urine averaged 1.03 ± 0.04 Modern and was not normally distributed, with some samples as low as 0.8 Modern. This was traced to plasticizer or plastic dust within the usual urine collection vessels, which was eliminated by alcohol rinses prior to use. AMS quantitation depends on thoughtful practices to maintain the isotope ratio of the sample as collected but can be applied to samples from most clinical sites.

16.11 AMS ANALYSIS OF REACTIVE METABOLITES

AMS sensitivity was first applied in biochemistry to the comparative quantitation of reactive metabolites, metabolites that are such strong electrophiles that they have undetectable concentrations in circulation. These metabolites are not quantifiable by LC separations of biological fluids but are rapidly bound to nucleophilic molecules and macromolecules depending on their site of reactive transformation. The first studies involved the quantitation deoxyribonucleic acid (DNA) adducts of food-borne genotoxins of the heterocyclic amine family in rodents (Turtellub et al., 1990). Further studies expanded to quantify reactive products bound to circulation proteins and DNA at exposure-relevant doses in rodents and humans (Dingely et al., 1998a; Lang et al., 1999; Garner et al., 1999; Cupid et al., 2004) as well as to tumor and healthy tissue in humans (Turteltaub et al., 1997; Mauthe et al., 1999; Dingely et al., 1999). Brown et al. (2005) provided an overview of these efforts. Many of these studies involved environmental exposures, whether industrial, nutritional, or natural, but macromolecular trapping of reactive drug metabolites were also quantified by AMS for tamoxifen (White et al., 1997; Martin et al. 1997, 2003), acetaminophen (Vogel et al., 2001), doxorubicin (DeGregorio et al., 2006), and adriamycin (Coldwell, et al. 2008). Specific analyses can be devised to add the sensitivity, precision, and accuracy of AMS to many particular metabolic studies, including protein binding (Dingley et al., 1998a), and specific receptor site reactions (Bennett et al., 1998). AMS is one of the few highly sensitive techniques to quantify reactive metabolites of compounds, and this remains one of the prominent uses of AMS in biomedical research environments.

16.12 SPECIES METABOLITE COMPARISON

A compound's metabolites may be potential pharmacological or toxicological agents themselves. Preclinical studies reveal mammalian metabolic pathways and identify potentially worrisome endpoints, while human cell or tissue cultures indicate whether the preclinic is likely predictive of the clinic. Nonetheless, determination of the complete set of circulating human metabolites of a drug candidate as early in development as possible is important to the pharmacology of the compound as well as its future suitability as a pharmaceutical. Quantitative clinical metabolite analyses are made during clinical ADME studies, once sufficient radiolabeled compound is available. However, obtaining whole-body metabolism for comparison among species earlier in the program can greatly affect the direction and speed of clinical development and toxiocological assurances.

The potential importance of metabolites in safety and toxicology of a drug candidate was long recognized in the pharmaceutical industry and was summarized by a report from interested parties published in 2002 (Baillie et al., 2002). The FDA and industry organizations then covened focused discussions that led to the 2008 Guidance on Metabolites in Safety Testing (FDA MIST, 2008) formalizing one approach to the issue from the many publications and opinion expressed in the intervening years. A recent issue of *Chemical Research in Toxicology* brought together commentary and examples of how this Guidance affects drug development (Guengerich, 2009). A critical analytical requirement in the Guidance is the need for comparative metabolite quantitation among humans and toxicology species. Contemporary bioanalytical tools have reached a high degree of specificity and sensitivity so that even minor metabolites are identified, even when their quantitation is of little likely importance. Safety assurance, however, comes from confirmation that the sum of all metabolites is complete compared to the quantified burden of compound derivatives in the matrix. Although there are creative liquid chromatography–tandem mass spectroscopy (LC–MS/MS) methods that address this requirement (Tiller et al., 2008), high-precision analysis of isotopic-labeled compounds across a wide dynamic range is the most facile, robust, and universal approach. AMS is particularly suited for comparative quantitation that has no dependence on species or matrix. AMS sensitivity means that metabolite quantitation may even occur prior to large-scale production of labeled compound for the ADME studies in later phases.

Metabolite quantitation in humans and two species of preclinical animals at $C_{\max}$ in plasma is shown in Figure 16.11. UPLC–AMS separated and quantified fractions of variable width that depended on metabolites expected from the animal species. UV absorbance (dashed lines in the figure) monitored resolution and provided a marker near 8 min for one metabolite spiked into the plasma samples before separation. AMS ^{14}C data is plotted as the fraction of the total eluted isotope signal in the trace on both linear and log scales for 18 compounds that produced "peaks" in one or more of the three species. Multiple metabolic differences are noted not only between the human and the animals but also between the two animal species. Specifically, the metabolites included as a spike, M_a, is disproportionately found in humans at more than 10% of the total ^{14}C content and only minimally, but distinctly, in species 2. Chromatographic conditions broadened the UV metabolite peak in species 1 with resolution also lost in ^{14}C, leaving only a hint of the metabolite presence. M_b is the only other metabolite exceeding 10% of total ^{14}C in human plasma in this trace. It is present at similar concentration in species 2, fulfilling the requirement for toxicological coverage in a preclinical species. Metabolites M_c and M_d appear in human plasma at and just less than 10% of total load but do not appear in the linear scale plots for the toxicology species. These may also require identification despite their high polarity and, hence, low likely toxicity. In the log scale data, both these metabolites are also present in the animals at very low levels, indicating, as with M_a, that they do naturally arise, but that they are either produced in much lower quantities or are cleared much faster. Urine data confirmed the former. Their presence in the toxicological species means that the metabolic pathways exist for these products but are controlled by some other element of physiochemistry. Other metabolites are seen in humans and in one or other species at comparable levels.

The data in Figure 16.11 confirmed that the major metabolite spiked into the plasma samples is disproportionate in humans and that, according to the MIST Guidance, further toxicology screens might include some other animal that does produce that product, or the chmical may be synthesized for direct dosing in a toxicology survey. Metabolite profiles in Figure 16.11 were obtained at $C_{\max}$ in plasma, but the MIST Guidance calls for a profile that reflects circulating exposure at steady state. One way to obtain such data is to use "AUC pooling" of samples drawn over the exposure period of an acute labeled dose given after chronic unlableled compound induced any enzymatic responses. (Hamilton et al., 1981; Raid et al., 1991; Hop et al., 1998 Cheung et al., 2005). Hop et al. (1998) showed excellent average agreement ($<5\%$) between pooled analysis and separately analyzed exposures in two dogs of multiple compounds, but Cheung et al. (2005) had rms standard errors between the methods of 10–30% among parent and metabolites of a potentially

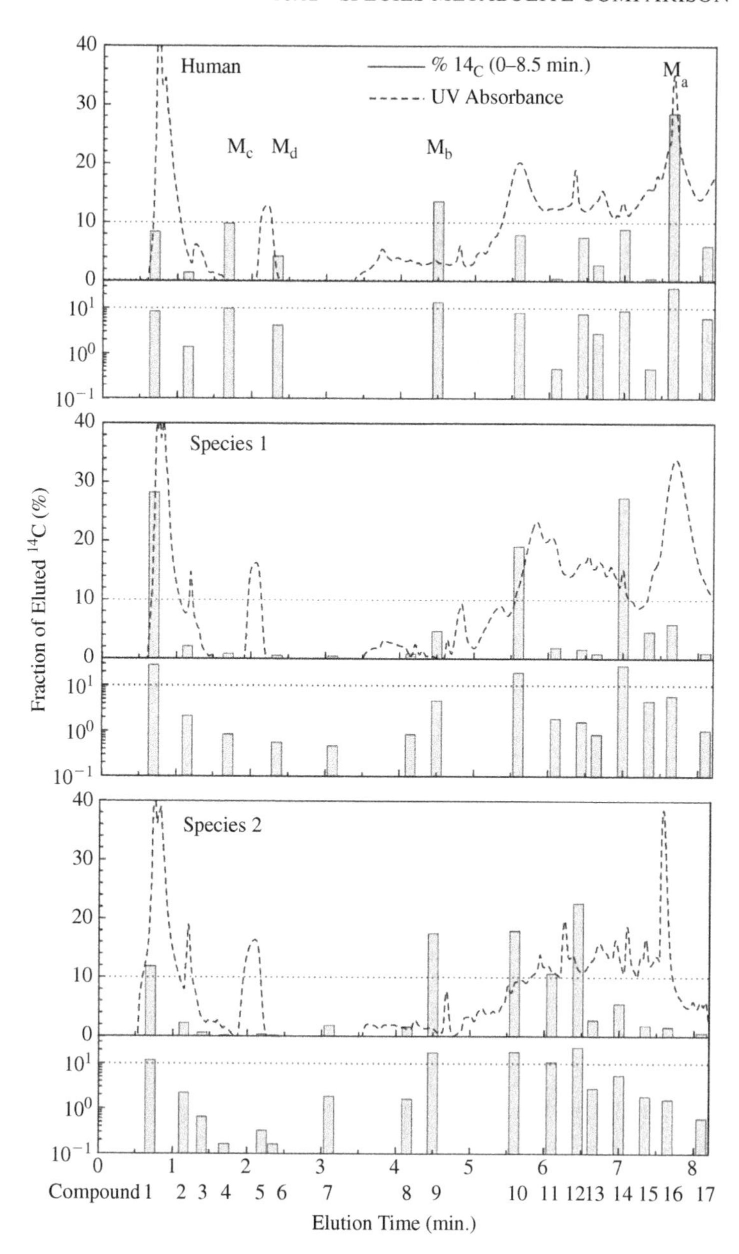

FIGURE 16.11 Metabolite compound concentrations of human (top) and two animal species plasmas after dosings with a ^{14}C-labeled compound are shown as fraction of the displayed total eluted ^{14}C signal along with UV absorbance (dashed line). The upper frames have a linear ^{14}C scale that is also given in log scale. A metabolite marker is seen in the UV trace near 8 min.

unstable compound over a wide period of analysis in both dogs and human. AMS provides a precise measure across a wide range of concentrations of a stable signal (the ^{14}C concentration) that may make the best use of this method in quantitative metabolite analysis.

16.13 NEW METABOLIC STUDIES ENABLED BY AMS

Precise measurement of all metabolites to attomole levels using very low amounts of ^{14}C promotes quantitation of the human metabolic response of potential therapeutics from the later stages of clinical development to much earlier phases of clinical trials, providing a wealth of human physiochemical information. This earlier measurement permits more confidence in decisions about the chemistries and forms that are taken through the full gamut of toxicology and clinical trials, reducing expensive surprises from arising in full human ADME studies. This strategy is now strengthened by the decision of at least one clinical research unit to accept human exposures to less than 1 μCi of ^{14}C-labeled compounds in humans without animal species whole-body dosimetry measurements.

AMS quantitation potentiates entirely new metabolic studies of very potent drugs in humans at very low doses. The specificity of the isotopic label distinguishes metabolism of biopharmaceuticals and other biological equivalents from endogenous "background" (Lappin et al., 2006). Finally, AMS sensitivity enables metabolism surveys of special populations who must be shielded from high chemical and/or isotopic doses, such as renally impaired subjects, pediatric populations, and even infants. Vuong et al. (2010) provide an example that covers all three classes of new studies: a very low dose of a biopharmaceutical to neonatal subjects. Figure 16.12 shows the plasma kinetic profile of the ^{14}C concentration after ^{14}C-ursodiol dosed gastrically at 8ng (37 Bq, 4 pg/kg) in three preterm infants. Wide variability is expected and seen in this small chort of a population that has not yet reached post-natal homeostasis. The 37 Bq (nCi) of ^{14}C is sufficient to follow this nanodose above LLOQ for at least 4 days in even the lowest responding subject. The LLOQ (six times the standard deviation of the predose blood draws) was 0.072 Modern or 0.3 pM ^{14}C, which corresponds to 0.4 pM for the ^{14}C-ursodiol that had a isomolar fraction of 80% (50 Ci/mol). The enterohepatic recycling of this natural acid is distinctly recognizable in two of the subjects by the return of signal at 24h. The elimination half-life for the low responder is 120 h.

No metabolic analysis was made on these neonate plasmas since this study sought to determine dose and isotopic levels required to quantitatively monitor neonatal cholestasis therapy with ursodiol. Plasma ursodiol is largely bound to protein and first-pass metabolism in the liver conjugates it with glycine and taurine. These conjugates excrete with other bile components into the intestine where they are deconjugated and reabsorbed. The efficiency of this cycle, even in neonates, is responsible for the very long clearance noted in Figure 16.12.

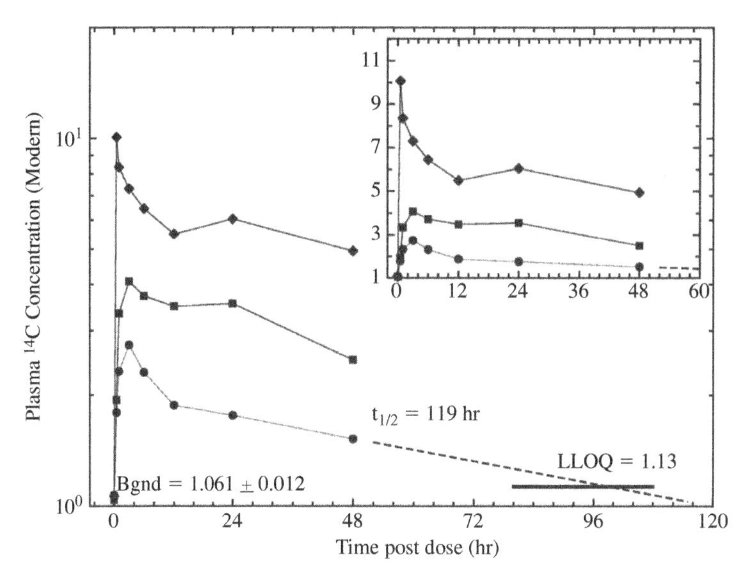

FIGURE 16.12 Plasma kinetic response of three preterm infants to an 8-ng, 37-Bq dose of ^{14}C-ursodiol is shown on a log scale with the linear scale shown in the inset. The lowest response remains above the LLOQ of AMS, 1.13 Modern, for 4 days postdose.

Fecal excretion occurs as lithocholic acid or chenodiol (Crosignani et al., 1996). Bile acids are transparent to UV, do not fluoresce, and are not distinctive electrochemically, so that LC quantitation requires solid-phase extraction and derivatization for detection (Nobilis et al., 2001). This measurement requires too large a plasma volume for use with neonates who have only 100 mL of plasma. The volume sensitivity of UPLC–AMS reduces the sample volume to tens of microliters by providing the direct quantitation of the isotopic label with minimal processing. The low responder on Figure 16.12 has 4 p*M* tracer ^{14}C in plasma at 24h postdose ([2 Modern-Bgnd] × 0.1 fmol/mg × 40 mg/mL), the majority of which is protein bound. Thus, UPLS–AMS with only a 37-Bq dose has potentially a dynamic range of 10–20 above the LLOQ for metabolite analysis, expected to be approximately 0.1 p*M* ^{14}C from Figures 16.7–16.9. This is sufficient for obtaining kinetic analyses of the well-studied metabolites for this compound. Metabolic studies in neonates may better generally use 100 Bq ^{14}C to widen this dynamic range for compounds that could have unexpected metabolites in this population. This is still only a factor of 3 increase in ^{14}C over a 2-kg neonate's natural ^{14}C content of 35 Bq and produces no potential risks from radiation above what a few days of life already provide.

This example of planned neonatal metabolic studies emphasizes the potential of UPLC–AMS to obtain important metabolic data from any population for ^{14}C-labeled compounds at any tracer or dose level. There is no analytical impediment to obtaining metabolite profiles under clinical restraints on population (e.g., metabolically or renally impaired) or specimen source (e.g., cerebral spinal fluid).

16.14 CONCLUSIONS

This chapter provides an introduction to the absolute biological quantitation offered by liquid chromatography, especially UPLC, with AMS quantitation of ^{14}C-labeled compounds. Precise and accurate quantitation is the primary goal of radioisotope molecular labeling, and AMS brings high sensitivity along with unprecedented accuracy and precision to ^{14} C quantification. This sensitivity reduces radioisotope exposures in clinical subjects to the point that no population need be excluded from quantitative clinical study. New insights into human physiochemistry are enabled by the quantitative recovery of simplified AMS processes that provide biological concentrations of all labeled metabolic products. This chapter should allow bioanalysts to understand fully the quantitative sensitivity and potential, as well as the limitations, of AMS as an isotope detector quantifying chromatographic separations of metabolites in biological matrices.

REFERENCES

Baillie T, Cayen M, Fouda H, Gerson RJ, Green JD, Grossman SJ, Klunk LJ, LeBlanc B, Perkins DG, Shipley LA. Drug metabolites in safety testing. Toxicol. Appl. Pharmacol. 182(3):188–196 (2002).

Barker J, and Garner RC. Biomedical applications of accelerator mass spectrometry—Isotope measurements at the level of the atom. Rapid Commum Mass Spectrom 1999;13(4):285–23.

Baxter M, Castle L, Crews HM, Rose M, Garner C, Lappin G, and Leong D. A sensitive method for the determination of chlorine-36 in foods using accelerator mass spectrometry. Food Additives and Contaminants Part a—Chem Anal Control Expos Risk Assess 2009;26(1):139–144.

Bennett CL, Beukens RP, Clover MR, Gove HE, Liebert RB, Litherland AE, Purser KH, and Sondheim WE. Radiocarbon dating using electrostatic accelerators: Negative-ions provide key. Science 1977;198 (4316):508–510.

Bennett JS, Bell DW, Buchholz BA, Kwok ESC, Vogel JS, and Morton TH. Accelerator mass spectrometry for assaying irreversible covalent modification of an enzyme by acetoacetic ester. In J Mass Spectrom 1998;180:185–193.

Bronk Ramsey C, and Hedges REM. Hybrid ion sources: Radiocarbon measurements form microgram to milligram. Nucl Instrum Methods Phys Res 1997;B 123:539–545.

Brown K, Dingley KH, Turteltaub KW. Accelerator mass spectrometry for biomedical research. Methods Enzymol 2005;402:423–434.

Buchholz, BA, Freeman S, Haack KW, and Vogel JS. Tips and traps in the C-14 bio-AMS Preparation laboratory. Nucl Instrum Methods Phys Res Sect B Beam Interact Mater Atoms 2000;172:404–408.

Buchholz BA, Fultz E, Haack KW, Vogel JS, Gilman SD, Gee SJ, Hammock BD, Hui X, Wester RC, and Maibach HI. HPLC-accelerator MS measurement of atrazine metabolites in human urine aftger dermal exposure. Anal Chem 1999;71 (16):3519–3525.

Cheung BWY, Cartier LL, Russile HQ, Sawchuk RJ. The application of sample pooling methods for determining AUC, AUMC and mean residence times in pharmacokinetic studies. Fund Clin Pharmacol 2005;19(3):347–354.

Chaiarappa-Zucca ML, Finkel RC, Martinelli RE, McAnianch JE, Nelson DO, Turteltaub KW. Measurement of beryllium in biological samples by accelerator mass spectrometry: Applications for studdying chronic beryllium disease. Chem Res Toxicol 2004;17(12):1614–1620.

Coldwell KE, Cutts SM, Ognibene TJ, Henderson PT, Phillips DR. Detection of Adriamycin-DNA adducts by accelerator mass spectrometry at clinically relevant Adriamycin concentrations. Nucleic Acids Res 2008;36:e100–110.

Colizza, K, Awad M, Kamel A. Metabolism, pharmacokinetics, and excretion of the substance preceptor antagonist CP-122,721 in humans: Structural characterization of the novel major circulating metabolite 5-trifluoromethoxy salicylic acid by high-performance liquid chromatography–tandem mass spectrometry and NMR spectroscopy. Drug Metab Dispos 2007;35(6):884–897.

Creek MR, Frantz CE, Fultz E, Haack K, Redwine K, Shen N, Turteltaub KW, Vogel JS. ^{14}C AMS quantification of biomolecular interactions using microbore and plate separations. Nucl Instrum Methods 1994;B92:454–458.

Creek, MR, Mani C, Vogel JS, Turteltaub KW. Tissue distribution and macromolecular binding of extremely low doses of C-14-benzene in B6C3F1 mice. Carcinogenesis 1997;18(12):2421–2427.

Crosignani A, Setchell KDR, Invernizzi P, Larghi A, Rodrigues CMP, Podda M. Clinical pharmacokinetics of therapeutic bile acids. Cliln Pharmacokinet 1996;30 (5):333–358.

Cupid BC, Lightfoot TJ, Russell D, Gant SJ, Turner PC, Dingley KH, Curtis KD, Leveson SH, Turteltaub KW, Garner RC. The formation of AFB(1)-macromolecular adducts in rats and humans at dietary levels of exposure. Food Chem Toxicol 2004;42(4):559–569.

Cuyckens F, Koppen V, Kembuegler R, Leclercq L. Improved liquid chromatography—Online radioactivity detection for metabolite profiling. J Chromatogr A 2008;1209 (1–2):128–135.

DeGregorio MW, Dingley KH, Wurz GT, Ubick E, Turteltaub KW. Accelerator mass spectrometry allows for celluar quantification of doxorubicin at femtomolar concentrations. Cancer Chemother Pharmacol 2006;57(3):335–342.

Denk E, Hillegonds D, Hurrell RF, Vogel J, Fattinger K, Hauselmann HJ, Kraenzlin M, Walczyk T. Evaluation of (41) calcium as a new approach to assess changes in bone metabolism: Effect of a bisphosphonate in tervention in postmenopausal women with low bone mass. J Bone Mineral Res 2007;22(10):1518–1525.

Denk E, Hillegonds D, Vogel J, Synal A, Geppert C, Wendt K, Fattinger K, Hennessy C, Berglund M, Hurrell RF, Walczyk T. Labeling the human skeleton with Ca-41 to assess changes in bone calcium metabolism. Anal Bioanal Chem 2006;386(6):1587–1602.

Dingley KH, Curtis KD, Nowell S, Felton JS, Lang NP, Turteltaub KW. DNA and protein adduct formation in the colon and blood of humans after exposure to a dietary-relvant dose of 2-amino-1methyl-6-phenylimidazo[4,5-*b*]pyridine. Cancer Epidemiol Biomakers Prev 1999;8(6):507–512.

Dingley KH, Freeman S, Nelson DO, Garner RC, Turteltuab KW. Covalent binding of 2-amino-3,8-dimethylimidazo[4,5-*f*] quinoxaline to albumin and hemoglobin at environmentally relevant doses—Comparison of human subjects and F344 rats. Drug Metab Dispos 1998a;26(8):825–828.

Dingley KH, Roberts ML, Velsok CA, Turteltaub KW. Attomole detection of 3H in biological samples using accelerator mass spectrometry: Application in low-dose, dual-isotope tracer studies in conjunction with C-14 accelerator mass spectrometry. Chem Res Toxicol 1998b;11(10)1217–1222.

Dingley KH, Ubick EA, Chiarappa-Zucca ML, Nowell-S, Abel S, Ebeler SE, Mitchell AE, Burns SA, Steinberg FM, Clifford AJ. Effect to dietary constituents with chemopreventive potential on adduct formation of a low dose of the heterocyclic amines PhIP and 1Q and phase II hepatic enzyhmes. Nutr Cancer In J 2003;46(2):212–221.

Dueker SR, Lin Y, Buchholz BA, Schneider PD, Lame MW, Segall HJ Vogel JS Clifford AJ. Long-term kinetic study of beta-carotene, using accelerator mass spectrometry in an adult volunteer. J Lipid Res 2000;41(11):1709–1800.

Ebeler SE, Dingley KH, Ubick E, Abel S, Mitchell AE, Burns SA, Steinberg FM, Clifford AJ. Animal models and analytical approaches for understanding the relationships between wine and cancer. Drugs Exper Clin Res 2005;31(1):19–27.

Fallon SJ, Guilderson TP, Brown TA. CAMS/LLNL ion source efficiency revisited. Nucl Instrum Methods Phys Sect B Beam Interact Mater Atoms 2007;259 (1):106–110.

Flarakos J, Liberman RG, Tannenbaum SR, Skiper PI. Intergration of continuous-flow accelerator mass spectrometry with chromatography and mass-selective detection. Anal Chem 2008;80(13):5079–5085.

Flarend RE, Hem SL, White JL, Elmore D, Sucknow MA, Rudy AC, Dandashli EA. In-vivo absorption of aluminum-containing vaccine adjuvants using A1-26. Vaccine 1997;15(12–13):1314–1318.

Garner RC, Goris I, Laenen AA, Vanhoutte E, Meuldermans W, Gregory S, Garner JV, Leong D, Whattam M, Calam A, Snel CA. Evaluation of accelerator mass spectrometry in human mass balance and pharmacokinetic study—Experience with 14C-labeled (*R*)-6-[amino(4-chlorophenyl)(1-methy-1H-imidazol-5-yl)]methyl-4-(3-chlorophenyl)-1-methy-2(1H)-quinolinone (R115777), a farnesyl transferase inhibitor. Drug Metab Disposl 2002;30(7):823–830.

Garner RC, Lightfoot TJ, Cupid BC, Russell D, Coxhead JM, Kutschera W, Priller A, Rom W, Steier P, Alexander DJ, Leveson SH, Dingley KH, Mauthe RJ, Turteltaub KW. Comparative biotransformation studies of MelQx and PhIP in animal models and humans. Cancer Lett 1999;143(2):161–165.

Gilman SD, Gee SJ, Hammock BD, Vogel JS, Haack K, Buchholz BA, Freeman SP, Wester RC, Hui X, Maibach HI. Analytical performance of accelerator mass spectrometry metabolites in human urine. Anal Chem 1998;70(16):3463–3469.

Guengerich FP. Introduction: Human metabolites in ssfety testing (MIST) issue. Chem Res Toxicol 2009;22(2):237–238.

Hamilton RA, Garnett WR, Kline BJ. Determination of mean valproic acid serum level by assay of a single pooled sample. Clin Pharmacol Ther 1981;29(3):408–413.

Hickenbottom SJ, Limke SL, Dueker SR, Y, Follett JR, Carkeet C, Buchholz BA, Vogel JS, Clilfford AJ. Dual isotope test for assessing beta-carotene cleavage to vitamin A in humans. Eur J Nutr 2002;41(4):141–147.

Ho CC, de Moura FF, Kim SH, Burri BJ, Clifford AJ. A minute dose of 14C-{beta}-carotene is absorbed and converted to retinoids in humans. J Nutr 2009;139 (8):1480–1486.

Hop C, Wang Z, Chen Q, Kwei G.Plasma-pooling methods to increase throughput for in vivo pharmacokinetic screening. J Pharm Sci 1998;87(7):901–903.

Kamen M. Tracers. Sci Am 1949;180(2):30–40.

Kautiainen A, Vogel JS, Turteltaub KW. Dose-dependent binding of trichloroethylene to hepatic DNA and protein at low doses in mice. Chem-Biol Interact 1997;106 (2):109–121.

Keck BD, Ognibene T, Vogel JS. Analytical validation of accelerator mass spectrometry for pharmaceutical development. Bioanalysis 2010;2(3):469–485.

Kim SH, Kelly PB, Clifford AJ. Accelerator mass spectrometry targets of submilligram carbonaceous samples using the high-throughput Zn reduction method. Anal Chem 2009;81(41):5949–5954.

Kustschera W. Progress in isotope analysis at ultra-trace level by AMS. Int J Mass Spectrom 2005;242(2–3):145–160.

Lang NP, Nowell S, Malfatti MA, Kulp KS, Knize MG, Davis C, Massengill J, Williams S, MacLeod S, Dingely KH, Felton JS, Turteltaub KW. In vivo human metabolism of 2-C-14 2-amino-1-methy-6-phenylimidazo-4,5-*b* pyridine (PhIP). Cancer Letth 1999;143(2):135–138.

Lappin G, Garner RC, Meyers T, Powell J, Varely P. Novel use of accelerator mass spectrometry for the quantification of low levels of systemic therapeutic recombinant protein. J Pharm Biomed Anal 2006;41(4):1299–1302.

Lappin G, Stevens L. Biomedical accelerator mass spectrometry: Recent applications in metabolism and pharmacokinetics. Expert Opin Drug Metab Toxico 2008;4 (8):1021–1033.

Lemke SL, Dueker SR, Follett JR, Lin YM, Carkeet C, Buchholz BA, Vogel JS, Clifford AJ. Absorption and retinol equivalence of beta-carotene in humans is influjenced by dietary vitaman A intake. J Lipid Res 2003;44(8):1591–1600.

Liberman RG, Skipper PL, Prakash C, Shaffer CL, Flarakos J, Tannenbaum SR. BEAMS Lab: Novel approaches to finding a balance between throughput and sensitivity. Nucl Instrum Methods Phys Res Sect Beam Interact Mater Atoms 2007;259(1):773–778.

Liberman RG, Tannenbaum SR, Hughey BJ, Shefer RE, Klinkowstein RE, Prakash C, Harriman SP, Skipper PL. An interface for direct analysis of C-14 in nonvolatile samples by accelerator mass spectrometry. Anal Chem 2004;76(2):328–334.

Lightfoot TJ, Coxhead JM, Cupid BC, Nicholson S, Garner RC. Analysis of DNA adducts by accelerator mass spectrometry in human breast tissue after administration of 2-amino-1-methyl-6phenylimidazo[4,5-*b*]pyridine and benzo[*a*]pyrene. Mutat Res2000;472(1–2):119–17.

Lin YM, Dueker SR, Follett JR, Fadel JG, Arjomand A, Schneider PD, Miller JW, Green R, Buchholz BA, Vogel JS, Phair RD, Clifford AJ. Quantitation of in vivo human folate metabolism. Am J Clin Nutr 2004;80(3):680–691.

Mani C, Freeman S, Nelson DO, Vogel JS, Turteltaub KW. Species and strain comparisons in the macromolecular binding of extremely low doses of C-14 benezene in rodents, using accelerator mass spectrometry. Toxicol Appl Pharmacol 1999;159(2):83–90.

Martin EA, Brown K, Gaskell M, Al-Azzawi F, Garner RC, Boocock DJ, Mattock E, Pring DW, Dingley K, Turteltaub KW, Smith LL, White INH. Tamoxifen DNA damage detected in human endometrium using accelerator mass spectrometry. Cancer Res 2003;63(23):8461–8465.

Martin EA, Carthew P, White INH, Heydon RT, Gaskell M, Mauthe RJ, Turteltaub KW, Smith LL. Investigation of the formation and accumulation of liver DNA adducts in mice chronically exposed to tamoxifen. Carcinogenesis 1997;18 (11):2209–2215.

Mauthe RJ, Dingley KH, Leveson SH, Freeman S, Turesky RJ, Garner RC, Turteltaub KW. Comparison of DNA-adduct and tissue-available dose levels of MeIQx in human and rodent colon following administration of a very low dose. Int J Cancer 1999;80(4):539–545.

McIntyre CP, Sylva SP, Roberts ML, Gas chromatograph-combustion system for C-14-accelerator mass spectrometry. Anal Chem 2009;81(15):6422–6428.

Middleton R., Survey of negative-ion sources for tandem accelerators. Nucl Instrum Methods 1974;122(1–2):35–43.

Middleton R. A review of ion sources for accelerator mass spectrometry. Nucl Instrum Methods Phys Res B Beam Interact Mater Atoms (Netherlands) 1984;233 (2):193–199.

Miyashita M, Presley JM, Buchholz BA, Lam KS, Lee YM, Vogel JS, Hammock BD. Attomole level protein sequencing by Edman degradation coupled with accelerator mass spectrometry. Proc Natl Acad Sci U.S.A. 2001;98(8):4403–4408.

Nassar AEF, Bjorge SM, Lee DY. On-line liquid chromatography-accurate radioisotope counting coupled with a radioactivity detector and mass spectrometer for metabolite identification in drug discovery and development. Anal Chem 2003;75(4) 785–790.

Nelson DE, Korteling RG, Stott WR. Carbon-14: Direct detection at natural concentrations. Science 1977;198(4316):507–508.

Nobilis M, Kunes J, Kopecky J, Kvetina J, Svoboda Z, Sladkova K, Vorterl J. High-performance liquid chromatographic determination of ursodeoxycholic acid after solid phase extraction of blood serum and detection-oriented derivatization. J Pharm Biomed Anal 2001;24(5–6):937–946.

Ognibene TJ, Bench G, Brown TA, Vogel JS. Ion-optics calculations and preliminary precision estimates of the gas-capable ion source for the 1-MV LLNL BioAMS spectrometer. Nucl Instrum Methods Phys Res Sect B Beam Interact Mater Atoms 2007;259(1):100–105.

Ognibene TJ, Bench G, Vogel JS, Peaslee GF, Murov S. A high-throughput method for the conversion of CO_2 obtained from biochemical samples to graphite in septa-sealed vials for quantification of C-14 via accelerator mass spectrometry. Anal Chem 2003;75(9):2192–2196.

Prakash C, Shaffer CL, Tremaine LM, Liberman RG, Skipper PL, Flarakos J, Tannenbaum SR. Applications of liquid chromatography—accelerator mass spectrometry (LC–AMS) to evaluate the metabolic profiles of a drug candidate in human urine and plasma. Drug Metab. Lett. 1, 226–231 (2007).

Riad LE, Chan KK, Sawchuk RJ. Determination of the relative formation and elimination clearance of 2 major carbamazepine metabolites in humans—A comparison between traditional and pooled sample analysis. Pharm Res 1991;8(4):541–543.

Rouwenhorst RJ, Jzn JF, Scheffers WA, Vandijken JP. Determination of protein-concentration by total organic-carbon analysis. J Biochem Biophys Methods 1991;22(2):119–128.

Schoenheimer R, Rittenberg D, Foster GL, Keston A, Ranter S. The application of the nitrogen isotope N-15 for the study of protein metabolism. Science 1938;88:599–600.

Schulze-König T, Dueker SR, Giacomo J, Suter M, Vogel JS, Synal HA. BioMICA-DAS: Compact next generation AMS system for pharmaceutical science. Nucl Instrum Methods 2010;B268:891–894.

Skog G. The single stage AMS machine at Lund University: Status report. Nucl Instrum Methods Phys Res Sect B Beam Interact Mater Atoms 2007;259(1):1–6.

Synal HA, Dobeli M, Jacab S, Stocker M, Suter M. Radiocarbon AMS towards its low-energy limits. Nucl Instrum Methods Phys Res Sect B Beam Interact Mater Atoms 2004;223:339–345.

Synal HA, Stocker M, Suter M. MICADAS: A new compact radiocarbon AMS system. Nucl Instrum Methods Phys Res Sect B Beam Interact Mater Atoms 2007;259(1):7–13.

Taylor RE, Long A, Kra RS. Radiocarbon after Four Decades: An Interdisciplinary perspective, Springer-Verlag, New York, 1992.

Tiller PR, Yu S, Castro-Perez J, Fillgrove KL, Baillie TA. High-throughput, accurate mass liquid chromatography/tandem mass spectrometry on a quadrupole time-of-flight system as a “first-line” approach for metabolite indentification studies. Rapid Commun Mass Spectrom 2008;22(7)1053–1061.

Turteltaub KW, Dingley KH, Curtis KD, Malfatti MA, Turesky RJ, Garner RC, Felton JS, Lang NP. Macromolecular adduct formation and metabolism of heterocyclic amines in humans and rodents at low doses. Cancer Lett 1999;143(2):149–155.

Turtelaub KW, Felton JS, Gledhill BL, Vogel JS, Southon JR, Caffee MW, Finkel RC, Nelson DE, Proctor ID, Davis JC. Accelerator mass-spectrometry in biomedical dosimetry—Relationship between low-level exposure and covalent binding of heterocyclic amine carcinogens to DNA. Proc Natl Acad Sci USA 1990;87 (14):5288–5292.

Turteltaub KW, Frantz CE, Creek MR, Vogel JS, Shen N, Fultz E. DNA-adducts in model systems and humans J Cell Biochem 1993;S17:138–148.

Turteltaub KW, Mauthe RJ, Dingley KH, Vogel JS, Frantz CE, Garner RC, Shen N. MeIQx-DNA adduct formation in rodent and human tissues at low doses. Mutat Res Fun Mol Mech Mutagen 1997;376(1–2):243–252.

US FDA. Centre for Drug Evaluation and Research. Guidance for industry: Bioanalytical method validation. UCM070107 (2001).

US FDA. Centre for Drug Evaluation and Research. Guidance for Industry: Safety Testing of Drug Metabolites. UCM065014 (2008).

US FDA. Guidance for industry: ICH Q2B validation of analytical procedures: methodology CDER/CBER. (1996).

Vogel JS. Rapid production of graphite without contamination for biomedical AMS. Radiocarbon 1992;34:344–350.

Vogel JS, Grant PG, Buchholz BA, Dingley K, Turteltaub KW. Attomole quantitation of protein separations with accelerator mass spectrometry. Electrophosesis 2001;22 (10):2037–2045.

Vogel JS, Love AH (2005). Quantitating isotopic molecular labels with accelerator mass spectrometry. Meth Enzm 402:402–422.

Vogel JS, Nelson DE, Southon JR. Background levels in an AMS system.Radiocarbon 1987;29:215–222.

Vogel JS, OgnibeneT, Palmblad M, Reimer P. Counting statistics and ion interval density in AMS. Radiocarbon 2004;46(3):1103–1109.

Vogel JS, Southon JR, Nelaon DE, Brown TA. Performance of catalytically condensed carbon for use in accelerator mass spectrometry. Nucl Instrum Methods 1984; B5:289–293.

Vuong LT, Lohstroh P, Blood A, Vasquez H, Vogel JS. Applying AMS in neonatal research and care. J Isotope Labelled Comp Radiopharm. 2010;53:352–354.

White INH, Martin EA, Mauthel RJ, Vogel JS, Turteltaub, KW, and Smith LL. Comparisons of the binding of C-14 radiolabelled tamoxifen or toremifene to rat DNA using accelerator mass spectrometry. Chem-Biol Interact 1997;106(2): 149–160.

Xu XM, Trumbore SE, Zheng SH, Southon JR, McDuffee KE, Luttgen M, Liu JC. Modifying a sealed tube zinc reduction method for preparation of AMS graphite targets: Reducing background and attaining high precision. Nucl Instrum Methods Phys Res Sect B Beam Interact Mater Atoms 2007;259(1):320–329.

Young GC, Corless S, Felgate CC, Colthup PV. Comparison of a 250kV single-stage accelerator mass spectrometer with a 5 MV tandem accelerator mass spectrometer—Fitness for purpose in bioanalysis. Rapid Commun Mass Spectrom 2008;22 (24):4035–4042.

Zhu MS, Zhao WP, Vazquez N, Mitroka JG. Analysis of low level radioactive metabolites in biological fluids using high-performance liquid chromatography with microplate scintillation counting: Method validation and application. J Pharm Biomed Anal 2005;39(1–2):233–245.

Zoppi U, Crye J, Song Q, Arjomand A. Performance evaluation of the new ams system at Accium Biosciences. Radiocarbon 2007;49(1):173–182.

17 Standard-Free Estimation of Metabolite Levels Using Nanospray Mass Spectrometry: Current Statutes and Future Directions

JING-TAO WU

Drug Metabolism and Pharmacokinetics, Millennium Pharmaceuticals, Inc., Cambridge, Massachusetts

17.1 INTRODUCTION

Metabolite quantitation plays an important role in drug discovery and development as the level of metabolites may potentially impact the efficacy and safety of a drug (Baillie et al., 2002; Leclercq et al., 2009; Smith and Obach, 2006). Even for metabolites with no biological activity at the clinical relevant dose, the amount of a specific metabolite formed often reveals a major metabolic pathway that may allow medicinal chemists to block some labile sites on the molecule or "soft spots" to metabolically stabilize the compound. The recent U.S. Food and Drug Administration (FDA) guidance on Safety Testing of Drug Metabolites further put forward the requirements for

Mass Spectrometry in Drug Metabolism and Disposition: Basic Principles and Applications, First Edition. Edited by Mike S. Lee and Mingshe Zhu.

metabolite characterization as part of the basis for a regulatory decision (FDA, 2008). In this guidance, a decision tree flowchart was proposed to determine whether any nonclinical toxicity studies are required for any specific metabolites. In this decision tree flowchart, all of the decision points are based on the quantity or exposure of the metabolites in humans relative to the exposure of the parent compound in humans and the exposure of the metabolites in animal species. Because of the importance of metabolite quantitation, various analytical technologies have been developed and are quickly evolving to determine metabolite exposure in humans and animals (Zhu et al., 2009).

If a synthetic standard of a specific metabolite is available, then the quantitation of the metabolite will be similar to the quantitation of the parent compound from a bioanalytical perspective. Unfortunately, in many cases, the synthetic standards of the metabolites are not available. This is not only common in the early discovery stage but also in the development stage because one compound may often form multiple, and in some cases, a large number of metabolites. The synthesis of every metabolite is resource intensive unless they are considered major metabolites based on exposure or specific pharmacological and toxicological reasons. This situation often forms a dilemma if metabolite exposure determination and metabolite standards synthesis are dependent on each other. Therefore, analytical technologies that enable the quantitation or semiquantitation of metabolites in the absence of synthetic standards are highly desirable.

17.2 CURRENT APPROACHES FOR METABOLITE QUANTITATION IN THE ABSENCE OF SYNTHETIC STANDARDS

A variety of approaches have been reported for the quantitation of metabolites in the absence of synthetic standards. If a radio-labeled parent compound is available, then it is the next best thing to having synthetic standards for the metabolites. A radio-labeled compound can be used directly for metabolite quantitation if the radio-labeled compound can be administered in the species of interest. Radio-labeled ADME (adsorption, distribution, metabolism, and excretion) studies in humans can provide information on mass balance and metabolite exposure in humans. Unfortunately, these studies are usually conducted at Phase II or III stages, which are often too late to address any metabolite-related safety issues in the clinic. Recently, accelerator mass spectrometry (AMS) technology has been used in the mass balance or metabolite profiling studies in humans (Young et al., 2001; Hah, 2009). Taking advantage of the high sensitivity offered by AMS, significantly reduced radioactivity (less than 100 nCi compared to approximately 50–100 μCi in a conventional human ADME study using a ^{14}C-labeled drug) is used, which allowed these mass balance and metabolite profiling studies to be conducted as early as part of first-in-human studies.

If the radio-labeled compound cannot be directly administered in the species of interest, the use of a calibrator approach has been reported by Yu et al.

(2007). In this work, metabolites generated in preclinical species dosed with the radio-labeled parent compound were quantified using radiometric detection, and the quantitative information of the metabolites was used to calibrate the relative mass spectrometric response of these metabolites in the preclinical species. By employing adequate chromatographic separation and matrix matching (add equal volume of blank human sample matrix to the animal sample matrix and vice versa), the levels of metabolites in human plasma can be estimated using the same calibration factor derived from preclinical species. Using carefully optimized experimental conditions, this method can provide reliable quantitative estimate of metabolites in humans. This method requires the availability of radio-labeled parent compound and an in vitro or in vivo preclinical biological system that can generate the metabolites. Obviously, this method does not apply to the quantitative estimation of human unique metabolites that can only be generated in vivo.

When a radio-labeled parent compound is not available, the alternative is to use a "universal response" detector, which would provide an equimolar response to the metabolites of diversified structures. Deng et al. (2004) demonstrated the feasibility of using a liquid chromatography–mass spectroscopy (LC–MS) system coupled with a chemiluminescent nitrogen detector for the quantitative estimation of metabolites in the absence of synthetic standards. The intensity of nitrogen-specific chemiluminescence is only proportional to the number of nitrogen atoms in a molecule regardless of the chemical structure. Therefore, the relative amount of a metabolite against the parent compound can be estimated based on chemiluminescence, which can be subsequently used to calibrate the response of the metabolite in mass spectrometric detection. The challenge of this method is to develop effective chromatographic separations to achieve the specificity since most of the endogenous components in the biological matrices contain nitrogen, and they have to be completely resolved from the metabolite to achieve reliable quantitation. In the feasibility test, only metabolites in urine but not in plasma were reported to be quantified successfully due to the complex nature of the plasma.

As an approximation to a universal response detector, an ultraviolet (UV) detection with a high concentration of in vivo sample or in vitro incubation can be used to calibrate the mass spectrometric response of the metabolites. Josephs et al. (2009) recently reported the use of high-performance liquid chromatography (HPLC)–UV detection to get area responses of a 30-μM in vitro microsome or hepatocyte incubation. The incubated sample was then diluted in matrix to create a single point calibration for mass spectrometric quantitation of the metabolites. The results from this method were successfully verified using buspirone and proprietary compounds for which the synthetic standards for their metabolites were available.

Another universal response detector that has been used to perform standard-free metabolite-level determination is quantitative nuclear magnetic resonance (NMR) spectroscopy. The concept of quantitation using NMR was established in the 1980s (Mackenzie, 1984). Malz and Jancke (2005) validated the use of

NMR for drug quantitatation in biological fluids. Espina et al. (2009) reported the use of ^{1}H-NMR to determine the concentration of biologically generated and isolated metabolites. These determined concentrations can then be used as standards for liquid chromatography tandem mass spectrometry (LC–MS/MS) (Espina et al., 2009).

Although all of the above methods have shown some success in standard-free metabolite quantitation, there is a strong demand to continue to explore analytical methodologies that offer improved specificity and sensitivity for the quantitation of metabolites in the absence of synthetic standards of the metabolites or radio-labeled parent compound. A new analytical methodology that offers a high level of specificity and sensitivity and has demonstrated initial success and good potential for standard-free metabolite quantitation is nanospray. This chapter will review the principles, initial applications, and future directions of nanospray for standard-free metabolite quantitation.

17.3 USE OF NANOSPRAY FOR STANDARD-FREE METABOLITE QUANTITATION

Electrospray ionization has become the predominant ionization method for coupling liquid chromatography (LC) with tandem mass spectrometry (MS/MS) and has enabled the use of LC–MS/MS for routine pharmacokinetic and drug metabolism studies. Besides ion suppression, a major limitation of the method is that the response factors differ significantly for compounds with diversified structures, and, thus, a reference standard of the analyte must be used for quantitative applications. Nanospray, or electrospray, at reduced flow rate (typically below 1 μL/min) was first reported by Wilm and Mann (1994, 1996). This technique quickly drew the attention of the research community because of its unique characteristics compared to conventional electrospray. In addition to its significantly reduced sample consumption, which has yielded sub-attomole sensitivity for intact proteins, a key attribute of this technique is the reduction of ion suppression. Because of these unique attributes, nanospray has been explored as an analytical tool in different areas of drug metabolism and pharmacokinetic-related applications including in vitro Caco-2 studies, microsomal stability studies, in vivo pharmacokinetic studies, and metabolite profiling studies (Van Pelt et al., 2003; Dethy et al., 2003). More recently, the use of a nanospray-based technique for the quantitation of metabolites in the absence of synthetic standards has emerged (Schmidt et al., 2003; Hop et al., 2005; Valaskovic et al., 2006; Ramanathan et al., 2007).

17.3.1 Nanospray and Equimolar Response

Although electrospray is a relatively simple technique from an instrumentation stand point, the mechanism involved in the ionization and the factors that affect the efficiency of a compound is not straightforward (Cole, 1997). There are many factors that affect how efficient a compound can be ionized in electrospray. The electrospray process generally involves the initial formation of charged droplets, which are converted into smaller and smaller droplets through a fission process called Columbic explosion. The gas-phase ions are formed from these small droplets by evaporation. A solvated species must migrate to the surface of a droplet to be transferred to the smaller droplets formed at the next stage. The chemical structure of a compound will largely affect its ability to be preferentially enriched at the droplet surface, which largely forms the basis of diversified responses of compounds in electrospray.

In nanospray, the initial droplets formed have a much smaller size than those formed in electrospray and result in increased total available surface area. Also, the diffusion time for solvated species to migrate to the droplet surface decreases in smaller droplets. In addition, the smaller sizes of the initial droplets in nanospray result in a reduced number of Columbic explosions required to form sufficiently small droplets suitable for ionization. All of these reduce the extent of preferential enrichment of any chemical species. Schmidt et al. (2003) first observed this equimolar phenomenon in nanospray at low flow rates using a nanospray emitter with a capillary tip with an outside diameter of less than 1 μm. Using a mixture of turanose and *n*-octyl-glucopyranoside at a 10:1 molar ratio, they observed the response ratio changed from approximately 2:1 at flow rates above 50 nL/min to approximately equimolar response (10:1) at flow rates of a few nanoliters/minute (Fig. 17.1). Turanose is very hydrophilic while *n*-octyl-glucopyranoside is hydrophobic and has high surface

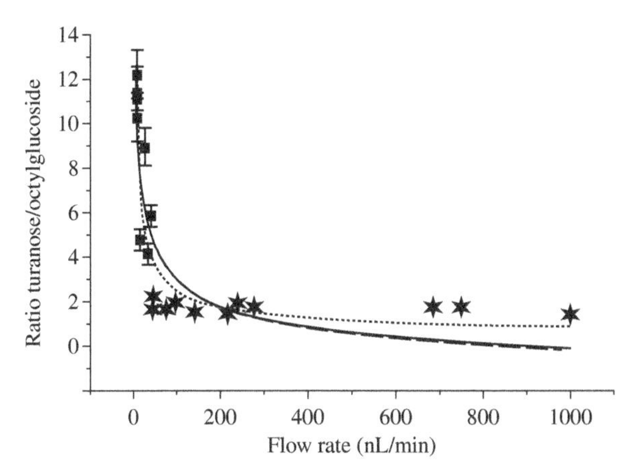

FIGURE 17.1 Mass spectrometric response ratio of turanose over *n*-octyl-glucopyranoside as a function of flow rates in nanospray. Turanose (10^{-5} mol/L) and *n*-octyl-glucopyranoside (10^{-6} mol/L) were prepared in neat solvent as a 10:1 (molar ratio) mixture. The curves represent the surface-to-volume ratio of initial droplets at different flow rates predicated using different theoretical models. [Reprinted from Schmidt et al. (2003) with permission.]

activity. Neither of these compounds is pre-charged in solution, and their response difference in electrospray is mostly determined by their surface activity. At sufficiently low flow rates, the role of surface activity in response diminished and equimolar response was observed. They further evaluated the effect of the precharge state on surface activity and electrospray response. Using a mixture of neurotensin (protonated in solution) and maltoheptaose (uncharged in solution), approximate equimolar response was also observed when the flow rate was reduced to a few nanoliters/minute in the nanospray mode.

17.3.2 Application of Nanospray in Estimating Metabolite Levels

The use of nanospray for equimolar response in a pharmaceutical setting was first reported by Hop et al. (2005). In this work, they used a commercially available nanospray device, the NanoMate, to study the response of a set of 25 compounds with 6 structurally distinct classes. Each of these compounds was prepared at a concentration of 1 μM in solvent. The responses from LC–MS and the NanoMate for these 25 compounds are shown in Figure 17.2. Compared to LC–MS data that showed responses varied approximately 21-fold across these compounds, the responses with NanoMate only varied approximately 2.2-fold. It is worth noting that nanospray not only generated more uniformed responses for compounds in the same class but also across the classes. Valaskovic et al. (2006) first reported the use of nanospray for standard-free estimation of metabolites in biological fluids. In this work, a pulled fused-silica nanospray tip with a diameter of 2 or 5 μm was used as the nanospray emitter, which could operate at a flow rate in the low nanoliter/minute range. A total of six drugs and their respective metabolites were spiked into reconstituted rat plasma extracts as an equimolar mixture. The relative responses of these parent and metabolite groups in both nanospray and electrospray mode were shown in Figure 17.3. Overall, nanospray generated much more uniformed responses among the parent compounds and their metabolites compared to electrospray. There also appeared to be a trend of increasing response uniformity when the flow rate further decreased within the nanospray mode. Of the six groups tested, the largest difference in responses observed was between cocaine and benzoylecogonine, which showed approximately a 2-fold difference. These encouraging results suggested the feasibility of nanospray operated at an ultra-low flow rate to quantify or at least semi-quantify metabolites in the absence of synthetic standards. Also, Valaskovic et al. (2006) proposed a practically useful calibration approach to quantify metabolites in the biological matrices. A flow diagram of this approach is shown in Figure 17.4. In this approach, a response correction factor for the metabolite against the parent compound was obtained from one study sample with ultra-low-flow nanospray. Since the purpose of this correction factor was only for mass spectrometric response but not for extraction recovery, a selective extraction method such as the liquid–liquid extraction method was used to

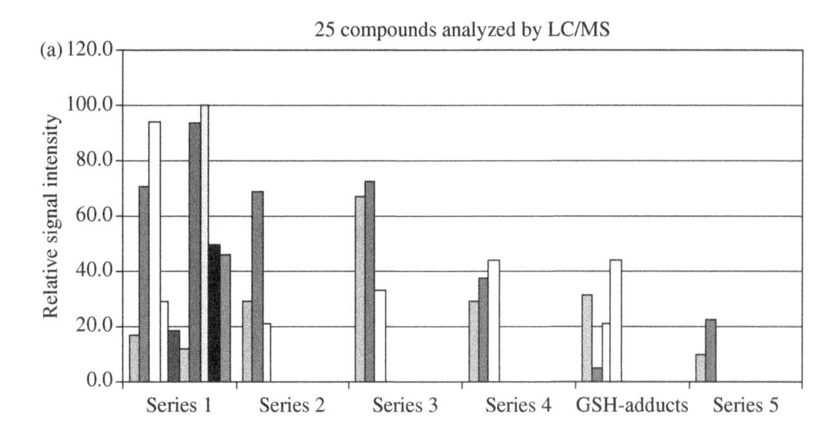

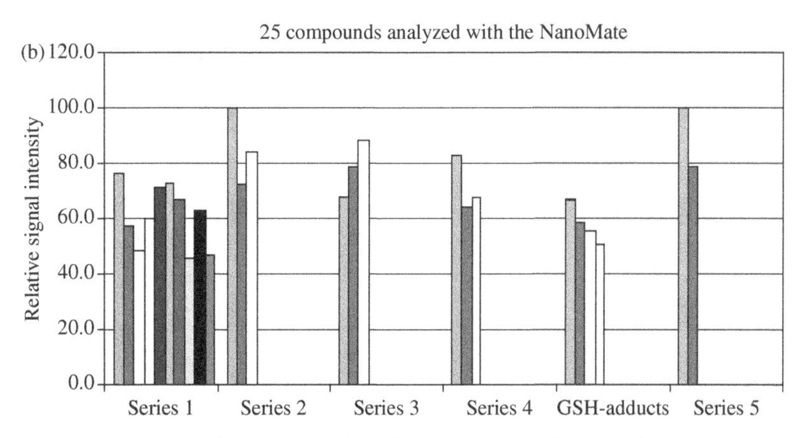

FIGURE 17.2 Comparison of relative signal intensity for 25 pharmaceutical compounds of different series spiked at the same molar concentration in neat solvent. (*a*) Compounds analyzed with LC–MS and (*b*) compounds analyzed by nanospray. [Reprinted from Hop et al. (2005) with permission.]

generate a cleaner sample for the nanospray. Once a response correction factor was obtained, all study samples were prepared with a protein precipitation method and analyzed with LC–MS followed by applying the response correction factor. Two assumptions in this approach were that extraction recovery for the metabolite was similar to the parent compound with protein precipitation and that the difference in matrix effects between samples prepared by liquid–liquid extraction and protein precipitation were negligible by employing good chromatographic separations.

The procedures described above provided initial feasibility for the estimation of metabolite levels in the absence of synthetic standards. One key drawback of the approach was that there was no differentiation among metabolites that were isomers as no chromatographic separation was used in this approach. Ramanathan et al. (2007) developed an improved methodology that

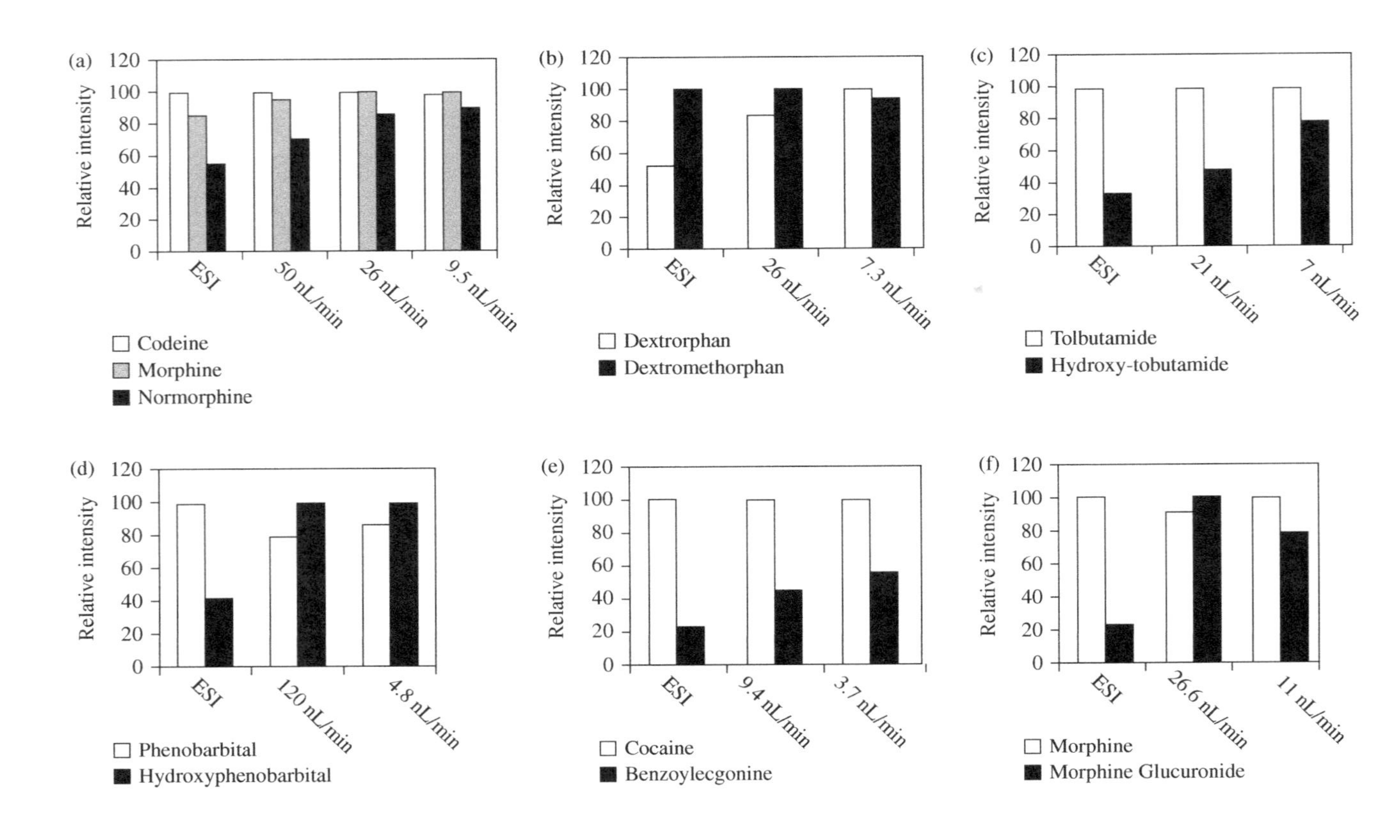

FIGURE 17.3 Comparison of relative intensity of parent compounds and metabolites in electrospray and nanospray at different flow rates. The compound and metabolites were spiked at the same molar concentration in rat plasma extracts. [Reprinted from Valaskovic et al. (2006) with permission.]

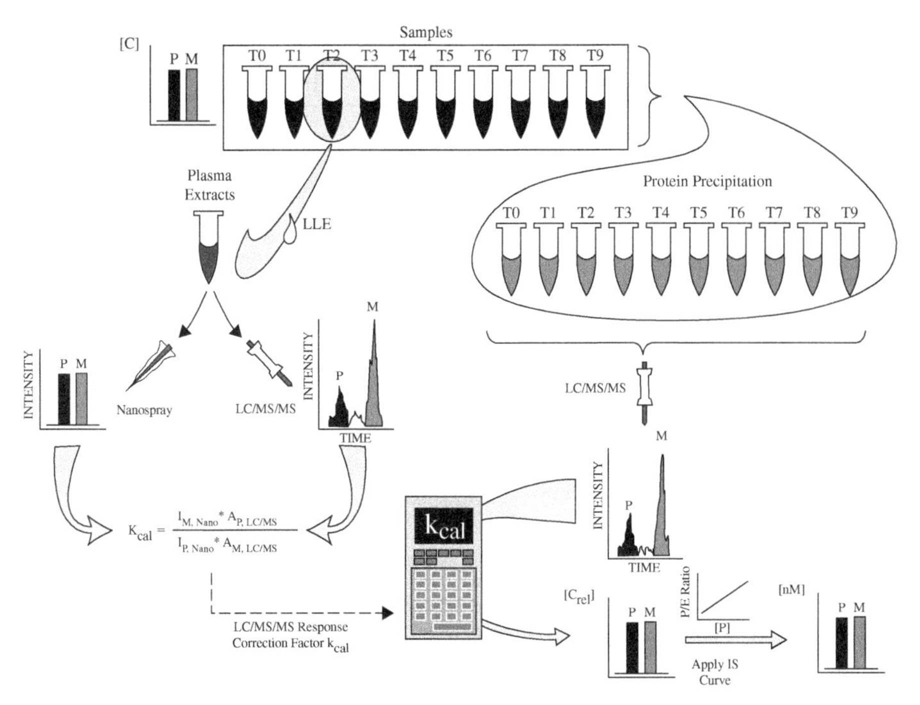

FIGURE 17.4 Schematic diagram of the calibration approach used in standard-free metabolite quantitation. [Reprinted from Valaskovic et al. (2006) with permission.]

incorporated HPLC separations before nanospray ionization. To compensate for the response difference from organic solvent composition in a gradient separation, a second HPLC pump was used to deliver a reverse gradient through postcolumn mixing so that the solvent composition at the nanospray emitter remained isocratic. A schematic diagram of this response normalized liquid chromatography nanospray ionization mass spectrometry is shown in Figure 17.5. Using this system, they demonstrated that the responses generated for vicriviroc and its metabolites quantitatively approached those generated using a radioactivity detector.

17.4 FUTURE DIRECTIONS

Although encouraging results have been reported by multiple groups in using nanospray for standard-free metabolite estimation, it is still a quickly evolving and highly controversial area. There is skepticism whether this equimolar response phenomenon is applicable to a wide range of structurally diversified compounds. In the author's opinion, considering equimolar response as a general property for nanospray under a variety of conditions may be over claiming. It must be recognized that equimolar response may only be achieved under some carefully optimized nanospray conditions. This may explain why

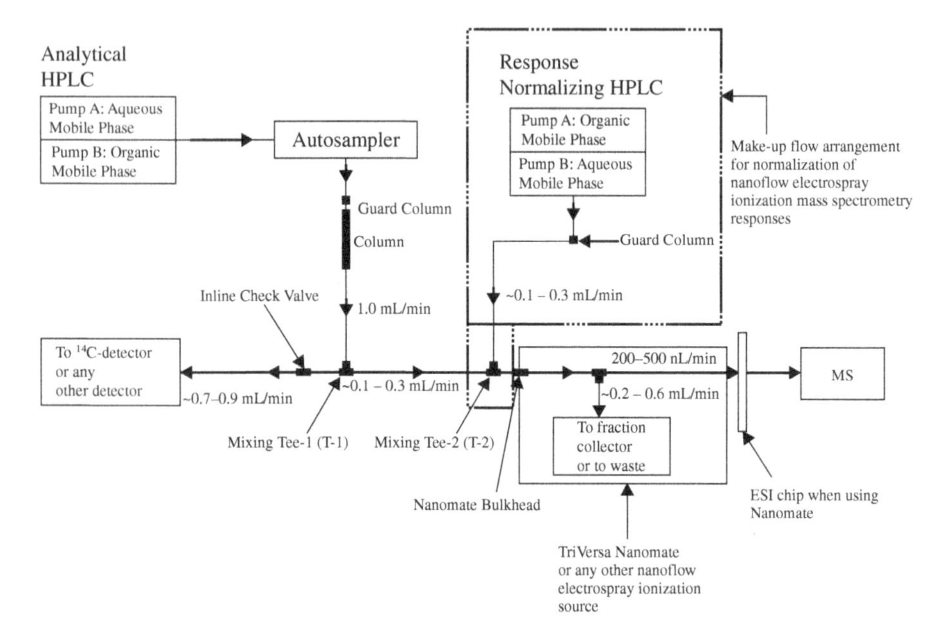

FIGURE 17.5 Diagram of the use of nanospray with response normalizing HPLC for standard-free metabolite quantitation. [Reprinted from Ramanathan et al. (2007) with permission.]

conflicting results exist among labs employing different operating conditions for the nanospray. It is very likely that true equimolar response can only be achieved at a region that matrix effects are not only reduced but virtually eliminated. The region may require nanospray to operate at a flow rate of low or even sub-nanoliter/minute range. This is not easily achievable with most of the commercial nanospray devices currently available. Therefore, improvements on instrumentation toward reliably operating in the low to sub-nanoliter/minute flow rate may be highly desirable before exploring a large chemical space to verify equimolar response.

As the technology stands today, nanospray under carefully optimized experimental conditions is a useful tool for obtaining meaningful information for initial and semiquantitative assessment of metabolites in drug discovery and early development. This includes applications to understand major metabolic pathway and "soft" spots during lead optimization and applications to understand the significance of any specific metabolites in terms of efficacy and toxicity. The information obtained can be used for initial decision making for further follow-up. It must be recognized that at this stage the technology is not robust enough to form the basis of any regulatory decision, such as accessing the criteria defined in the FDA guidance for Safety Testing of Drug Metabolites. However, this stance may change as the technology is still quickly evolving. Developing a set of experimental parameters specially optimized for equimolar response with a commercially available nanospray platform and implement the platform for a large chemical space is the path forward to fully assess the utility of this technology and to bring this technology to wide acceptance by the industry and the regulatory agencies.

REFERENCES

Baillie TA, Cayen MN, Fouda H, Gerson RJ, Green JD, Grossman SJ, Klunk LJ, LeBlanc B, Perkins DG, Shipley LA. Drug metabolites in safety testing. Toxicol Appl Pharmacol 2002;182:188–196.

Cole RB, Ed. Electrospray Ionization Mass Spectrometry, Fundamentals, Instrumentation and Applications. Wiley, New York, 1997.

Deng Y, Wu J-T, Zhang H, Olah TV. Quantitation of drug metabolites in the absence of pure metabolite standards by high-performance liquid chromatography coupled to a chemiluminescence nitrogen detector and mass spectrometer. Rapid Commun Mass Spectrom 2004;18:1681–1685.

Dethy J-M, Ackermann BL, Delatour C, Henion JD, Schultz GA. Demonstration of direct bioanalysis of drugs in plasma using nanoelectrospray infusion from a silicon chip coupled with tandem mass spectrometry. Anal Chem 2003;75:805–811.

Espina R, Yu L, Wang J, Tong Z, Vashishtha S, Talaat R, Scatina J, Mutlib A. Nuclear magnetic resonance spectroscopy as a quantitative tool to determine the concentration of biologically produced metabolites: Implications in metabolites in safety testing. Chem Res Toxicol 2009;22:299–310.

Hah SS. Recent advances in biomedical applications of accelerator mass spectrometry. J Biomed Sci 2009;16:54, pp. 111–124.

Hop CECA, Chen Y, Yu LJ. Uniformity of ionization response of structurally diverse analyte using a chip-based nanoelectrospray ionization source. Rapid Commun Mass Spectrom 2005;19:3139–3142.

Josephs JL, Luk CE, Grubb M, Yang Y, Zhang H, Cai H, Langish R, Shipkova P, Sanders M, Humphreys WG. An integrated approach to in vitro and in vivo metabolite quantitation based on high resolution full scan MS Data. Oral presentation. Presented at the 57th Annual Conference of Am Soc Mass Spectrom, Philadelphia, June 2009.

Leclercq L, Cuyckens F, Mannens GSJ, de Vries R, Timmerman P, Evans DC. Which human metabolites have we mist? Retrospective analysis, practical aspects, and perspectives for metabolite identification and quantification in pharmaceutical development. Chem Res Toxicol 2009;22:280–293.

Mackenzie IS. Theoretical aspects of quantitative NMR. Anal Proc 1984;21:500–502.

Malz F, Jancke H. Validation of quantitative NMR. J Pharm Biomed Anal 2005;38:813–823.

Ramanathan R, Zhong R, Blumenkrantz N, Chowdhury SK, Alton KB. Response normalized liquid chromatography nanospray ionization mass spectrometry. J Am Soc Mass Spectrom 2007;18:1891–1899.

Schmidt A, Karas M, Dulcks T. Effect of different solution flow rates on analyte ion signals in nano-ESI MS, or: when does ESI turn into nano-ESI? J Am Soc Mass Spectrom 2003;14:492–500.

Smith DA, Obach RS. Metabolites and safety: What are the concerns, and how should we address them? Chem Res Toxicol 2006;19:1570–1579.

U. S. Food and Drug Administration (FDA). Guidance for Industry: Safety Testing of Drug Metabolites. FDA Center for Drug Evaluation of Research (CDER), Rockville, MD, 2008.

Valaskovic GA, Utley L, Lee MS, Wu J-T. Ultra-low flow nanospray for the normalization of conventional liquid chromatography/mass spectrometry through equimolar response: Standard-free quantitative estimation of metabolite levels in drug discovery. Rapid Commun Mass Spectrom 2006;20:1087–1096.

Van Pelt CK, Zhang S, Fung E, Chu I, Liu T, Li C, Korfmacher WA, Henion J. A fully automated nanospray tandem mass spectrometric method for analysis of Caco-2 samples. Rapid Commun Mass Spectrom 2003;17:1573–1578.

Wilm MS, Mann M. Electrospray and Taylor-cone theory, Dole's beam of macromolecules at last? Int J Mass Spectrom Ion Process 1994;136:167–180.

Wilm MS, Mann M. Analytical properties of the nanoelectrospray ion source. Anal Chem 1996;68:1–8.

Young G, Ellis W, Ayrton J, Hussey E, Adamkiewicz B. Accelerator mass spectrometry (AMS): Recent experience of its use in a clinical study and the potential future of the technique. Xenobiotica 2001;31:619–632.

Yu C, Chen CL, Gorycki FL, Neiss TGA. A rapid method for quantitatively estimating metabolites in human plasma in the absence of synthetic standards using a combination of liquid chromatography/mass spectrometry and radiometric detection. Rapid Commun Mass Spectrom 2007;21:497–502.

Zhu M, Zhang D, Zhang H, Shyu WC. Integrated strategies for assessment of metabolite exposure in humans during drug development: Analytical challenges and clinical development considerations. Biopharm Drug Dispos 2009;30:163–184.

18 Profiling and Characterization of Herbal Medicine and Its Metabolites Using LC–MS

ZEPER ABLIZ, RUIPING ZHANG, PING GENG, DONGMEI DAI, JIUMING HE, and JIAN LIU

Key Laboratory of Bioactive Substances and Resource Utilization of Chinese Herbal Medicine, Ministry of Education, Institute of Materia Medica, Chinese Academy of Medical Sciences and Peking Union Medical College, Beijing, 100050 China

18.1 INTRODUCTION

Chinese herbal medicine has developed over thousands of years. The vast human pharmacological information and experience that have been accumulated form an integrated theory system. However, Chinese herbal medicine is such a complex system that a large number of chemical constituents and metabolites, in vivo and in vitro, need both qualitative and quantitative analysis. More importantly, the chemical compositions and metabolites that exist in various chemical processes are closely related to the therapeutic effects

Mass Spectrometry in Drug Metabolism and Disposition: Basic Principles and Applications, First Edition. Edited by Mike S. Lee and Mingshe Zhu.

of Chinese herbal medicine (Chen et al., 2006; Chung et al., 2004; Lee, 2000). Nowadays, with the trend of returning to nature and especially Chinese herbal medicine and natural products as the most important resource for screening lead compounds, scientists in all countries show great interest in the research that involves elucidation of the material foundation of Chinese herbal medicine. In all, there is a great desire to obtain the information about the chemical composition of Chinese herbal medicine.

The characteristics of multiconstituent, multitarget effect mechanisms have restricted the revelation of the active material basis of Chinese herbal medicine, which makes the quality control of herbal medicine a problem. Therefore, study of analytical methods of the chemical constituents in Chinese herbal medicine is not only a significant challenge that pharmaceutical analysts face but also a key link in the investigation of the material basis of Chinese herbal medicine. Generally, there are three types of chemical analysis of Chinese herbal medicine: (1) the analysis of chemical constituents in Chinese medicine materials; (2) the analysis of chemical constituents in a prescription; and (3) the analysis of the original drug and its metabolites in body fluids (such as blood serum, tissue, urine, etc.) or in vitro bacteria/cell metabolism. The analysis of various trace-level constituents and metabolites in Chinese herbal medicines and their pharmaceutical preparations is critical to the research of such complex mixture systems.

Along with the development of modern analytical techniques, liquid chromatography–mass spectrometry (LC–MS) with high sensitivity and selectivity has become the most powerful tool for the separation and analysis of chemical constituents in Chinese herbal medicine (Cai et al., 2002; Yang et al., 2009). Moreover, for the application of tandem mass spectrometry (MS/MS) and multistage mass spectrometry (MS^n), LC–MS/MS can provide more plentiful and efficient structural information to establish a rapid and efficient analytical system. Online analysis of Chinese herbal medicine cannot only avoid the complicated, cumbersome, and time-consuming separation but also promptly provide structure information through a proper scan mode (neutral loss scan, parent ion scan, and product ion scan) of the original drug and its metabolites, which are trace and difficult to identify (Chang et al., 2000). LC–MS/MS has unique advantages and plays an increasingly important role in the analysis of active ingredients and their metabolites in Chinese herbal medicine.

Because many of the chemical constituents or active ingredients in Chinese herbal medicine are not yet identified, studies on metabolism of Chinese herbal medicine are less reported (Wang et al., 2005, 2008; Tang, 2002; Zhang et al., 2005; Zhao et al., 2007; Wei et al., 2007; Wu et al., 2008). The fingerprint analysis of Danshen injection fluid and its raw materials (roots and rhizoma of *Salvia miltiorrhiza*) by high-performance liquid chromatography (HPLC)–MS was described (Zhang et al., 2005). As a result, 11 major chromatographic peaks were characterized by their MS spectra and comparison with the reference standards. In addition, rat plasma was analyzed by HPLC–MS after intravenous administration of Danshen injection fluid at different time intervals to explore the in vivo metabolism of the major active constituents. The comparison of chemical fingerprints with metabolic fingerprints indicated that polyphenolic acids were significant for biological activity of Danshen injection fluid. It might be concluded that chemical fingerprints combined with metabolic fingerprints is a useful means to control the quality and to clarify the possible mechanism of action of herbal products. In another study (Wu et al., 2008), HPLC Fourier transform ion cyclotron (FTICR) mass spectroscopy (HPLC–FTICR–MS) was developed to identify active compounds and their metabolites after oral administration of an herbal extract of *Epimedium koreanum* Nakai to rats. By contrasting the analytical results obtained from the herbal extract with those obtained from biological specimens, the profile of flavonoid biotransformation in *Epimedium* was obtained, and the metabolic pathways of the main components, in rats, were described. It was found that the main flavonoids of *Epimedium*, such as epimedin A, epimedin B, and epimedin C, metabolize to baohuoside I, in accordance with the terminal metabolite of icariin. Therefore, it can be proposed that the quality of *Epimedium* should be evaluated using multi-indexical quantitative method, not by individual icariin.

The routine research of active fractions or active components from Chinese herbal medicine includes repeated extraction, separation, and bioactivity screening. The whole process is tedious, laborious, and expensive. Clearly, it would be difficult to predict the fate of these active components when they undergo the study of efficacy, pharmacology, and metabolism. Perhaps another point of view would be to consider how the human body reacts to a drug and the subsequent analysis of metabolism of active fractions or active components. This perspective may allow for novel studies on therapeutic basis of Chinese herbal medicine and result in breakthroughs in medicine.

There are three situations for the in vivo process of Chinese herbal medicine: (1) Most of the components in complex chemical systems are absorbed as prototype into systemic circulation. Thus, in vivo absorption spectrum of the active components provides evidence for therapeutic basis of Chinese herbal medicine. (2) Most components of the active extracts undergo biotransformation during the absorption process. So their prototypes are hardly detected in biological fluids. This pathway always occurs with plant extracts of which most components are polar compounds and undergo phase II metabolism directly in body. The information accumulated from the analysis of the raw plant mixture and biological specimens is combined with that of the metabolic pathway, and then the transforming relationship between prototype and metabolite is found step by step. (3) There are metabolic transforming relationships between the components in the active fractions. These components always undergo phase I metabolism, including oxidation, reduction, hydrolysis, and hydration, to produce metabolites that are identical to the original prototypes. In this situation, it is hardly determined whether the constituents detected in the biofluid are prototypes or metabolites that just transform to some prototypes through metabolism. Due to complexity of the active mixture, there are many difficulties using the isotope labeling method. The transforming relationship of a mixture is speculated based on metabolism of the single component at present. If the transforming relationships between each component are not considered, then it would be feasible to find material basis of herbal medicine by regarding the in vivo constituents as the possible active components.

Compared with the metabolism of the usual chemical medicine, the metabolism of herbal medicine is more difficult due to the objective fact mentioned above. However, such a complex system still has its own rules since most metabolites remain the skeleton structure or substructure of the original compounds. So metabolites may have the similar mass spectrometric behaviors with the prototype such as losing the same neutral fragments or forming the same characteristic ions. In recent years, the systematic investigation of the mass spectrometric behavior and the relationship between the fragmentation difference and the structural features of natural products with different skeleton types was investiaged using electrospray ionization (ESI)–MS/MS in positive and/or negative ion modes, as well as positive ion fast atom bombardment (FAB)–MS techniques (Liang et al., 2002; Xiang et al., 2002; Liu et al., 2004, 2009; Li et al., 2005; Ablajan et al., 2006; Cui et al., 2004). The discoveries included a series of novel and specific fragmentation reactions, characteristic rules of fragmentation behavior, and different effects of positive and negative ionization, which can indicate explicitly the minor structural differences of similar compounds. The notable diversity of fragmentation behavior that arises from structural differences becomes a powerful diagnostic tool for identification of novel natural product structures. Meanwhile the information obtained provides the basis for the online identification of trace components in the complex mixture. Furthermore, many kinds of bioactive crude extracts from various Chinese herbal medicines were analyzed by online LC–MS/MS, LC–ultraviolet (UV)–MS and ESI–MS^n techniques (Cui et al., 2004; Li et al., 2006). The rapid and efficient analytical method was developed based on a complementary strategy that combined mass spectrometry and the fragmentation rules with chromatographic analysis. This strategy proved to be suitable for the structural characterization of the main and trace-level constituents with different skeleton types, such as alkaloid, saponin, and flavonoid, in various crude extracts. On the basis of the investigation cited above, a method for the fast profiling of the constituents of flavonol in an active herbal extract and their metabolites in biological fluids was established (Geng et al., 2007). *Gossypium herbaceam* L. is an ethnic medicine to treat mental retardation by the Uygur people in Xinjiang, China. An active extract obtained from *Gossypium herbaceam* L., named AB-8-2, mainly contains flavonols. Investigations on AB-8-2-related therapeutic agents are ongoing, and the identification of AB-8-2 components, and the profiling of its metabolites, will offer a better understanding of its mechanism of action. The research strategy that obtains complementary information from mass spectrometric behavior and metabolic pathway to support the structure elucidation of prototypes and metabolites was proposed and employed for profiling the integral metabolism of flavonols of AB-8-2.

18.2 CHARACTERIZATION OF CHEMICAL CONSTITUENTS IN CHINESE HERBAL MEDICINE

18.2.1 Systematic Identification Method for Flavonols

Structure information on the flavonol compounds (Table 18.1), which includes the type of aglycone, the attachment points of the substituents to the aglycone, the sequence of the glycan part, and the interglycosidic linkages, could be obtained from the MS/MS spectra. The results of our experiments combined with publication data (Cuyckens and Claeys, 2005; Ferreres et al., 2004; Cuyckens et al., 2001) were applied to establish the systematic identification method for flavonol constituents.

To describe the fragmentations observed in product ion spectra, the nomenclature proposed by Domon and Costello (1988) was adopted (Scheme 1).

Figure 18.1 contains the product ion spectrum that features a high abundance radical aglycone ion $[Y_0{-}H]^{-\bullet}$ indicated by 3-*O*-glycosylation,

TABLE 18.1 Structures of Flavonols Studied

Compound	R_1	R_2	R_3	R_4	R_5	MW (Da)
Isoquercetrin	Glc	H	H	OH	H	464
Quercimeritrin	H	Glc	H	OH	H	464
Quercetin-3′-glucoside	H	H	H	Oglc	H	464
Quercetin-3-glucuronide	GlcUA	H	H	OH	H	478
Rutin	Rutinose	H	H	OH	H	610
Quercetin	H	H	H	OH	H	302
Isorhamnetin	H	H	H	OCH_3	H	316
Tamarixetin	H	H	H	OH	CH_3	316
Astragalin	Glc	H	H	H	H	448
8-Methoxyl-kaempferol-7-rh Amnoside	H	Rha	OCH_3	H	H	462
Kaempferol-3-(6″-p-coumaroyl)-glucoside	(6″-p-coumaroyl)-glucoside	H	H	H	H	594
Kaempferol	H	H	H	H	H	286

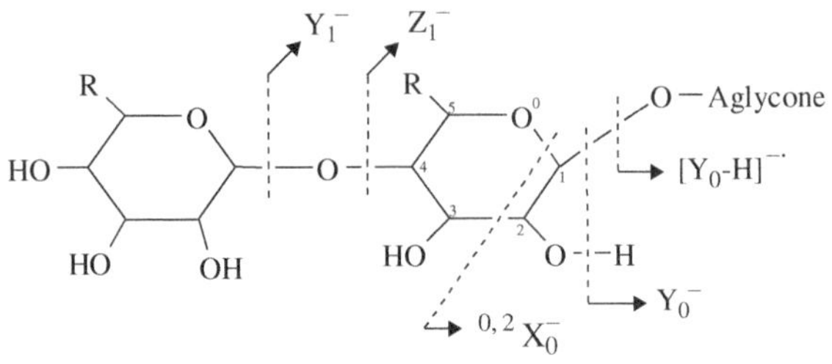

SCHEME 18.1 Ion nomenclature used for flavonol glycosides (Domon and Costello, 1988).

which is consistent with previous results (Ablajan et al., 2006; Cuyckens and Claeys, 2005). In addition, a cross-ring product ion $^{0,2}X_0^-$ could be observed in the MS/MS spectra of flavonol 3- and 7-*O*-glycosides, but not in flavonol 3′-*O*-glycosides. Therefore, the $[Y_0{-}H]^{-\bullet}$ and $^{0,2}X_0^-$ ions were used as diagnostic ions to differentiate 3-*O*, 7-*O*, and 3′-*O*-glycosyl flavonols. A high abundance $[Y_0\text{-}H]^{-\bullet}$ ion along with the observation of the $^{0,2}X_0^-$ ion characterized 3-*O*-glycosylation (Fig. 18.1*a* and 18.1*d*); 7-*O*-glycosylation yielded no obvious $[Y_0\text{-}H]^{-\bullet}$ ion, but the $^{0,2}X_0^-$ ion was present (Fig. 18.1*b*), and neither the $[Y_0\text{-}H]^{-\bullet}$ nor the $^{0,2}X_0^-$ ion was present in the spectrum of quercetin-3′-glucoside (Fig. 18.1*c*). Furthermore, the fragmentation of rutin was similar to that of isoquercetrin (Fig. 18.1*a*), and featured the loss of an intact disaccharide moiety (rutinose). For quercetin-3-glucuronide, only the aglycone ion Y_0^- was observed, while the two diagnostic ions were absent.

Flavonol-di-*O*-glycoside and flavonol-mono-*O*-diglycoside could easily be distinguished from each other. In the MS^2 spectrum of the $[M\text{-}H]^-$ ion, the Y_1^- ion formed by loss of a terminal residue was the base peak for the former, while the Y_0^- ion was dominant for the latter (Ferreres et al., 2004). As for the interglycosidic linkage on flavonoid dihexosides, such as rutinose [rhamnosyl-(1→6)-glucose] and neohesperidose [rhamnosyl-(1→2)-glucose] (Ma et al., 2001), the common rules were stated that the observation of both the Y_1^- and the Z_1^- ions with high relative abundance pointed to the 1→2 linkage, whereas in the 1→6 case those two ions were in very low abundance or absent (Ferreres et al., 2004; Cuyckens et al., 2001). These rules were well proven in structural elucidation of flavonol glycosides.

18.2.2 Online Structural Characterization of Constituents in AB-8-2

Most of the pure compounds (Table 18.1) originally exist in *Gossypium herbaceam* L. The common biogenetic origin increased the confidence for the identification of the constituents in the active fraction AB-8-2. According to the systematic identification method of flavonols, constituents in AB-8-2 were profiled in one chromatographic run with data-dependent full-scan MS^n (Table 18.2).

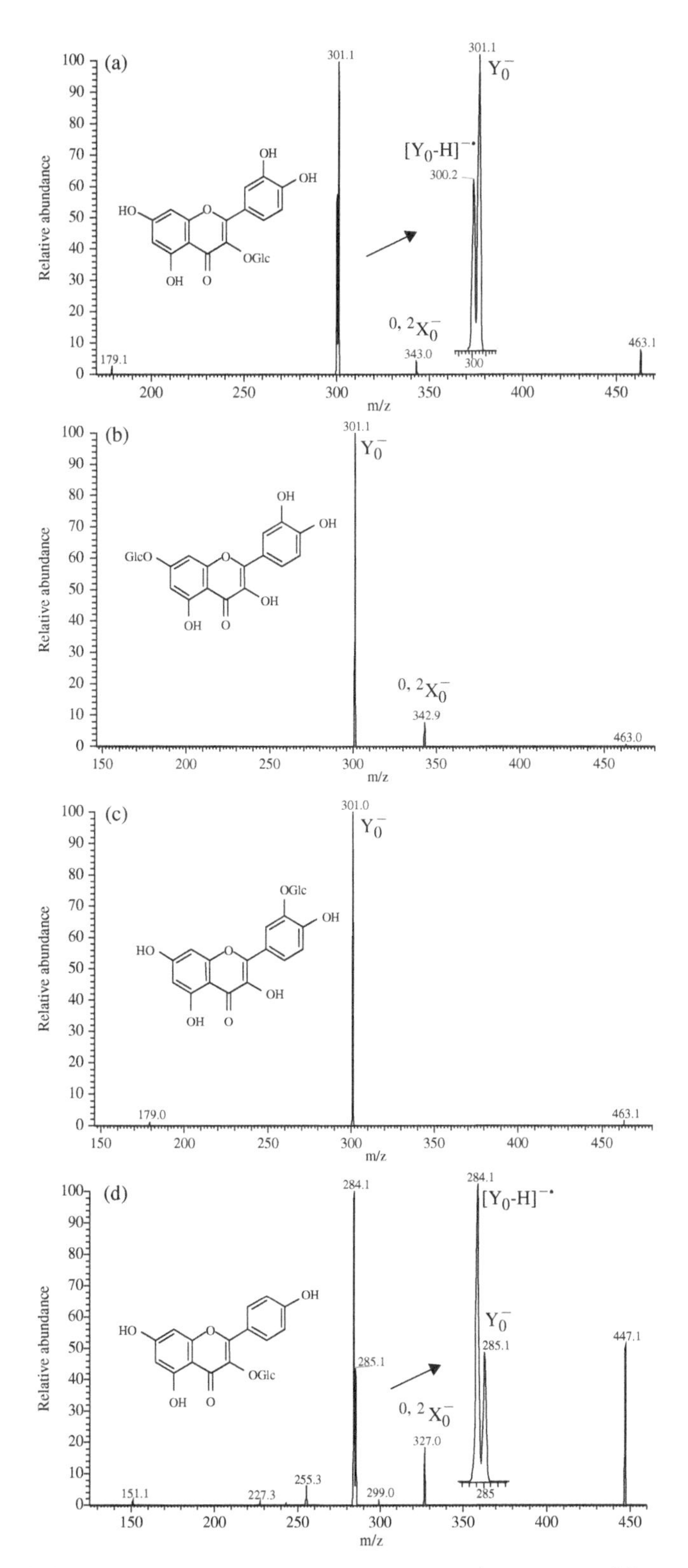

FIGURE 18.1 Comparison between $[M-H]^-$ product ion spectra of flavonol mono-*O*-glycosides: (*a*) isoquercetrin, (*b*) quercimeritrin, (*c*) quercetin-3′-glucoside, and (*d*) astragalin.

TABLE 18.2 HPLC–MS^n Data Of Parent Constituents in AB-8-2[a]

Component[b]	t_R (min)	[M-H]−(m/z)	ESI–MS^n Data (m/z, Relative Abundances, %)	Identity[c]	Location[d]
1	2.69	595	MS^2[595]: 595(2), 475(8), 463(26), 445(15), 343(3), 301(18), 300(100), 271(8), 255(4)	Q-3-Hex-Pen (1→2)	AB
2	2.75	625	MS^2[625]: 625(2), 505(1), 463(100), 462(14), 343(1), 301(16) MS^3[625→463]: 463(9), 343(5), 301(100), 300(43), 271(1)	Q-Hex-Hex	AB
3	4.59	625	MS^2[625]: 505(1), 463(100), 462(21), 343(1), 301(13) MS^3[625→463]: 463(6), 343(2), 325(2), 301(100), 300(53)	Q-Hex-Hex	AB
4	6.10	609	MS^2[609]: 489(8), 447(100), 285(3) MS^3[609→447]: 447(23), 284(100) MS^4[609→447→284]: 255(100)	K-Hex-Hex	AB
6	9.81	625	MS^2[625]: 607(2), 463(100), 445(3), 301(5), 300(3) MS^3[625→463]: 301(100), 300(14) MS^4[625→463→301]: 179(100)	Q-Hex-Hex	AB
7	11.08	625	MS^2[625]: 463(100), 301(6), 300(1) MS^3[625→463]: 301(100), 300(22) MS^4[625→463→301]: 301(72), 273(2), 257(11), 193(2), 179(100), 151(41)	Q-Hex-Hex	AB
8	11.48	623	MS^2[623]: 477(100), 461(58), 460(9), 315(23) MS^3[623→477]: 477(13), 462(100), 314(35), 299(37) MS^3[623→461]: 446(6), 315(100) MS^4[623→477→299]: 299(19), 271(100)	MQ-Hex-dHx	AB, iv
9	11.48	595	MS^2[595]: 475(6), 463(13), 445(16), 427(1), 355(1), 343(2), 301(21), 300(100), 273(1), 271(9), 255(4), 179(1) MS^3[595→445]: 445(31), 355(100) MS^4[595→445→355]: 355(46), 337(100)	Q-3-Hex-Pen (1→2)	AB
10	12.27	595	MS^2[595]: 595(1), 577(1), 475(7), 463(31), 445(8), 343(3), 301(20), 300(100), 273(1), 271(10), 255(4), 179(1) MS^3[595→300]: 300(100), 271(70), 255(38), 254(1), 179(5)	Q-3-Hex-Pen (1→2)	AB

11	12.42	623	MS^2[623]: 608(3), 477(2), 459(26), 357(4), 314(100), 299(46) MS^3[623 → 314]: 299(100)	MQ-3-Hex-dHx (1 → 2)	AB
12	12.86	609	MS^2[609]: 609(3), 343(10), 301(95), 300(100), 271(10), 255(5), 179(2) MS^3[609 → 343]: 343(77), 297(100)	Rutin	AB
13	13.61	463	MS^2[463]: 343(8), 301(100) MS^3[463 → 301]: 301(100), 179(10), 151(5)	Quercimeritrin	AB, iv
14	13.97	463	MS^2[463]: 343(2), 301(100), 300(57), 179(1) MS^3[463 → 301]: 301(70), 300(2), 273(13), 271(2), 257(11), 255(2), 239(2), 229(2), 193(1), 179(100), 151(52), 107(1) MS^4[463 → 301 → 179]: 169(10), 151(100)	Isoquercetrin	AB, iv
18	15.53	447	MS^2[447]: 447(21), 419(1), 357(1), 327(13), 285(54), 284(100), 257(1), 255(5) MS^3[447 → 284]: 284(14), 255(100), 227(10)	K-3-Hex	AB, iv, ig
19	15.53	433	MS^2[433]: 343(1), 301(41), 300(100) MS^3[433 → 300]: 300(100), 271(70), 255(44), 254(2), 243(1), 179(8), 151(2)	Q-3-Pen	AB, iv, ig
20	15.81	505	MS^2[505]: 463(43), 445(5), 343(2), 301(84), 300(100), 273(1) MS^3[505 → 463]: 301(74), 300(100) MS^3[505 → 300]: 300(100), 271(67), 257(2), 255(33), 179(50), 151(18)	Q-3-acetylHex	AB
22	16.32	447	MS^2[447]: 447(40), 327(16), 299(1), 285(78), 284(100), 257(1), 255(7) MS^3[447 → 284]: 284(15), 255(100), 227(11)	Astragalin	AB, iv, ig
25	16.98	505	MS^2[505]: 505(3), 463(50), 445(4), 301(96), 300(100), 271(2), 179(1)	Q-3-acetylHex	AB, iv

(*Continued*)

TABLE 18.2 (Continued)

Component[b]	t_R (min)	[M-H]−(m/z)	ESI−MS^n Data (m/z, Relative Abundances, %)	Identity[c]	Location[d]
			MS^3[505→463]: 301(100), 300(69)		
			MS^3[505→300]: 300(100), 271(77), 255(44), 179(39), 151(14)		
26	17.02	447	MS^2[447]: 447(11), 327(13), 285(100), 284(36), 255(1)	K-7-Hex	AB
			MS^3[447→285]: 285(100), 257(8), 256(7), 241(1), 151(2)		
27	17.10	477	MS^2[477]: 477(50), 462(4), 449(14), 387(2), 357(19), 315(33), 314(100), 300(2), 299(22), 285(1), 271(2)	MQ-3-Hex	AB, iv, ig
			MS^3[477→314]: 314(7), 299(100), 286(23), 285(50), 271(40), 257(6), 243(10)		
			MS^4[477→314→271]: 271(100), 243(11)		
28	17.29	417	MS^2[417]: 417(20), 327(7), 285(36), 284(100)	K-3-Pen	AB, iv
			MS^3[417→284]: 284(28), 255(100), 227(9)		
31	17.91	505	MS^2[505]: 505(1), 463(78), 445(7), 343(1), 301(54), 300(100), 299(1), 271(1), 179(1)	Q-3-acetylHex	AB, iv
			MS^3[505→445]: 300(100)		
38	18.82	463	MS^2[463]: 301(100), 300(2)	Quercetin-3′-glucoside	AB, iv
			MS^3[463→301]: 301(83), 273(4), 179(100), 151(36)		
			MS^4[463→301→179]: 169(100), 151(100)		
40	19.75	477	MS^2[477]: 477(1), 315(100), 314(3), 300(7)	MQ-3′-Hex	AB, iv
			MS^3[477→315]: 315(12), 300(100)		
			MS^4[477→315→300]: 300(100), 271(49), 255(28), 151(1)		
42	21.08	461	MS^2[461]: 461(33), 433(44), 432(9), 299(100), 298(78), 271(5)	MK-Hex	AB
			MS^3[461→299]: 299(100), 271(80)		
			MS^3[461→433]: 433(100), 271(3)		

43	21.87	461	MS^2[461]: 461(45), 433(21), 341(6), 315(2), 299(100), 298(38) MS^3[461 → 299]: 299(100), 271(34)	MK-Hex	AB
46	22.79	461	MS^2[461]: 461(20), 446(7), 315(100), 314(52), 300(19), 299(4) MS^3[461 → 315]: 315(23), 300(100), 299(4), 287(4), 272(1), 256(1), 181(6), 166(1) MS^4[461 → 315 → 300]: 300(3), 272(100), 256(8), 255(2), 239(3), 166(1)	8-methoxyl-kaempferol-7-rhamnoside	AB, iv, ig
47	23.39	593	MS^2[593]: 447(20), 307(3), 285(100), 284(8) MS^3[593 → 447]: 327(25), 285(100) MS^3[593 → 285]: 285(100), 257(6), 151(1)	Kaempferol-3-(6″-*p*-coumaroyl)-glucoside	AB, iv, ig
48	24.38	593	MS^2[593]: 447(19), 327(1), 307(3), 285(100), 284(9), 257(2) MS^3[593 → 285]: 285(100), 267(1), 257(7), 241(2), 151(1) MS^3[593 → 447]: 447(7), 285(100), 284(8)	Kaempferol-3-(6″-*p*-coumaroyl)-hexoside	AB, iv
49	24.53	301	MS^2[301]: 301(67), 273(15), 257(10), 239(1), 229(2), 193(4), 179(100), 151(48) MS^3[301 → 179]: 151(100) MS^4[301 → 179 → 151]: 107(100)	Quercetin	AB

(*Continued*)

TABLE 18.2 (Continued)

Component[b]	t_R (min)	[M-H]−(m/z)	ESI−MS^n Data (m/z, Relative Abundances, %)	Identity[c]	Location[d]
52	29.57	607	MS^2[607]: 607(3), 579(1), 461(5), 307(4), 299(100), 298(1), 284(17) MS^3[607→299]: 299(100), 284(59) MS^3[607→284]: 284(100), 255(17)	4′-methyl-kaempferol-3-(6″-*p*-coumaroyl)-β-D-glucoside	AB, iv

[a] t_R, retention time; Q, quercetin; M, methyl; MQ, methylquercetin; MK, methylkaempferol; Hex, hexose; dHx, deoxyhexose; Pen, pentose; AB, AB-8-2.

[b] All the parent constituents in Table 18.1 and metabolites in Table 18.2 were sorted and numbered uniformly by the retention time under the same chromatographic condition.

[c] "Q-3-Hex-Pen (1→2)" indicates flavonol mono-*O*-diglycoside with the interglycosidic linkages in bracket; "Q-Hex-Hex" indicates flavonol-di-*O*-glycoside. Methyl in **42** and **43** was not located on hydroxyl, probably on the phenyl group.

[d] "AB, iv" indicates the corresponding component was detected both in AB-8-2 and rat bile after intravenous administration; "AB, iv, ig" indicates the corresponding component was detected in AB-8-2 and rat bile after intravenous and oral administration.

The availability of reference compounds facilitated the identification of rutin (component **12**), quercimeritrin (**13**), isoquercetrin (**14**), astragalin (**22**), quercetin-3′-glucoside (**38**), 8-methoxylkaempferol-7-rhamnoside (**46**), kaempferol-3-(6″-*p*-coumaroyl)glucoside (**47**), quercetin (**49**), isorhamnetin (**53**), and tamarixetin (**54**) by their fragmentation behavior and retention times.

However, nuclear magnetic resonance (NMR) data were sometimes needed to provide complementary information to validate the chemical structure determined by MS/MS. NMR/LC–MS parallel dynamic spectroscopy (PDS) (Dai et al., 2009) combined with an incomplete separation strategy was proposed and developed for the simultaneous structure identification of constituents in AB-8-2.

The overview of the NMR/LC–MS PDS is shown in Figure 18.2. Through the co-analysis of visualized MS and NMR data with signal amplitude co-variation in the NMR/LC–MS PDS spectra, the intrinsic correlation between retention time (t_R), mass/charge (m/z), and chemical shift (δ) data of the same individual constituent in the LC fractions can be found. As a consequence, the complementary spectral information was obtained from mixture spectra for unambiguous structure identification of individual constituents in crude extracts.

The LC–UV chromatogram of AB-8-2 is shown in Figure 18.3*a*. Figure 18.3*b* was the composition profiles of 12 constituents, which was reconstructed by plotting the HPLC–UV areas of the 12 constituents in the series of fractions. AB-8-2 was incompletely separated and 12 constituents were eluted into different fractions and processed different concentration changing profiles. The LC–MS and ^{1}H NMR data of the series of fractions were acquired separately. After data processing with the routines compiled in MATLAB, the signal amplitude covariation of the extracted ion chromatograms (XIC) and ^{1}H NMR signals were visualized in parallel to produce the NMR/LC–MS PDS spectrum of AB-8-2, as shown in Figure 18.4.

Because the signal amplitude is proportional to concentration, the spectroscopic signals of the same constituent covaried in the series of the artificial fractions, which manifested as sharing the same fraction ranges and intensity changing profiles. Based on this fact, the XIC and ^{1}H NMR signals that were derived from an individual constituent in the series of fractions can be correlated and assigned. In other words, this visualization tool can discover the intrinsic correlation between LC–MS and NMR data of the same constituent from mixture spectra. Thus, the use of NMR/LC–MS PDS was featured.

In Figure 18.4, the fraction axis (vertical axis) resembles retention time axis in a typical chromatogram, and every line represents the XIC and ^{1}H NMR spectrum of each fraction. The series of spectra are similar to the multiple spectra collected over chromatographic peaks from online LC–NMR–MS analysis. In addition, the XICs of different molecular ions are displayed with a different color, which was helpful to distinguish the isomers and coeluted

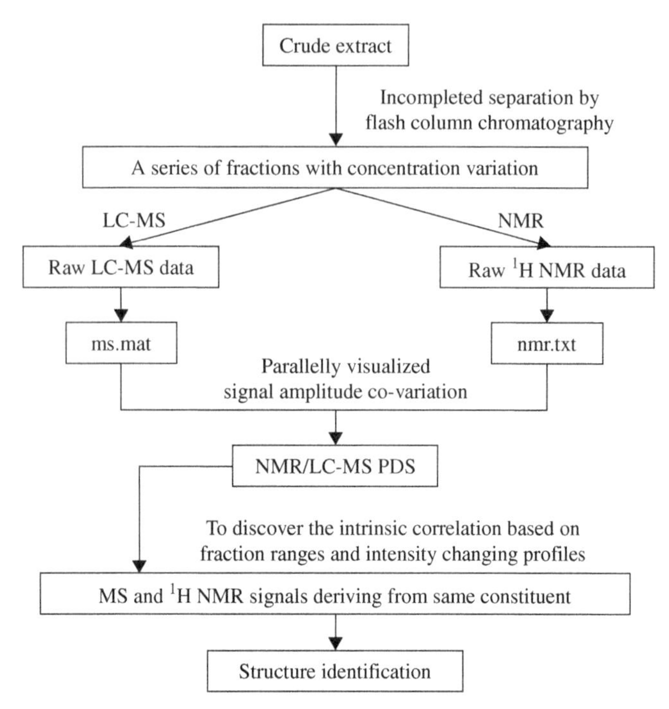

FIGURE 18.2 Overview of the NMR/LC–MS parallel dynamic spectroscopy (NMR/LC–MS PDS) combined with an incomplete separation strategy for simultaneous structure identification of natural products in crude extract.

constituents. For example, three peaks at t_R 13.6, 13.8, and 14.1 min were observed in the "green" XIC of the $[M-H]^-$ ion at m/z 463, corresponding to three isomers existing in AB-8-2.

As a result, the complementary LC–MS and ^{1}H NMR data of the 12 constituents in the active crude extract were correlated and recovered successfully for simultaneous structure identification by using the NMR/LC–MS PDS method without the need for completed separation and isolation. Five quercetin-based (constituents 1, 2, 3, 7, and 8) and five kaempferol-based compounds (constituent 4, 5, 6, 9, and 10), besides quercetin (constituent 11) and kaempferol (constituent 12), were identified. The NMR/LC–MS PDS technique with the incomplete separation strategy played a more important role in the structure identification of coeluted isomers, such as the coelution of constituents 1, 2, and 3. In this case, the chromatographic separation conditions should be carefully optimized in the combination NMR techniques. Moreover, besides the recovered XICs and ^{1}H NMR signals, relevant spectral information, such as LC–MS/MS data and one-dimensional total correlation spectroscopy (TOCSY) spectra, were collaborated through index of the recovered spectroscopic signals for the unambiguous structure identification.

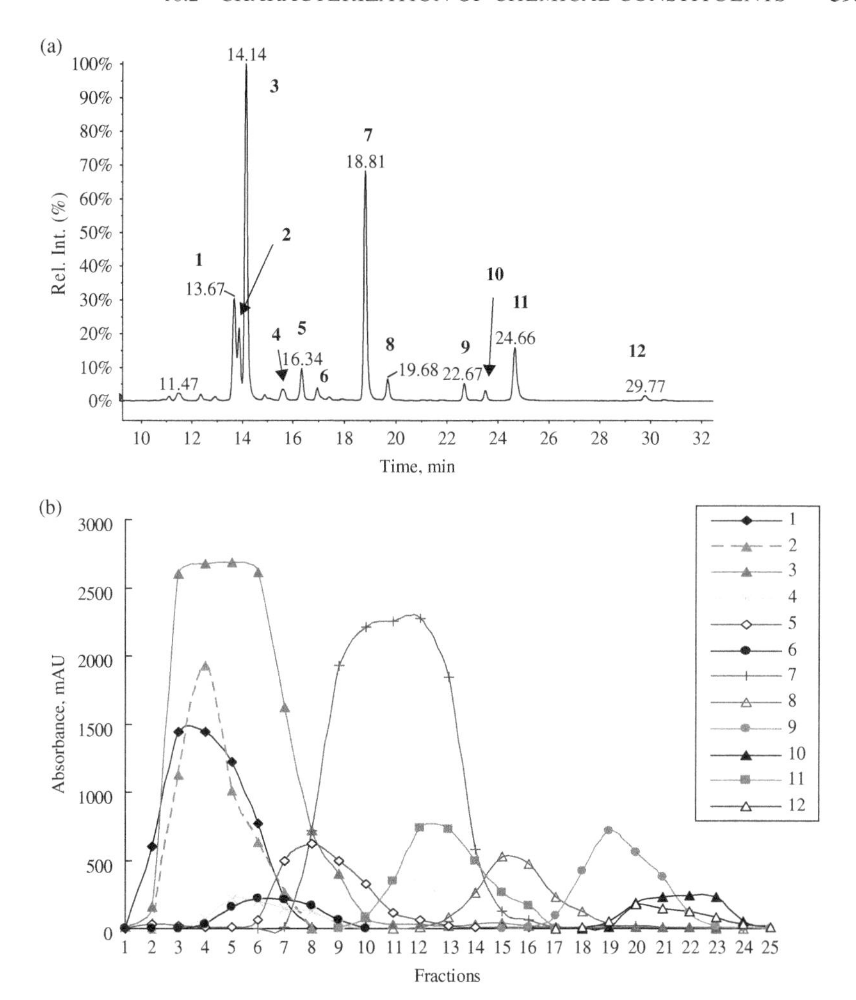

FIGURE 18.3 (a) HPLC chromatogram of AB-8-2 at λ 370 nm. Constituents are numbered in order by their retention times, which account for 0.1–34.9% of the crude extract. (*b*) The composition profiles of 12 constituents in the series of fractions from AB-8-2 reconstructed from HPLC analysis (λ 370 nm), displaying the concentration changing profile of each constituent and the composition of each fraction. A different numbering system is adopted in this part, not in accordance with those in Tables 18.2 and 18.3.

The results demonstrated that NMR/LC–MS PDS achieved the similar function of online LC–NMR–MS analysis in offline mode by such a "virtual combination approach." Moreover, the offline mode demonstrated advantages that address some of the compatibility problems that exist with direct combination NMR techniques. Furthermore, various NMR experiments can be performed as needed without the restriction of acquiring time and mode. In conclusion, NMR/LC–MS PDS combined with an incomplete separation

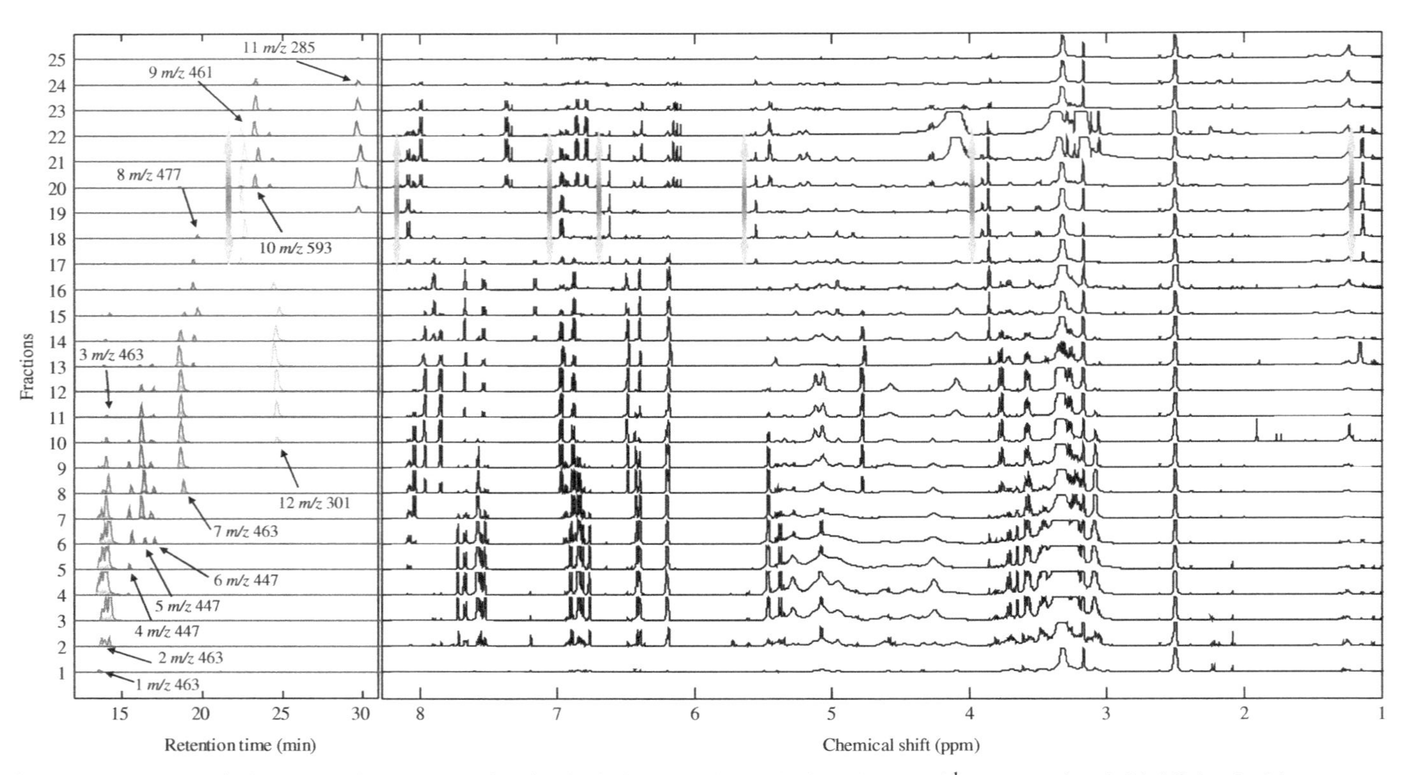

FIGURE 18.4 NMR/LC–MS PDS spectrum of active herbal extract AB-8-2. Six columns of ^{1}H NMR signals highlighted with gray arrows at left side were correlated to the column of $[M-H]^-$ ion at m/z 461 at t_R 22.6 min based on the covariation among their fraction ranges and intensity changing profiles, which were assigned to constituent **9**.

TABLE 18.3 HPLC–MS^n Data of Metabolites in Rat Bile

Component	t_R (min)	$[M\text{-}H]^-$ (m/z)	ESI–MS^n Data (m/z, Relative Abundances, %)	Identity[a]	Location
5	8.98	477	MS^2[477]: 301(100) MS^3[477 → 301]: 273(100)	Q-GlcUA	ig
15	14.36	609	MS^2[609]: 609(2), 577(5), 477(16), 459(13), 357(2), 315(100), 314(75), 300(25), 299(16), 271(3) MS^3[609 → 300]: 300(100), 271(82) MS^3[609 → 477]: 357(100)	MQ-3-Hex-Pen (1 → 2)	iv
16	14.65	477	MS^2[477]: 301(100) MS^3[477 → 301]: 301(100), 273(9), 257(11), 179(89), 151(42)	Q-GlcUA	ig
17	14.97	609	MS^2[609]: 609(3), 477(14), 459(2), 357(14), 315(100), 314(17), 300(46), 299(18), 271(4) MS^3[609 → 315]: 315(13), 300(100)	MQ-3-Hex-Pen (1 → 2)	iv
21	16.14	491	MS^2[491]: 315(100), 300(3) MS^3[491 → 315]: 315(52), 300(100) MS^4[491 → 315 → 300]: 300(100), 272(46)	MQ-GlcUA	ig
23	16.68	477	MS^2[477]: 477(49), 462(3), 449(17), 387(1), 357(17), 342(1), 329(2), 315(37), 314(100), 301(9), 300(21), 286(3) MS^3[477 → 314]: 314(4), 299(100), 286(21), 285(40), 271(30), 257(4), 243(7) MS^4[477 → 314 → 271]: 271(100), 243(17), 227(4)	MQ-3-Hex	ig, iv
24	16.83	461	MS^2[461]: 285(100) MS^3[461 → 285]: 285(100), 267(2), 257(11), 229(2)	K-GlcUA	ig
29	17.57	491	MS^2[491]: 315(100), 300(9) MS^3[491 → 315]: 315(34), 300(100) MS^4[491 → 315 → 300]: 300(100), 271(34), 255(7)	MQ-GlcUA	ig
30	17.74	543	MS^2[543]: 463(100), 381(4), 301(7) MS^3[543 → 463]: 343(8), 301(100)	Q-Hex-S	iv

(Continued)

TABLE 18.3 (Continued)

Component	t_R (min)	$[M\text{-}H]^-$ (m/z)	ESI–MS^n Data (m/z, Relative Abundances, %)	Identity[a]	Location
32	18.2	491	MS^2[491]: 357(1), 315(100) MS^3[491 → 315]: 315(100), 300(47)	MQ-GlcUA	ig
33	18.32	651	MS^2[651]: 651(3), 609(100), 591(30), 519(14), 501(3), 357(8), 315(52), 314(22), 300(22), 299(15), 271(4) MS^3[651 → 609]: 609(1), 477(2), 417(1), 357(1), 315(100), 314(4), 300(47), 299(9) MS^4[651 → 609 → 315]: 315(1), 300(100)	Acetyl MQ-3-Hex-Pen(1 → 2)	iv
34	18.41	447	MS^2[447]: 447(37), 432(3), 419(14), 357(10), 315(37), 314(100), 299(32), 271(11), 175(4) MS^3[447 → 314]: 299(100), 271(3) MS^4[447 → 314 → 299]: 299(100), 271(64), 243(2)	MQ-3-Pen	iv, ig
35	18.5	461	MS^2[461]: 461(13), 285(100) MS^3[447 → 285]: 285(100), 257(8), 256(7), 241(1), 151(2)	K-GlcUA	ig
36	18.7	557	MS^2[557]: 477(100), 381(37), 301(16) MS^3[557 → 477]: 301(100) MS^4[557 → 477 → 301]: 301(100), 179(6), 151(6)	Q-GlcUA-S	ig
37	18.79	491	MS^2[491]: 315(100) MS^3[491 → 315]: 315(47), 300(100)	MQ-GlcUA	ig
39	18.97	477	MS^2[477]: 301(100) MS^3[477 → 301]: 301(100), 179(79), 151(20)	Q-GlcUA	ig
41	20.38	519	MS^2[519]: 519(73), 504(8), 491(29), 459(5), 387(1), 357(25), 342(5), 315(33), 314(100), 299(65), 285(1), 271(5) MS^3[519 → 314]: 314(1), 299(100), 286(3), 285(10), 271(11) MS^4[519 → 314 → 299]: 299(100), 271(57), 243(1)	MQ-acetylHex	iv
44	22.19	475	MS^2[475]: 299(100), 284(2) MS^3[475 → 299]: 299(31), 284(100) MS^4[475 → 299 → 284]: 284(100), 255(58)	MK-GlcUA	ig

45	22.28	381	MS^2[381]: 381(94), 335(10), 333(6), 317(27), 301(100), 299(5), 209(8)	Q-S	ig
			MS^3[381 → 301]: 273(100), 257(7), 207(5)		
			MS^4[381 → 301 → 273]: 245(100), 179(16)		
50	25.33	475	MS^2[475]: 475(5), 299(100)	MK-GlcUA	ig
			MS^3[475 → 299]: 299(87), 284(100)		
			MS^4[475 → 299 → 284]: 284(100), 255(60)		
51	26.67	585	MS^2[585]: 525(1), 505(100), 409(1), 381(1), 301(3)	acetyl Q- Hex-S	iv
			MS^3[585 → 505]: 343(8), 301(100)		
			MS^4[585 → 505 → 301]: 301(100), 179(2)		
53	30.10	315	MS^2[315]: 315(100), 314(1), 300(75)	Isorhamnetin	ig
			MS^3[315 → 300]: 300(100), 271(2)		
54	30.46	315	MS^2[315]: 315(74), 300(100)	Tamarixetin	ig
			MS^3[315 → 300]: 300(100), 272(5), 271(4), 151(4)		
55	34.75	395	MS^2[395]: 315(100)	MQ-S	ig
			MS^3[395 → 315]: 315(85), 300(100), 287(4), 284(1), 151(7)		
			MS^4[395 → 315 → 300]: 300(100), 271(34), 151(62)		
56	35.26	365	MS^2[365]: 365(1), 285(100)	K-S	ig
			MS^3[365 → 285]: 285(100), 257(13), 241(10), 169(1), 151(29)		
57	36.94	395	MS^2[395]: 315(100)	MQ-S	iv
			MS^3[395 → 315]: 315(100), 300(32)		
58	36.94	381	MS^2[381]: 301(100)	Q-S	iv, ig
			MS^3[381 → 301]: 301(77), 273(15), 257(13), 193(1), 179(100), 151(48), 107(1)		
			MS^4[381 → 301 → 179]: 179(17), 151(100)		

[a] GlcUA, glucuronyl unit; S, SO_3.

strategy has the potential to expedite the structure identification of natural products in crude extracts.

18.3 PROFILING THE INTEGRAL METABOLISM OF HERBAL MEDICINE

18.3.1 Analysis of Parent Constituents and Metabolites in Rat Bile

The mass spectrometric fragmentation in combination with a metabolic pathway was explored in profiling components in bile samples. The in vivo metabolic reactions of flavonols were mainly methylation, glucuronidation, and sulfation (Mullen et al., 2004; Day et al., 2000; Williamson et al., 2000). The neutral losses of 15, 80, and 176 Da observed in MS/MS spectra were used to characterize methyl, sulfate, and glucuronide conjugate, respectively. In comparison with the blank sample, a total of 31 quercetin-based and 12 kaempferol-based compounds were identified in rat biles after oral and intravenous (IV) administration of AB-8-2. All the components were sorted and numbered by the retention time under the same chromatographic condition. The parent constituents and metabolites are listed in Tables 18.2 and 18.3, respectively.

As the first report of in vivo acetylation on aglycone of flavonol glycosides, component **33** was used as an example to illustrate the analytical procedure. It was only detected in the bile after IV administration. The MS^n spectra of **33** are shown in Figure 18.5. In the MS^2 spectrum of the $[M-H]^-$ ion at *m/z* 651, the existence of an acetyl group was confirmed by the ions at *m/z* 609 and 591, which correspond to the elimination of ketene (42 Da) and acetic acid (60 Da), respectively. The other product ion at *m/z* 519 (loss of 132 Da) indicated the presence of a pentose residue and also demonstrated that the acetyl group was not located on the pentose residue. In the MS^3 spectrum of the ion at *m/z* 609, the product ion at *m/z* 315 (base peak) was formed by the loss of 294 Da (162 + 132) and indicated a disaccharidic residue. The additional loss of a pentose residue (*m/z* 477) suggested that the pentose residue was located in a terminal position (i.e., hexosyl-pentose). Component **33** underwent a similar fragmentation pathway to rutin, with the *m/z* 314 ($[Y_0-H]^{-\bullet}$) and 357 ($^{0,2}X_0^-$) ions indicated a 3-*O*-glycosylation. Due to the biogenetic features of this plant, the ion at *m/z* 300 in the MS^4 spectrum of *m/z* 315 is proposed to be the radical aglycone ion of quercetin generated by loss of a $\cdot CH_3$ radical.

Acetylation on a sugar moiety is commonly encountered in natural products. However, the possibility that acetylation occurred on the aglycone must be considered with metabolites. In addition to being the $^{0,2}X_0^-$ ion shown in Figure 18.5*b*, the *m/z* 357 ion could originate by a loss of 162 Da (a hexose residue) from the acetyl-containing ion at *m/z* 519 (Fig. 18.6), which indicated that acetylation occured on the aglycone moiety of **33**.

The interglycosidic linkage of component **33** was determined by tracing its parent form in AB-8-2. Component **33** was presumed to be formed via

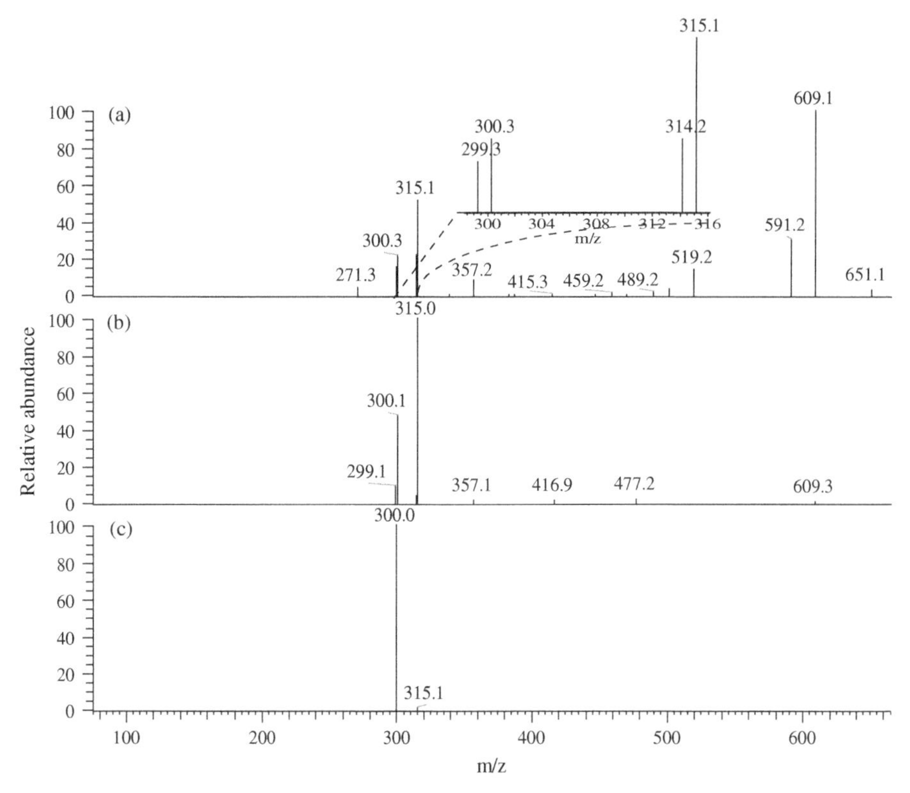

FIGURE 18.5 LC–MS^n spectra of **33**. (*a*) MS^2 spectrum of $[M-H]^-$ ion at *m*/*z* 651, (*b*) MS^3 spectrum of ion at *m*/*z* 609, and (*c*) MS^4 spectrum of ion at *m*/*z* 315.

acetylation or methylation, or both. Therefore, the probable molecular weight (MW) of the parent form could be speculated as 610, 638, or 596. No corresponding peak was found in the XIC for *m*/*z* 637. However, peaks of **4** and **12** and peaks of **9** and **10** were observed, as putative parents of **33**, in XICs for *m*/*z* 609 and 595, respectively (Fig. 18.7). Apparently, **4** and **12** (rutin) could not be parents of **33** because of the kaempferol nucleus of **4** and the rutinose sugar moiety of **12** (as shown in Table 18.2). Thus, **33** could only be derived from **9** and/or **10** via both acetylation and methylation with the glycan remaining intact. Based on the fragmentation of **9** and **10** listed in Table 18.2, the two glycosyl units of **33** were involved in a 1→2 interglycosidic linkage.

A combination of the structural features discussed above allowed for the proposal of an acetylated methylquercetin-3-*O*-[pentosyl-(1→2)]-hexoside structure for component **33**. To our knowledge, this is the first report of in vivo acetylation of flavonol. Component **51** was identified as another in vivo acetylated product by similar reasoning. Its fragmentation data are listed in Table 18.3.

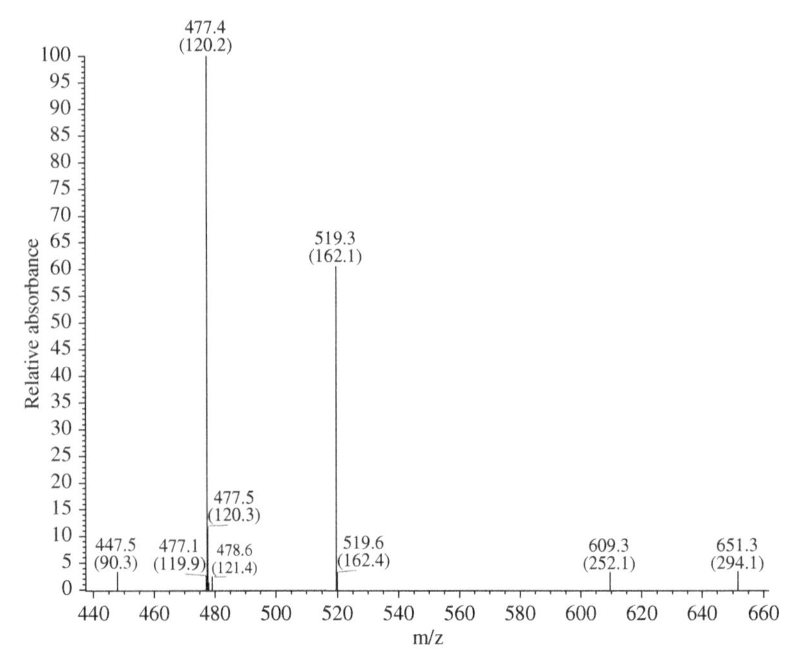

FIGURE 18.6 Precursor ion spectrum of *m/z* 357 (**33**), with the mass difference of corresponding precursor ions shown in parentheses.

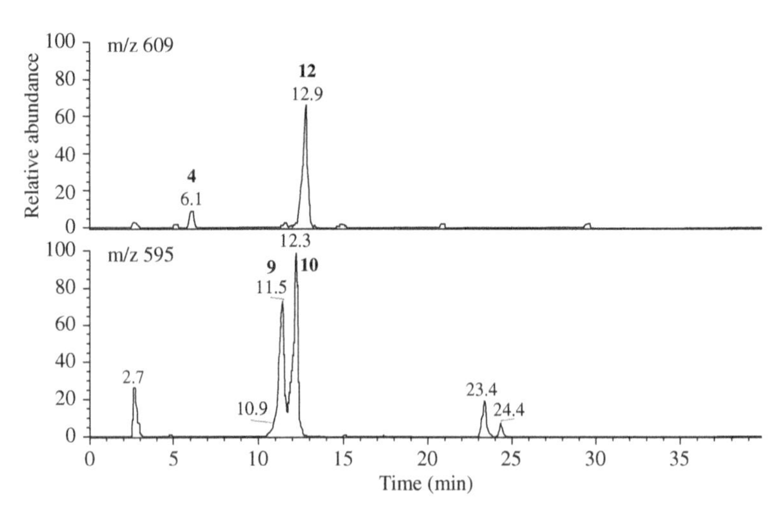

FIGURE 18.7 LC–MS analysis of AB-8-2 in negative ion mode. Extracted ion chromatograms for *m/z* 609 and 595.

Some metabolites could be isomeric forms of parent constituents, and the isomers with MW 478 were chosen as an example of the reasoning procedure. Figure 18.8 shows the XIC of the ion at *m/z* 477 in three analytes and in blank bile. Components **5**, **16,** and **39** were identified as quercetin

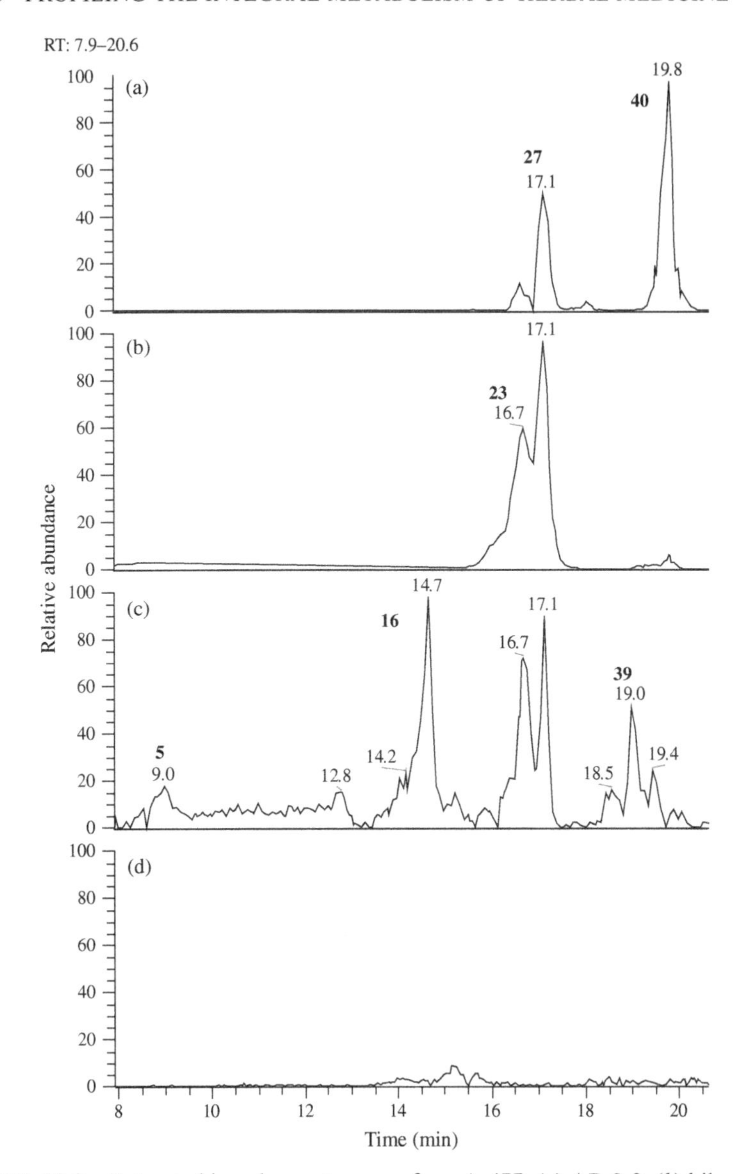

FIGURE 18.8 Extracted ion chromatograms for m/z 477: (*a*) AB-8-2, (*b*) bile after IV administration, (*c*) bile after oral administration, and (*d*) blank bile.

monoglucuronides based on their similar fragmentation behavior to quercetin-3-glucuronide. The fragmentation behavior of **40** (Fig. 18.9) was similar to that of quercetin-3′-glucoside (Fig. 18.1*c*), and it was thus proposed to be methyl quercetin-3′-hexoside. Components **23** and **27** were designated as methyl quercetin-3-hexoside by the same reasoning. Comparing the XIC for m/z 477 in the three analytes shown in Figure 18.8, the relative signal

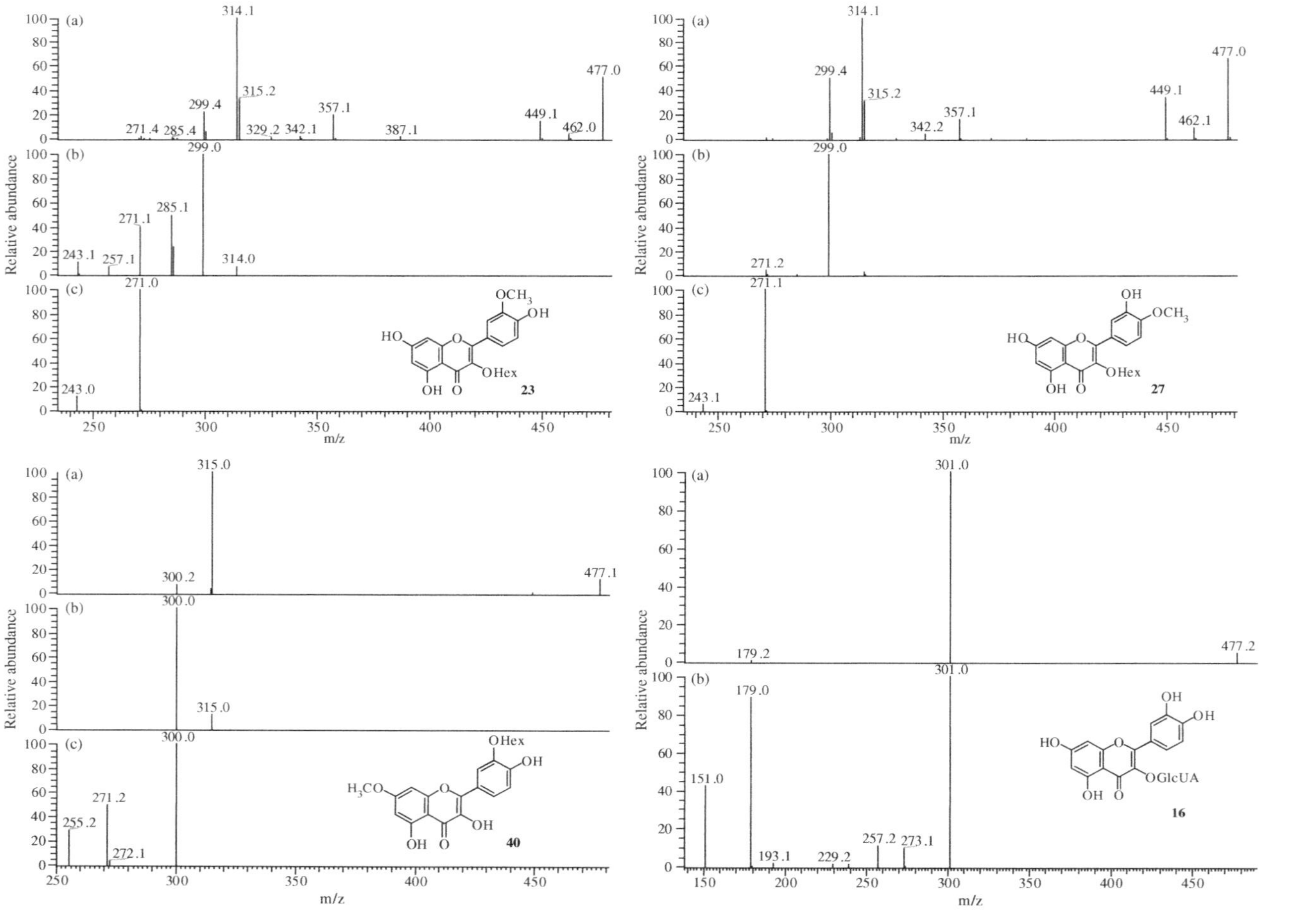

FIGURE 18.9 MS^n ($n = 2$–4) negative ion spectra of isomers with MW 478. Component **23**: (*a*) MS^2 [m/z 477], (*b*) MS^3 [m/z 477→314], and (*c*) MS^4 [m/z 477→314→271]. Component **27**: (*a*) MS^2 [m/z 477], (*b*) MS^3 [m/z 477→314], and (*c*) MS^4 [m/z 477→314→271]. Component **40**:

intensity of **40** was greater than that of **23** and **27** in AB-8-2. The opposite results were observed with the bile samples where **23** and **27** gave much stronger peaks than **40**. The difference in metabolic fate could be attributed to different methyl-substituted sites. Quercetin can be *O*-methylated in vivo by catechol-*O*-methyl-transferase (COMT) (van der Woude et al., 2006; Zhu et al., 1994; Hollman and Arts, 2000), as meant that methylation occurs on the 3′- or 4′-OH in the B ring. Thus it can be deduced that component **23** and **27** are 3′- or 4′-methylquercetin-3-hexosides. Methyl substitution on the B ring for **40** was, however, unlikely, owing to its decrease and even disappearance in the biological samples. Thus the methyl group of **40** was assigned to the 7-*O*-position. As the polarity of isorhamnetin (3′-methylated quercetin) was stronger than that of tamarixetin (4′-methylated quercetin), it is logical to extrapolate that the same polarity order remained after they were glycosylated with a hexose sugar on the same positon. Therefore, the respective retention times under the same chromatographic conditions led us to assign the structure of component **23** as 3′-methylquercetin-3-hexoside and that of **27** as 4′-methylquercetin-3-hexoside.

18.3.2 Integral Metabolic Characteristics of Flavonols in AB-8-2

Based on the analysis performed in our laboratory and on literature results (Day et al., 2000; van der Woude et al., 2004; O′Leary et al., 2003; Moon et al., 2001), the metabolic pathway of flavonols with a quercetin nucleus in AB-8-2 is exhibited in Figure 18.10. The in vivo acetylation on aglycone of flavonol glycosides is reported for the first time. The integral metabolic characteristics of

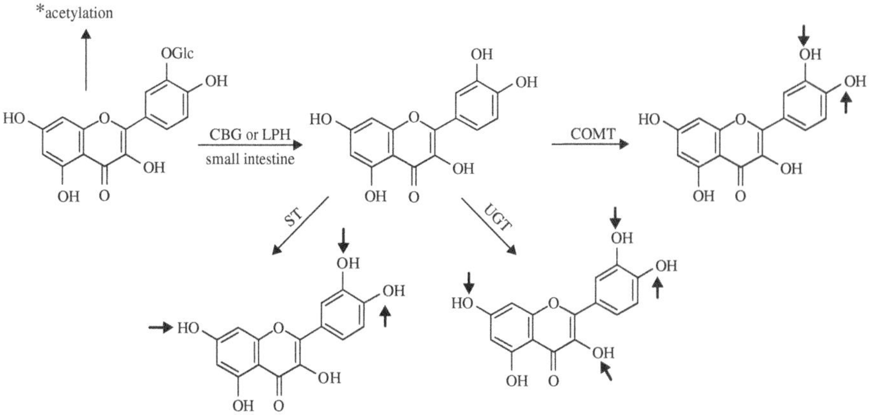

FIGURE 18.10 Metabolic pathway of flavonols with quercetin nucleus (quercetin-3′-glucoside is used as an example). The heavy arrows indicate the position of conjugation. *Novel metabolic reaction taken place on the aglycone of flavonols in vivo. COMT, catechol-*O*-methyltransferase; UGT, UDP-glucuronosyltransferase; ST, sulfotransferase; LPH, lactase-phlorizin hydrolase; CBG, cytosolic β-glucosidase.

flavonols in AB-8-2 were profiled by categorizing and analyzing the components in rat bile after oral and IV administration.

Components identified only in the bile after oral dosing are listed in Table 18.3. These include monoglucuronide conjugates of quercetin and kaempferol (e.g., **5**, **16**, **24**, and **39**), methylquercetin monoglucuronides (e.g., **21**, **29**, **32**, and **37**), methylkaempferol monoglucuronides (e.g., **44** and **50**), glucurono-sulfoconjugate of quercetin **36**; quercetin sulfate **45**; kaempferol sulfate **56**; methylquercetin sulfate **55**; and two methylated quercetins, isorhamnetin (**53**) and tamarixetin (**54**). All the above phase II metabolites were formed following deglycosylation of flavonol glycosides.

Metabolites in the bile after IV dosing were mainly produced by methylation (e.g., **15**, **17**, and **41**) and sulfation (e.g., **30**). They were characterized by the presence of a glycosyl unit. No glucuronide conjugates were detected from bile after IV administration, which is in sharp contrast to what was seen in bile after oral administration.

As shown in Table 18.3, only **23**, **34**, and **58** were detected in both bile samples, indicating the great distinction in metabolic fate for flavonol components with and without intestinal metabolism.

The components **8**, **13**, **14**, **25**, **28**, **31**, **38**, **40**, **48**, and **52** (Table 18.2) were the parent constituents recovered in bile after IV administration but disappeared in bile after oral administration. These components remained as the parent forms of flavonol glycosides, which indicated that no deglycosylation and glucuronidation occurred to them, owing to the avoidance of intestinal metabolism for the IV administration route.

Component **18**, **19**, **22**, **27**, **46**, and **47** (Table 18.2) were present in all three analytes. We speculated that **18**, **22** (astragalin, Table 18.1), **46** and **47** were glycosides of kaempferol or methoxylkaempferol and that they remained as the parent form because of the lack of a catechol group in the B ring, which was the necessary structure for methylation in vivo. For **19**, as the catechol group in the B ring was susceptible to COMT, we speculated that methylated products with MW 448 should be present in the bile samples. The peak of **34** was found in the XIC for the $[M-H]^-$ ion at m/z 447, and it was inferred to be the methylated product of **19** by comparing the fragmentation behavior of these two compounds. Unlike **13** (quercimeritrin) and **14** (isoquercetrin), **19** could only be partly metabolized, which indicated that the pentose moiety may contribute to its metabolic stability. Component **27** was deduced to be methylquercetin-3-*O*-hexoside, which was a component in AB-8-2, and which may also originate from other ingredients (e.g., isoquercetrin, component **14** in Table 18.2) via a metabolic reaction mediated by COMT. Regardless of its origin, there is no doubt that **27** was a stable derivative of quercetin after intestinal and hepatic metabolism.

The contrast between the two bile samples indicated that deglycosylation and subsequent glucuronidation occurred during the absorption process in the intestinal tract (O'Leary et al., 2003; Nemeth et al., 2003; Day et al., 2003; Crespy et al., 1999). Metabolites formed by the catechol-*O*-methylation reaction were extensively identified in the bile samples no matter the administration route, which was in accordance with the liver being an active site for the methylation of flavonoids (Crespy et al., 2003).

18.3.3 Analysis of the Metabolites of AB-8-2 in Rat Urine

The rat urine sample after oral administration of AB-8-2 was analyzed by using the method established. In one analytical run, 51 constituents including 35 di-glucuronidated metabolites, 12 monoglucuronidated metabolites and 4 flavonol aglycones were characterized. The main aglycone of metabolites was quercetin or kaempferol (Table 18.4).

The incomplete chromatographic separation of the isomers in the rat urine and the similarities of the MS^n spectra of the coeluted components resulted in difficulties in the structural elucidation of metabolites in rat urine. These coeluted components were distinguished on the basis of the metabolic pathway and biogenic feature. The metabolite **M11** was used as an example to illustrate the analytical procedure.

The retention time of **M11** was at 9.49 min in XIC for m/z 653 (Fig. 18.11). In the MS^2 spectrum of $[M-H]^-$ ion at m/z 653 (Fig. 18.12*a*), the intensive peaks at m/z 477 and m/z 301 produced by loss of glucuronide (176 Da) from the ions at m/z 635 and 477 suggested **M11** to be a diglucuronide conjugates of flavonol. Meanwhile, in the high mass range of the ions at m/z 477 and 301, the weak ions at m/z 491 and 315 were formed by the loss of hexose (162 Da) from the ion at m/z 653 and 477. Both the ions at m/z 315 and 301 presented in the MS^3 spectrum of the ion at m/z 477 (Figure 18.12*b*), which was originated by the loss of hexose (162 Da) and glucuronide (176 Da) from the ion at m/z 477. The aglycone of **M11** was confirmed as quercetin because of the characteristic A ring ions at m/z 179 and 151 in the MS^4 spectrum of the ion at m/z 301 (Fig. 18.12*c*). Because it was impossible that the two groups (162 Da, 176 Da) connected to quercetin simultaneously, the ion at m/z 477 was a mixture composed of two kinds of aglycone with glycosyl (301+176, 315+162). From the above evidence, **M11** was a mixture composed of two isomers, the structures of which were Q-GlcUA-GlcUA and MQ-Hex-GlcUA.

In contrast with various metabolites in the bile samples, the metabolic end products of flavonols by renal excretion were mainly glucuronide conjugates of flavonol, in accordance with the literature results (Graf et al., 2005). Moreover, some flavonol diglycosides of AB-8-2 probably remained one glycoside in the structure during the absorption process in intestinal tract, then excreted in urine as secondary glycosides conjugated with glucuronic acid. Some coeluated glucuronide methyl ester conjugated metabolites were also identified in the urine sample, which may be produced in the process of sample preparation.

The metabolic characteristics of flavonols were profiled by systematic analysis of substantial components in AB-8-2 and rat bile after oral and IV administration. Liver and intestine, as the two main sites for flavonol metabolism, dominated the different metabolic pathways. Deglycoslylation

TABLE 18.4 Metabolites Identified in Rat Urine

Metabolite	t_R (min)	$[M\text{-}H]^-$ (m/z)	Identity[a]
M1	2.79	639	Q-Hex-GlcUA
M2	2.79	653	Q-GlcUA-GlcUA
M3	2.79	667	MQ-GlcUA-GlcUA
M4	4.44	639	Q-Hex-GlcUA
M5	5.11	653	Q-GlcUA-GlcUA
M6	6.61	637	K-GlcUA-GlcUA
M7	7.43	667	MQ-GlcUA-GlcUA
M8	7.87	667	MQ-GlcUA-GlcUA
M9	8.21	639	Q-Hex-GlcUA
M10	8.67	653	Q-GlcUA-GlcUA
M11	9.49	653	Q-GlcUA-GlcUA; MQ-Glc-GlcUA
M12	9.59	637	K-GlcUA-GlcUA
M13	10.39	667	MQ-GlcUA-GlcUA; Q-GlcUA-MGlcUA
M14	11.04	637	K-GlcUA-GlcUA
M15	11.04	667	MQ-GlcUA-GlcUA; Q-GlcUA-MGlcUA
M16	11.69	667	MQ-GlcUA-GlcUA; Q-GlcUA-MGlcUA
M17	11.45	653	Q-GlcUA-GlcUA
M18	11.69	637	K-GlcUA-GlcUA
M19	11.96	637	MQ-dHx-GlcUA
M20	12.31	667	MQ-GlcUA-GlcUA
M21	12.38	653	Q-GlcUA-GlcUA
M22	12.70	681	MQ-GlcUA-MGlcUA
M23	12.89	667	MQ-GlcUA-GlcUA
M24	13.40	639	Q-Hex-GlcUA
M25	13.62	667	MQ-GlcUA-GlcUA; Q-GlcUA-MGlcUA
M26	13.78	653	Q-GlcUA-GlcUA
M27	14.10	653	Q-GlcUA-GlcUA
M28	14.27	681	MQ-GlcUA-MGlcUA
M29	14.54	477	Q-GlcUA
M30	14.87	667	MQ-GlcUA-GlcUA; Q-GlcUA-MGlcUA
M31	15.10	477	Q-GlcUA
M32	15.32	681	MQ-GlcUA-MGlcUA
M33	15.38	667	MQ-GlcUA-GlcUA
M34	16.52	667	MQ-GlcUA-GlcUA
M35	16.75	461	K-GlcUA
M36	16.93	667	MQ-GlcUA-GlcUA; Q-GlcUA-MGlcUA
M37	17.17	461	K-GlcUA
M38	17.55	667	MQ-GlcUA-GlcUA; Q-GlcUA-MGlcUA
M39	17.70	491	MQ-GlcUA
M40	17.83	681	MQ-GlcUA-MGlcUA
M41	17.92	491	MQ-GlcUA
M42	18.35	491	MQ-GlcUA
M43	19.09	491	MQ-GlcUA
M44	20.80	491	MQ-GlcUA

Metabolite	t_R (min)	$[M\text{-}H]^-$ (m/z)	Identity[a]
M45	23.10	505	MQ- MGlcUA
M46	23.80	505	MQ- MGlcUA
M47	24.60	505	MQ- MGlcUA
M48	25.12	301	quercetin
M49	29.99	285	kaempferol
M50	30.48	315	isorhamnetin
M51	30.77	315	tamarixetin

[a]GlcUA, glucuronyl unit; S, SO_3; MGlcUA, methyl-glucuronopyranosyl unit.

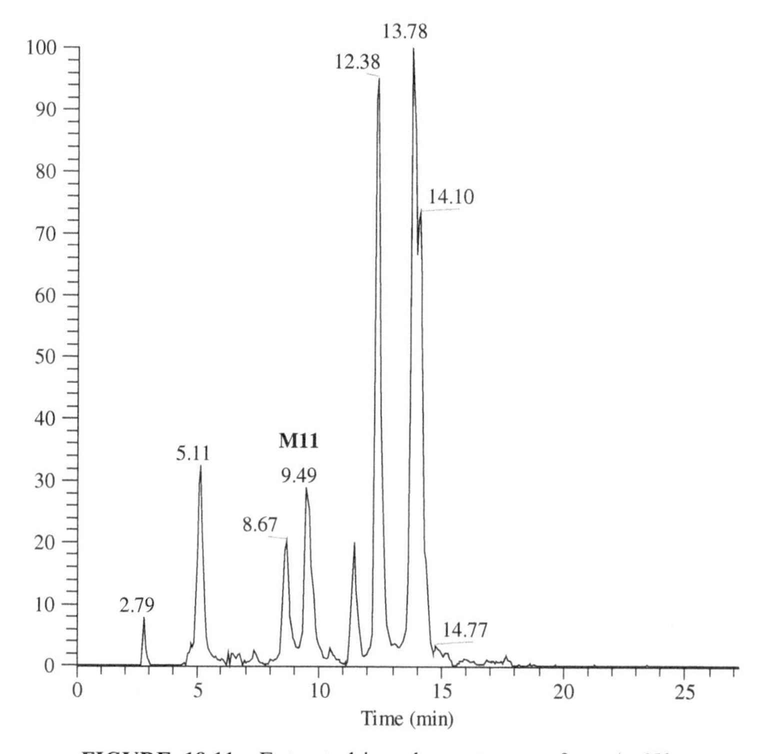

FIGURE 18.11 Extracted ion chromatogram for m/z 653.

and subsequent conjugation with glucuronic acid occurred in the small intestine, and the primary metabolites were glucuronidated aglycone or methyl aglycone. Some flavonol glycosides administered intravenously, avoiding the intestinal metabolism, could remain as the intact forms. Catechol-*O*-methylation was the most obvious reaction for flavonols in hepatic metabolism. By profiling the constituents in AB-8-2 and metabolites in bile, a view on the

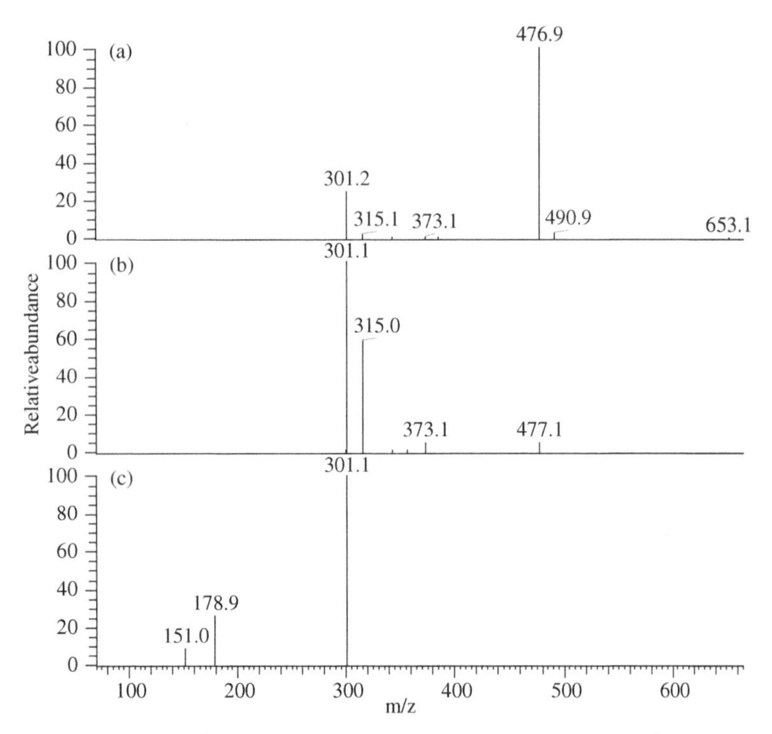

FIGURE 18.12 (*a*) MS^2 spectrum of $[M-H]^-$ ion at *m/z* 653, (*b*) MS^3 spectrum of ion at *m/z* 477, and (*c*) MS^4 spectrum of ion at *m/z* 301.

biotransformation of the constituents in AB-8-2 was obtained, which helped to judge which components in the active fraction merited further development.

18.4 CONCLUSIONS

The determination of the metabolism profile of medicinal herbs and plant extracts is very difficult due to their inherent complexity. In the present work, a rapid analytical method using LC–MS^n techniques for flavonol component screening in crude herbal extract and biosamples was established. Mass spectrometric fragmentation behavior and metabolic pathway complement each other in structural identification and correlating the metabolites and their parent forms. This information provides a basis for research that involves the metabolism profile of complex mixtures.

LC–MS is a powerful tool for the analysis of Chinese herbal medicines including identification of active ingredients, compositional analysis, and metabolic profiling. However, structural verification with the complementary information from other spectrometric methods or hyphenated techniques is needed along with comparison with reference compounds.

Over recent decades, a number of lead compounds and new natural products derived from medicinal herbs have been successfully isolated and identified, and great efforts have been made in chemical and pharmacological studies of Chinese herbs. The scientific basis of the majority of Chinese medicinal materials has remained poorly understood. The metabolite profile of herbal medicine is important for screening its active constituents and, thus, provide a valuable contribution to the drug discovery process and elucidation of the underlying mechanism of action. Owing to the rapid technical advances and increasing availability of the instrumentation of LC–MS technique, the profiling of the integral metabolism and characterization of parent constituents and metabolites will greatly prompt the development and modernization of Chinese herbal medicines.

ACKNOWLEDGMENT

This research was supported by the National Natural Science Foundation of China (Nos. 20475064 and 20775091) and the Key Project of Knowledge Innovation Program of the Chinese Academy of Sciences (No. KGCX2-SW-507).

REFERENCES

Ablajan K, Abliz Z, Shang XY, He JM, Zhang RP, Shi JG. Structural characterization of flavonol 3,7-di-*O*-glycosides and determination of the glycosylation position by using negative ion electrospray ionization tandem mass spectrometry. J Mass Spectrom 2006;41:352–360.

Cai Z, Lee F, Wang X, Yu W. A capsule review of recent studies on the application of mass spectrometry in the analysis of Chinese medicinal herbs. J Mass Spectrom 2002;37:1013–1024.

Chang Y, Abliz Z, Wang MZ. New techniques of tandem mass spectrometry and its application in the study of drug metabolism. Acta Pharm Sin 2000;35:73–78.

Chen P, Li C, Liang SP, Song GQ, Sun Y, Shi YH, Xu SL, Zhang JW, Sheng SQ, Yang YM, Li M. Characterization and quantification of eight water-soluble constituents in tubers of *Pinellia ternata* and in tea granules from the Chinese multiherb remedy Xiaochaihu-tang. J Chromatogr B 2006;843:183–193.

Chung VQ, Tattersall M, Cheung HT. Interactions of a herbal combination that inhibits growth of prostate cancer cells. Cancer Chemother Pharmacol 2004;53:384–390.

Crespy V, Morand C, Besson C, Cotelle N, Vezin H, Demigne C, Remesy C. The splanchnic metabolism of flavonoids highly differed according to the nature of the compound. Am J Physiol Gastrointest Liver Physiol 2003;284:G980–G988.

Crespy V, Morand C, Manach C, Besson C, Demigne C, Remesy C. Part of quercetin absorbed in the small intestine is conjugated and further secreted in the intestinal lumen. Am J Physiol Gastrointest Liver Physiol 1999;277:G120–G126.

Cui LJ, Abliz Z, Xia M, Zhao LY, Gao S, He WY, Xiang Y, Liang F, Yu SS. On-line identification of phenanthroindolizidine alkaloids in a crude extract from *Tylophora atrofolliculata* by liquid chromatography combined with tandem mass spectrometry. Rapid Commun Mass Spectrom 2004;18:184–190.

Cuyckens F, Claeys M. Determination of the glycosylation site in flavonoid mono-*O*-glycosides by collision-induced dissociation of electrospray-generated deprotonated and sodiated molecules. J Mass Spectrom 2005;40:364–372.

Cuyckens F, Rozenberg R, de Hoffmann E, Claeys M. Structure characterization of flavonoid *O*-diglycosides by positive and negative nano-electrospray ionization ion trap mass spectrometry. J Mass Spectrom 2001;36:1203–1210.

Dai DM, He JM, Sun RX, Zhang RP, Haji AA, Abliz Z. Nuclear magnetic resonance and liquid chromatography–mass spectrometry combined with an incompleted separation strategy for identifying the natural products in crude extract. Anal Chim Acta 2009;632:221–228.

Day AJ, Bao Y, Morgan MR, Williamson G. Conjugation position of quercetin glucuronides and effect on biological activity. Free Radic Biol Med 2000;29:1234–1243.

Day AJ, Gee JM, DuPont MS, Johnson IT, Williamson G. Absorption of quercetin-3-glucoside and quercetin-4′-glucoside in the rat small intestine: The role of lactase phlorizin hydrolase and the sodium-dependent glucose transporter. Biochem Pharmacol 2003;65:1199–1206.

Domon B, Costello CE. A systematic nomenclature for carbohydrate fragmentations in FAB-MS/MS spectra of glycoconjugates. Glycoconj J 1988;5:397–409.

Ferreres F, Llorach R, Gil-Izquierdo A. Characterization of the interglycosidic linkage in di-, tri-, tetra- and pentaglycosylated flavonoids and differentiation of positional isomers by liquid chromatography/electrospray ionization tandem mass spectrometry. J Mass Spectrom 2004;39:312–321.

Geng P, Zhang RP, Haji AA, He JM, Qu K, Zhu HB. Fast profiling of the integral metabolism of flavonols in the active fraction of *Gossypium herbaceam* L. using liquid chromatography/multi-stage tandem mass spectrometry. Rapid Commun Mass Spectrom 2007;21:1877–1888.

Graf BA, Mullen W, Caldwell ST, Hartley RC, Duthie GG, Lean ME, Crozier A, Edwards CA. Disposition and metabolism of [2-14C] quercetin-4′-glucoside in rats. Drug Metab Dispos 2005;33:1036–1043.

Hollman, PC H, Arts ICW. Flavonols, flavones and flavanols—nature, occurrence and dietary burden. J Sci Food Agric 2000;80:1081–1093.

Lee KH. Research and future trends in the pharmaceutical development of medicinal herbs from Chinese medicine. Public Health Nutr 2000;3:515–522.

Li B, Abliz Z, Fu GM, Tang MJ, Yu SS. Characteristic fragmentation behavior of some glucuronide-type triterpenoid saponins using electrospray ionization tandem mass spectrometry. Rapid Commun Mass Spectrom 2005;19:381–390.

Li B, Abliz Z, Tang MJ, Fu GM, Yu SS. Rapid structural characterization of triterpenoid saponins in crude extract from *Symplocos chinensis* using liquid chromatography combined with electrospray ionization tandem mass spectrometry. J Chromatogr A 2006;1101:53–62.

Liang F, Li LJ, Abliz Z, Yang YC, Shi JG. Structural characterization of steroidal sapoins by electrospray ionization and fast-atom bombardment tandem mass spectrometry. Rapid Commun Mass Spectrom 2002;16:1168–1173.

Liu Y, He J, Zhang R, Shi J, Abliz Z. Study of the characteristic fragmentation behavior of hydroquinone glycosides by electrospray ionization tandem mass spectrometry with optimization of collision energy. J Mass Spectrom 2009;44:1182–1187.

Liu YZ, Liang F, Cui LJ, Xia M, Zhao LY, Yang YC, Shi JG, Abliz Z. Multi-stage mass spectrometry of furostanol saponins combined with electroapray ionization in positive and negative ion modes. Rapid Commun Mass Spectrom 2004;18:235–238.

Ma YL, Cuyckens F, Van den Heuvel H, Claeys M. Mass spectrometric methods for the characterisation and differentiation of isomeric *O*-diglycosyl flavonoids. Phytochem Anal 2001;12:159–165.

Moon JH, Tsushida T, Nakahara K, Terao J. Identification of quercetin 3-*O*-β-D-glucuronide as an antioxidative metabolite in rat plasma after oral administration of quercetin. Free Radic Biol Med 2001;30:1274–1285.

Mullen W, Boitier A, Stewart AJ, Crozier A. Flavonoid metabolites in human plasma and urine after the consumption of red onions: Analysis by liquid chromatography with photodiode array and full scan tandem mass spectrometric detection. J Chromatogr A 2004;1058:163–168.

Nemeth K, Plumb GW, Berrin JG, Juge N, Jacob R, Naim HY, Williamson G, Swallow DM, Kroon PA. Deglycosylation by small intestinal epithelial cell β-glucosidases is a critical step in the absorption and metabolism of dietary flavonoid glycosides in humans. Eur J Nutr 2003;42:29–42.

O'Leary KA, Day AJ, Needs PW, Mellon FA, O'Brien NM, Williamson G. Metabolism of quercetin-7- and quercetin-3-glucuronides by an in vitro hepatic model: The role of human β-glucuronidase, sulfotransferase. Biochem Pharmacol 2003;65:479–491.

Tang HM. A study on material base of acorus Tatarinowii Schott for cerebral disease. Li shi zhen Med and Mater Med Res 2002;13:1–2.

van der Woude H, Boersma MG, Alink GM, Vervoort J, Rietjens IM. Consequences of quercetin methylation for its covalent glutathione and DNA adduct formation. Chem Biol Interact 2006;160:193–203.

van der Woude H, Boersma MG, Vervoort J, Rietjens IM. Identification of 14 quercetin phase II mono- and mixed conjugates and their formation by rat and human phase II in vitro model systems. Chem Res Toxicol 2004;17:1520–1530.

Wang XJ, Sun WJ, Sun H, Zhou DX, Lu HT, Wang P, Liu L. Effects of formula compatibility of Yin Chen Hao Tang on the constituents absorbed into blood of rat. Chin J Nat Med 2008;6:43–47.

Wang YL, Liang YZ, Chen BM, He YK, Li BY, Hu QN. LC-DAD-APCI-MS-based screening and analysis of the absorption and metabolite components in plasma from a rabbit administered an oral solution of Dang gui. Anal Bioanal Chem 2005;383:247–254.

Wei YJ, Li P, Shu B, Song Y. Identification of chemical and metabolic components of Fu Fang Dan Shen prescription by high performance liquid chromatography–electrospray ion trap mass spectrometry. Chin J Anal Chem 2007;35:13–18.

Williamson G, Day AJ, Plumb GW, Couteau D. Human metabolic pathways of dietary flavonoids and cinnamates. Biochem Soc Trans 2000;28:16–22.

Wu CS, Sheng YX, Zhang YH, Zhang JL, Guo BL. Identification and characterization of active compounds and their metabolites by high-performance liquid chromatography/Fourier transform ion cyclotron resonance mass spectrometry after oral administration of a herbal extract of *Epimedium koreanum* Nakai to rats. Rapid Commun Mass Spectrom 2008;22:2813–2824.

Xiang Y, Abliz Z, Li LJ, Huang X, Yu SS. Study on structural characteristic features of phenanthriondolizidine alkaloids by fast atom bombardment with tandem mass spectrometry. Rapid Commun Mass Spectrom 2002;16(17):1668–1674.

Yang M, Sun J, Lu Z, Chen GT, Guan SH, Liu X, Jiang BH, Ye M, Guo DA. Phytochemical analysis of traditional Chinese medicines (TCM) using liquid chromatography coupled with mass spectrometry. J Chromatogr A 2009;1216:2045–2062.

Zhang JL, Cui M, He Y, Yu HL, Guo DA. Chemical fingerprint and metabolic fingerprint analysis of Danshen injection by HPLC-UV and HPLC-MS methods. J Pharm Biomed Anal 2005;36:1029–1035.

Zhao X, Qiang GH, Guo Y, Zhao XF, Liu Q, Zheng XH. The new metabolites of notoginseng in compound Dan Shen Dripping pills in human plasma. J of Chin Mass Spectrom Soc 2007;28:24–26.

Zhu BT, Ezell EL, Liehr JG. Catechol-*O*-methyltransferase-catalyzed rapid O-methylation of mutagenic flavonoids. Metabolic inactivation as a possible reason for their lack of carcinogenicity in vivo. J Biol Chem 1994;269:292–299.

19 Liquid Chromatography Mass Spectrometry Bioanalysis of Protein Therapeutics and Biomarkers in Biological Matrices

FUMIN LI

Bioanalytical Department, Covance Laboratories, Madison, Wisconsin

QIN C. JI

Bioanalytical Sciences, Analytical Research & Development, Bristol-Myers Squibb, Princeton, New Jersey

Mass Spectrometry in Drug Metabolism and Disposition: Basic Principles and Applications, First Edition. Edited by Mike S. Lee and Mingshe Zhu.

19.1 GENERAL INTRODUCTION

Liquid chromatography coupled with tandem mass spectrometry (LC–MS/MS) technology is routinely used for the bioanalysis of small organic molecules in biological matrices. There is an increased and accelerated interest in the application of LC–MS/MS for the quantitative analysis of proteins. In proteomic research, while the majority of the work focuses on qualitative analysis, growing emphasis is being placed on quantitative applications (Motoyama and Yates, 2008; Aebersold and Mann, 2003, Linscheid et al., 2009; Elliott et al., 2009). Furthermore, LC–MS/MS is an emerging analytical platform for protein bioanalysis in applications such as biomarker validation and toxicokinetic (TK)/pharmacokinetic (PK) determinations in preclinical and clinical studies of pharmaceutical research and development (Ji et al., 2004, 2008; Yang et al., 2007; Heudi et al., 2008). There is a large body of literature that focuses on large-scale protein characterization applications using LC–MS/MS (Yates et al., 2009). Indeed, proteomics has been one of the main drivers for the implementation of novel analytical technologies, which can achieve faster, more sensitive, and more selective analyses. There is no doubt that methodologies and technologies applied in proteomics can be adapted to the bioanalysis of proteins in drug discovery and development programs. However, the quantitative bioanalysis of protein therapeutics and protein biomarkers presents unique challenges and can be vastly different from proteomics. The content of this chapter is mainly devoted to the absolute quantitation of protein drugs and biomarkers in biological matrices by LC–MS/MS. As discussed in a review (Ji et al., 2004), there are two primary strategies used in the LC–MS/MS quantitative analysis of proteins in biological matrices: (1) protein quantitation by direct mass spectrometric detection of intact proteins and (2) protein quantification through mass spectrometric detection of a representative peptide segment generated from the enzymatic digestion or chemical cleavage of the protein analyte.

Similar to the method development process of LC–MS assays for small molecules, the main aspects of the assay development process are sample preparation, chromatographic separation, and mass spectrometry detection. The goal of sample preparation is to separate the analyte of interest from the rest of the matrix components as much as possible. If a representative peptide(s) is used for quantitation, then sample preparation can take place either before and/or after the enzymatic digestion. Several strategies of sample preparation to enhance the assay sensitivity are reviewed extensively in this chapter. The chromatographic method development depends heavily on the sample preparation methodologies, the strategy for protein analysis, and the targeted analyte(s) for measurement. The endpoint of chromatographic development is to provide a reliable LC method that can sufficiently resolve the analytes of interest from matrix components that may not be removed during sample preparation. The LC method should be capable of providing consistent peak integration especially at the lower limit of quantitation (LLOQ). Elimination/reduction of injection carryover from the column and/or autosampler is another important consideration for the LC development. Selective reaction monitoring (SRM), widely used for the small-molecule mass spectrometry detection, can be effectively employed for the detection of small proteins, and peptides derived from enzymatic digestion of large proteins. It is worth noting that multiple charges carried by these precursor ions enhance the collision energy applied in the collision-activated dissociation (CAD) process in the collision cell of triple quadrupole mass spectrometers.

Discussions in this chapter also include several aspects that are of particular importance to evaluation of the assay performance. For the large protein analysis using representative peptides, the digestion efficiency reviewed is critical for assay throughput, sensitivity, and accuracy and precision. The pros and cons of different types of internal standard (ISTD) used are discussed extensively. At the end of the chapter, assay development strategy based on the various stages of the drug development and biomarker validations are addressed. The application of an LC–MS/MS assay may result in extended protocols for studies that involve the investigation of the mechanism action and safety profile of therapeutic protein drugs.

19.2 PROTEIN QUANTITATION BY LC–MS/MS

Advances in biotechnology and the advent of "personalized medicine" have triggered the development of a whole new class of drugs, mainly biomolecules. The emergence of protein-based therapeutics, in terms of number and frequency, is indicative of the paradigm shift in drug discovery and development in the last two decades (Leader et al., 2008). As described in the publication by Leader and co-workers in 2008, the U.S. Food and Drug Administration (FDA) has approved more than 130 proteins and peptides for clinical use, and the number of candidates currently under development certainly dwarfs this figure. At least 14 monoclonal antibody (mAb) drugs are currently on the market in various therapeutic areas, including oncology, inflammation, infectious disease, and cardiovascular diseases. These numbers will likely increase in the years to come as more than 100 mAb drugs were in clinical development in 2007 (Mascelli et al., 2007).

Pharmacokinetics and pharmacodynamics (PK/PD) are essential in the drug discovery and development process. The ability to obtain accurate PK information in a timely manner for a drug candidate can influence the progress the drug development. This ability highlights the critical nature of quantitative bioanalysis especially in a good laboratory practice (GLP) environment to support the Investigational New Drug (IND)- or New Drug Application (NDA)-enabling filings. The process of quantitative bioanalytical method development and validation has been well established to generate PK data to support studies on absorption, distribution, metabolism, and elimination (ADME) for small molecules (Ackermann et al., 2008). Compared to small-molecule drugs, protein therapeutics offers high target-binding specificity (Jones et al., 1986). It is possible to optimize and engineer key characteristics of the protein drugs such as the half-life, biodistribution, and affinity (Weiner, 2006). Large biomolecules are chemically and biologically different from small-molecule drugs. The bioanalytical method development and validation for small molecules cannot be directly applied to macromolecule analysis in biological matrices. Nevertheless, the vast knowledge and experience accumulated in small-molecule bioanalysis can help in the method development and validation for quantitative determination of large biomolecules in support of drug discovery and development.

Protein quantitation in complex biological samples has traditionally been conducted with immunoassays (DeSilva et al., 2003; Lobo et al., 2004). Although immunoassays are both sensitive (nano- to picomolar detection limits) and amenable to high-throughput formats, these approaches are usually time consuming to develop. For example, the method development and validation of an enzyme-linked immunosorbant assay (ELISA) may take up to 5 months. Further, the binding of antibodies or antigens to other components in a complex matrix is common and can affect the assay selectivity or even render inaccurate results such as a false positive or a false negative. An ELISA process developed during the preclinical stages of drug development may not be transferred directly to analyze clinical samples due mainly to the differences in matrix specificity between animal and human subjects.

As a result, there is an emerging trend to research and develop alternative approaches for sensitive, selective, and high-throughput bioanalytical methods for protein drug quantitation in complex biological matrices to support drug discovery and development programs. LC–MS/MS is a mainstream technology used to obtain drug metabolism and pharmacokinetics (DMPK) information for small-molecule drugs. The platform has the benefits of good specificity, reproducibility, high throughput, and wide dynamic range, making it the workhorse for absolute quantitation of low-molecular-weight pharmaceuticals (Jemal et al., 2006, 2010). LC–MS/MS is widely used for protein characterization, and is increasingly applied to protein quantification (Anderson and Hunter, 2006; Ong and Mann, 2005). Moreover, LC–MS has the potential to determine protein drugs and their metabolite(s) simultaneously. The time to develop an LC–MS method could be as short as 2 weeks (Ji et al., 2009). All of these unique advantages of LC–MS make it a suitable alternative for protein quantitation in biological matrices.

19.3 PROTEIN QUANTITATION USING INTACT PROTEINS BY LC–MS/MS

The quantitation of small-sized therapeutic proteins or peptides in plasma can be conducted with LC–MS/MS detection of the intact proteins. Electrospray ionization (ESI) is the most widely used ionization technique for quantitative LC–MS/MS analysis of proteins or peptides. There are some occasions where atmospheric pressure chemical ionization (APCI) was used to circumvent matrix effects (Volosov et al., 2001). ESI typically generates multiply-charged protein or peptide ions, depending upon the number of basic charges in the polypeptide backbone. For small proteins, high abundance peaks of multiply charged molecular ions usually falls into the normal mass-to-change ratio m/z range of most quadrupole mass spectrometers. Quadrupole mass spectrometry is broadly used for the bioanalysis of small molecules because of the assay selectivity and sensitivity. It has been proved that direct measurement of relatively small biomolecules by ESI with quadrupole mass spectrometry is practical as discussed in the following paragraphs. Quantitative peptide analysis by LC–MS/MS is not the focus of this chapter. Readers are directed to reviews (Careri and Mangia, 2003; John et al., 2004; Van den Broek et al., 2008) on the quantitative bioanalysis of peptides.

The following are two detailed examples about quantitation of relatively small proteins of using an intact protein LC–MS/MS detection strategy. An LC/ESI–MS/MS method (Becher et al., 2006) was developed to determine EPI-hNE4, a 56-amino acid recombinant protein, in both human and monkey plasma. The method was demonstrated as capable of achieving an impressive sensitivity similar to that obtained from an ELISA assay at 5 ng/mL with 40 fmol loaded to the column. Protein precipitation with acetonitrile was used to remove the bulk of the plasma proteins. In addition, EPI-hNE4 in plasma samples was initially dissociated from possible antibodies with a small amount of acidified acetonitrile. Good intra- and interassay precision (CV in between 8.5 and 8.3%), and accuracy (90.6–102.1%), as well as good linearity (R^2 = 0.9982) were obtained for intact EPI-hNE4 in plasma by LC–MS/MS.

In another application, Ji and co-workers (2003) developed an LC–MS/MS strategy for the analysis of rk5, a small protein of 10,464 amu, in plasma by measuring rk5 as an intact protein. Plasma samples were processed by solid-phase extraction (SPE) to remove interference compounds. A N^{15} isotopically labeled version of the rK5 protein was used as its internal standard (ISTD), which enhanced the assay precision and accuracy. This assay was validated following the FDA's *Guidance for Industry-Bioanalytical Method Validation* (FDA, 2000). Of 18 replicates of QC samples from three consecutive runs, the intraassay CV was between 0.6 and 3.8%, and the interassay CV was between 1.7 and 3.2%. The interassay mean accuracies, expressed as percent of the theoretical value, were between 101.5 and 104.7%. An LLOQ of 99.2 ng/mL, using a sample volume of 50 μL corresponding to 380 fmol (3.97 ng) of rK5 injected onto the analytical column (assuming 100% extraction recovery), was obtained. Good linearity with a coefficient of determination (R^2) ranging from 0.9972 and 0.9994 was obtained between 99.2 and 52 920.0 ng/mL. The assay has proven to be rugged and specific and has been employed to generate data in support of preclinical studies.

As has been demonstrated, LC–MS/MS quantitation of intact proteins in biological matrix is a viable option for proteins of relatively low molecular weight. LC–MS/MS was demonstrated to be selective, high throughput, sensitive (down to a few nanograms/milliliter), and have a wide dynamic range (i.e., at least three orders of magnitude) for intact protein quantitation in plasma. Due to the presence of large quantities of endogenous proteins and other compounds in biological matrix, sample cleanup [e.g., protein precipitation, liquid–liquid extraction (LLE), and SPE] are essential for small organic molecules bioanalysis. Protein analytes are physiochemically more similar to the endogenous proteins present in biological fluids than small-molecule analytes. The effective extraction of the protein of interest from the bulk of endogenous proteins represents a key challenge for a successful quantitative protein analysis by LC–MS/MS. It is worth noting that LLE does not seem to be a good method for the sample preparation of proteins. Consistent with our experience, no LLE methods were found in the literature for protein extraction. Low recovery was often observed when LLE was used. Due to their polar nature, most proteins would stay in the aqueous phase. For relatively small-sized protein therapeutics, the advantage of using protein precipitation for sample preparation is that they may not precipitate as easily as large biomolecules found in the matrix. Thus, protein precipitation (Becher et al., 2006) is an attractive sample preparation technique for LC–MS/MS analysis of small-sized protein therapeutics in biological matrices. However, precautions need to be taken to consider potential sample loss due to protein binding between therapeutic agents and endogenous antibodies. As shown in Figure 19.1 (Becher et al., 2006), a dissociation agent effectively disrupted the binding of EPI-hNE4 with a high-affinity rabbit polyclonal antiserum EPI-hNE4 spiked in human plasma. A 20-fold increase in concentration was observed in a TK study when the dissociation agent was applied (Becher et al., 2006). SPE can be an effective sample preparation approach prior to the LC–MS/MS detection of proteins (Ji et al., 2003). It has been demonstrated that binding between the antidrug antibody and the drug analyte may not be completely disrupted under the

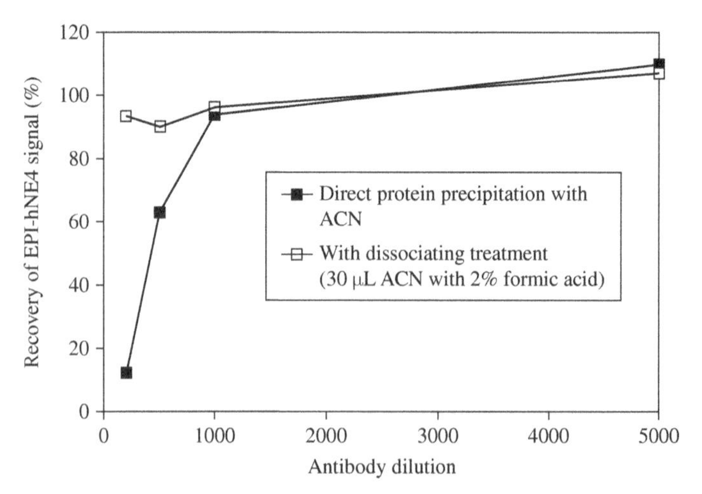

FIGURE 19.1 Influence of antibodies on the LC–MS/MS signal of EPIhNE4: Recovery of the EPI-hNE4 signal with and without dissociating sample processing when dilution of antidrug body is decreased from 1/5000 to 1/200. [Reprinted from Becher et al. (2006) with permission of the American Chemical Society.]

normal procedure of solid-phase extraction. The addition of dissociation agents may be required if the total protein drug concentration needs to be measured (Ji et al., 2007).

With the use of ESI, proteins tend to have an envelope of peaks with different charge states while only one peak can be effectively monitored using SRM detection. This distribution of protein ion intensity to multiple peaks leads to substantial loss in sensitivity, which inevitably constrains the achievable LLOQ of an LC–MS/MS assay. This feature can present a significant challenge for the development of an efficient LC method to separate large protein molecules from complex matrix components. The addition of ion-paring reagents such as trifluoroacetic acid (TFA) can improve the LC separation of biomolecules; however, its use results in the loss of mass spectrometric signal. As a general rule, the intact protein analysis strategy may not be ideal for quantification by LC–MS/MS of large proteins.

19.4 PROTEIN QUANTITATION USING REPRESENTATIVE PEPTIDES BY LC–MS/MS

The LC–MS/MS of intact protein approach may not be practical for large proteins, such as monoclonal antibodies (mAbs), mainly due to limitations of mass spectrometry (e.g., mass range and resolution) and the electrospray ionization process (multiply charged nature). Protein quantitation can be conducted by subjecting the proteins to enzyme digestion, commonly trypsin, followed by monitoring a representative tryptic peptide by LC–MS/MS. Denaturation, reduction, and alkylation are generally performed prior to proteolytic digestion. This strategy is based on the hypothesis that the proteolytically generated representative peptides can truly represent the intact proteins for quantitation.

In other words, the LC–MS/MS response of a selected peptide can be used to determine the quantity of their corresponding protein(s). The selection of representative peptides from the therapeutic proteins is of paramount importance because the selection can influence assay selectivity, sensitivity, matrix effects, and recovery. The selected peptide should be unique and distinct from endogenous peptides and, to this effect, may be compared to the peptides generated and recorded in the human GenBank. The selected peptides need to exhibit good ionization efficiency, which directly relates to the LC–MS/MS assay sensitivity. The peptide MS fragmentation pattern is also an important consideration because it can alter assay selectivity and sensitivity. For instance, the N-terminal end of the proline residue is a dominant MS/MS cleavage site, where extensive and predicable fragmentation paths are expected (De Hoffman et al., 2001). The presence of basic amino acids (e.g., arginine, lysine, histidine, or proline) at or close to either end of the peptide induces the formation of fragment ions that contain both the N- and the C-terminus. Any possible chemical modification from the selected peptides needs to be avoided as well.

Peptides that contain amino acid residues that are prone to chemical modification (Brönstrup, 2004), such as methionine, tryptophan, or cysteine, should not be considered.

The use of proteolytic digestion coupled with mass spectrometry detection was first developed by Barr and colleagues in 1996 to quantitate apolipoprotein A-1. Similar approaches were employed to the quantification of prostate-specific antigen (Barnidge et al., 2004), intermediate abundance proteins (Lin et al., 2006), C-reactive protein (Kuhn et al., 2004), and apolipoproteins (Kay et al., 2007). Yang and co-workers (2007) developed an LC–MS/MS assay for the quantification of somatropin and a therapeutic human monoclonal antibody using a protein analog (i.e., bovine fetuin) as an ISTD and a two-dimensional solid-phase extraction (2D-SPE) for the cleanup of the plasma digest. The plasma samples were first digested with trypsin overnight, and appropriate tryptic peptides for both analytes of interest and the ISTD were chosen. The representative peptides were selected for the analysis based on factors such as assay selectivity and sensitivity. The combination of reversed-phase SPE and cation exchange SPE greatly reduced the complexity of the tryptic peptide mixtures from the digestion of plasma, and thus, provided a clean LC–MS/MS background and enhanced selectivity. Good accuracy and precision, within ±15%, were established for somatropin from 1 to 1000 μg/mL using its representative peptide for quantitation. Good linearity was demonstrated as well, which showed the effectiveness of an analog protein as an ISTD to compensate for LC–MS/MS assay variability starting from trypsin digestion. When applied to a rat PK study of a proprietary therapeutic monoclonal antibody, the results from the LC–MS/MS method showed good agreement for the data obtained from an ELISA assay in Figure 19.2 (Yang et al., 2007).

Recently, Heudi and co-workers (2008) developed a method for the quantification of a therapeutic protein that combines proteolytic digestion with isotope dilution mass spectrometry using the isotope-labeled version of the protein as the ISTD. The ISTD was labeled with each threonine containing four ^{13}C atoms and one ^{15}N atom. Serum samples that contain the therapeutic protein were spiked with the labeled internal standard before being subjected to overnight trypsin digestion. SPE based on mixed-mode cation exchange was applied to clean up the serum digest. Tryptic peptides were eluted in a basic methanol–ethanol solvent mixture, which was evaporated to dryness and reconstituted prior to LC–MS/MS analysis. The dry-down and reconstitution helped to concentrate the samples and dissolve the tryptic peptides in an LC–MS-compatible solvent [i.e., 0.1% formic acid in a water : methanol mixture (v : v, 1 : 1)). The method showed good linearity between 5.00 and 1000 μg/mL for the mAb drug candidate with good assay accuracy and precision. As indicated in Figure 19.3 (Heudi et al., 2008), similar PK profiles were obtained from an ELISA and the LC–MS/MS method using the mAb's representative tryptic peptide (TGPFDYWGQGTLVTVSSASTK). However, the LC–MS/MS measurements resulted in ~60% higher exposure compared to those from ELISA. The discrepancy was attributed to LC–MS/MS that measured the

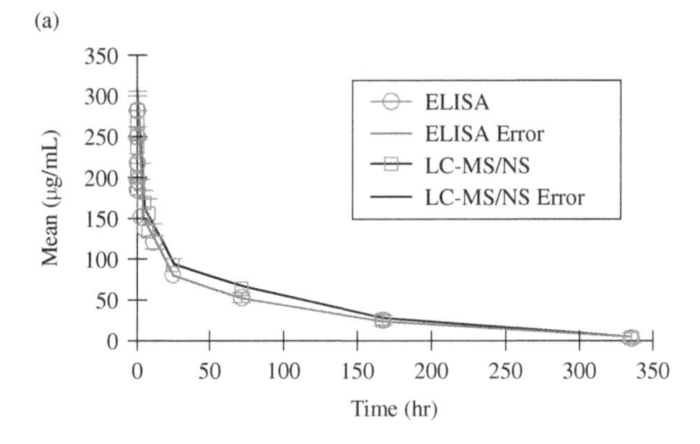

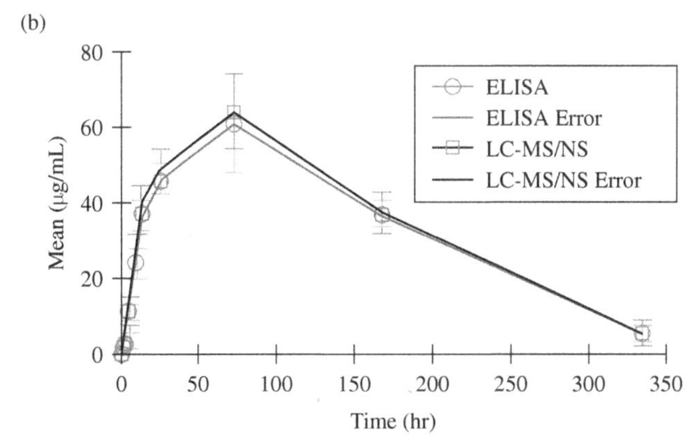

FIGURE 19.2 Rat pharmacokinetic profiles of mAbX obtained from ELISA and LC–MS bioassay: (*a*) intravenous (IV) profile and (*b*) intraperitoneal (IP) profile. The LC–MS assay results showed good agreement with the ELISA results. [Reprinted from Yang et al. (2007) with permission of the American Chemical Society.]

total concentration of the target proteins, both active and inactive. In comparison, ELISA measures only active or free proteins in the samples. The results point out the unique advantage of ELISA that provides a more accurate assessment of the effective exposure. Although LC–MS/MS provides a higher exposure by also measuring inactive forms of the therapeutic proteins, it is capable of generating PK profiles, similar to those from the ELISA assay. Above all, ELISA and LC–MS/MS can be employed in tandem as complementary technologies for the quantitative bioanalysis of proteins in biological matrix.

As has been demonstrated, enzymatic digestion followed by monitoring a representative peptide by LC–MS/MS is a good alternative to ELISA for protein quantitation in biological matrix (Yang et al., 2007; Heudi et al., 2008). The similar time profiles were often observed from both LC–MS/MS and ELISA methods with the possibility that a higher area under the curve (AUC)

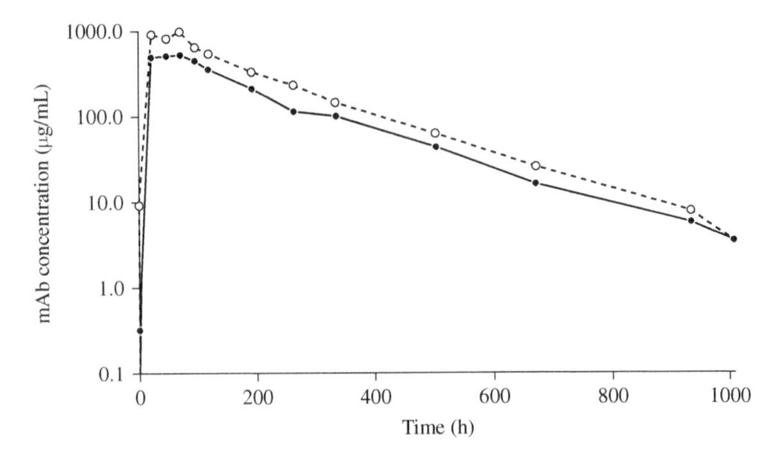

FIGURE 19.3 Comparison of mean concentration time profiles of mAb between (- - -) LC–MS/MS and (the solid line) ELISA in marmoset serum. [Reprinted from Heudi et al. (2008) with permission of the American Chemical Society.]

might be observed from the LC–MS/MS measurement. The use of either an isotopically labeled version or an analog protein of the target protein as the ISTD was shown to sufficiently correct for variability that includes enzymatic digestion, subsequent sample preparation, and LC–MS/MS analysis. However, almost all reported enzymatic digestions were conducted overnight, presumably >12 h, which greatly limits the method development and sample analysis throughput.

Previous reports (Arsene et al., 2008; Russell et al., 2001) have demonstrated that the addition of organic solvent to an enzymatic cleavage mixture can increase the reaction rate. Arsene and colleagues completed a tryptic digestion of recombinant human growth hormone (rhGH) under an hour by conducting the digestion in 80% acetonitrile (ACN) (Arsene et al., 2008). Another group demonstrated the efficacy of adding organic solvent to digestions for qualitative proteomics, obtaining 68% sequence coverage for a tryptic digest of myoglobin in 80% ACN in less than 5 min. It is important to note that these experiments were carried out in controlled systems, where simple protein mixtures were dissolved in solvents.

Recently, accelerated trypsin digestion was achieved in the presence of organic solvent for protein quantitation in complex matrix (i.e., human plasma) (Li et al., 2009). Myoglobin was used as the model protein and somatropin was served as the analog protein internal standard. The digestion was accomplished in 30 min in human plasma, and 50% methanol (by volume) was found to be optimal in terms of tryptic peptide yield for this assay (Fig. 19.4; Li et al., 2009). The increase in the enzymatic reaction rate with the addition of organic solvent likely resulted from the partial denaturation of the substrates, which allows trypsin quick access to cleavage sites that are otherwise buried deep within the

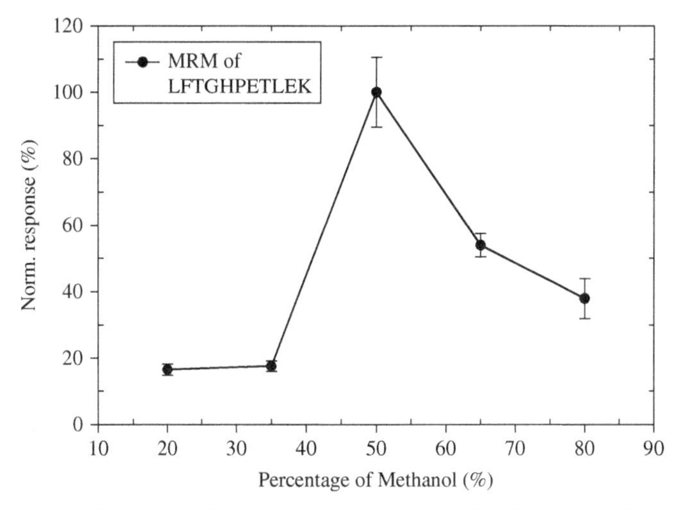

FIGURE 19.4 Effect of methanol concentration on the digestion of myoglobin in human plasma ($n = 3$). [Reprinted from Li et al. (2009) with permission of *Rapid Communication Mass Spectrometry*.]

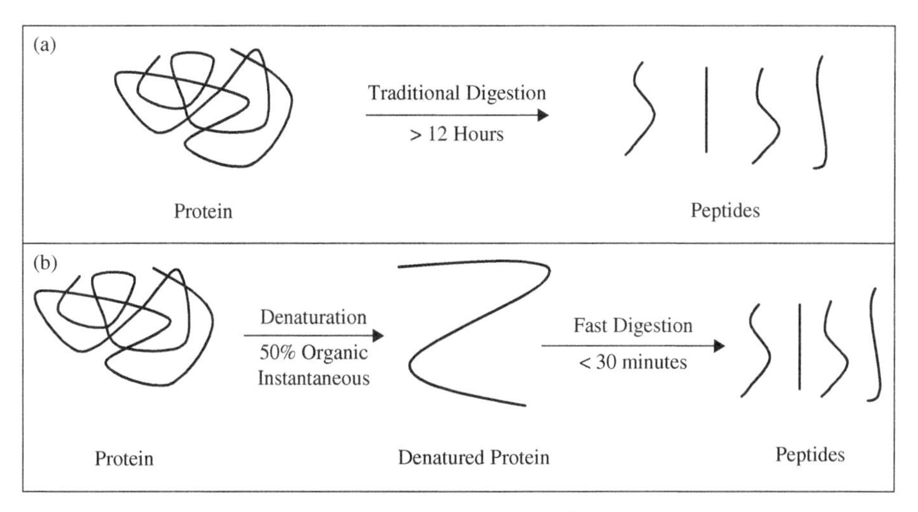

FIGURE 19.5 Schematic illustration of the enzymatic (trypsin) digestion of a traditional overnight digestion in (*a*) aqueous buffer and (*b*) an accelerated digestion in the presence of organic solvent.

native protein conformation. In contrast, under aqueous or traditional overnight digestion conditions, the enzyme approaches external substrate sites until the protein loses its conformational structure due to cleavage, at which point the enzyme can access other sites for further digestion. The process is depicted in Figure 19.5 (Li et al., 2009).

A total of 10 myoglobin tryptic peptides were identified from both the accelerated digestion protocol (i.e., 30 min) and the overnight digestion, representing 72% sequence coverage of myoglobin. The results demonstrated that the rapid digest protocol is qualitatively equivalent to an overnight digestion in terms of amino acid sequence coverage. Compared to a typical protein-to-enzyme ratio of 100 : 1 in the traditional overnight digestion, a lower protein-to-enzyme ratio of ~7 : 1 is critical in generating a higher tryptic peptide yield within 30 min. It is understandable that a high protein-to-enzyme ratio and a high-quality (often expensive) enzyme are needed in the proteomic research because peptides from enzyme self-digestion may cause interferences, especially for protein identification work. For the bioanalysis of protein quantitation in plasma samples, it is likely that the generation of an interference results from matrix is much higher than that from enzyme self digestion. A low protein-to-enzyme ratio should be beneficial in protein bioanalysis, as it greatly accelerates the digestion speed. The LC–MS/MS method was selective for measuring both myoglobin and somatropin tryptic peptides from accelerated digestion (Fig. 19.6; Li et al., 2009). Somatropin, a human growth hormone, was used as the protein analyte to establish a linear calibration curve from 2.00

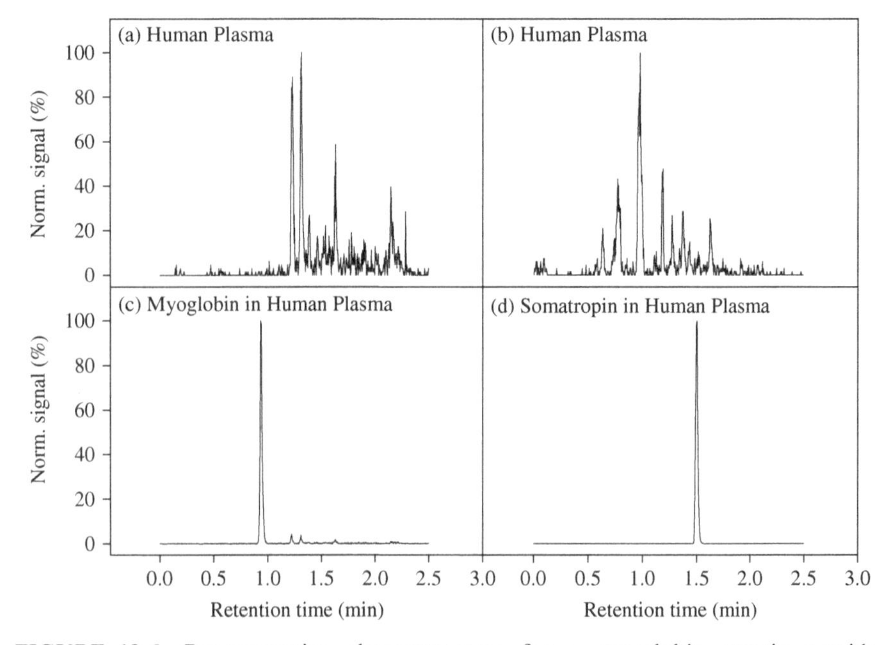

FIGURE 19.6 Representative chromatograms for a myoglobin tryptic peptide, LFTGHPETLEK, obtained from (*a*) human plasma blank and (*b*) myoglobin spiked in human plasma, and for a somatropin tryptic peptide, LFDNAMLR, obtained from (*c*) human plasma blank and (*d*) somatropin spiked in human plasma. [Figure (*a*) and (*c*) reprinted from Li et al. (2009) with permission of *Rapid Communication Mass Spectrometry*.]

to 1000 μg/mL in human plasma while myoglobin was used as the internal standard.

The amount and type of the organic solvent used is important to the related assay. Excessive organic solvent could reduce the enzyme activities. In a further study, Li and co-workers (2009, in preparation) quantitatively characterized accelerated trypsin digestion of targeted proteins in human plasma in the presence of 50% methanol. The accelerated tryptic digestion was compared with the regular digestion of myoglobin in human plasma for 8 time points ranging from 15 min. to 24 h. The results demonstrated that the accelerated digestion proceeded to almost the same degree of completeness in 15 min as seen after 24 h in an aqueous buffer.

Overall, it is both practical and feasible to develop LC–MS/MS methods for the quantitation of large therapeutic proteins, including monoclonal antibodies (mAbs) by monitoring representative peptides resulting from enzymatic digestion. Accelerated digestion can offer the advantage of shortened sample preparation time for proteins in complex biological matrices such as plasma and serum. Thus, sample preparation will no longer be a rate-limiting step for protein bioanalysis that involves enzymatic digestion. The ultimate benefit is the fast turn-around time for protein sample analysis by LC–MS/MS to support drug development programs.

19.5 CONSIDERATION OF INTERNAL STANDARD FOR PROTEIN QUANTITATION

In quantitative bioanalysis in complex biological fluids, the incorporation of an ISTD is essential to achieve accurate, precise, and reproducible results because methods typically employ various sample preparation steps that can introduce variability. An ideal ISTD is the stable isotopically labeled version of the analyte of interest. The labeled version is physically and chemically similar to the target analyte so that variability derived from sample preparation to LC–MS/MS analysis can be tracked. This situation is the preferred approach for the quantitative bioanalysis of small-molecule drugs.

As for protein bioanalysis, the stable isotopically labeled version of the target protein is the ideal ISTD for LC–MS/MS assays too. For example, an isotope-labeled monoclonal antibody version was used for the quantification of an mAb drug candidate by monitoring a pair of representative peptides for the analyte and the ISTD (Heudi et al., 2008). The labeled ISTD was spiked to the serum samples prior to trypsin digestion so that the enzymatic digestion process variation can be compensated for should there be any variability between samples. The linearity was established between 5 and 1000 μg/mL with the accuracy and precision of five QCs meeting the acceptance criteria of method validation. It was noted that peak height ratio, instead of peak area ratio, of its representative peptide to the corresponding isotope-labeled peptide was used for concentration calculation. In another study, a N^{15} isotopically labeled version of a small protein, 10,464 amu, was used as the ISTD for the quantitation in plasma by measuring the intact protein directly with LC–MS/MS (Ji et al., 2003, 2005). Good intraassay and interassay CVs and good interassay accuracies demonstrated the ruggedness of the LC–MS/MS assay utilizing an isotopically labeled protein as the ISTD.

The benefit of using isotope-labeled full-length proteins as ISTD is obvious for protein quantification using representative peptides because they can track the variability of enzymatic digestion and sample extraction. An additional benefit arises from the flexibility of method development. For instance, the representative peptide can be chosen and changed, or several representative peptides can be monitored within the same run. Nevertheless, the approach of labeling proteins with stable isotopes as the ISTD cannot be readily realized since the production of isotopically labeled proteins is a challenging and complicated process (Kippen et al., 1997). Thus, the use of labeled proteins for every assay development is not practical. Instead, isotopically labeled versions of the tryptic peptides are added for LC–MS/MS protein quantitation after protein digestion (Kirkpatrick et al., 2005; Gerber et al., 2007; Brönstrup, 2004). However, the labeled peptides do not follow the sample preparation steps such as enzymatic digestions or immunoprecipitation or albumin depletion.

Representative or analog protein ISTDs have been used in the LC–MS/MS protein quantitation. Czerwenka and co-workers (2007) used analog proteins from other species for the quantitation of β-lactoglobulin by LC–MS. Recombinant proteins that differ by a few amino acids were used as the analog protein standard (Tubbs et al., 2006). Other analog proteins were used as ISTD as well. A mouse monoclonal antibody specific to human sEGFR was used to monitor the overall sample preparation process from immunoprecipitation to tryptic digestion to LC–MS/MS analysis.

Bovine fetuin (Yang et al., 2007) was used as an analog protein ISTD for the quantification of somatropin and a proprietary monoclonal antibody drug (mAbX). Unlike an isotope-labeled peptide, bovine fetuin was added at the beginning of the sample preparation, which compensated for variability that might occur during the sample processing steps, including protein digestion. Figure 19.7 (Yang et al., 2007) is a comparison between the use of bovine fetuin as the analog ISTD and the use of stable isotope-labeled peptide as ISTD for the quantification of mAbX, where linear regression with R^2 of 0.9943 and 0.9886 were obtained, respectively. These results indicate that the protein analog ISTD (bovine fetuin) performed similarly or better than an isotope-labeled peptide ISTD.

Li and co-workers (2009) demonstrated the effectiveness of an analog protein ISTD for protein quantitation employing accelerated tryptic digestion. Somatropin was used as the analyte of interest while myoglobin served as its analog protein ISTD to correct for the variability during the enzymatic digestion and sample preparation process. The solid circles in Figure 19.8 (Li et al., 2009) were obtained experimentally from seven calibrators, whereas the line is from a linear regression of the experimental data. The curve is linear over the range of 2.00–1000 μg/mL for somatropin with an R^2 of 0.9985.

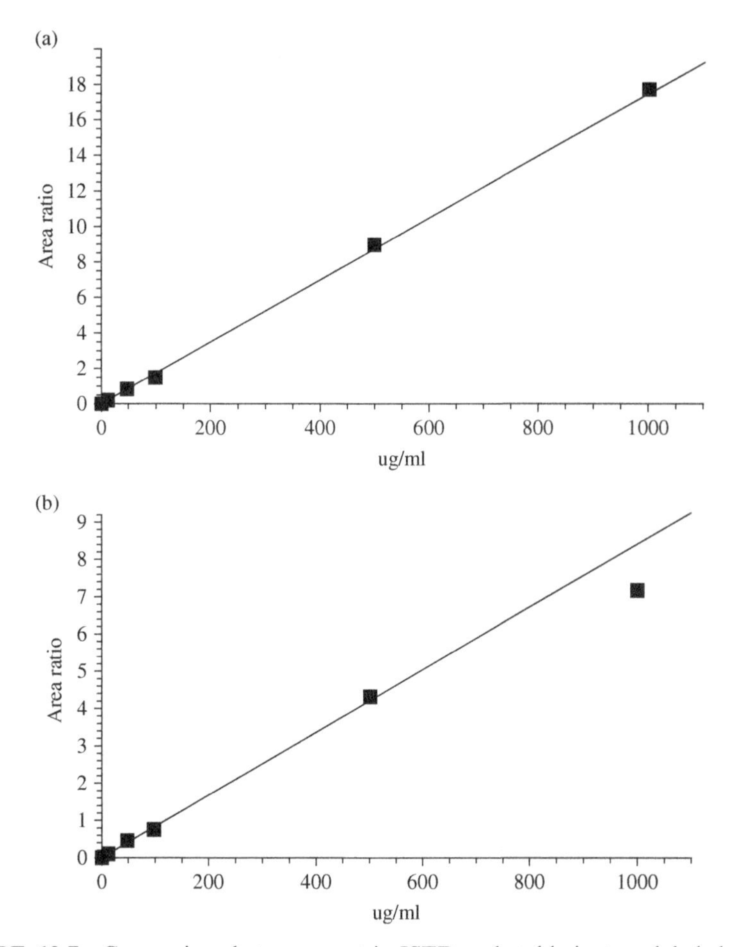

FIGURE 19.7 Comparison between protein ISTD and stable-isotope-labeled peptide ISTD: (*a*) Use of bovine fetuin as ISTD for the mAbX assay. The r2 of the linear regression was 0.9943. (*b*) Use of stable-isotope-labeled peptide as ISTD for the mAbX assay. The r2 of the linear regression was 0.9886. [Reprinted from Yang et al. (2007) with permission of the American Chemical Society.]

Chemical modification, in which functional groups on proteins are either added or blocked, is a good alternative to stable isotope-labeled protein ISTD. The active research of comparative proteomic studies has led to the use of various strategies that include isotope-coded affinity tagging (Gygi et al., 1999), acid-labile isotope coded extractants (Qiu et al., 2002), phosphoserine isotope-coded affinity tag that labels phosphorylated residues (Goshe et al., 2002) or nicotinyl-*N*-hydroxysuccinimide (Münchbach et al., 2000), and quantitation using enhanced signal tags (Beardsley and Reilly, 2003) for N-terminus labeling. This approach could be adapted for the creation of ISTD for plasma

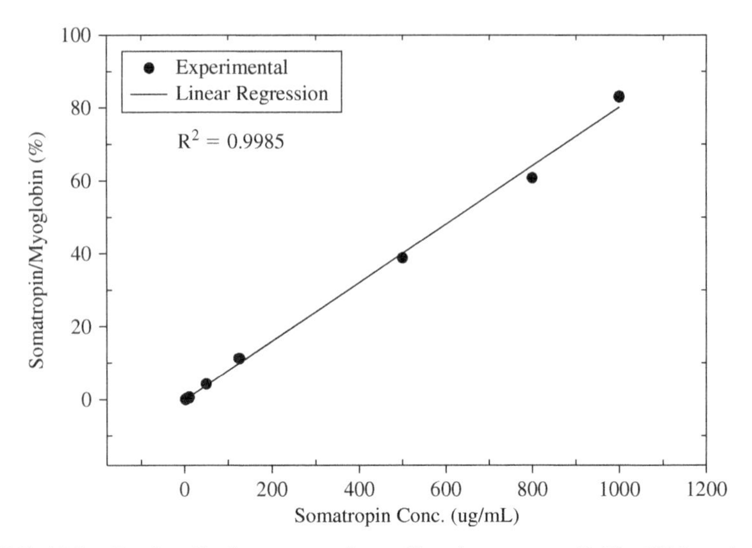

FIGURE 19.8 Fresh-spiked somatropin calibration curve (2.00–1000 μg/mL) in human plasma using myoglobin as the ISTD. [Reprinted from Li et al. (2009) with permission of *Rapid Communication Mass Spectrometry*.]

or protein assays (Becher et al., 2006; Ji et al., 2009). An alkylated version of the target protein, EPI-hNE4, was employed as its ISTD using iodoacetamide to block the sulfhydryl groups of protein (Becher et al., 2006). The modification leads to a mass shift of 57 amu for each alkylated sulfhydryl group. The alkylated ISTD was purified by high-pressure liquid chromatography (HPLC) and spiked into the plasma sample prior to sample preparation. EPI-hNE4 and the alkylated EPI-hNE4 showed similar retention time, as indicated in Figure 19.9 (Becher et al., 2006). Good intra- and interassay precision (CV in between 8.5 and 8.3%), and accuracy (90.6–102.1%) were obtained for EPI-hNE4 validation with the use of the alkylated ISTD. Ji and co-workers (2009) reported an approach of using tryptic peptides generated by purified therapeutic monoclonal antibody and labeled with *d*(2)-formaldehyde as ISTD while those derived from proteins in cynomolgus monkey serum were labeled with *d*(0)-formaldehyde. This provides a fast, cost-effective, and standardized approach to generate internal standards for any surrogate peptides that are used to quantify the therapeutic monoclonal antibody in biological matrices and allow a fast development and validation within 1–2 weeks.

19.6 MATRIX EFFECT, MATRIX SUPPRESSION/ENHANCEMENT, AND RECOVERY

In quantitative bioanalysis in support of drug discovery and development, the majority of the experiments are carried out in complex biological matrices,

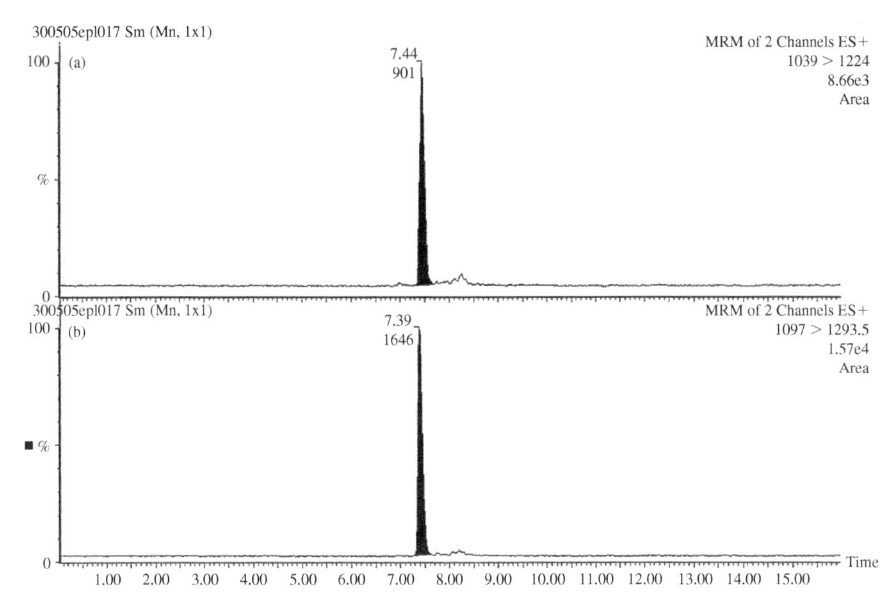

FIGURE 19.9 (*a*) Chromatographic separation of 50 ng/mL EPI-hNE4 on a Zorbax SB-C18 with a gradient mobile phase. (*b*) Chromatographic separation of IS (alkylated EPI-hNE4) in the same conditions. [Reprinted from Becher et al. (2006) with permission of the American Chemical Society.]

including plasma, serum, and urine. Therefore, some level of sample preparation/cleanup is necessary to extract the analyte of interest from the bulk of endogenous components. All sample preparation procedures have the potential to result in the loss of analyte. On the other hand, the coextracted matrix components may cause matrix suppression/enhancement and matrix effect if an appropriate ISTD is not used, leading to possibly inaccurate results. Therefore, recovery and matrix effects are two important parameters to consider during LC–MS/MS assay development.

Matrix effect and matrix suppression/enhancement are terms in bioanalysis that are often used interchangeably. However, the matrix effect is a more specific terminology in bioanalysis and is primarily used to describe whether an assay's performance can be reproduced for individual subject samples, regardless of the difference in matrices. If the response ratio of an analyte to its ISTD can be kept consistent under the same concentration level in different matrices, then no adverse matrix effect can be claimed for the bioanalytical assay. On the other hand, matrix suppression/enhancement is more explicit to evaluate whether an analyte signal is affected by coeluted matrix components even though they are not directly measured by mass spectrometry. The suppression/enhancement can be ascribed to an analyte's ionization efficiency, which is impacted, either negatively or positively, by the underlying matrix components.

An assay can have a minimal matrix effect even if substantial matrix suppression/enhancement is observed, provided that labeled ISTDs are employed. An isotopically labeled ISTD can compensate for the ionization variability for the analyte of interest. In comparison, matrix suppression/enhancement can affect assay sensitivity and linearity.

Matrix effect in complex biological matrices is an important factor to consider during LC–MS/MS assay development for protein quantitation via monitoring either intact proteins or representative peptides. Matrix effects from six different lots of plasma were investigated for the quantitation of intact EPI-hNE4 using protein precipitation prior to LC–MS/MS analysis (Becher et al., 2006). Minimal matrix effects were noted. In a separate study, it was demonstrated that there was no matrix effect for the quantitation of rK5 in plasma by measuring the intact rK5 via LC–MS/MS (Ji et al., 2003, 2005, 2007). It is worth noting that SPE was used to extract rK5 from plasma samples and stable-labeled ISTDs were used.

Severe matrix suppression/enhancement was noted for the quantitation of an mAb by monitoring a representative peptide followed by trypsin digest by LC–MS/MS (Heudi et al., 2008). Their results indicated that sample preparation/cleanup and chromatography were not sufficient to separate the underlying matrix components from the tryptic peptide of interest. More severe matrix suppression/enhancement could be observed when tryptic peptides are analyzed by LC–MS/MS than intact proteins. Mammalian proteomes can comprise more than 20,000 different proteins without counting posttranslational modifications and sequence and splicing variants (Aebersold and Mann, 2003). Up to 1 million distinct peptides can be generated from a proteolytic digestion of such a mixture following standard protocols of bottom-up proteomics. The number of peptides will increase if missed cleavage sites, due to the inevitable imperfections in enzyme activity, are considered. Such a huge number of endogenous peptides make it excessively challenging to cleanup targeted tryptic peptides from analytes of interest, which inevitably leads to more severe matrix ion suppression/enhancement. The results illustrate the necessity of either developing thorough sample cleanup strategies and/or chromatography to reduce the complexity of the digested samples to minimize this ion suppression/enhancement effects.

Sample preparation efficiency is referred to recovery. The absolute recovery directly impacts assay sensitivity (i.e., LLOQ) of a method. Recovery can be measured precisely for small molecules by spiking a known amount of the analyte into blank matrix extracts. Recovery can be calculated by comparison of the response obtained from the postspiking samples with the response from extracted samples at the same concentration level(s). Becher and co-workers (2006) evaluated the overall extraction recovery for EPI-hNE4 in both human and monkey plasma of an LC–MS/MS assay using protein precipitation as the sample preparation. The recovery for EPI-hNE4 was determined to be between 44.1 and 48.8%, while the recovery for the ISTD was ~43.4%. Ji and co-workers (2003) used SPE to extract rK5, a small protein drug, from plasma samples, followed by direct analysis of intact rK5 by LC–MS/MS. The peak area ratio, analyte/ISTD, from recovery controls was compared to that obtained from extracted QC samples. The mean extraction recovery of rK5 was between 72 and 85%, which was sufficient to achieve accurate, precise, and reproducible results at the LLOQ. In an LC–MS/MS method development for the quantitation of an mAb (Heudi et al., 2008), digested samples were cleaned up by SPE and tryptic peptides were subsequently monitored. The overall method recovery was estimated to be only 14%, where the recovery of the SPE purification is 58%. About 32% loss was due to the sample digestion step of the method. In a similar study (Yang et al., 2007), an mAb drug candidate and its analog protein standard were spiked into plasma, which was submitted to tryptic digestion. Representative tryptic peptides from the mAb and ISTD were used for quantitation. A two-dimensional (2D) SPE was utilized to clean up resulting digested samples. The recovery for the 2D-SPE was about 57% across low and high concentrations.

19.7 SENSITIVITY ENHANCEMENT VIA IMMUNOCAPTURE/ PURIFICATION

Quantitation of serum proteins poses analytical challenges because proteins are present at vastly different concentrations, up to 10 orders of magnitude (Anderson and Anderson, 2002). For example, human serum albumin (HSA) accounts for nearly 50% of the net protein contents in human serum. In comparison, the protein drugs are typically circulating at relatively low nanogram/milliliter concentrations. Additionally, protein therapeutic drugs are physicochemically similar to the endogenous components of the human proteome. The abundant endogenous components will inevitably interfere with the analysis of low-concentration target protein drugs. The interference ultimately causes matrix suppression and selectivity issues for LC–MS/MS based quantitative bioanalytical assays, leading to poor assay precision and accuracy. To detect and quantify low abundant proteins of interest, antibody enrichment/capture of specific proteins or peptides has been developed prior to LC–MS/MS analysis. Proteins can be enriched up to 1000-fold by antibodies (Whiteaker et al., 2007), which makes limits of quantification below 1 ng/mL readily achievable.

Stable isotope standards with capture by antipeptide antibodies (SISCAPA) were developed for quantitation of peptides in complex digests (Anderson and Andeson, 2002). Antipeptide antibodies, immobilized on nanoaffinity columns, are used to enrich specific peptides and their corresponding stable-isotope-labeled ISTD. Measured by either selected ion monitoring (SIM) or selected reaction monitoring (SRM) ESI mass spectrometry, an average 120-fold enrichment of the antigen peptide relative to others was demonstrated with simple binding and elution from the affinity columns. SISCAPA is specifically designed to enrich tryptic peptides of interest after enzymatic digestion, which is suited for multiprotein assays. Immunoaffinity LC–MS/MS is the most widely recognized strategy for the characterization of low abundant proteins. Nevertheless, the strategy may not be ideal for target quantification of protein therapeutics in biological matrix because tryptic peptides from endogenous proteins with similar properties as the ones from target proteins are likely to be enriched by the antipeptide antibodies. In essence, the interference could be enriched as well.

Protein enrichment via specific antibodies prior to LC–MS/MS seems to be better suited for quantitative analysis of protein therapeutics in biological matrix. Immunoaffinity purification of Myl3, a 23-kDa isoform of the subunits of myosin, was performed on serum samples (Berna et al., 2007). The enriched Myl3 was digested on-bead with trypsin to release a representative peptide for LC–MS/MS analysis with a corresponding stable-isotope-labeled ISTD. The method was validated to measure Myl3 in rat serum. A LLOQ of 0.073 n*M* Myl3 in serum was achieved. Myl3 interday accuracy and precision was within 7.6 and 11.1%, while interday accuracy and precision did not exceed 12.9 and 13.2%, respectively. The same group (Berna and Ackermann, 2009) reported the use of immunoprecipitation in 96-well ELISA format (IPE) and microwave-assisted protein digestion to perform LC–MS/MS protein analyses within a single day. Lower limits of quantification of 100 pg/mL (NTproBNP) and 0.95 ng/mL (Myl3) were achieved. One of the major advantages of immunoprecipitation in 96-well format over traditional immunoprecipitation is that the preparation can be processed in parallel instead of individually. Therefore, throughput in the sample enrichment step can be greatly increased.

Dubois and co-workers (2008) proposed an immunocapture/mass spectrometry assay to quantitate Erbitux (cetuximab), an mAb used for the treatment of colorectal cancer. Soluble epidermal growth factor receptor (sEGFR), the cellular target of Erbitux, was covalently bound to the M-280 tosyl-activated magnetic beads. The functionalized beads were added to serum samples to capture Erbitux and its analog ISTDs. Erbitux was released from the beads under acidic conditions, reduced, and digested by strypin. The resulting peptide mixtures were analyzed by LC–MS/MS as shown in Figure 19.10 (Dubois et al., 2008). Since elution volume was set at 50 μL using a sample size of 500 μL, the sensitivity of Erbitux was increased 10-fold due to immunocapture. The assay achieved an LLOQ of 20 ng/mL for a three-order of magnitude of dynamic calibration curve range. The introduction of an analog ISTD at the immunoprecipitation step helped control the assay variability below 20%. This result clearly demonstrates the practical utility of immunocapture to enrich low abundance protein therapeutics in biological matrices.

The potential issues with bead-based immunocapture methods are saturation, nonspecific adsorption, and nonspecific protein binding, which through different mechanisms lead to nonlinear response of analyte at high and low analyte concentrations. Bead saturation at high protein concentrations can be solved by narrowing the curve range. Bead pretreatment with addition of a

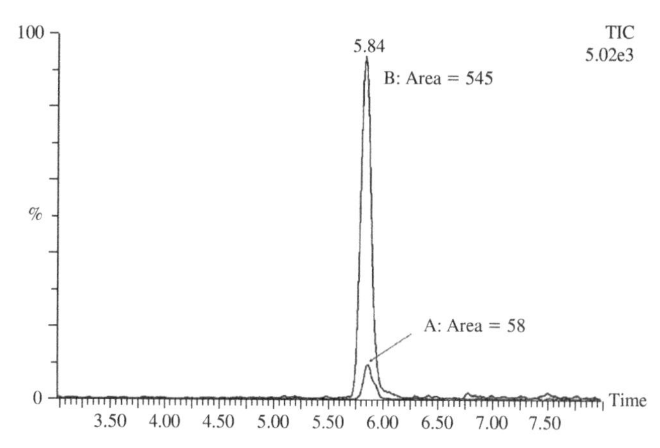

FIGURE 19.10 Reconstituted LT3 signals after trypsin digestion from (*a*) a sample of Tris-HCl 50 m*M* buffer (1 m*M* CaCl2 and 100 μg/mL BSA) spiked with 200 ng/mL Erbitux and (*b*) a sample of human serum spiked with 200 ng/mL Erbitux submitted to immunocapture. Chromatographic separation was performed with a gradient mobile phase [Reprinted from Dubois et al. (2008) with permission of the American Chemical Society.]

known amount of standard proteins will alleviate the nonspecific adsorption at low protein concentrations.

It is obvious that LC–MS/MS assay sensitivity can be proportionally increased by enriching the protein of interest using immunocapture. The distinctive advantage of immunoprecipitation is that structural modifications are taken into consideration via the capture of the active form of the protein drugs in the biological matrix. However, this at the same time could be the drawback of immunoprecipitation where a specific antibody or antigen to the protein of interest needs to be obtained.

19.8 SENSITIVITY ENHANCEMENT VIA DEPLETION OF ABUNDANT PROTEINS

Novel protein biomarker discovery has gained momentum in drug development in hopes that the markers can provide diagnostic or prognostic information of a disease. However, biomarker discovery poses a major technical challenge (Bodovitz and Joos, 2004) because the dynamic range of concentrations of proteins in human serum spans greater than 10 orders of magnitude (Anderson and Anderson, 2002). The removal of high abundance proteins has been utilized for the identification of low abundant proteins in proteomic analysis (Chromy et al., 2004). Enhanced detection and identification of low abundant peptides and proteins by mass spectrometry have been reported (Björhall et al., 2005; Taylor, 2005). A variety of different strategies have been used to deplete abundant proteins prior to LC–MS/MS analysis. Cibacron Blue, a dye-modified affinity resin, binds to albumin and is a commonly used method of removing albumin. This dye can also bind other proteins, such as nicotinamide adenine dinucleotide (NAD+), flavin adenine dinucleotide (FAD), adenosine triphosphate (ATP) (Prestera et al., 1992; Thresher and Swaisgood, 1990). This nonspecific interaction can result in the removal of low abundance proteins of interest (Gianazza and Arnaud, 1982). Shores and Knapp (2007) identified 46 unique proteins in cerebrospinal fluid (CSF) using a 50-kDa molecular weight cutoff filter, which is less favorable compared to 135 proteins identified from analysis without depletion. The results strongly indicate the nonspecific removal of high-molecular-weight proteins by filter does not offer clear advantages and could lead to significant loss of important proteins. In addition, Ji and co-workers (2007) demonstrated that a significant amount of albumin could be depleted through solid-phase extraction sample preparation.

More specific and complete depletion of targeted abundant proteins to enable detection of low abundant proteins have been demonstrated with individual antibody methods. For instance, an immunodepletion strategy based on use of polyclonal antibodies such as the commercially available multiple affinity removal system (MARS) has shown rapid and efficient depletion of serum proteins. The system depletes the six most abundant proteins in human plasma (i.e., albumin, IgG, IgA, transferrin, haptoglobin, and antitrypsin). These six most abundant proteins represent approximately 85% of the total protein mass in human serum or plasma (Echan et al., 2005). Tissue plasminogen activator (tPA), 10–60 ng/mL in human plasma, was detected by 2D gel followed by MARS depletion of the six most abundant proteins (Cho et al., 2005). With the use of the more sensitive LC–MS/MS platform, a detection limit of tPA down to low nanogram/milliliter concentration range should be practical.

Depletion of the most abundant serum proteins is an attractive option that can facilitate the detection of low abundance proteins. This strategy is mostly applied in qualitative proteomics, where the focus is placed on the identification. To support drug discovery and development, quantitative bioanalysis of targeted protein therapeutics in complex biological matrices takes center stage. In addition to having the capability of detection of low concentrations of drug candidates, any depletion method should enable consistent and reproducible removal of high abundance proteins and minimize nonspecific removal of the analytes of interest. This capability is crucial to achieving precise and accurate quantitative bioanalysis.

Hagman et al. (2008) described a method for quantifying monoclonal antibodies in human serum and its validation in cynomolgus monkey serum. The serum was depleted of albumin prior to analysis to reduce sample complexity and enhance sensitivity. A peptide from the CDR3-region of its heavy chain was chosen as the representative tryptic peptide to quantify the entire mAb. Albumin depletion efficiency and specificity in human serum were evaluated using five commercially available albumin depletion kits. Figures 19.11*a* and 19.11*b* (Hagman et al., 2008) contain the sodium dodecyl sulfate polyacrylamide

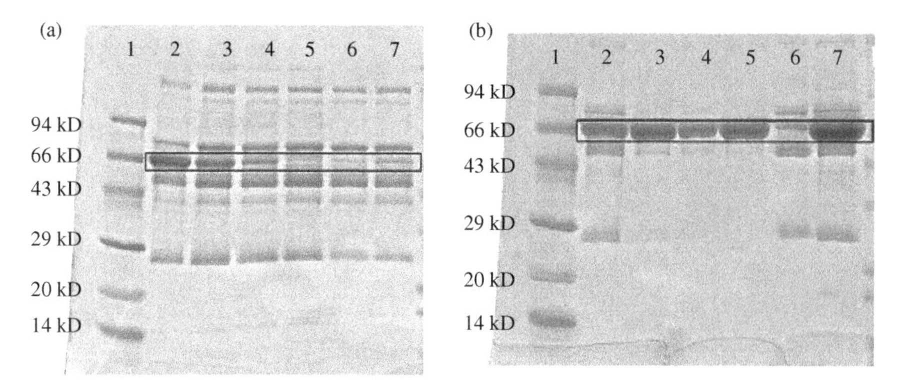

FIGURE 19.11 (*a*) SDS-PAGE of albumin-depleted human serum. In lane 1, mixtures of reference proteins were separated. Crude serum was separated in lane 2. In lane 3, albumin-depleted serum using the Montage albumin depletion kit was separated. Serum depleted with the Vivapure Anti-HSA depletion kit was separated in lane 4. In lane 5, serum depleted using the ProteoExtract albumin removal kit was separated. Serum depleted with the Aurum Affi-Gel Blue mini kits was separated in lane 6. Finally, serum depleted using the Enchant albumin depletion kit was separated in lane 7. (*b*) SDS-PAGE of eluted proteins from albumin depletion column. In lane 1, mixtures of reference proteins were separated. Crude serum was separated in lane 2. In lane 3, the proteins with affinity for the Montage albumin depletion kit column were separated. Proteins eluting from the Vivapure Anti-HSA depletion kit were separated in lane 4. In lane 5, the proteins with affinity for the ProteoExtract albumin removal kit columns were separated. Proteins eluting from the Aurum Affi-Gel Blue minikit columns were separated in lane 6. In lane 7, proteins with affinity for the Enchant albumin depletion kit columns were separated. [Reprinted from Hagman et al. (2008) with permission of the American Chemical Society.]

gel electrophoresis (SDS-PAGE) of the albumin-depleted human serum and eluted proteins from the albumin depletion column. The results illustrate that nondye affinity has higher specificity compared to dye-based affinity (Ramström et al., 2005). Approximately 50% of the total serum protein was removed prior to LC–MS/MS analysis based on the results. Although the selected human monoclonal antibody analyte has high sequence homology with IgG, no interference was observed at the retention time of the targeted tryptic peptides as shown in Figure 19.12*a* (Hagman et al., 2008). An LLOQ of 2.00 μg/mL using 50 μL of serum was achieved (Fig. 19.12*b*; Hagman et al., 2008).

Following are some aspects that limit the broad application of the depletion kits. The potential issue of employing this albumin depletion strategy is the limited sample size, typically less than 50 μL, which can be treated by the depletion kits. This sample size prevents the use of a large sample volume to obtain assay sensitivity enhancement. The depletion kits are relatively expensive. Perhaps the most significant drawback of the immunodepletion methods is nonspecific removal of plasma proteins of interest that may be bound to the removed proteins (Anderson and Hunter, 2006; Liu et al., 2006).

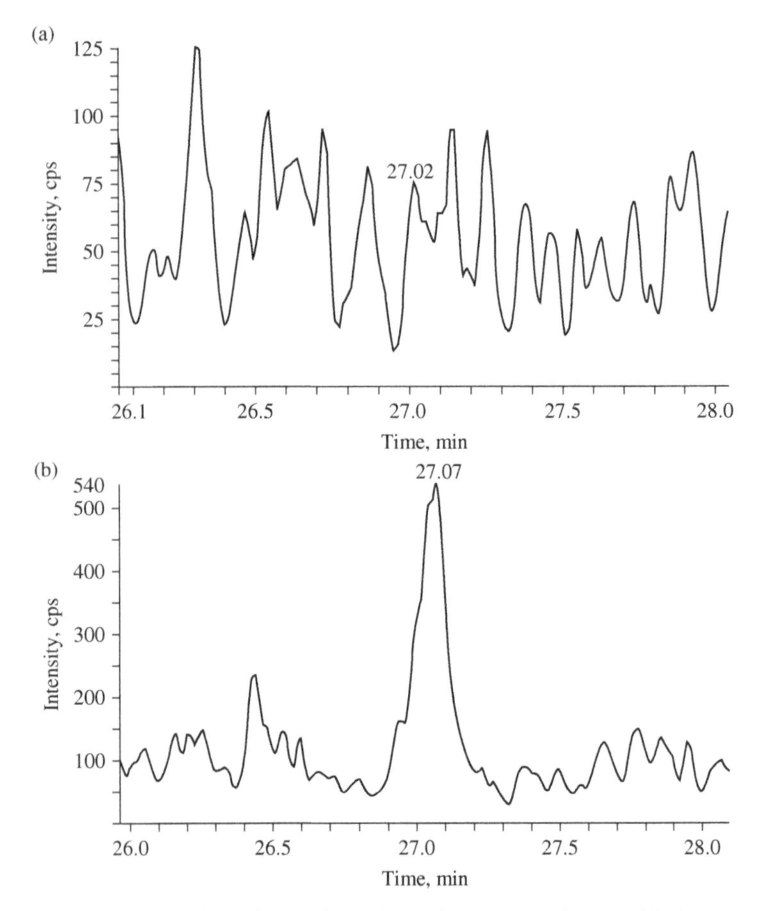

FIGURE 19.12 Transition of the triply charged representative peptide ion monitored as a function of time. (*a*) LC–MS/MS chromatogram of the blank human serum sample. (*b*) LC–MS/MS chromatogram of the 2-μg/mL HmAb human serum sample.[Reprinted from Hagman et al. (2008) with permission of the American Chemical Society.]

19.9 PRACTICAL ASPECTS OF LC–MS ASSAY FOR PROTEINS IN DRUG DEVELOPMENT

19.9.1 "Fit-for-Purpose" Assay Development Strategy

As discussed in previous sections, LC–MS/MS assays for protein quantitation can be a valuable alternative to immunoassays. A "fit-for-purpose" strategy can be applied for selection of either immuno or LC–MS/MS assays to meet the requirement at different drug development stages. The selection of a drug candidate for further development in the drug discovery and preclinical development stage requires a highly specific bioanalytical assay that is readily available and does not require significant investment. An immunoassay may not be a good choice to fit for this purpose because the development of antibody and antigen reagents can be both expensive and time consuming. An immunoassay may not have the flexibility to accommodate rapid sequence changes of the protein compounds being evaluated. Each protein analyte may require a unique antigen or antibody reagent. The number of samples needing to be analyzed may not justify the investment for the development of an immunoassay during the early discovery stage. LC–MS/MS can be the assay of choice to fit this purpose because of its shorter development cycle. Additional cost saving may be achieved using peptide ISTD and simple sample preparation. An additional reason for choosing LC–MS/MS assays over immunoassays is that LC–MS/MS can often provide the needed assay accuracy/precisions and sensitivity. As the (protein) drug development advances further, an ISTD of the stable isotopically labeled version of the protein may become necessary to improve assay ruggedness. A more refined sample preparation procedure may also be needed to enhance assay performance with regard to recovery, matrix effect, sensitivity, and accuracy/precision.

The number of samples that need to be analyzed likely increases as the drug development proceeds to clinic stages. When an immunoassay is well developed and characterized, sample analysis can be sensitive, fast, and cost effective. Similarly, the considerations for selecting an LC–MS/MS assay or an immunoassay could follow the same approach at different stages of biomarker development, validation, and application.

19.9.2 LC–MS/MS Assay for Pegylated Proteins, Protein Homologs, and Posttranslational Modified Proteins

While small-molecule drug compounds often have a well-defined molecular structure, protein drugs or protein biomarkers are often a group of structurally related compounds that have the same therapeutic functions or biological responses. For example, a protein drug could be an ensemble of protein homologs. Pegylated proteins will have a range of different sized poly(ethylene glycol) (PEG) polymers attached. Additionally, proteins may have posttranslational modification in vivo. When a representative peptide from the protein digestion is used for the protein quantitation, the LC–MS/MS will measure all related protein with different PEG chain length as demonstrated in a study published by Yang and co-workers (2009). For protein homologs and post-translational modified proteins, depending on the selection of the representative peptide in the assay, molecules that consist of the representative peptide will be included in the content of the measured concentration. Representative peptide approach can have the advantage of including all related proteins for quantitation, which may otherwise be excluded if detection was purely based on the molecular weight of the whole protein. The posttranslational modification of a protein is an issue frequently encountered in biomaker bioanalysis. An LC–MS/MS assay may detect proteins with the posttranslational modification regardless of whether it is related to the intention of the measurement or not.

Another issue for biomarker measurement is underestimated protein concentration in the case of polymorphism due to the genetic diversity of different populations. However, these issues also exist even if an immunoassay is used. In this case, simultaneous measurement of two or more representative peptides from the same protein by LC–MS/MS may provide better understanding of the assay performance (Hoofnagle and Wener, 2009).

19.9.3 Total and Free Protein Concentration Measurement

Not all the proteins are detected as the measurable concentrations in immunoassays, depending on the antigen or antibody reagent used in the assay. The measured protein concentrations from LC–MS/MS assays, however, are generally considered total drug concentrations. The total drug concentration could provide valuable information complementary to the concentrations obtained by immunoassays. The sample preparation process, especially during enzymatic digestion, should disrupt potential protein bindings when a representative peptide is measured as the representative of the whole protein. Strong binding between the protein analytes and antibodies can be dissociated in the sample extraction process when an intact protein is used for the concentration measurement (Ji et al., 2007). In addition, for small protein analytes, dialysis methods can be used to separate free and bound proteins prior to the detection of the protein in each dialysis chamber to obtain free protein concentration, which could be used as the surrogate "active" drug concentration for pre-clinical studies (Ji et al., 2005). Other sample preparations can be incorporated to define the protein concentrations measured prior to the LC–MS, or LC–MS/MS measurements of the treated samples.

19.9.4 Protein Metabolism

Progress in protein quantitation is the result of merging the knowledge of LC–MS/MS bioanalysis of small molecules and mass spectrometry application in proteomic researches. Up until now, the evaluation of the metabolism of the protein drugs has not been at the forefront of drug development. Because of this, no information is available regarding potential impact of the metabolites of protein drugs from both efficacy and safety points of view. It is still debatable whether it would be worthwhile to investigate the biotransformation of protein drugs using LC–MS/MS technology. In addition, some of the questions related to drug metabolism that have been raised in the development of small-molecule drugs could be raised in the drug development of large molecules in the future.

19.10 CONCLUSIONS

An increase in protein drug candidates within drug development pipelines combined with an increase in the use of protein biomarkers for decision making in clinical practice have prompted bioanalytical scientists to search for an analytical tool capable of direct quantitative measurement of proteins. LC–MS/MS certainly fits this criterion very well. The assay capability and performance largely depend on the method development with regard to the following aspects: (1) The capability of LC–MS/MS detection of the small protein or the representative peptide; (2) selection of representative peptides; (3) the efficiency of the protein digestion in biological matrices; (4) sample cleanup prior to and/or after the digestion; and (5) selection of the ISTDs. A fit-for-purpose strategy can be used to balance the investment in assay development and the intended use of the assay during the drug development process from drug discovery through development. Advances in mass spectrometric technology and separation sciences continually provide an array of tools that will facilitate the development of LC–MS/MS assays for protein bioanalysis. Clearly, the advantage of obtaining quantitative protein information in the biological matrix by LC–MS/MS will ultimately help accelerate the pace of protein therapeutics discovery and development.

ACKNOWLEDGMENTS

The authors would like to thank Dr. Mark E. Arnold, Dr. Mohammed Jemal, and Dr. Anne-Francoise Aubry for their comprehensive review of this chapter and helpful discussions.

REFERENCES

Ackermann BL, Berna MJ, Eckstein JA, Ott LW, Chaudhary AK. Current applications of liquid chromatography/mass spectrometry in pharmaceutical discovery after a decade of innovation. Annu Rev Anal Chem 2008;1:357–396.

Aebersold R, Mann M. Mass spectrometry-based proteomics. Nature 2003;422 (6928):198–207.

Anderson L, Hunter CL. Quantitative mass spectrometric multiple reaction monitoring assays for major plasma proteins. Mol Cell Proteomics 2006;5(4):573–588.

Anderson NL, Anderson NG. The human plasma proteome: History, character, and diagnostic prospects. Mol Cell Proteomics 2002;1(11):845–867.

Arsene CG, Ohlendorf R, Burkitt W, Pritchard C, Henrion A, O'Connor G, Bunk DM, Güttler B. Protein quantification by isotope dilution mass spectrometry of proteolytic fragments: Cleavage rate and accuracy. Anal Chem 2008;80(11):4154–4160.

Barnidge DR, Goodmanson MK, Klee GG, Muddiman DC. Absolute quantification of the model biomarker prostate-specific antigen in serum by LC-MS/MS using protein cleavage and isotope dilution mass spectrometry. J Proteome Res 2004;3(3):644–652.

Barr JR, Maggio VL, Patterson DG Jr, Cooper GR, Henderson LO, Turner WE, Smith SJ, Hannon WH, Needham LL, Sampson EJ. Isotope dilution—Mass spectrometric quantification of specific proteins: Model application with apolipoprotein A-I. Clin Chem 1996;42(10):1676–1682.

Beardsley RL, Reilly JP. Quantitation using enhanced signal tags: A technique for comparative proteomics. J Proteome Res 2003;2(1):15–21.

Becher F, Pruvost A, Clement G, Tabet JC, Ezan E. Quantification of small therapeutic proteins in plasma by liquid chromatography–tandem mass spectrometry: Application to an elastase inhibitor EPI-hNE4. Anal Chem 2006;78(7):2306–2313.

Berna M, Ackermann B. Increased throughput for low-abundance protein biomarker verification by liquid chromatography/tandem mass spectrometry. Anal Chem 2009;81(10):3950–3956.

Berna MJ, Zhen Y, Watson DE, Hale JE, Ackermann BL. Strategic use of immunoprecipitation and LC/MS/MS for trace-level protein quantification: Myosin light chain 1, a biomarker of cardiac necrosis. Anal Chem 2007;79(11):4199–4205.

Björhall K, Miliotis T, Davidsson P. Comparison of different depletion strategies for improved resolution in proteomic analysis of human serum samples. Proteomics 2005;5(1):307–317.

Bodovitz S, Joos T. The proteomics bottleneck: Strategies for preliminary validation of potential biomarkers and drug targets. Trends Biotechnol 2004;22(1):4–7.

Brönstrup M. Absolute quantification strategies in proteomics based on mass spectrometry. Expert Rev Proteomics 2004;1(4):503–512.

Careri M, Mangia A. Analysis of food proteins and peptides by chromatography and mass spectrometry. J Chromatogr A 2003;1000(1–2):609–635.

Cho SY, Lee EY, Lee JS, Kim HY, Park JM, Kwon MS, Park YK, Lee HJ, Kang MJ, Kim JY, Yoo JS, Park SJ, Cho JW, Kim HS, Paik YK. Efficient prefractionation of low-abundance proteins in human plasma and construction of a two-dimensional map. Proteomics 2005;5(13):3386–3396.

Chromy BA, Gonzales AD, Perkins J, Choi MW, Corzett MH, Chang BC, Corzett CH, McCutchen-Maloney SL. Proteomic analysis of human serum by two-dimensional differential gel electrophoresis after depletion of high-abundant proteins. J Proteome Res 2004;3(6):1120–1127.

Czerwenka C, Maier I, Potocnik N, Pittner F, Lindner W. Absolute quantitation of beta-lactoglobulin by protein liquid chromatography–mass spectrometry and its application to different milk products. Anal Chem 2007;79(14):5165–5172.

De Hoffman E, Charette J, Stroobant V. Mass Spectrometrys Principles and Applications, 2nd ed. Wiley, New York, 2001.

DeSilva B, Smith W, Weiner R, Kelley M, Smolec J, Lee B, Khan M, Tacey R, Hill H, Celniker A. Recommendations for the bioanalytical method validation of ligand-binding assays to support pharmacokinetic assessments of macromolecules. Pharm Res 2003;20(11):1885–1900.

Dubois M, Fenaille F, Clement G, Lechmann M, Tabet JC, Ezan E, Becher F. Immunopurification and mass spectrometric quantification of the active form of a chimeric therapeutic antibody in human serum. Anal Chem 2008;80(5):1737–1745.

Echan LA, Tang HY, Ali-Khan N, Lee K, Speicher DW. Depletion of multiple high-abundance proteins improves protein profiling capacities of human serum and plasma. Proteomics 2005;5(13):3292–3303.

Elliott MH, Smith DS, Parker CE, Borchers C. Current trends in quantitative proteomics. J Mass Spectrom 2009;44(12):1637–1660.

Food and Drug Administration (FDA). Guidance for Industry-Bioanalytical Method Validation; U.S. Department of Health and Human Services, Center for Drug Evaluation and Research (CDER), Center for Veterinary Medicine (CVM), May 2000, available: http://www.fda.gov/cder/guidance/index.htm.

Gerber SA, Kettenbach AN, Rush J, Gygi SP. The absolute quantification strategy: Application to phosphorylation profiling of human separase serine 1126. Methods Mol Biol 2007;359:71–86.

Gianazza E, Arnaud P. Chromatography of plasma proteins on immobilized Cibacron Blue F3-GA. Mechanism of the molecular interaction. Biochem J 1982;203 (3):637–641.

Goshe MB, Veenstra TD, Panisko EA, Conrads TP, Angell NH, Smith RD. Phosphoprotein isotope-coded affinity tags: Application to the enrichment and identification of low-abundance phosphoproteins. Anal Chem 2002;74(3):607–616.

Gygi SP, Rist B, Gerber SA, Turecek F, Gelb MH, Aebersold R. Quantitative analysis of complex protein mixtures using isotope-coded affinity tags. Nat Biotechnol 1999;17(10):994–999.

Hagman C, Ricke D, Ewert S, Bek S, Falchetto R, Bitsch F. Absolute quantification of monoclonal antibodies in biofluids by liquid chromatography–tandem mass spectrometry. Anal Chem 2008;80(4):1290–1296.

Heudi O, Barteau S, Zimmer D, Schmidt J, Bill K, Lehmann N, Bauer C, Kretz O. Towards absolute quantification of therapeutic monoclonal antibody in serum by LC-MS/MS using isotope-labeled antibody standard and protein cleavage isotope dilution mass spectrometry. Anal Chem 2008; 80(11):4200–4207.

Hoofnagle AN, Wener MH. The fundamental flaws of immunoassays and potential solutions using tandem mass spectrometry. J Immunol Methods 2009;347(1–2):3–11.

Jemal M, Ouyang Z, Xia YQ. Systematic LC-MS/MS bioanalytical method development that incorporates plasma phospholipids risk avoidance, usage of incurred sample and well thought-out chromatography. Biomed Chromatogr 2010;24:2–19.

Jemal M, Xia YQ. LC-MS development strategies for quantitative bioanalysis. Curr Drug Metab 2006;7(5):491–502.

Ji C, Li W, Ren XD, El-Kattan AF, Kozak R, Fountain S, Lepsy C. Diethylation labeling combined with UPLC/MS/MS for simultaneous determination of a panel of monoamine neurotransmitters in rat prefrontal cortex microdialysates. Anal Chem 2008;80(23):9195–9203.

Ji C, Sadagopan N, Zhang Y, Lepsy C. A universal strategy for development of a method for absolute quantification of therapeutic monoclonal antibodies in biological matrices using differential dimethyl labeling coupled with ultra performance liquid chromatography–tandem mass spectrometry. Anal Chem 2009;81(22):9321–9328.

Ji QC, Rodila R, El-Shourbagy TA. A sample preparation process for LC-MS/MS analysis of total protein drug concentrations in monkey plasma samples with antibody. J Chromatogr B Analyt Technol Biomed Life Sci. 2007;847(2):133–141.

Ji QC, Rodila R, Gage EM, El-Shourbagy TA. A strategy of plasma protein quantitation by selective reaction monitoring of an intact protein. Anal Chem 2003; 75(24):7008–7014.

Ji QC, Rodila R, Gage EM, El-Shourbagy TA. Mass spectrometric approaches for protein quantitation in drug development. Am Pharm Rev 2004;7(3):84–89.

Ji QC, Rodila R, Morgan SJ, Humerickhouse RA, El-Shourbagy TA. Investigation of the immunogenicity of a protein drug using equilibrium dialysis and liquid chromatography tandem mass spectrometry detection. Anal Chem 2005;77(17):5529–5533.

John H, Walden M, Schäfer S, Genz S, Forssmann WG. Analytical procedures for quantification of peptides in pharmaceutical research by liquid chromatography–mass spectrometry. Anal Bioanal Chem 2004;378(4):883–897.

Jones PT, Dear PH, Foote J, Neuberger MS, Winter G. Replacing the complementarity-determining regions in a human antibody with those from a mouse. Nature 1986;321:522–525.

Kay RG, Gregory B, Grace PB, Pleasance S. The application of ultra-performance liquid chromatography/tandem mass spectrometry to the detection and quantitation of apolipoproteins in human serum. Rapid Commun Mass Spectrom 2007;21(16): 2585–2593.

Kippen AD, Cerini F, Vadas L, Stöcklin R, Vu L, Offord RE, Rose K. Development of an isotope dilution assay for precise determination of insulin, C-peptide, and proinsulin levels in non-diabetic and type II diabetic individuals with comparison to immunoassay. J Biol Chem 1997;272(19):12513–12522.

Kirkpatrick DS, Gerber SA, Gygi SP. The absolute quantification strategy: A general procedure for the quantification of proteins and post-translational modifications. Methods 2005;35(3):265–273.

Kuhn E, Wu J, Karl J, Liao H, Zolg W, Guild B. Quantification of C-reactive protein in the serum of patients with rheumatoid arthritis using multiple reaction monitoring mass spectrometry and 13C-labeled peptide standards. Proteomics 2004;4(4):1175–1186.

Leader B, Baca QJ, Golan, DE. Protein therapeutics: A summary and pharmacological classification. Nat Rev Drug Discov 2008;7:21–39.

Lee MS. Strategies for Drug Discovery Using Mass Spectrometry. Wiley, Hoboken, NJ, 2006.

Li F, Schmerberg CM, Ji QC. Accelerated tryptic digestion of proteins in plasma for absolute quantitation using a protein internal standard by liquid chromatography/tandem mass spectrometry. Rapid Commun Mass Spectrom 2009;23(5):729–732.

Li F, Schmerberg CM, Ji QC. Characterization of methanol assisted accelerated tryptic digestion for absolute protein quantitation in human plasma using LC-MS/MS. Manuscript in preparation.

Lin S, Shaler TA, Becker CH. Quantification of intermediate-abundance proteins in serum by multiple reaction monitoring mass spectrometry in a single-quadrupole ion trap. Anal Chem 2006;78(16):5762–5767.

Linscheid MW, Ahrends R, Pieper S, Kühn A. Liquid chromatography–mass spectrometry-based quantitative proteomics. Methods Mol Biol 2009;564:189–205.

Liu T, Qian WJ, Mottaz HM, Gritsenko MA, Norbeck AD, Moore RJ, Purvine SO, Camp DG 2nd, Smith RD. Evaluation of multiprotein immunoaffinity subtraction for plasma proteomics and candidate biomarker discovery using mass spectrometry. Mol Cell Proteomics 2006;5(11):2167–2174.

Lobo ED, Hansen RJ, Balthasar JP. Antibody pharmacokinetics and pharmacodynamics. J Pharm Sci 2004;93(11):2645–2668.

Mascelli MA, Zhou H, Sweet R, Getsy J, Davis HM, Graham M, Abernethy D. Molecular, biologic, and pharmacokinetic properties of monoclonal antibodies:

Impact of these parameters on early clinical development. J Clin Pharmacol 2007;47:553–565.

Mayr BM, Kohlbacher O, Reinert K, Sturm M, Gröpl C, Lange E, Klein C, Huber CG. Absolute myoglobin quantitation in serum by combining two-dimensional liquid chromatography–electrospray ionization mass spectrometry and novel data analysis algorithms. J Proteome Res 2006;5(2):414–421.

Motoyama A, Yates JR 3rd. Multidimensional LC separations in shotgun proteomics. Anal Chem 2008;80(19):7187–7193.

Münchbach M, Quadroni M, Miotto G, James P. Quantitation and facilitated de novo sequencing of proteins by isotopic N-terminal labeling of peptides with a fragmentation-directing moiety. Anal Chem 2000;72(17):4047–4057.

Ong SE, Mann M. Mass spectrometry-based proteomics turns quantitative. Nat Chem Biol 2005;1(5):252–262.

Prestera T, Prochaska HJ, Talalay P. Inhibition of NAD(P)H:(quinone-acceptor) oxidoreductase by cibacron blue and related anthraquinone dyes: A structure-activity study. Biochemistry 1992;31(3):824–833.

Qiu Y, Sousa EA, Hewick RM, Wang JH. Acid-labile isotope-coded extractants: A class of reagents for quantitative mass spectrometric analysis of complex protein mixtures. Anal Chem 2002;74(19):4969–4979.

Ramström M, Hagman C, Mitchell JK, Derrick PJ, Håkansson P, Bergquist J. Depletion of high-abundant proteins in body fluids prior to liquid chromatography Fourier transform ion cyclotron resonance mass spectrometry. J Proteome Res 2005;(2):410–416.

Russell WK, Park ZY, Russell DH. Proteolysis in mixed organic-aqueous solvent systems: Applications for peptide mass mapping using mass spectrometry. Anal Chem 2001; 73(11):2682–2685.

Shores KS, Knapp DR. Assessment approach for evaluating high abundance protein depletion methods for cerebrospinal fluid (CSF) proteomic analysis. J Proteome Res 2007;6(9):3739–3751.

Taylor PJ. Matrix effects: The Achilles heel of quantitative high-performance liquid chromatography–electrospray–tandem mass spectrometry. Clin Biochem 2005; 38(4):328–334.

Thresher WC, Swaisgood HE. Characterization of specific interactions of coenzymes, regulatory nucleotides and cibacron blue with nucleotide binding domains of enzymes by analytical affinity chromatography. J Mol Recognit 1990;3(5–6):220–228.

Tubbs KA, Kiernan UA, Niederkofler EE, Nedelkov D, Bieber AL, Nelson RW. Development of recombinant-based mass spectrometric immunoassay with application to resistin expression profiling. Anal Chem 2006;78(10):3271–3276.

Van den Broek I, Sparidans RW, Schellens JH, Beijnen JH. Quantitative bioanalysis of peptides by liquid chromatography coupled to (tandem) mass spectrometry. J Chromatogr B Analyt Technol Biomed Life Sci 2008;872(1–2):1–22.

Volosov A, Napoli KL, Soldin SJ. Simultaneous simple and fast quantification of three major immunosuppressants by liquid chromatography–tandem mass-spectrometry. Clin Biochem 2001;34(4):285–290.

Weiner LM. Fully human therapeutic monoclonal antibodies. J Immunother 2006; 29(1):1–9.

Whiteaker JR, Zhao L, Zhang HY, Feng LC, Piening BD, Anderson L, Paulovich AG. Antibody-based enrichment of peptides on magnetic beads for mass-spectrometry-based quantification of serum biomarkers. Anal Biochem 2007;362(1):44–54.

Yang Z, Hayes M, Fang X, Daley MP, Ettenberg S, Tse FL. LC-MS/MS approach for quantification of therapeutic proteins in plasma using a protein internal standard and 2D-solid-phase extraction cleanup. Anal Chem 2007;79(24):9294–9301.

Yang Z, Ke J, Hayes M, Bryant M, Tse FL. A sensitive and high-throughput LC-MS/MS method for the quantification of pegylated-interferon-alpha2a in human serum using monolithic C18 solid phase extraction for enrichment. J Chromatogr B Analyt Technol Biomed Life Sci 2009;877(18–19):1737–1742.

Yates JR, Ruse CI, Nakorchevsky A. Proteomics by mass spectrometry: approaches, advances, and applications. Annu Rev Biomed Eng 2009;11:49–79.

20 Mass Spectrometry in the Analysis of DNA, Protein, Peptide, and Lipid Biomarkers of Oxidative Stress

STACY L. GELHAUS, and IAN A. BLAIR
Centers for Cancer Pharmacology and Excellence in Environmental Toxicology, University of Pennsylvania, Philadelphia, Pennsylvania

Mass Spectrometry in Drug Metabolism and Disposition: Basic Principles and Applications, First Edition. Edited by Mike S. Lee and Mingshe Zhu.

20.1 INTRODUCTION

Research in the area of oxidative stress, which spans the fields of chemistry, toxicology, pharmacology, and biology, is becoming increasingly important. Oxidative stress research focuses on diet and nutrition, cardiovascular disease, neurodegenerative diseases, carcinogenesis, and aging. In all of these areas, the current approach for early detection and prevention has been to develop biomarkers specific to a particular disease or disease progression. Oxidative stress primarily results from the formation of reactive oxygen species (ROS), which are constantly generated in vivo by a variety of endogenous processes such as normal mitochondrial aerobic respiration, phagocytosis of bacteria- or virus-containing-cells, and peroxisome-mediated degradation of fatty acids (Blair, 2008). Increased oxidative stress results from increases in ROS formation during inflammation, radiation, and the metabolism of hormones, drugs, and environmental toxins. In many instances, these increases overwhelm protective pathways and increase ROS-mediated damage. Glutathione (GSH), one of the major defense mechanisms against oxidative stress, levels decrease with corresponding increases in glutathione disulfide (GSSG) concentration. Also, induction of many enzymes such as cyclooxygenase (COX) and lipoxygenase (LO), which contribute to the oxidative metabolism of xenobiotics and endogenous compounds occurs. This leads to ROS-mediated damage of deoxyribonucleic acid (DNA), proteins, and lipids. Some of these biomarkers are endogenous, such as the formation of 8-oxo-7,8-dihydro-2′-deoxyguanosine (8-oxo-dGuo) and the F_2-isoprostanes (isoPs), both derived from ROS modifications. Some are response biomarkers, such as the formation of the heptanone-etheno-2′-deoxyguanosine DNA- adduct (HεdGuo) that arises from lipid peroxidation in cigarette smokers. Exposure biomarkers, not discussed in this chapter, are those biomarkers related to exposure to an environmental chemical and, thus, contain some moiety of the chemical or its metabolite. The most common biological samples used to monitor these biomarkers of oxidative stress include blood, serum, urine, nasal and lung lavages, and biopsied tissue and cells (Winnik and Kitchin, 2008). Most important to the research in the area of biomarkers is the ability to develop accurate, precise, sensitive, and selective biomarker assays. To this end, liquid chromatography–mass spectrometry (LC–MS) has been extensively used because of its ability to meet most, if not all, of these requirements of method development.

The most common instrumentation for the analysis of biomarkers includes microbore and capillary reversed-phase chromatography coupled to a triple-stage quadrupole (TSQ) mass spectrometer or ion trap, with an atmospheric pressure ionization source such as electrospray ionization (ESI), nanospray ionization (NSI), or atmospheric pressure chemical ionization (APCI). Ion trap mass spectrometers provide higher sensitivity in full-scan mode, which is useful for product ion identification of a metabolite; however, TSQs are used most often due to improved sensitivity for quantification in multiple reaction monitoring (MRM) mode. Unlike a linear ion trap, TSQs also provide constant neutral loss scan capabilities. This scan provides a list of analytes from a class of compounds that loses a specific neutral part of the molecule, for instance, a glucuronide conjugate. In some instances, such as the targeted lipidomic profiling method described by Lee and Blair (Lee et al., 2005c), a chiral normal phase, instead of reversed-phase, is used for the separation of various eicosanoids and other metabolites of lipid peroxidation. For a more in-depth discussion of instrumentation, please refer to the extensive information in Part II of this book.

One of the most important analytical techniques one should incorporate into any LC–MS method is the use of stable isotope dilution. There is no other better internal standard then the heavy labeled version of the compound of interest. The use of either [^{13}C] and/or [^{15}N] labels are preferred as deuterium ([^{2}H]) incorporation alters the retention time of the internal standard. On reversed-phase columns, a [^{2}H]-labeled analog will elute earlier than the protium form and on normal phase columns the analog will elute later. This elution order is due to slight differences in their interactions with the stationary phase. Stable isotope dilution methods provide the assurance that the internal standard experiences the same environment during sample preparation as well as during ionization in the source. This method allows for accurate and precise quantification of the analyte of interest. This technique is becoming increasingly popular; however, many labeled internal standards, especially those of novel biomarkers, are not commercially available and must be synthesized.

Each section of this chapter is divided into the biomarker compound class (i.e., nucleic acids, proteins, and lipids) and further divided into endogenous biomarkers, which are typically formed by direct oxidative action on the compound, or response biomarkers, which may be a combination thereof. For example, when smokers are compared with nonsmokers, there is increased ROS-mediated formation. In addition, there is also increased lipid hydroperoxide-mediated DNA damage resulting in the formation of HεdGuo. The results from the numerous studies described in this chapter are meant to give an overview of the accomplishments within the years 2005–2009. Thus, this chapter is not intended to be comprehensive.

20.2 DNA BIOMARKERS OF OXIDATIVE STRESS

20.2.1 Background

One of the first thoughts that come to mind when one hears the phrase "oxidative stress" is DNA damage, specifically oxidation of DNA, which is typically implicated in carcinogenesis, neurodegenerative and autoimmune diseases, and aging. DNA oxidation can result from reaction with transition metals, reactive oxygen species (ROS) released during the metabolism of xenobiotics, and also from ROS formation during normal cellular functions such as respiration and inflammation (Ames et al., 1993). Because guanine (Gua)

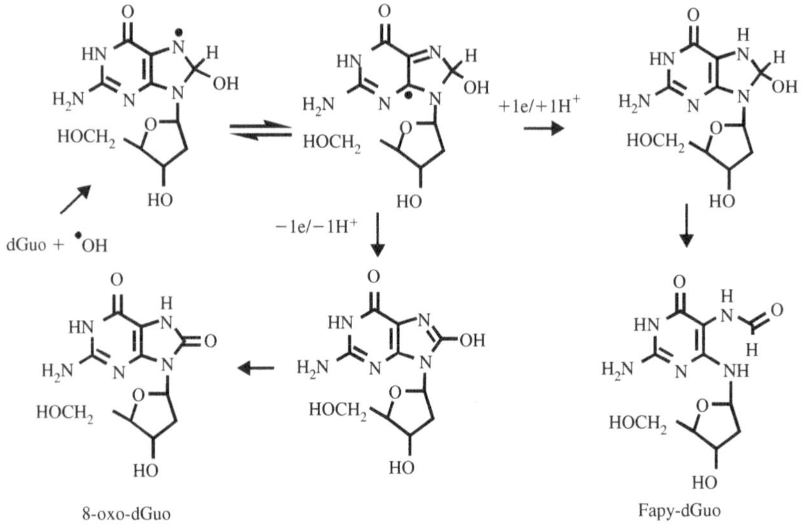

FIGURE 20.1 Hydroxyl-radical-mediated formation of 8-oxo-dGuo and Fapy-dGuo.

has the lowest oxidation potential of all nucleic acid bases, it is the base most commonly oxidized (Armitage, 1998; Burrows and Muller, 1998). Research in this area has also determined that the sequence of nucleic acids modulates the oxidation potential of Gua in DNA. For example, several Gua residues stacked as 5′-GG or GGG oxidize more easily than a single Gua alone. Primary oxidation of Gua results in the formation of 7,8-dihydro-8-oxo-2′-deoxyguanosine (8-oxo-dGuo). 8-oxo-dGuo can be formed from the one electron oxidation of the C8-OH radical where as formamidopyrimidines-dGuo (Fapy-dGuo), another common product of oxidative stress, is formed from the one electron reduction of the C8-OH radical (Figure 20.1). In spite of the enormous literature on the analysis of 8-oxo-dGuo, the exact mechanism of formation during oxidative stress still remains to be elucidated. An alternative pathway, originally proposed by the Frelon et al. (2003), involves the oxidation of DNA by singlet oxygen. The finding that oxidative stress induced by dimethylsulfoxide (DMSO) causes 8-oxo-dGuo formation lends further credence to this concept because DMSO is such a good scavenger of hydroxyl radicals. Therefore, increased concentrations of DMSO should decrease hydroxyl radical-mediated 8-oxo-dGuo formation rather than cause an increase (Mangal et al., 2009).

20.2.2 Oxidative Damage to DNA Bases: 8-Oxo-dGuo

The analysis of 8-oxo-dGuo as an index of oxidative DNA damage using MS-based methodology has been fraught with numerous methodological problems. The current state-of-the-art methodology involves the use of a comet assay to measure 8-oxo-dGuo lesions. Previous LC–MRM/MS methods have reported 8-oxo-dGuo levels 10 times higher than those of the comet assay (Gedik and Collins, 2005). Even those MS-based assays, which use immunoaffinity and stable isotope dilution, were unable to match the low levels reported in the comet assay, and thus indicate that artifactual oxidation of 2′-deoxyguanosine occurred during the sample preparation process. Unfortunately, the comet assay also has limitations and does not provide an accurate quantitative readout, is not linear, and hOGG1, human 8-oxo-guanine glycosylase, is not specific for the 8-oxo-dGuo lesion as hOGG1 is capable of excising Fapy derivatives (Dizdaroglu et al., 2008).

A recent paper from our laboratory has described a combination of immunoaffinity purification, stable isotope dilution LC–MS, together with a new method of sample preparation to obtain accurate basal levels of 8-oxo-dGuo in lung, liver, and cervical epithelial carcinoma cell lines (Mangal et al., 2009). In this study, human bronchoalveolar, H358, cells were treated with $KBrO_3$ as a positive control and methyl methanesulfonate (MMS) as a negative control. A linear dose–response was observed for 8-oxo-dGuo formation in H358 cells treated with increasing concentrations of the $KBrO_3$ ($r = 0.962$) from 0.05 mM to 2.50 mM, while there was no increase when cells were treated with MMS. Good correlation with the comet assay was observed at low levels of oxidative damage; however, the comet assay became nonlinear with higher levels of $KBrO_3$ while the stable isotope dilution LC–MRM/MS method continued to provide a direct linear correlation. Analyses were performed using a Phenomenex Luna C18 (2) (250 mm × 4.6 mm ID, 5 μm) column coupled to an API 4000 equipped with a positive electrospray ionization source. Stable isotope dilution was also employed for accurate quantification of 8-oxo-dGuo. MRM experiments were conducted by using the m/z 284 (MH^+) → m/z 168 [MH^+-2′-deoxyribose+H] transition for 8-oxo-dGuo and m/z 289 (MH^+) → m/z 173 [MH^+-2′-deoxyribose+H] transition for [$^{15}N_5$]-8-oxo-dGuo (Mangal et al., 2009).

As mentioned earlier, other LC–MS methods that use stable isotope dilution and immunoaffinity purification have been developed. What sets the method developed in our laboratory apart from the rest is the detailed attention to sample preparation (Mangal et al., 2009). The early recognition that increased levels of 8-oxo-dGuo were due to artifactual formation during sample preparation led to specific adjustments in this method. In this work, several DNA isolation techniques were compared to determine which method provided the lowest levels of artifactual oxidation. DNAzol, a proprietary reagent from Invitrogen that contains guanidine thiocyanate, was used for DNA extraction in conjunction with Chelex-treated buffers. These precautions were used to prevent Fenton chemistry-mediated generation of ROS and artifactual oxidation of DNA bases. Deferoxamine was also added to all buffers in order to complex any residual transition metal ions that remained after Chelex treatment (Mangal et al., 2009). Basal levels of oxidative damage in H358 cells were determined to be 2.2 ± 0.4 8-oxo-dGuo bases/10^7 dGuo or 5.5 ± 1.0 8-oxo-dGuo/10^8 bases. The observed levels are in good correlation with the levels determined by the comet assay and 10 times lower than any previous level determined by LC–MS (Mangal et al., 2009).

20.2.3 Oxidative Damage to DNA Bases: Formamidopyrimidines

Fapys are structurally unique as they are derived from purines and can be regarded as purines with an open imidazole ring. However, the pyrimidine is connected to the sugar moiety at the exocyclic amino group attached to the C6 of the pyrimidine (Dizdaroglu et al., 2008). These DNA adducts were identified almost 50 years ago; however, the relevance of these lesions has only recently come to light as their biological importance and potential contribution to mutagenicity has been overshadowed by 8-oxo-Gua. Published studies indicate that FapyGua and Fapy-adenine (FapyAde) can be formed from a variety of biological insults such as ionizing radiation, H_2O_2/metal ions, hypoxanthine and xanthine oxidase, photosensitization, stimulated neutrophils, ultraviolet (UV) radiation, and redox-cycling (Akman et al., 1992; Aruoma et al., 1989a, 1989b; Birincioglu et al., 2003; Blakely et al., 1990; Boiteux et al., 1992; Dizdaroglu et al., 1991; Frelon et al., 2003; Zastawny et al., 1995). FapyGua and FapyAde have been analyzed in cell culture and human and animal tissue. In most cases the levels of FapyGua have been comparable or greater than 8-oxo-Gua. Although oxygen was previously thought to completely inhibit the formation of Fapy, studies have shown that oxygen levels merely alter the balance between 8-oxo-purines and Fapy (Aruoma et al., 1989b; Frelon et al., 2003; Nackerdien et al., 1991). Increased levels of reducing agents such as GSH and ascorbic acid increase the levels of formamidopyrimidines (Dizdaroglu et al., 2008).

Formamidopyrimidines are repaired through the base excision repair (BER) process that involves various glycosylases. Many eukaryotic glycosylases excise FapyGua and FapyAde. A DNA glycosylase that removes N7-Me-FapyGua from DNA was first identified in various strains of *Escherichia coli.* This enzyme, which was also found in mammalian cells, purified and named formamindopyrimidine DNA glycosylase (Fpg) can also excise 8-oxo-Gua from DNA. One study showed that hOGG1 efficiently excised FapyGua when correctly paired with cytosine from synthetic 2′-deoxyoligonucleotides (Krishnamurthy et al., 2008). A general survey of glycosylases that includes Fpg, MutY, Nth, Nei, Neil, yOgg1, hoGG1, Ntg1, and Ntg2 revealed that FapyAde and FapyGua are physiological substrates of most known prokaryotic and eukaryotic DNA repair enzymes (Dizdaroglu et al., 2008). In some cases, the efficiency of glycosylases, such as Ogg1, to abstract FapyAde and FapyGua from DNA, is similar to that for removal of 8-oxo-Gua. This suggests that Ogg1 is not specific for 8-oxo-Gua. Numerous biological studies have shown that FapyGua is formed as abundantly as 8-oxo-Gua and has similar mutagenic properties in mammalian cells. While Fapy formation is an important area of study, many methods for analysis are based on gas chromatography (GC)–MS instead of LC–MS (England et al., 1998; Jaruga et al., 2008). Certainly, more attention will need to be given to this endogenous marker of oxidative stress.

20.2.4 Lipid-Hydroperoxide-Derived Genotoxins

In addition to direct DNA damage, oxidative stress can induce lipid-peroxidation followed by lipid hydroperoxide-mediated formation of bifunctional electrophiles that can covalently modify DNA. During oxidative stress there is a decrease in GSH and a corresponding increase in GSSG, reactive oxygen, and nitrogen species and an increase in COX-2 and LOs. DNA adducts resulting from nucleophilic attack by bifunctional electrophiles such as 4-hydroperoxy-2(*E*)-nonenal (HPNE), 4-hydroxy-2(*E*)-nonenal (HNE), 4-oxo-2(*E*)-nonenal (ONE), *trans*, *trans*-2,4-decadienal (DDE), 9,12-dioxo-10 (*E*)-dodecenoic acid (DODE) and 4,5-epoxy-2(*E*)-decenal (EDE), malondialdehyde (MDA), and acrolein are all possible and several have been characterized by NMR and LC–MS/MS (Blair 2005; Lee and Blair, 2001). DNA repair enzymes will excise the DNA adducts, most likely through base excision repair (BER) pathways, so the DNA adduct may potentially be detected in vivo as urinary adducts, thus providing a non-invasive measure of oxidative stress implicated in various disease pathologies (Sharma and Farmer, 2004). However, analysis of urinary DNA adducts of lipid-hydroperoxide-derived bifunctional electrophiles has not been particularly successful.

Analysis of DNA adducts derived from lipid hydroperoxide bifunctional electrophiles has relied primarily on reversed-phase ESI–MS (Blair, 2005). Some of the relatively hydrophobic bifunctional electrophiles that result from lipid hydroperoxide decomposition cannot be analyzed under these conditions because they are poorly ionized. These electrophiles can be converted to their oxime derivatives to improve ESI efficiency; however, this conversion results in syn- and anti-oxime isomers and extremely complex LC chromatograms (Lee and Blair, 2000). Therefore, normal-phase LC–APCI/MS has proved to be much more successful for the analysis of lipid-hydroperoxide-derived bifunctional electrophiles (Lee et al., 2001, 2005b; Williams et al., 2005).

20.2.5 Lipid-Hydroperoxide-Derived DNA Adducts

Metabolism of arachidonic acid by COX-2, 15-LO, or oxidation by ROS all lead to the formation of 15(*S*)-hydroperoxyeicosatetraenoic acid (HPETE). 15(*S*)-HPETE undergoes homolytic decomposition to form HPNE (Blair, 2008). It is also important to note that 15-LO-derived 13(*S*)-hydroperoxy-9*Z*,11*E*-octadecadienoic acid (HPODE) from linoleic acid (LA) also undergoes decomposition to form HPNE (Blair, 2008). A formal loss of water (2-electron oxidation) converts HPNE to ONE, whereas a 2-electron reduction results in the formation of HNE (Blair, 2008). ONO is formed enzymatically from ONE through reduction by aldo-keto reductases (AKRs). ONE is highly reactive, and upon translocation to the nucleus it readily forms HεdGuo adducts. ONE also forms adducts with 2′-deoxyadenosine (dAdo), and 2′-deoxycytidine (dCyd). Previous studies determined that ONE is much more efficient than HNE at modifying DNA through the formation of Hε- adducts (Lee et al., 2005b). HεdGuo DNA adducts have been found in the DNA of rat intestinal epithelial (RIE) cells that stably express COX-2 (Lee et al., 2005d).

In 2006, Williams et al. (2006) provided the first in vivo example of COX-2-mediated formation of lipid-hydroperoxide-derived ONE-DNA adducts in mammalian tissue. In this study, a quantitative comparison was made between C57BL/6J and C57BL/6JAPC^{min} mice using stable isotope dilution LC–MS/MS. Min mice are one of the most common models of colorectal cancer and in addition to developing large numbers of polyps, the up-regulation of COX-2 has been well characterized (Williams et al., 1996). Transitions were monitored for carboxynonanone-εdCyd (CεdCyd) and CεdGuo and heptanone-etheno-dCyd, -Guo, and dAdo (HεdCyd, HεdGuo, HεdAdo) together with their corresponding [^{15}N]-heavy-labeled internal standards. Potential artifact formation of heptanone-etheno-DNA-adducts was examined by adding [^{13}C^{15}N]-labeled DNA to the tissue sample before extraction was conducted. Amounts of the three [^{13}C^{15}N]-labeled Hε-DNA- adducts were all $<0.05\%$ of the relative [^{15}N]-internal standard; thus artifactual adduct formation is not an issue in the sample preparation and analysis (Williams et al., 2006). Overall, HεdGuo was the most abundant adduct found in tissue where HεdAdo and HεdCyd were the least abundant, and suggests that they are more readily repaired. However, HεdCyd was the most abundant adduct found in calf thymus DNA after treatment with 15(*S*)-HPETE. No carboxynonanone-etheno adducts were detected. These data suggest that HεdGuo and HεdCyd may be good markers for the detection of colorectal cancer (Lee et al., 2005d). Previous studies have shown that environmental chemicals such as vinyl fluoride and chloroethylene oxide form the same unsubstituted ε-DNA adducts (Barbin and Bartsch, 1986; Swenberg et al., 1999). Therefore, unsubstituted ε-DNA adducts do not arise solely from lipid peroxidation and indicate that the Hε-DNA adducts are true biomarkers of lipid peroxidation.

Much discussion centers on the formation of lipid hydroperoxide-derived DNA adducts from the oxidation of arachidonic acid; however, oxidation of linoleic acid also leads to the formation of bifunctional electrophiles that form DNA adducts (Lee et al., 2005a). 9, 12-Dioxo-10(*E*)-dodecenoic acid (DODE) was shown to arise from homolytic decomposition of 13(*S*)-HPODE. Formation of this bifunctional electrophile is proposed to occur through a similar pathway as ONE and HPNE (Lee et al., 2005a) through the intermediate formation of 12-oxo-9(*Z*)-dodecenoic acid and 9-hydroperoxy-12-oxo-10(*E*)-dodecenoic acid. It is possible that peroxidation of linoleic acid may lead to the modification of DNA and proteins in a manner similar to ONE. A report from Kawai et al. (2004) described the formation of etheno adducts when 13-HPODE was allowed to decompose in the presence of 2′-deoxynucleosides or DNA. Studies performed by Lee et al. (2005a) described the synthesis of multigram quantities of DODE starting from 1,8-octanediol using a furan homologation procedure. This procedure led to LC–MS studies that confirmed synthetic DODE produced the same CεdGuo adduct as was formed when 13 (*S*)-HPODE was allowed to decompose in the presence of dGuo (Lee et al., 2005a). Initial nucleophilic attack is thought to occur from the exocylic N^2 amino group of Gua at the C-12 aldehyde of DODE to form an unstable carbinolamine intermediate. This step is followed by intramolecular Michael addition of the pyrimidine N1 of dGuo to C-11 of the resulting α-,β-unsaturated ketone. Subsequent dehydration then gives the CεdGuo-adduct.

20.2.6 DNA Adducts from Other Aldehydes and Base Propenals

M_1dG (Zhou et al., 2005a) is a major endogenous, peroxidation-derived adduct of dGuo formed in reactions with either MDA (a product of lipid peroxidation, DNA peroxidation, and prostaglandin synthesis) or base propenals. The major source of M_1dG is thought to arise from base propenals as a result of ROS-mediated damage to the DNA sugar backbone (Zhou et al., 2005b). M_1dG is a substrate for the nucleotide excision repair (NER) pathway, which accounts for its presence in human urine as the 2′-deoxynucleoside. Knutson et al. (2008) determined that 6-oxo-M_1dG is the primary metabolite of M_1dG. Only M_1dG and 6-oxo-M_1dG were detected in urine and feces of male Sprague–Dawley rats given [^{14}C]-M_1dG. In this study, samples were separated and fractions were collected using high performance liquid chromatography (HPLC) and then sent for further analysis by accelerator MS (AMS) (Knutson et al., 2008). The AMS platform features the acceleration of ions to very high kinetic energy levels, which enables the separation of rare isotopes from abundant neighboring isotopes. This method is very useful for the detection and quantitation of long-lived isotopes such as ^{10}Be, ^{36}Cl, ^{26}Al, and ^{14}C. Using AMS, it was determined that 49% of the total ^{14}C was recovered in the urine and 51% in the feces (Knutson et al., 2008). Analysis of HPLC fractions revealed that 40% of [^{14}C]-M_1dG (2.0 nCi/kg, ~2.5 ng/rat) had been further metabolized to [^{14}C]-6-oxo-M_1dG over a period of 24 h in urine. This level decreased to 25% when the [^{14}C]-M_1dG dosage was decreased to 0.5–54 pCi/kg. The fact that 25–40% of the [^{14}C]-M_1dG administered to the rats by intravenous (I.V.) was metabolized to 6-oxo-M_1dG and cleared in urine indicates that this metabolism may occur under physiological conditions (Knutson et al., 2008). Our laboratory is currently involved with the development of these biomarkers oxidation in human urine; however, it is likely that the levels will be too low to assess the amount of base propenal- and MDA-mediated DNA damage (Hoberg et al., 2004). Despite this fact, the idea that lipid hydroperoxides-derived DNA adducts may undergo further metabolism before they are excreted into urine is intriguing. Although M_1dG is not a specific marker of lipid peroxidation, this adduct may serve as a useful biomarker of endogenous DNA damage that results from increased oxidative stress. This adduct has also been detected in the liver DNA of humans and in the DNA of circulating human leukocytes (Chaudhary et al., 1994; Rouzer et al., 1997).

Recent studies have demonstrated that acrolein, a bifunctional electrophile produced through lipid peroxidation and found in cigarette smoke and car exhaust, reacts with dGuo to form cyclic 1,N^2-propano-dGuo adducts. Cyclic 1,N^2-propano-dGuo adducts are also readily formed by crotonaldehyde, also known at β-methacrolein (Zhang et al., 2007). Depending on the direction of the initial Michael addition, two pairs of stereoisomers of (6*R*/*S*)-3-(2′-deoxyribos-1′-yl)-5,6,7,8-tetrahydro-6-hydroxypyrimido[1,2-*a*]purine-10(3*H*) one (α-OHAcr-dGuo) and (8*R*/*S*)-3-(2′-deoxyribos-1′-yl)-5,6,7,8-tetrahydro-8-hydroxypyrimido[1,2-*a*]purine-10(3*H*)one (γ-OH-Acr-dGuo) are formed. In particular α-OH-Acr-dGuo is mutagenic in human cells and predominantly induces G→T transversions (Liu et al., 2005; Nath and Chung, 1994; Nath et al.,1996, 1998); In 2006, Feng et al. (2006) reported that acrolein-DNA adducts are preferentially formed at p53 mutational hotspots, thus implicating a role in DNA repair inhibition. A 2007 study by the Hecht group at the University of Minnesota has shown that α-OH-Acr-dGuo and γ-OH-Acr-dGuo were found in the lung tissue of former and current smokers at relatively high levels compared to benzo[*a*]pyrene (B[*a*]P)-derived DNA adducts (Zhang et al., 2007). Acrolein is thought to occur in quantities up to 10,000 times higher than B[*a*]P. Thus, acrolein-dGuo DNA adducts could serve as an exposure biomarker of cigarette smoke and other environmental pollutants such as vehicle exhaust.

20.3 PROTEIN AND PEPTIDE BIOMARKERS OF OXIDATIVE STRESS

20.3.1 Introduction

Adduction of proteins and peptides by xenobiotics was first discussed over 60 years ago by the Millers who were pioneers in xenobiotic metabolism (Miller and Miller, 1947). Unfortunately, little progress was made in this area for some time due to the limitations of the available technology and a shift emphasis to DNA damage. Interest in protein modifications began to resurface in the 1980s with significant improvements in technology.

The relationship between binding and toxicity is not simple. Protein modification does not necessarily correlate with toxicity, and it has taken researchers such as Farmer, Liebler, Tannenbaum, Ehrenberg, and others to show this (Ehrenberg et al., 1974; Farmer et al., 1980; Green et al., 1984). Over the last 30 years, oxidative damage to proteins has been shown to play a critical role in many disease pathologies related to inflammation and oxidative damage. Some of these include cancer, asthma and other pulmonary diseases, neurodegenerative diseases, diabetes, and metabolic diseases. In fact, electrophilic targets are even implicated in cellular mechanisms of response. Electrophilic adduction of certain proteins acts as a sensor to oxidative stress. For instance, Keap1, is a thiol-rich protein that mediates the transcription factor Nrf2. Upon modification of the cysteine groups, Keap1 dissociates from Nrf2 and allows Nrf2 to translocate to the nucleus and activate numerous phase II antioxidant genes (Potter et al., 1973; Roberts et al., 1990; Streeter et al., 1984). Other potential sensors include phosphatases, peroxiredoxins, and thioredoxin (Baillie, 2006; Evans and Baille, 2005; Evans et al., 2004; Schroder and Kaufman, 2005). To this end, researchers have begun to globally analyze the electrophilic modifications of proteins in specific cellular fractions such as the cytoplasm, mitochondria, the endoplasmic reticulum (ER) and nucleus. The degree of protein modification and level of specificity is unclear. Furthermore, the role of protein structure in this modification process on a proteome-wide scale has received greater attention. Mass spectrometry analyses have been used to characterize protein modifications by dozens of electrophiles. In most cases, a target protein was incubated with electrophiles to generate adducts, which were then analyzed by MS of either the intact protein or a proteolytic digest. Mapping of adducts to specific sequence locations has been achieved by MS/MS or by peptide mass fingerprinting and chemical inference. Studies revealed that protein modification is selective and specific, as only a subset of sites is available for reaction. The reactivity of solvent-exposed nucleophiles (e.g., cysteine, histidine, and lysine) is expected, but solvent exposure is not necessarily the only factor in adduction site selectivity (Liebler, 2008).

20.3.2 Protein Adducts from Lipid-Hydroperoxide-Derived Bifunctional Electrophiles

Protein biosynthesis is modulated during oxidative stress (Yan et al., 2005) and in addition many proteins are oxidatively modified. Unfortunately, the amount of oxidized proteins compared to total cell or tissue protein is relatively small. Initially, analytical techniques were not capable of differentiating specific modifications from the background noise. However, Liebler and others have used affinity enrichment tools to make global protein analysis more feasible (Liebler, 2008). Previous techniques included immunoblotting and radiolabeling; however, these techniques are limited by the availability, sensitivity, and specificity of antibodies against adducts and the specific activity of the radiolabel. In some instances AMS was used to detect protein adducts in tissues treated with [^{14}C]-labeled compounds. Use of the label provided some enrichment of protein adducts (Williams et al., 2002). A more current variation combines two-dimensional (2D) gel electrophoresis and antibodies directed to specific protein targets. The work done by Petersen et al. represents an example of this enrichment approach, and HNE adduction of heat shock proteins Hsp72 and Hsp 90 was reported Carbone et al. 2004, 2005. In vivo adduct formation was detected with 2D gel electrophoresis followed by immunoblotting with anti-HNE Michael adduct antibodies (Kapphahn et al., 2006; Meier et al., 2005, 2007). The 2D gel method represented a step in the right direction; however, the number of oxidative proteins captured in this step was still limited. Better enrichment techniques along with shotgun proteomics by

LC–MS was used by Liebler (2008). The antibodies had limitations in affinity and specificity; therefore, biotinylating agents were used instead (Liebler, 2008). Three compounds used in their initial studies were *N*-iodoacetyl-*N*-biotinylhexylenediamine (IAB), (+)-biotinyl-iodoacetamidyl-3,6-dioxaoctanediamine (PEO-IAB), and 1-biotinamido-4-(4′-[maleimidoethylcyclohexane] carboxamido)butane (BMCC). These tags penetrate cells and react readily with proteins in multiple cellular compartments. Tagging proteins alkylated by electrophiles allows for easy capture and analysis by LC–MS. Their tag simply consisted of the electrophile—linker and biotin. These two electrophile tags were chosen because they form protein adduct through mechanistically distinct pathways. PEO-IB displays *sn-2* electrophilic chemistry analogous to toxicologically relevant electrophiles, such as aliphatic epoxides, alkyl halides, and episulfonium ions. BMCC reacts with cysteine thiols by a Michael addition, much like quinones and α,β-unsaturated carbonyls (many of which result from lipid peroxidation). These two mechanisms mimic the most relevant reaction chemistry electrophiles (Dennehy et al., 2006). One can easily see how this type of analysis can be applied using an electrophile such as HNE and others produced through lipid peroxidation.

Proteins in subcellular fractions were incubated under near-native conditions with electrophile. Previous optimization experiments indicated that incubation with 200 μ*M* PEO-IAB or 10 μ*M* BMCC for 30 min was sufficient to achieve detectable biotin labeling across the molecular weight range of proteins without appearing to saturate all targets (Dennehy et al., 2006). Following incubation with electrophiles, protein fractions were treated with DTT (dithiothreitol) to neutralize unreacted electrophile and digested with trypsin to enrich for adducted peptides with immobilized neutravidin on agarose beads (Fig. 20.2). The capture of only the labeled peptides ensures the LC–MS analysis of an enriched set and provides high-quality mass spectra of primarily adducted peptides. If the tags are captured before digestion, then the protein adducts, which are of much lower abundance, are consequently lost in the analysis of the many untagged peptides. However, several good spectra are still necessary to identify the modified protein with confidence (Dennehy et al., 2006).

A total of 897 cysteine residues on 539 proteins were reproducibly adducted and captured. Most interesting was the resulting thiol modification selectivity. Only 90 proteins and 125 peptides were modified by both electrophiles (Dennehy et al., 2006). These results are indicative of selective alkylation rather than random adduction and also selective thiol chemistry of the two electrophiles. Alkylation by the Michael acceptor, BMCC, was selective for cysteines located next to arginine or lysine residues and also indicated some structural chemical selectivity. Most interestingly, of all the proteins modified, 80% were only adducted at a single cysteine residue (Dennehy et al., 2006). Results also correlated with previously suspected sites known to be susceptible to alkylation. The study by Liebler et al. (2008) represented the first global

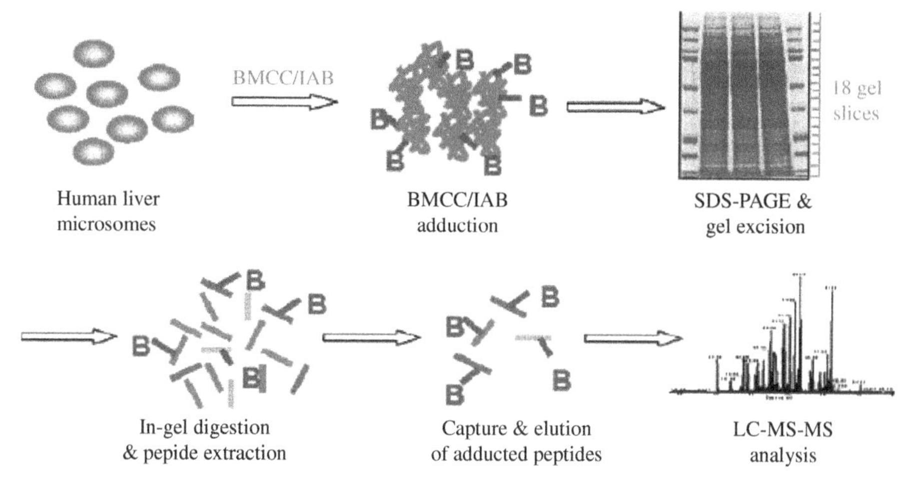

FIGURE 20.2 Adduction of proteins with the biotin-labeled reactive electrophiles 1-biotinamido-4-(4′-[maleimidoethylcyclohexane]-carboxamido)-butane (BMCC) and *N*-iodoacetyl-*N*-biotinylhexylenediamine (IAB). Adduction is followed by SDS PAGE gel separation followed by gel digestion and peptide extraction. The protein–electrophile adducts are then captured by biotin-avidin chromatography and analyzed by LC–MS/MS.

characterization of proteins and protein sites targeted by electrophiles (Dennehy. et al., 2006).

The endoplasmic reticulum (ER) is of particular interest because many of the proteins residing in the ER are involved in the metabolism of xenobiotics. The identification of targets associated with toxic outcomes may increase specificity of toxicity predictions. In this work, done in human liver microsomes, Shin et al. 2007 identified 376 modified peptides and 263 proteins modified by the iodoacetamide IAB and the *N*-alkylmaleimide BMCC labeling method. Again, only 20% of the proteins were modified by both electrophiles. These nonselective targets may make good general biomarkers of oxidative stress. Following the same pattern, protein adduction was found to be selective and reproducible, and approximately 25% of the targets reacted with both electrophiles (Shin et al., 2007). In their study in the mitochondria, 1693 cysteine thiol targets were mapped to 809 proteins (Wong and Liebler, 2008). Forty-four of the adducted proteins were found to be apoptosis-related and adduct site specificity on these targets differed between the two probes.

Protein modifications by HNE, ONE, and MDA have been characterized on specific proteins, but there are few global characterizations, which could potentially provide the "big picture" necessary to determine the role oxidative stress plays in disease etiology. Most recently, Condreanu et al. (2009) conducted a study on protein damage in RKO, human colorectal carcinoma, cells treated with either 50 or 100 μM HNE. Michael adducts of HNE were biotinylated by reaction with biotinamidohexanoic acid hydrazide and captured with streptavidin. The biotinylated hydrazide reacts with the residual carbonyl formed through the Michael addition of HNE to the protein nucleophile. The captured proteins were resolved by denaturing gel electrophoresis and then digested with trypsin before being analyzed by LC–MS/MS on a Thermo Finnigan LTQ (Codreanu. et al., 2009). Of the 1500+ proteins identified, 417 displayed a statistically significant increase in adduction with increasing HNE exposure concentration. Since the false-positive rate can be quite high due to nonspecific binding in affinity capture methods, Liebler et al. (2008) took extensive measures in their data analyses. A nonlabeled approach was used to quantify proteins as a function of HNE exposure concentration. Statistical analyses were applied to identify target proteins that demonstrated concentration dependent adduction. In addition, 18 biotin hydrazide-HNE-adducted peptides were detected by specific capture biotin antibodies and high-resolution LC–MS/MS on an LTQ-Orbitrap, which enabled detection of HNE exposure levels down to 1 μM (Codreanu et al., 2009). A full scan was run at a resolution of 60,000 and was followed by 5 data-dependent MS/MS scans. Proteins were only considered HNE adducted if they appeared in the treated sample and not the control (Codreanu et al., 2009).

A downside of using the biotin–linker tag is the unknown reactivity between the tag and a protein of interest. Postlabeling alternatives include Click chemistry or the Staudinger ligation, which add the biotin–linker during sample processing (Kolb and Sharpless, 2003; Saxon et al., 2000). A last consideration of global protein adduct detection is adduct stability. For instance, quinone adducts may reoxidize, which allows for potential cross-linking and epoxides. The epoxides can undergo ring opening to form esters that may be prone to hydrolysis under certain conditions. Burcham et al. (2003) reported the reaction of acrolein adducts with the amino group of Tris buffers, and thus, prevents antibody-based detection.

One of the misuses of MS-based analysis of protein adducts is the determination of relative reactivities of different nucleophilic sites. Unfortunately, the inference of reactivity from MS spectral or signal intensities is often complicated by the different ionization efficiencies of different peptide sequences that bear the same adduct. A particular sequence may or may not have a strong MS signal or MS/MS spectrum of the peptide adduct depending upon its ability to ionize and also if there are any potential interferences in the sample matrix. Several methods have been developed to assess competing reactivities. One method involves measuring the rates of reaction at the sites and then comparing these rates to the kinetic constants. However, the measurement of adduction kinetics presents a challenging problem, largely because of the difficulty of absolutely quantifying individual adducts. Liebler recently reported an approach that employed differential stable isotope tagging of peptides in digests of adducted proteins (Mason and Liebler, 2003). Derivatization of the N-termini of tryptic peptides with [$^{12}C_6$]- or [$^{13}C_6$]-phenylisocyanate (PIC) results in light and heavy derivatives that are easily distinguished by MS and MS/MS. To extract kinetic parameters from a time-course study with electrophile-treated protein, the sample from the last time point is digested and labeled with the heavy label PIC. All other time points are digested and labeled with the light label PIC (Orton and Liebler, 2007; Szapacs et al., 2006). Each light-labeled sample is spiked with an equal portion of the heavy-labeled sample as a reference standard, and then, the ratios of light to heavy signals extracted from MS/MS data are plotted as a function of reaction time. The plots are fitted to appropriate first-order exponential curves to derive k_{obs} values, which can be used directly to compare reactivities and different protein sites. Application of this approach to measure k_{obs} for HNE adduction of several histidine and lysine residues in human serum albumin revealed a range of three orders of magnitude in the measured constants (Szapacs et al., 2006). The most reactive target of HNE was His-242, which lies in a hydrophobic binding pocket in the IIa subdomain known to bind fatty acids and lipophilic xenobiotics. The reactivity of the His-242 imidazole was further enhanced by electrostatic effects of nearby lysine residues, which reduce the estimated pK_a of the imidazole nitrogen to less than 1.0. Another method to accurately measure reactivities was developed in our laboratory (Ciccimaro et al., 2009). For a known modification on a specific protein, light- and heavy-labeled peptide sequences with tryptic overhangs and the modification of interest can be synthesized or purchased. Heavy-labeled peptides with tryptic overhangs are digested with the protein of interest and provide an internal standard for stable isotope dilution methodology. Light and heavy peptides can also be run together at varying concentrations of light peptide to develop a standard curve, which may be used for quantification. This method allows for quantitative analysis of a specific modification at a specific site on a protein and takes into account interferences that may alter ionization efficiency (Ciccimaro et al., 2009; Shah. et al., 2009).

20.3.3 Oxidized Methionine, Histidine, and Tyrosine

Amino acid residues are potential targets of free radical oxidation and nitration. Carbonyl derivatives of proteins may be formed by the interaction of protein amino acid side chains, mainly cysteine, histidine, and lysine residues with reactive aldehydes, such as HNE and ONE generated by peroxidation of PUFAs (polyunsaturated fatty acids). Amino acid and peptide biomarkers of oxidative stress are typically focused on specific proteins related to disease pathology. For instance, the oxidation of histidine and methionine are typically discussed in β-amyloid plaque formation and HNE-derived histidine adducts are the main focus of modifications on low-density lipoprotein (LDL) (Annangudi et al., 2008). However, there are several specific examples of general biomarkers of oxidative stress that include endogenous histidine containing dipeptides such as carnosine and anserine as well as the very stable *o,o′*-dityrosine. These will be discussed below.

Metal-catalyzed oxidation has been suggested to occur on histidines 13 and 14 of amyloid β peptides. Inoue et al. (2006) used LC–MS/MS to characterize these sites of oxidation on tryptic amyloid β (Aβ) peptides. These peptides are derived from the amyloid precursor protein (APP). Normal catabolism of APP is known to involve cleavage by β-secretase and γ-secretase, to produce Aβ peptides with carboxyl-terminal heterogeneity. Evidence has emerged that supports a role for metal ions and ROS in the pathogenesis of Alzheimer's disease (AD); therefore, oxidized Aβ peptides may serve as a biomarker of AD in cerebrospinal fluid (CSF) (Bush, 2003; Castellani et al., 2004; Huang et al., 1999; Huang et al., 2000; Multhaup et al., 2002). Several studies have shown that the oxidation of Aβ occurs through the formation of Cu^{I} and subsequent generation of ROS from molecular oxygen, which depend on the presence of three N-terminal histidines, His^{6}, His^{13}, and His^{14} (Barnham et al., 2004; Huang et al., 1999, 2000). There is also evidence that methionine, specifically, Met^{35}, is involved in plaque formation. When Met^{35} was replaced with norleucine, oxidative stress and the neurotoxicity induced by $A\beta_{1\text{-}42}$ were both attenuated (Butterfield and Kanski, 2002). It has been reported that Aβ isolated from AD brain had lost approximately 1 mol of histidine/mol of peptide, but the oxidation product was not characterized (Atwood et al., 2000). Inuoe et al. (2006) characterized the oxidation products of $A\beta_{1\text{-}16}$ and $A\beta_{1\text{-}40}$ using metal-catalyzed oxidation (MCO) to mimic what may occur in vivo. After MCO, samples underwent a trypsin digest and were analyzed by LC–ESI–MS/MS to establish the exact sites of oxidation. Oxidation occurred only under conditions where Cu^{II} and ascorbic acid were present simultaneously and up to 85% of the native $A\beta_{1\text{-}16}$ peptide was consumed. Under these conditions ascorbic acid converted Cu^{II} to Cu^{I}, necessary for the one-electron reduction of molecular oxygen. This work clearly demonstrates that ascorbic acid/Cu^{II}-mediated ROS generation exclusively occurs at His^{13}, His^{14} and that Met^{35} oxidation does not occur in $A\beta_{1\text{-}40}$ under these conditions (Inoue et al., 2006).

Endogenous histidine-containing dipeptides such as carnosine (β-alanyl-l-histidine, CAR), homocarnisone (γ-aminobutyryl-L-histidine, HCAR), and anserine (*N*-β-alanyl-3-methyl-L-histidine, ANS) react with toxic aldehydes such as HNE and acrolein and are biomarkers of lipid peroxidation. CAR and ANS are carbonyl quenching and behave as detoxifying agents for cytotoxic aldehydes, by reacting with HNE and other aldehydes in biological systems through a "sacrificial" mechanism that mimics the preferred HNE addition sites in proteins (Orioli et al., 2005). Orioli et al. (2007) developed the first validated LC–MS/MS method for the simultaneous analysis of carnosine, homocarnosine, and anserine for biological matrices. This method was demonstrated in rat skeletal muscle. Unfortunately, this method does not include stable isotope dilution as only H-Tyr-His-OH is utilized as an internal standard. In a 2007 follow-up paper, Orioli et al. (2007) extended their work to monitor free and protein-bound histidine residues modified by reactive carbonyl species in urine. In this work, His-1,4-dihydroxynonane (His-DHN), DHN-mercapturic acid (DHN-MA), His-HNE, and carnosine-HNE (CAR-HNE) were simultaneously detected in the urine of Zucker obese rats. Again, H-Tyr-His-OH was used as an internal standard (Orioli et al., 2007). This work highlights the utility of a panel of biomarkers to assist with the analysis of oxidative stress-related disease states.

Dityrosine is a fluorescent molecule that forms as a result of normal posttranslational processing, but alterations in the level of dityrosine formation is also used as a biomarker of increased oxidative stress (Orhan et al., 2005). The crosslinked structure confers resistance to proteolysis; therefore, tyrosine cannot be reused in de novo protein synthesis. This biomarker is also stable in environments with oxygen or pH changes. Dityrosine has been used as an important biomarker for oxidatively modified proteins during UV and γ-irradiation, aging, and exposure to free radicals, nitrogen dioxide, peroxynitrite, and lipid hydroperoxides. Dityrosine formation is initiated with tyrosyl radical formation followed by the *ortho*-to-*ortho* (to the OH groups) cross-linkage of two tyrosyl radicals. Increased dityrosine levels have been associated with varying pathologies such as eye cataracts, atherosclerosis, acute inflammation, and Alzheimer's disease (Hensley et al., 1998; Kato et al., 2000; Leeuwenburgh et al.,1997, 1999b). Several groups have analyzed this oxidized protein biomarker in urine using stable isotope dilution LC–MS/MS with either ESI or APCI sources. Like IsoPs, dityrosine has not been correlated to specific disease pathology; rather dityrosine is used as a general biomarker of protein damage caused by increases in oxidative stress. Concentrations of dityrosine were found to be 100-fold higher in low-density lipoproteins isolated from atherosclerotic lesions compared to normal LDL (Heinecke, 2002). Recent studies performed with rats demonstrated that dityrosine is secreted into the urine rather than recycled into proteins (Leeuwenburgh et al., 1997, 1999a, 1999b). To this end, Orhan et al. (2005) have developed a stable isotope dilution LC–APCI–MS/MS method for the quantification of *o,o'*-dityrosine in human urine. A TSQ and ion trap were compared in terms of sensitivity, specificity, and reproducibility. The results of this comparison illustrates that the TSQ has 2.5-fold higher sensitivity (limit of detection 0.01 μM) compared to the ion trap. The heavy-labeled internal standard *o,o'*-[$^2H_6$]-dityrosine was added to increase sensitivity and specificity and was used in both methods. The TSQ-based method employed an APCI source in positive ion mode superseded by separation on a Phenomenex Intersil ODS-2 (5 μm, 150 mm × 4.6 mm) at a flow rate of 0.8 mL/min. The urine of 14 smoker subjects was analyzed for *o, o'*-dityrosine and normalized to creatinine values (Orhan et al., 2005). MRM experiments were used for quantification of the analyte and internal standard with the following transitions, m/z 361→316 (dityrosine) and m/z 367→319 ([$^2H_6$]-dityrosine). The calibration was linear over a range of 0.01–10 μM with an limit of detection of 0.01 μM and a limit of quantification of 0.03 μM. Urinary concentrations of dityrosine in smoker subjects were 0.08 ± 0.01 μM, corresponding to 10.1 ± 0.4 μmol/mol creatinine.

Finally, nitro-tyrosine is emerging as an important oxidatively modified amino acid, particularly when present in proteins. Nitrotyrosine has been suggested to arise through activation of myeloperoxidase (MPO) during oxidative stress (Citardi et al., 2006). Furthermore, MPO-catalyzed oxidation of high-density lipoprotein (HDL) and apolipoprotein A-I (apoA-I) resulted in selective inhibition in ABCA1-dependent cholesterol efflux from macrophages.

A dramatic selective enrichment in nitrotyrosine content was observed within apoA-I recovered from serum and human atherosclerotic lesions. The analysis of serum from sequential subjects demonstrates that the nitrotyrosine contents of apoA-I were markedly higher in individuals with cardiovascular disease. An increase in nitrotyrosine was observed in the atherosclerotic lesions and plasma of apolipoprotein A-I-deficient mice (Parastatidis et al., 2007). LC–MS/MS studies confirm that nitration of apoA-I occurs specifically on tyrosine-192 (Shao et al., 2005). Similarly, LC–MS/MS methods were used to identify tyrosine residues 292 and 422 at the carboxyl terminus of the β chain as the principal sites of fibrinogen nitration in vivo (Parastatidis et al., 2008). Stable isotope dilution LC–MRM/MS methodology will likely be employed extensively in the future to assess the role of protein tyrosine nitration in the oxidative stress associated with cardiovascular disease.

20.3.4 Lipid Hydroperoxide-Derived GSH Adducts

GSH is the major low-molecular-weight thiol in mammalian cells with intracellular concentrations ranging from 4 to 10 m*M* (Wu. et al., 2004). Approximately 85–90% of GSH is found in the cytosol while the remaining GSH is located primarily in the nucleus, mitochondria, and peroxisomes (Lu, 2000). Aside from its involvement in endogenous eicosanoid formation, GSH serves as reducing equivalents in biosynthetic reactions, as a cofactor in the reduction of ROS and lipid hydroperoxides by GSH peroxidase, and GSH S-transferases (GSTs). GSH also plays a key role in the reduction of reactive intermediates derived from arylamines and in the conjugation of reactive intermediates to form S-substituted adducts through its nucleophilic cysteine residue (Brigelius-Flohe, 1999; Cohen.and Hochstein, 1963; Dickinson.and Forman, 2002; Evans and Baillie, 2005; Evans. et al., 2004; Lam,and Austen,. 2002; Murphy and Zarini, 2002; Rock et al., 2000; Soberman and Christmas, 2003; Turesky, 2004; Wang and Ballatori, 1998). When extensive metabolism of reactive intermediates occurs, formation of the intermediate-GSH-adduct may result in depletion of GSH-levels and cause reactive intermediate adduct formation on free sulfhydryl groups of proteins. Reactive intermediates may form ribonucleic acid (RNA) adducts or also be transported into the nucleus and react with DNA. In some cases these intermediates are long-lived and cross into the plasma and form protein adducts. Protein adducts of hemoglobin and serum albumins provide a good measure of oxidative insult and exposure to environmental chemicals (Kolman et al., 2002; Lee et al., 2005d; Swenberg et al., 2001; Yocum et al., 2005).

GSH adducts can be readily detected by LC–ESI/MS/MS in the positive ion mode with either the full-scan mode to search for anticipated conjugates or the constant neutral loss scanning for 129 Da (γ-glutamyl moiety) to detect GSH-derived metabolites followed by collision induced dissociation (CID) and MS/MS analysis of MH^+ (Baillie and Davis, 1993; Levsen et al., 2005; Nikolic et al., 1999). More recently, negative ion ESI methodology in combination with the precursor ion scan mode of *m/z* 272 (M-H-H_2S) was shown to provide a more general method for the detection of GSH adducts in cellular incubations (Dieckhaus et al., 2005). The characterization of GSH adducts from drug candidates allows potential sites and mechanisms of bioactivation to be determined (Evans and Baillie, 2005; Evans et al., 2004). Such studies can be coupled with structural modification of the drug candidate to block sites susceptible to metabolic activation and formation of a reactive intermediate. This strategy results in the development of drug candidates that are less likely to bind to cellular macromolecules and, as a result, have a reduced potential to cause adverse events if ultimately used in clinical studies (Evans and Baillie, 2005; Evans. et al., 2004).

Typically, GSSG levels are maintained at approximately 1% of GSH levels through GSH reductase-mediated reduction with NADPH (Dickinson and Forman, 2002). Cellular concentrations of GSSG are also regulated through its removal by ATP-dependent transporters (Homolya et al., 2003; Suzuki and Sugiyama, 1998). When GSH levels are rapidly depleted, there is a concomitant increase in GSSG levels (Cook et al., 2004; Ketterer, 1988; Turesky, 2004). Under these conditions, thiol exchange with sulfhydryl groups on intracellular proteins can occur to form mixed disulfides (Dickinson and Forman, 2002). Surprisingly, little is known about specific mechanisms and/or enzyme(s) that catalyze formation of protein–SSG mixed disulfides. The disulfides are relatively long-lived and they appear to be involved in GSH-mediated cell signaling in an analogous manner to phosphorylation (Shelton et al., 2005). Glutaredoxins are specific catalysts of protein–SSG de-glutathionlyation and have been suggested to function in a similar role to the phosphatases that control protein phosphorylation (Fernandes and Holmgren, 2004; Shelton et al., 2005). The ability of protein thiols, mixed disulfides, and GSH to rapidly redox cycle when cells are isolated has significantly inhibited our ability to thoroughly understand their interrelationship during oxidative stress. However, the availability of new LC–MS-based methodology in which thiols are rapidly covalently modified, thus preventing artifactual oxidation, has enhanced our ability to conduct such studies (Zhu et al., 2008).

Oxidative stress also leads to the up-regulation of NF-κB pathways such as the induced formation of COX-2 (Farrow and Evers, 2002; Jang and Surh, 2005; Rahman et al., 2004; Tanabe and Tohnais, 2002; Wu, 2005). As described in the previous section on DNA response biomarkers, COX-2 converts arachidonic and linoleic acids to 15(*S*)-HPETE and 13(*S*)-HPODE, respectively. These products of lipid peroxidation undergo homolytic decomposition to HPNE and eventually result in the formation of the bifunctional electrophiles ONE and HNE (Lee et al., 2001; Schneider et al., 2002; Williams et al., 2005). Aside from the formation of DNA and protein adducts, ONE is capable of forming a novel endogenous GSH adduct that is readily excreted into the cellular milieu (Jian et al., 2005a). This novel endogenous GSH adduct was subsequently characterized as a novel thiadiazabicyclo-ONE-GSH-adduct (TOG) that arose from GST-mediated addition to the α,β-unsaturated ketone

(Jian et al., 2007). HNE is also capable of forming a GSH adduct, which arose from a GST-mediated Michael addition of GSH to the α,β-unsaturated aldehydes (Carini et al., 2004; Murphy and Zarini, 2002; Uchida, 2003). Structural analogs of ONE, including DODE and DOOE, are formed from the carboxylate terminus of linoleic-acid-derived 13(*S*)-HPODE and AA-derived 5(*S*)-HPETE, respectively (Jian et al., 2005b; Lee. et al., 2005a). Therefore, TOG represents a prototypic member of a family of endogenous GSH adducts that can arise from both nonenzymatic (ROS) and enzymatic (COX and LO) lipid peroxidation. These endogenous GSH adducts and their metabolites offer promise as potential new biomarkers of oxidative stress.

During apoptotic induction studies in EaHy.926 (A549 and HUVEC hybrid) endothelial cells, ONE was observed to unexpectedly form the intracellular, monomeric GSH adduct with an MH^+ of 420 (Jian et al., 2005a). Significantly, this adduct was the major product when ONE was treated with 40-fold excess of GSH in the presence of equine or rat GST. The enzymatically generated ONE-GSH adduct exhibited an MH^+ of 426.1717 with high-resolution ESI–MS/MS. Proton nuclear magnetic resonance (NMR) revealed the GSH adduct to be a pyrrole derivative, which exists as a mixture of two molecular forms in solution. These data were consistent with a molecular structure of (2*S*, 7*R*)-7-[*N*-(carboxymethyl)carbamoyl]-5-oxo-12-pentyl-9-thia-1,6-diazabicyclo [8.2.1]trideca-10(13),11-diene-2-carboxylic acid. The thiadiazabicyclo-ONE-GSH-adduct (TOG) possesses an 11-membered macrocyclic ring that has never been previously observed. LC–MS^n analysis of TOG revealed a stable product ion at *m*/*z* 280, which permitted a highly specific and quantitative stable isotope dilution LC–SRM/MS method using the 420 (MH^+)→280 (MH^+-$CONHCH_2CO_2H$-$CONH_2$). TOG was also discovered to form if cells were treated with Fe^{II} (500 μ*M*), which is able to initiate oxidative stress in endothelial cells. Care was taken to ensure that ONO did not arise from the interconversion of HNE and ONE and that the only pathway of formation was through homolytic lipid hydroperoxide decomposition in endothelial cells. ONE-derived ONO formation most likely occurs through an AKR-mediated pathway. Significant levels of ONO were detected, indicating that the biological contributions attributed to HNE may need to be evaluated due to the fact that HNE and ONO are isomeric. TOG may just be the first of a new class of ONE-GSH adducts that provide potential biomarker of both enzymatic and nonenzymatic lipid-hydroperoxide-mediated intracellular oxidative stress similar to that suggested for the HNE-GSH adducts, which are most likely involved in cellular signal transduction (Jian et al., 2007). Both DODE and DOOE are also capable of forming TOG-like structures. Once these structures have been fully characterized it will be possible to determine which PUFA-derived lipid hydroperoxides are involved in the induction of intracellular oxidative stress, thus providing quantitative biomarkers of this phenomenon.

Cellular redox status itself can be considered a biomarker of oxidative stress. The redox potential of the 2GSH/GSSG couple, which is determined by the Nernst equation, provides a means to assess cellular redox status. To this end, we have developed a stable isotope dilution LC–MS method to analyze GSH and GSSG. Accurate quantification of GSH has always been troublesome due to the ease with which GSH is oxidized to GSSG or reduced protein disulfides during sample preparation. To overcome this challenge derivatization of GSH with 4-fluoro-7-sulfamoylbenzofurazan (ABD-F) was employed. Derivatization of the GSH thiol group was rapid, quantitative, and occurred at room temperature. Concentrations of GSH and GSSG were determined in two monocyte/macrophage RAW 267.4 cell lines with or without 15-LO expression (R15LO and RMock, respectively) after treatment with ONE. R15LO cells can synthesize higher levels of 15(*S*)-HPETE compared to RMock cells; however, it was also determined that R15LO cells had higher concentrations of GSH and thus increased resistance to ONE-mediated GSH depletion. Consequently, R15LO cells had lower reduction potentials at all concentrations of ONE. GSSG concentrations were higher in R15LO cells after ONE treatment when compared with the ONE-treated RMock cells. Thus, 15(*S*)-HPETE may modulate the activity of GSH reductases or the transporters involved in the removal of GSH (Zhu et al., 2008). This stable isotope dilution method is sensitive and robust and can be readily adapted to the quantification of GSH and GSSG in tissue and plasma samples.

20.4 LIPID BIOMARKERS OF OXIDATIVE STRESS

20.4.1 Introduction

There is intense interest in the analysis of ROS-derived IsoPs because of their utility as biomarkers of oxidative stress. Enzymatic pathways of eicosanoid formation are regioselective and enantioselective, whereas ROS-mediated eicosanoid formation proceeds with no stereoselectivity. Many of the eicosanoids are also present in only picomolar concentrations in biological fluids. Such low levels present a formidable analytical challenge because methodology is required that can separate enantiomers and diastereomers with high sensitivity and specificity. However, API methodologies, such as ESI, APCI, and electron capture (EC) APCI have revolutionized our ability to analyze endogenous eicosanoids. LC separations of eicosanoids can now be readily coupled with API ionization, collision-induced dissociation (CID) and tandem MS (MS/MS). Targeted eicosanoid analyses can be conducted with LC–MRM/MS. Several examples of targeted eicosanoid lipid analysis using conventional LC–ESI/MS have been discussed and some new data on the analysis of eicosanoids using chiral LC–ECAPCI/MS has been presented.

20.4.2 Isoprostanes

Arachidonic acid is an important fatty acyl component of the lipidome that is present as the free fatty acid in the plasma as well as the esterified form in sterol lipids, and at the *sn*-2 position of glycerolipids and glycerophospholipids

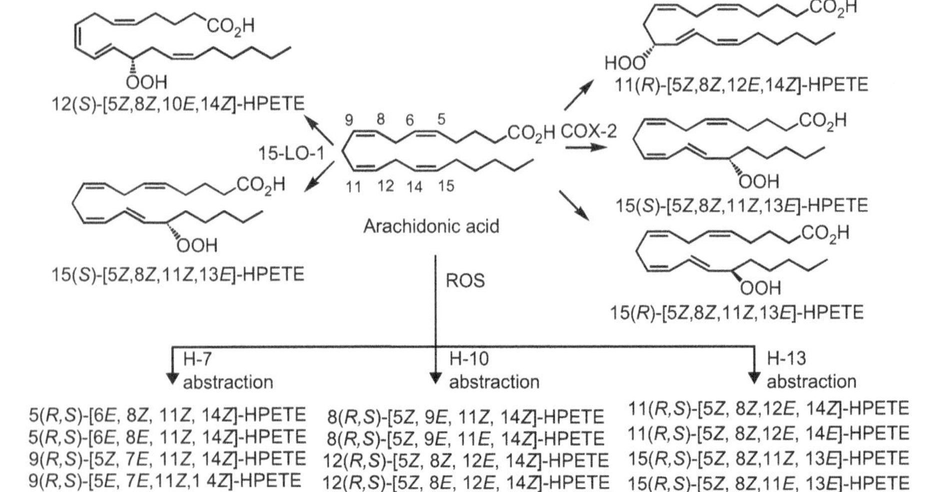

FIGURE 20.3 ROS-, COX-2-, and 15-LO-1-mediated HPETE formation. [Reprinted with permission from Mesaros et al. (2009).]

(Farooqui et al., 2001; Phinney et al., 1990; Roberts et al., 1996; Zhou and Nilsson, 2001). IsoPs are oxidation products of arachidonic acid, which were originally discovered as artifacts present in stored plasma samples (Morrow et al., 1990). The action of hydroxyl radicals on esterified arachidonic acid results in formation of the IsoP classes D, E, F, and J. These IsoPs are structural analogs of the corresponding COX-derived prostaglandins (PGs) (Milne et al., 2008; Morrow et al., 1992; 1994). Hydroxyl-radical-mediated abstraction of *bis*-allylic hydrogen atoms at C-7, C-10, and C-13 of esterified arachidonic acid generates four intermediate allylic radicals (Fig. 20.3), which each add two molecules of molecular oxygen to form a complex mixture of esterified endoperoxide hydroperoxides. The endoperoxide hydroperoxides are subsequently reduced to give four classes of esterified F_2 IsoPs (III, IV, V, and VI) and are released as their free (unesterified) forms by various lipases (Chaitidis et al., 1998; Stafforini et al., 2006; Tselepis and John, 2002). Each F_2 IsoP class contains a cyclopentane ring with two side chains in a *cis*-configuration on adjacent carbon atoms, which distinguishes them from the *trans*-configuration of the side chains attached at C-8 and C-12 of COX-derived $PGF_{2\alpha}$. In addition to the chiral side-chain carbon atoms, each F_2 IsoP class has three other chiral centers allowing for the potential formation of 64 isomers (Fig. 20.4). There are currently two nomenclatures used for IsoPs. One system, which was developed by Taber et al. (1997), uses the carbon atom of the initially formed peroxyl radical in order to determine each class. A second system, which was developed by Rokach et al. (1997), is based on the omega-carbon as a starting reference. The Rokach nomenclature system is used in this section. IsoPs were validated as reliable biomarkers of oxidative stress in a multilaboratory collaboration (Kadiiska et al,. 2005a; 2005b). The studies

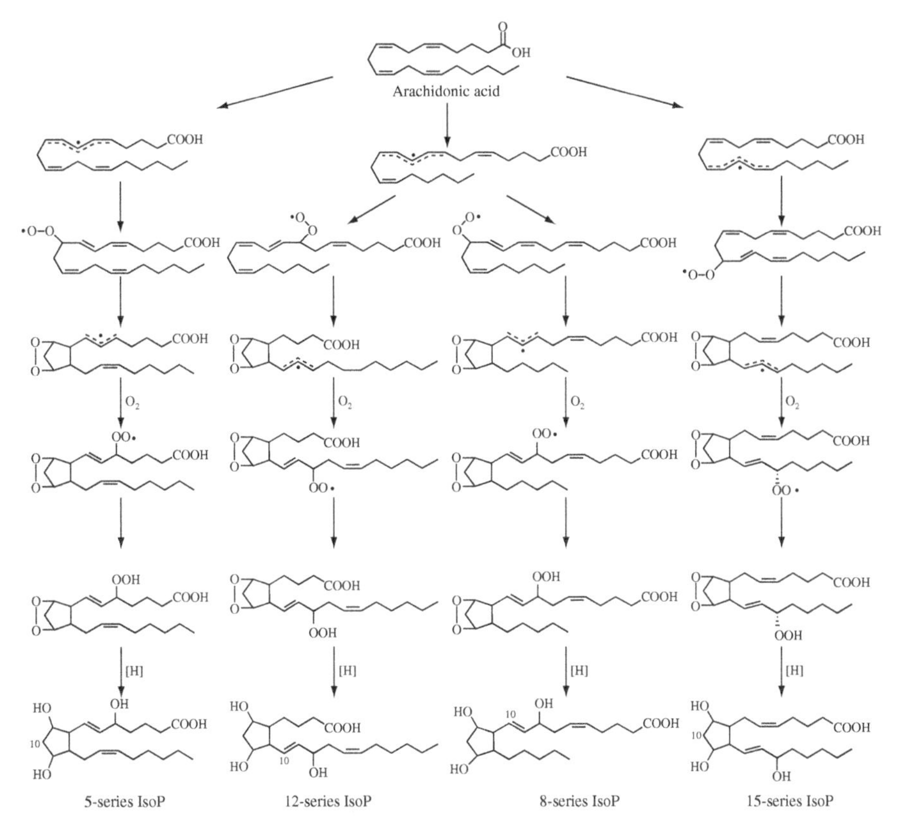

FIGURE 20.4 F_2 IsoP formation from arachidonic acid. [Reprinted with permission from Milne and Morrow (2008).]

focused on 8-*iso*-PGF$_{2\alpha,}$ (class III) and 8,12-*iso*-PGF$_{2\alpha}$, VI (class VI). 8-*iso*-PGF$_{2\alpha}$ also known as 8-epi-PGF$_{2\alpha}$ and iPF$_{2\alpha}$-III, is the most studied of the IsoP isomers. Significant elevations of both isomers were shown to occur in the urine of rats treated with carbon tetrachloride (Kadiiska et al., 2005a).

Remarkably, since the first reports of esterified IsoPs, the interest in the detection of the IsoPs in urine had generated almost 400 papers (Morrow et al., 1992, 1994). LC–MS methods used for IsoP analysis employ primarily reversed-phase LC coupled with positive ESI/MS. 8-*iso*-PGF$_{2\alpha}$ is the most frequent IsoP analyzed. There are relatively few reports that detect more than one IsoP in a single LC–MS analysis. Each class of IsoP forms a specific product ion during LC–MS/MS analysis making it possible to differentiate the four classes (Lawson et al., 1998). The characteristic product ion for class III and class IV is m/z 193 and m/z 115, respectively. However, there are potentially 16 class III and 16 class VI IsoP isomers; therefore, efficient LC separations are essential. Furthermore, class VI isomers also have a product ion at m/z 193 that is approximately 10 times less intense as the one from class III.

The first publication that reported the use of LC–MS for quantification of IsoPs in urine used reversed-phase LC coupled with ESI/MS. The method used only 1 mL urine and the clean-up procedure using solid phase extraction (SPE) columns gave quantitative recovery of the IsoP. The chromatographic runs and SPE purification methods are short, and this results in a very easy and user-friendly procedure (Li et al., 1999). A comprehensive review by Tsikas et al. describes sample preparation techniques and compares GC–MS methods with the most recent LC–MS/MS methods (Tsikas et al., 2003). The focus is primarily on 8-*iso*-$PGF_{2\alpha}$ and highlights the difficulty to detect only one IsoP isomer without immunoaffinity chromatography preparation. The large concentration differences for the various IsoP classes in urine are also addressed.

Among the recent publications that detail IsoP quantification in urine, Yan et al. (2007) have reported the detection of three class III isomers and two class VI isomers. Concentrations determined by LC–MS were then correlated with those obtained by enzyme-linked immunosorbent assay (ELISA) measurements. Interestingly, the amount reported by LC–MS was approximately half that found by ELISA. The report by Yan et al. is not the only example of this problem, which most likely results from the cross reactivity of the antibody with other IsoPs in the urines. LC–MS/MS methods are potentially more accurate than ELISA-based methodology because they can separate the individual isomers and reduce interference from the same class of IsoP.

Several groups have used immunoaffinity chromatography for the purification of eicosanoids prior to analysis (Sircar and Subbaiah, 2007; Tsikas et al., 2003). Immunoaffinity purification can be regarded as a "gold standard" for the purification of endogenous compounds when conducted under rigorous conditions (Oe et al., 2004). Immunoaffinity columns for 8-*iso*-$PGF_{2\alpha}$ are commercially available. When these columns were coupled with stable isotope dilution LC–MS/MS methodology, excellent sensitivity and specificity could be obtained. Furthermore, relatively small urine volumes (0.3 mL) were required (Sircar and Subbaiah, 2007). The specificity of this methodology is particularly useful for analyses by GC–MS when LC–MS/MS instruments are not available. However, care has to be taken to ensure there is no carry over from sample to sample. Columns should be checked routinely for residual endogenous eicosanoids between analytical runs by independent analysis of the internal standard. Unfortunately, the method is somewhat limited for the IsoPs because there are very few antibodies available for other structural analogs.

We have recently applied stable isotope dilution chiral LC–ECAPCI/MRM/MS methodology to the analysis of multiple IsoPs in human urine after liquid–liquid extraction and PFB derivatization. This methodology has made it possible to show a significant increase in 8-*iso*-$PGF_{2\alpha}$ (class III) and 8,12-*iso*-$PGF2_{\alpha}$-VI (class VI) in the urine of smokers compared with nonsmokers (Fig 20.5). The concentration of 8-*iso*-$PGF_{2\alpha}$ in a typical smoker's urine sample was 0.6 ng/mg creatinine and 0.1 ng/mg of creatinine in a typical nonsmoker's sample (data not shown). 8, 12-*iso*-$PGF2_{\alpha}$-VI had a concentration 1.9 ng/mg

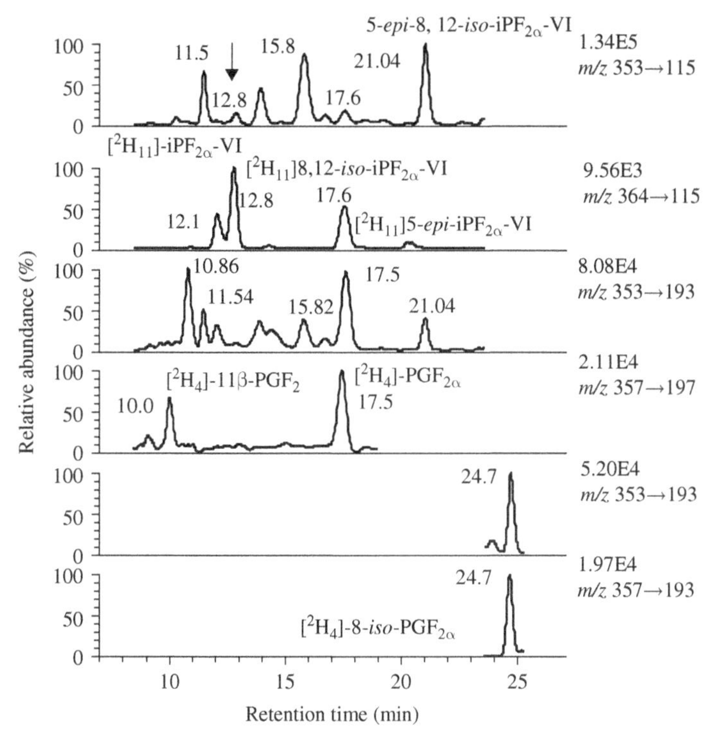

FIGURE 20.5 Typical LC–ECAPCI/MS chromatogram of IsoPs in a 3-mL smoker's urine sample spiked with a mixture of synthetic heavy isotope analog ISs (1 ng of each). [Reprinted with permission from Mesaros et al. (2009).]

creatinine in a typical smoker's urine sample and 0.2 ng/mg creatinine in a typical non-smoker's sample.

A number of years ago we reported that another hydroxylated eicosanoid (20-hydroxyeicostetraenoic acid, 20-HETE) is excreted primarily as a glucuronide conjugate in the urine of human subjects. This finding, which has since been confirmed by a number of other groups, arises primarily as a result of UGT1A1- and UGT1B7-mediated glucuronidation of 20-HETE (Little et al., 2004; Prakash et al., 1992; Rivera et al., 2004; Sacerdoti et al., 1997; Watzer et al., 2000). Interestingly, there are two studies, which were reported at scientific meetings, suggesting that IsoPs can be excreted in the urine as glucuronide conjugates (Sasaki et al., 2002). It is also noteworthy that after infusion of radiolabeled 8-*iso*-$PGF_{2\alpha}$ into a human subject, 42% of the radioactivity was excreted into the urine as polar metabolites (possibly glucuronides) (Roberts et al., 1996). Over the last several years, highly specific and sensitive methodology has been developed for the analysis of eicosanoids based on the use of chiral-normal-phase LC coupled with ECAPCI/MS (Lee and Blair, 2007; Lee et al., 2005c; Singh et al., 2000). This methodology has now been applied to the analysis of IsoPs in order to determine whether they can be excreted in the urine of tobacco smokers and nonsmokers as glucuronide conjugates. It was found that levels of glucuronidated IsoPs were similar to the non-conjugated IsoPs. Additional studies will be required in order to fully understand the relevance of this finding to analysis of IsoPs as biomarkers of oxidative stress.

20.4.3 Hydroxyeicosatetraenoic Acids (HETEs)

Since our finding that 20-HETE is excreted in the urine primarily as a glucuronide metabolite (Prakash et al., 1992), several studies have shown that it is excreted in higher concentrations in hypertensive subjects when compared with normal control subjects (McGiff and Quilley, 2001). The relevance of these findings to oxidative stress in the formation of 20-HETE remains to be fully elucidated. However, it is noteworthy that a single nucleotide polymorphism in the cytochrome P450 (P450) *4F2* gene but not the P450 *4A11* gene was associated with increased 20-HETE excretion and increased blood pressure (Ward et al., 2008). In addition, 12(*S*)-HETE was found to be increased in the urine of diabetic subjects when compared with controls (Suzuki et al., 2003). This finding suggests that a specific 12-LO is up-regulated in diabetes and that the resulting increased production of 12(*S*)-HPETE could induce cellular oxidative stress in a similar manner to 15(*S*)-HPETE (Zhu et al., 2008). However, there have been no follow-up studies to determine whether 12(*S*)-HETE biosynthesis is up-regulated in oxidative stress or whether this biosynthesis arises from a pathway that is specific to the pathophysiology of diabetes. Clearly, more studies on the relevance of urinary HETE excretion and the role of glucuronidation are required. The availability of our recently developed highly sensitive stable isotope dilution chiral LC–ECAPCI/MS method for the analysis of urinary 12(*S*)-HPETE and 20-HETE (Fig 20.6) will facilitate these studies (Mesaros et al., 2009).

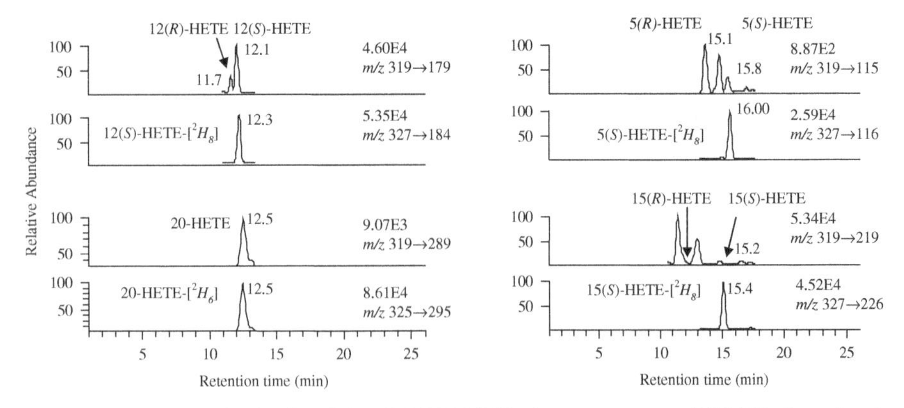

FIGURE 20.6 Typical LC-ECAPCI/MRM/MS chromatogram of HETEs in a 3-mL smoker's urine sample spiked with a mixture of synthetic heavy isotope analog ISs (1 ng of each). [Reprinted with permission from Mesaros et al. (2009).]

20.5 CREATININE: THE COMMON DENOMINATOR

Creatinine is a cyclic breakdown product of creatine and creatine phosphate. Creatinine is produced in muscle at a fairly constant rate depending on muscle mass (Fig 20.7). The rate of nonenzymatic formation of creatinine from creatine is nearly constant, and the creatine constantly diffuses out of the tissues into the blood. Creatinine is then filtered out of the blood by the kidneys into the urine. Although a small amount of creatinine is actively secreted by the kidneys, there is very little reabsorption in the tubules. Levels of creatine and guanidinoacetic acid in blood and urine are normally very low so that urinary creatinine levels provide a useful index of renal excretion. Therefore, intra- and interdifferences in urine volume can be normalized, and biomarker concentrations can be reported in terms of creatinine as the denominator rather than the urine volume. Typically, creatinine is measured by a colorimetric assay, which provides only a precise but not very accurate measurement as creatinine levels are underestimated by approximately 20% (Fig. 20.8). Therefore, the accuracy of the biomarker assay is often compromised when a creatinine normalization factor is employed. This problem has become recognized by researchers in the biomarker field and LC–MS is starting to be employed for the analysis of urinary creatinine. However, it is a very polar molecule that tends to elute in the void volume of C18 columns. Therefore, we have begun to use cyano-columns, which retain creatine, creatinine, and guanidinoacetic acid. Stable isotope dilution methodology allows for the quantification of creatinine and its two upstream products to give a more accurate assessment of urinary excretion for normalization of urinary biomarker measurements.

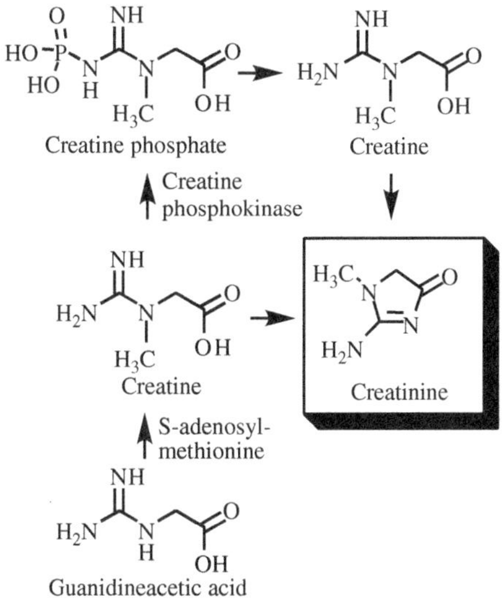

FIGURE 20.7 Formation of creatine phosphate, creatinine phosphate, and creatinine.

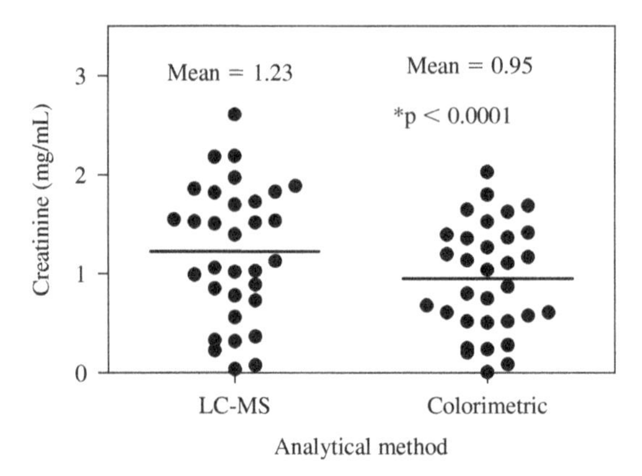

FIGURE 20.8 Analysis of urinary creatinine obtained by LC–MS and a colorimetric assay in the same urine samples.

20.6 SUMMARY AND CONCLUSIONS

LC–MS has provided researchers an invaluable tool to analyze oxidative metabolites and adducts on lipids, nucleic acids, and proteins. Many of these oxidative metabolites go on to form adducts such as the HεdGuo from ONE and 1-N^2-propano-dGuo from ACR. For the most part, these modifications are used as individual biomarkers of specific disease states. Only recently has there been an emphasis on the formation of a panel of biomarkers for specific disease states or environmental exposures. For instance, our laboratory has focused on exposure and response biomarkers of cigarette smoke. Many of these reactive compounds have been implicated in cancers, pulmonary diseases, and cardiovascular diseases, to name a few. A panel of biomarkers that range from oxidized DNA adducts to IsoPs has provided a tool to better understand different patterns of disease progression and biological outcomes.

Advances in LC–MS methods, whether in sample preparation, such as the inhibition of artifactual formation of 8-oxo-dGuo, or in the instrumentation, such as electron-capture APCI for chiral lipids, and nanospray LC–MS for the shotgun analysis of proteins will provide the necessary tools that can be used for future biomarker studies. The most important immediate change is to define the relationship between specific oxidative modifications and changes in function related to toxicity. Furthermore, the availability of reliable nanoflow ultraperformance liquid chromatography (UPLC) promises to provide substantial increases in sensitivity and specificity for biomarker analyses. The coupling of UPLC with new and more sensitive TSQ instrumentation will significantly enhance the ability to make such determinations. The enhanced sensitivity will permit much smaller plasma and urine samples to be analyzed. This performance will allow smaller volumes of urine and plasma to be collected and stored, and allow for larger and much needed population studies to be conducted.

ACKNOWLEDGMENTS

We acknowledge the support of NIH grants R01CA091016, R01CA130038, P30ES013508, U01ES016004, and 5F32ES016683.

REFERENCES

Akman SA, Doroshow JH, Burke TG, Dizdaroglu M. DNA base modifications induced in isolated human chromatin by NADH dehydrogenase-catalyzed reduction of doxorubicin. Biochemistry 1992;31:3500–3506.

Ames BN, Shigenaga MK, Hagen TM. Oxidants, antioxidants, and the degenerative diseases of aging. Proc Natl Acad Sci USA 1993;90:7915–7922.

Annangudi SP, Deng Y, Gu X, Zhang W, Crabb JW, Salomon RG. Low-density lipoprotein has an enormous capacity to bind (*E*)-4-hydroxynon-2-enal (HNE): Detection and characterization of lysyl and histidyl adducts containing multiple molecules of HNE. Chem Res Toxicol 2008;21:1384–1395.

Armitage B. Photocleavage of nucleic acids. Chem Rev 1998;98:1171–1200.

Aruoma OI, Halliwell B, Dizdaroglu M. Iron ion-dependent modification of bases in DNA by the superoxide radical-generating system hypoxanthine/xanthine oxidase. J Biol Chem 1989a;264:13024–13028.

Aruoma Ol, Halliwell B, Gajewski E, Dizdaroglu M. Damage to the bases in DNA induced by hydrogen peroxide and ferric ion chelates. J Biol Chem 1989b;264:20509–20512.

Atwood CS, Scarpa RC, Huang X, Moir RD, Jones WD, Fairlie DP, Tanzi RE, Bush AI. Characterization of copper interactions with alzheimer amyloid beta peptides: Identification of an attomolar-affinity copper binding site on amyloid beta1-42. J. Neurochem 2000;75:1219–1233.

Baillie TA. Future of toxicology-metabolic activation and drug Design: Challenges and opportunities in chemical toxicology. Chem Res Toxicol 2006;19:889–893.

Baillie TA, Davis MR. Mass spectrometry in the analysis of glutathione conjugates. Biol Mass Spectrom 1993;22:319–325.

Barbin A, Bartsch H. Mutagenic and promutagenic properties of DNA adducts formed by vinyl chloride metabolites. IARC Sci Publ 1986;70:345–358.

Barnham KJ, Haeffner F, Ciccotosto GD, Curtain CC, Tew D., Mavros C, Beyreuther K, Carrington D, Masters CL, Cherny, RA, Cappai R, Bush AI. Tyrosine gated electron transfer is key to the toxic mechanism of Alzheimer's disease beta-amyloid. FASEB J 2004; 18:1427–1429.

Birincioglu M, Jaruga P, Chowdhury G, Rodriguez H, Dizdaroglu M, Gates KS. DNA base damage by the antitumor agent 3-amino-1,2,4-benzotriazine 1,4-dioxide (tirapazamine). J Am Chem Soc 2003;125:11607–11615.

Blair IA. The role of oxidative stress in cancer and toxicology. In: Encyclopedia of Mass Spectrometry, Caprioli RM, Gross ML Eds., Elsevier, New York, 2005, pp. 283–307.

Blair IA. DNA adducts with lipid peroxidation products. J Biol Chem 2008;283: 15545–15549.

Blakely WF, Fuciarelli, AF, Wegher, BJ, Dizdaroglu, M. Hydrogen peroxide–induced base damage in deoxyribonucleic acid. Radiat Res 1990;121:338–343.

Boiteux S, Gajewski E, Laval J, Dizdaroglu M. Substrate specificity of the *Escherichia coli* Fpg protein (formamidopyrimidine-DNA glycosylase): Excision of purine lesions in DNA produced by ionizing radiation or photosensitization. Biochemistry 1992;31:106–110.

Brigelius-Flohe R. Tissue-specific functions of individual glutathione peroxidases. Free Radic Biol Med 1999;27:951–965.

Burcham PC, Fontaine FR, Petersen DR, Pyke SM. Reactivity with Tris(hydroxymethyl)aminomethane confounds immunodetection of acrolein-adducted proteins. Chem Res Toxicol 2003;16:1196–1201.

Burrows CJ, Muller JG. Oxidative nucleobase modifications leading to strand scission. Chem Rev 1998;98:1109–1152.

Bush AI. The metallobiology of Alzheimer's disease. Trends Neurosci 2003;26:207–214.

Butterfield DA, Kanski J. Methionine residue 35 is critical for the oxidative stress and neurotoxic properties of Alzheimer's amyloid beta-peptide 1-42. Peptides 2002;23:1299–1309.

Carbone DL, Doorn JA, Kiebler Z, Ickes B.R, Petersen DR. Modification of heat shock protein 90 by 4-hydroxynonenal in a rat model of chronic alcoholic liver disease. J Pharmacol Exp Ther 2005;315:8–15.

Carbone DL, Doorn JA, Kiebler Z, Sampey BP, Petersen DR. Inhibition of Hsp72-mediated protein refolding by 4-hydroxy-2-nonenal. Chem Res Toxicol 2004;17:1459–1467.

Carini M, Aldini, G, Facino, R.M. Mass spectrometry for detection of 4-hydroxy-trans-2-nonenal (HNE) adducts with peptides and proteins. Mass Spectrom Rev 2004;23:281–305.

Castellani RJ, Honda K, Zhu X, Cash AD, Nunomura A, Perry G, Smith MA. Contribution of redox-active iron and copper to oxidative damage in Alzheimer disease. Ageing Res Rev 2004;3:319–326.

Chaitidis P, Schewe T, Sutherland M, Kuhn H, Nigam S. 15-Lipoxygenation of phospholipids may precede the sn-2 cleavage by phospholipases A2: Reaction specificities of secretory and cytosolic phospholipases A2 towards native and 15-lipoxygenated arachidonoyl phospholipids. FEBS Lett. 1998;434:437–441.

Chaudhary AK, Nokubo M, Reddy GR, Yeola SN, Morrow JD, Blair IA, Marnett LJ. Detection of endogenous malondialdehyde-deoxyguanosine adducts in human liver. Science 1994;265:1580–1582.

Ciccimaro E, Hanks SK, Yu KH, Blair IA. Absolute quantification of phosphorylation on the kinase activation loop of cellular focal adhesion kinase by stable isotope dilution liquid chromatography/mass spectrometry. Anal Chem 2009; 81: 3304–3313.

Citardi MJ, Song W, Batra PS, Lanza DC, Hazen SL. Characterization of oxidative pathways in chronic rhinosinusitis and sinonasal polyposis. Am J Rhinol 2006;20:353–359.

Codreanu SG, Zhang B, Sobecki SM, Billheimer DD, Liebler DC. Global analysis of protein damage by the lipid electrophile 4-hydroxy-2-nonenal. Mol Cell Proteomics 2009;8:670–680.

Cohen G, Hochstein P Glutathione peroxidase: The primary agent for the elimination of hydrogen peroxide in erythrocytes. Biochemistry 1963;2:1420–1428.

Cook JA, Gius D, Wink DA, Krishna MC, Russo A, Mitchell JB. Oxidative stress, redox, and the tumor microenvironment. Semin Radiat Oncol 2004;14:259–266.

Dennehy MK, Richards KA, Wernke GR, Shyr Y, Liebler DC. Cytosolic and nuclear protein targets of thiol-reactive electrophiles. Chem Res Toxicol 2006;19:20–29.

Dickinson DA, Forman HJ. Cellular glutathione and thiols metabolism. Biochem Pharm 2002;64:1019–1026.

Dieckhaus CM, Fernandez-Metzler CL, King R, Krolikowski PH, Baillie TA. Negative ion tandem mass spectrometry for the detection of glutathione conjugates. Chem Res Toxicol 2005;18:630–638.

Dizdaroglu M, Kirkali G, Jaruga P. Formamidopyrimidines in DNA: Mechanisms of formation, repair, and biological effects. Free Radic Biol Med 2008; 45: 1610–1621.

Dizdaroglu M, Rao G, Halliwell B, Gajewski E. Damage to the DNA bases in mammalian chromatin by hydrogen peroxide in the presence of ferric and cupric ions. Arch Biochem Biophys 1991;285:317–324.

Ehrenberg L, Hiesche KD, Osterman-Golkar S, Wenneberg I. Evaluation of genetic risks of alkylating agents: Tissue doses in the mouse from air contaminated with ethylene oxide. Mutat Res 1974;24:83–103.

England TG, Jenner A, Aruoma OI, Halliwell B. Determination of oxidative DNA base damage by gas chromatography–mass spectrometry. Effect of derivatization conditions on artifactual formation of certain base oxidation products. Free Radic Res 1998;29:321–330.

Evans DC, Baillie TA. Minimizing the potential for metabolic activation as an integral part of drug design. Curr Opin Drug Discov Devel 2005;8:44–50.

Evans DC, Watt AP, Nicoll-Griffith DA, Baillie TA. Drug-protein adducts: An industry perspective on minimizing the potential for drug bioactivation in drug discovery and development. Chem Res Toxicol 2004;17:3–16.

Farmer PB, Bailey E, Lamb JH, Connors TA. Approach to the quantitation of alkylated amino acids in haemoglobin by gas chromatography mass spectrometry. Biomed Mass Spectrom 1980;7:41–46.

Farooqui AA, Yi OW, Lu XR, Halliwell B, Horrocks LA. Neurochemical consequences of kainate-induced toxicity in brain: Involvement of arachidonic acid release and prevention of toxicity by phospholipase A(2) inhibitors. Brain Res Brain Res Rev 2001;38:61–78.

Farrow B, Evers BM. Inflammation and the development of pancreatic cancer. Surg Oncol 2002;10:153–169.

Feng Z, Hu W, Hu Y, Tang MS. Acrolein is a major cigarette-related lung cancer agent: Preferential binding at p53 mutational hotspots and inhibition of DNA repair. Proc Natl Acad Sci USA 2006;103:15404–15409.

Fernandes AP, Holmgren A. Glutaredoxins: Glutathione-dependent redox enzymes with functions far beyond a simple thioredoxin backup system. Antioxid Redox Signal 2004;6:63–74.

Frelon S, Douki, T, Favier, A, Cadet, J. Hydroxyl radical is not the main reactive species involved in the degradation of DNA bases by copper in the presence of hydrogen peroxide. Chem Res Toxicol 2003;16:191–197.

Gedik CM, Collins A Establishing the background level of base oxidation in human lymphocyte DNA: Results of an interlaboratory validation study. FASEB J 2005;19:82–84.

Green LC, Skipper PL, Turesky RJ, Bryant MS, Tannenbaum SR. In vivo dosimetry of 4-aminobiphenyl in rats via a cysteine adduct in hemoglobin. Cancer Res 1984;44:4254–4259.

Heinecke JW. Oxidized amino acids: Culprits in human atherosclerosis and indicators of oxidative stress. Free Radic Biol Med 2002;32:1090–1101.

Hensley K, Maidt ML, Yu Z, Sang H, Markesbery WR, Floyd RA. Electrochemical analysis of protein nitrotyrosine and dityrosine in the Alzheimer brain indicates region-specific accumulation. J Neurosci 1998;18:8126–8132.

Hoberg AM, Otteneder M, Marnett LJ, Poulsen HE. Measurement of the malondialdehyde-2'-deoxyguanosine adduct in human urine by immuno-extraction and liquid chromatography/atmospheric pressure chemical ionization tandem mass spectrometry. J Mass Spectrom 2004;39:38–42.

Homolya L, Varadi A, Sarkadi B. Multidrug resistance-associated proteins: Export pumps for conjugates with glutathione, glucuronate or sulfate. Biofactors 2003;17:103–114.

Huang X, Cuajungco MP, Atwood CS, Hartshorn MA, Tyndall JD, Hanson GR, Stokes KC, Leopold M, Multhaup G, Goldstein LE, Scarpa RC, Saunders AJ, Lim J, Moir RD, Glabe C, Bowden EF, Masters CL, Fairlie DP, Tanzi RE, Bush AI. Cu (II) potentiation of alzheimer abeta neurotoxicity. Correlation with cell-free hydrogen peroxide production and metal reduction. J Biol Chem 1999;274:37111–37116.

Huang X, Cuajungco MP, Atwood CS, Moir RD, Tanzi RE, Bush AI. Alzheimer's disease, beta-amyloid protein and zinc. J Nutr 2000;130:1488S-1492S.

Inoue K, Garner C, Ackermann BL, Oe T, Blair IA. Liquid chromatography/tandem mass spectrometry characterization of oxidized amyloid beta peptides as potential biomarkers of Alzheimer's disease. Rapid Commun Mass Spectrom 2006;20:911–918.

Jang JH, Surh YJ. Beta-amyloid-induced apoptosis is associated with cyclooxygenase-2 up-regulation via the mitogen-activated protein kinase-NF-kappaB signaling pathway. Free Radic Biol Med 2005;38:1604–1613.

Jaruga P, Kirkali G, Dizdaroglu M. Measurement of formamidopyrimidines in DNA. Free Radic Biol Med 2008;45:1601–1609.

Jian W, Arora JS, Oe T, Shuvaev VV, Blair IA. Induction of endothelial cell apoptosis by lipid hydroperoxide-derived bifunctional electrophiles. Free Radic Biol Med 2005a;39:1162–1176.

Jian W, Lee SH, Arora JS, Silva Elipe MV, Blair IA. Unexpected formation of etheno-2′-deoxyguanosine adducts from 5(S)-hydroperoxyeicosatetraenoic acid: Evidence for a bis-hydroperoxide intermediate. Chem Res Toxicol 2005b;18:599–610.

Jian W, Lee SH, Mesaros C, Oe T, Elipe MV, Blair IA. A novel 4-oxo-2(*E*)-nonenal-derived endogenous thiadiazabicyclo glutathione adduct formed during cellular oxidative stress. Chem Res Toxicol 2007;20:1008–1018.

Kadiiska MB, Gladen BC, Baird DD, Germolec D, Graham LB, Parker CE, Nyska A, Wachsman JT, Ames BN, Basu S, Brot N, FitzGerald GA, Floyd RA, George M, Heinecke JW, Hatch GE, Hensley K, Lawson JA, Marnett LJ, Morrow JD, Murray DM, Plastaras J, Roberts LJ, Rokach J, Shigenaga MK, Sohal RS, Sun J, Tice RR, Van Thiel DH, Wellner D, Walter PB, Tomer KB, Mason RP, Barrett JC. Biomarkers of oxidative stress study II: are oxidation products of lipids, proteins, and DNA markers of CCl4 poisoning? Free Radic Biol Med 2005a;38:698–710.

Kadiiska MB, Gladen BC, Baird DD, Graham LB, Parker CE, Ames BN, Basu S, FitzGerald GA, Lawson JA, Marnett LJ, Morrow JD, Murray DM, Plastaras J, Roberts LJ, Rokach J, Shigenaga MK, Sun J, Walter PB, Tomer KB, Barrett JC, Mason RP. Biomarkers of oxidative stress study III. Effects of the nonsteroidal anti-inflammatory agents indomethacin and meclofenamic acid on measurements of oxidative products of lipids in CCl4 poisoning. Free Radic Biol Med 2005b; 38:711–718.

Kapphahn RJ, Giwa BM, Berg KM, Roehrich H, Feng X, Olsen TW, Ferrington DA. Retinal proteins modified by 4-hydroxynonenal: Identification of molecular targets. Exp Eye Res 2006;83:165–175.

Kato Y, Wu X, Naito M, Nomura H, Kitamoto N, Osawa T. Immunochemical detection of protein dityrosine in atherosclerotic lesion of apo-E-deficient mice using a novel monoclonal antibody. Biochem Biophys Res Commun 2000;275:11–15.

Kawai Y, Uchida K, Osawa T. 2′-deoxycytidine in free nucleosides and double-stranded DNA as the major target of lipid peroxidation products. Free Radic Biol Med 2004;36:529–541.

Ketterer B. Protective role of glutathione and glutathione transferases in mutagenesis and carcinogenesis. Mutat Res 1988; 202:343–361.

Knutson CG, Skipper PL, Liberman RG, Tannenbaum SR, Marnett LJ. Monitoring in vivo metabolism and elimination of the endogenous DNA adduct, M1dG {3-(2-deoxy-beta-**D**-erythro-pentofuranosyl)pyrimido[1,2-alpha]purin-10(3*H*)-one}, by accelerator mass spectrometry. Chem Res Toxicol 2008;21:1290–1294.

Kolb HC, Sharpless KB. The growing impact of click chemistry on drug discovery. Drug Discov Today 2003;8:1128–1137.

Kolman A, Chovanec M, Osterman-Golkar S. Genotoxic effects of ethylene oxide, propylene oxide and epichlorohydrin in humans: Update review (1990–2001). Mutat Res 2002;512:173–194.

Krishnamurthy N, Haraguchi K, Greenberg MM, David SS. Efficient removal of formamidopyrimidines by 8-oxoguanine glycosylases. Biochemistry 2008;47:1043–1050.

Lam BK, Austen KF. Leukotriene C4 synthase: a pivotal enzyme in cellular biosynthesis of the cysteinyl leukotrienes. Prostaglandins Other Lipid Mediat. 2002;68–69: 511–520.

Lawson JA, Li H, Rokach J, Adiyaman M, Hwang SW, Khanapure SP, FitzGerald GA. Identification of two major F2 isoprostanes, 8,12-iso- and 5-epi-8, 12-iso-isoprostane F2alpha-VI, in human urine. J Biol Chem 1998;273:29295–29301.

Lee SH, Blair IA. Characterization of 4-oxo-2-nonenal as a novel product of lipid peroxidation. Chem Res Toxicol 2000;13:698–702.

Lee SH, Blair IA. Oxidative DNA damage and cardiovascular disease. Trends Cardiovasc. Med 2001a;11:148–155.

Lee SH, Blair IA. Targeted chiral lipidomics analysis by liquid chromatography electron capture atmospheric pressure chemical ionization mass spectrometry (LC-ECAPCI/MS). Methods Enzymol 2007;433:159–174.

Lee SH, Elipe MVS, Arora JS, Blair IA. Dioxododecenoic acid: A lipid hydroperoxide-derived bifunctional electrophile responsible for etheno DNA adduct formation. Chem Res Toxicol 2005a;18:566–578.

Lee SH, Oe T, Arora JS, Blair IA. Analysis of Fe-II-mediated decomposition of a linoleic acid-derived lipid hydroperoxide by liquid chromatography/mass spectrometry. J Mass Spectrom. 2005b;40:661–668.

Lee SH, Oe T, Blair IA. Vitamin C-induced decomposition of lipid hydroperoxides to endogenous genotoxins. Science 2001;292: 2083–2086.

Lee SH, Williams MV, Blair IA. Targeted chiral lipidomics analysis. Prostaglandins Other Lipid Mediat 2005c;77:141–157.

Lee SH, Williams MV, Dubois RN, Blair IA. Cyclooxygenase-2-mediated DNA damage. J Biol Chem 2005d;280:28337–28346.

Leeuwenburgh C, Hansen PA, Holloszy JO, Heinecke JW. Hydroxyl radical generation during exercise increases mitochondrial protein oxidation and levels of urinary dityrosine. Free Radic Biol Med 1999a;27:186–192.

Leeuwenburgh C, Hansen PA, Holloszy JO, Heinecke JW. Oxidized amino acids in the urine of aging rats: potential markers for assessing oxidative stress in vivo. Am J Physiol 1999b;276:R128–R135.

Leeuwenburgh C, Rasmussen JE, Hsu FF, Mueller DM, Pennathur S, Heinecke JW. Mass spectrometric quantification of markers for protein oxidation by tyrosyl radical, copper, and hydroxyl radical in low density lipoprotein isolated from human atherosclerotic plaques. J Biol Chem 1997;272:3520–3526.

Levsen K, Schiebel HM, Behnke B, Dotzer R, Dreher W, Elend M, Thiele H. Structure elucidation of phase II metabolites by tandem mass spectrometry: An overview. J Chromatogr A 2005;1067:55–72.

Li H, Lawson JA, Reilly M, Adiyaman M, Hwang SW, Rokach J, FitzGerald GA. Quantitative high performance liquid chromatography/tandem mass spectrometric analysis of the four classes of F(2)-isoprostanes in human urine. Proc Natl Acad Sci USA 1999;96:13381–13386.

Liebler DC. Protein damage by reactive electrophiles: targets and consequences. Chem Res Toxicol 2008;21:117–128.

Little JM, Kurkela M, Sonka J, Jantti S, Ketola R, Bratton S, Finel M, Radominska-Pandya A. Glucuronidation of oxidized fatty acids and prostaglandins B1 and E2 by human hepatic and recombinant UDP-glucuronosyltransferases. J Lipid Res. 2004;45:1694–1703.

Liu X, Lovell MA, Lynn BC. Development of a method for quantification of acrolein-deoxyguanosine adducts in DNA using isotope dilution-capillary LC/MS/MS and its application to human brain tissue. Anal Chem 2005;77:5982–5989.

Lu SC. Regulation of glutathione synthesis. Curr Top Cell Regul 2000;36: 95–116.

Mangal D, Vudathala D, Park JH, Lee SH, Penning TM, Blair IA. Analysis of 7,8-dihydro-8-oxo-2′-deoxyguanosine in cellular DNA during oxidative stress. Chem Res Toxicol 2009;22(5):788–97.

Mason DE, Liebler DC. Quantitative analysis of modified proteins by LC-MS/MS of peptides labeled with phenyl isocyanate. J Proteome Res 2003;2:265–272.

McGiff JC, Quilley J. 20-Hydroxyeicosatetraenoic acid and epoxyeicosatrienoic acids and blood pressure. Curr Opin Nephrol Hypertens 2001;10:231–237.

Meier BW, Gomez JD, Kirichenko OV, Thompson JA. Mechanistic basis for inflammation and tumor promotion in lungs of 2,6-di-*tert*-butyl-4-methylphenol-treated mice: Electrophilic metabolites alkylate and inactivate antioxidant enzymes. Chem Res Toxicol 2007;20:199–207.

Meier BW, Gomez JD, Zhou A, Thompson JA. Immunochemical and proteomic analysis of covalent adducts formed by quinone methide tumor promoters in mouse lung epithelial cell lines. Chem Res Toxicol 2005;18:1575–1585.

Mesaros C, Lee SH, Blair IA. Targeted quantitative analysis of eicosanoid lipids in biological samples using liquid chromatography–tandem mass spectrometry. J Chromatogr B 2009;877:2736–2745.

Miller EC, Miller JA. The presence and significance of bound aminoazo dyes in the livers of rats fed para-dimethylaminoazobenzene. Cancer Res 1947;7:468–480.

Milne GL, Yin H, Morrow JD. Human biochemistry of the isoprostane pathway. J Biol Chem 2008;283:15533–15537.

Morrow JD, Awad JA, Boss HJ, Blair IA, Roberts LJ. Non-cyclooxygenase-derived prostanoids (F2-isoprostanes) are formed in situ on phospholipids. Proc Natl Acad Sci USA 1992;89:10721–10725.

Morrow JD, Harris TM, Roberts LJ. Noncyclooxygenase oxidative formation of a series of novel prostaglandins: analytical ramifications for measurement of eicosanoids. Anal. Biochem 1990;184:1–10.

Morrow JD, Minton TA, Mukundan CR, Campbell MD, Zackert WE, Daniel VC, Badr KF, Blair IA, Roberts LJ. Free radical-induced generation of isoprostanes in vivo. Evidence for the formation of D-ring and E-ring isoprostanes. J Biol Chem 1994;269:4317–4326.

Multhaup G, Scheuermann S, Schlicksupp A, Simons A, Strauss M, Kemmling A, Oehler C, Cappai R, Pipkorn R, Bayer TA. Possible mechanisms of APP-mediated oxidative stress in Alzheimer's disease. Free Radic Biol Med 2002;33:45–51.

Murphy RC, Zarini S. Glutathione adducts of oxyeicosanoids. Prostaglandins Other Lipid Mediat 2002;68-69:471–482.

Nackerdien Z, Kasprzak KS, Rao G, Halliwell B, Dizdaroglu M. Nickel(II)- and cobalt (II)-dependent damage by hydrogen peroxide to the DNA bases in isolated human chromatin. Cancer Res 1991;51:5837–5842.

Nath RG, Chung FL. Detection of exocyclic 1,N2-propanodeoxyguanosine adducts as common DNA lesions in rodents and humans. Proc Natl Acad Sci USA 1994;91:7491–7495.

Nath RG, Ocando JE, Chung FL. Detection of 1,N2-propanodeoxyguanosine adducts as potential endogenous DNA lesions in rodent and human tissues. Cancer Res 1996;56:452–456.

Nath RG, Ocando JE, Guttenplan JB, Chung FL. 1,N2-Propanodeoxyguanosine adducts: Potential New Biomarkers of smoking-induced DNA damage in human oral tissue. Cancer Res 1998;58:581–584.

Nikolic D, Fan PW, Bolton JL, van Breemen RB. Screening for xenobiotic electrophilic metabolites using pulsed ultrafiltration–mass spectrometry. Comb Chem High Throughput Screen 1999;2:165–175.

Oe T, Lee SH, Arora JS, Blair IA. Reactivity and selectivity of lipid hydroperoxide-derived modifications to peptides. Abstr Pap Am Chem Soc 2004;228:U391.

Orhan H, Coolen S, Meerman JH. Quantification of urinary *o,o′*-dityrosine, a biomarker for oxidative damage to proteins, by high performance liquid chromatography with triple quadrupole tandem mass spectrometry. A comparison with ion-trap tandem mass spectrometry. J Chromatogr B 2005;827:104–108.

Orioli M, Aldini G, Benfatto MC, Facino RM, Carini M. HNE Michael adducts to histidine and histidine-containing peptides as biomarkers of lipid-derived carbonyl stress in urines: LC-MS/MS profiling in Zucker obese rats. Anal Chem 2007;79:9174–9184.

Orioli M, Aldini G, Beretta G, Facino RM, Carini M LC-ESI-MS/MS determination of 4-hydroxy-*trans*-2-nonenal Michael adducts with cysteine and histidine-containing peptides as early markers of oxidative stress in excitable tissues. J Chromatogr B 2005;827:109–118.

Orton CR, Liebler DC. Analysis of protein adduction kinetics by quantitative mass spectrometry: competing adduction reactions of glutathione-*S*-transferase P1-1 with electrophiles. Chem Biol Interact 2007;168:117–127.

Parastatidis I, Thomson L, Burke A, Chernysh I, Nagaswami C, Visser J, Stamer S, Liebler DC, Koliakos G, Heijnen HF, FitzGerald GA, Weisel JW, Ischiropoulos H. Fibrinogen beta-chain tyrosine nitration is a prothrombotic risk factor. J Biol. Chem 2008;283:33846–33853.

Parastatidis I, Thomson L, Fries DM, Moore RE, Tohyama J, Fu X, Hazen SL, Heijnen HF, Dennehy MK, Liebler DC, Rader DJ, Ischiropoulos H. Increased protein nitration burden in the atherosclerotic lesions and plasma of apolipoprotein A-I deficient mice. Circ Res 2007;101:368–376.

Phinney SD, Odin RS, Johnson SB, Holman RT. Reduced arachidonate in serum phospholipids and cholesteryl esters associated with vegetarian diets in humans. Am J Clin Nutr 1990;51:385–392.

Potter WZ, Davis DC, Mitchell JR, Jollow DJ, Gillette JR, Brodie BB. Acetaminophen-induced hepatic necrosis. 3. Cytochrome P-450-mediated covalent binding in vitro. J Pharmacol Exp Ther 1973;187:203–210.

Prakash C, Zhang JY, Falck JR, Chauhan K, Blair IA. 20-Hydroxyeicosatetraenoic acid is excreted as a glucuronide conjugate in human urine. Biochem Biophys Res Commun 1992;185:728–733.

Rahman I, Marwick J, Kirkham P. Redox modulation of chromatin remodeling: Impact on histone acetylation and deacetylation, NF-kappaB and pro-inflammatory gene expression. Biochem Pharm 2004;68:1255–1267.

Rivera J, Ward N, Hodgson J, Puddey IB, Falck JR, Croft KD. Measurement of 20-hydroxyeicosatetraenoic acid in human urine by gas chromatography-mass spectrometry. Clin Chem 2004;50:224–226.

Roberts LJ, Moore KP, Zackert WE, Oates JA, Morrow JD. Identification of the major urinary metabolite of the F2-isoprostane 8-*iso*-prostaglandin F2alpha in humans. J Biol Chem 1996;271:20617–20620.

Roberts SA, Price VF, Jollow DJ. Acetaminophen structure-toxicity studies: In vivo covalent binding of a nonhepatotoxic analog, 3-hydroxyacetanilide. Toxicol Appl Pharmacol 1990;105:195–208.

Rock CL, Lampe JW, Patterson RE. Nutrition, genetics, and risks of cancer. Annu Rev Public Health 2000;21:47–64.

Rokach J, Khanapure SP, Hwang SW, Adiyaman M, Lawson JA, FitzGerald GA. Nomenclature of isoprostanes: A proposal. Prostaglandins 1997;54:853–873.

Rouzer CA, Chaudhary AK, Nokubo M, Ferguson DM, Reddy GR, Blair IA, Marnett LJ. Analysis of the malondialdehyde-2'-deoxyguanosine adduct pyrimidopurinone in human leukocyte DNA by gas chromatography/electron capture/negative chemical ionization/mass spectrometry. Chem Res Toxicol 1997;10:181–188.

Sacerdoti D, Balazy M, Angeli P, Gatta A, McGiff JC. Eicosanoid excretion in hepatic cirrhosis. Predominance of 20-HETE. J Clin Invest 1997;100:1264–1270.

Sasaki DM, Yuan Y, Gikas K, Kanai K, Taber D, Morrow JD, Roberts LJ, Callewaert DM. Enzyme immunoassays for 15-F2T isoprostane-M, an urinary biomarker for oxidant stress. Adv Exp Med Biol 2002;507:537–541.

Saxon E, Armstrong JI, Bertozzi CR. A "traceless" Staudinger ligation for the chemoselective synthesis of amide bonds. Org Lett 2000;2:2141–2143.

Schneider C, Boeglin WE, Prusakiewicz JJ, Rowlinson SW, Marnett LJ, Samel N, Brash AR. Control of prostaglandin stereochemistry at the 15-carbon by cyclooxygenases-1 and -2. A critical role for serine 530 and valine 349. J Biol Chem 2002;277:478–485.

Schroder M, Kaufman RJ. ER stress and the unfolded protein response. Mutat Res 2005;569:29–63.

Shah SJ, Yu KH, Sangar V, Parry SI, Blair IA. Identification and quantification of preterm birth biomarkers in human cervicovaginal fluid by liquid chromatography/tandem mass spectrometry. J Proteome Res 2009;8:2407–2417.

Shao B, Bergt C, Fu X, Green P, Voss JC, Oda MN, Oram JF, Heinecke JW. Tyrosine 192 in apolipoprotein A-I is the major site of nitration and chlorination by myeloperoxidase, but only chlorination markedly impairs ABCA1-dependent cholesterol transport. J Biol Chem 2005;280:5983–5993.

Sharma RA, Farmer PB. Biological relevance of adduct detection to the chemoprevention of cancer. Clin Cancer Res 2004;10:4901–4912.

Shelton MD, Chock PB, Mieyal JJ. Glutaredoxin: Role in reversible protein s-glutathionylation and regulation of redox signal transduction and protein translocation. Antioxid Redox Signal 2005;7:348–366.

Shin NY, Liu Q, Stamer SL, Liebler DC. Protein targets of reactive electrophiles in human liver microsomes. Chem Res Toxicol 2007;20:859–867.

Singh G, Gutierrez A, Xu K, Blair IA. Liquid chromatography/electron capture atmospheric pressure chemical ionization/mass spectrometry: Analysis of pentafluorobenzyl derivatives of biomolecules and drugs in the attomole range. Anal Chem 2000;72:3007–3013.

Sircar D, Subbaiah PV. Isoprostane measurement in plasma and urine by liquid chromatography–mass spectrometry with one-step sample preparation. Clin Chem 2007;53:251–258.

Soberman RJ, Christmas P. The organization and consequences of eicosanoid signaling. J Clin Invest 2003;111:1107–1113.

Stafforini DM, Sheller JR, Blackwell TS, Sapirstein A, Yull FE, McIntyre TM, Bonventre JV, Prescott SM, Roberts LJ. Release of free F2-isoprostanes from esterified phospholipids is catalyzed by intracellular and plasma platelet-activating factor acetylhydrolases. J Biol Chem 2006;281:4616–4623.

Streeter AJ, Bjorge SM, Axworthy DB, Nelson SD, Baillie TA. The microsomal metabolism and site of covalent binding to protein of 3′-hydroxyacetanilide, a nonhepatotoxic positional isomer of acetaminophen. Drug Metab Dispos 1984;12:565–576.

Suzuki H, Sugiyama Y. Excretion of GSSG and glutathione conjugates mediated by MRP1 and cMOAT/MRP2. Semin Liver Dis 1998;18:359–376.

Suzuki N, Hishinuma T, Saga T, Sato J, Toyota T, Goto J, Mizugaki M. Determination of urinary 12(*S*)-hydroxyeicosatetraenoic acid by liquid chromatography–tandem mass spectrometry with column-switching technique: Sex difference in healthy volunteers and patients with diabetes mellitus. J. Chromatogr B Analyt Technol Biomed Life Sci 2003;783:383–389.

Swenberg JA, Bogdanffy MS, Ham A, Holt S, Kim A, Morinello EJ, Ranasinghe A, Scheller N, Upton PB. Formation and repair of DNA adducts in vinyl chloride- and vinyl fluoride-induced carcinogenesis. IARC Sci Publ 1999;150:29–43.

Swenberg JA, Koc H, Upton PB, Georguieva N, Ranasinghe A, Walker VE, Henderson R. Using DNA and hemoglobin adducts to improve the risk assessment of butadiene. Chem Biol Interact 2001;135–136:387–403.

Szapacs ME, Riggins JN, Zimmerman LJ, Liebler DC. Covalent adduction of human serum albumin by 4-hydroxy-2-nonenal: Kinetic analysis of competing alkylation reactions. Biochemistry 2006;45:10521–10528.

Taber DF, Morrow JD, Roberts LJ. A nomenclature system for the isoprostanes. Prostaglandins 1997;53:63–67.

Tanabe T, Tohnai N. Cyclooxygenase isozymes and their gene structures and expression. Prostaglandins Other Lipid Mediat 2002;68-69:95–114.

Tselepis AD, John CM. Inflammation, bioactive lipids and atherosclerosis: Potential roles of a lipoprotein-associated phospholipase A2, platelet activating factor-acetylhydrolase. Atheroscler Suppl 2002;3:57–68.

Tsikas D, Schwedhelm E, Suchy MT, Niemann J, Gutzki FM, Erpenbeck VJ, Hohlfeld JM, Surdacki A, Frolich JC. Divergence in urinary 8-*iso*-PGF(2alpha) (iPF(2alpha)-III, 15-F(2t)-IsoP) levels from gas chromatography-tandem mass spectrometry quantification after thin-layer chromatography and immunoaffinity column chromatography reveals heterogeneity of 8-*iso*-PGF(2alpha). Possible methodological, mechanistic and clinical implications. J Chromatogr B 2003;794:237–255.

Turesky RJ. The role of genetic polymorphisms in metabolism of carcinogenic heterocyclic aromatic amines. Curr Drug Metab. 2004;5:169–180.

Uchida K. 4-Hydroxy-2-nonenal: A product and mediator of oxidative stress. Prog Lipid Res 2003;42:318–343.

Wang W, Ballatori N. Endogenous glutathione conjugates: Occurrence and biological functions. Pharmacol Rev 1998;50:335–356.

Ward NC, Tsai IJ, Barden A, van Bockxmeer FM, Puddey IB, Hodgson JM, Croft KD. A single nucleotide polymorphism in the CYP4F2 but not CYP4A11 gene is associated with increased 20-HETE excretion and blood pressure. Hypertension 2008;51:1393–1398.

Watzer B, Reinalter S, Seyberth HW, Schweer H. Determination of free and glucuronide conjugated 20-hydroxyarachidonic acid (20-HETE) in urine by gas chromatography/negative ion chemical ionization mass spectrometry. Prostaglandins Leukot. Essent. Fatty Acids 2000;62:175–181.

Williams CS, Luongo C, Radhika A, Zhang T, Lamps LW, Nanney LB, Beauchamp RD, Dubois RN. Elevated cyclooxygenase-2 levels in Min mouse adenomas. Gastroenterology 1996;111:1134–1140.

Williams KE, Carver TA, Miranda JJ, Kautiainen A, Vogel JS, Dingley K, Baldwin MA, Turteltaub KW, Burlingame AL. Attomole detection of in vivo protein targets of benzene in mice: Evidence for a highly reactive metabolite. Mol Cell Proteomics 2002;1:885–895.

Williams MV, Lee SH, Blair IA. Liquid chromatography/mass spectrometry analysis of bifunctional electrophiles and DNA adducts from vitamin C mediated decomposition of 15-hydroperoxyeicosatetraenoic acid. Rapid Commun Mass Spectrom 2005;19:849–858.

Williams MV, Lee SH, Pollack M, Blair IA. Endogenous lipid hydroperoxide-mediated DNA-adduct formation in Min mice. J Biol Chem 2006;281:10127–10133.

Winnik WM, Kitchin KT. Measurement of oxidative stress parameters using liquid chromatography–tandem mass spectroscopy (LC-MS/MS). Toxicol Appl Pharmacol 2008;233:100–106.

Wong HL, Liebler DC. Mitochondrial protein targets of thiol-reactive electrophiles. Chem Res Toxicol. 2008;21:796–804.

Wu G, Fang YZ, Yang S, Lupton JR, Turner ND. Glutathione metabolism and its implications for health. J Nutr 2004;134:489–492.

Wu, K.K. Control of cyclooxygenase-2 transcriptional activation by pro-inflammatory mediators. Prostaglandins Leukot Essent Fatty Acids 2005;72:89–93.

Yan W, Byrd GD, Ogden MW. Quantitation of isoprostane isomers in human urine from smokers and nonsmokers by LC-MS/MS. J Lipid Res 2007;48: 1607–1617.

Yan Y, Weaver VM, Blair IA. Analysis of protein expression during oxidative stress in breast epithelial cells using a stable isotope labeled proteome internal standard. J. Proteome Res 2005;4:2007–2014.

Yocum AK, Yu K, Oe T, Blair IA. Effect of immunoaffinity depletion of human serum during proteomic investigations. J Proteome Res 2005;4:1722–1731.

Zastawny TH, Doetsch PW, Dizdaroglu M. A novel activity of *E. coli* uracil DNA *N*-glycosylase excision of isodialuric acid (5,6-dihydroxyuracil), a major product of oxidative DNA damage, from DNA. FEBS Lett 1995;364:255–258.

Zhang S, Villalta PW, Wang M, Hecht SS. Detection and quantitation of acrolein-derived 1,N2-propanodeoxyguanosine adducts in human lung by liquid chromatography–electrospray ionization–tandem mass spectrometry. Chem Res Toxicol 2007;20:565–571.

Zhou L, Nilsson A. Sources of eicosanoid precursor fatty acid pools in tissues. J Lipid Res 2001;42:1521–1542.

Zhou X, Taghizadeh K, Dedon PC. Chemical and biological evidence for base propenals as the major source of the endogenous M1dG adduct in cellular DNA. J Biol Chem 2005a;280:25377–25382.

Zhou X, Taghizadeh K, Dedon PC. Chemical and biological evidence for base propenals as the major source of the endogenous M1dG adduct in cellular DNA. J Biol Chem 2005b;280:25377–25382.

Zhu P, Oe T, Blair IA. Determination of cellular redox status by stable isotope dilution liquid chromatography/mass spectrometry analysis of glutathione and glutathione disulfide. Rapid Commun. Mass Spectrom 2008;22:432–440.

21 LC–MS in Endogenous Metabolite Profiling and Small-Molecule Biomarker Discovery

MICHAEL D. REILY, PETIA SHIPKOVA, and SERHIY HNATYSHYN

Bioanalytical and Discovery Analytical Sciences, Applied and Investigational Metabonomics, Bristol-Myers Squibb, Princeton, New Jersey

Mass Spectrometry in Drug Metabolism and Disposition: Basic Principles and Applications, First Edition. Edited by Mike S. Lee and Mingshe Zhu.

21.1 INTRODUCTION

A full appreciation of the complexity of life and how living things interact with their environment requires a fundamental understanding of many biological processes. At a molecular level, these biological processes involve knowledge of the interrelations of a wide variety of biochemical components. Entirely within an organism, these processes include a complicated array of genetic deoxyribonucleic acid (DNA) transcripts, proteins, and endogenous metabolites that constantly change in response to the requirements of survival and perpetuation. Of course, living organisms also have to interact with their environment. These can range from seemingly simple interactions, such as chemotaxis in bacteria, to highly complex interactions, as in the symbiotic relationships that exist in higher plants and animals. Thus, processes in living organisms can only be thoroughly understood when the exact makeup of these various molecular compartments and how they change can be assessed in a quantitative way. Unlike much of this volume, this chapter intends to focus on the use of liquid chromatography–mass spectrometry (LC–MS) to measure endogenous metabolites. However, the term endogenous here must be kept in context. For example, many of the small molecules that course through our veins and tissues come from the environment, and, thus, changes in these molecules are inextricably connected with our lifestyles and behaviors. The metabolome, a term first defined as the "total compliment of metabolites in a cell" (Tweeddale et al., 1998), was subsequently defined more restrictively as "only of those native small molecules (definable, non-polymeric compounds) that are participants in general metabolic reactions and that are required for the maintenance, growth and normal function of a cell" and explicitly excludes xenobiotics and xenobiotic metabolites since these are not required for the normal function of the cell (Beecher, 2003). More recently, the metabolome has been suggested to include all small molecules introduced and modified by diet, medication, environmental exposure, and coexisting organisms (Dunn, 2008). From an analytical perspective, this broader definition of the metabolome is the most common interpretation, and the relevance of any particular component of the measured metabolome must be ascertained in the context of the study at hand. Regardless of the exact definition, as the substrates and products of biochemical and enzymatic reactions, the metabolome reflects the overall global biochemical state of an organism, and changes in the functional genome, transcriptome, and proteome can be inferred based on changes in the metabolome. A comprehensive measurement of changes in the metabolome in response to external stressors can lead to the deduction of the relationship between the perturbation and effected biochemical pathways and, ultimately, to the discovery of biomarkers that report upon the perturbation.

21.2 MEASURING THE METABOLOME

The study of the time-dependent changes in the metabolome as a function of physiological stimulation or genetic modification is known as metabonomics

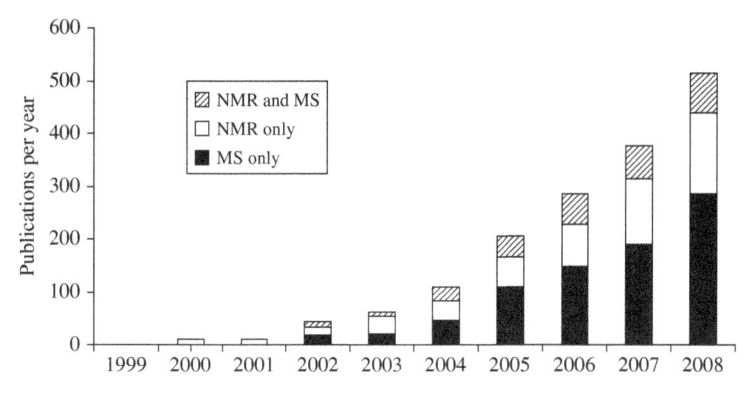

FIGURE 21.1 Metabonomics publications over the last 2 decades broken down by primary analytical discipline.

(Nicholson et al., 1999). A related term that was more recently defined but has gained substantially in popularity, metabolomics (Fiehn et al., 2000; Fiehn, 2001), is the measurement of the entire metabolome under a given set of conditions. Although subtle differences in these definitions can be (and have been) argued, for the purposes of discussion in this chapter the two terms are synonymous. The published systematic practice of the correlation of analytical observables in biofluids to human health dates back to Hippocrates (Kouba et al., 2007) and the first truly modern vestiges of metabonomics date back nearly 40 years. Hippocrates had the instruments of his own senses—sight, smell, and taste, replaced in the modern era with the use of pattern recognition for the analysis of complicated gas chromatography (GC) (Pauling et al., 1971) and gas chromatography mass spectrometry (GC-MS) (Jellum et al., 1971) data in the context of human disease. Since these early reports, our understanding of biology, analytical instrumentation, and methods to analyze analytical data have improved dramatically. The recent exponential growth in the number of literature reports on metabonomics applications speaks to the current interest level in this technology. If these applications are organized by analytical discipline, mass spectrometry clearly emerges as the most popular analytical platform (Fig 21.1).

21.3 ANALYTICAL APPROACHES

There are three main approaches within the field of metabonomics: fingerprinting, nontargeted metabonomics, and targeted metabonomics. These approaches differ philosophically in their ease of execution and information content, but each has practical utility and unique analytical work flows. However, regardless of the approach, any successful metabonomics study requires a high-quality data set that can be translated into a biochemical snapshot that reflects the temporal state of an organism's metabolome.

This chapter will focus on variations of LC–MS applications suitable for each of these approaches, along with mention of complimentary techniques as appropriate. Mass spectrometry, through isotopic distribution, fragmentation, and adduct formation, produces many ions for each metabolite. Thus, a typical LC–MS data set may contain tens of thousands of peaks (not including noise) that represent only hundreds of compounds. An important theme in the application of mass spectrometry to metabonomics is to properly convolute these individual ions into molecular species, and they will be discussed at length later in this chapter. One big advantage of LC–MS over nuclear magnetic resonance (NMR) or even GC–MS approaches is flexibility. The combination of stationary phases, mobile phases, ionization modes, and detection schemes is almost limitless. The downside of this flexibility is that there are few standardized approaches. For example, there are no extensive libraries of reference spectra or established deconvolution methods as there are for GC–MS (Tautenhahn et al., 2007). While NMR has unique advantages such as ease of sample preparation and data collection and linear response with high dynamic range, LC–MS provides several orders of magnitude higher sensitivity, which is essential for a comprehensive metabolome characterization. While GC–MS often rivals the sensitivity of LC–MS, this technique requires extensive sample preparation and data processing is far from trivial. Finally, accurate mass capabilities and MS^n-enhanced structural elucidation, explain the continued success of LC–MS-based metabonomics.

21.3.1 Fingerprinting Methods

The fingerprinting approach involves generating rich analytical data sets on complex biological samples, typically minimally prepared peripheral biofluids or cell media, which allow differentiation of subjects or samples based on spectroscopic, chromatographic, or spectrometric data. These analytical fingerprints can contain thousands of individual data points that can be easily reduced to numerical representations of the concentration components that comprise the sample. These "analytical fingerprints" reflect the composition of the matrix under investigation and, thus, can be treated as large multivariate data sets. With the use of widely available statistical tools such as principal component analysis (PCA), these data sets can be used to identify samples that fall outside of control or normal samples and provide direct information on individuals that should be scrutinized further. Classification models can be constructed as surrogates for the biological endpoints (Fig 21.2) with the use of more sophisticated statistical analysis and the incorporation of known biological endpoints such as cell density, histopathology, or clinical chemistry parameters. The advantage of this approach is that annotation of the analytical data is not required and, therefore, is only rate limited by the analytical technique. This type of analysis is useful for establishing differentiation between individuals or groups and has applications in approaches such as in vivo screening of similar drug candidates (Robosky et al., 2002; Dieterle et al.,

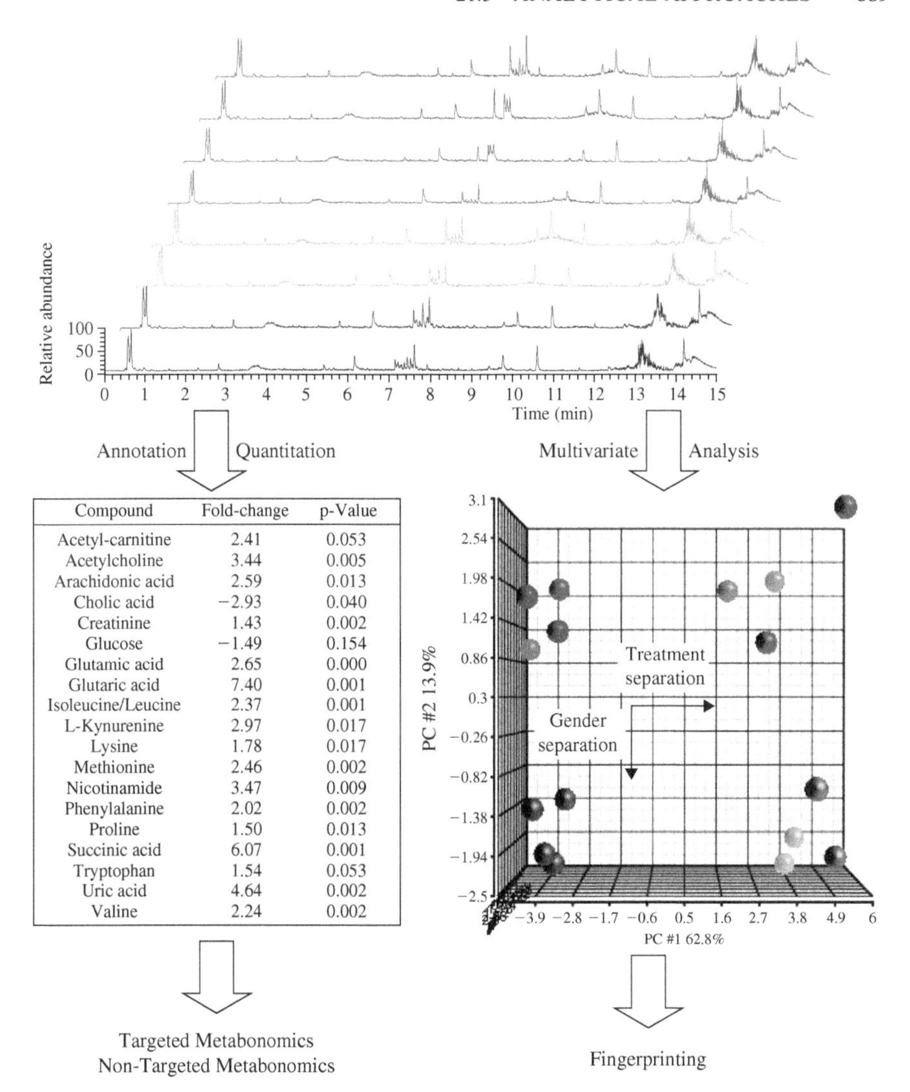

Compound	Fold-change	p-Value
Acetyl-carnitine	2.41	0.053
Acetylcholine	3.44	0.005
Arachidonic acid	2.59	0.013
Cholic acid	−2.93	0.040
Creatinine	1.43	0.002
Glucose	−1.49	0.154
Glutamic acid	2.65	0.000
Glutaric acid	7.40	0.001
Isoleucine/Leucine	2.37	0.001
L-Kynurenine	2.97	0.017
Lysine	1.78	0.017
Methionine	2.46	0.002
Nicotinamide	3.47	0.009
Phenylalanine	2.02	0.002
Proline	1.50	0.013
Succinic acid	6.07	0.001
Tryptophan	1.54	0.053
Uric acid	4.64	0.002
Valine	2.24	0.002

FIGURE 21.2 Metabonomics work flow. PCA plot of serum data obtained from fed and fasted male and female rats (n = 4). The four treatment groups are clearly separated into clusters; female fed (upper left), male fed (lower left), female fasted (upper right) and male fasted (lower right).

2006; Robertson et al., 2007), prescreening animals prior to extensive toxicological evaluations, and quality control of cell media.

The utility of an analytical method for fingerprinting is related to both its level of detail and reproducibility. The latter should be manifested not only within a given sample set but preferably over long time periods and on different platforms and allows for the comparison of data sets acquired in different laboratories months or years apart. Inasmuch as reproducibility is affected by analytical and sample variability, fingerprinting techniques should minimize and simplify sample handling procedures (Robertson et al., 2005). Examples of analytical fingerprinting methods that have been reported include NMR spectroscopy (Robosky et al., 2002; Shockcor and Holmes, 2002; Coen et al., 2008), infrared (IR) and Raman spectroscopy (Ellis and Goodacre, 2006; Elliott et al., 2007; Brown et al., 2009; Pistorius et al., 2009), capillary electrophoresis–MS (CE–MS) (Allard et al., 2008; Garcia-Perez et al., 2008), atmospheric pressure desorption–MS (Chen et al., 2007a), direct injection (DI–MS) (Brown et al., 2005; Hansen and Smedsgaard, 2007; Beckmann et al., 2008; Madalinski et al., 2008), LC–MS (Wilson et al., 2005; Gika et al., 2008c; Plumb et al., 2008; Michopoulos et al., 2009), and GC–MS (Jonsson et al., 2005; Vallejo et al., 2009). In general, hyphenated techniques such as LC–MS and GC–MS provide highly detailed analytical data, but introduce sources of variability at the chromatographic level, such as column deterioration, temperature, and matrix effects. Further sources of variability that can affect these and DI–MS techniques include ion suppression within the MS instrument itself. As a result, special care must be taken to account for and minimize analytical variability in order for these methods to be considered as robust fingerprinting tools (Zelena et al., 2009). In contrast, remarkable reproducibility for spectroscopic methods such as NMR (Keun et al., 2002) and IR (Chen et al., 2007b; Pistorius et al., 2009), which can generate data on minimally manipulated samples, make them superior fingerprinting approaches.

21.3.2 Nontargeted Metabonomics

As with analytical fingerprinting, nontargeted metabonomics makes use of ultra-rich analytical data sets but goes a step beyond and seeks to assign as many individual chromatographic or spectroscopic peaks as possible. This approach produces a comprehensive list of metabolites with absolute or relative quantification and has been referred to as metabolomics (Fiehn, 2001) or metabolic profiling (Brown et al., 2005). The changes in metabolome components can be mapped to specific pathways and provide biomarkers and/or mechanistic information about a process under study. Since much of the data generated with this approach is unassigned, it is important that methods used for this approach are amenable to subsequent identification of unknown components. These requirements essentially limit the nontargeted metabonomic analytical techniques to NMR spectroscopy and MS-based approaches. Of these, NMR is highly quantitative and very reproducible both longitudinally and across laboratories (Keun et al., 2002), although it suffers from poor sensitivity and, thus, has been largely limited to proton NMR (Reily and Lindon, 2005). As has been noted, MS has staked a claim as a preeminent analytical technique for metabonomics applications. Unlike NMR, which is primarily used for the analysis of whole biofluids, MS is normally used for highly sensitive and selective analyses in conjunction with separation techniques. The most widely used hyphenated MS systems employ GC, LC, or CE as a front-end separation technique. General MS methodologies for metabonomics have been recently reviewed (Villas-Boas et al., 2005; Dunn, 2008;

Lokhov and Archakov, 2009) as have the specific hyphenated techniques of LC–MS (Lu et al., 2008b; Vogeser and Seger, 2008), GC–MS (Fiehn 2008; Pasikanti et al. 2008a, 2008b), and CE–MS (Monton and Soga, 2007; Garcia-Perez et al., 2008). For nontargeted profiling approaches, a high-resolution (>10,000) accurate mass (<5 ppm) full scan is the preferred detection approach. Orbitrap and Fourier transform ion cyclotron resonance FTICR mass spectrometers offer excellent measurement accuracy (<2 ppm) and a wide resolution range, however, high resolution is obtained at the expense of the duration of duty cycle. Therefore, the number of scans may be insufficient for a typical ultra-high-performance liquid chromatography (UHPLC) peak, which may be only 1–2 s wide (discussed in detail later in the chapter).

Because the primary objective of nontargeted metabonomics is essentially to create a list of molecular concentrations, the ability to assign and measure as many analytes as possible is absolutely essential in order to provide meaningful data. It is well recognized that in a typical metabonomics study, detected peaks that correspond to real molecules and correlate with biological endpoints, often remain unannotated. In such cases, a common practice is to label these unknown peaks and attempt to identify them. This "identification of unknowns" is an important bottleneck in this field since it is labor intensive and often requires expertise from multiple analytical disciplines. However, the potential payoff for success is novel insight into biochemical mechanism, or at the very minimum expansion of the number of annotatable metabolites. The ultra-high sensitivity and powerful molecular identification capabilities of MS is particularly well suited for this application.

Finally, it should be pointed out that nontargeted metabonomics always provides an analytical fingerprint, but the inverse is only true if the analytical data are readily annotated with unambiguous molecular identifications. For example, vibrational spectroscopies such as IR can provide a highly reproducible and detailed spectrum that reflects the molecular composition of a complex mixture, but assignment of IR bands to individual molecular species can generally not be made. Thus, IR is a good metabolomic fingerprinting method but cannot generally be used for nontargeted metabonomics.

21.3.3 Targeted Metabonomics

This approach differs from those discussed above in that measured analytes have been selected a priori, usually to address certain specific biological questions within a study. Furthermore, since the exact analytes are known, it is possible and desirable to measure their absolute concentrations with appropriate use of internal standards and calibration curves. The lines between targeted metabolomics and traditional LC–MS-based assay development are arguably ill defined. Certainly, no one would equate an MS assay developed to measure alanine with targeted metabonomics. However, a multiplex assay to measure 13 or 14 organic acids in a urine sample probably qualifies. Often, targeted approaches can be developed for specific compound classes such as lipids, bile acids, or amino acids that are pertinent to a particular study. For example, in our laboratory, we routinely measure 13 bile acids using a targeted metabonomics approach. Sample preparation and chromatographic methods are easier to develop with this approach and produce good separation of the desired analytes, which, when combined with the use of appropriate standards, can facilitate absolute quantitation. The approach is not limited to molecules with similar physicochemical properties, however; targeted metabolite assays often result from the desire to more accurately quantify specific sets of molecules observed in an untargeted study. The drawback to a targeted metabonomics approach is, of course, that serendipitous observations are not enabled and so unanticipated changes and the discovery of novel biomarkers or mechanisms may not be achieved.

Many MS-based techniques lend themselves to a targeted approach. Multiple reaction monitoring (MRM) using tandem quadrupole mass spectrometry (TQMS) allows highly selective monitoring of precursor ions and product ions of each metabolite and enables high sensitivity over a broad dynamic range without the need for isotope labeled-standards. This approach was used to accurately quantify 100 metabolites in each sample of plant extracts (Sawada et al., 2009). The MRM approach is preferred when a limited number of analytes (<50) are to be measured, where as a full-scan, high resolving power system (e.g., time-of-flight, FTICR, or Orbitrap) has advantages with larger target analyte lists (Lu et al., 2008a). Methods based on DI–MS have been developed to work specifically with kits that contain many isotopically labeled standards to account for potential significant ion suppression associated with the introduction of complex mixtures directly into the MS source (Altmaier et, al. 2008). For creating MS/MS fragmentation libraries, it has been shown that ion traps can generate universal MS/MS spectra, which can be reproduced across different laboratories and different instruments and, therefore, are amendable to library generation analogous to electron impact (EI) fragment libraries used for GC–MS. (Josephs and Sanders, 2004)

21.4 EXPERIMENTAL DESIGN

Designing a successful metabonomics experiment is truly a multidisciplinary exercise, and one must consider details of the in-life portion, sample collection, sample preparation, analytical data generation, data processing, analysis, and interpretation. The in-life portion involves study subjects selection, acclimatization, dosing or treatment, and sample collection and has been discussed in detail elsewhere (Robertson et al., 2002). The following sections will focus on the remaining steps in this process as it pertains to LC–MS.

21.4.1 Sample Selection

Almost any biological fluid or tissue can be used for LC–MS-based metabonomic investigations. In the plant world, extracts of pulverized plant tissue are normally used (Sumner et al., 2003; Hagel and Facchini, 2008) and for cellular metabonomics, culture media or cell extracts are used (Teng et al., 2008). In animals, many LC–MS studies have been reported on blood plasma or serum and urine. Additionally, more exotic fluids have been reported as well, such as breath condensate (Carraro et al., 2008; Effros 2008), milk (Boudonck et al., 2009; German, 2009), saliva (Yan et al., 2008), and cerebrospinal fluid (Wishart et al., 2008). While sample selection will largely depend upon the investigation at hand, the vast majority of studies in animals will use peripheral fluids that are easily obtained and available in sufficient quantity, (e.g., blood plasma or serum and urine). These are often the only samples available in the clinic and so from a pharmaceutical perspective, identification of biomarkers in these matrices is attractive from a translational (animal-to-human) standpoint. While urine is a popular metabonomics biofluid, the very high inorganic salt content can be problematic with respect to ionization suppression and could lead to lower sensitivity and lack of reproducibility for small polar compounds that are often the target analytes in metabonomics. In addition, urine concentration can be highly variable and can often be contaminated with fur, food, and feces in animal studies. On the other hand, as a waste fluid, the composition of urine can vary widely with pathophysiological condition and, therefore, can be a sensitive measure of adverse events in toxicology studies. Additionally, urine samples can be collected serially and essentially noninvasively, enabling individuals to serve as their own controls and, therefore, reducing both animal and drug requirements (Robertson et al., 2002). Blood fractions are the standard for comparison with clinical laboratory and pharmacokinetic analyses. As such, plasma or serum collections are almost always included in clinical and preclinical protocols. This is an important point since metabonomics is frequently added as an adjunct to a larger study with traditional endpoints (e.g., clinical chemistry or histopathology) and its inclusion is facilitated by not requiring major departures from standard laboratory procedures. Furthermore, as blood composition is tightly (homeostatically) controlled, there are many endogenous components that will be constant from sample to sample and small changes in composition are likely to be significant. However, there are limitations (particularly for small animals) on how often blood can be drawn, and so the advantages of serial collection may be limited compared with urine.

21.4.2 Sample Preparation

Fingerprinting and nontargeted metabonomics benefit from an unbiased view of all compound classes. In these approaches, the endogenous metabolites that might be of interest are not known a priori, and, therefore, sample preparation techniques that preserve as much of the metabolome as possible are essential. There are a number of reports in the literature that describe tissue profiling, but those techniques are generally more specific for particular tissues (i.e., brain, liver, muscle, etc.) and/or expected metabolites of interest. In general, for fatty tissues such as adipose, brain, and even liver, liquid–liquid extraction (LLE) is the method of choice for the separation of polar water-soluble components from the lipophilic components (Kohler et al., 2009). A recent review (Villas-Boas et al., 2005) highlights the advantages and disadvantages of many different sample quenching approaches for cells, fungi, plants, and animal tissues. The following discussion will focus on sample preparation for two most commonly used matrices for biomedical metabonomic studies, urine, and blood plasma/serum.

21.4.2.1 Sample Preparation for Urine Urine is directly compatible with reverse-phase and hydrophilic interaction liquid chromatography (HILIC) LC–MS analyses, and, therefore, minimal sample preparation is required. The easiest approach involves dilution followed by centrifugation for removal of precipitates. There are literature reports that cite protein precipitation or solid-phase extraction (SPE) approaches [Villas-Boas et al., (2005) and references within], but we have found that protein precipitation does not add a significant benefit since the protein levels in urine are relatively low. Even under optimized conditions for small polar molecules, SPE suffers from a significant breakthrough in the loading step, which results in the loss of a large number of endogenous compounds (Shipkova et al., 2007). Table 21.1 contains the recovery amounts for succinic, glutaric, and pimelic acids using a C18 SPE approach. Fractions were collected during the loading, washing, and elution steps of the SPE process. Acidification of the urine and elution buffers aided retention of the higher molecular weight organic acids but a significant amount of the sample was still lost in the loading and washing steps. This approach would still be valid if sample fractionation is required, for example, where removal of components such as nonpolar xenobiotic or endogenous molecules, peptides, or proteins is necessary. The loading and the washing portions would retain all of the small, very polar molecules, while different eluants with

TABLE 21.1 Percent Recovery of Three Common Endogenous Dicarboxylic Acids from a C18 Solid-Phase Extraction Column, Calculated Based on Total Area of Extracted Ion Chromatogram in Each of Three Collected Fractions (Load, Wash and Elute)[a]

Dilution	SPE Steps	Succinic Acid	Glutaric Acid	Pimelic Acid
Water	Load	49	82	80
	Wash	11	18	20
	Elute	40	0	0
Aqueous 0.1% FA	Load	14	1	0
	Wash	56	48	0
	Elute	29	51	100

[a]Samples were diluted 10-fold in either water or 0.1% formic acid (FA).

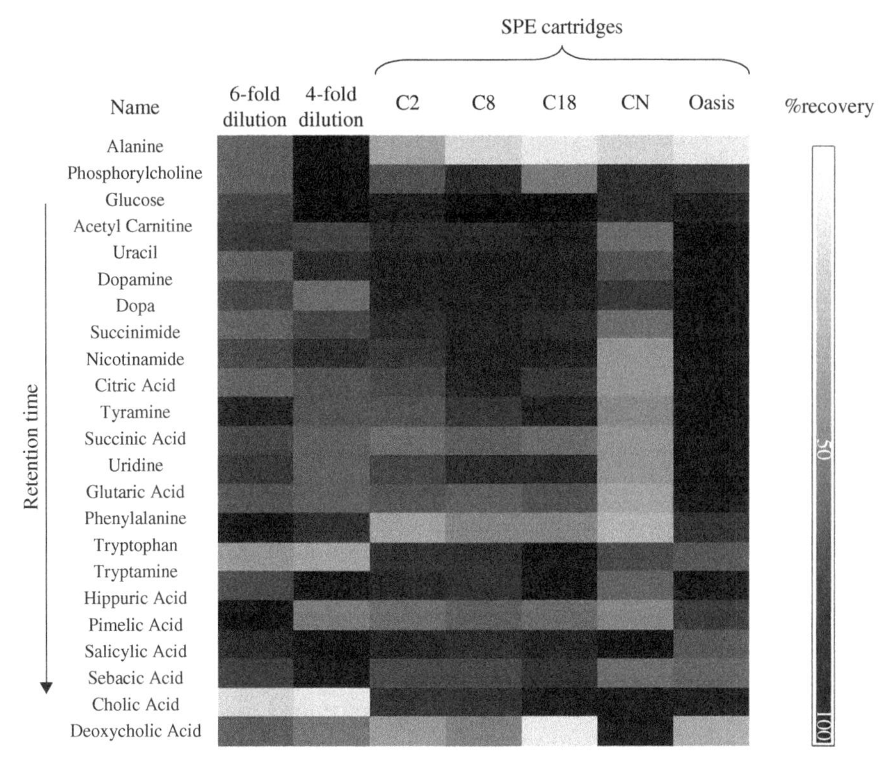

FIGURE 21.3 Recovery of stable-labeled standards spiked in urine. Color indicates recovery, according to the scale on the right. Despite the better recovery of many metabolites with some solid-phase extraction (SPE) cartridges, expediency often dictates a "dilute-and-shoot" approach (see text). OASIS SPE cartridges were obtained from Waters Corp. and used per manufacturer instructions.

increasing organic content would contain the less polar components. To illustrate the overall comparison between extraction efficiency for various SPE cartridges *versus* dilution for urine samples, a study was conducted where 20 stable-label standards were spiked into urine and their recovery was calculated. A heat map of the observed recoveries for 6-fold, 4-fold dilution, and five different SPE cartridges is shown in Figure 21.3. The best recovery of all analytes was attained with direct dilution, and our preferred strategy features a 4-fold direct dilution with water. A complicating factor for urine LC–MS analysis is the need for concentration normalization. As a waste fluid, urine concentrations can vary significantly from sample-to-sample due to normal physiology (water is almost always provided *ad libitum* in clinical and nonclinical studies) as well as pathophysiological effects. This complication, of course, is true for any analytical or clinical chemistry approach using urine as a matrix and is commonly dealt with by normalizing to creatinine. Creatinine is used primarily because under normal physiological conditions its total excretion is expected to be constant for a given muscle mass. In reality, creatinine production can also be highly variable (Miller et al., 2004), and so care must be taken to ensure that the constant creatinine production assumption is valid for a given study. A 5-fold or higher difference in 24-h urine volumes from rats or mice within a study is not unusual (Warrack et al., 2009). In principle, when 24-h collections are made, normalizing to total urine volume should be a valid approach. However, that has not been the experience for reasons that are not completely understood, but which may have to do with sample collection artifacts (Warrack et al., 2009). Like many clinical chemistry assays, it is possible to normalize to creatinine after data analysis (Warrack et al., 2009), either measured independently or in the metabonomics experiment itself. Another approach, independent of the metabonomic measurements themselves involves the normalization to osmolality (Warrack et al., 2009, Shipkova et al., 2010). Osmolality, measured by freezing-point depression, reflects total osmolite concentration; inasmuch as most of the components in urine are osmolites, it follows that this would be a reasonable external concentration calibrant. Osmolality normalization can be performed after data collection or during sample preparation. In cases where large differences in urine concentration are expected, each sample can be diluted to a constant osmolality prior to data-collection. With such concentration normalization significant differences in ionization and matrix suppression from samples with wildly different concentrations would be avoided. Finally, a post-data-acquisition normalization to total useful signal has also been previously reported (Warrack et al., 2009) as is currently our preferred approach.

21.4.2.2 Sample Preparation—Plasma/Serum For LC–MS analyses, both plasma and serum are suitable matrices and can be processed identically. For plasma, the choice of anticoagulant is a consideration. Citrate is problematic as it interferes with the assessment of the endogenous levels of citric acid. High levels of citrate are also likely to cause matrix suppression, whereas ethylenediamine tetraacetic acid (EDTA) and heparin are well tolerated. Prior to establishing a sample preparation protocol for plasma and serum, various protein precipitation, as well as SPE, approaches were investigated (Shipkova et al., 2007). The "dilute-and-shoot" approach used for urine is not feasible for plasma and serum due to their very high protein content. Similar to our investigation of urine sample preparation, 20 stable isotope-labeled standards were spiked into rat plasma and the overall recovery was calculated. The results are outlined in Figure 21.4 and indicate that the two best options were either C8 SPE or protein precipitation with cold (−20°C) methanol, 0.1% formic acid. The recovery of all detected components in plasma were also investigated (Fig. 21.5). Since these results are comparable and because it is far simpler and cheaper, our current protocol features protein precipitation with 2–3 volumes of −20°C methanol containing 0.1% formic acid. A recent publication (Bruce

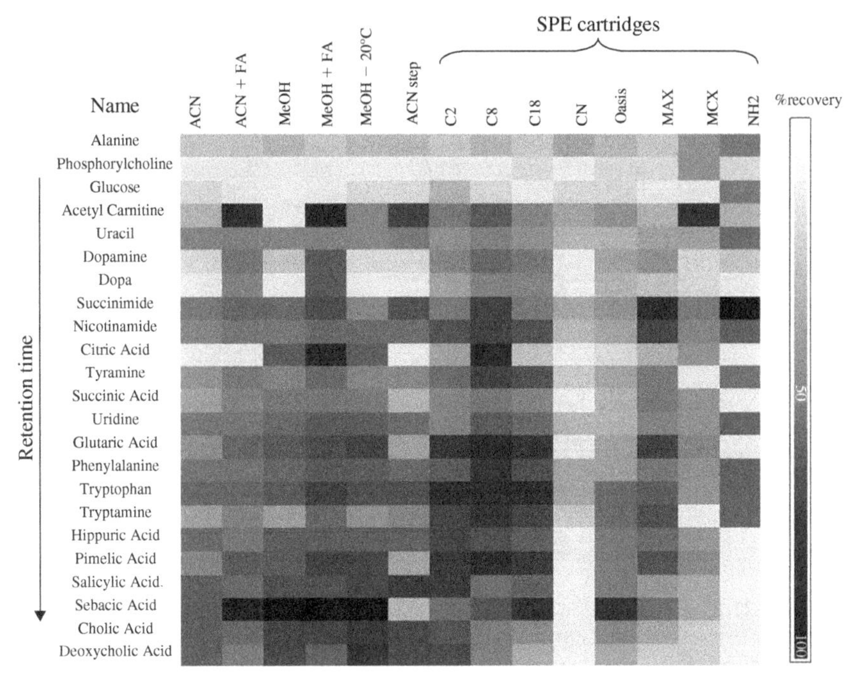

FIGURE 21.4 Recovery of stable-labeled standards spiked in plasma. Color indicates recovery, according to the scale on the right. Abbreviations: ACN = acetonitrile; FA = 0.1% formic acid; MeOH = methanol; SPE= solid-phase extraction. OASIS SPE cartridges were all obtained from Waters Corp. and used per manufacturer instructions.

et al., 2008) investigates in great detail the effect of the methanol or acetonitrile content for protein precipitation approaches.

21.4.3 Chromatography and Mass Spectral Detection

An LC–MS chromatogram is essentially a three-dimensional data set with separate chromatographic, mass-to-charge (m/z) and ion count (intensity) dimensions. The combination of chromatography and mass spectral approaches can take on many incarnations and are dependent on the nature of the metabonomics investigation. Because of the complexity of most metabonomics samples, the final solution is often a compromise between good science and expediency since the ideal analytical conditions for 1000+ metabolome components is an unrealistic goal. In this section, we will discuss each of these elements separately with an emphasis on biomedical metabonomic applications.

21.4.3.1 Chromatography Although the final detection is based on the m/z ratios, optimized chromatographic separation is crucial to obtain high-quality data since it is directly related to peak definition and subsequent data

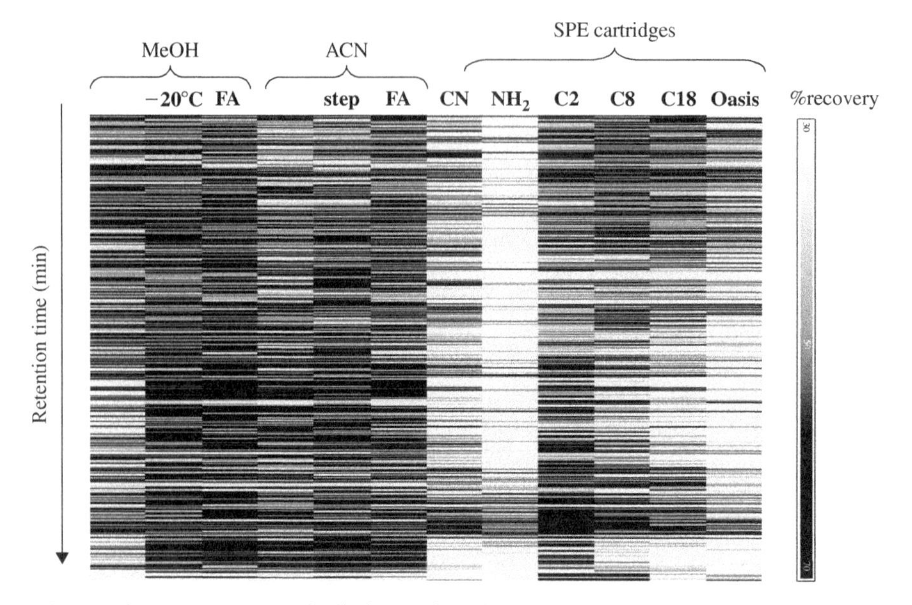

FIGURE 21.5 Recovery of all detected endogenous components in plasma. Color indicates recovery, according to the scale on the right. All recoveries above 75% are shown in red (good recovery) and all recoveries below 25% are shown in white (poor recovery). Abbreviations: ACN = acetonitrile; FA = 0.1% formic acid; MeOH = methanol; SPE= solid-phase extraction; step = stepwise addition of acetonitrile. OASIS SPE cartridges were all obtained from Waters Corp. and used per manufacturer instructions. Best overall recovery was observed using a methanol precipitant with 0.1% formic acid.

processing. The separation of components based on their chromatographic properties provides two major benefits. First and foremost, robust chromatographic separation minimizes matrix effects, such as ion suppression, charge competition, and other possible interactions between coeluting matrix components. Matrix effects cannot be completely eliminated since it is analytically impossible to separate every single matrix component within a reasonable analytical time frame. However, good chromatographic separation can dramatically decrease signal variability that results from these types of artifacts. Matrix effects from biological fluids are well-known problems, especially with electrospray ionizaion (ESI) LC–MS analyses (Dams et al., 2003; Moller et al., 2004). Second, good chromatographic separation minimizes the potential of coelution of isobaric components. Often, components that coelute would be undistinguishable in the mass spectrometer since they have identical elemental compositions and therefore *m/z* ratios. A real-world example of coelution of isobaric components involves the analysis of betaine and valine. The former is an endogenous molecule involved in glycine, serine, and threonine metabolism, and the latter is a common amino acid involved in protein synthesis and degradation. Both have identical elemental composition ($C_5H_{11}NO_2$) and similar reverse-phase chromatographic properties but play very different roles in the metabolome. Similarly, various isobaric pairs can be found between different amino acid derivatives, bile acids, and phospholipids.

For nontargeted metabonomics, a chromatographic separation that is amenable to various different components—from small polar molecules to larger lipophilic material—is desirable. Reverse-phase chromatography is the most commonly used approach and is compatible with UHPLC, which has quickly become the chromatographic method of choice due to its superior resolution and speed of analysis compared to high-performance liquid chromatography (HPLC) (Gika et al., 2008a; Plumb et al., 2008; Michopoulos et al., 2009; Zelena et al., 2009). Recently, a number of different reverse-phase UHPLC columns and gradients were evaluated based on overall number of endogenous plasma components detected (Luk, 2008). The study indicated that the Acquity BEH C18, 150-mm with 1.7-μm particle size provided the best compound resolution, retention and subsequent detection. Our typical nontargeted chromatography includes this column, along with a very shallow, high aqueous gradient to promote separation of small polar compounds, followed by a steeper increase in organic solvent to elute the less polar components (Shipkova et al., 2007, 2010). HILIC has gained in popularity over the last few years (Gika et al., 2008b; Kamleh et al. 2008) and may provide an excellent orthogonal separation approach to traditional reverse-phase chromatography. HILIC would appear to be well suited for metabonomics since much of the metabolome is comprised of small polar compounds. However, the current prevalence of metabonomics reports that utilize reverse-phase chromatography suggests that this technology has not yet been fully adopted. Based on our own experiences, this relatively slow adoption could be due to the limited robustness and/or stability of available HILIC phases compared to the incumbent reverse phases such as C18. Perhaps as this technology matures, a more important role in metabonomics studies will be assumed. Alternatives to LC include CE, GC, and direct infusion-which are beyond the scope of this chapter. For targeted applications, even when there are a large number of compounds of interest, the analyte structures are known, and, therefore, optimization of chromatography is possible. For example, if a particular study is focused on fatty acids and bile acids, the gradient and/or column should be adjusted to optimize for these. Although these molecules can be captured within a broad nontargeted chromatographic approach, the shallow aqueous gradient designed for polar components has a negative effect on the peak shape and elution profiles of fatty acids and bile acids in particular. These molecules benefit from an elution with a strong organic content (typically $> 50\%$) and can be well separated within a short time and, therefore, are well suited for a custom chromatographic approach as opposed to a general nontargeted.

21.4.3.2 MS Parameters and Data Collection The most commonly used LC–MS ionization techniques are ESI and atmospheric pressure chemical ionization (APCI). While ESI provides a much broader coverage of the metabolome, certain classes of endogenous metabolites, such as cholesterol and related sterols, have a much better APCI response. The vast majority of literature methods for nontargeted LC–MS metabonomics analyses use ESI in both positive and negative mode, and APCI is either used as an orthogonal ionization approach or for targeted analyses.

The choice of mass spectral detection is mainly determined by the nature of the experiment—targeted *versus* nontargeted metabonomic analyses. As discussed above, triple quadrupole MRM-based approaches are often used for targeted analyses due to a low duty cycle and low limits of detection. For nontargeted analyses, high-resolution accurate mass is essential for identification and annotation since the potential targets of interest are not known a priori. For such analyses, time-of-flight (TOF), hybrid quadrupole-time-of-flight (Q-TOF), FTICR, or Orbitraps can be used. Additionally, the data acquisition rate significantly impacts the ability to deconvolute closely eluting peaks. The scan rate required for the evaluation of LC–MS data increases with chromatographic overlap and narrower chromatographic peaks afforded by modern UHPLC systems, which can be 1–2 s across. We have determined that a minimum of five scans across a peak are required for obtaining accurate and reproducible peak shapes. We have chosen the hybrid linear ion trap (LTQ) Orbitrap platform (Thermo Fisher Scientific) for our nontargeted metabonomics analyses, and we are currently evaluating the utility of the faster scanning Exactive system (Thermo Fisher Scientific), which uses the same ion detection scheme as the Orbitrap. Most TOF instruments also offer high data acquisition speeds and therefore better peak definition while retaining high resolution and accuracy.

21.4.4 Quality Controls

For studies that include large numbers of samples it is important to implement a quality control (QC) paradigm with periodic insertion of appropriate samples so that instrument performance can be assessed. A pooled sample strategy has been shown to provide QC samples that can be used to assess the stability of many of the endogenous metabolites measured within a study (Sangster et al., 2006). Data from these QC samples can be used to assess spectral quality by the automated evaluation of several parameters such as charge state consistency, total signal intensity, and signal-to-noise ratio (Purvine et al., 2004). The use of pooled samples is discussed further in the section on quantifiability filtering below.

21.5 DATA PROCESSING AND ANALYSIS

The goal of LC–MS-based metabonomics is to convert measured signals into either a reproducible pattern for fingerprinting or into a quantitative assessment of metabolites that may be descriptive of physiological responses. As described above, LC–MS is a three-dimensional data set that always includes: (1) mass/charge (m/z) number, (2) chromatographic retention time, and (3) signal intensity. The data-handling task for LC–MS metabonomics can be divided into two main categories: processing and analysis. The first of these includes the multistep extraction of signals followed by assimilation of this extracted information into a reduced data set. The subsequent analysis category includes annotation of the processed data, verification that responses are concentration dependent (quantifiability filtering), and variety of statistical methods that may be employed for the interpretation (Want et al., 2005; Katajamaa and Oresic, 2007).

LC–MS data presents some unique challenges for processing and analysis. These are discussed briefly, and approaches to deal with them are described in more detail in the sections that follow. First, the majority of measured signal intensities in an LC–MS profile are proportional to the concentration of the corresponding component (Radulovic et al., 2004). However, numerous signals that do not arise from the analyte itself such as chemical noise or spurious peaks can contaminate the data. There are also many factors related to the analytes that may cause variability in signal intensity such as ion suppression. Second, matrices typically used for metabonomics contain hundreds of components and individual analytes that elute closely with each other and therefore require postacquisition computational deconvolution. Third, each molecule generates multiple ions that arise from different isotopomers, fragments, and adducts, which requires convolution of these many factors back to the molecular level. Finally, annotation of the reduced data must be made with a high level of confidence despite the presence of many similar and even isobaric components. Thus, the primary task of data processing and analysis is to filter out artifacts, extract and annotate relevant signals from the raw LC–MS data, and separate the true biological variability from that introduced by sample preparation, instrumentation, and processing (Katajamaa and Oresic, 2007).

21.5.1 Anatomy of an LC–MS Profile

A typical LC–MS profile is as a collection of mass spectra successively recorded during the duration of the chromatographic experiment. Each mass spectrum consists of a number of data points, with each representing one ion recorded by the detector in the form of mass-to-charge ratio (m/z) and intensity (ion counts) pair, associated with a retention time Fig. 21.6*a*. Recent instrumentation improvements have resulted in increased sensitivity and resolution but also increased complexity of the spectra. As a result, processing high-resolution LC–MS profiles presents a number of challenges. First, there can be more than 1 million signals in a single LC–MS profile, rendering manual processing futile. Second, as many as 96% of all the recorded signals can correspond to noise (Fleming et al., 1999). This noise can be chemical (e.g., solvents, column bleed, source contaminants, etc.) or electronic. Indeed, a simple visual analysis of a total ion chromatogram (TIC) of a complex sample such as urine yields only a few distinct peaks with the vast majority of analytes overshadowed by background signals and noise (Fig. 21.6*b*). Third, a large number of signals may correspond to a single chemical entity in the sample.

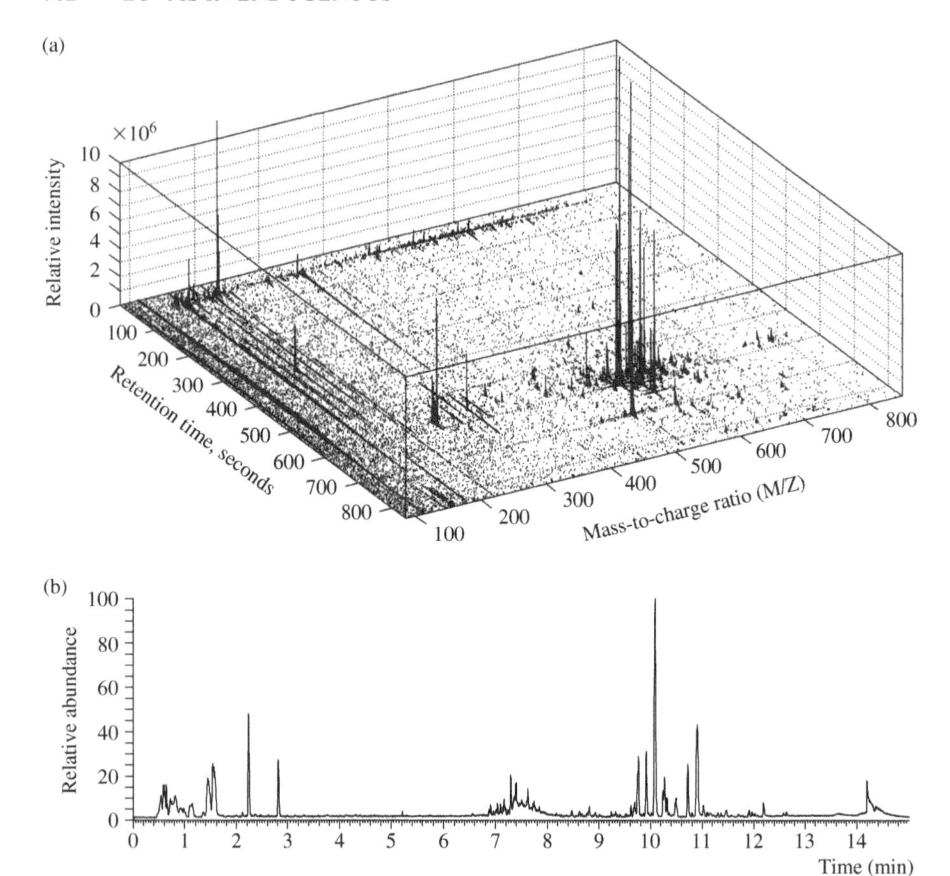

FIGURE 21.6 LC–MS profiles of extracted human plasma. Data were collected on Thermo Fisher FTMS in negative mode ESI using sample preparation and reverse-phase chromatography. (*a*) three-dimensional image and (*b*) total ion current chromatogram for the same sample.

Each component in a sample will consist of a population of distinct species due to naturally occurring isotopic forms (Jaitly et al., 2009). Further compounding this problem, gas-phase ions will often react in the electrospray source to form adducts, multimers, and fragments. As an example, the mass spectrum at the apex of the HPLC peak from d5-hippuric acid is represented by multiple related signals, which arise from coeluting isotopic contributions, adducts, multimers, and other spectral features (Fig. 21.7) (Hnatyshyn et al., 2008).

21.5.2 Processing

The interpretation of LC–MS metabonomics data set from the complicated situation described above to something useful requires several steps as depicted in Figure 21.8. While the exact method of execution of these processing steps is

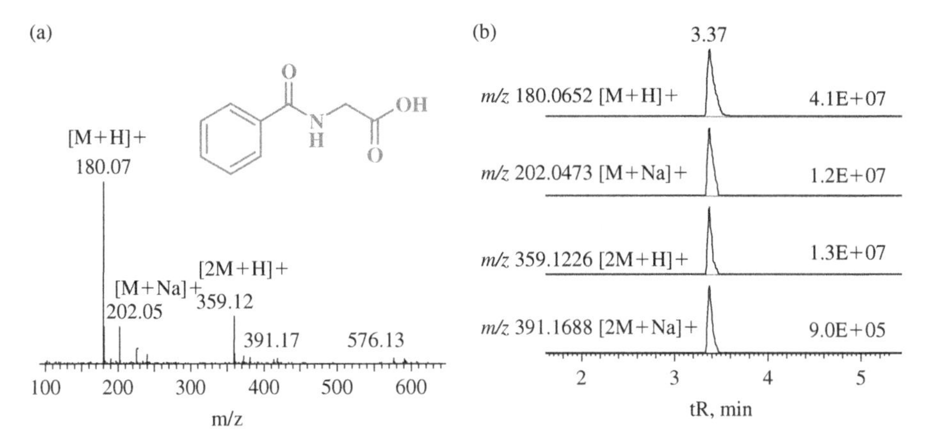

FIGURE 21.7 Molecular and related ions for hippuric acid. Data were collected in negative mode electrospray using reverse-phase chromatography. (*a*) The mass spectrum of hippuric acid, collected at the apex of chromatographic peak. (*b*) Ion chromatograms extracted with a 5.0-ppm *m*/*z* window showing chromatographic profiles of ions related to hippuric acid.

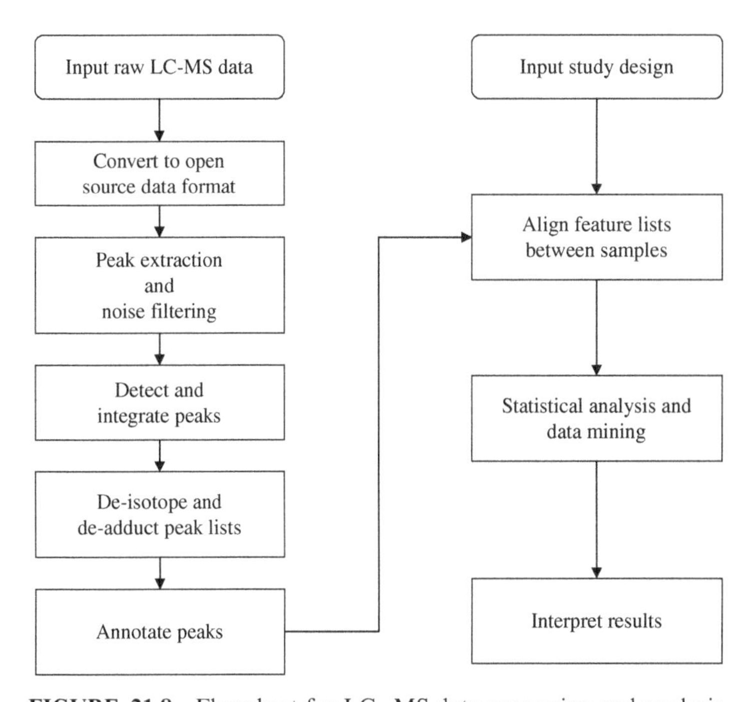

FIGURE 21.8 Flowchart for LC–MS data processing and analysis.

somewhat arbitrary, there are common unavoidable transformations that are present in almost all reported processing pipelines (Listgarten and Emili, 2005; Katajamaa and Oresic, 2007; Sumner et al., 2007; Lu et al., 2008b), which are discussed below.

21.5.2.1 Conversion of Raw Data Often, data processing tools first require conversion of the raw data to an open-source data structure to allow for the extraction of relevant signals and elimination of noise. Some instrument manufacturers allow direct access to the raw data files through dynamically linked libraries (DLLs) or a component object model (COM), but all provide software to convert proprietary raw data to various open file formats, such as American Standard Code of Information Interchange (ASCII) (Katajamaa and Oresic, 2007), netCDF (Katajamaa and Oresic, 2007), mzData, AnalysisXML, and mzXML (Pedrioli et al., 2004; Sturm et al., 2008) files.

21.5.2.2 Noise Filtering The sources of noise can be divided into two major categories: random and chemical (Hastings et al., 2002; Andreev et al., 2003). Either source of noise can mask or mimic the signals and cause false-positive or -negative identification of sample components. Random noise may be spurious or normally distributed and is mainly attributed to characteristics of the mass spectrometer, which includes thermal noise, ion source instabilities, and noise that originates from many other parts of the LC–MS system that are in the direct contact with the sample (Hastings et al., 2002; Andreev et al., 2003; Kast et al., 2003; Winfield et al., 2005). Chemical noise is a common feature of ESI spectra and is usually attributed to the solvent clusters generated by the ionization process which survive ion desolvation (Kast et al., 2003). Other sources of chemical noise can include solvent impurities, column bleed, and analytes from previous injections. Removal of chemical noise is a complex task that may include baseline correction, blank subtraction, and evaluation of the quality of extracted ion chromatograms.

Methods to remove noise from chromatograms include peak sharpening (e.g., back-folding algorithms), which attempt to enhance signal intensity among the noise (Biller and Biemann, 1974), component detection algorithm (CODA) (Winfield et al., 2005), match filtering algorithms (Andreev et al., 2003; Kast et al., 2003; Wang et al., 2006), and peak filters based on the Orphan Survival Strategy (P-BOSS) algorithm (Morohashi et al., 2007). Another approach that can be used for removal of chemical noise is background subtraction, originally described 15 years ago (Goodley and Imatani, 1993), is of limited use for typical metabonomics samples because the background is normally not constant over the duration of the chromatographic analysis. A more suitable extension of this approach is to build a background function compiled from LC–MS profiles collected from the blank injections, where the intensity of background signals from the blank can be evaluated at any (m/z, t_R) point and compared with the corresponding signal from the sample (Fig. 21.9).

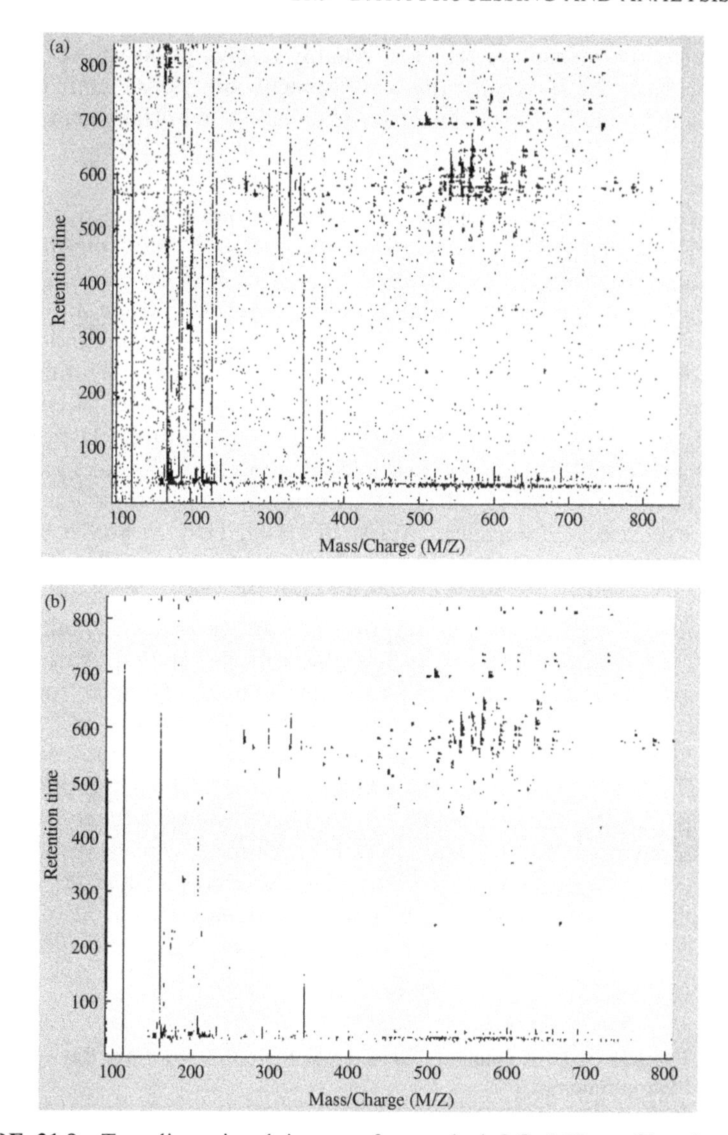

FIGURE 21.9 Two-dimensional image of a typical LC–MS profile of extracted plasma: (*a*) before and (*b*) after noise removal.

21.5.2.3 Peak Detection Peak picking often goes hand in hand with noise filtering since, if one can accurately identify all of the noise, the remaining components must be analyte-related and vice versa. Peak detection can be performed directly in two dimensions (Stolt et al., 2006; Vivo-Truyols et al., 2005; Hastings et al., 2002) or at the level of extracted ion chromatograms. Additional peak optimization and filtering may be applied in some cases (Hartley et al., 1987; Hamilton and Paul, 1990; Idborg et al., 2004). While most of the methods will produce reasonable peak lists for well-defined signals, there is currently no method that can reliably resolve peaks at every level of signal intensity. As a result, the development of robust, unbiased, and automated methods for peak detection remains an important area of active research.

21.5.2.4 Grouping-Related Signals A noise-free list of detected peaks from an LC–MS chromatogram on a protein-precipitated blood plasma sample typically contains 5,000–10,000 entries. Such a list can be used as input for outlier identification or fingerprinting analysis using multivariate statistics as is. However, this list also includes signals that correspond to multiple charged species, isotopic clusters, adducts, multimers, and fragments, creating a situation where many of the derived variables are not independent of each other. Indeed, most of the analytes in the sample will have multiple related peaks. A significant reduction of the list is possible by converting all peaks to a charge state of one while removing all peaks associated with isotopic envelopes of higher charge states (Zhang and Marshall, 1998; Hermansson et al., 2005; Kaur and O'Connor, 2006; Sturm et al., 2008; Jaitly et al., 2009). This list of quasi-molecular ions can be further reduced to a list of components by grouping together peaks that correspond to various adducts, multimers, or fragments. In certain cases, deducing the isotopic pattern can facilitate metabolite identification. An example of this is in the complex metabolism of bromoethylamine (BEA) and its metabolites (Shipkova et al., 2010) see Table 21.2. Advanced adduct deconvolution algorithms are present in some commercial software packages such as Mass Hunter (Agilent) or Sieve (Thermo Fisher) that allow the user to specify a custom list of known adduct formulas. A more comprehensive list of adducts and multimers can be produced by repetitive study of specific biological samples where coeluting peaks are tested for their chemical relationship to a base peak (Hnatyshyn et al., 2008); see Table 21.3. The final output of this process is an analyte peak table that should

TABLE 21.2 Commonly Detected Isotopic Patterns from Rat Plasma Samples

Pattern	Number of Assignments
C, O, N, Na, P, F, I, H	2803
C, O, N, S, (Mg), H	51
C, O, N, K, H	23
C, O, N, Fe, H	35
C, O, N, Ca, H	22
C, O, N, Ni, H	16
C, O, N, Cl, Cl, H	2
C, O, N, Cl, H	7
Unknown[a]	841

[a] Insufficient number of peaks to classify pattern.

TABLE 21.3 Commonly Detected Adducts from Rat Plasma Samples

Adduct	% Assignment
$[M+H]^+$	100
$[M+Na]^+$	12.1
$[M\text{-}H_2O+H]^+$	8.3
$[2M+H]^+$	4.7
$[M+NH_4]^+$	3.8
$[2M+Na]^+$	3.1
$[M\text{-}2(H_2O)+H]^+$	2.5
$[M+C_2H_3N+H]^+$	2.1
Other adducts	> 2

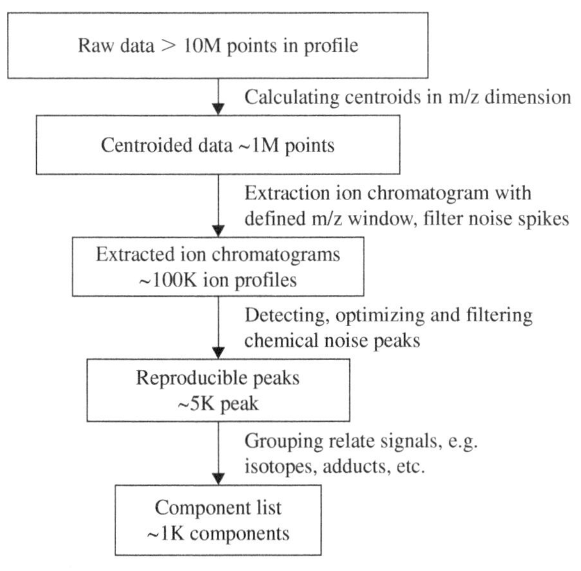

FIGURE 21.10 Summary of the data reduction achieved by convolution of all analyte-related ions.

contain only one peak for each measured analyte, which is derived from all ions related to that analyte, and which is defined as a component in the LC–MS profile. Thus, the ultimate data reduction can be achieved by converting LC–MS profiles into component lists. A typical data reduction ultimately converts a raw data file with 10 million points into a table of intensities of ~ 1,000 components, as illustrated in Figure 21.10.

21.5.2.5 Alignment Chromatographic alignment is required to account for slight retention time differences between sample runs. Alignment may be performed as a first step before noise filtering, signal extraction, and grouping;

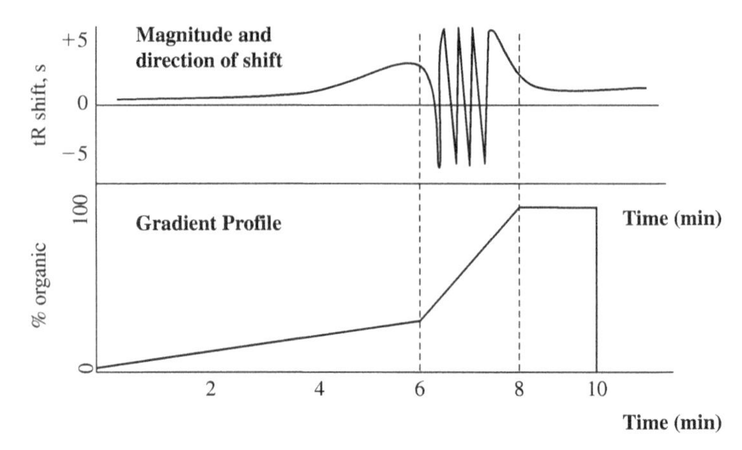

FIGURE 21.11 Simulated sample-to-sample chromatographic peak shifts and elution order for a typical reverse-phase gradient. Peak retention shifts are random inside the indicated area and therefore not predictable.

however, this process is a very difficult and computationally expensive task due to nonuniform sampling rates for mass spectra during chromatographic separation and unpredictable fluctuations in the chromatographic system itself. Such fluctuation may occur due to instrument drift, changes in temperature of the column or mobile phase, the chemistry of the sample, column history, or the responsiveness of the mobile phase mixing system and always results in peak shifts of some magnitude (Fig. 21.11). Alternatively, alignment can be applied at the peak-list level (either quasi-molecular ions or components). This approach is less computationally intensive and can benefit from the general observation that peaks with the same m/z do not change elution order for a given set of chromatographic conditions. Although LC–MS profiles contain many peaks that have obvious correspondence across a data set (Fig. 21.12*a*), the complexity of the alignment process is defined by the number of closely eluting peaks with the same m/z window, which may not have adequate correspondence (Fig. 21.12*b*), or are not fully resolved. Figure 21.12*c* represents a real-life example of extracted ion chromatograms for urine samples from a normal (top trace) and polyuric animal (bottom trace). These chromatograms illustrate the difficulty in transposing peak assignments between samples run within the same experiment.

Methods for LC–MS data alignment are based on the computational techniques that include binning, clustering, and warping or combinations thereof (Aberg et al., 2009; Szymanska et al., 2007) as well as continuous profile model (Listgarten and Emili, 2005; Listgarten et al., 2007) and selected peak alignment (Zhang X et al., 2005).

21.5.2.6 Data Normalization Normalization is a statistical or experimental procedure that is used to remove systematic variability within a set of measurements that is unrelated to treatment. Statistical approaches to normalization are based on the use of defined error models for the complete data set

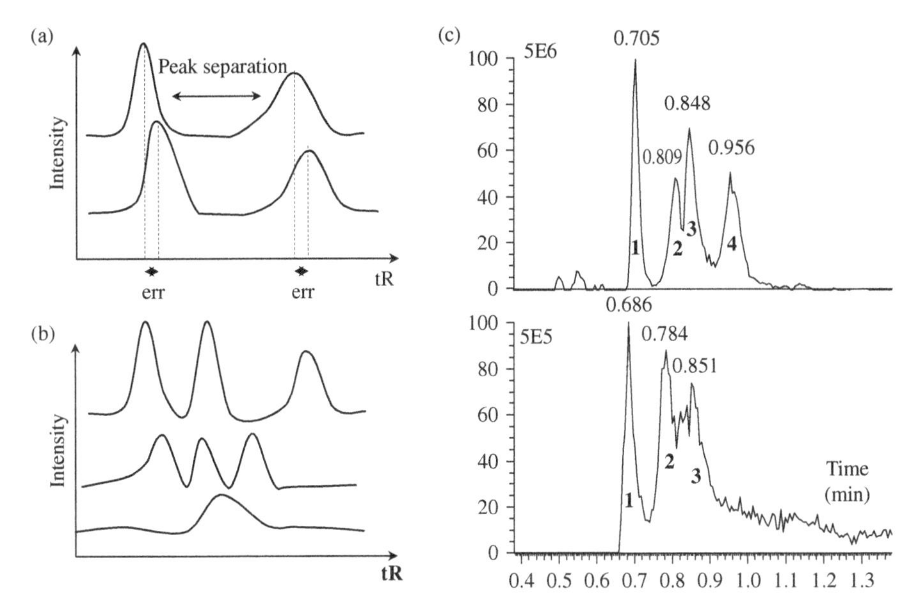

FIGURE 21.12 Hypothetical and experimental illustrations of the correspondence problem in peak alignment. (*a*) hypothetical example of an obvious retention time shift for all components, which is corrected with a simple factor; (*b*) hypothetical example of an ambiguous alignment problem, and (*c*) experimental example of partial resolution of a sample that contains four different bile acids. The assigned peaks in the chromatogram in the top trace and are not readily transferred to the more compressed chromatography in the bottom trace.

and the determination of the best scaling factors for each sample to fit with the error model (Katajamaa and Oresic, 2007). Experimental normalization is based on the introduction of single or multiple internal standards to compensate for instrumental error (Katajamaa and Oresic, 2005). As discussed above in the sample preparation section, normalization prior to data collection can be beneficial for samples of significantly different concentrations. MS total useful signal (MSTUS) and osmolality have been reported to be suitable normalization approaches (Warrack et al., 2009).

21.5.2.7 Quantifiability Filtering Although progress is being made and there are distinct exceptions, LC–MS as applied to metabonomics is still not a quantitative science (Listgarten and Emili, 2005). Peak detection and quantification, even if done optimally, does not guarantee linearity of the peak signal relative to analyte concentration (Wang et al., 2003). This observation is true even if the analyte in question behaves well in isolation or in a different matrix. The uncertainty is largely due to the high likelihood of coelution and the resulting ionization suppression and renders dangerous the assumption that a given analyte will behave similarly across similar samples. These effects can be mitigated to some extent through the judicious use of internal standards and even more convincingly with artificial isotopomers of target analytes (e.g., ^{13}C

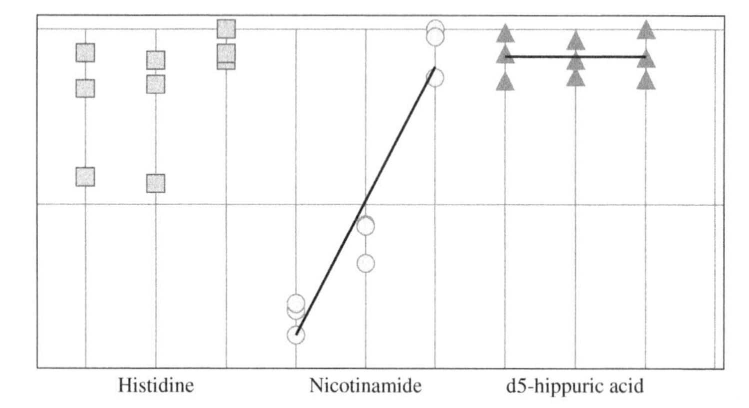

FIGURE 21.13 Quantifiability filtering based on responses of individual analytes measured in the LC–MS profiles from serial-diluted pooled samples. Each component is evaluated for a positive slope before attempts are made to interpret changes in its signal in test samples. Shown are examples of one quantifiable metabolite (nicotinamide, showing the expected increase in response with increase in concentration), one metabolite that cannot be reliably quantified (histidine) due to lack of concentration-dependant response and the internal standard (d5-hippuric acid) which shows the expected constant response.

or ^{2}H metabolite analogs). However, such adjunctive studies are time consuming and expensive and force practitioners to compromise, particularly in nontargeted metabonomics studies. A reasonable approach is to include serial dilution of pooled samples from which a general idea of the concentration response of a particular analyte can be obtained. These pooled QCs should be analyzed at the beginning and end of each sample set. Before intensity changes of a specific component are interpreted, it should be confirmed that its intensity increases with concentration in the dilution series. It is not uncommon to see a lack of concentration dependence for signals from particular analytes, and occasionally an inverse correlation is observed (the peak intensity goes down with increasing concentration) (Fig. 21.13). Such components cannot be used for biological interpretation.

21.5.3 Data Analysis

The desired output of data processing is often represented as a table of grouped (convoluted) peaks. This format is manageable in size and each entry corresponds to a single analyte in the sample and its relative intensity. Such a data set is suitable for fingerprinting (Dettmer et al., 2007) but requires annotation for targeted or untargeted metabonomics evaluation. Analysis of the data at this point with multivariate statistical approaches can be used to differentiate the samples to: (1) focus the next phase of analysis on a subset of the samples, (2) identify biological outliers, or (3) identify aberrant data sets. Untargeted or targeted metabonomics applications require further data analysis steps described in the following sections.

21.5.3.1 Data Annotation Annotation of components in a data set is a significant challenge and currently represents a major bottleneck in metabonomics analysis. Approaches that have been applied toward the routine annotation of components measured in a sample include: (1) deriving the molecular formula from accurate mass measurements; (2) use of databases to identify biologically relevant structures from molecular formulas; (3) the use of libraries of LC–MS/MS fragmentation data; (4) traditional structure elucidation approaches that use MS^n and NMR on isolated or semi-isolated samples; (5) confirmation of putative identification with an orthogonal technique; and (6) prediction of expected metabolites based on other genomic, transcriptomic, or proteomic data. Some of these approaches are described in more detail below.

Elemental Composition In principle, accurate mass can be used for the de novo structure elucidation of small molecules since only certain elemental compositions will be consistent with a mass measured with very high accuracy. Indeed, when accurate mass measurements are available, the elemental composition determination of monoisotopic ions is the first step of characterization because it provides a simple, efficient, and automatable way to search chemical and metabolic databases (Werner et al., 2008). Highly accurate mass determinations (<5ppm) can routinely be made with modern mass spectrometers using time of flight and cyclotron-based detection. However, even in cases where sub-part-per-million accuracy is achieved, many chemically possible formulas are obtained, and incorporation of additional information, such as isotope distribution, is required in order to reduce the number of possible elemental compositions (Kind and Fiehn, 2006). Caution should be exercised, however, since the isotope ratio can be affected by interference and saturation effects due to both the matrix and experimental conditions (Werner et al., 2008) and have been reported to be up to 20% of predicted values (Grange et al., 2006). This approach can be extended to include the grouping of other related species and physicochemical properties to reduce the number of possible formulas even further (e.g., valence rules, isotopic patterns, nitrogen rule, ring plus double bode equivalent) (Kind and Fiehn, 2007; Werner et al., 2008). Even if a single elemental composition can be determined, many isomeric possibilities are still plausible, even within a biological context. For example, a ChemSpider search on $C_5H_8O_4$ yeilds 156 unique molecules. A similar search of the Madison Metabolomics Consortium Database (www.mmcd.nmrfam.wisc.edu) yields 12 unique entities that have biological relevance. Ten of these molecules contain a CO_2 group that will be eliminated as a dominant fragment in an MS/MS experiment. Despite such important isobaric redundancies, there is much potential in the exploitation of mass spectral measurements to identify the primary structure of molecules.

Libraries and Database Resources for LC–MS Excellent compendiums of Web-based metabolomics resources that cover libraries, databases, and other useful tools have recently been published (Go, 2009; Tohge and Fernie, 2009). The following discussion will therefore be biased to the authors' familiarity with respect to data analysis rather than comprehensive in nature.

Gas chromatography–MS has long enjoyed a database-oriented peak identification approach based on similarity matching to electron-impact (EI) fragment mass spectra available from the National Institute of Standards and Technology (NIST), John Wiley & Sons, and recently, the metabolomics-specific Golm database (Hummel et al., 2007). The LC–MS analog to this approach would rely on parent and fragmentation spectra generated with LC–MS/MS. Such libraries are difficult to standardize, primarily because fragmentation and corresponding mass spectra are highly dependent upon conditions and instrument types (Jansen et al., 2005). Nonetheless, LC–MS/MS data for large numbers of compounds has become available in accessible formats (Halket et al., 2005). For example, METLIN (http://metlin.scripps.edu/about.php) (Smith et al. 2005), in addition its metabolite database (see below), currently has over 4600 MS/MS spectra on almost 900 metabolites acquired with different collision energies. Similarily, MassBank (http://www.massbank.jp/index.html) and NIST (Want et al., 2005; Dettmer et al., 2007) have extensive libraries of LC–MS and LC–MS/MS data. Although it seems unlikely that a single, universal set of conditions is forthcoming for LC–MS/MS for metabonomics applications, newly available MS/MS libraries will prove to be useful for the assignment of some unknowns where accurate mass by itself is insufficient.

Numerous databases are available for metabonomic researchers (Dettmer et al., 2007; Go, 2009) and fall into four categories: spectral, metabolite, pathway analysis, and information management. Most metabolite databases also contain spectral data. These databases are quite useful when used in combination for the identification of unknowns. For example, accurate mass or molecular formula information can be used to query spectral/metabolite databases such as METLIN (Smith et al., 2005), MMCD (Cui et al., 2008), or the human metabolome database, HMDB (Wishart et al., 2009). Potential leads can be investigated within a biological context from pathway databases such as KEGG (http://www.genome.jp/kegg/) or BioCyc (http://biocyc.org/). An interesting tool, BioSpider (http://www.biospider.ca), has recently been reported that generates dossiers on metabolites by collating data from a variety of online databases (Knox et al., 2007).

21.5.3.2 Statistical Analysis and Reporting Methods for statistical analysis of metabonomics data sets include a variety of supervised and unsupervised multivariate techniques (Holmes et al., 2000) as well as univariate analysis strategies. These chemometric approaches have been recently reviewed (Holmes and Antti, 2002; Robertson et al., 2007), and a thorough discussion of these is outside the scope of this chapter. Perhaps the best known of the unsupervised multivariate techniques is principle component analysis (PCA) and is widely used for the detection of outliers, visualizing trends in data, sample classification, and a qualitative assessment of overall data quality (Lu et al., 2008c). This approach can reduce a metabonomics data set with thousands of variables to a few principle components that describe the majority of variation in a data set (Want et al., 2005). Most commonly available statistical packages support PCA. Other tools for sample classification based on unsupervised pattern recognition include hierarchical clustering analysis and independent component analysis (Dettmer et al., 2007).

Supervised pattern recognition allows inclusion of group information to define a statistical model that describes the relationship between independent variables (e.g., intensity of spectral features) and dependent response (e.g., histopathology, toxicity scores, etc.). When the dependent response is known, supervised methods such as partial least-square discriminate analysis, linear discriminate analysis, soft-independent method of class analogy, and analysis of variance are preferred (Dettmer et al., 2007). In addition to these linear methods, a family of methods capable of nonlinear classification and regression were adapted from the field of artificial intelligence and include artificial neural networks, self-organizing maps, genetic algorithms, and evolutionary programming (Want et al., 2005; Dettmer et al., 2007; Listgarten et al., 2007; Lu, et al., 2008c). These methods are also useful for the detection of patterns and the building predictive models.

The successful application of multivariate statistical methods is highly dependent on the quality of prior processing, and, thus, incomplete data reduction (e.g., presence of isotopic peaks and adducts) may lead to a multicolinearity problem. In this situation, many of the measured variables come from the same molecule and therefore, are not independent of each other. This can result in the regression coefficient estimates to behave erratically in response to small changes in the data as well as increase the standard errors of the affected coefficients (Listgarten and Emili, 2005).

Once the data is properly reduced, univariate statistics can be used on lists of components (annotated or not). In the situation when variance in the data is distributed normally and is equal between groups, the simple Student's *t* test may be sufficient to evaluate the differences in individual metabolites between data sets. Mann-Whitney or Kruskal-Wallis nonparametric tests can be used for calculating p values when the variance is unknown. The magnitude of the difference between groups can be calculated as a fold–change between group averages. For molecules that survive the quantifiability filter, a fold–change is a convenient way to assess biochemical changes without the requisite of rigorous quantitation. These types of calculations are easily implemented using Microsoft Excel Visual Basic scripts. Table 21.4 contains the results of such statistical calculations on annotated components in rat plasma obtained from fed and fasted groups of animals (Donald Robertson, personal communication). Perhaps lacking the glamour of fancy multivariate statistical approaches, these types of data presentations are the most useful for biological interpretation of metabonomics results.

TABLE 21.4 Data Presentation from Targeted or Nontargeted Metabonomics Analysis

	Male control vs. Treated		Female Control vs. Treated	
Compound Name	Fold–Change	P–Value	Fold–Change	P–Value
Acetyl-carnitine	1.28	0.307	0.85	0.540
Acetylcholine	0.81	0.417	0.59	0.164
Arachidonic acid	1.40	0.308	1.40	0.376
Arginine	0.78	0.200	0.88	0.580
Betaine	1.01	0.942	0.89	0.678
Carnitine	0.49	0.017	0.49	0.010
Chenodeoxycholic acid	70% samples below detection limit[a]		18.85	0.051
Cholic acid	30.36	0.009	6.69	0.225
Choline	1.05	0.450	1.58	0.027
Citrulline	0.90	0.368	0.85	0.313
Creatine	3.22	0.005	4.09	0.002
Creatinine	1.05	0.654	1.22	0.457
Cytidine	1.34	0.002	1.39	0.287
Deoxycholic acid	70% samples below detection limit[a]		22.47	0.036
Pantothenic acid	0.75	0.281	0.65	0.244
Glucose	0.76	0.541	0.88	0.676
Glutamic acid	1.07	0.586	0.95	0.776
Glutamine	0.97	0.754	1.01	0.914
Glutaric acid	2.14	0.005	3.11	0.063
Glycocholic acid	1.29	0.737	3.75	0.093
Glycoursodeoxycholic acid	5.61	0.129	35.22	0.080
Histidine	0.91	0.718	0.84	0.600
Isoleucine/leucine	1.07	0.653	0.85	0.593
Kynurenine	1.20	0.487	1.88	0.197
Lysine	1.29	0.201	0.91	0.699
Methionine	1.01	0.887	1.07	0.791
Methyl-histidine	2.36	0.009	2.99	0.004
Nicotinamide	0.86	0.591	0.84	0.656
Phenylalanine	1.11	0.512	0.92	0.814
Proline	0.49	0.001	0.42	0.007
Sebacic acid	0.83	0.189	0.91	0.719
Succinic acid	0.76	0.293	0.70	0.353
Sulfoglycolithocholic acid	2.60	0.190	3.66	0.049
Taurine	1.33	0.027	1.64	0.002
Taurochenodeoxycholic acid	0.62	0.383	2.92	0.096
Taurocholic acid	0.62	0.164	3.80	0.148
Threonine	1.55	0.097	1.00	0.990
Tryptophan	0.84	0.411	0.67	0.239
Tyrosine	0.96	0.828	1.09	0.786
Uric acid	1.06	0.805	1.16	0.633
Ursodeoxycholic acid	70% samples below detection limit[a]		10.52	0.041
Valine	1.32	0.138	0.94	0.852

[a] Significant numbers of samples within a group with values below the detection limit preclude fold-change estimation.

21.5.3.3 Dealing with Xenobiotics In nontargeted metabonomics, it is crucial to reliably distinguish endogenous metabolites from xenobiotic components in order to identify perturbations in normal biochemical processes and postulate mechanisms of action. However, the xenobiotic components are often not easily recognized in the large and complex pool of small-molecule endogenous metabolites. In addition to the dosed drug substance, biological fluids can contain various drug metabolites, impurities associated with the parent molecule, and various formulation components. In drug studies, samples from control animals dosed only with vehicle are necessary to identify any exogenous components associated with the dosing formulations, food effects, and analytical artifacts. However, the vehicle control samples do not compensate for drug metabolites and often there is limited information on drug biotransformation products and dynamics. Even when this information is available, due to the large variations between species and the complexities of drug metabolism and distribution, it is very difficult or impossible to correctly predict formation of all metabolites. Hence, there is a need for a reliable xenobiotic metabolite detection and identification step as part of any metabonomics study. A variety of traditional drug metabolite identification tools, such as predicted metabolite searches, information-dependant acquisition (Xia et al., 2003; Li et al., 2007), and mass defect filter (Zhang et al., 2003; Zhu et al., 2006; Bateman et al., 2007) can be employed for typical drug molecules. These tools become less effective when the dosed molecules are small reactive species, as the generated metabolites are usually conjugates with little structural similarity to the dosed compound. A recent publication discusses parallel dosing of stable isotope-labeled molecules as an approach to overcome such problems (Shipkova et al., 2010).

21.6 CONCLUSION

Metabonomics provides a window on how external influences perturb an organism's numerous biochemical pathways as measured through changes in many endogenous small molecules. At a minimum, these molecular changes report directly or indirectly on the nature of the perturbation. In ideal circumstances, with rigorous correlation to accepted endpoints, the identified species can become small-molecule biomarkers that can serve as easy-to-measure surrogates of the endpoint and can, in turn, be used to facilitate monitoring systems under study (Weckwerth and Morgenthal, 2005). As an example, a panel of urinary markers can substitute for more labor-intensive histopathological assessment within a drug discovery and development setting. Across the spectrum of drug discovery, such biomarkers can be extremely valuable. In early discovery, where resources are at a premium, a biomarker can mean the difference between detecting or missing a critical efficacious or toxic effect. In development, because of the conservation of the metabolome across species, endogenous metabolite biomarkers are easily translated from a preclinical to clinical setting. Although the industry has just begun to experience the impact that mass spectrometry will make in the rapidly growing field of metabonomics, it is clear that LC–MS technology will play an important role in this field for years to come.

REFERENCES

Aberg KM, et al. The correspondence problem for metabonomics datasets. Anal Bioanal Chem 2009;394:151–162.

Allard E, et al. Comparing capillary electrophoresis–mass spectrometry fingerprints of urine samples obtained after intake of coffee, tea, or water. Anal Chem 2008;80:8946–8955.

Altmaier E, et al. Bioinformatics analysis of targeted metabolomics—Uncovering old and new tales of diabetic mice under medication. Endocrinology 2008;149: 3478–3489.

Andreev VP, et al. A uiversal denoising and peak picking algorithm for LC-MS based on matched filtration in the chromatographic time domain. Anal Chem 2003;75:6314–6326.

Bateman KP, et al. MSE with mass defect filtering for in vitro and in vivo metabolite identification. Rapid Commun Mass Spectrom 2007;21:1485–1496.

Beckmann M, et al. High-throughput, nontargeted metabolite fingerprinting using nominal mass flow injection electrospray mass spectrometry. Nature Protocols 2008;3:486–504.

Beecher CWW. The human metabolome In Metabolic Profiling: Its Role in Biomarker Discovery and Gene Function, Harrogan GG, Ed. Boston, Kluwer Academic, 2003, pp. 311–319.

Biller JE, Biemann K. Reconstructed mass spectra, a novel approach for the utilization of gas chromatography–mass spectrometer data. Anal Lett 1974;7:515.

Boudonck KJ, et al. Characterization of the biochemical variability of bovine milk using metabolomics. Metabolomics 2009;5:1–12.

Brown KL, et al. Raman spectroscopic differentiation of activated versus non-activated T lymphocytes: An in vitro study of an acute allograft rejection model. J Immunol Methods 2009;340:48–54.

Brown SC, et al. Metabolomics applications of FT-ICR mass spectrometry. Mass Spectrom Rev 2005;24:223–231.

Bruce SJ, et al. Evaluation of a protocol for metabolic profiling studies on human blood plasma by combined ultra-performance liquid chromatography/mass spectrometry From extraction to data analysis. Anal Biochem 2008;372:237–249.

Carraro S, et al. Metabolomics in exhaled breath condensates. Am J Respir Crit Care Med 2008;177:236.

Chen H, et al. Neutral desorption sampling coupled to extractive electrospray ionization mass spectrometry for rapid differentiation of biosamples by metabolomic fingerprinting. J Mass Spectrom 2007a;42:1123–1135.

Chen W, et al. Application of FTIR and metabolomics analysis in high-throughput screening strains. Huagong Xuebao/J Chem Ind Eng (China) 2007b;58:2336–2340.

Coen M, et al. NMR-based metabolic profiling and metabonomic approaches to problems in molecular toxicology. Chem Res Toxicol 2008;21:9–27.

Cui Q, et al. Metabolite identification via the Madison Metabolomics Consortium Database. Nature Biotechnol 2008;26:162–164.

Dams R, et al. Matrix effect in bio-analysis of illicit drugs with LC-MS/MS: Influence of ionization type, sample preparation, and biofluid. J Am Soc Mass Spectrom 2003;14:1290–1294.

Dettmer K, et al. Mass spectrometry-based metabolomics. Mass Spectrom Rev 2007;26:51–78.

Dieterle F, et al. Application of metabonomics in a compound ranking study in early drug development revealing drug-induced excretion of choline into urine. Chem Res in Toxicol 2006;19:1175–1181.

Dunn WB. Current trends and future requirements for the mass spectrometric investigation of microbial, mammalian and plant metabolomes. Phys Biol 2008;5:1–24.

Effros RM. Metabolomics in exhaled breath condensates. Am J Respir Crit Care Med 2008;177:236.

Elliott GN, et al. Soil differentiation using fingerprint Fourier transform infrared spectroscopy, chemometrics and genetic algorithm-based feature selection. Soil Biol and Biochem 2007;39:2888–2896.

Ellis DI, Goodacre R. Metabolic fingerprinting in disease diagnosis: Biomedical applications of infrared and Raman spectroscopy. Analyst 2006;131:875–885.

Fiehn O. Combining genomics, metabolome analysis, and biochemical modelling to understand metabolic networks. Compar Funct Genomics 2001;2:155–168.

Fiehn O. Extending the breadth of metabolite profiling by gas chromatography coupled to mass spectrometry. Trends Anal Chem 2008;27:261–269.

Fiehn O, et al. Metabolite profiling for plant functional genomics. Nature Biotechnol 2000;18:1157–1161.

Fleming CM, et al. Windowed mass selection method: A new data processing algorithm for liquid chromatography–mass spectrometry data. J Chromatogr A 1999;849:71–85.

Garcia-Perez I, et al. Metabolic fingerprinting with capillary electrophoresis. J Chromatogr A 2008;1204:130–139.

German JB. Genomics and milk. Austral J Dairy Technol 2009;64:94–101.

Gika HG, et al. Evaluation of the repeatability of ultra-performance liquid chromatography-TOF-MS for global metabolic profiling of human urine samples. J Chromatogr B Analy Technol Biomed Life Sci 2008a;871:299–305.

Gika HG, et al. Hydrophilic interaction and reversed-phase ultra-performance liquid chromatography TOF-MS for metabonomic analysis of Zucker rat urine. J Separ Sci 2008b;31:1598–1608.

Gika HG, et al. Liquid chromatography and ultra-performance liquid chromatography–mass spectrometry fingerprinting of human urine. Sample stability under different handling and storage conditions for metabonomics studies. J. Chromatogr A 2008c;1189:314–322.

Go EP. Database resources in metabolomics: An overview. J Neuroimmune Pharmacol 2009;5:18–30.

Goodley P, Imatani K. Background-subtraction software to improve LC-MS chromatographic and MS data. Ame Lab 1993;25:36B–36D.

Grange AH, et al. Determination of ion and neutral loss compositions and deconvolution of product ion mass spectra using an orthogonal acceleration time-of-flight mass spectrometer and an ion correlation program. Rapid Commun Mass Spectrom 2006;20:89–102.

Hagel JM Facchini PJ. Plant metabolomics: Analytical platforms and integration with functional genomics. Phytochem Rev 2008;7:479–497.

Halket JM, et al. Chemical derivatization and mass spectral libraries in metabolic profiling by GC/MS and LC/MS/MS. J Exp Botany 2005;56:219–243.

Hamilton JC, Paul JG. Mixture analysis using factor analysis. II: Self-modeling curve resolution. J Chemometri 1990;4:1–13.

Hansen MAE, Smedsgaard J. Automated work-flow for processing high-resolution direct infusion electrospray ionization mass spectral fingerprints. Metabolomics 2007;3:41–54.

Hartley TF, et al. Computerized chromatographic peak detection using the trigg tracking signal. An application devised for use with an online analog to digital converter connected between an amino acid analyser and a personal computer. Biomed Chromatogr 1987;2:104–109.

Hastings CA, et al. New algorithms for processing and peak detection in liquid chromatography/mass spectrometry data. Rapid Commun Mass Spectrom 2002;16:462–467.

Hermansson M, et al. Automated quantitative analysis of complex lipidomes by liquid chromatography/mass spectrometry. Anal Chem 2005;77:2166–2175.

Hnatyshyn S, Sanders M, Shipkova P, Luk E, Warrack B, Reily M, Poster 375: Automated Mass Spectra Interpretation Approach to Data Reduction for LC-MS Metabonomics Analysis, Poster presented at the American Society for Mass Spectrometry Conference, June 5, Denver, CO, 2008

Holmes E, Antti H. Chemometric contributions to the evolution of metabonomics: Mathematical solutions to characterising and interpreting complex biological NMR spectra. Analyst 2002;127:1549–1557.

Holmes E, et al. Chemometric models for toxicity classification based on NMR spectra of biofluids. Chem Res Toxicol 2000;13:471–478.

Hummel J, et al. The golm metabolome database: A database for GC-MS based metabolite profiling. Topics Curr Genet 2007, 18:75–95.

Idborg H, et al. Multivariate approaches for efficient detection of potential metabolites from liquid chromatography/mass spectrometry data. Rapid Commun Mass Spectrom 2004;18:944–954.

Jaitly N, et al. Decon2LS: An open-source software package for automated processing and visualization of high resolution mass spectrometry data. BMC Bioinformatics 2009;10:87.

Jansen R, et al. LC-MS/MS systematic toxicological analysis: Comparison of MS/MS spectra obtained with different instruments and settings. Clin Biochemi 2005;38:362–372.

Jellum E, et al. Screening for metabolic disorders using gas-liquid chromatography, mass spectrometry, and computer technique. Scand J Clin Lab Investig 1971;27:273–285.

Jonsson P, et al. High-throughput data analysis for detecting and identifying differences between samples in GC/MS-based metabolomic analyses. Anal Chem 2005;77:5635–5642.

Josephs JL, Sanders M. Creation and comparison of MS/MS spectral libraries using quadrupole ion trap and triple-quadrupole mass spectrometers. Rapid Commun Mass Spectrom 2004;18:743–759.

Kamleh A, et al. Metabolomic profiling using Orbitrap Fourier transform mass spectrometry with hydrophilic interaction chromatography: A method with wide applicability to analysis of biomolecules. Rapid Commun Mass Spectrom 2008;22:1912–1918.

Kast J, et al. Noise filtering techniques for electrospray quadrupole time of flight mass spectra. J Am Soc Mass Spectrom 2003;14:766–776.

Katajamaa M, Oresic M. Processing methods for differential analysis of LC/MS profile data. BMC Bioinformatics 2005;6:179–191.

Katajamaa M, Oresic M. Data processing for mass spectrometry-based metabolomics. J Chromatogr A 2007;1158:318–328.

Kaur P, O'Connor PB. Algorithms for automatic interpretation of high resolution mass spectra. J Am Soc Mass Spectrom 2006;17:459–468.

Keun HC, et al. Analytical reproducibility in (1)H NMR-based metabonomic urinalysis. Chem Res Toxicol 2002;15:1380–1386.

Kind T, Fiehn O. Metabolomic database annotations via query of elemental compositions: Mass accuracy is insufficient even at less than 1 ppm. BMC Bioinformatics 2006;7:234.

Kind T, Fiehn O. Seven Golden Rules for heuristic filtering of molecular formulas obtained by accurate mass spectrometry. BMC Bioinformatics 2007;8:105.

Knox C, et al. BioSpider: A web server for automating metabolome annotations. Pacific Symp Biocompu 2007;1:145–156.

Kohler M, et al. Characterization of lipid extracts from brain tissue and tumors using Raman spectroscopy and mass spectrometry. Anal Bioanal Chem 2009;393:1513–1520.

Kouba E, et al. Uroscopy by Hippocrates and Theophilus: Prognosis versus diagnosis. J Urol 2007;177:50–52.

Li AC, et al. Complete profiling and characterization of in vitro nefazodone metabolites using two different tandem mass spectrometric platforms. Rapid Commun Mass Spectrom 2007;21:4001–4008.

Listgarten J, Emili A. Statistical and computational methods for comparative proteomic profiling using liquid chromatography–tandem mass spectrometry. Mol Cell Proteomics 2005;4:419–434.

Listgarten J, et al. Difference detection in LC-MS data for protein biomarker discovery. Bioinformatics 2007;23:e198–e204.

Lokhov PG, Archakov AI. Mass spectrometry methods in metabolomics. Biochem (Moscow) Suppl Ser B: Biomed Chem 2009;3:1–9.

Lu W, et al. Analytical strategies for LC-MS-based targeted metabolomics. J Chromatogr B Anal Technol Biomed Life Sci 2008a;871:236–242.

Lu X, et al. LC-MS-based metabonomics analysis. J Chromatogr B Anal Technol Biomed Life Sci 2008b;866:64–76.

Lu X, et al. LC-MS-based metabonomics analysis. J Chromatogr B 2008c;866: 64–76.

Luk C.E, et al. Evaluation of columns and gradients for LC/MS-based non-targeted metabonomics. Paper presented at the 56th American Society for Mass Spectrometry Conference, Denver, CO. 2008.

Madalinski G, et al. Direct introduction of biological samples into a LTQ-orbitrap hybrid mass spectrometer as a tool for fast metabolome analysis. Anal Chem 2008;80:3291–3303.

Michopoulos F, et al. UPLC-MS-based analysis of human plasma for metabonomics using solvent precipitation or solid phase extraction. J Proteome Res 2009;8:2114–2121.

Miller RC, et al. Comparison of specific gravity and creatinine for normalizing urinary reproductive hormone concentrations. Clin Chem 2004;50:924–932.

Moller K, et al. Investigation of matrix effects of urine on a molecularly imprinted solid-phase extraction. J Chromatogr B Anal Technol Biomed Life Sci 2004;811:171–176.

Monton MRN, Soga T. Metabolome analysis by capillary electrophoresis–mass spectrometry. J Chromatogr A 2007;1168:237–246.

Morohashi M, et al. Model-based definition of population heterogeneity and its effects on metabolism in sporulating *Bacillus subtilis*. J Biochem 2007;142:183–191.

Nicholson JK, et al. "Metabonomics". Understanding the metabolic responses of living systems to pathophysiological stimuli via multivariate statistical analysis of biological NMR spectroscopic data. Xenobiotica 1999;29:1181–1189.

Pasikanti KK, et al. Development and validation of a gas chromatography/mass spectrometry metabonomic platform for the global profiling of urinary metabolites. Rapid Commun Mass Spectrom 2008a;22:2984–2992.

Pasikanti KK, et al. Gas chromatography/mass spectrometry in metabolic profiling of biological fluids. J Chromatogr B Anal Technol Biomed Life Sci 2008b;871: 202–211.

Pauling L, et al. Quantitative analysis of urine vapor and breath by gas-liquid partition chromatography. Proc Nat Acad Sci 1971;68:2374–2376.

Pedrioli PGA, et al. A common open representation of mass spectrometry data and its application to proteomics research. Nature Biotechnol 2004;22:1459–1466.

Pistorius AMA, et al. Monitoring of biomass composition from microbiological sources by means of FT-IR spectroscopy. Biotechnol Bioeng 2009;103:123–129.

Plumb RS, et al. A rapid simple approach to screening pharmaceutical products using ultra-performance LC coupled to time-of-flight mass spectrometry and pattern recognition. J Chromatogr Sci 2008;46:193–198.

Purvine S, et al. Spectral quality assessment for high-throughput tandem mass spectrometry proteomics. OMICS A J Integrative Biol 2004;8:255–265.

Radulovic D, et al. Informatics platform for global proteomic profiling and biomarker discovery using liquid chromatography–tandem mass spectrometry. Mol Cell Proteomics 2004;3:984–997.

Reily MD, Lindon JC. NMR spectroscopy: Principles and instrumentation. In. Metabonomics in Safety Assessment, Robertson DG, Holmes E, Nicholson JK. Taylor & Francis, New York, 2005.

Robertson DG, et al. Metabonomic technology as a tool for rapid throughput in vivo Toxicity Screening. In Comprehensive Toxicology, Vanden Heuvel JP, Perdew GJ, Mattes WB, Greenlee WF, Eds. Elsevier Science, 2002, pp. 583–610.

Robertson DG, et al. Metabonomics in preclinical drug development. Expert Opin Drug Metab Toxicol 2005;1:363–376.

Robertson DG, et al. Metabonomics in pharmaceutical discovery and development. J Proteome Res 2007;6:526–539.

Robosky LC, et al. In vivo toxicity screening programs using metabonomics. Combinatorial Chem High Throughput Screen 2002;5:651–662.

Sangster T, et al. A pragmatic and readily implemented quality control strategy for HPLC-MS and GC-MS-based metabonomic analysis. Analyst 2006;131:1075–1078.

Sawada Y, et al. Widely targeted metabolomics based on large-scale MS/MS data for elucidating metabolite accumulation patterns in plants. Plant Cell Physiol 2009;50:37–47.

Shipkova P, et al. Sample preparation approaches and data analysis for metabonomic profiling of plasma samples. Paper presented at the 55th American Society for Mass Spectrometry Conference, Indianapolis, IN, 2007.

Shipkova P, et al. Urinary metabolites of 2-bromoethanamine identified by stable isotope labelling: evidence for carbamoylation and glutathione conjugation. Xenobiotica 2010;40:862–873.

Shockcor JP, Holmes E. Metabonomic applications in toxicity screening and disease diagnosis. Curr Topics Med Chem 2002;2:35–51.

Smith CA, et al. METLIN: A metabolite mass spectral database. Ther Drug Monitor 2005;27:747–751.

Stolt R, et al. Second-order peak detection for multicomponent high-resolution LC/MS data. Anal Chem 2006;78:975–983.

Sturm M, et al. OpenMS–An open-source software framework for mass spectrometry. BMC Bioinformatics 2008;9:163.

Sumner LW, et al. Plant metabolomics: Large-scale phytochemistry in the functional genomics era. Phytochemistry 2003;62:817–836.

Sumner LW, et al. Proposed minimum reporting standards for chemical analysis: Chemical Analysis Working Group (CAWG) Metabolomics Standards Initiative (MSI). Metabolomics 2007;3:211–221.

Szymanska E, et al. Evaluation of different warping methods for the analysis of CE profiles of urinary nucleosides. Electrophoresis 2007;28:2861–2873.

Tautenhahn R, et al. Annotation of LC/ESI-MS mass signals. Lecture Notes in Computer Science (including subseries Lecture Notes in Artificial Intelligence and Lecture Notes in Bioinformatics) 2007;4414:371–380.

Teng Q, et al. A direct cell quenching method for cell-culture based metabolomics. Metabolomics 2009;5:199–208.

Tohge T, Fernie AR. Web-based resources for mass-spectrometry-based metabolomics: A user's guide. Phytochemistry 2009;70:450–456.

Tweeddale H, et al. Effect of slow growth on metabolism of *Escherichia coli*, as revealed by global metabolite pool ("metabolome") analysis. J Bacteriol 1998;180:5109–5116.

Vallejo M, et al. Plasma fingerprinting with GC-MS in acute coronary syndrome. Anal Bioanal Chem 2009;394:1–8.

Villas-Boas SG, et al. Mass spectrometry in metabolome analysis. Mass Spectrom Rev 2005;24:613–646.

Vivo-Truyols G, et al. Automatic program for peak detection and deconvolution of multi-overlapped chromatographic signals: Part II: Peak model and deconvolution algorithms. J Chromatogr A 2005;1096:146–155.

Vogeser M, Seger C. A decade of HPLC-MS/MS in the routine clinical laboratory—Goals for further developments. Clin Biochem 2008;41:649–662.

Wang SC, et al. Parametric studies of matched filters to enhance the signal-to-noise ratios of LC-MS-MS peaks. Analy Chim Acta 2006;556:201–207.

Wang W, et al. Quantification of proteins and metabolites by mass spectrometry without isotopic labeling or spiked standards. Anal Chem 2003;75:4818–4826.

Want EJ, et al. The expanding role of mass spectrometry in metabolite profiling and characterization. ChemBioChem 2005;6:1941–1951.

Warrack BM, et al. Normalization strategies for metabonomic analysis of urine samples. J Chromatogr B. Anal Technol Biomed Life Sci 2009;877:547–552.

Weckwerth W, Morgenthal K. Metabolomics: From pattern recognition to biological interpretation. Drug Discov Today 2005;10:1551–1558.

Werner E, et al. Mass spectrometry for the identification of the discriminating signals from metabolomics: Current status and future trends. J Chromatogr B Analy Technol Biomed Life Sci 2008;871:143–163.

Wilson ID, et al. High resolution "ultra performance" liquid chromatography coupled to oa-TOF mass spectrometry as a tool for differential metabolic pathway profiling in functional genomic studies. J Proteome Res 2005;4:591–598.

Winfield SR, et al. System and methods for non-targeted processing of chromatographic data. Icoria, 2005; U. S. patent application number 20050143931/A1.

Wishart DS, et al. HMDB: A knowledgebase for the human metabolome. Nucleic Acids Res 2009;37:D603–D610.

Wishart DS, et al. The human cerebrospinal fluid metabolome. J Chromatogr B Anal Technol Biomed Life Sci 2008;871:164–173.

Xia YQ, et al. Use of a quadrupole linear ion trap mass spectrometer in metabolite identification and bioanalysis. Rapid Commun Mass Spectrom 2003;17:1137–1145.

Yan SK, et al. A metabonomic approach to the diagnosis of oral squamous cell carcinoma, oral lichen planus and oral leukoplakia. Oral Oncol 2008;44:477–483.

Zelena E, et al. Development of a robust and repeatable UPLC—MS method for the long-term metabolomic study of human serum. Anal Chemi 2009;81:1357–1364.

Zhang H, et al. A software filter to remove interference ions from drug metabolites in accurate mass liquid chromatography/mass spectrometric analyses. J Mass Spectrom 2003;38:1110–1112.

Zhang X, et al. Data pre-processing in liquid chromatography–mass spectrometry-based proteomics. Bioinformatics 2005;21:4054–4059.

Zhang Z, Marshall AG. A universal algorithm for fast and automated charge state deconvolution of electrospray mass-to-charge ratio spectra. J Am Soc Mass Spectrom 1998;9:225–233.

Zhu M, et al. Detection and characterization of metabolites in biological matrices using mass defect filtering of liquid chromatography/high resolution mass spectrometry data. Drug Metab Dispos 2006;34:1722–1733.

Appendix

Following tables are summaries of common biotransformation reactions and resultant drug metabolites. Mass shifts and molecular formula changes of common metabolites from the parent drugs and mass defect shifts of common metabolites from either the parent drugs or corresponding mass defect templates are also included. The concepts and applications of mass defect filer templates are described in references: Zhu M, Ma L, Zhang D, Ray K, Zhao W, Humphreys WG et al. Detection and characterization of metabolites in biological matrices using mass defect filtering of liquid chromatography/high resolution mass spectrometry data. Drug Metab Dispos 2006;34:1722–33.

Table A1 Metabolites with Similar Structures to the Parent Drugs

Mass Shift (Da)	Molecular Formula Change	Mass Defect Shift (mDa)	Biotransformation Reaction[a]
+48	$+O_3$	−0.0153	$RSH \rightarrow RSO_3H$ $3 \times (RH \rightarrow ROH)$
+34	+2OH	+0.0054	$RCH{=}CHR' \rightarrow RCH(OH){-}CH(OH)R'$
+32	$+O_2$	−0.0102	$2 \times (RH \rightarrow ROH)$ $RSR' \rightarrow RSO_2R'$
+30	$+OCH_2$	+0.0106	$RH \rightarrow ROH \rightarrow ROCH_3$
	$+O_2, -H_2$	−0.0259	$RCH_3 \rightarrow RCOOH$ Quinone formation
+18	$+H_2O$	+0.0106	$RCN \rightarrow RCONH_2$ $RCH{=}CHR' \rightarrow RCH_2CH(OH)R'$ Epoxide hydrolysis
+16	+O	−0.0051	$RH \rightarrow ROH$ Epoxidation *S*- or *N*-oxidation
	$+O_2, -CH_4$	−0.0415	$RCH_2CH_3 \rightarrow RCOOH$

Mass Spectrometry in Drug Metabolism and Disposition: Basic Principles and Applications, First Edition. Edited by Mike S. Lee and Mingshe Zhu.

Table A1 (Continued)

Mass Shift (Da)	Molecular Formula Change	Mass Defect Shift (mDa)	Biotransformation Reaction[a]
+15	$+O_2$, $-NH_3$	−0.0367	$RCH_2NH_2 \rightarrow RCOOH$
+14	$+CH_2$	+0.0157	Methylation
	$+O$, $-H_2$	−0.0208	$RCH_2R' \rightarrow RCOR'$ $RCH_2OH \rightarrow RCOOH$
+4	$+H_4$	+0.0313	$2 \times (RCH{=}CHR' \rightarrow RCH_2CH_2R')$
+3	$+OH$, $-N$	−0.0004	Reductive cleavage of isoxazole followed by hydrolysis of imine to carbonyl
+2	$+H_2$	+0.0157	$RCOR' \rightarrow RCH(OH)R'$ $RCH{=}CHR' \rightarrow RCH_2CH_2R'$
	$+O$, $-CH_2$	−0.0208	$RCH_3 \rightarrow ROH$
+1	$+O$, $-NH$	−0.0160	$RCH_2NH_2 \rightarrow RCH_2OH$
	$+H$	+0.0078	*N*-oxide → *N*-hydroxy
0	$+O$, $-CH_4$	−0.0364	$R\text{-}CH(OH)CH_3 \rightarrow RCOOH$
−1	$+O$, $-NH_3$	−0.0317	$RCHNH_2R' \rightarrow RCOR'$
−2	$-H_2$	−0.0157	$RCH_2CH_2R' \rightarrow RCH{=}CHR'$ $RCH(OH)R' \rightarrow RCOR'$ $RNH\text{-}NH_2 \rightarrow RN{=}NH$
−4	$-H_4$	−0.0313	$RCH_2NH_2 \rightarrow RCN$ $RCH_2CH_2CH_2CH_2R' \rightarrow$ $RCH{=}CH\text{-}CH{=}CHR'$
−12	$+O$, $-C_2H_4$	−0.0364	$RCH_2COCH^3 \rightarrow RCOOH$
−14	$+H_2$, $-O$	+0.0208	$RCOOH \rightarrow RCH_2OH$ $RNO_2 \rightarrow RNHOH$
	$-CH_2$	−0.0157	De-methylation
−16	$-O$	+0.0051	$RSOR' \rightarrow RSR'$ $RNHOH \rightarrow RNH_2$
	$+O$, $-S$	+0.0330	$RNHNHR'C{=}S \rightarrow RNHNHR'C{=}O$
−18	$-H_2O$	−0.0106	$RCH_2CHOHR \rightarrow RCH{=}CHR$ $RCH{=}N\text{-}OH \rightarrow R\text{-}CN$
−28	$-CO$	+0.0051	$RCOR' \rightarrow RR'$
	$-C_2H_4$	−0.0313	$RNHCH_2CH_3 \rightarrow RNH_2$
−29	$+H$, $-NO$	0.0098	$RNO \rightarrow RH$
−30	$+H_2$, $-O_2$	+0.0259	$RNO_2 \rightarrow RNH_2$
	$-CH_2O$	−0.0106	$ROCH_3 \rightarrow RH$
−44	$-CO_2$	+0.0102	$RCOOH \rightarrow RH$
−45	$+H$, $-NO_2$	+0.0149	$RONO_2 \rightarrow ROH$

[a]The parent drugs can be used as mass defect filter templates for searching for these oxidative metabolites.

Table A2 Metabolites Formed via Dealkylation, Hydrolysis or Other Metabolic Reactions, Whose Molecular Masses Are Significantly Smaller Than the Parent Drugs

	Biotransformation Reaction	Filter Template[a]
Dealkylation	$RCH_2XCH_2R' \rightarrow RCH_2XH + R'CHO$ (X = NH, *O* or *S*) (RCH_2XH or $R'CHO$ > 50 Da),	RCH_2XH, $R'CHO$,
Cleavage of ester, amide or ether	$RC(=O)OCH_2R' \rightarrow RCOOH + R'CH_2OH$	$RCOOH$, $R'CH_2OH$
	$RC(=O)NHCH_2R' \rightarrow RCOOH + R'CH_2NH_2$	$RCOOH$, $R'CH_2NH_2$
	$RSC(=O)R' \rightarrow RSH + R'COOH$	RSH, $R'COOH$
	$RCH_2OR' \rightarrow RCOOH + R'OH$	$RCOOH$, $R'OH$
Reduction of azo and disulfide	RN=NR"	RNH_2, $R'NH_2$
	$RS\text{-}SR' \rightarrow RSH + R'SH$	RSH, $R'SH$

[a]Any expected metabolites that have significantly lower nominal masses than the parent drugs can be used as core structure filter templates for searching for these metabolites and their derivatives from metabolic cleavage reactions.

Table A3 Conjugated Metabolites, Whose Molecular Masses Are Significantly Greater Than the Parent Drugs

Biotransformation Reaction[a]	Mass defect shift (mDa)	Mass shift (Da)	Molecular formula change
GSH conjugation	68.2	+305.682	+10C+15H+3N+6O+S
Glucuronidation	32.1	+176.0321	+6C+8H+7O
Glycine conjugation	21.5	+57.0215	+2C+3H+N+O
Methylation	15.7	+14	+C+2H
N-acetylcysteine conjugation	14.7	+161.0147	+6C+8H+N+3O+S
Acetylation	10.6	+42.0106	+2C+2H+O
Systeine conjugation	4.1	+119.0041	+3C+5H+N+2O+S
Taurine conjugation	4.1	+107.0041	+2C+5H+N+2O+S
Sulfate conjugation	−43.2	+80.0432	+3O+S

[a]The mass defect filter template for each class of conjugates in the positive ion mode is the mass of the protonated parent drug + the mass shift listed. For example, the diclofenac–GSH adduct filter template is the protonated diclofenac (296.0245) + the mass shift of GSH adducts (305.682). Accordingly, the mass defect filter for detecting diclofenac-GSH adducts is designed based on the filter template.

Table A4 Metabolites Formed via Typical Dehalogenation Reactions

Mass shift (Da)	Molecular formula change	Mass defect difference from RCH3 (mDa)	Mass difference from RCH_3	Biotransformation reaction	Filter template
−2	−F, +OH	−0.0051	+16	$RCH_2F \rightarrow RCH_2OH$	RCH_3
−18	−F, +H	0	0	$RCH_2F \rightarrow RCH_3$	
−18	−Cl, +OH	−0.0051	+16	$RCH_2Cl \rightarrow RCH_2OH$	
−20	−HF	−0.0157	−2	$RCH_2F \rightarrow R{=}CH_2$	
−34	−Cl, +H	0	0	$RCH_2Cl \rightarrow RCH_3$	
−36	$-F_2$, $+H_2$	0	0	$RCHF_2 \rightarrow RCH_3$	
−36	−HCl	−0.0157	−2	$RCH_2Cl \rightarrow R{=}CH_2$	
−40	−2HF	−0.0313	−4	2 x ($RCH_2F \rightarrow R{=}CH_2$)	
−58	$-F_3$, $+H_3$	0	0	$RCF_3 \rightarrow RH_3$	
−62	−Br, +OH	−0.0051	+ 16	$RCH_2Br \rightarrow RCH_2OH$	
−68	$-Cl_2$, $+H_2$	0	0	$RCHCl_2 \rightarrow RCH_3$	
−72	−2HCl	−0.0313	−4	2 x ($RCH_2Cl \rightarrow R{=}CH_2$)	
−78	−Br, +H	0	0	$RCH_2Br \rightarrow RCH_3$	
−80	−HBr	−0.0157	−2	$RBrCH_3 \rightarrow R{=}CH_2$	
−156	$-Br_2$, $+H_2$	0	0	$RCHBr_2 \rightarrow RCH_3$	
−160	−2HBr	−0.0313	−4	2 x ($RBrCH_3 \rightarrow R{=}CH_2$)	

Index

Mass Spectrometry in Drug Metabolism and Disposition: Basic Principles and Applications, First Edition. Edited by Mike S. Lee and Mingshe Zhu.

Printed in the USA/Agawam, MA
May 3, 2019

702516.002